Long-Term Durability of Polymeric Matrix Composites

Kishore V. Pochiraju • Gyaneshwar P. Tandon
Gregory A. Schoeppner
Editors

Long-Term Durability of Polymeric Matrix Composites

Springer

Editors
Kishore V. Pochiraju
Department of Mechanical Engineering
Stevens Institute of Technology
Castle Point of Hudson
Hoboken, NJ 07030, USA
kishore.pochiraju@stevens.edu

Gyaneshwar P. Tandon
Multi-Scale Composites
and Polymers Division
University of Dayton Research Institute
Dayton, OH 45469-0060, USA
Gyaneshwar.Tandon.ctr@wpafb.af.mil

Gregory A. Schoeppner
US Air Force, Air Force Research
Laboratory Materials and
Manufacturing Directorate
Wright-Patterson AFB
Dayton, OH 45433, USA
greg.schoeppner@wpafb.af.mil

ISBN 978-1-4419-9307-6 e-ISBN 978-1-4419-9308-3
DOI 10.1007/978-1-4419-9308-3
Springer New York Dordrecht Heidelberg London

Library of Congress Control Number: 2011935374

Printed on acid-free paper

Springer is part of Springer Science+Business Media (www.springer.com)

Preface

Polymers and composites contribute to substantial weight and cost savings in aerospace, marine, ground vehicle, wind power, and infrastructure applications. However, the development of a durable and damage tolerant composite structure is a major challenge, particularly for applications with elevated temperature or harsh operating environments. Meeting this challenge requires advances on several fronts: understanding key requirements of various application domains; innovations in polymer chemistry, material science and nanotechnology for the development of high performance constituent materials; methodologies for tailoring high-performance composite architectures; understanding mechanisms and mechanics of degradation and damage evolution; reliable model-based life predictions; and enabling design practices and validation methods.

Our motivation for this book was to compile a comprehensive knowledgebase and a compendium of analytical and experimental techniques for engineers and materials scientists designing composite structures with long-term durability and damage tolerance requirements. The book presumes that the reader has a basic understanding of composite materials and their performance. The authors of the book, drawn from academia, defense and government research laboratories, and leading practitioners from the industry, provide both scientific and practical perspectives for assessing and predicting long-term durability of the structures. As editors, we have tried to balance the rigorous scientific treatment of the problem with empirical methodologies, modeling/simulation with experimental techniques, and analytical methods with design and validation practices. We collaborated with the contributing authors to provide up to date research and methodologies available at the time of writing. As a field in which scientific theories are just beginning to provide adequate understanding of aging and degradation mechanisms, this book provides the composites community with a firm basis which can pave the way for newer developments that improve the long-term durability and damage tolerance of composite structures. The mechanism based durability and life prediction mechanisms will eventually mature to replace the pervasive empiricism in this field.

The contents of this book span from fundamental physical, chemical and engineering principles required for the evaluation and prediction of the long-term durability of polymeric composites structures to the design and application domain requirements. The sixteen chapters of the book can be classified into five related topics. The first three chapters of the book deal with behavior of constituent materials – resin, additives and modifiers, and the fiber-matrix interfaces. The chapters in this section present elements of chemistry of polymer resins along with the changes that occur when resins are exposed to environmental factors, a survey of the broad range of protection solutions available for composites, and thermodynamic and kinetics-based models used to develop the processing-interphase and interphase-property linkages.

The mechanisms of the response of the structures, damage tolerance, impact damage and physical aging are the subjects of the second section. The damage tolerance of composite structures is intimately linked to the morphology and extent of flaws or damage. Evolution of damage is treated as the formation of a multitude of crack surfaces within a composite that permanently change its response. Impact is a commonly occurring source of threat to composites that can produce a “seed” damage state, which inherently controls the subsequent durability and damage tolerance of the affected structure. Fundamentals of aging, its impact on the mechanical response, and environmental effects on physical aging behavior are discussed. The four chapters of the third section of the book deal with degradation due to environmental factors – moisture and temperature. Determination of diffusivity and moisture uptake in a polymer composite is a key step in the accurate prediction of moisture-induced degradation. The combined influences of damage and stress on moisture diffusion and the prediction of strength degradation in composites due to moisture ingress are illustrated. An understanding of the effects of thermo-oxidation in high-temperature PMCs for structural components subjected to arbitrary service environments is critical to life performance predictions. A comprehensive mechanism-based evaluation of thermo-oxidation is presented as three chapters of this section. These chapters describe the fundamental oxidation chemical reactions in the polymer, the morphological and material behavior changes due to oxidation, oxidation-damage coupling, experimental characterization techniques and a predictive simulation framework. This mechanisms based treatment of the problem provides a scientific basis for simulating long-term thermo-oxidative stability of composite structures as opposed to the empirical weight loss measurements.

The last two sections of the book are devoted to design and validation techniques for achieving long-term durability and application-specific durability requirements. In Sect. 4, the structural design and durability issues of composite joints provide the knowledge base and techniques for designers to address long-term durability concerns at the design stage. Issues to be considered in the design of composites used in primary and secondary composite structures including corrosion prevention measures associated with joining composite and metallic components are described. The unique requirements for developing load spectra for accelerated full-scale durability testing for both composite and combined composite/metallic structures are also discussed. Structural joints are where the durability, or lack of durability, is often most evident in a structure's response to combined environmental effects and mechanical loads. The fifth section addresses application related factors for aerospace structures, engines and civil infrastructure. Since economic considerations often require the use of military and commercial aircraft beyond their original design service lives, it is necessary to understand aircraft in-service induced aging and damage to ensure the airworthiness and structural integrity of these airframes.

The reader of this book may be particularly interested in how the long-term durability needs require considerations at several length and time scales. While the consequences of the durability (or lack of) are felt at the structural scale as illustrated by the complete tear-down analysis of an aircraft structure, the origins of durability are in the design of the polymer molecule at the nano-scale, the microstructure and micro-mechanics of the composite, and complex interactions of the interfaces at the meso-scales. Reliable predictions of performance degradation over long-term use and the modes of eventual failure require rigorous and validated models at several length and time scales along with homogenization theories dealing with scale transition. Validation of such models needs effective and controlled acceleration of degradation processes so long-term effects can be reproduced in laboratory environments within reasonable observation times. We hope that the discussion of the five facets of the problem and their interrelationships, as presented in this book, will inspire researchers – both theoreticians and experimentalists – from the fields of thermo-mechanics, materials, and polymer science to work together and formulate structural durability and life prediction frameworks.

Finally, we thank all the authors for their efforts and valuable writings and all our colleagues who served as reviewers for their comments and opinions.

Hoboken, NJ — Kishore V. Pochiraju
Dayton, OH — Gyaneshwar P. Tandon
Dayton, OH — Gregory A. Schoeppner

Contents

1 **Resin Systems and Chemistry: Degradation Mechanisms and Durability** 1
Jeffrey A. Hinkley and John W. Connell

2 **Additives, Nanocomposites, and Barrier Coatings** 39
Alexander B. Morgan

3 **Interphases in Composites** 77
Ranga Pitchumani

4 **Modeling of Response of Composite Materials with Damage** 121
Ramesh Talreja

5 **Impact Damage of Composite Structures** 143
Hyonny Kim, John C. Halpin, and Gabriela K. DeFrancisci

6 **Moisture-Induced Degradation** 181
Samit Roy

7 **Physical Aging in Glasses and Composites** 237
Gregory B. McKenna

8 **Mechanisms and Kinetics of Organic Matrix Thermal Oxidation** ... 311
Xavier Colin and Jacques Verdu

9 **Characterization of Thermo-Oxidation in Laminated and Textile Composites** 345
G.P. Tandon

10 **Modeling Thermo-Oxidative Aging and Degradation of Composites** 383
Kishore V. Pochiraju

11 **Composite Structures Durability Design and Substantiation** 427
Gregory A. Schoeppner and Matthew S. Tillman

12 **Durability of Structural Joints** 483
Carl Q. Rousseau and Endel V. Iarve

13 **Durability and Aging of Composite Aircraft Structures** 513
John Tomblin, Lamia Salah, and Dan Hoffman

14 **Durability of Polymer Composites in Power and Propulsion Applications** 549
John C. Thesken, Cheryl L. Bowman, and James K. Sutter

15 **Durability of Composites in Aircraft Engine Applications** 597
Rajiv A. Naik

16 **Composite Durability and Aging in Civil Infrastructures** 625
Dean C. Foster

Index 657

Chapter 1
Resin Systems and Chemistry: Degradation Mechanisms and Durability

Jeffrey A. Hinkley and John W. Connell

Abstract This chapter focuses on the chemistry of polymer resins and the changes that occur when resins are exposed to environmental factors that can cause degradation. The exposures that are considered are elevated temperatures (with or without oxygen); contact with water and other fluids; radiation; and mechanical loads.

Some general observations about the types of effects to be expected for each exposure condition are outlined first. Then, the chemistries of the various classes of resins that are in current use as matrices for high-performance composites are described. Separate sections treat epoxies, bismaleimides, PMR-type thermosets, phenyethynyl-terminated imides, and high-temperature thermoplastics. For each resin type, the formulation and curing are briefly discussed. This discussion is followed by a review of the available literature on mechanisms of long-term degradation as determined by spectroscopy, chromatography, and other analytical techniques. Consequences to constituent properties are also described: increases or decreases in glass transition temperatures; shrinkage and cracking; and changes in mechanical stiffnesses, strengths, and toughnesses.

1.1 Introduction

In choosing a polymer–matrix composite material for a particular application, a number of factors need to be weighed. Among these are mechanical requirements, fabrication method (e.g., press-molding, resin infusion, filament winding, tape layup), and use conditions. Primary among the environmental exposures encountered in aerospace structures are moisture and elevated temperatures, but certain applications

J.A. Hinkley (✉)
Advanced Materials and Processing Branch, Mail Stop 226, NASA Langley Research Center, 6A W. Taylor St., Hampton, VA 23681, USA
e-mail: jeffrey.a.hinkley@nasa.gov

K.V. Pochiraju et al. (eds.), *Long-Term Durability of Polymeric Matrix Composites*,
DOI 10.1007/978-1-4419-9308-3_1, © Springer Science+Business Media, LLC 2012

Table 1.1 Classes of matrix resins

	Typical Tg, °C			MIL-HDBK-17 test recommendations		
Resin type	Dry	Wet	Processing temperature, °C	Default elevated temperature	Default range	Typical use temp, °C
Vinyl ester	130	120	20–80	–	–	20
Epoxy	199	140	20–180	104	23–135	20–100
Cyanate ester	232		135–200	–	–	100
Bis maleimide	296	210	177–204	177	23–232	200
Polyarylene ethers	290	–	250	–	–	160
Polyimides				288	23–315	
PMR	338	–	316			232
Thermoplastic	340		350			177

may require resistance to other fluids and solvents, alkaline agents, thermal cycling, radiation, or rapid, localized heating (e.g., lightning strike). In this chapter, the main classes of polymer resin systems found in aerospace composites will be discussed. Within each class, their responses to environmental factors and the associated degradation mechanisms will be reviewed. Table 1.1 lists some common resin systems along with their processing and nominal long-term use temperatures. Exactly what is meant by "long-term" is not always specified. For airframe structure and aircraft engines, it should probably mean tens of thousands of hours [21].

General statements about aging behavior of particular backbone structures (e.g., "Structure A is more stable than Structure B") must be tempered with the knowledge that degradation is affected by (1) polymer molecular weight and end groups, (2) purity and defects, (3) physical form (e.g., bulk, composite, or powder), and (4) property examined (e.g., weight loss, microcracking, strength or toughness loss) [71]. In addition, there is a potential gap between academic studies on purified systems and performance metrics on composites from commercial prepreg.

1.1.1 General Effects of Thermo-Oxidative Exposure

In the following sections, detailed information on chemical changes for various resin chemistries will be presented where it is available. It is possible to make some generalized observations even in the absence of detailed mechanisms, however.

The first distinction to be made is between physical and chemical aging. A glassy polymer (thermoplastic or thermoset) or the glassy phase of a semicrystalline polymer is typically in a kinetically frozen nonequilibrium state. If the material is annealed below its glass transition temperature (Tg), it evolves toward equilibrium. This evolution is characterized by decreases in volume, enthalpy, and creep rate and the material generally becomes stiffer and more brittle [162]. These changes can be reversed by heating the polymer above Tg and quenching. Because no chemical changes are taking place, the material can be "rejuvenated" multiple times in the laboratory by repeatedly cycling above Tg [113]. Of course, it is not practical to do this on an article in service.

In addition to physical aging, thermal exposure can cause chemical aging. The curing reactions of thermoset resins may continue, leading to increased cross-link density, evidenced by rises in Tg and often embrittlement. On the other hand, the strength of an initially under-cured resin may actually increase during this stage due to post-curing. It may be worth mentioning that some elastomers, although they are not usually matrix resins, may undergo reversion, where a *decrease* in cross-link density brought about by heat causes them to soften or even liquefy [46]. Aging of composites at higher temperatures and longer times may lead to weight loss in the resin due to chain fragmentation and volatilization of the resulting small molecules. In the presence of oxygen, it is not unusual to see an initial small weight *gain* due to the formation of oxidation products [2] followed by faster weight loss. Oxidative weight loss is usually indicative of extensive chemical changes in the resin, leading to embrittlement, shrinkage, and cracking. When cracking occurs, weight loss commonly accelerates due to the increased surface area in contact with oxygen [34].

1.1.2 General Effects of Exposure to Water and Other Solvents

Water and other small molecules in polymers usually cause a decrease in Tg known as plasticization. The effect has been fitted to equations originally used to predict the Tgs of copolymers and blends [25]. Usually the glassy modulus is not much affected. In some cases where the small molecule interacts strongly with the matrix, there is a density and modulus *increase* in the glassy phase, a phenomenon called antiplasticization (even though Tg still decreases).

The amount of water sorbed by polymers varies widely. At low degrees of absorption, the diffusion into the polymer obeys Fick's law and produces at most a small degree of reversible swelling. Larger degrees of solvent swelling can lead to what is known as Case II sorption, in which a swollen layer with a very sharp boundary moves into the material with time. Even in the absence of that effect, certain materials, especially epoxies, may exhibit irreversible water sorption/desorption/resorption. This is usually attributed to cracking arising from swelling stresses [167], but the existence of nanoscopic inhomogeneities that can retain water has also been proposed [151, 179].

Chemical changes due to reaction with water are not common in composite matrix resins, but they can occur under certain circumstances. One example is the hydrolysis of polyimides at high pH or in the presence of superheated steam [148]. Another example is the degradation of the matrix resin or of fiber/matrix interfaces when a graphite composite is galvanically coupled to an active metal, usually in the presence of salt water [108]. This degradation may be the result of hydroxyl ions or peroxy-species evolved at the graphite cathode.

1.1.3 General Effects of Exposure to Radiation

Ultraviolet radiation is effectively blocked by carbon fibers, but can still produce severe surface degradation of films, coatings, and composite matrix resins. Weathering by combined UV/air/water exposure removes substantial quantities of composite surfaces [91]. Other forms of ionizing radiation (electrons, protons, gamma rays) produce free radicals via homolytic scission of covalent bonds. These in turn lead to simultaneous polymer chain scission (decrease in molecular weight), cross-linking, and evolution of small molecules, chiefly H_2. In the presence of oxygen, chain scission tends to dominate and properties such as Tg and toughness decrease dramatically. In general, all-aromatic polymers are more radiation-resistant than those containing aliphatic structures [142].

1.1.4 General Effects of Mechanical Loads

Although creep and fatigue damage will be considered in more detail in Chap. 4, it is worth mentioning that time-dependent behavior in carbon–fiber and glass–fiber composites is controlled by the matrix. In particular, matrix cracking (in response to thermal stresses) and interply delamination depend on matrix toughness. Interestingly, while static strength and toughness are directly related to matrix ductility, fatigue thresholds vary less [115].

1.2 Epoxies

Because they cure readily from liquid or tacky oligomers without the release of volatiles, epoxy resins are very popular starting materials for composites. For adhesives and lower-temperature structural resins, a common monomer is the diglycidyl ether of bisphenol A, DGEBA, shown in Fig. 1.1. In aerospace and other high-temperature applications, it is helpful to use a resin with higher functionality such as the tetrafunctional TGMDA (tetraglycidyl methylene dianiline, Fig. 1.2), the tri-functional triphenyl methane derivative (Fig. 1.3) or a multifunctional novolac epoxy (Fig. 1.4). All these various epoxy components tend to be miscible, and the mixture can be formulated to give the desired film-forming ability, prepreg tack and drape, and cured Tg. Additional components may be incorporated in order to modify toughness, moisture resistance, and other properties.

While commodity epoxy adhesives and coatings may be cured with a variety of catalysts and curing agents, (e.g., anhydrides, aliphatic polyamines) most

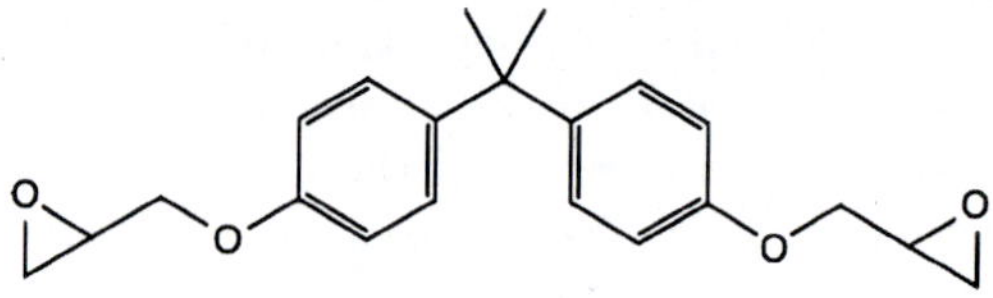

Fig. 1.1 DGEBA

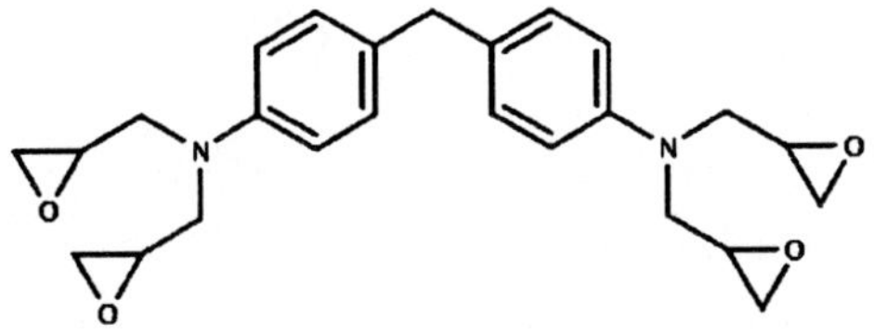

Fig. 1.2 TGMDA

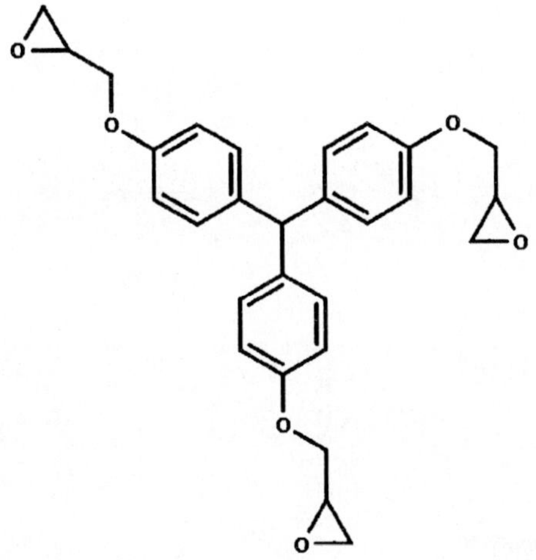

Fig. 1.3 Trifunctional epoxy

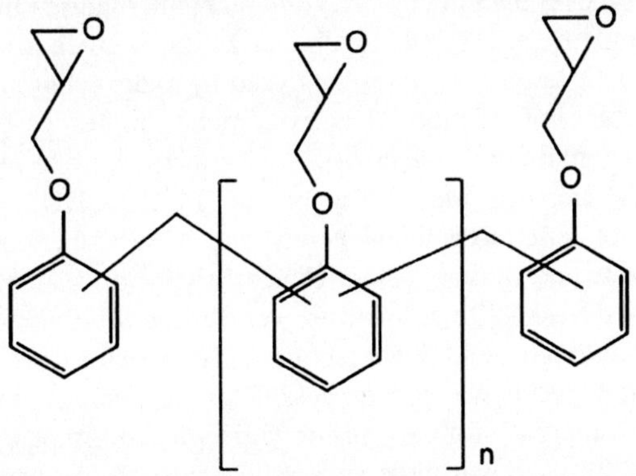

Fig. 1.4 Novolac epoxy. Typically 3-4 epoxy groups per molecule; meta and para links

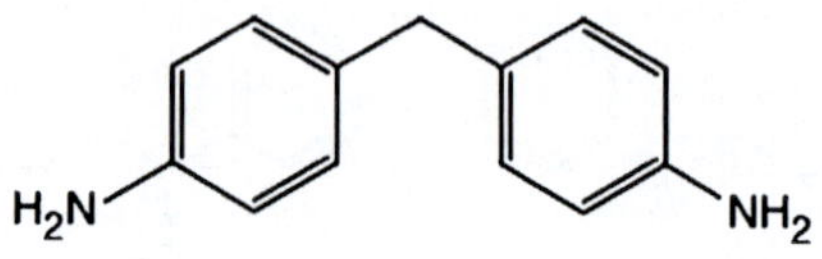

Fig. 1.5 MDA

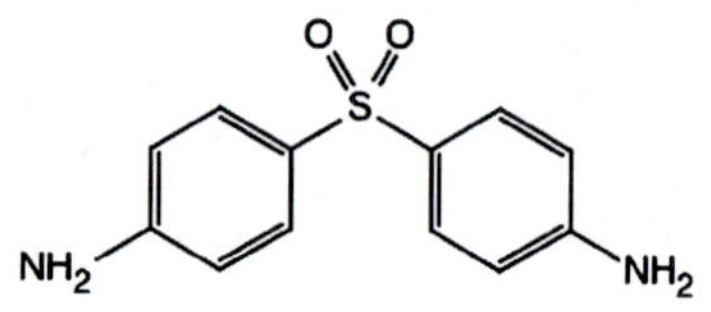

Fig. 1.6 DDS

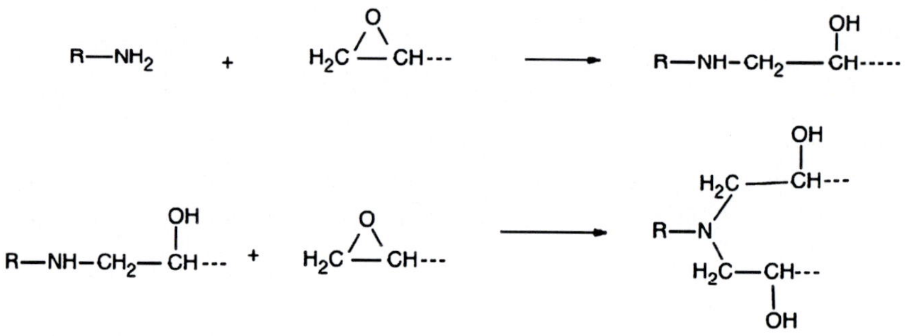

Fig. 1.7 Epoxy-amine reactions

high-performance matrix resins are cured with aromatic diamines such as MDA (methylene dianiline, Fig. 1.5) or DDS (diamino diphenyl sulfone, Fig. 1.6). The reaction of epoxy groups with primary and secondary aromatic amine produces cross-links of the types shown in Fig. 1.7. At higher temperatures and when epoxy is present in stoichiometric excess, homopolymerization to ether structures (Fig. 1.8) may occur [133]. This latter reaction is catalyzed by hydroxyl groups, so it tends to occur in the later stages of reaction. Thus, even in the simplest case, a cured epoxy-amine system will contain multiple cross-link chemistries, some of them cyclic.

Early aerospace prepregs were based on the TGMDA/DDS chemistry [110]. The combination of a tetrafunctional epoxy and a tetrafunctional amine cross-linker led to a network of very short chains, giving a desirable high Tg and high modulus. These resins, however, were very brittle. Furthermore, the addition of functionalized liquid rubber additives (such as ATBN, amine-terminated butadiene/acrylonitrile) – a strategy that was very successful for toughening of epoxy adhesives – was not an effective approach for these resins with high cure temperatures and limited inherent ductility [30]. A much more successful approach has been to incorporate thermoplastics, preferably with amine end groups that can react directly into the curing resin [78]. The properties of the resulting matrix material depend on the solubility of the

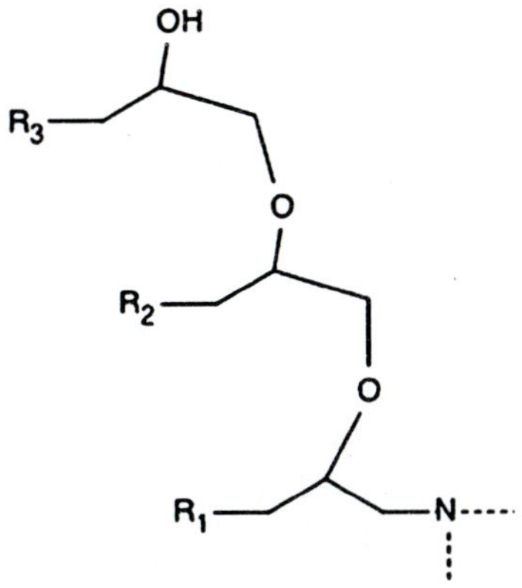

Fig. 1.8 Polyether formation via homopolymerization

thermoplastic and on the reaction conditions, but in general, toughness can be improved by at least a factor of 2 over that of the base epoxy without a loss in room temperature modulus or hot/wet properties. It should be noted that, in contrast to rubber toughening, which usually involves no more than 15% rubber, thermoplastic toughening can use large amounts of the additive. In fact, a "phase-inverted" morphology, where the thermoplastic becomes the continuous phase, may give optimum toughness [89]. In an additional toughening approach, a current state-of-the-art graphite/epoxy contains spherical particles of an amorphous polyamide at the ply–ply interface; in the finished composite laminate, the particle-toughened interlayer is about 30 μm thick, or perhaps 15% by volume [118]. Modern resins are thus designed to be inhomogeneous on several length scales in order to capture the desired combination of properties (chiefly good laminate compression strength along with interlaminar toughness for damage resistance and damage tolerance).

It may be mentioned that there have been efforts to modify epoxy–matrix composites with various nanoscale additives, although they do not seem to be commercial (e.g. [150]). Some modifiers might be expected to slow water permeation [87] but strength improvements have been generally modest so far [182] and long-term data are lacking.

1.2.1 Thermal Aging

A typical aerospace epoxy, Hexcel 8552, is a 180°C curing system "designed to operate in environments up to 121°C." With a dry Tg of 200°C and a wet Tg of 154°C, it is usually evaluated for aircraft service at 82°C. Data for a range of epoxy systems, shown in Table 1.2, indicate that resin moduli fall off by about 25–30% at this temperature under moisture-saturated conditions [39]. This softening adversely affects composite compression strength, so a temperature of 120°C probably already represents an accelerated exposure relative to a practical use temperature.

Table 1.2 Temperature/moisture effects on epoxy modulus

		Neat resin		
		Tensile modulus, GPa		
System		Room temperature, RT	82°C	121°C
3502	Dry	3.8	3.1	2.8
	Wet	3.5	2.6	1.9
2220-3	Dry	3.0	2.5	2.1
	Wet	3.1	2.1	0.9
8551-7	Dry	3.1	2.4	2.2
	Wet	2.8	2.3	0.8

The composites handbook (MIL-HDBK-17) recommends an aircraft service environment for testing purposes of 85% RH. The corresponding test temperature or Material Operating Limit is not specified, but is often 28°C below the Tg.

Wolfrum and coworkers [172] attempted to rapidly evaluate thermo-oxidative effects in graphite/epoxy laminates using exposures at up to 200°C for 450 days. Under these conditions, oxidation is much more severe at the composite surface, although changes in the infrared spectrum were observed even 500 μm below the surface. These were interpreted as oxidative destruction of the epoxy resin. Specifically, the absorption band at 1,510 cm^{-1} assigned to an aromatic CH vibration, but sensitive to attached aliphatic structures [44] decreased almost to zero. Simultaneously, peaks assigned to carbonyl (1710) and amide (1,668 cm^{-1}) groups increased. Though not shown by Wolfrum, the –OH and $-CH_2-$ peaks correspondingly drop sharply. This is consistent with the mass spectroscopy results indicating release of 2-propanol, acetic acid, etc. At 200°C, aromatic groups are being eliminated by oxidation [44] or volatilization of aniline, N,N-dimethylaniline, etc. The weight loss of a 16-ply composite under these conditions is over 3%, but is concentrated at the surface. Still, it is probable that elevated temperature exposure, even in the absence of oxygen, will lead to chain-breaking [15, 156].

It is important to reiterate that the detailed structure of an epoxy network is dependent upon the stoichiometry, impurities in the monomers, and cure temperature [66].

1.2.2 Radiation Effects

Ultraviolet (UV) radiation, including sunlight at sea level, is very injurious to epoxy–matrix composites. UV does not penetrate beyond the near-surface layer so paints or other coatings can provide adequate protection. Unprotected laminates erode rather quickly. Kumar et al. [91] showed that a 177°C-cured laminate lost the equivalent of over half a ply's thickness and underwent severe microcracking when exposed to only 500 h of alternating UV/condensation exposure. Giori et al. [63], in a simulation of space exposure, detected fairly high yields of CO, CO_2 and H_2 from TGMDA-based composites that were exposed to 200–400 nm UV in vacuum. The CO_2 was attributed to oxidized species formed during fabrication; the H_2 is considered indicative of generalized free radical processes. In this study compressive or flexural strengths of the composites were not greatly affected. In the same study, electron radiation in vacuo at 10^9 rads produced a measureable increase in Tg for one of the resins (Narmco 5208) but a decrease for the other (Fiberite 934). Fox et al. [58] exposed 934 resin and composite to 10^{10} rads of 1 MeV electrons, corresponding to a "worst case" 30-year exposure in geosynchronous earth orbit. The decline in Tg and production of low-molecular-weight products was attributed to polymer chain scission. This had the effect of making the resin more brittle, especially at cryogenic temperature, and led to decreases in composite transverse strength. Comparison of neat resin to composite properties was complicated by the significant finding that the neat resin cured differently, at least in this study. The report mentions in passing that 934 contained a minor amount of BF_3 complex (a homopolymerization catalyst). Noting that the flexural strength as a function of absorbed dose is the same for either electron or gamma radiation, Egusa [53] concluded that their effects are equivalent. Ultimate strain of TGMDA/DDS resin fell by over a factor of 2 at 150 MGy dose. Takeda et al. [158] studied a TGMDA/DDS system with either carbon or glass cloth reinforcement. The decreases in composite Mode I interlaminar fracture toughness that they observed upon irradiation are consistent with resin embrittlement.

Bellenger and Verdu [14] proposed the interesting idea that internal stresses (valence distortions) caused by the curing reaction and by "cool down" from the cure temperature may promote bond-breaking under oxidative conditions. This idea explained the larger degree of network degradation (decrease in Tg) when photo oxidation was conducted deep in the glassy region, where thermal residual stresses would be highest.

1.2.3 Moisture and Solvent Effects

In cured epoxy resins, there are a variety of polar groups (hydroxyls, amines, sulfone) that would be expected to have an affinity for water. In fact, it is difficult to desorb the last traces of absorbed moisture by heating at moderate temperatures (60°C). Zhou and Lucas [180] suggest that this water is doubly hydrogen-bonded to hydroxyls, whereas Zinck and Gerard [183] infer (by studying different cross-linkers) that tertiary amines are responsible. How much water is strongly bound seems to be a function of the sample history.

Many investigators have posited therefore that water may exist in two states in epoxies, for example, bound/free; dissolved/clustered; in nanovoids/microvoids [128]. The existence of more than one kind of water absorption site is not of merely academic interest, since hysteresis in the absorption/desorption curves will affect the nature of the steady-state distribution (and hence the internal stresses) when a composite part undergoes cyclic temperature/humidity exposure [161]. Although details differ, these models call attention to the fact that at least for some resins at some temperature/moisture conditions, plateau moisture uptake values may depend on sample history [157] even in the absence of gross cracking, cavitation, or failure at fiber/matrix interfaces [168]. Moisture hysteresis may be less prominent in high-Tg tetrafunctional resins than in the commodity DGEBA materials. Lee and Peppas [94] found no evidence of cracking or irreversible changes in TGMDA/DDS exposed to water at 70°C, even though that system absorbs almost 7% by weight [181]. Nogueira [114] studied the effect of cure schedule on water uptake in a DDS-cured tetrafunctional blend containing excess epoxy. Even with 0.3 mm thin resin plaques, moisture uptake did not reach equilibrium in several cases, suggesting that ongoing damage processes may be occurring beyond 1,000 h. Reversibility was not examined in that study. In amine-cured epoxies there does not seem to be any evidence for actual hydrolysis of chemical bonds. In contrast, epoxy adhesives cured with dicyandiamide or anhydride may hydrolyze [5, 48, 176], causing permanent damage to the network.

Effects of moisture exposure on mechanical properties are illustrated in some early work by Augl [8, 9]. He studied composite properties of Narmco 5,208 on two different carbon fibers. Thin (2.0 mm) unidirectional specimens were exposed to 80% relative humidity (RH) at 75°C for 21 days, absorbing 0.64% moisture. This is not long enough to reach moisture saturation or to achieve a uniform through-thickness water concentration profile, but Augl pointed out that in his flexural testing, the outer surfaces were the most highly stressed. Although the hot/wet rigidity of the neat resin at 175°C was half the dry value after the 80% RH exposure, the room-temperature resin modulus was practically unaffected. Still, the room temperature composite flex strength dropped about 12% after the humid exposure and the failure mode changed. Interestingly, a composite made with a higher-modulus, but unsized, fiber seemed to retain more strength, implying that the undercured epoxy sizing may have been more moisture-sensitive than the bulk matrix. In Augl's experiments, all moisture effects were completely reversed upon re-drying the specimens.

Humidity decreases resin modulus and strength (at least at elevated temperatures), but the increased ductility in water-plasticized epoxies might be expected to improve toughness. This is in fact the case: some neat resin castings of a bisphenol-A epoxy cured with an aromatic diamine doubled in toughness when stored in laboratory air (50% humidity, 25°C) relative to the as-prepared state [32]. Ordinarily, higher resin toughness would be expected to translate to higher composite (e.g., interlaminar) toughness. The situation is not quite so simple, though, given the large role played by fiber/matrix adhesion in both the initiation and propagation of delaminations [47, 88]. Russell and Street [140] found for 3501-6/AS1 that wet composite G_{Ic} was only about 10% higher than dry. Wet G_{IIc}, on the other hand, was marginally lower. The toughness trends with temperature were opposite for the neat resin and composite, a discrepancy that the authors attributed to the roles of residual stresses and constraint effects.

G_{IIc} of a composite that is toughened with thermoplastic particles in the interply regions decreased about 20% when the laminate was wet [102]. Fatigue thresholds for Mode II interlaminar crack growth were also about 25% lower in wet specimens. This was attributed to a change in crack path toward failure at the particle/epoxy interface. The matrix itself did become more ductile, and creep tests on $[\pm 45]_{2s}$ specimens of this same graphite/epoxy system with about 1% water showed about a 20% increase in creep compliance at 26°C and much more at higher temperatures [174]. The creep rate was directly related to the proximity of the wet Tg.

Sala [59] studied carbon fabric/epoxy laminates that absorbed 4.8% moisture, whereupon static moduli of matrix-dominated (±45) specimens decreased by nearly 50%. When loaded in fatigue, wet (±45) specimens underwent a much slower stiffness degradation than dry ones, although fatigue failure stresses were similar. Wet (0/90) specimens had about 30% higher fatigue failure stress.

There is not a lot of information available on effects on matrix resins of other fluids besides water. In general, it would be expected that solvent sensitivity would depend strongly on the detailed chemistry of the network [84] as well as on its cross-link density. Curliss and Carlin [42] found decreases in modulus and strength when both AS4/3501-6 and toughened epoxy systems were exposed to hot JP-4. Although they attributed the loss in strength to degradation of the fiber/matrix interface, there was also a measureable decrease in Tg due to preferential sorption of certain fuel components. The degradation in 0° flex strength amounted to 25%. Hooper et al. [79] and Hooper and Subramanian [80] showed that absorption of jet fuel tended to toughen AS4/3501-6. Almen et al. [3] show only minor effects on 977-2 and 3501-6 composites due to immersion in methyl ethyl ketone, although the specimen thickness was not reported.

1.2.4 Effects of Stress and Time

Like other polymers, composite matrix resins exhibit time-dependent behavior. This applies to stiffness, strength, and toughness properties. For example, Ashcroft et al. [7] modeled cracking in an epoxy using a viscoelastic fracture model. The implication of this is that subcritical crack growth could occur over a very long time. Absorbed moisture would affect the time-dependent (e.g., creep) behavior of a resin, but there is also a more subtle effect: applied stress also increases both the diffusion constant and the ultimate water pickup [49, 55]. Residual stresses due to cure and "cool down" may cause similar effects; this may explain why ±45 coupons absorbed more fuel than did unidirectional coupons under the same conditions [43].

An interesting new approach to damage tolerant composites has been demonstrated: a "self-healing" epoxy matrix [117]. At this point mostly a laboratory curiosity, such a material system when damaged would release an encapsulated adhesive into small cracks to seal them or slow their growth [120]. The more general research area of remendable polymers (which includes thermoplastics and certain epoxy-like materials designed with thermally reversible cross-links) has been reviewed by Bergman [15].

1.3 Bismaleimides

Bismaleimide (BMI) resins get their name from monomeric structures that are synthesized by condensing maleic anhydride with an aromatic diamine, followed by cyclodehydration to the bis-imide [152]. A common component is built on 4, 4′-diaminodiphenyl methane (Fig. 1.9). Maleimides can undergo free-radical polymerization with a catalyst or thermally without a catalyst, but the highly cross-

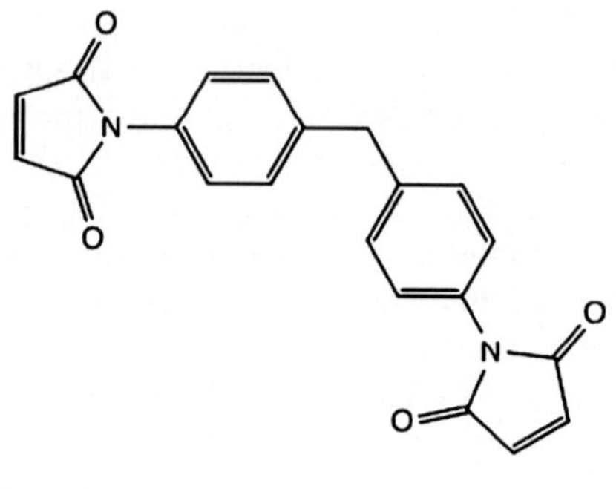

Fig. 1.9 BMI monomer

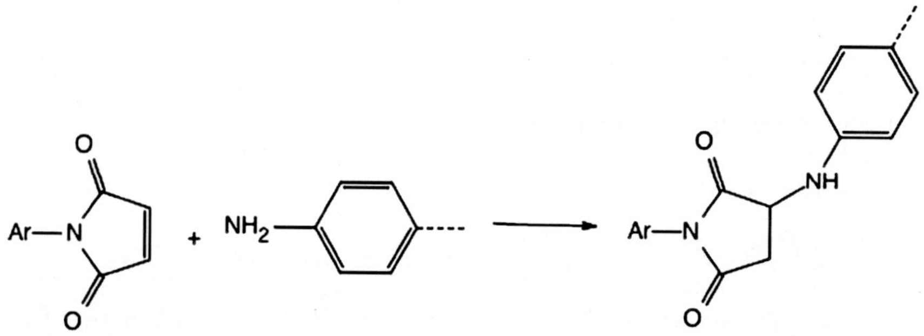

Fig. 1.10 Michael addition

linked product arising from homopolymerization of this low-molecular-weight monomer is extremely brittle and, therefore, not very useful. Commercial BMI matrix resins are formulated with other components: chain extenders, reactive diluents, catalysts, and tougheners; their cured structures thus may vary widely and are probably very complicated.

One method of chain extension for monomers like that in Fig. 1.9 employs the Michael addition with aromatic diamines (Fig. 1.10). Another uses Diels-Alder chemistry (Fig. 1.11) to copolymerize either divinyl benzene [155] or a monomeric or oligomeric bis-propenyl compound [154] with the maleimide. Recent generations of BMI prepregs contain substantial percentages (~30%) of propenylphenoxy or allylphenyl or allylphenol comonomers that improve toughness; a typical example is diallyl bisphenol A (Fig. 1.12). The latter becomes incorporated via an "ene"-type reaction (Fig. 1.13). The ene reaction is complete early in the cure cycle, and further reaction of maleimide and allyl double bonds occurs in the 200–250°C range [111]. At higher temperatures and longer times, there is spectroscopic evidence of etherification via the hydroxyl functionality. Morgan et al. call this continued cross-linking, but the observed increase in Tg could also be caused by intramolecular reactions that increase the structures' rigidities.

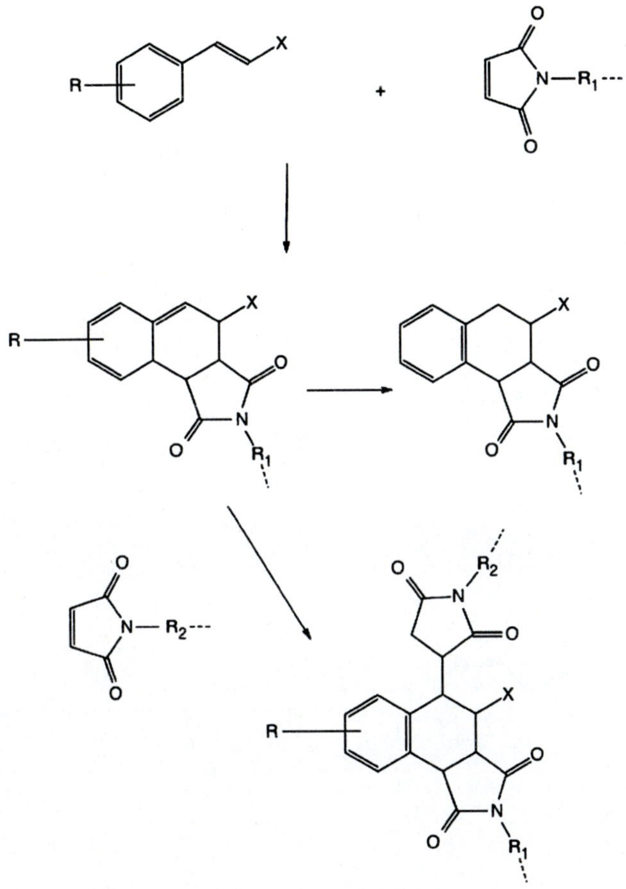

Fig. 1.11 Diels-Alder-type extension

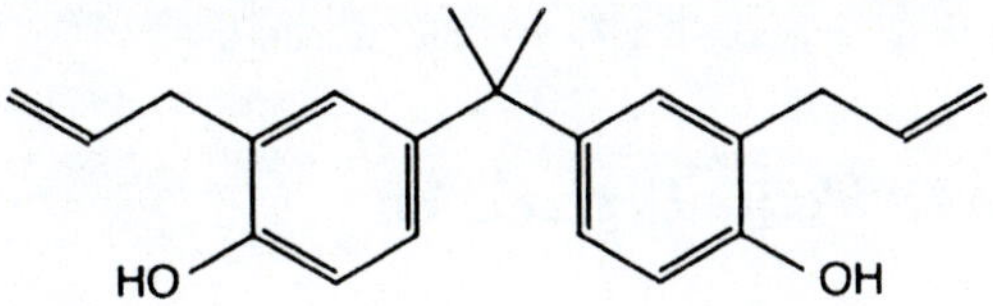

Fig. 1.12 Diallyl bisphenol A

Because they offer epoxy-like processing with some improvement in thermal stability, BMIs are also blended and co-reacted with other thermosets, such as epoxies and/or cyanate esters [51]. BMIs have also been toughened by the addition of thermoplastics, including maleimide-capped sulfone polymers, although at a sacrifice of prepreg tack and drape [153]. An RTM version of the commercial 5250-4 is said to contain, in addition to the diaminodiphenylmethane-derived monomer, the BMI of toluene diamine [97, 98], probably as a viscosity modifier.

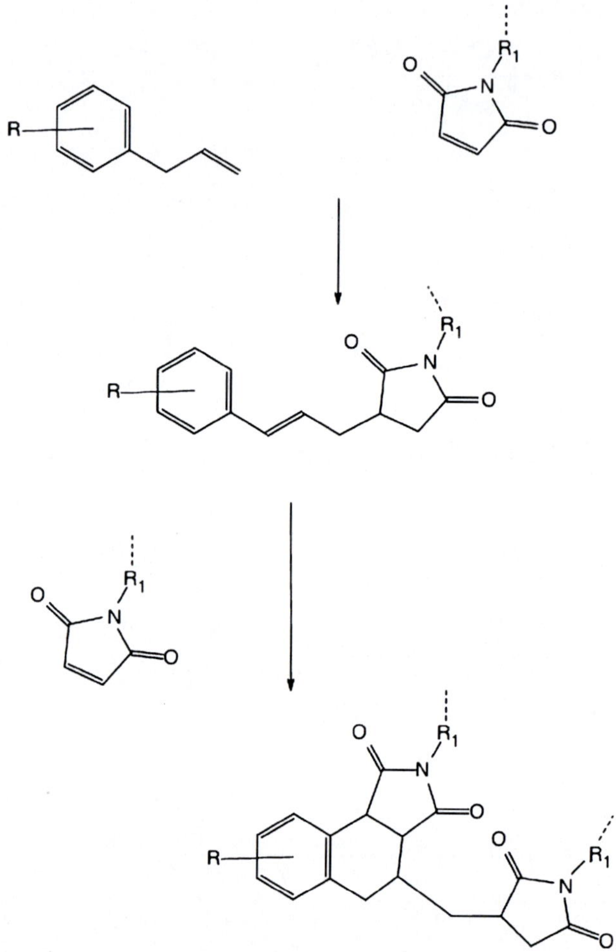

Fig. 1.13 Chain extension and cross-linking via "ene" reactions

1.3.1 Thermo-Oxidative Degradation

Working with cross-plied laminates of BASF X5260, Pederson et al. [125] showed that isothermal aging in air at 177°C produced a weight loss of 1.5% and increased Tg from 250°C to 300°C, which suggests significant chemical changes at this temperature during the 2,232-h aging period. At the same time, the resistance to interlaminar cracking decreased by 50%. Sixteen-ply quasi-isotropic specimens formed microcracks (up to 50/cm) under the influence of thermal degradation and residual stress. The data for crack density as a function of time and ply depth fit the assumption that the reaction was diffusion-controlled. Similarly, Ohno et al. [119] reported that the surface layers of BMI composites blacken as cracking proceeds. It is difficult to do much detailed chemical analysis on a cross-linked resin, but XPS seemed to indicate that oxidation of carbon to carbonyl (occurring initially at phenol and isopropylidene groups) produced colored species and eventually loss of small molecules (e.g., benzoquinone). Dao et al. [45] interpreted their FTIR data as showing that even the initial cure conditions (at 170°C or higher) contributed to surface oxidation. –C–OH groups arising from oxidation of –CH as well as the vinyl groups left by incomplete cure are thought to provide sites for continued oxidative attack. These oxidation processes lead to easily measureable weight gains in the early stages of aging [33]. After that, the surface composition stabilizes as material is lost.

At accelerated aging temperatures, resin from entire surface plies may be vaporized while the interior of a 2-mm-thick laminate is little affected. In spite of this, DMA measurements could detect increases in Tg. Increases in Tg in a few thousand hours at more typical use temperatures have also been seen [99]. Morgan et al. [111] associated this post-curing with "a more brittle mechanical response" but no mechanical data were shown. Boyd and Chang [23], however, did show that aging at 205°C in nitrogen produced both an increase in Tg and a decrease in composite compression after impact strength, which certainly suggests chemical changes unrelated to oxidation.

1.3.2 Moisture and Solvent Effects

Like epoxies, commercial BMI resins absorb several percent of water [11]. A manufacturer's data sheet quotes a dry Tg of 300°C and a wet Tg of 200°C (Cycom 5250-4). Room temperature properties (modulus, strength) are not greatly affected by moisture [40]. When tested wet at 82°C, however, about a 15% lower interlaminar shear strength and a 20% lower composite in-plane shear modulus were recorded [56].

Unlike epoxies, BMIs contain hydrolysable links. Upon water aging a 90°C, Bao and Yee [11] saw changes in the infrared spectrum that they ascribed to hydrolysis of imides to amides and carboxylic acids. This change was coupled to continuing long-term water uptake as the network relaxed.

A related degradation mechanism is the hydrolysis that occurs when BMI composites are galvanically coupled to metals. Woo [175] showed that BMI was more susceptible to galvanic degradation than epoxies, and with Rommel et al. [137] blamed hydroxyl ion produced at the cathode. Taylor [160], however, compared galvanic results to added caustic and blamed peroxide and superoxide radicals.

Another difference from epoxies is that BMI's have use temperatures above 100°C. Rapid heating of moisture-saturated coupons ("thermal spiking") can cause blistering and other permanent damage.

In addition to water, Falcone and Dow [56] soaked BMI coupons in JP-4, hydraulic fluid, deicing fluid, methylene chloride, and MEK. At room temperature, only methylene chloride (paint stripper) produced obvious damage in 4 days. Jet fuel did affect the 82°C properties substantially (producing over 20% decrease in shear modulus and strength).

1.3.3 Mechanical and Time-Dependent Effects

A principal mechanism observed upon long-term thermal cycling is the formation of microcracks and their eventual propagation as delaminations. Ju and Morgan [83] showed that cycling from room temperature to 250°C produced 45% more cracks than room temperature to 200°C. They attribute this to continued post-curing. It is interesting that 0.8% prestrain leads to measureable cracking after only two cycles to −196°C; the mechanical prestrain had little effect on the extent of cracking in 25–250°C cycling. The most severe condition included both elevated and cryogenic excursions, especially when combined with humidity. Since moisture-induced swelling should ameliorate ply-level "cool down" stresses, a more localized effect of water at the fiber/matrix interface is implicated. Alternatively, the resin itself may be degraded by 100% humidity at 100°C.

1.4 PMR and Related Thermosets

The PMR (Polymerization of Monomeric Reactants) approach to addition polyimides was first proposed in 1972. By using monomers in a low-boiling solvent such as methanol, it is relatively easy to make a flexible prepreg. The original composition, PMR-15, incorporated monomethyl ester of 5-norbornene-2,3-dicarboxylic acid (nadic ester or NE), 4,4′-methylene dianiline (MDA), and dimethylester of 3,3′,4,4′-benzophenone tetracarboxylic acid (BTDE), formulated to give a polyimide molecular weight of about 1,500 [146]. The nominal structure is shown in Fig. 1.14. Since the original publication, modifications have been proposed to the diamine (e.g., RP-46 [123]), to the dianhydride (e.g., "PMR-II" [145]) and even to the endcap [31], but PMR-15 is still a commercial product and has received a great deal of study.

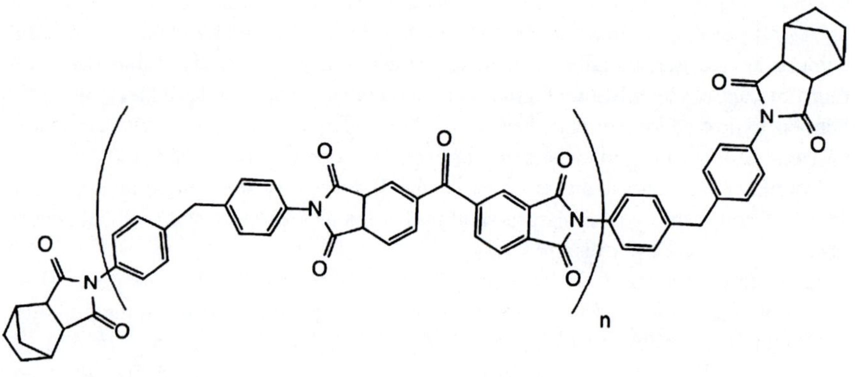

Fig. 1.14 Nominal PMR-15 oligomer (n=2.09)

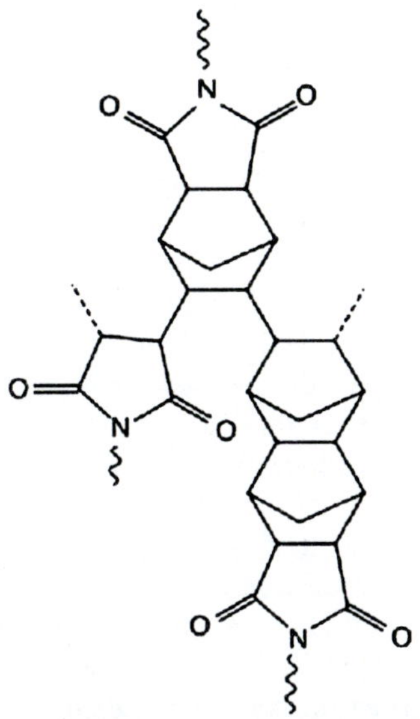

Fig. 1.15 Proposed cross-link structures

1.4.1 Cure and Thermooxidative Aging

During the manufacture of a PMR part, loss of solvent, imidization, and some cross-linking may occur simultaneously, depending on the cure schedule. Imidization is fairly straightforward, and according to IR measurements, can be completed at temperatures as low as 165°C [82]. Model compound studies (Meador et al. [107]) suggest that the initial reaction between monomers proceeds via base-catalyzed anhydride formation rather than by direct displacement from the ester-acid. The bisnadimide of MDA (i.e., Fig. 1.1 with n = 0) constitutes 35–50% of the oligomers [107]. It might be expected that the product of the thermal reaction of such short oligomers would be very highly cross-linked. The nadimide-terminated polyimide oligomer undergoes an endo- to exo-isomerization at temperatures above 160°C [109] to give a mixture of isomers. The model compound N-phenyl nadimide polymerizes at 285°C to give a product with degree of polymerization n = 6–7 [121]. The kinetics depend on pressure, suggesting a role of cyclopentadiene (CP). Maleimides produced in the reverse Diels-Alder reaction may play a role in initiating the addition polymerization. Although some authors draw structures with CP copolymerized into the backbone via a double bond, few olefinic carbons are actually detected in the cured resin [173]. Cyclopentadiene may instead be reacting with the nadic groups yielding end groups with multiple bridges. The cross-link that might result is shown schematically in Fig. 1.15.

One factor that distinguishes much of the literature on this class of materials is a focus on applications at 316°C (600°F) even though the original paper [146] showed that a composite made from a similar composition lost 18% of its weight in 600 h at this temperature. The curing conditions were not reported in that paper, but the constant push for higher-temperature performance [20] has led to aggressive post-cure conditions: conventionally 16–96 h at 316°C, but in one case up to 100 h at 371°C [12]. Scola [143], in spite of measureable surface erosion and weight loss, found no evidence of pyrolysis in the interior of unidirectional composites. In fact, the elevated-temperature flex strength increased, perhaps due to a 25°C increase of the resin Tg during aging. Thus, in contrast to epoxies, where (wet) Tg limits the use temperature and chemical aging is rarely a major concern, the high use temperatures expected of these addition polyimides produce simultaneous curing and degradation, with the result that, for example, short beam shear strength increases for the first ~2,000 h, then falls rather precipitously [124]. Weight loss accelerates at about the same point.

In thermosets in general, a high degree of cross-linking is associated with thermo-oxidative stability. Comparisons among PMR-15, 30, 50, and 75, however, showed that the higher the molecular weight was between endcaps, the less weight was lost during 188°C aging [164]. This improved thermo-oxidative performance came at a very minor sacrifice in cured Tg. Leung et al. [95] also varied PMR prepolymer molecular weight. They interpreted their data in terms of cross-link density only, but it could equally well be inferred that the (aliphatic) cross-link structure itself was less stable than the imide backbone with respect to weight loss. Propose a scheme whereby oxidation of the nadic methylene bridge leads to aromatized structures (quinones and hydroquinones) although these were not detected directly. Their scheme maintains the links formed during the initial polymerization, so it does not account for volatilization of small fragments. It is clear though, that early in the aging or oxidation process, the nadimides are transformed, and that this chemical change is accompanied by weight loss, even in nitrogen [20].

1.4.2 Moisture and Solvent Effects

Due to the emphasis on high-temperature applications, there are relatively few papers on moisture effects in PMR-type resins. Moisture can, however, be expected to have three kinds of effects: (1) matrix plasticization; (2) retention of water after hygrothermal cycling, leading to delamination, blistering, or void growth upon thermal spiking; and (3) hydrolysis of imide linkages.

As an example of the first effect, moisture saturation had no effect on compression strength of Celion/PMR laminates at room temperature, but at elevated temperatures, the moisture effect was substantial [130]. Strength was down about 50% at 300°C. This may be an underestimate, since some drying out may have occurred as the test temperature was reached; the heat-up rate is not reported.

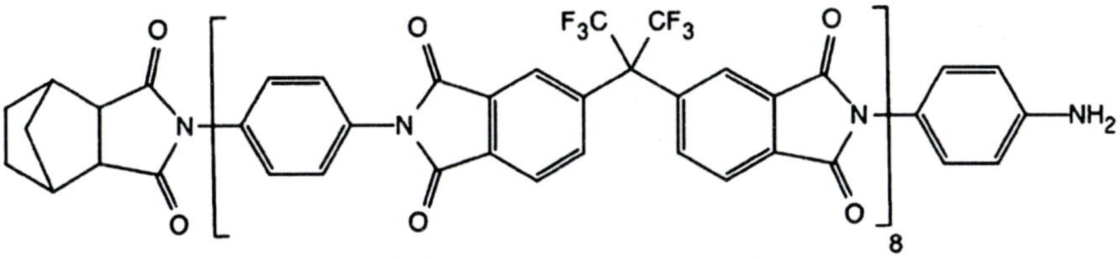

Fig. 1.16 Imidized structure of AFR700B as formulated

Shimokawa [147] saw a similar 50% strength drop at 260°C in wet specimens. Post-impact compression failure strain is apparently improved by moisture, perhaps due to better matrix ductility [129].

The second mechanism, void growth, is illustrated by data of Morgan et al. [112] on AFR700B (Fig. 1.16). Moisture contents above 0.4 wt% lead to blistering at around 400°C (wet Tg~340°C).

Finally, although no difference was seen between PMR oxidation in dry air and ambient air [132], hydrolysis of imides occurs above a threshold water content and leads to significant losses in toughness at rather moderate temperatures, for example, 80°C immersed [68].

1.4.3 Radiation

Bowles et al. [19] exposed both graphite/epoxy and C6000/PMR-15 to 6×10^9 rads of electron radiation in high vacuum near room temperature. DMA showed a decrease in the Tg of the epoxy (attributed to chemical degradation) whereas Tg of the PMR-15 composite increased about 11°C. Vyhnal [165] reported on LDEF-exposed composites and concluded that the mechanical properties were unchanged. The latter results may have been compromised somewhat by differences in the quality of the exposed and control panels.

1.4.4 Time and Stress Effects

Aging (air, 288°C, 1,000 h) degrades the tensile strength of the resin "dramatically" [138]. Matrix embrittlement can cause cracking, and cracks, in turn, provide additional paths for air ingress [22]. Thus even though oxidative degradation is largely a surface phenomenon, stresses due to the layup, fiber architecture [139], and mechanical loads [83] can interact with thermo-oxidative exposure to affect composite lifetimes.

In contrast to the behavior of thermoplastics, interlaminar fracture toughness of PMR is not greatly affected by the fatigue waveform [163] in spite of the fact that the matrix does show nonlinear creep at 250°C [101].

1.5 Phenylethynyl-Terminated Imides (PETI)

Aromatic polymers containing phenylethynyl (PE) groups were first reported in the mid-1970s [69]; a review was published in 2000 [38]. Reaction of the PE group (Fig. 1.17) leads to both chain extension and cross-linking, and does not evolve significant volatiles when conducted under pressure. With the proper choices of monomers and molecular weight, oligomers can, therefore, have low melt viscosity – low enough to be suitable for transfer molding – but high Tg and reasonable toughness when cured. NASA holds many patents on PE aromatic polyimides; these include the composition of matter for PETI-5, AFR-PE-4, and PETI-330 [37, 72]. They have emerged as versatile matrix resins for high-temperature composites on aerospace vehicles. NASA invested significant resources during the 1990s on the development of PETI-5 for the High Speed Civil Transport. More recently, the U.S. Air Force has invested in the development of the AFR-PE-4 matrix resin [65] and generated a composites design-allowable database [103].

The resin transfer molding (RTM) resins come as powders in the imide form and are free of unreacted monomers or solvent. Prepreg materials can be obtained in two forms: as a mixture of monomers in alcohol (referred to as the PMR approach), or as an amide acid oligomer in a polar aprotic solvent such as N-methyl-2-pyrrolidinone (NMP). The mixture-of-monomers prepreg contains unreacted aromatic diamine that can be hazardous to handle since most aromatic diamines have varying degrees of toxicity [6, 141]. The U.S. Environmental Protection Agency has ruled that exposure to at least one diamine, methylene dianiline, presents a significant risk to humans as a carcinogen. The amide acid oligomer form does not contain unreacted aromatic diamine but it does contain NMP, which also must be handled carefully.

The nature of the cross-linking chemistry is not known in detail, but there may be an analogy to the ethynyl (acetylene)-terminated resins described by Lee et al. [93]. Gel permeation chromatography, electron spin resonance, kinetic and mass spectroscopic results on ethynyl model compounds, are consistent with radical

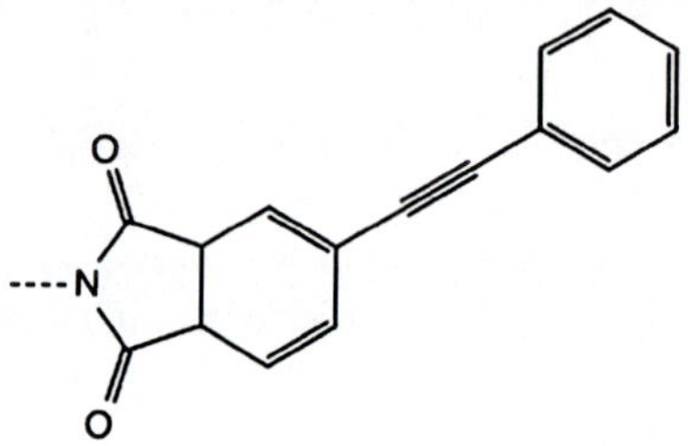

Fig. 1.17 End group from phenylethynyl phthalic anhydride

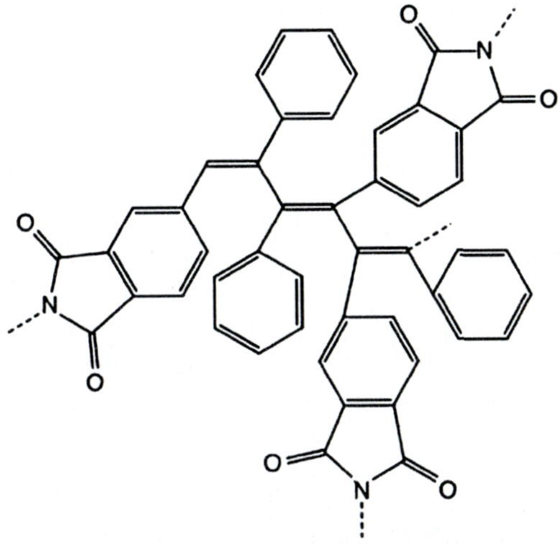

Fig. 1.18 Putative polyene cross-link formed in early stage of phenylethynyl cure

polymerization to give polyene structures. Li and Morgan [96], citing Fang et al. [57], show a similar structure arising from the phenylethynyl endcap (Fig. 1.18). Using NMR on labeled oligomers, Roberts et al. [135] detected some polyenes as well as aromatic groups including substituted benzenes, especially after post-curing. Interestingly, the product distribution seemed to depend on the oligomer molecular weight, perhaps due to concentration and viscosity effects. The average degree of polymerization of the end groups in the usual range of oligomers may be as low as 2.5 [75], due to steric restrictions and radical chain termination. Thus a substantial amount of the curing actually amounts to chain extension. This explains the high elongations and fracture toughness of the cured neat resins and the damage tolerance (compression strength after impact) of composites. It was thought that the phenylethynyl group reacted without volatile evolution, but recent TGA-mass spectroscopy studies have indicated that under ambient pressure small amounts of volatiles are released during cure [62]. The volatiles appear to emanate from degradation of the phenylethynyl groups; benzene and substituted benzenes and vinyl compounds evolve near the cure temperature (300°C). The amount of weight lost is small (1%), but it can lead to significant void formation in composite laminates. Volatilization of the decomposition products is suppressed by pressure so that void-free composites are routinely fabricated at pressures of 0.7–1.4 MPa.

1.5.1 Thermo-Oxidative Aging

In addition to changes in backbone chemistry [67], reactive oligomers can undergo post-curing reactions. These reactions may occur even at temperatures below Tg [26].

Many features of the degradation of PETI resins are very similar to those of the corresponding linear polymers [76], but the cross-link structure may be more susceptible to oxidation than the backbone is: weight loss of cured PETI oligomers at high temperatures may be slightly greater than in the corresponding linear polymers [74, 85, 159]. In the second of these studies, the difference did not look significant when highly accelerated aging was extrapolated to probable use temperatures. Takekoshi and Terry assert that resins from phenylethynyl imides perform "substantially better" than, for example, nadic-terminated ones. Connell et al. [35] showed that laminates from cured PETI oligomers with a formulated molecular weight of ~750 g/mol (not including the endcap) retained ~75% of room-temperature short-beam shear strength and open-hole compression strength after 1,000 h at 288°C in air. The mechanism of the strength losses was not examined. Accelerated aging at up to 343°C for 100 h led to weight losses of less than 2%; cracking was confined to surface plies. Exposure to 371°C, which is close to Tg, produced much larger mass losses [10].

1.5.2 Moisture and Solvent Effects

Composites made from these polyimides pick up measureable amounts of moisture. Amarillas [4] reported a wet Tg 90°C below that of the dry material for AFR-PE-4 composite. A PETI polyetherimide exposed to 85% relative humidity/90°C gained about 1.60% water based on the matrix [27]. This led to substantial loss of transverse strength. Han and Nairn [68] showed that PETI-5 composite was considerably more resistant to water-induced microcracking than K3B thermoplastic composite.

Cross-linking should also improve the solvent resistance relative to a thermoplastic. Adhesive bonds employing PETI-5 were virtually unaffected by a hydraulic fluid soak [29]. MEK produced crazing in some cured neat resin films, but the most noticeable effect of various solvents was plasticization evidenced by slight decreases in modulus and consistent increases in elongation at break [70]. Interestingly, a 1-week water boil seemed to decrease the neat resin failure strain measured at 177°C.

1.5.3 Radiation

Generally, phenylethynyl imides have an aromatic backbone to maximize thermal stability; this also gives them radiation resistance. After thermal curing, ethynyl and phenylethynyl resins contain a significant population of free radicals, and these radicals are stable, most likely due to resonance stabilization (Hill and Connell unpublished ESR results 1994). The exposure to various types of radiation may, however, cause these radicals to further react inducing cross-linking and/or chain extension. In one experiment, both molded ethynyl-containing neat resin and carbon fiber reinforced composite specimens were exposed to electron radiation at ambient temperature. Modulus and mechanical properties measured at elevated temperature increased, and toughness as determined by elongation decreased [36]. These mechanical property changes are similar to those observed after isothermal aging in air.

1.5.4 Time and Stress

Strengths of AFR-PE-4 composites drop about 30–60% between room temperature and 343°C [169]. Composites also undergo microcracking when they are cycled between 177°C and cryogenic temperatures [13]. Isothermal treatment of IM7/PETI-5 at −184°C for 555 h produced unexplained strength increases in some cases, despite the appearance of damage that was visible on polished edges in the form of matrix pitting [170].

1.6 Thermoplastics

A wide variety of thermoplastics such as nylons, polypropylene, and polycarbonate are common matrix materials for chopped-fiber, injection-moldable composites. For high-performance applications in transportation and aerospace, continuous-fiber unidirectional tape and fabric are chosen, and the matrix choices quickly narrow to aromatic polymers [104] containing ether, ketone, and sulfone groups (and sometimes thioether). Figure 1.19 shows a generic structure. Also in this class are thermoplastic imides, shown in Fig. 1.20. Table 1.3 gives some well-known examples. Melt viscosities of all of these materials tend to be high, so successful composite

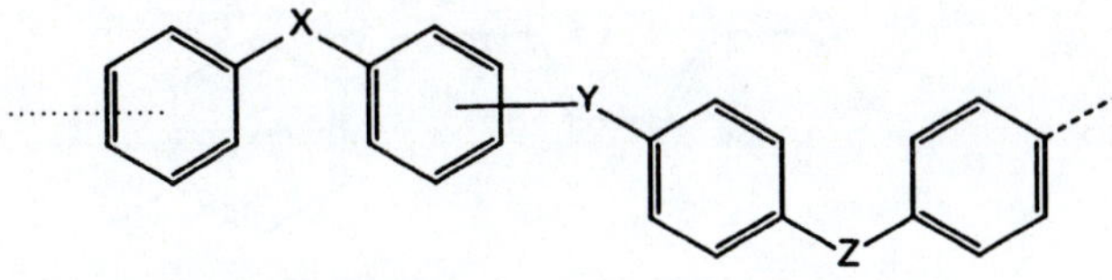

Fig. 1.19 Generic aromatic polymer

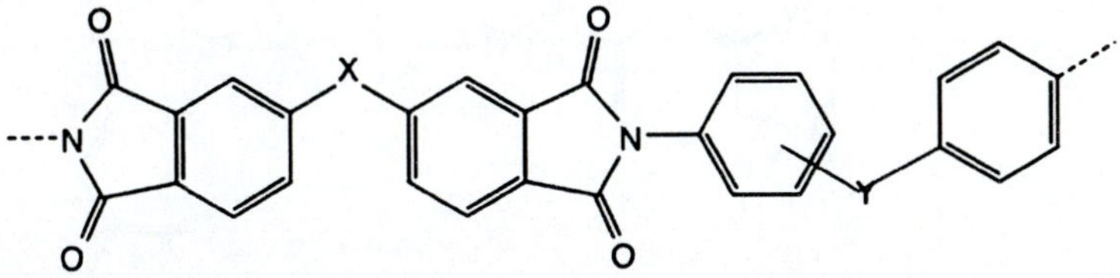

Fig. 1.20 Generic linear polyimide

Table 1.3 Representative thermoplastics

Polymer	Structure	Tg [T_m]
Polyphenylene sulfide (PPS)	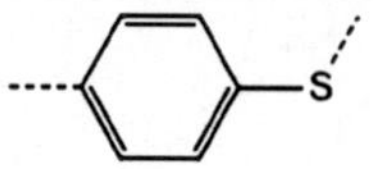	90 [285]
Poly(ether ether ketone) (PEEK)	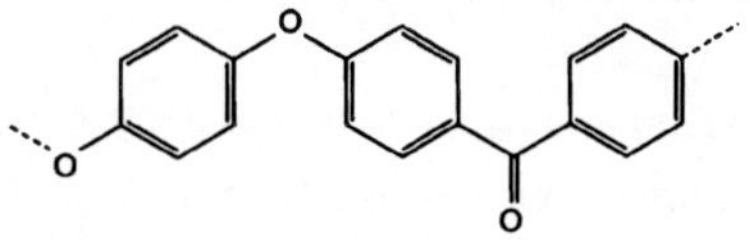	145 [343]
Polysulfone (PSF)	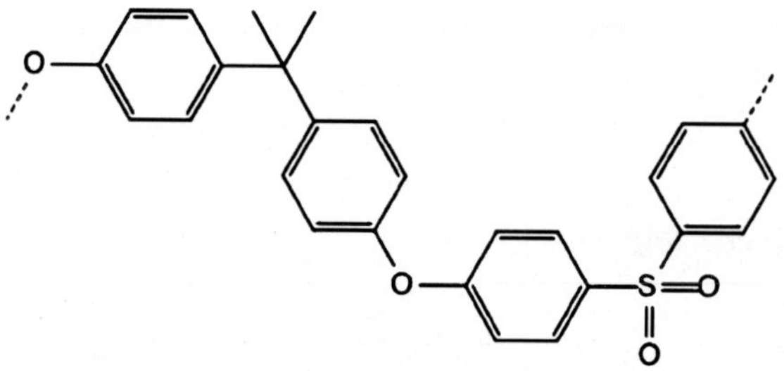	190
Ultem polyimide	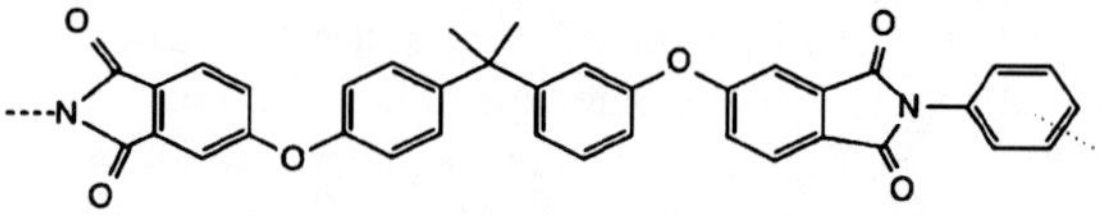	215
Polyethersulfone (PES)	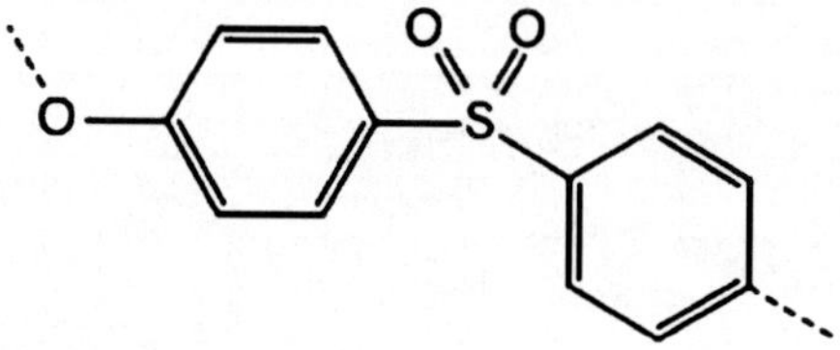	230
K3B	Proprietary	245
Polyamide-imide (PAI)	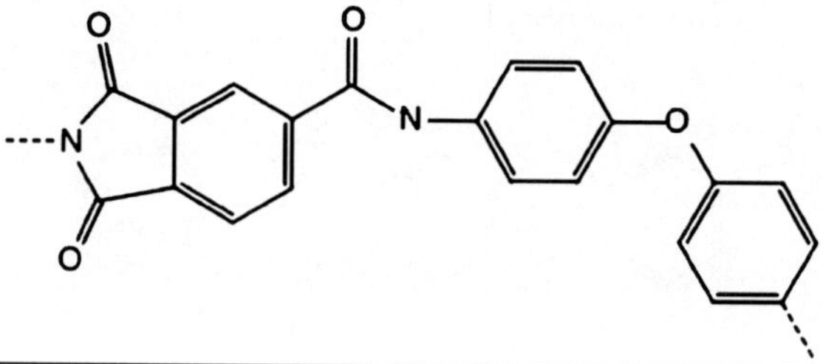	280

processing relies on keeping molecular weights low. In semicrystalline materials, crystallites can function as cross-links to improve toughness and solvent resistance and raise the use temperature. In linear amorphous materials, the molecular weight range that can be used in composite fabrication may be quite small [81] so chain extension/cross-linking via reactive end groups or pendant groups is helpful. The chemistries and degradation of these materials are taken up in other sections of this chapter. Another approach to processing a linear polyimide is illustrated by DuPont's Avimid N, (Tg = 352°C) which is prepregged from a mixture of monomers, probably the fluorinated anhydride 2,2-bis-(3,4-dicarboxyphenyl)hexafluoropropane dianhydride and the diamines *p*-phenylenediamine and *m*-phenylenediamine.

1.6.1 Thermo-Oxidative Aging

Because both the use temperatures and the processing temperatures of the thermoplastics tend to be higher than those of epoxies, oxidation tends to play a significant role in loss of properties during aging. Even relatively brief exposures to processing in air at Tg + 100°C [24] produced measureable changes in Tg and melt rheology [127].

Many studies of the pyrolysis of thermoplastics exist (e.g. [54, 126]) but the degradation mechanisms in air and at more realistic use temperatures are different from those that occur during rapid heating in vacuo. In general, exposure of thermoplastics to high temperatures and oxidation will lead to simultaneous backbone rearrangements, cross-linking or branching and chain scissions. The latter would eventually lead to the volatilization of small fragments. To give an example, Kuroda et al. [92] heated a soluble polyimide powder in air or in vacuum and were able to quantify scission and cross-linking reactions by employing both gel permeation chromatography and measurements of the elongation of small films in the rubbery state. They also tracked changes in Tg. Although mechanical properties of the glassy polymer were not measured, it is known [60] that reductions in molecular weight lead eventually to a ductile–brittle transition and it is also known that the concentration of cross-links affects toughness directly [73].

Weight loss, on the other hand, while it is certainly easy to measure, especially for resin films and powders, might not correlate with changes in mechanical properties [77]. This is because neither cross-linking nor changes in backbone chemistry are necessarily closely tied to reactions that produce small molecules, especially under accelerating conditions. By the time a linear polyimide has experienced a weight loss of 2–3% at 400°C, the remaining polymer is chemically very different from the starting material [52].

When used to study long-term aging, weight loss is also not very sensitive. For example, Abate et al. [1] could not detect "appreciable" weight loss in their polyarylene ether ketones during 25,000 h at 270°C, an aging temperature that was above Tg, and thus far above any reasonable use temperature for that material.

Parvatareddy et al. [122] aged IM8/ITX (a semicrystalline thermoplastic matrix with a Tg of ~185°C) for 6–9 months at 150°C leading to a 48% drop in composite 90° flex strength. Comparison of N_2 and air aging suggested that a combination of physical and chemical effects caused the simultaneous rises in resin Tg and modulus. The concomitant 1.7% weight loss in the composite is good evidence that some chemical events were occurring.

Physical aging is a known problem for amorphous thermoplastics, but it also occurs in semicrystalline materials like PPS [90]. Physical aging of PEI below Tg increased the yield stress and modulus [136] but produced no discernable embrittlement; composite specimens were stronger after aging.

1.6.2 Moisture and Solvent Effects

Because of the high processing and use temperatures of most thermoplastic composites, effects of water have not received a lot of attention. Zhang and Piggott [178] showed that graphite/PEEK absorbed about 0.5% water based on the matrix weight fraction; failure strain decreased by ~25%, which they attribute to the matrix and *not* to the interface. Bismarck et al. [16] also saw a decrease in transverse strength. The change was irreversible on drying, so may relate to crystallinity [17]. Polyphenylene sulfide (Tg = 90°C) undergoes additional crystallization during hot/wet aging [100]. Wang and Springer [166] found a small (probably insignificant) increase in the interlaminar fracture toughness of graphite/PEEK with moisture. On the other hand, GIIc decreased about 20% upon moisture saturation [144]. While salt water had little effect on polysulfone, steam at 123°C did degrade it [105], causing a change in tensile failure mode from ductile to brittle. A thick graphite composite degraded after 23 h in the pressure cooker.

Shin et al. [149] detected the hydrolysis of a thermoplastic polyimide, Avimid K3B, when it was exposed to thermal spiking or pressurized steam. Burcham et al. [28] concluded that water immersion was unrealistically severe for K3B and they also assumed that degradation of the fiber/matrix interface was responsible for the loss in interlaminar toughness that they observed. Han and Nairn [68], however, found that even humid exposures at 80°C decreased the toughness of K3B enough to cause spontaneous microcracking of the composite under the influence of residual stress.

As was the case for BMIs, aromatic polyimides may be degraded by aqueous base. A structure–property study by Croall and StClair [41] concluded that films containing the pyromellitic diimide residue were more susceptible to attack than polymers made from dianhydrides that contain bridging groups.

Because of interest in ICI's APC composite for aircraft, there were a number of studies of immersion of PEEK in aircraft fluids (e.g. [42]). Although PEEK absorbs measureable amounts of JP-4 and hydraulic fluid, the Tg and mechanical properties are not greatly affected. Stronger solvents such as MEK and paint stripper plasticize the matrix and weaken the composite. There is some evidence that aromatic compounds such as phenol are extracted from jet fuel [43].

So-called environmental stress cracking (crazing of stressed polymers in the presence of weak solvents or of surface-active agents) requires careful attention in applications of commodity polymers. Aromatic polyimides seem to be susceptible to stress cracking [50, 177] although it has not received a lot of study.

1.6.3 Radiation Effects

All-aromatic thermoplastics are considered to be inherently radiation-resistant. Bonin et al. [18], for example, found little change in the properties of graphite-PEEK exposed to 1 MGy nuclear radiation. Exposures to UV [134] or electron radiation [64] can cause polymer chain cleavage as evidenced by the production of benzene and other aromatics in addition to H_2 and CH_4. At 10^9 rads, Giori saw over 20% decrease in flex strength in a polysulfone composite. The layup was not reported, but one expects that the change in matrix-dominated properties might be even larger. The linear polysulfone seemed more radiation-sensitive in terms of mechanical properties than were highly cross-linked (tetrafunctional) epoxies.

1.6.4 Stress and Time

It is known that stress can enhance the transport and net uptake of solvent in neat PEEK [171]. Pritchard and Randles [131] found that diffusion and sorption of aircraft fluids in a composite were strongly dependent on layup and specimen thickness, "probably because of the effect of residual stress." The reversibility of moisture/temperature cycling in a graphite thermoplastic was highly dependent on internal stress [61].

Studying the effects of water immersion on K3B, Kim et al. [86] found decreases in the fracture toughness of the neat resin and spontaneous microcracking in composites. No significant changes in Tg occurred, and the authors suggested that plasticization by water may have increased the rate of physical aging; stress gradients due to nonuniform water absorption could possibly initiate damage.

The interlaminar toughness of thermoplastics can be quite high compared to that of typical epoxies. Under fatigue loading, however, the Mode-II toughness is degraded: the threshold strain-energy release rate is much smaller than the static value [116].

References

1. Abate L, Blanco I, Motta O, Pollicino A, Recca A (2002) 'The isothermal degradation of some polyetherketones: a comparative kinetic study between long-term and short-term experiments' Polymer Degradation and Stability 75 465–471
2. Akay M, Spratt GR 2008 Evaluation of thermal ageing of a carbon fibre reinforced bismalemide Composites Science and Technology Volume: 68 Issue: 15–16 Pages: 3081–3086
3. Almen GR, Byrens RM, Mackenzie PD, Maskell RK, McGrail PT and Sefton, MS 1989 "977 – a Family of New Toughened Epoxy matrices", 34 Int'l SAMPE Symposium May 8–11, 1989 pp 259–270
4. Amarillas M 2009 'Wet Tg Determination of High Temperature Composites Using DMA' abstract submitted to North American Thermal Analysis Society, https://www.egr.msu.edu/apps/natas09/?action=list accessed 8/13/09
5. Antoon MK and Koenig JL, (1981) 'Irreversible Effects of Moisture on the Epoxy Matrix in Glass-Reinforced Composites' Journal of Polymer Science: Polymer Physics Edition, Vol. 19, 197–212
6. National Research Council 1981 'Aromatic Amines: An Assessment of the Biological and Environment Effects' Conmittee on Amines, Board on Toxicology and Environmental Health Hazards, Assembly of Life Sciences, National Academies Press Washington DC
7. Ashcroft IA, Abdel Wahab MM and Crocombe AD (2003) 'Predicting Degradation in Bonded Composite Joints Using a Semi-Coupled Finite-Element Method' Mechanics of Advanced Materials and Structures 10:3, 227–248
8. Augl JM 1977 'The effect of moisture on carbon fibre reinforced epoxy composite mechanical property changes' NSWC/WOL/TR-76-149
9. Augl JM, Berger AE 1976 'Moisture effect on carbon epoxy composites Proceedings of the 8th national SAMPE technical conference' Seattle, WA, SAMPE, Covina p. 383–427
10. Bain S, Ozawa H and Criss JM 2006 'Development of a Cure/Postcure Cycle for PETI-330 Laminates Fabricated by Resin Transfer Molding' High Performance Polymers 18 991–1001
11. Bao LR, Yee AF 2002 'Effect of temperature on moisture absorption in a bismaleimide resin and its carbon fiber composites' Polymer Vl 43 IS 14 P 3987–3997
12. Baugher AH Espe MP Goetz JM Schaefer J and Pater RH 1997 'Cross-Linked Structures of Nadic-End-Capped Polyimides at 371°C' Macromolecules, 30, 6295–6301
13. Bechel, VT; Arnold, F 2006 'Permeability of polymer composites for cryogenic applications' Collection of Technical Papers – 47th AIAA/ASME/ASCE/AHS/ASC Structures, Structural Dynamics and Materials Conference v 7, p 5092–5102
14. Bellenger V, Verdu J 1985 'Oxidative Skeleton Breaking In Epoxy-Amine Networks' Journal of Applied Polymer Science Vl 30 Is 1 pp. 363–374
15. Bergman, SD 2008 'Mendable polymers' Journal of Materials Chemistry Volume: 18 Issue: 1 Pages: 41–62
16. Bismarck A, Hofmeler M, Dorner G 2007 'Effect of hot water immersion on the performance of carbon reinforced unidirectional poly(ether ether ketone) (PEEK) composites: Stress rupture under end-loaded bending' Composites Part A-Applied Science and Manufacturing Volume 38 Issue 2 Pages 407–426
17. Boinard, E Pethrick, RA MacFarlane, CJ 2000 'The influence of thermal history on the dynamic mechanical and dielectric studies of polyetheretherketone exposed to water and brine' Polymer Vl 41 IS 3 p 1063–1076
18. Bonin HW Miedema I Bui, VT 2008 'Advanced Polymer Composites for the Fabrication of Spent Nuclear Fuel Disposal Containers' Nuclear Technology Vol 164 Is 2 P 286–304
19. Bowles DE Tompkins SS Sykes GF 1986 'Electron Radiation Effects on the Thermal Expansion of Graphite Resin Composites' Journal of Spacecraft and Rockets, v 23, n 6, p 625–629

20. Bowles KJ 1988 'A thermally modified polymer matrix composite material with structural integrity to 371 °C' 20th Intl SAMPE Technical conference September 27–29, 1988 pp 552–561
21. Bowles KJ, McCorkle L, Ingrahm L 1998 'Comparison of graphite fabric reinforced PMR-15 and Avimid N composites after long-term isothermal aging at various temperatures' Journal of Advanced Materials Vol 30 Iss 1 P 27–35
22. Bowles, KJ 1999 'Durability of Graphite-Fiber-Reinforced Pmr-15 Composites Aged at Elevated Temperatures' J Compos Technol Research Vol 21 no 3 pp 127–132
23. Boyd, JD; Chang GEC 1993 'Bismaleimide composites for advanced high-temperature applications' International SAMPE Symposium and Exhibition (Proceedings), v 38, n 1, p 357–365
24. Brauman SK and Chow AW 1993 'Processing of Poly(ethersulfone)/Graphite Fiber Composites: Thermooxidation, Rheology, and Consolidation' Polymer Composites, February, Vol. 14, No. 1 pp 42–50
25. Brewis DM, Comyn J, Shalash RJA, Tegg JL 1980 'Interaction of Water with Some Epoxide Adhesives' Polymer Volume: 21 Issue: 3 Pages: 357–360
26. Bullions TA, McGrath JE, Loos AC 2003a 'Thermal-oxidative aging effects on the properties of a carbon fiber-reinforced phenylethynyl-terminated poly(etherimide)' Composites Science and Technology 63 1737–1748
27. Bullions, TA Loos, AC McGrath, JE 2003b 'Moisture sorption effects on and properties of a carbon fiber-reinforced phenylethynyl-terminated poly(etherimide)' Journal of Composite Materials Vl 37 Is 9 P 791–809
28. Burcham LJ Vanlandingham MR Eduljee RF Gillespie Jr JW 1996 "Moisture effects on the behavior of graphite/polyimide composites" Polymer Composites, v 17, n 5, p 682–690
29. Cano RJ and Jensen BJ 1997 'Effect of molecular weight on processing and adhesive properties of the phenylethynyl-terminated polyimide LARC-PETI-5' Journal of Adhesion, v 60, n 1–4, p 113–123
30. Chen TK, Jan YH 1995 'Effect of Matrix Ductility on the Fracture-Behavior of Rubber-Toughened Epoxy-Resins' Polymer Engineering and Science Volume: 35 Issue: 9 Pages: 778–785
31. Chuang KC Bowman CL Tsotsis TK Arendt CP 2003 '6 F-Polyimides with phenylethynyl endcap for 315–370 C Applications' High Performance Polymers VL 15 IS 4 p 459–472
32. Clarke JA 1985 "An ageing effect in cured epoxy castings relevant to fracture toughness testing" Polymer communications v 26, April pp 113–15
33. Colin X Marais C Verdu J (2001) 'Thermal Oxidation Kinetics for a Poly(bismaleimide)' Journal of Applied Polymer Science, Vol. 82, 3418–3430
34. Colin X, Mavel A, Marais C, Verdu J 2005 'Interaction between cracking and oxidation in organic matrix composites' Journal of Composite Materials Volume: 39 Issue: 15 Pages: 1371–1389
35. Connell JW Smith Jr JG, Hergenrother PM Criss JM 2003 'High Temperature Transfer Molding Resins: Laminate Properties of PETI-298 and PETI-330' High Perform. Polymers 15: 375–394
36. Connell JW, Hergenrother PM and Havens SJ, 1989 "Thermosetting Resins from Blends of Ethynyl Terminated Materials", High Performance Polymers, 1(2) 119
37. Connell JW Smith JG and Hergenrother PM 2002 "Composition of and method for making high performance resins for infusion and transfer molding processes" U.S. Patent 6,359,107, assigned to The United States of America as represented by the Administrator of the National Aeronautics and Space Administration
38. Connell JW, Smith JG, Hergenrother PM (2000) "Oligomers and Polymers Containing Phenylethynyl Groups" Journal of Macromolecular Science-Reviews in Macromolecular Chemistry and Physics Volume: C40 Issue: 2–3 Pages: 207–230

39. Coquill SL and Adams DF 1989 Mechanical Properties of Several Neat Polymer Matrix Materials and Unidirectional Carbon Fiber-Reinforced Composites NASA Contractor Report NASA-CR-181805 May 1985 - Oct. 1989
40. Costa ML Rezende MC And De Almeida SFM 2005 'Strength of Hygrothermally Conditioned Polymer Composites with Voids' Journal of Composite Materials, Vol. 39, No. 21 p 1943–1961
41. Croall CI and StClair TL 1992 'Mechanical properties of polyimide films after exposure to high pH' Journal of Plastic Film and Sheeting, v 8, n 3, p 172–190
42. Curliss DB, Carlin DM 1990 'The Effect of Jet Fuel Exposure on Advanced Aerospace Composites II: Mechanical Properties'. Final rept. May-Dec 89 Wright Research and Development Center Wright-Patterson AFB OH Defense Technical Information Center Accession Number : ADA227529
43. Curliss DB, Carlin, DM, Arnett, MS 1990 "The effect of jet fuel absorption on advanced aerospace thermoset and thermoplastic composites", National SAMPE Symposium and Exhibition (Proceedings), v 35, n pt 1 pp 332–345
44. Dao B, Hodgkin J, Krstina J, Mardel J, Tian W (2006) Accelerated Aging Versus Realistic Aging in Aerospace Composite Materials. II. Chemistry of Thermal Aging in a Structural Composite Journal of Applied Polymer Science, Vol. 102, 3221–3232
45. Dao B; Hodgkin J; Krstina J; Mardel J; Tian W 2007 'Accelerated ageing versus realistic ageing in aerospace composite materials. III. The chemistry of thermal ageing in bismaleimide based composites' Journal of Applied Polymer Science, v 105, n 4, p 2062–2072
46. Datta RN, Huntink NM, Datta S, Talma AG 2007 Rubber vulcanizates degradation and stabilization Rubber Chemistry and Technology Volume: 80 Issue: 3 JUL-AUG Pages: 436–480
47. Davies R, Pomies F Carlsson LA 1996 "Influence of Water and Accelerated Aging on the Shear Fracture Properties of Glass/Epoxy Composite" Applied Composite Materials 3: 71–87
48. Deneve B, Shanahan MER 1995 'Physical and Chemical Effects in An Epoxy-Resin Exposed To Water-Vapor' Journal Of Adhesion V. 49 Is. 3–4 pp 165–176
49. Derrien K and Gilormini P 2007 'The effect of applied stresses on the equilibrium moisture content in polymers' Scripta Materialia 56 297–299
50. Dillard DA Hinkley JA Johnson WS and StClair TL 1994 "Spiral Tunneling Cracks Induced by Environmental Stress Cracking in LaRCTM-TPI Adhesives" Journal of Adhesion Volume: 44 Issue: 1–2 Pages: 51–67
51. Dinakaran K, Kumar RS, Alagar M 2003 Preparation and characterization of bismaleimide-modified bisphenol dicyanate epoxy matrices Journal of Applied Polymer Science Volume: 90 Issue: 6 Pages: 1596–1603 Published: NOV 7
52. Dine-Hart RA, Parker DBV, Wright WW 1971 British Polymer Journal 'Oxidative degradation of a polyimide film: II. Studies using hydrazine hydrate' British Polymer Journal Volume 3 Issue 5, Pages 226–234
53. Egusa S 'Mechanism of radiation-induced degradation in mechanical properties of polymer matrix composites' Journal of Materials Science 23 (1988) 2753–2760
54. Ehlers GFL; Fisch KR; Powell WR 1969 'Thermal Degradation Of Polymers With Phenylene Units in the Chain- 1, 2, 3' Source: J Polym Sci Part A-1 Polym Chem, v 7, n 10, p 2931–2981
55. Fahmy AA and Hurt JC 1980 'Stress Dependence of Water Diffusion in Epoxy Resin', Polymer Composites, v1 no 2 pp 77–80
56. Falcone A and Dow MB 1995 'The Effects Of Aircraft Fuel And Fluids on the Strength Properties of Resin Transfer Molded (RTM) Composites' Third NASA Advanced Composites Technology Conference, Volume 1, Part 1, p 399–413 Document ID:19950022626 NASA Technical Report Server Accessed 6/25/2009
57. Fang X, Xie X-Q, Simone CD, Stevens MP, Scola DA 2000 'A Solid-State 13 C NMR Study of the Cure of 13 C-Labeled Phenylethynyl End-Capped Polyimides' Macromolecules 2000, 33, pp 1671–1681

58. Fox DJ, Sykes, Jr. GF, Herakovich CT 1987 'Space Environmental Effects on Graphite-Epoxy Compressive Properties and Epoxy Tensile Properties' NASA-TM-89297; NAS 1.15:89297; CCMS-87-11; VPI-E-87-13 177p Available From:CASI
59. G. Sala 2000 'Composite degradation due to fluid absorption' Composites: Part B 31 (2000) 357–373
60. Gardner RJ, Martin JR 1979 'Humid Aging Of Plastics - Effect of Molecular-Weight on Mechanical-Properties and Fracture Morphology of Polycarbonate' Journal of Applied Polymer Science V 24 Iss: 5 Pages: 1269–1280
61. Gates TS Grayson MA 1999 'On the use of accelerated aging methods for screening high temperature polymeric composite materials' AIAA-1999-1296 AIAA/ASME/ASCE/AHS/ ASC Structures, Structural Dynamics, and Materials Conference and Exhibit, 40th, St. Louis, MO, Apr. 12–15, 1999, Collection of Technical Papers. Vol. 2
62. Ghose S, Cano RJ Watson KA Britton SM Jensen BJ Connell JW Herring HM and Linberry QJ (2009) "High Temperature VARTM of Phenylethynyl Terminated Imides" High Performance Polymers, 21(5), 000
63. Giori C, Yamauchi T, Rajan K and Melt R 1983 Mechanisms of Degradation of Graphite Composites in a Simulated Space Environment AIAA 21st Aerospace Sciences Meeting January 10–13, 1983/Reno, Nevada AIAA-83-0590
64. Giori, C Yamauchi, T 1984 'Effects of Ultraviolet And Electron Radiations on Graphite-Reinforced Polysulfone and Epoxy-Resins' Journal of Applied Polymer Science Vl 29 Is 1 P 237–249
65. Gray RA Magato JR Vannucci R. Dillingham, G, 2006, "High-temperature adhesive development" 38th SAMPE Fall Technical Conference: Global Advances in Materials and Process Engineering, November 6, 2006 - November 9, 2006 Soc. for the Advancement of Material and Process Engineering
66. Gupta A, Cizmecioglu M, Coulter D, Liang RH, Yavrouian A, Tsay FD and Moacanin J, 1983 'The Mechanism of Cure of Tetraglycidyl Diaminodiphenyl Methane with Diaminodiphenyl Sulfone' Journal of Applied Polymer Science, Vol. 28, P 1011–1024
67. Hale WF, Farnham AG, Johnson RN, and Clendinning RA (1967) 'Poly(aryl Ethers) by Nucleophilic Aromatic Substitution. 11. Thermal Stability' Journal of Polymer Science: Part A-1 VOL. 5, 2399–2414
68. Han M-H and Nairn JA (2003) "Hygrothermal Aging of Polyimide Matrix Composite Laminates" Composites Part A, 34, 979–986
69. Hedberg FL and Arnold FE (1976) "Phenylethynyl-Pendant Polyphenylquinoxalines Curable By An Intramolecular Cycloaddition Reaction" Journal Of Polymer Science Part A-Polymer Chemistry Vol 14 Iss 11 P 2607–2619
70. Hergenrother PM Smith Jr, JG 1994 "Chemistry and properties of imide oligomers end-capped with phenylethynylphthalic anhydrides" Polymer, v 35, n 22, p 4857–4864
71. Hergenrother PM 2003 The Use, Design, Synthesis, and Properties of High Performance/ High Temperature Polymers: An Overview High Performance Polymers v15 p 3–45
72. Hergenrother; PM and Smith, Jr. JG 1996 "Imide oligomers endcapped with phenylethynyl phthalic anhydrides and polymers therefrom" U.S. Patent 5,567,800 assigned to The United States of America as represented by the Administrator of the National Aeronautics and Space Administration
73. Hinkley JA and Campbell FJ (1983) "Fracture Toughness of Irradiated Polyethersulfone" J. Mat. Sci. Lett. 2, 267–271
74. Hinkley JA Jensen, BJ 1995 'Relative thermooxidative stability of a phthalimide against a phenylethynylphthalimide terminated polyimide' High Performance Polymers, v 7, n 1, p 1–9
75. Hinkley JA, Jensen BJ 1996 'Crosslinking in phenylethynyl-terminated polyimides' High Performance Polymers, v 8, n 4, p 599–605
76. Hinkley JA, Proctor DA 2000 'Oxidative aging of cured polymer from phenylethynyl terminated imide oligomers' Journal Of Advanced Materials Volume: 32 Issue: 4 Pages: 35–41

77. Hinkley JA and Yue JJ 1995 'Oxidative Aging of Thermoplastic Polyimide Films', J. Appl. Polym. Sci. 57, 1539–1543
78. Hodgkin JH, Simon GP, Varley RJ 1998 Thermoplastic toughening of epoxy resins: a critical review Polymers For Advanced Technologies Volume: 9 Issue: 1 Pages: 3–10
79. Hooper, SJ Subramanian, R Toubia RF 1991 'Effects of moisture absorption on edge delamination. Part II. An experimental study of jet fuel absorption on graphite-epoxy' in ASTM Special Technical Publication, n 1110, p 107–125
80. Hooper, SJ, Subramanian, R 1993 'Effects of water and jet fuel absorption on mode I and mode II delamination of graphite/epoxy' in ASTM Special Technical Publication, n 1156, p 318–340, 1993
81. Hou TH, Johnston NJ, St Clair TL 1995 'IM7/LARCTM-IA polyimide composites' High Performance Polymers, Vol. 7, No. 1, 105–124
82. Jordan, K Iroh, JO 1996 'Effect of isothermal aging on the imidization of PMR-15' Polymer Engineering and Science VL 36 IS 20 P 2550–2555
83. Ju J and Morgan RJ 2004 'Characterization of Microcrack Development in BMI-Carbon Fiber Composite under Stress and Thermal Cycling' Journal of Composite Materials, Vol. 38, No. 22/p 2007–2024
84. Kaplan ML 1991 'Solvent Penetration in Cured Epoxy Networks' Polymer Engineering and Science, May 1991, Vol. 31, No. 10, 689
85. Kiefer R, Yue JJ; Hinkley JA 1995 'Kinetic mapping of oxidative weight loss in polyimide composites' Journal of Advanced Materials, v 26, n 3, p 55–59
86. Kim HW Grayson MA Nairn JA 1995 "The Effect of Hygrothermal Aging on the Microcracking Properties of Some Carbon Fiber/Polyimide Laminates" Advanced Composite Letters, Vol. 4, No. 5, 185–188
87. Kim J-K, Hu C, Woo RSC and Sham M-L 2005 'Moisture barrier characteristics of organoclay–epoxy nanocomposites' Composites Science and Technology Volume 65, Issue 5 Pages 805–813
88. Kim K-Y, Ye L 2004 'Interlaminar fracture toughness of CF/PEI composites at elevated temperatures: roles of matrix toughness and fibre/matrix adhesion' Composites: Part A 35 477–487
89. Kinloch AJ, Yuen ML and Jenkins SD 1994 Thermoplastic-toughened epoxy polymers Journal of Materials Science Volume 29, Number 14
90. Krishnaswamy, RK Geibel, JF Lewis, BJ 2003 'Influence of semicrystalline morphology on the physical aging characteristics of poly(phenylene sulfide)' Macromolecules VL 36 IS 8 p 2907–2914
91. Kumar BG, Singh RP And Nakamura T 2002 'Degradation of Carbon Fiber-reinforced Epoxy Composites by Ultraviolet Radiation and Condensation' Journal of Composite Materials, Vol. 36, No. 24 pp. 2713–2733
92. Kuroda S-I and Mita I, 1989, 'Degradation Of Aromatic Polymers – II The Crosslinking During Thermal and Thermo-Oxidative Degradation of a Polyimide' Eur. Polym. J. Vol. 25, No. 6, pp. 611–620
93. Lee CY-C, Goldfarb IJ, Helminiak, TE, Arnold FE (1983) 'Air Force Review of Acetylene Terminated Resin Technology' National SAMPE Symposium and Exhibition (Proceedings), p 699–710
94. Lee MC and Peppas NA (1993) 'Water Transport in Graphite/Epoxy Composites' Journal of Applied Polymer Science, Vol. 47, 1349–1359
95. Leung CL, Ghaffarian R, Leung KC 1997 'Thermo-oxidative stability of polyimides - I. Effect of prepolymer molecular weights' Polymer Degradation And Stability Vol 58 Iss 1–2 P 11–14
96. Li Y, Morgan RJ 2006 'Thermal Cure of Phenylethynyl-Terminated AFR-PEPA-4 Imide Oligomer and a Model Compound' Journal of Applied Polymer Science, Vol. 101, 4446–4453
97. Li, Y 2004 'Synthesis and Cure Characterization of High Temperature Polymers', PhD Dissertation, Materials Science and Engineering, Texas A & M University

98. Link PE 2007 'High Temperature Degradation of 5250-4 Polymer Resin', Master's Thesis, Air Force Institute of Technology Graduate School of Engineering and Management (AFIT/EN) published as report no. AFIT/GAE/ENY/07-J12 14 June 2007
99. Loechet, C and Rothschild R 1992 Personal communication cited in Morgan et al. 1997
100. Ma C-CM, Lee C-L and Tai N-H 1992 'Chemical Resistance of Carbon Fiber-Reinforced Poly(ether ether ketone) and Poly(phenylene sulfide) Composites' Polymer Composites, Vol. 13, No. 6 pp 435–440
101. Marais, C Villoutreix G 1998 'Analysis and modeling of the creep behavior of the thermostable PMR-15 polyimide' Journal of Applied Polymer Science VL 69 IS 10 P 1983–1991
102. Matsuda S, Hojo M, Ochiai S 1999 'Effect of water environment on mode II delamination fatigue in interlayer-toughened CFRP' JSME International Journal Series A-Solid Mechanics and Material Engineering Volume: 42 Issue: 3 Pages: 421–428
103. McConnell VP (2009) "Resins for the hot zone, part I: Polyimides" High Performance Composites July Issue http://www.compositesworld.com/articles/resins-for-the-hot-zone-part-i-polyimides.aspx accessed 9/17/2009
104. McGrail PT (1996) "Polyaromatics" Polymer International v 41 pp 103–121
105. McKenna GB, Bradley GW Dunn HK, and Statton WO 1979 'Degradation Resistance of Some Candidate Composite Biomaterials' Journal of Biomedical Materials Research, Vol. 13, 783–798
106. Meador MAB, Johnston JC, Cavano PJ 1997 'Elucidation of the cross-link structure of nadic-end-capped polyimides using NMR of C-13-labeled polymers' Macromolecules Volume 30 Issue 3 Pages 515–519
107. Milhourathammadi, A Gaudemer, F Merienne, C Gaudemer, A 1994 'Reactivity Of Fragments Present In The Polyimide Prepolymer Pmr-15 - Thermal-Isomerization And Oxidation Of Imide Derivatives Of Norborne (Ne) And Methane Dianiline (MDA)' Journal Of Polymer Science Part A-Polymer Chemistry VL 32 IS 8 P 1593–1597
108. Miriyala SK, Tucker WC, Rockett TJ, Brown R, 1992 Degradation of carbon reinforced polymer composites under galvanic coupling conditions AIAA-1992-2553 in AIAA/ASME/ASCE/AHS/ASC Structures, Structural Dynamics and Materials Conference, 33 rd, Dallas, TX, Apr 13–15, 1992, Technical Papers. Pt. 5 (A92-34522 13–39). Washington, DC, American Institute of Aeronautics and Astronautics, p 3036–3045
109. Mison, P Sillion, B 1999 'Thermosetting oligomers containing maleimides and nadimides end-groups' Advances in Polymer Science VL 140 P 137–179
110. Morgan RJ and Mones ET (1987) 'The Cure Reactions, Network Structure, and Mechanical Response of Diaminodiphenyl Sulfone-Cured Tetraglycidyl 4,4′Diaminodiphenyl Methane Epoxies' Journal of Applied Polymer Science, Vol. 33, 999–1020
111. Morgan RJ, Shin EE, Rosenberg B and Jurek A 1997 'Characterization of the cure reactions of bismaleimide composite matrices' Polymer Vol. 38 No. 3, pp 639–646
112. Morgan, RJ Shin EE Lincoln JE Zhou J Drzal LT Wilenski MS Lee A Curliss D 1998 'Durability Characterization Of Bismaleimide And Polyimide-Carbon Fiber Composites' International SAMPE Symposium and Exhibition (Proceedings), v 43, n 1, p 106–119
113. Nicholson LM, Whitley KS, Gates TS 2001 'The combined influence of molecular weight and temperature on the physical aging and creep compliance of a glassy thermoplastic polyimide' Mechanics of Time-Dependent Materials Volume: 5 Issue: 3 Pages: 199–227
114. Nogueira P, Ramirez C, Torres A, Abad MJ, Cano J, Lopez J, Lopez-Bueno I, Barral L (2001) 'Effect of Water Sorption on the Structure and Mechanical Properties of an Epoxy Resin System' Journal of Applied Polymer Science, Vol. 80, 71–80
115. O'Brien TK (1988) Fatigue Delamination Behavior of PEEK Thermoplastic Composite Laminates Journal of Reinforced Plastics and Composites, Vol. 7, No. 4, 341–359
116. O'Brien TK, Murri GB, and Salpekar SA 1987 "Interlaminar Shear Fracture Toughness And Fatigue Thresholds For Composite Materials" NASA Technical Memorandum 89 157, USAAVSCOM Technical Memorandum 87-B-9 August 1987

117. O'Brien TK; White SR 2008 'Assessment of Composite Delamination Self-Healing Via Micro-Encapsulation' American Society for Composites 23rd Annual Technical Conference; 9–11 Sep. 2008; Memphis, TN Available From CASI or from NASA Aeronautics and Space Database; accessed 23 Jul 2009 from http://www2.sti.nasa.gov
118. Odagiri N Kishi, H and Nakae T 1991 Torayca T800/3900-2 toughened epoxy prepreg system: toughening concept and mechanisms in Proceeding of the American Society for Composites 6th Technical Conference Technomic, Lancaster, PA page 43
119. Ohno S, Lee M-H, Lin KY, Ohuchi FS 2000 Thermal degradation of IM7:BMI5260 composite materials: characterization by X-ray photoelectron spectroscopy Materials Science and Engineering A293 88–94
120. Pang JWC, Bond IP 2005 'Bleeding composites - damage detection and self-repair using a biomimetic approach' Composites Part A-Applied Science and Manufacturing Volume: 36 Issue: 2 Pages: 183–188
121. Panigot, MJ Waters JF Varde U Sutter, JK. Sukenik CN 1992 'Benzonorbornadiene end caps for PMR resins' Macromolecules, v 25, n 2, p 530–534
122. Parvatareddy H Wang JZ Dillard DA Ward TC Rogalski, ME 1995 'Environmental Aging of High-Performance Polymeric Composites – Effects On Durability' Composites Science And Technology Vl 53 Is 4 pp 399–409
123. Pater RH 1991 "The 316°C and 371°C Composite properties of an Improved PMR Polyimide: LaRC RP46" Proc. 36th Int. SAMPE Symp. and Exhibition Closed Papers Publication p 78
124. Pater RH Whitley K Morgan C And Chang A 1991 'Crosslinking-Property Relationships in PMR Polyimide Composites. Part I' Polymer Composites Vol. 12, No. 2 pp 126–132
125. Pederson, CL Gillespie, JW McCullough, RL Rothschilds, RJ Stanek, SL 1995 'The Effect Of Isothermal Aging On Transverse Crack Development In Carbon-Fiber-Reinforced Cross-Ply Laminates' Polymer Composites Vl 16 Is 2 Pp 154–160
126. Perng, L.H. 1999 'Thermal cracking characteristics of PEEK under different environments by the TG/FTIR technique' Journal of Polymer Science, Part A: Polymer Chemistry, v 37, n 24, p 4582–4590
127. Phillips, R Glauser, T Manson, JAE1997 'Thermal stability of PEEK/carbon fiber in air and its influence on consolidation' Polymer Composites Vl 18 IS 4 pp 500–508
128. Popineau S, Rondeau-Mouro C, Sulpice-Gaillet C, Shanahan MER (2005) 'Free/bound water absorption in an epoxy adhesive' Polymer 46 10733–10740
129. Postlewaite J McLaren D 1985 'Development Of Design Data For Propulsion Pmr-15 Composites' Source: NASA Conference Publication 2385, p 335–338
130. Poveromo, Leonard M. 1985 'Polyimide Composites – Application Histories' in High Temperature Polymer Matrix Composites, NASA Conference Publication 2385 Proceedings of a conference held at NASA Lewis Research Center Cleveland, Ohio March 16–18. 1983
131. Pritchard, G Randles, SJ 1995 'The Interaction Of Liquids With Polyaryl Ether Ether Ketone Composites' Plastics Rubber And Composites Processing And Applications Vl 24 Is 4 P175–180
132. Putthanarat S, Tandon GP and Schoeppner GA 2008 'Influence of aging temperature, time, and environment on thermo-oxidative behavior of PMR-15: nanomechanical characterization' Journal of Materials Science Volume 43, Number 20/October, Pages 6714–6723
133. Pyun E, Sung CSP 1991 Network structure in diamine-cured tetrafunctional epoxy by UV-visible and fluorescence spectroscopy Macromolecules 24 (4), pp 855–861
134. Rivaton, A Gardette, JL 'Photodegradation of polyethersulfone and polysulfone' Polymer Degradation And Stability 1999 Vl 66 Is 3 P 385–403
135. Roberts CC Apple TM Wnek GE 2000 'Curing chemistry of phenylethynyl-terminated imide oligomers: Synthesis of C-13-labeled oligomers and solid-state NMR studies' Journal Of Polymer Science Part A-Polymer Chemistry Vl 38 Is 19 Pp 3486–3497

136. Robertson CG, Monat JE, Wilkes GL 1999 ‘Physical aging of an amorphous polyimide: Enthalpy relaxation and mechanical property changes’ Journal of Polymer Science Part B-Polymer Physics Volume 37 Issue 15 Pages 1931–1946
137. Rommel, ML Postyn, AS Dyer, TA 1993 ‘Accelerating Factors In Galvanically Induced Polyimide Degradation’ SAMPE Journal Vl 29 is 2 p 19–24
138. Ruggles-Wrenn, MB Broeckert, JL 2009 ‘Effects of Prior Aging at 288 degrees C in Air and in Argon Environments on Creep Response of PMR-15 Neat Resin’ J Appl Polym Sci VL 111 IS 1 P 228–236
139. Rupnowski, P Gentz, M Sutter JK Kumosa, M 2004 ‘Mechanical response of a woven graphite/polyimide composite to in-plane shear dominated loads at room and elevated temperatures Acta Materialia VL 52 IS 19 P 5603–5613
140. Russell AJ and Street KN, 1985 ‘Moisture and temperature effects on the mixed-mode delamination fracture of unidirectional graphite/epoxy’ in WS Johnson, ed. Delamination and Debonding of Materials, ASTM STP 876 American Society for Testing and Materials, Philadelphia PA pp 349–370
141. Sands JM, Fink BK, McKnight SH, Newton CH, Gillespie Jr. JW and Palmese GR 2001 “Environmental issues for polymer matrix composites and structural adhesives” Clean Products and Processes Volume 2, Number 4, pages 228–235
142. Sasuga T; Hagiwara M 1987 ‘Radiation Deterioration Of Several Aromatic Polymers Under Oxidative Conditions’ Polymer, v 28, n 11, p 1915–1921
143. Scola, DA Vontell, JH 1991 ‘Mechanical-Properties And Mechanism Of The Degradation Process Of 316-Degrees-C Isothermally Aged Graphite Fiber Pmr-15 Composites’ Polymer Engineering And Science VL 31 IS 1 P 6–13
144. Selzer, R Friedrich, K 1995 ‘Influence Of Water-Uptake On Interlaminar Fracture Properties Of Carbon-Fiber-Reinforced Polymer Composites’ Journal Of Materials Science Vl 30 Is 2 P 334–338
145. Serafini TT Vannucci RD Alston WB 1976 ‘Second generation PMR polyimides’ Natl. Symp. and Exhibition of the Soc. for the Adv. of Mater. and Process Eng., 21st Los Angeles, CA, United States
146. Serafini TT, Delvigs P, Lightsey GR 1972 ‘Thermally Stable Polyimides From Solutions Of Monomeric Reactants’ Journal Of Applied Polymer Science Volume: 16 Issue: 4 Pages: 905–
147. Shimokawa T, Hamaguchi Y, Katoh H 1999 ‘Effect of moisture absorption on hot/wet compressive strength of T800H/PMR-15 carbon/polyimide’ Journal Of Composite Materials Volume: 33 Issue: 18 Pages: 1685–1698
148. Shin EE Morgan RJ, Zhou J “Hydrolytic Degradation Mechanisms and Kinetics of Polyimides for Advanced Composites” SAMPE 2000 - Long Beach, CA May 21–25, 2000
149. Shin, EE Morgan, RJ Zhou, JM Lincoln, J Jurek, R Curliss, DB 2000 Hygrothermal durability and thermal aging behavior prediction of high-temperature polymer-matrix composites and their resins Journal Of Thermoplastic Composite Materials VL 13 IS 1 pp 40–57
150. Siddiqui NA, Woo RSC, Kim JK, Leung CCY, Munir A 2007 ‘Mode I interlaminar fracture behavior and mechanical properties of CFRPs with nanoclay-filled epoxy matrix’ Composites Part A-Applied Science And Manufacturing Volume: 38 Issue: 2 Pages: 449–460
151. Soles CL Chang FT Bolan BA Hristov HA Gidley DW Yee AF 1998 Contributions of the Nanovoid Structure to the Moisture Absorption Properties of Epoxy Resins Journal of Polymer Science: Part B: Polymer Physics, Vol. 36, 3035–3048
152. Stenzenberger HD 1990 “Chemistry and properties of addition polyimides” in Wilson, D, Stenzenberger HD and Hergenrother PM, Polyimides, Blackie, Glasgow and London, Published in USA by Chapman and Hall, NY
153. Stenzenberger HD, Herzog M, Koenig P and Roemer W 1989 ‘Bismaleimide resins: past, present, and future’ 34th international SAMPE symposium May 8–11, pp 1877–1888
154. Stenzenberger HD; Koenig P; Herzog M; Roemer W. 1987 Bismaleimide Resins Improved Novel Toughening Modifiers For BMI Resins. Into SAMPE Symp Exhib, v 32, p 44–58

155. Street, SW 1980 'V-378A, A New Modified Bis Maleimide Matrix Resin For High Modulus Graphite' National SAMPE Symposium and Exhibition (Proceedings), p 366–375
156. Stutz H 'Lifetime Assessment of Epoxies by the Kinetics of Thermal Degradation' Journal of Applied Polymer Science, Vol. 91, 1881–1886 (2004)
157. Suh D-W, Ku M-K, Nam J-D, Kim B-S, Yoon, S-C 2001 "Equilibrium Water Uptake of Epoxy/Carbon Fiber Composites in Hygrothermal Environmental Conditions" Journal of Composite Materials, Vol. 35, No. 03 pp 264–278
158. Takeda N Tohdoh M And Takahashi K (1995) 'Interlaminar fracture toughness degradation of radiation-damaged GFRP and CFRP composites' Adv Composite Mater vol4, no 4 pp 343–354
159. Takekoshi T and Terry JM, 1994 'High-Temperature Thermoset Polyimides Containing Disubstituted Acetylene End-Groups' Polymer Volume: 35 Issue: 22 Pages: 4874–4880
160. Taylor, SR Wall, FD Cahen, GL 1996 'The detection and analysis of electrochemical damage in bismaleimide/graphite fiber composites Journal Of The Electrochemical Society VL 143 IS 2 P 449–458
161. Tenney DR and Unnam J 1978 'Analytical Prediction of Moisture Absorption in Composites' J. Aircraft Vol.15, no. 3
162. Truong VT and Ennis BC 1991 'Effect Of Physical Aging On The Fracture-Behavior Of Cross-Linked Epoxies' Polymer Engineering And Science VL 31 IS 8 pp 548–557
163. Uematsu, Y; Kitamura, T; Ohtani, R. 'Creep-fatigue interaction in delamination crack propagation of advanced CFRPs at high temperatures' in ASTM Special Technical Publication, v 1302, p 110–132 Proceedings of the 1995 Symposium on High Temperature and Environmental Effects on Polymeric Composites, November 13, 1995
164. Vannucci, RD. 1987 'PMR Polyimide Compositions For Improved Performance At 371 Degree C' SAMPE quarterly, v 19, n 1, p 31–36
165. Vyhnal RF 1993 'The Effects Of Long-Duration Space Exposure On The Mechanical Properties Of Some Carbon-Reinforced Resin Matrix Composites p 941–955 in LDEF: 69 Months in Space. Part 3: Second Post-Retrieval Symposium Publication Date: Apr 1, 1993 NASA. Langley Research Center
166. Wang Q Springer GS 1989 'Moisture absorption and fracture toughness of PEEK polymer and graphite fiber reinforced PEEK' Journal of Composite Materials, v 23, n 5, p 434–447
167. Weitsman Y.J.*, Guo Ya-Jun 2002 'A correlation between fluid-induced damage and anomalous fluid sorption in polymeric composites' Composites Science and Technology 62 889–908
168. Weitsman YJ, Y-J 2002 "A correlation between fluid-induced damage and anomalous fluid sorption in polymeric composites" Composites Science and Technology 62 889–908
169. Whitley KS and Collins TJ 2006 'Mechanical Properties of T650-35/AFR-PE-4 at Elevated Temperatures for Lightweight Aeroshell Designs' AIAA 2006–2202 47th AIAA/ASME/ASCE/AHS/ASC Structures, Structural Dynamics, and Materials Conference
170. Whitley KS and Gates TS 2002 'Thermal/Mechanical Response And Damage Growth In Polymeric Composites At Cryogenic Temperatures' AIAA-2002-1416 43 rd AIAA/ASME/ASCE/AHS/ASC Structures, Structural Dynamics, and Materials Conference
171. Wolf, CJ Fu, H 1996 'Stress-enhanced transport of toluene in poly aryl ether ether ketone (Peek)' Journal Of Polymer Science Part B-Polymer Physics Vl 34 IS 1 P 75–82
172. Wolfrum J, Eibl S, Lietch L 2009 Rapid evaluation of long-term thermal degradation of carbon fibre epoxy composites Composites Science and Technology 69 523–530
173. Wong AC 1981 'Nuclear Magnetic-Resonance Study Of Norbornene End-Capped Polyimides 2 – Solution And Solid-State C-13 Study Of 2NE-MDA And PMR Polymerization' Macromolecules 14 832–836
174. Woo EM 1994 'Moisture Temperature Equivalency In Creep Analysis Of A Heterogeneous-Matrix Carbon-Fiber Epoxy Composite' Composites Volume: 25 Issue: 6 Pages: 425–430

175. Woo, EM Chen, JS Carter, CS 1993 'Mechanisms Of Degradation Of Polymer Composites By Galvanic Reactions Between Metals And Carbon-Fiber' Polymer Composites Vl 14 is 5 p 395–401
176. Xiao GZ, Shanahan MER 1998 'Irreversible Effects of Hygrothermal Aging on DGEBA/DDA Epoxy Resin' Journal of Applied Polymer Science, Vol. 69, 363–369
177. Yang, ACM. and Brown, HR 1988 'Solvent-Induced Local Deformation Zones In Polyimide Films Adhered To A Rigid Substrate' Journal of Materials Science, v 23, n 1, p 65–71, Jan 1988
178. Zhang L and Piggott MR 2000 'Water Absorption and Fiber-Matrix Interface Durability in Carbon-PEEK' Journal of Thermoplastic Composite Materials; v13 n2 p 162–172
179. Zheng Q and Morgan RJ 1993 Synergistic Thermal-Moisture Damage Mechanisms of Epoxies and Their Carbon Fiber Composites Journal of Composite Materials v. 27 no. 15: 1465–1478
180. Zhou J and Lucas JP (1999) 'Hygrothermal effects of epoxy resin. Part I: the nature of water in epoxy' Polymer 40 5505–5512
181. Zhou J, Lucas JP (1999) Hygrothermal effects of epoxy resin. Part II: variations of glass transition temperature Polymer 40 5513–5522
182. Zhou YX, Pervin F, Jeelani S, Mallick PK 2008 'Improvement in mechanical properties of carbon fabric-epoxy composite using carbon nanofibers' Journal Of Materials Processing Technology Volume: 198 Issue: 1–3 Pages: 445–453
183. Zinck P and Gerard J-F 2008 'Polyepoxide-water interactions: Influence of the chemical structure of the network' Polymer Degradation and Stability 93, 1231–1237

Chapter 2
Additives, Nanocomposites, and Barrier Coatings

Alexander B. Morgan

Abstract Material use is often governed by the properties it brings to an application, but in some cases those desirable properties are rapidly degraded by their intended use environment. Heat, abrasion caused by part-on-part wear, and particulate impact can damage material properties, especially polymer–fiber-reinforced composites. To ensure that the benefits of polymer composites can be utilized in these extreme environments, protection is needed. The simplest form of protection is the use of additives to the polymer matrix, such as antioxidants, thermal stabilizers, and flame retardants. Of newer interest is the use of nanocomposite technology, which provides enhanced thermal and mechanical durability, which sometimes brings multifunctional performance to the composite. Barrier coatings represent an engineering solution to protect the composite part, but newer research focuses on incorporation of the barrier coating during composite fabrication so that the protection is engineered to be a covalently bound part of the polymer rather than a post-fabrication add-on coating produced via painting or adhesive bonding. This chapter provides a survey of the broad range of protection solutions available for composites, with an emphasis on approaches that yield thermal and/or abrasion protection in polymer composites.

2.1 Introduction

Synthetic polymers as a class of materials are relatively new (100+ years old) in comparison to metals, ceramics, and natural materials (1,000+ years old). Polymers in use with fiber reinforcement are even newer and, so not surprisingly, as these very new materials with attractive properties (ease of processing and forming, light

A.B. Morgan (✉)
Advanced Polymers Group, Multiscale Composites and Polymers Division,
University of Dayton Research Institute, Dayton, OH 45469-0160, USA
e-mail: Alexander.Morgan@udri.udayton.edu

K.V. Pochiraju et al. (eds.), *Long-Term Durability of Polymeric Matrix Composites*,
DOI 10.1007/978-1-4419-9308-3_2, © Springer Science+Business Media, LLC 2012

weight, corrosion resistance) get more used, they eventually come into environments where they replace the older metals, ceramics, and natural materials. However, that replacement may not always be successful; sometimes the polymer composite may solve one problem, but have some other failing which prevents it from being completely successful in the final application. Specifically, a polymer composite may be superior in corrosion resistance, but not at all suitable to replace a metal in a very hot and wet environment. So, the materials scientist is left with a dilemma: Either go back to the original material and try to engineer a solution, or try to protect the polymer composite against the damage it is suffering in its end-use application. This is, unfortunately, easier said than done, due to growing trends pushing the further use of polymer composites, as well as a lack of solutions to protect the polymer composite against the many types of damage it may encounter. The lack of solutions is often due to the complexity of the material science problem in the first place, as both scientific issues and regulatory issues may predetermine what the material scientist can use. To better explain this point, some background on trends pushing for polymer composite use is needed, along with a general discussion of polymer composite protection design philosophy. Once these are defined, it becomes possible to understand and develop protection for these composites.

2.1.1 Polymer Composite Trends

Polymer composites as a class of materials continue to grow in use for a variety of reasons. Usually it is one particular property that drives their use, but in some cases it is multiple benefits that lead to their increasing use. In building construction, polymer composites have increased in use due to benefits in environmental resistance (no wood rot or termite damage), while in vehicles, composite growth is driven by light weight (fuel savings) and a resistance to environmental damage (no rust or corrosion). These factors leading to composite selection are usually driven by economics in that the composite is replacing something due to lower cost, or lower total maintenance costs, but more often in the case of vehicles and aerospace, the drive is almost strictly economics from fuel savings provided by the lightweight composites. At the time of this writing (June 2009), oil prices had dropped from their (2008) $100 per barrel price to $71–73 per barrel, but even at these lower prices, airlines, military air forces, and even mass ground transport (bus, truck) are still looking at unacceptable fuel efficiency. Plus, the cost of oil continues to be erratic, making it difficult to plan economically and budget for fuel costs. Thus, the drive to incorporate more polymer composites for fuel savings will continue, but the replacement of metal and ceramics is not so straightforward.

Metals and ceramics do have key advantages, which is why they have been in use for so long. Specifically, they tend to be very durable in most extreme environments (heat resistance and mechanical durability for metal; heat and corrosion for ceramics) where they are used as engine components, heat shields on aircraft, or external hulls exposed to the environment (UV, rain erosion, sand erosion).

But since these components are major parts of vehicles today, this is where all the weight savings are located, and so if you are to replace a metal or ceramic with a polymer composite, the composite must be suited to resist the environmental damage that metals and ceramics handle with ease. As will be explained later in this chapter (and has been explained in other chapters in this book), this is not an easy task due to the chemical structure of the polymers themselves. Carbon-based polymers are prone to oxidation and heat damage, tend not to do well against erosion due to their relative softness, and therefore require protection if they are to be used in extreme environments. So one must consider protection schemes in material and component design for polymer composites if they are to be used more to replace metals and ceramics, or find some way to incorporate the material properties of metals and ceramics directly into the composite structure. First we will discuss protection schemes, but later it will be clear that for practical protection to occur, we must look at the structure–property relationships of metals and ceramics and find a way to incorporate them into polymer + metal or polymer + ceramic hybrid composites that bring practical protection while not compromising the reasons for selecting the polymer composite in the first place. Before those protection schemes for polymer composites can be implemented, though, we first need to understand how polymer composites degrade.

2.2 Degradation of Polymer Composites: Reaction to Heat, Fire, Electrical, and Wear

2.2.1 Degradation vs. Decomposition

The terms "degradation" and "decomposition" are sometimes used interchangeably when discussing polymeric composites, but they should not be used this way, as they have distinctly separate mechanisms. While they are related in that decomposition can lead to degradation of properties, or degradation can precede decomposition, the two are quite different, and how one protects against them may involve different schemes.

Degradation is a loss of properties caused by a physical change in the polymer. The properties and physical changes that occur can be quite broad in behavior, so some specific examples are needed to describe this further. For a polymer exposed to heat which causes the polymer to melt, flow, or soften (but not break chemical bonds), the physical, mechanical, and electrical properties of that material may degrade [1]. Likewise, for a polymer exposed to great amounts of mechanical stress, the polymer may begin to buckle and crack, leading to a degradation in mechanical integrity.

Decomposition occurs when the chemical structure of the polymer itself is changed by a particular stimulus. Going back to heat damage, decomposition would be caused by enough heat being put into the polymer such that the polymer bonds break and change, either lowering polymer molecular weight, which leads to irreversible property degradation, or worse, that leads to small molecule pyrolysis and polymer ignition.

To better explain these two terms and to understand how to protect a polymer composite against degradation and decomposition, three typical types of extreme environment damage are discussed below. Ultimately though, damage is caused by energy from external stimuli either causing a physical change in the matrix polymer, or the energy exciting the bonds of the polymer such that the bonds break, which results in chemically induced physical damage to the polymer composite. While the exact stimuli that induce this type of bond cleavage and loss of properties may differ, ultimately it can be said that any type of energy which can break chemical bonds will lead to polymer degradation and decomposition in a composite.

2.2.2 *Heat and Fire Damage*

Heat and fire damage to polymeric materials is probably the most well-understood of the polymer degradation/decomposition mechanisms, although there is still plenty that is not fully understood or elucidated yet, especially for specific polymer systems and complex systems such as polymer composites. Heat damage is a typical precursor to fire damage in that heat causes the polymer to melt and soften first, which causes degradation of mechanical properties. The heat source causing the damage to a composite may be a hot object (engine, radiant energy source), or it may be a fire which causes heat damage through convection and radiation, or through those means and direct thermo-oxidative decomposition of the polymer composite if it is in direct contact with the open flame.

In regards to heat damage, it is quite common for composites to begin to fail structurally once the matrix resin temperature goes above the glass transition temperature (T_g) of the polymer. T_g is the temperature at which the polymer chains slip past each other from a solid phase into a rubbery or semi-liquid phase, and the polymer can greatly lose stiffness and modulus above this temperature. T_g is often well below the decomposition temperature of the polymer, which is the temperature at which enough heat energy has entered into the polymer to cause bonds to begin to break and the polymer to fragment. If the fragments are of a low enough molecular weight, they can immediately begin to pyrolyze, which leads to rapid loss of mechanical properties as the matrix resin vaporizes; more often, if enough mass has been pyrolyzed, this decomposition of the polymer serves as the source for polymer ignition which leads to fire propagation. However, some polymers will cross-link and form a carbonaceous residue (char) rather than pyrolyzing small fragments; these materials will still lose mechanical integrity, but will not easily ignite as the polymer decomposes. The mechanisms and pathways from this type of heat damage are complex and often polymer-specific (see Chap. 1 for some details), but other references provide some general reaction schemes to consider [2–6].

The previous paragraph is a rather simplistic explanation of the phenomenon of heat damage on a polymer composite, but the reality is more complex. Depending upon how fast a polymer goes from solid to liquid to gas (or solid to char), the mechanical degradation of the composite can rapidly lead to mechanical failure, which can exacerbate the heat and fire damage to the composite. Specifically, fiber reinforcement can fail and delaminate, which in turn exposes more surface area to additional heat/fire damage, which further accelerates the rate of polymer decomposition and burn-through. In drastic cases, the burning polymer can lead to rapid structural failure so that fire spreads rapidly, yielding significant losses. One notable example is the Norwegian minesweeper Orkla, which was an all-composite vessel that caught fire and rapidly sank after the fire spread through the ship and parts of the composite structure began to fail [7]. Some polymers degrade in such a way that they drip and form pools of molten flaming materials which cause rapid growths in heat release. Chief among these polymers are polyolefins and polyurethanes; the latter is well-known to cause pool-fire-type conditions that can rapidly release large amounts of heat that in turn can cause other nearby objects to ignite, leading to flashover fire conditions [8, 9]. Figures 2.1–2.3 attempt to better illustrate some of these phenomena relating to polymer decomposition caused by heat damage and how different types of systems affect flammability or the ability to propagate further heat/fire damage.

In Fig. 2.1, the typical burning behavior for a polymer is shown. As the material is heated, it begins to go from solid to liquid and can flow. As further heat is applied the material will decompose (thermal decomposition) (Fig. 2.1), and then materials will pyrolyze away from the surface, thus leaving the condensed phase (solid polymer/fuel) and entering into the gas phase where they can mix with oxygen and be combusted. Once ignition occurs, all oxygen is consumed at the flame front and pyrolysis products coming off from the condensed phase are continually carried to the flame front through convection and other fire-induced thermal flows. This process will continue until all of the polymer fuel that can be pyrolyzed is consumed, or the material is extinguished.

In Fig. 2.2, an example of how cross-linking and char formation to mitigate flame propagation for a polymer is shown. As the polymer is exposed to heat, it will begin to decompose, but instead of forming smaller fragments and flowing/melting under heat exposure, it will instead solidify further. The now highly cross-linked material can then turn to char which slows down the rate of fuel release/pyrolysis products into the gas phase for combustion. This phenomenon creates a low heat release material since more and more of the potential carbon fuel is bound up in a form that cannot be pyrolyzed further. In effect, char formation keeps the polymer in the condensed phase rather than allowing it to enter into the vapor phase (Fig. 2.2). The formation of this char can greatly lower heat release, as shown in Fig. 2.3.

In summary, heat damage to a polymer composite can at a minimum lead to a structural failure, but at its worst can lead to catastrophic fire loss, especially if the polymer composite is part of a major structure in a mass transport application or part of a building structure. Understanding all the possible fire-risk scenarios is beyond the scope of this chapter, but heat damage to a polymer composite is

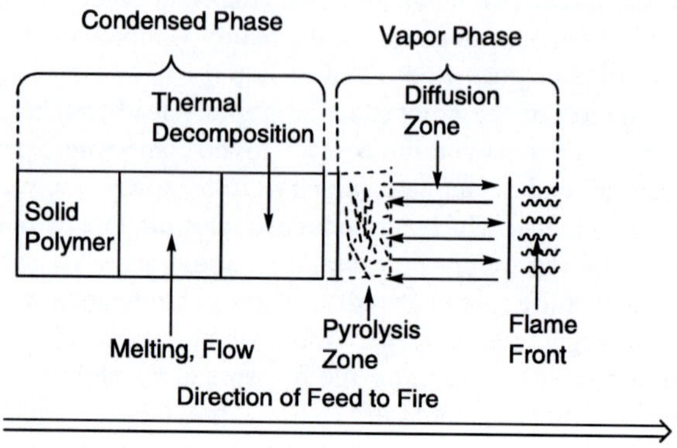

Fig. 2.1 Physical heat-induced decomposition behavior for non-charring polymers

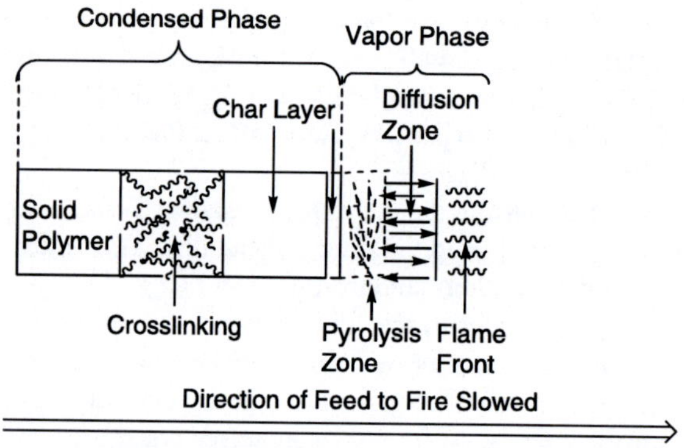

Fig. 2.2 Physical heat-induced decomposition behavior for charring polymers

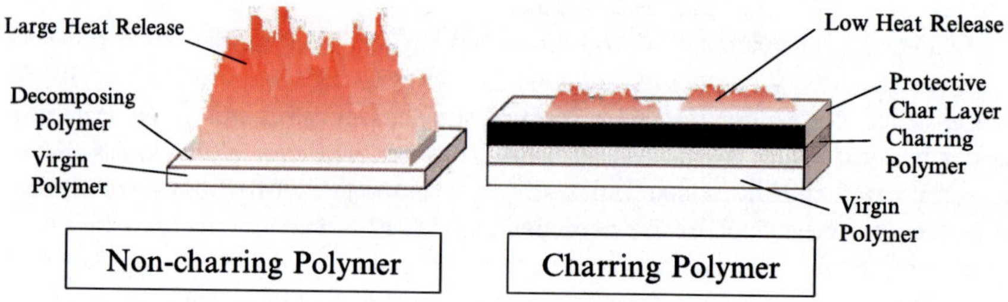

Fig. 2.3 Differences in heat release caused by non-charring and charring polymers [10]

something that the material scientist must pay attention to, as heat damage is probably the most likely type of damage that a polymer composite can encounter outside of its original design parameters. Knowing that the extent of heat damage is related to the composite matrix polymer chemical structure, we can focus on protecting that polymer from heat damage. Methods of heat protection will be discussed in detail later in this chapter, but it should be made clear that if one can protect the polymer from getting too hot in the first place, all possible risks of structural failure and ignition are completely eliminated.

2.2.3 Electrical Damage

Electrical damage to composites primarily occurs when electrical arcing such as might be caused by circuit failure or lightning strike hits the composite and results in decomposition and degradation of the polymer. Electrical damage does not typically lead to just degradation of properties alone since most polymer composites are typically insulators, although if enough electrical current is dumped into a composite, it may heat to the point that it begins to soften. While the mechanism of damage is very similar to that of heat-induced damage, the timescale during which damage occurs is typically much shorter. The electrical arc rapidly decomposes the polymer by putting energy into the polymer bonds which then break via homolytic bond cleavage (single electron reactions), and then these unstable radicals react with other parts of the polymer to form char, or rapidly undergo depolymerization which results in localized matrix polymer vaporization. In some cases, the local electrical damage leads to rapid heating of the composite, which can lead to both electric and heat-induced damage to the composite. Worst-case scenarios will lead ignition of the composite, and this now-burning polymer will cause damage to nearby objects.

What makes electrical damage different from heat damage is that typically the electrical damage is localized, while most heat damage is spread over the entire composite or a large area of radiative/convective heat exposure. Specifically, damage occurs at the point of electrical impact. One example of electrical damage is lightning strike. Lightning strikes are very fast and very intense, and continue to be an area of concern for composite protection, especially with the increased use of composites for aircraft wings/fuselages and wind turbine blades. Damage from electrical failures (such as shorting electrical components in electronic enclosures) is much less intense, but is also localized and occurs very quickly. Since the damage occurs so quickly and will result in polymer decomposition, the best solution is similar to that of heat protection – make a change to the material so that the electrical damage never occurs. In this case, the best approach is to make the composite electrically conductive, or set up conductive pathways in the composite so the electrical charge goes through the composite, rather than being absorbed and causing damage.

2.2.4 Wear and Abrasion Damage

Mechanical wear or abrasion on a polymer composite typically causes degradation, not decomposition. However, if the motion causing the abrasion is fast enough and lasts long enough, the polymer can be heated to the point that decomposition will occur. Indeed, sometimes the friction can be intense enough that mechanochemistry can occur which will result in chemical changes at the surface of the abrading materials [11]. Still, wear and abrasion typically cause a physical change to the composite, and not a chemical one. With wear and abrasion phenomena, either the polymer is worn away, leaving bare fiber reinforcement which loses mechanical integrity, or the fiber is worn away as well leading to structural weakness in the entire part. Like electrical damage, damage from abrasion and wear is localized; it occurs at the spot where the composite is in contact with another object. So with degradation being the main concern for this type of damage, the best ways to prevent damage are to prevent wear through engineering changes, or to modify the properties of the composite itself such that it can tolerate wear damage or has a harder surface to resist this type of damage. However, modification of the composite properties requires an understanding of the abrasion/wear on the composite surface, which can be complex, depending upon polymer structure and the media that is abrading/wearing out the composite surface. Some polymers are soft and rip away from abrasion while others are brittle and slowly chip away. With rain erosion, the mechanism of degradation begins with high-velocity water impacts first causing crack sites that the water seeps into and then parts of the composite spall (explosively pop off) as the water expands in the composite although the force from the water impact can lead to additional composite delamination [12]. So some understanding of what exactly is causing the damage is required before designing the protective system, but fundamentally this type of damage is due to physical changes at the macroscale and microscale for the polymer composite, not at the molecular scale.

2.2.5 *Other Damage Types*

For polymeric materials, degradation and decomposition can occur through other means besides the three discussed above. Ultraviolet (UV) radiation can result in severe polymer decomposition either by chemical bond cleavage or by creating free-radical sites along the polymer backbone. These sites lead to chemical changes in the polymer when this radical reacts with the rest of the polymer or with oxygen. These decompositions can in turn lead to degradation of properties for the composite. Oxidation of a polymer is discussed in other parts of this book (Chaps. 1, 5, and 13), and while this type of decomposition is often accelerated by heat, all polymeric materials suffer oxidation damage over time, even at room temperature or lower, although the kinetics of such oxidation are greatly reduced at lower temperatures. Such oxidation leads to significant chemical changes in the polymer structure, which in turn degrades the polymer properties in ways both minor (discoloration) and major (embrittlement and mechanical failure). UV and non-heat-induced oxidation damage is certainly something that the material science engineer needs to consider, but it is not the emphasis of this chapter. Additional review articles on this topic can be obtained elsewhere [13].

2.3 Active and Passive Protection Schemes

So now that the various types of damage have been described, we can begin to discuss how a material scientist should provide protection against these types of damage. After considering the type of damage that will occur and whether one is protecting against degradation or decomposition, the scientist must then choose from one of two general protection schemes: active or passive.

2.3.1 Active Protection

This type of protection scheme primarily uses engineering solutions or active labor (regular inspection by maintenance personnel) either to prevent damage to a material or to monitor for damage and then replace/repair the damaged part as it occurs. It is typically used in a reactive mode when a polymer composite is put into an application and damage is expected. For engineering solutions with active protection, part of the apparatus in which the composite is located is used to protect the composite against damage. This can be as simple as a heat shield/plate which protects the composite against thermal and abrasion damage, or complex venting and moving parts which either keep the composite cool or move erosion surfaces away from the composite. Even with this approach, there is frequently a need to monitor the composite through nondestructive evaluation (NDE) testing to make sure that no significant damage has occurred. The constant NDE work can add a lot of cost to the part in terms of long-term maintenance and mapping out the limits of use on the composite before it has to be replaced. Still, despite the higher costs associated with an active protection approach, it can be very easy to implement during component design, or to monitor and mitigate damage after the composite part is in use. However, this approach, while easy to implement, can easily be defeated. Should an unforeseen scenario occur, damage to the composite may be caused when the damage stimuli (UV, heat, abrasion, etc.) bypass the engineering protection.

2.3.2 Passive Protection

This approach is similar to active protection in that engineering solutions can be used, but instead this approach utilizes specific material design and chemical modification to resist damage which might occur, rather than actively testing and trying to prevent damage with through constant composite repair and replacement. This is the approach used for most polymeric materials, especially for oxidation, UV-induced, and flammability-induced damage. This is typically achieved through the use of antioxidants, UV stabilizers, or flame-retardant additives, but other forms of passive protection can be employed. As mentioned in the description of active protection, engineering solutions can sometimes be considered passive protection because the heat or abrasion shield provides protection for the composite, but again, engineering solutions can be defeated. For passive protection, the design philosophy is to set a particular damage level for the composite and then test (often destructively) its ability to resist damage up to the set level. If the test specimen passes, then it is assumed that the composite part will do fine in the application that correlates to that test, and no additional monitoring of the composite is needed. However, NDE is still sometimes needed to see whether damage has occurred, or to set a lifetime for the part, as even passive protection schemes will wear out. What makes passive protection so attractive for composite durability enhancement is that the protection is always available and can be tailored to provide durability to a polymer that normally could not be used for a particular application. Sometimes this approach is quite cost-effective, but admittedly, it can result in the materials scientist playing a "balance of properties" game for the material as the researcher makes a modification which enhances one property at the expense of another.

For polymer composites today, both active and passive protection schemes are used, but this chapter focuses on the fundamentals and philosophies of design for the passive approach. Active protection schemes are application-specific and are driven by engineering designs, which are too detailed to try to cover in this chapter. Passive protection, though, has some underlying universal fundamentals that can be used to develop a damage-resistant composite. For passive protection there are three universal approaches: additives, polymer nanocomposite technology, and integrated barrier coatings. These three approaches will be discussed in turn in the next three sections.

2.4 Additive Approaches

The number of additive solutions for composite protection is so extensive that it is impossible to cover all of them in this chapter, especially since some of the chemistries for protection are polymer-specific. Where appropriate, some specific chemistries will be described to illustrate how the additive provides protection in general to the polymer, but for more detailed discussions on these additives, the reader is encouraged to read the other more comprehensive books on the subject [14–16].

For polymer composite applications, choosing the right additive can be a daunting decision. Most types of additives in use today have been optimized for a particular polymer chemistry and end-use performance, so that what works for one polymer may not work at all for another. Luckily, most polymer manufacturers have technical service researchers who deliberately formulate their polymeric materials to have the optimal balance of protective additives in it for certain applications. So if a material scientist was interested in purchasing a polyolefin for outdoor applications, he or she can contact the polymer company and just buy a polyolefin already formulated for that end-use application: it already has the necessary antioxidants, UV stabilizers, and other additives needed for passive protection. This is particularly true for thermoplastic materials: one cannot buy a pure polymer, but must specify what the polymer needs to do and then buy the thermoplastic polymer. For thermoset resins, though, preformulated systems with additives included are not as widespread, and they are almost nonexistent for high-temperature thermosets such as polyimides. The primary reason for this is economics, but also, no one has really developed (high-temperature) antioxidants or UV stabilizers for high-temperature aerospace polymers.

So in the end, material scientists manufacturing fiber-reinforced composite can save themselves a lot of time by just selecting the right additive-filled polymer. To select the right polymer, though, the material scientist will need to know the end-user requirements in detail and be willing to interact with the vendor selling the preformulated material. Should no suitable system be available to the material scientist, then the scientist can begin to consider formulating their own additive package to protect the composite, based upon the damage it needs to resist and the measurement metrics of damage resistance needed to commercialize/implement the composite successfully. For fiber-reinforced thermoplastics and some commodity thermoset resins, this approach will work very well, but for high-temperature thermoset resins and some high-melt-temperature thermoplastic polymers, the scientist will have to consider one of the other mechanisms of protection such as polymer nanocomposite technology or barrier coatings described later in this chapter.

2.4.1 UV Stabilizers and Antioxidants

Damage to a polymer initiated by UV or oxidation in general occurs through free-radical reactions. This begins with initiation reactions, followed by propagation reactions in which the radicals rearrange and form more stable radicals which in turn propagate additional initiation reactions. The process ends when the radicals terminate into a more stable product, or are stopped from further propagation by either the UV inhibitor or antioxidant. Some generic reaction schemes to illustrate this are shown in Fig. 2.4. In this figure the three main free-radical reaction types (initiation, propagation, termination) are shown, with two example chemical reactions representative for that reaction type given in the figure. Metals can also catalyze oxidation damage in a polymer, depending upon the polymer chemical structure and the propensity of the metal to undergo one-electron transfer reactions. Again, all of these reactions will be governed by the actual chemical structure and physical form (crystalline, amorphous) of the polymer, but the general scheme of reaction holds regardless of polymer chemistry and form.

There are three ways in which antioxidants and UV stabilizers work: by performing radical scavenging, decomposing peroxides that may form, or deactivating the metal that may be catalyzing the damage. Due to the variety of specific chemical reactions that may occur, antioxidant and UV stabilizer chemistry is tailored to specific polymer systems to address that system's decomposition mechanism and to be compatible with the polymer. Additives which are not fully compatible with

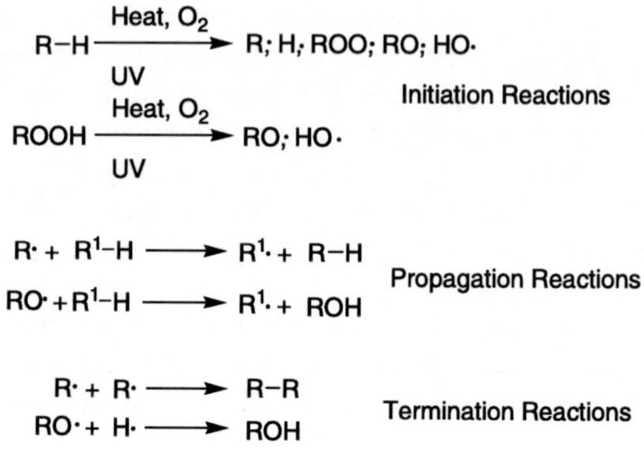

Fig. 2.4 Initiation, propagation, and termination reactions in UV/oxidation damage

the polymer matrix may work very well as an antioxidant/UV stabilizer, but will migrate out of the polymer over time (known as "blooming"), thus making the polymer vulnerable to damage.

Additives that are typically effective as antioxidants and UV stabilizers include hindered phenols, aromatic amines, hindered amines (only those which have a tendency to form N-oxides), hydroxylamines, benzofuranones, divalent sulfur compounds, phosphorus (III) compounds (phosphines and phosphites, with phosphites most common), and multi-dentate metal ligands (such as ethylenediaminetetracetic acid – EDTA). Just about every class of polymer in use today can be matched with a commercially available additive that has been optimized for that polymer. The exception to this would be high-temperature polymers (those processed above 300°C); no current commercial antioxidants or UV stabilizers are thermally stable enough to be processed at that temperature.

2.4.2 *Flame-Retardant Additives*

The chemistry of flame-retardant additives is highly varied and is optimized not only for specific polymer chemistries, but also to address flammability effects (flame spread, dripping, smoke release, etc.) that are required to be addressed in a specific regulatory fire test. The science of flame retardants as a class of additives is almost solely driven by fire-safety regulations, which address specific fire-risk scenarios. So while an antioxidant or UV inhibitor may be optimized for a particular polymer, and once in place can be used in all applications of that polymer, the same cannot be said of a flame-retardant additive. What helps one polymer pass one particular fire-safety test may not work for that same polymer (or a different polymer) in a different fire-safety test. Therefore, to select the proper flame-retardant additive, one must consider the polymer and the end-use application at the same time.

Flame-retardant chemistry includes such classes of compounds as halogenated organics, char formers, cross-linking compounds, mineral fillers, intumescent packages, phosphorus compounds, nitrogen-based compounds, and even certain metal and boron compounds. This chapter cannot cover all the flame-retardant chemistries available, but they are extensively covered elsewhere in review books and articles [17–29]. However, most flame-retardant additives achieve fire protection for a polymer through one or more of three broad mechanisms: vapor phase radical inhibition, endothermic cooling, or condensed phase char formation. Extensive details on these mechanisms can be found in the review papers cited for flame retardants in this chapter, but the general mechanisms will be described here. There will also be some discussion on vapor and condensed phase flame retardancy here, which can be seen in Figs. 2.1 and 2.2.

2.4.2.1 Vapor Phase Radical Inhibition

Flame retardants that act as vapor phase radical inhibitors begin by pyrolyzing with the polymer fuel and fragmenting into stable radicals, which help prevent free-radical propagation reactions in a flame front. They specifically form stable intermediates which give off less heat than the normal free-radical reactions that make up combustion, which helps pull heat out of the flame. Once enough radicals and heat have been removed, the free-radical cascade reaction that is combustion will cease. Examples of vapor phase radical inhibitors include organochlorine and organobromine compounds, as well as some phosphorus compounds. This principle of action is how Halon fire-extinguishing agents work, but for flame-retardant additives that work in this manner, the additive can only provide self-extinguishment behavior if there is enough free-radical scavenger to put out the flame. In large fires, these types of flame retardants can be completely overwhelmed. Finally, these types of flame retardants, especially halogenated ones, will cause more smoke to be formed during combustion due to the inhibition of clean combustion in the vapor phase. This is a design issue that material scientists will need to consider for composites used in mass human transport (bus, rail, ship, aircraft) applications.

2.4.2.2 Endothermic Cooling

This class of flame retardants is primarily condensed phase in its activity, but there is also some vapor phase action. When this flame retardant is heated, it decomposes endothermically to cool the condensed phase (thus preventing further heat-induced decomposition and pyrolysis) and typically releases a nonflammable gas. This nonflammable gas dilutes the total amount of fuel in the vapor phase which either prevents/delays ignition or keeps heat release low, allowing for self-extinguishment once the external flame is removed from the polymer. The types of additives that fall into this class are typically mineral fillers, including hydroxides such as aluminum hydroxide (also known as alumina trihydrate) and magnesium hydroxide, as well as carbonates such as hydromagnesite. Organic carbonates can also be used as flame retardants, but they are not as effective as mineral filler systems. This is because the mineral filler flame retardants bring one additional benefit to the condensed phase: they dilute the total amount of fuel in the condensed phase, since after they release their nonflammable gas and cool the condensed phase, they are typically inorganic oxides, which cannot be burned further. Those oxides have not only replaced flammable polymer fuel, but sometimes will fuse together and form protective ceramic barriers as well.

2.4.2.3 Condensed Phase Char Formation

Materials that chemically react with a thermally decomposing polymer by forming newer, more stable bonds as old ones are broken by heat are referred to as char-forming flame retardants. Additives in this class typically include phosphorus flame retardants and intumescent fire-protection packages. Some metals can catalyze additional char formation, and sometimes additives which just slow down fuel pyrolysis rates through major increases in molten polymer viscosity (such as fumed silica and nanocomposites) can cause char to form. Specifically, the radicals in the condensed phase have more residence time in the condensed phase before volatilizing and therefore react with themselves rather than pyrolyzing and combusting. These polymers, by binding up polymeric fuel into highly cross-linked graphitic or glassy-carbon char, keep heat release rate low since there is very little mass being pyrolyzed for combustion (Fig. 2.3). Some flame-retardant additives (phosphorus in particular) can have some vapor phase activity while they promote char formation, but the dominance of condensed phase or vapor phase in the case of phosphorus is polymer dependent [17, 18, 24]. Finally, the polymer itself can include functional groups which activate under fire conditions to form char [30], but in some cases the cross-linking groups can release heat into the condensed phase, making flammability worse even though char yields are higher [31].

In a fire situation, polymer decomposition will occur. The flame retardant is only there to retard the growth of the fire, not to prevent it from occurring. In this aspect, flame-retardant additives are truly passive and only activate when enough heat is present to decompose the polymer. While the flame retardant may cause the polymer to self-extinguish quickly or char to the point that flames can no longer damage the composite, some damage will already have occurred during ignition and heat exposure. Therefore, with fire, one will always have polymer decomposition and degradation, but the question with the use of flame-retardant additives is how much damage has occurred. Did the flame retardant allow the structure to be saved so that it could be repaired? Did it just provide enough time for nearby people to evacuate the area and get to safety? How the flame retardant is used, and its ability to narrow the amount of damage that occurred or slow down the total destruction of the composite, depends upon the fire-safety regulation in use for that material. The use of flame retardants *does not* confer universal fire safety.

Just because something is flame retarded does not mean it is fire-safe. Certification of fire safety is the realm of fire-safety engineers who set the fire-safety codes which govern safe use of a polymeric composite material. Materials engineers seeking to understand and develop multifunctional polymeric composite materials where fire is a risk should take time to understand the regulations, and how those regulations affect flame-retardant choice and fire-safe material design. If the end use of the composite requires a high enough level of fire safety so that the composite can be repaired, multiple layers of defense may be needed to protect it. If the composite is sacrificial and must simply provide enough time for people to escape, then the use of additives may be enough. If one wants to save a structure, though, more robust flame-retardancy and fire-protection schemes will be needed. Flame-retardant additive use and fire safety of composites can be very tricky to implement into a composite on top of all the other product performance requirements. So if flame-retardant additives cannot be used due to strict product requirements, then the material scientist must consider using an inherently low heat release (low flammability) polymer or other approaches, such as polymer nanocomposites and integrated barrier systems.

2.5 Polymer Nanocomposites

In polymer nanocomposites, the minor (in concentration) reinforcing component is sub-micron in scale and is so evenly dispersed through the polymer that the major component of the final composite is not bulk polymer, but interfacial polymer. In effect, a polymer nanocomposite is a material in which interfacial polymer properties, not the bulk polymer + reinforcing article, dictate performance. Because of this distinction, and the fact that some nanoscale fillers can bring additional property enhancements (such as thermal, mechanical, and electrical), polymer nanocomposites are one of the fastest growing material science areas to date [32, 33]. However, polymer nanocomposites are not new technology as they have been around for quite a while hidden in common material forms. Only recent advances in polymer nanoscale science have revealed these materials to really be polymer nanocomposites (nanoscale fillers in a polymer matrix). Historical examples of polymer nanocomposites include rubber tires (carbon blacks), paints (organoclays as rheological modifiers), and the original technology which led to polymer nanocomposites as a field, polymer + clay hybrids produced by Toyota R&D 20 years ago [34, 35].

While there are numerous literature reviews on what a nanocomposite is, how it is prepared, and some of its uses [36–40], the technology is best described as a route to multifunctional materials. Specifically, through control of polymer structure (conversion of bulk polymer to interfacial polymer) and use of nanoscale filler structure–property relationships, it becomes possible to create a material that can serve more than one function and, in effect, replace several materials. For example, rather than using a conductive paint or embedded aluminum mesh to provide an

EMI shield on a composite part, one can use conductive nanofillers (such as carbon nanotubes and nanofibers) to provide an integrated EMI shield as well as enhanced mechanical properties in the final composite. Or, with the use of thermally conductive nanofillers, the polymer could replace a metal in a complex composite part assembly holding a heat-generating component, thus saving weight and possibly assembly costs while still allowing for heat dissipation.

2.5.1 Polymer Nanocomposite Protection Schemes

Despite the great promise and actual results achieved with polymer nanocomposite technology, preparing a successful polymer nanocomposite is not an easy task. One cannot simply drop nanofiller into a polymer and create a nanocomposite through simple mixing. Care must be taken to design the nanocomposite before synthesis so that the nanofiller will have the right interface with the polymer and will obtain uniform dispersion through processing. Only if good and uniform nanoparticle dispersion is achieved will one obtain enhanced properties [33–43]. Due to the all-interfacial nature of a polymer nanocomposite, the polymer degradation and decomposition mechanisms change in comparison to those of the bulk polymer. Because the polymer is all-interfacial, the nanofiller can have tremendous effects on the decomposition of the polymer under UV, abrasion, mechanical damage, or heat/fire damage scenarios, and the effects can be beneficial or detrimental.

Polymer nanocomposites greatly help the polymer resist heat and fire damage, from a typical delay in onset of depolymerization to large reductions in heat-release rate under fire [39, 40]. The mechanism for clay nanocomposite flammability reduction appears to be that the polymer pyrolysis rate is greatly slowed by the formation of a clay-rich layer, which increases molten resin viscosity and serves as a barrier against mass loss/fuel release [44]. For carbon nanotube and nanofiber nanocomposites, the mechanism of damage/heat release reduction is the formation of a nanoparticle network which also increases molten polymer viscosity, which in turn reduces mass loss/fuel release rates [45, 46]. These nanocomposites can also be combined with traditional flame-retardant additives to obtain synergistic flame-retardant effects, so that less additive is needed to obtain a regulatory passing flame performance [39, 40]. There are some cases of polymer nanocomposite + flame-retardant-additive antagonism, but these cases are very additive-specific, usually involving flame-retardant additives that help polymers pass tests by accelerating melt flow away from flames. Since polymer nanocomposites increase molten resin viscosity under fire conditions, the nanocomposites serve as "anti-drip" additives which leads to their antagonistic effect with melt-promoting flame-retardant additives.

For UV and oxidation damage, the results are mixed. There are reports that some nanocomposites accelerate UV degradation and UV-induced oxidation, but the cause is not clear. In some instances, metals in the nanofillers (such as the metals in the clay structure or absorbed onto the clay surface) accelerate the UV degradation [47, 48]. In other cases, it may be that the large surface area of the clay nanofiller causes it to absorb the UV stabilizers from the matrix, thus removing that additive protection for the polymer from the system [49]. For carbon nanotube nanocomposites, however, there appear to be some enhancements in UV stability [50, 51]. Finally, it is possible to create titanium oxide (TiO_2)-rich nanocomposite films that are UV and X-ray opaque (but visibly transparent) and can provide protection to an underlying polymer substrate [52]. So for UV protection, more studies are needed, as well as careful selection of nanofiller to provide protection to the polymer. For oxidation damage protection there have not been as many studies, but in one extreme case the clay nanocomposite provided significant protection against an aggressive oxygen plasma by forming a clay-rich protective barrier on the surface of the polymer [53], much like the flame-retardancy mechanism for clays in the case of fire [44].

Other types of damage to polymer nanocomposites have not yet been well-studied. Abrasion damage presumably would be reduced with a nanocomposite, but this may be nanofiller-dependent, with graphene and graphite nanocomposites providing lower friction and less surface wear versus other fillers which toughen the surface and make it harder for the abrading object to wear away the polymer. Electrical damage presumably would be mitigated through the use of conductive nanofillers (such as carbon nanotubes and nanofibers), which create a conductive path through the composite, thus allowing high-intensity electricity to pass through the composite. While this mechanism of electrical protection has been proposed, the author could find no published papers proving this protective capability against lighting strike or electrical arcing.

While polymer nanocomposites have shown impressive ability to resist damage, they may not always be appropriate for polymer composite use. Cost of nanofillers is still relatively high compared to traditional fillers and additives, and nanofillers often cause viscosity changes in the polymer, which make processing and synthesis of the polymer difficult. Therefore, it may be more appropriate to utilize nanofiller technology in protection barrier approaches, thus putting the nanofiller where it is needed to protect the composite. This approach should reduce costs as well as solve some of the processing issues associated with polymer nanocomposite technology.

2.6 Barrier Coatings

One way to provide protection to many materials is to apply a protective coating, such as a paint or a plasma-deposited layer, after the part is manufactured. For fire protection, even very thin layers of silica deposited onto a polymer surface can lower heat release as well as delay ignition of the polymer [54, 55]. Intumescent barriers can provide superb fire and heat protection to a polymer composite or steel by creating oxidation-resistant/low thermal conductivity carbon foam [56–58]. There have even been recently published results regarding the use of carbon nanotubes to create "buckypapers" (a nonwoven fabric of nanotubes) which help

Fig. 2.5 Polymer nanocomposite top layer protecting underlying polymer protection to underlying polymer + fiber composite

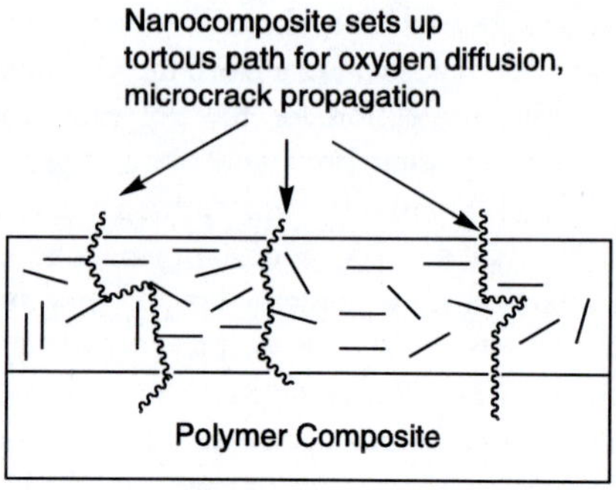

provide fire protection to composites [59]. Barrier-coating protection can also be more of a solid barrier than a true coating, such as ceramic tiles, insulation blankets, abrasion shields, and metal plates for electrical conduction/multi-threat protection. These protection systems, put in place after the part is made, are often referred to as "engineering solutions" in that one uses engineering to solve the problem rather than to address the flaw of the material itself. While all of these protection systems are effective and are in use today, they all have one key problem: because they are easily applied, they can be easily defeated. If the shield comes off, or the coating is broken or scratched, then protection is lost and the underlying polymer composite is subject to damage. Therefore, the use of an integrated protective coating to the polymer composite may be a superior approach.

2.6.1 *Integrated Protective Coatings for Composites*

When additive approaches are not acceptable and post-fabrication engineering solutions fail, it may be worthwhile to consider an integrated protective coating. This system borrows from additive, nanocomposite, and protective coating approaches, but puts the protection where it is needed in the final composite and makes sure that it is fully integrated into the part. In effect, the protective layer is designed to be fully compatible with the polymer composite and is created during the composite manufacture process. What follows in this section is an example of this type of protection conducted as a research project at UDRI, and so presents a particular case study on the integrated coating concept. More specifically, it represents an example of applying nanocomposite technology to provide protection for a composite. This example is not meant to be the definitive example of integrated coating technology, but instead a teaching example of how this problem can be looked at and potentially solved with polymer nanocomposites and the integrated coating approach.

In Figs. 2.5 and 2.6 are two generic examples of this approach. Figure 2.5 shows the nanocomposite approach applied as an outer barrier to the composite, while Fig. 2.6 shows a more complex protection scheme with multiple layers copolymerized.

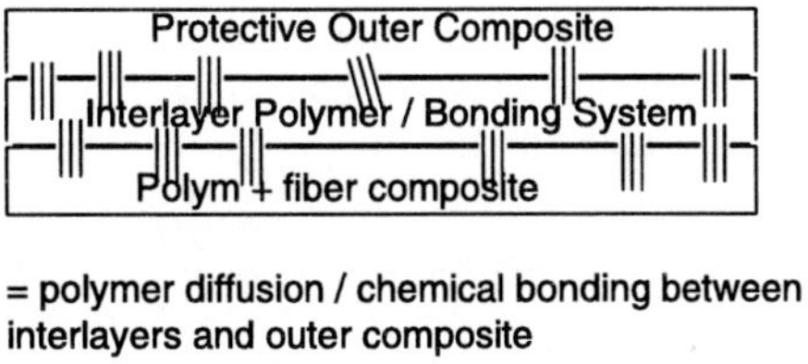

Fig. 2.6 Protective outer composite with interlayer bonding system to provide composite from oxidation damage

In both cases, the protection system is copolymerized with the base composite. Potentially, this allows for the underlying composite to remain lower in cost and maintain maximum mechanical integrity while the protective layer provides a barrier against polymer decomposition and degradation. The approach should be optimal for polymeric systems that are created from monomer (such as thermoset resin systems or thermoplastics that can be polymerized from liquid monomers), but it could work for fully polymerized thermoplastics through melt bonding/ specialized injection overmolding. For the purposes of this chapter, we will focus on the thermoset example and provide data for how this approach works in an aerospace grade polyimide. Note that the data provided in this chapter are meant to be as general as possible, while showing some actual results from fabricated composites utilizing the protective coating approach. We expect to publish the results in full in a peer-reviewed journal in the future.

The design approach to create a protective layer for composites follows three steps:

2.6.1.1 Step 1: Design Protective Coating System

In this step, we consider polymer chemistry (generally the bulk composite matrix polymer) and viscosity. We also consider what type of composite reinforcement, multifunctionality, service environment, and manufacturing process will be required. If we think about thermal protection for aerospace composites, we can focus on a polyimide with carbon fiber to be prepared by an autoclave process.

2.6.1.2 Step 2: Formulate Protective Coating System

In this step, we select the appropriate nanoparticles/surfactants which will yield thermal oxidative protection for dispersion into the polymer. The types of nanoparticles or fillers appropriate for this role are inorganic particles and fillers. The reason for this is that inorganic materials (typically metal and nonmetallic oxides) are already in their highest oxidation state and cannot be further damaged by high-temperature oxidation reactions. Through the use of ultrafine particle size alumina, fumed silica, carbon nanofibers, or exfoliated graphite, such layers can be prepared and co-cured with a polyimide + carbon fiber composite. However, one would need to use high loadings of such fillers to make the top barrier more inorganic in character and more likely to provide thermal protection. At this point, one can develop multiple formulations with varied levels of inorganic content to create a graded structure, but for the purposes of the example in this chapter, we will start with a monolayer of highly filled polyimide to provide protection just at the surface of the composite.

2.6.1.3 Step 3: Fabricate Coating Film/Composite System

In this final step, a film of desired thickness (similar to a film adhesive) is applied either to the tool surface of the composite mold or to the composite preform. Nanoparticles + resin will diffuse into the fiber perform, forming a gradient structure, or at least ensuring that the polymer chains entangle sufficiently throughout the top layer and the underlying composite that when the polymers are fully cured, the top film and the composite are fully bonded at the molecular level. Diffusion of resin and nanofillers into the top ply of the composite will be determined by resin viscosity, and so if it is desired to minimize diffusion of nanoparticles into the composite, one can increase polymer viscosity to affect this change.

So we utilized an aromatic polyimide (RTM grade polyimide) typically used in aerospace applications along with

- Alumina (Al_2O_3, median 1.5 μm particle size)
- Fumed silica (an organic treated grade)
- Vapor grown carbon nanofibers (high heat treated grade)
- Exfoliated graphite flake
- Multi-wall carbon nanotubes

These were all chosen as commercial off-the-shelf materials that can be provided in multi-kilogram to multiton amounts, so if they are successful, rapid scale-up and commercialization are possible.

First the materials were prepared into resin pellets for thermogravimetric analysis (TGA) screening. This was done by subjecting a fully cured resin pellet to TGA isothermal holds at 260°C (500°F) and 316°C (600°F) for 4 h under air. TGA results (Table 2.1) suggest that alumina alone reduces mass loss the most, but some of the other systems show good results at 260°C and so should be scaled up for additional testing. However, the reduction in mass loss provided by the alumina is not as impressive when one considers the fact that this particular sample is only 40 wt% polymer, and so other samples with significantly less inorganic content may provide more robust thermal protection. SEM analysis showed that in all cases, the dispersion of nanoparticles and alumina was uniform, so we can rule out nonuniform dispersion (nanoparticle agglomeration) as the reason for the discrepancies in results.

With these data, the following polyimide + alumina and/or nanoparticles were scaled up and filmed for composite fabrication. Films were laid up with carbon fabric to make an 8 ply 6″ square laminate, and the polyimide + alumina/

Table 2.1 TGA data for polyimide + inorganic systems

Sample	Weight loss after 4 h hold at 260°C (TGA)%	Weight loss after 4 h hold at 316°C (TGA)%
Polyimide (PI) control	0.22	0.76
PI + 60 wt% Al_2O_3	0.14	0.39
PI + 50 wt% Al_2O_3 + 10 wt% fumed silica	0.12	2.31
PI + 40 wt% Al_2O_3 + 10 wt% fumed silica + 5 wt% carbon nanofiber	0.17	1.08
PI + 50 wt% Al_2O_3 + 5 wt% exfoliated graphite	0.13	0.28
PI + 40 wt% Al_2O_3 + 10 wt% fumed silica + 5 wt% exfoliated graphite	0.12	0.93
PI + 50 wt% Al_2O_3 + 0.5 wt% carbon nanotubes	0.12	0.38
PI + 5 wt% carbon nanofiber	0.32	0.68

nanoparticle protective films consisted of 1 ply of the 8-ply stack applied on the tool side of the composite layup. All composites were then fully cured via autoclave processing, and the resulting panels were cut and machined for additional testing after autoclave cure.

- Polyimide + 60 wt% Al_2O_3
- Polyimide + 50 wt% Al_2O_3 + 10 wt% fumed silica
- Polyimide + 40 wt% Al_2O_3 + 10 wt% fumed silica + 5 wt% carbon nanofiber
- Polyimide + 50 wt% Al_2O_3 + 5 wt% exfoliated graphite
- Polyimide + 40 wt% Al_2O_3 + 10 wt% fumed silica + 5 wt% exfoliated graphite

Upon removal from the autoclave, it was observed that all the samples, with the exception of the polyimide + alumina sample, had some surface defects (not voids, but dents and very small cracks) on the protective ply/barrier layer. Optical microscopy revealed that all samples except the alumina-only sample had very thick top layers so that this resin area was not well-reinforced with carbon fiber. This makes sense in light of the known fact that fumed silica, carbon nanofibers, and exfoliated graphite are all known to increase polymer viscosity [60, 61]. Therefore, the increase in resin viscosity meant that during the autoclave processing, the resin in the top protective film did not fully migrate into the top layer of carbon fabric, thus resulting in a thicker outer layer on these composites. The alumina + polyimide sample did not show this effect, and indeed, during processing of the polyimide precursors with alumina, it was noted that the high levels of alumina did not affect the polymer viscosity much at all. With that in mind we can understand why the polyimide + alumina-only sample gave a good integrated top protective ply (Fig. 2.7) as compared to one of the polyimide + nanofiller systems (Fig. 2.8). When studying these figures, one needs to look closely at the top line of the panel (admittedly difficult to see in Fig. 2.7) and that thickness before it reaches the darker curve which is the top of the carbon fiber bundle. These pictures are micrographs taken edge-on in the composite so that one is looking at the top protective layer and how thick it is on top of the

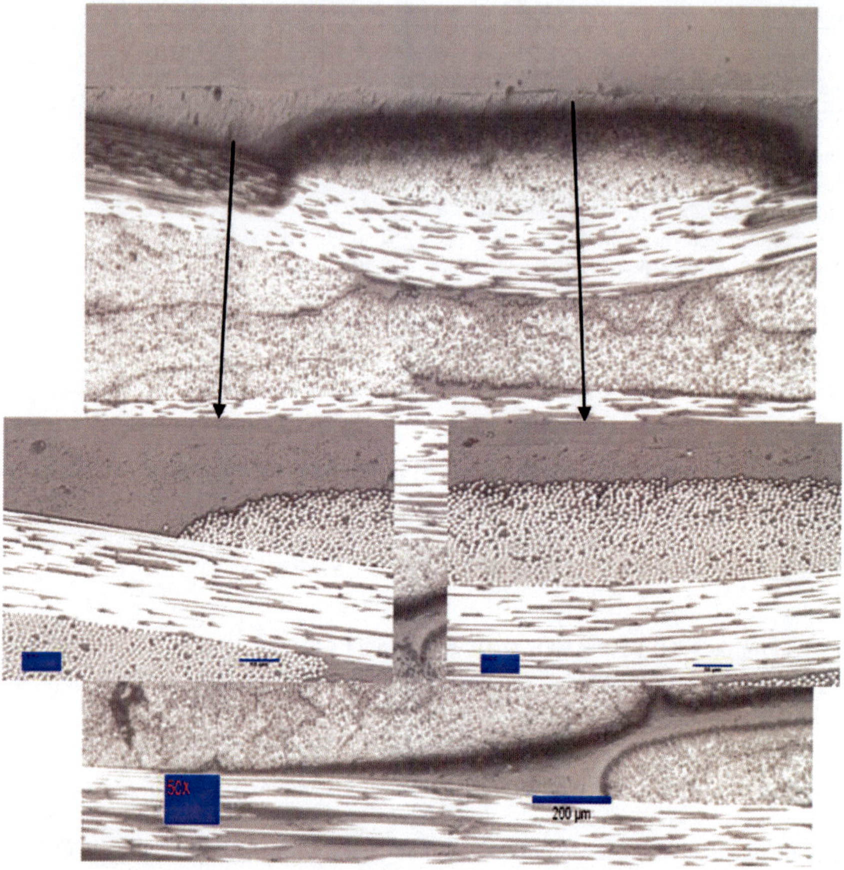

Fig. 2.7 Polyimide + 60 wt% Al_2O_3 micrographs

carbon fabric. In Fig. 2.7, the top layer is very thin, indicating that the resin-rich layer on this panel is very thin. For the other material (Fig. 2.8), the top layer is very thick, meaning that there is a large resin-rich layer which has no fiber reinforcement. This is important to keep in mind when looking at the thermal oxidative stability data and the panel skins after aging later in this section.

The panels with the protective plies were then subjected to thermal aging studies in calibrated ovens at 316°F for 200 h in air and then weighed for mass loss. The panels were all placed "protective laminate side up" in the ovens to minimize the effects of surface area, and the measured mass loss was normalized to surface area of the samples exposed to air to better understand the ability of the protective coatings to reduce the mass loss rate. The results from this study (Table 2.2) reflect the results in the TGA data (Table 2.1) in that the alumina sample gave the best performance. This makes sense in light of the fact that the top protective layer in

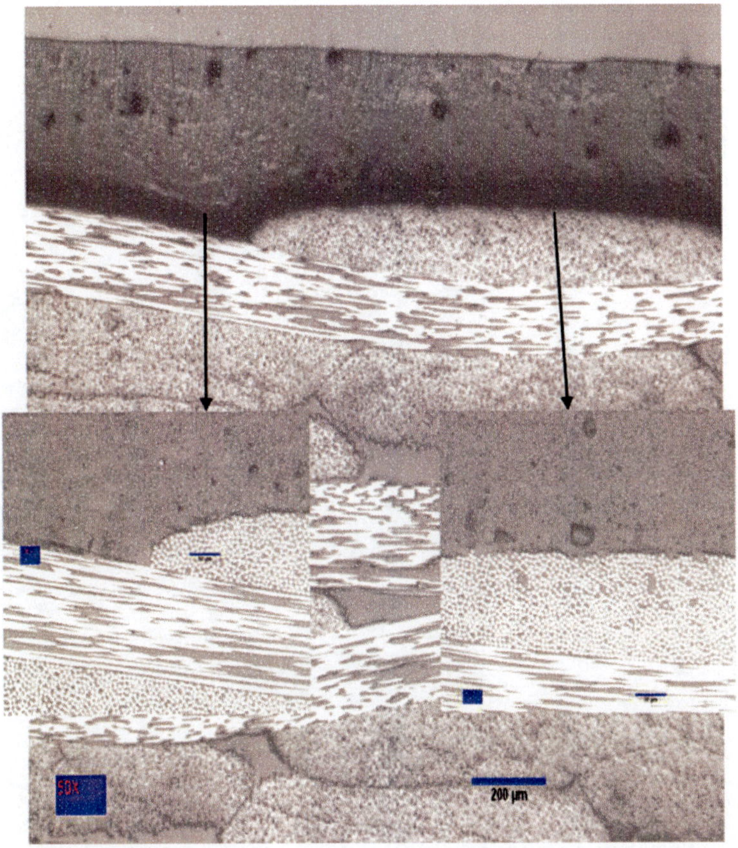

Fig. 2.8 Polyimide + 40 wt% Al_2O_3 + 10 wt% fumed silica + 5 wt% carbon nanofiber micrographs

Table 2.2 Polyimide + protective coating weight loss data at 316°F aging

Formulation	Wt loss-47 h (g/cm^2)	Wt loss-100 h (g/cm^2)	Wt loss-148 h (g/cm^2)	Wt loss-200 h (g/cm^2)	Coated side calculated
Polyimide (PI) control	0.083	0.116	0.138	0.180	N/A
PI + 60 wt% Al_2O_3	0.055	0.083	0.102	0.133	0.086
PI + 50 wt% Al_2O_3 + 10 wt% fumed silica	0.061	0.102	0.129	0.184	0.188
PI + 40 wt% Al_2O_3 + 10 wt% fumed silica + 5 wt% carbon nanofibers	0.061	0.012	0.148	0.213	0.246
PI + 50 wt% Al_2O_3 + 5 wt% exfoliated graphite	0.069	0.099	0.124	0.174	0.168
PI + 40 wt% Al_2O_3 + 10 wt% fumed silica + 5 wt% exfoliated graphite	0.068	0.109	0.138	0.196	0.212

Fig. 2.9 Surface of PI + 40 wt% Al_2O_3 + 10 wt % fumed silica + 5 wt% carbon nanofiber after 200 h @ 316°C. Sample is a 1″ square specimen

this sample is fully integrated into the composite structure (compare Fig. 2.7 to Fig. 2.8) and as polymer oxidizes away, it will leave an inorganic-rich top layer which can provide additional protection. Some of the other samples which were more effective in pellet form performed quite poorly in these tests, but post-aging analysis showed that the thick top layer was the reason for this. Specifically, the thicker top layers were more likely to crack open due to a lack of fiber reinforcement, and were also prone to polymer shrinkage during oxidation (which exacerbates the cracking), which fits with the known mechanism of polyimide oxidation, aging, and microcrack formation (Chaps. 9, 10 in this book). Examples of the extreme amounts of microcrack formation can be seen in Fig. 2.9, and the lack of microcracking can be seen in Fig. 2.10 for the polyimide + alumina sample.

With these preliminary results, we can state that the integrated protective coating concept works in principle, but more work is needed. Process work is needed to address resin viscosity increases created by the nanofibers, to see if it is possible to prevent the formation of thick top layers found in this study. Also, process work is needed to determine how much more alumina can be put into the top protective layer to further improve thermal oxidative protection. Fundamental research is also needed to see if it is possible to create fully inorganic top protective plies that

Fig. 2.10 Surface of PI + 60 wt% Al_2O_3 after 200 h @ 316°C. Sample is a 1″ square specimen

copolymerize with polyimides without interfering with composite mechanical properties, so that mass loss under thermal aging will approach zero. The data set at this time is rather limited, but the proof-of-concept experiment strongly suggests that this approach has merit, and indeed, research programs are underway to push this concept further and to develop not only better systems than the ones displayed here, but also systems that can be commercialized and used for composite protection in 2–5 years.

2.7 Future Trends and Unsolved Problems for Composite Protection

Material durability is not a new problem, but since polymeric composites are a relatively new class of materials, there is still much to learn despite all we know today. This final section of the chapter shares some likely trends in research on these materials as they continue to replace metals and ceramics, as well as some other unsolved problems for composites that need to be addressed as research in the area progresses.

2.7.1 Passive + Active Protection

It is expected that the use of nanocomposites or additives alone will not be enough for composite durability as these materials go into extreme environments. Indeed, with flame-retardant polymer nanocomposites, it is the combination of flame-retardant additives and polymer nanocomposites that have led to better fire-safe materials with improved mechanical properties and cost performance profiles. So with polymer composites in need of protection from erosion, electrical, and heat damage, we will likely see traditional means of protection (passive protection) combined with additives and nanocomposite approaches. The approach described in this section is one of multiple lines of defense. No carbon-based material can withstand very high temperatures and oxidation for long, so while the approaches of nanofillers and protective barriers greatly improve the durability of polymer composites exposed to heat and air, heat shields may still be needed. With the combination of engineering solutions and enhanced composite durability through nanocomposites, the parts may increase their usable lifetime several times over, thus saving on maintenance costs, or allowing for lighter-weight engineering solutions to be used, which improves the entire system. As existing materials are improved through additive, nanocomposite, and protective barrier approaches, engineers will get more savvy about taking advantage of these property enhancements first to make step changes in performance, and later major redesigns of systems. Most materials are inserted into new applications this way, by step change improvement in existing applications, so this trend is likely to be observed in the next decade.

2.7.2 Multifunctional Materials

"Multifunctional materials" simply means that one material does the job of many. It resists heat, conducts electricity and heat, has all the right mechanical, weight, and color requirements, is tough in all environments, parks the car, slices bread, and is dirt cheap too! One gets the idea that a multifunctional material is something truly impressive and can be used in many different ways, but really there is much to be done in this field. This field of study is certainly ongoing, and polymer + carbon nanotube/nanofiber nanocomposites are likely to be the first real example of a multifunctional material. But still, these materials are not perfect. While they may be electrically and thermally conductive, they are not conductive *enough* to replace metals outright. Then there is the cost issue: nanotubes and nanofibers are expensive, and so while effective in low loadings, at $100–200 per kg optimistically (sometimes $1,000 s per kg), one has to decide if the improvements are really justified by the cost. One definite unsolved problem for nanotubes is how to functionalize them without compromising electrical conductivity or properties. Most side-wall functionalization of the nanotubes results in loss of electrical conductivity, or worse, shortening of the tubes and loss of mechanical reinforcement [62]. Another unsolved issue is how to create an electrically and thermally conductive polymeric material that is transparent. The need for optical windows that can provide EMI shielding (or dissipate static discharge), energy for heads-up displays (computerized visual displays on transparent viewing surfaces), and the ability to resist erosion/wear/impact is still unsolved by any single material. Today it is accomplished with a multilayer system which adds cost and maintenance issues to the final system. For multifunctional materials to achieve actual success, one must first define an application that really can use such a material, and then design to that application. No one will say "Aha! I have the universal multifunctional material that everyone can use," because it cannot be achieved. Actual applications and needs (technology pull), not just discoveries in the lab pushing the technology out, will drive this field.

2.7.3 Multicomponent Systems

As suggested earlier in this section, polymeric composites are likely to be used in concert with existing materials and solutions, but getting these materials to work well together is still unsolved. This is partly an engineering design issue (How can engineering keep the part out of harm's way?) and partly a material selection issue (Which materials will minimize damage and work well with the polymer composite?). Most importantly, how each material influences the degradation and decomposition of the other has yet to be fully studied. If we are to go to metal + polymer, ceramic + polymer, or ceramic + metal + polymer hybrids, we need to understand at a macro, micro, and molecular scale how these materials interact in a range of environments. While there have been great strides in understanding the oxidation failure behavior of fiber-reinforced composites (see Chaps. 9,10), there is still much to learn, especially in regards to chemical decomposition pathways. Only by understanding the actual chemical species generated can the material scientist or polymer chemist develop a new means of defense. For truly complex structures, understanding how a metal catalyzes degradation (or char formation) in a polymer is important, as is understanding how changes in one structure lead to structural failure in another dissimilar material. In effect, it is an unknown problem which needs a multidisciplinary solution. It requires mechanical engineering, materials science, chemistry, and physics to come together on a problem of composite failure in an extreme environment and work in tandem toward a unifying understanding. Once that understanding is achieved, then one can design new materials or systems that survive the extreme environment. Much of this will have to rely upon experimental results and not just theory and modeling, although modeling is very important in understanding the behavior and in designing the right experiments to run. This unsolved problem is likely to continue to slow down composite insertion into many applications, but it is not insurmountable and can be addressed with careful study.

2.7.4 *Interfacial Fundamentals*

With both polymer nanocomposites and multi-material engineered structures, one of the great unknowns continues to be understanding the fundamental science (chemical, physical) at the interface between these dissimilar materials. The interfacial interactions can be complex, and without an understanding of these interactions, the material scientist has to think empirically at best, or outright guess at worst, to design a system that either takes advantage of these interactions or is not negatively affected by them. Since these interactions are complex, they will be explained in turn.

2.7.4.1 Polymer Nanocomposite Interface Issues

While polymer nanocomposites have been successfully commercialized, limitations on the interface between the nanoparticle and polymer have limited further commercialization of these materials. The interfacial issue is key to ensuring that a nanocomposite is formed in the first place, as without a well-designed interface to bridge the gap between nanoparticle and polymer, the two materials will typically separate, yielding expensive filled materials (microcomposites). The interfacial issues are nanoparticle-dependent; what works for a layered silicate is not always applicable to a carbon nanofiber/nanotube or colloidal inorganic particle.

For layered silicates and other layered inorganics, the main interfacial issue still to be addressed is the thermal stability of the organic treatment. Most inorganic layered materials are only miscible in polar polymer matrices, so organic treatments are needed to make the organophobic nanoparticle organophilic.

For layered silicates such as clays, the typical interface is an alkyl ammonium treatment put onto the clay surface through ion exchange. This treatment is chosen because of its low cost and well-established use to date in making organically treated clays. However, this treatment is thermally unstable above 180–200°C [63], and once the alkyl ammonium reaches this temperature it thermally decomposes to yield an amine and an alpha olefin via the Hoffman degradation reaction [64, 65]. The decomposition of this organic treatment causes the clay to return to organophobic status, which leads to polymer de-intercalation in the clay galleries (interlayer area) and subsequent phase separation. With sufficient thermal decomposition, the well-constructed nanocomposite will return to a microcomposite structure and all nanocomposite benefits will be lost [66–69]. With this in mind, one must consider the use of more thermally stable organic treatments for the layered silicate, such as imidazolium [70–73] or phosphonium compounds [74]. While these organic treatments are effective, they are not commercially available at this time, although more sources of imidazolium compounds are becoming available due to the increase in ionic liquid availability. However, not just any imidazolium or phosphonium compound can be used; one must ensure that long enough alkyl chains exist to enable intercalation of the polymer chains and miscibility with the polymer matrix as well as any other interactions that may be needed to obtain a good nanocomposite structure [75–78]. Indeed, in some cases alkyl ammonium can do a better job than imidazolium in rapidly dispersing into the polymer matrix during processing [79–82]. However, even though alkyl ammonium dispersion can occur very quickly before the alkyl ammonium degrades sufficiently, if the polymer is kept at elevated temperatures for long periods of time (thermal aging), the alkyl ammonium will degrade and the clays will collapse, resulting in a microcomposite structure [83–86]. The thermal instability issues of alkyl ammoniums will need to be addressed for these materials to become more accepted and further commercialized.

On a final note, the alkyl ammonium discussion above is relevant to cation-exchangeable layered materials such as montmorillonite, laponite, hectorite, and similar layered silicates. For anionic-exchangeable layered materials (such as double-layered hydroxides), the organic group used to prepare the interface can vary greatly due to a wide range of commercially available anionic surfactants. The layered nanoparticles that fall into this category have only recently begun to be studied, but so far they do not appear to have the thermal instability issues that cationic-exchanged layered materials have [38, 87–91].

For carbon nanotubes and nanofibers, the interface is created in a very different way. An interface is often needed since the graphitic surface chemistry of these rod-like carbon nanoparticles does not lend itself to good interaction between the polymer and the nanoparticle. While dispersion of the nanoparticle can be achieved without additional interface, there are far fewer mechanical benefits, since as the polymer chains slip past each other, they have no good way to interact with the nanotube other than pi–pi bond interactions if aromatic groups are present, and even then these interactions are relatively weak compared to actual covalent bonds or polymer chain entanglements [37, 92]. What typically happens in the absence of an interface between polymer and nanotube/nanofiber is that the fibers "pull out" when the polymer breaks [93, 94], much like how fibers without good "sizings" pull out of composites when they break and fail. To achieve an interface between nanoparticle and polymer, in this case, one must covalently react a functional group onto the surface of the nanoparticle. For carbon nanotubes, this is done either at the end of the tube or on the wall of the tube itself, with reactions that either partially break open the graphitic structure of the wall or take advantage of defect sites on the nanotube surface [95–98].

For carbon nanofibers, a graphitic edge structure already exists and this is likely where reactions occur, although reactions on the wall of the nanofiber are also possible. The edge structure already exists due to the stacked "cup" (or coiled ribbon) structure of the nanofiber structure as opposed to the rolled sheet structure of the nanotube. These covalent reactions, once achieved, are permanent and tend to have very good stability; the treatment is unlikely to decompose during processing as sometimes happens with alkyl-ammonium-treated layered silicates. However, in so reacting with the wall of the nanotube, one weakens the properties of the nanotube. By disrupting the continuous sp2 carbon network, electrical conductivity can be lost, and the tube itself can become mechanically weakened. This is a great problem for single-wall carbon nanotubes (SWNT) but less of a problem for multi-wall nanotubes (MWNT); only the outer wall of an MWNT is damaged during functionalization, and the inner core maintains its properties. There have been some breakthroughs in creating interfaces for nanotubes which are not covalently bonded to the side wall of the tube [99, 100], but these technologies need to be studied further to ensure that they are stable to all processing conditions and really bring a benefit to nanocomposite properties. So for nanotubes and nanofibers, the interfacial issues remain without a satisfactory solution. Covalently functionalized MWNT or nanofibers can be used, but they do have high costs, which prevents their widespread use. The other issue, and the main one to address, is that the fundamentals of what type of interfacial chemistry structure and amount of interface on the nanotube/nanofiber surface are necessary to achieve good nanocomposite properties have not been fully elucidated, and so the material scientist trying to utilize these types of nanoparticles for protective schemes will have to spend more time doing basic research before implementing these materials into a solution.

There are plenty of chemical routes for functionalization of colloidal nanoparticles available, from direct covalent functionalization of the surface to encapsulation technology. What remains unknown is what exactly the interface should be and how much should be present. For example, if too much interface greatly clouds the surface of the nanoparticle, some nanoparticle benefits could be lost (especially important to conductive nanoparticles and nanoparticles with electro-optic effects). Likewise, if there is not enough interface, the nanoparticles will aggregate and form microcomposites and no benefits will be obtained. Perhaps it can best be said that at this time we do not know when an organic interface is needed, nor do we know when it is not needed, and we definitely do not know what the structure of that interface should be to maintain good dispersion and gain benefit of nanoparticle structure–property relationships. So much is still unknown because this class of nanofillers is still quite new in polymer nanocomposite research, and there is very little published work on the subject. Some materials, like polyhedral oligomeric silsesquioxanes (POSS) [101] are somewhat understood, but other systems, especially those in which nanoparticle shape and structure play a role in properties, remain elusive [102]. Finally, there is the uncertain issue of when it is better to mix in a nanoparticle to obtain properties, or to try to form the nanoparticle in situ during polymerization and obtain a true polymer + inorganic hybrid [103, 104]. It may be that the best properties can only be obtained through a hybrid structure rather than through a filled-system approach.

2.7.4.2 Engineered Structure Interfacial Issues

The interfacial issues with engineered structures are predominantly physical, since the interfacial area between dissimilar materials is typically macroscale, and as such, chemical effects are usually minimized. While engineers can look at material properties and try to match them based upon their physical properties, most of the database properties to choose from either are limited in temperature range, or are not well-described as regards all the properties over the lifetime of the part. As an example, mismatch in coefficient of thermal expansion between a thermal ceramic shield and a polymer composite will cause the shield to come off during use, or worse, to break if constrained as one part presses against the other due to differences in thermal expansion. How does one engineer for this problem? Is it simply a case of putting an absorbing material (a negative CTE material) between the other two materials, or does the structure have to include engineered "air gaps" between parts to allow for expansion? The design principles to prevent damage to a composite are not always clear, but what is more, fundamental engineering rules to prevent negative interactions between materials do not appear to be very clear at this time. Some of the other chapters in this book suggest that "rules and tools" exist to help minimize these negative interactions and unforeseen failures in new applications for old materials, but much work needs to be done.

2.7.5 Life Cycle and Environmental Challenges

Because of recent interest in improving energy efficiency, polymer composites are making inroads into many applications where their light weight can result in energy savings (less fuel consumption) for conventional propulsion technology. However, as these materials are used more often, the questions of their life cycle and environmental impact come forward. More and more, materials used in part and vehicle construction need to be recycled at the end of their use. For thermoplastics, this is relatively easy; the materials can be reground and melted into new parts. If they cannot be reused in this manner, they can at least be incinerated easily for energy recovery. However, for thermosets, the material in most fiber-reinforced composite applications, recycling is not currently an option, at least not in commercially viable scales. Thermoset materials cannot be reground and reformed, and would need to be incinerated or sent to landfill. The incineration of commodity thermoset composites will yield energy generation, but high-performance composites (such as polyimides) present an incineration challenge in that they tend to be char formers, and so might absorb more energy than they would generate under incineration conditions. Plus, many of these composites have expensive fibers present as their reinforcement, so recovering the fiber while burning the resin is not something that can be easily done. Therefore, these materials present a real environmental and recycling challenge as they continue to grow in use. Recycling technology, either chemical (depolymerization and recapture of monomers) or physical (separation of polymer from fiber so fiber can be reused) is either in its infancy or nonexistent for thermoset polymer nanocomposites, and this presents a real unsolved problem for these materials. If the composites are damaged, replacement is easy, but what do you do with the damaged part? Along with preventing damage, figuring out what to do with the damaged composite at the end of its lifetime is of equal importance, and is something that researchers should begin considering when developing new materials.

2.7.6 Final Thoughts and Conclusions

As polymer composites continue to replace metals and ceramics, protection will be needed. As indicated throughout the chapter, there are many possible sources of protection to choose from. All the protection schemes available fall into active or passive schemes, but ultimately passive schemes are the ones most likely to succeed from a total cost and reliability perspective, although they will have a higher initial investment cost. Passive schemes of additives, polymer nanocomposites, and barrier coatings are available today, but each has its own limitations and appropriateness, depending upon the level of protection required. As long as one focuses utilizing on protection scheme/chemistry which addresses the mechanism of damage to the composite, it is very likely that the material scientist will enable the composite to replace a metal or ceramic successfully. Admittedly, there is still much work to do, but the goals are achievable, and it is highly probable that polymer composites designed to survive extreme environments will be discovered and commercialized.

Acknowledgments The author would like to thank the editors of this book for giving him an opportunity to write this chapter. Funding for the Integrated Coating System and Inorganic–Organic Hybrid work discussed in this chapter was provided by the Ohio Department of Development 3rd Frontier Research Commercialization Program (ODOD Grant Tech 09-007) and Air Force Research Laboratory "Aerospace Organic Matrix Composite Materials/Hybrids for Extreme Environments" (AFRL RXBC FA8650-05-D-5052) program, respectively.

References

1. ASTM E176, "Standard Terminology of Fire Standards," in *Annual Book of ASTM Standards,* Vol. 4.07, American Society for Testing and Materials, West Conshohocken, PA.
2. Hirschler, M. M.; Morgan, A. B. "Thermal Decomposition of Polymers" in "Society for Fire Protection Engineering Handbook" Eds. DiNenno, P. J.; Drysdale, D. D.; Beyler, C. L.; Walton, W. D.; Custer, R. L. P.; Hall, J. R.; Watts, J. M. National Fire Protection Association, Quincy, MA 2008, ISBN 978-0-87765-821-4.
3. "Fire Retardancy of Polymeric Materials" ed. Grand, A. F.; Wilkie, C. A. Marcel Dekker, Inc. New York. 2000. ISBN 0-8247-8879-6.
4. "Thermal decomposition, combustion and fire-retardancy of polyurethanes – a review of the recent literature" Levchik, S. V.; Weil, E. D. *Polym. Int.* 2004, *53*, 1585–1610.
5. "Thermal decomposition, combustion and flame-retardancy of epoxy resins – a review of the recent literature" Levchik, S. V.; Weil, E. D. *Polym. Int.* 2004, *53*, 1901–1929.
6. "Flammability" Tewarson, A. Chapter 42 in "Physical Properties of Polymers Handbook, Mark J. E. ed. AIP Press, NY 1996. pp 577–604.
7. "Screening Tests for Fire Safety of Composites for Marine Applications" Sorathia, U.; Long, G.; Gracik, T.; Blum, M.; Ness, J. *Fire Mater.* 2001, *25*, 215–222.
8. "Effects of Polyurethane Mattress Foam Properties and Geometry on Small and Large-Scale Fire Test Results" Hurd, M.; Weckman, E.; Enniful, E. Torvi, D. *Proceedings of Fire and Materials 2007 Conference*, January 29–31, 2007, San Francisco, CA. Interscience Communications.

9. "The piloted transition to flaming in smoldering fire retarded and non-fire retarded polyurethane foam" Putzeys, O. M.; Fernandez-Pello, A. C.; Rein, G.; Urban, D. L. *Fire and Materials*, 2008, *32*, 485–499.
10. Figure provided by Dr. Jeffrey W. Gilman – US Department of Commerce, National Institute of Standards and Technology. Public Domain figure not subject to copyright.
11. http://en.wikipedia.org/wiki/Mechanochemistry. Accessed July 7 2011.
12. Phelps, A. W. "Materials selection for wear resistance," Chapter 41, Handbook of Materials Selection, M. Kutz, ed., Wiley, New York, 2002, p. 1275.
13. "Degradation issues of polymer materials used in railway field" Ito, M,; Nagai, K. *Polym. Degrad. Stab.* 2008, *93*, 1723–1735 and extensive references therein.
14. "Kirk-Othmer Encyclopedia of Chemical Technology, 5th Ed." John Wiley & Sons, Hoboken, NJ, 2007. ISBN 13-978-0471484943. Sections on Antioxidants, Polymers, and Flame Retardants.
15. Handbook of Polymer Degradation, 2nd edition, Hamid, S. H., Ed. CRC Press, 2000 ISBN 978-0824703240.
16. Plastics Additives – An A-Z Reference (Pritchard, G.), Springer-Verlag, 2007. ISBN 978-0412727207.
17. "Fire Retardancy of Polymeric Materials, 2nd Ed." Wilkie, C. A.; Morgan, A. B. Eds. CRC Press. Baco Raton, FL, USA. ISBN 978-1420083996, December 2009.
18. "Advances in fire retardant materials" Horrocks, A. R.; Price, D. Eds. Woodhead Publishing Ltd., England. ISBN 978-1-84569-262-9.
19. "Commercial Flame Retardancy of Thermoplastic Polyesters – A Review." Weil, E. D.; Levchik, S. *J. Fire Sci.* 2004, *22*, 339–350.
20. "Commercial Flame Retardancy of Unsaturated Polyester and Vinyl Resins: Review" Weil, E. D.; Levchik, S. *J. Fire Sci.* 2004, *22*, 339–350.
21. "Thermal decomposition, combustion and flame-retardancy of epoxy resins – a review of the recent literature" Levchik, S. V.; Weil, E. D. *Polym. Int.* 2004, *53*, 1901–1929.
22. "Flame Retardants in Commercial Use or Development for Textiles" Weil, E. D.; Levchik, S. V. *J. Fire Sci.* 2008, *26*, 243–281.
23. "Flame Retardants in Commercial Use or in Advanced Development in Polycarbonates and Polycarbonate Blends" Levchik, S. V.; Weil, E. D. *J. Fire Sci.* 2006, *24*, 137–151.
24. "A Review of Recent Progress in Phosphorus-based Flame Retardants" Levchik, S. V.; Weil, E. D. *J. Fire Sci.* 2006, *24*, 345–364.
25. "Flame Retardants for Polystyrenes in Commercial Use or Development" Weil, E. D.; Levchik, S. V. *J. Fire Sci.* 2007, *25*, 241–264.
26. "Fire retardant polymers: recent developments and opportunities" Bourbigot, S.; Duquesne, S. *J. Mater. Chem.* 2007, *17*, 2283–2300.
27. "Flame Retardants in Commercial Use or Development for Polyolefins" Weil, E. D.; Levchik, S. V. *J. Fire Sci.* 2008, *26*, 5–42.
28. "Fire Properties of Polymer Composite Materials" Eds. Mouritz, A. P.; Gibson, A. G. Springer-Verlag, The Netherlands, 2006. ISBN 978-1-4020-5355-9.
29. "Flame retardancy of silicone-based materials" Hamdani, S.; Longuet, C.; Perrin, D.; Lopez-cuesta, J-M.; Ganachaud, F. *Polym. Degrad. Stab.*, 2009, *94*, 465–495.
30. XPS characterization of Friedel-Crafts cross-linked polystyrene" Wang, J.; Du, J.; Yao, H.; Wilkie, C. A. *Polym. Degrad. Stab.* 2001, *74*, 321–326.
31. "Synthesis and Testing of Nonhalogenated Alkyne-Containing Flame Retarding Polymer Additives" Morgan, A. B.; Tour, J. M. *Macromolecules* 1998, *31*, 2857–2865.
32. "Twenty Years of Polymer-Clay Nanocomposites" Okada, A.; Usuki, A. *Macromol. Mater. Eng.* 2007, *291*, 1449–1476.
33. "Polymer Nanocomposites" Krishnamoorti, R.; Vaia, R. A. *J. Polym. Sci.: Part B: Polym. Phys.* 2007, *45*, 3252–3256.
34. "Synthesis of nylon 6-clay hybrid" Usuki, A.; Kojima, Y.; Kawasumi, M.; Okada, A.; Fukushima, Y.; Kurauchi, T.; Kamigaito, O. *J. Mater. Res.* 1993, *8*, 1179–1184.

35. "The chemistry of polymer-clay hybrids" Okada, A.; Usuki, A. *Mat. Sci. Eng. C.* 1995, *3*, 109–115.
36. "Polymer/layered silicate nanocomposites: a review from preparation to processing" Ray, S. S.; Okamoto, M. *Prog. Polym. Sci.* 2003, *28*, 1539–1641.
37. "Polymer Nanocomposites Containing Carbon Nanotubes" Moniruzzaman, M.; Winey, K. I. *Macromolecules* 2006, *39*, 5194–5205.
38. "Synthetic, layered nanoparticles for polymeric nanocomposites (PNCs)" Utracki, L. A.; Sepehr, M.; Boccaleri, E. *Polym. Adv. Technol.* 2007, *18*, 1–37.
39. "Flame retarded polymer layered silicate nanocomposites: a review of commercial and open literature systems" Morgan, A. B. *Polym. Adv. Technol.* 2006, *17*, 206–217.
40. "Flame Retardant Polymer Nanocomposites" Edited by Alexander B. Morgan and Charles A. Wilkie. Book published by John Wiley & Sons, Hoboken, NJ 2007. ISBN 978-0-471-73426-0
41. "How Nano are Nanocomposites?" Schaefer, D. W.; Justice, R. S. *Macromolecules* 2007, *40*, 8501–8517.
42. "Polymer Nanocomposites with Prescribed Morphology: Going beyond Nanoparticle-Filled Polymers" Vaia, R. A.; Maguire, J. F. *Chem. Mater.* 2007, *19*, 2736–2751.
43. "Polymer nanotechnology: Nanocomposites" Paul, D. R.; Robeson, L. M. *Polymer* 2008, *49*, 3187–3204.
44. "A study of the flammability reduction mechanism of polystyrene-layered silicate nanocomposite: layered silicate reinforced carbonaceous char" Gilman, J. W.; Harris, R. H.; Shields, J. R.; Kashiwagi, T.; Morgan, A. B. *Polym. Adv. Technol.* 2006, *17*, 263–271.
45. "Relation between the viscoelastic and flammability properties of polymer nanocomposites" Kashiwagi, T.; Mu, M.; Winey, K.; Cipriano, B.; Raghavan, S. R.; Pack, S.; Rafailovich, M.; Yang, Y.; Grulke, E.; Shields, J.; Harris, R.; Douglas, J. *Polymer* 2008, *49*, 4358–4368.
46. "Nanoparticle networks reduce the flammability of polymer nanocomposites" Kashiwagi, T.; Du, F.; Douglas, J. F.; Winey, K. I.; Harris, R. H.; Shields, J. R. *Nature Materials* 2005, *4*, 928–933.
47. "Photo-oxidative degradation of polyethylene/montmorillonite nanocomposite" Qin, H.; Zhao, C.; Zhang, S.; Chen, G.; Yang, M. *Polym. Degrad. and Stab.* 2003, *81*, 497–500.
48. "Photo-oxidation of Polypropylene/Montmorillonite Nanocomposites. 1. Influence of Nanoclay and Compatibilizing Agent" Morlat, S.; Mailhot, B.; Gonzalez, D.; Gardette, J-L. *Chem. Mat.* 2004, *16*, 377–383.
49. "Photooxidation of ethylene-propylene-diene/montmorillonite nanocomposites" Morlat-Therias, S.; Mailhot, B.; Gardette, J-L.; Da Silva, C.; Haidar, B.; Vidal, A. *Polym. Degrad. Stab.* 2005, *90*, 78–85.
50. "Photo-oxidation behaviour of polyethylene/multi-wall carbon nanotube composite films" Dintcheva, N. Tz.; La Mantia, F. P.; Malatesta, V. *Polym. Degrad. Stab.* 2009, *94*, 162–170.
51. "Polymer/carbon nanotube composites: Influence of carbon Nanotubes on EVA photodegradation" Morlat-Therias, S.; Fanton, E.; Gardette, J-L.; Peeterbroeck, S.; Alexandre, M. Dubois, P. *Polym. Degrad. Stab.* 2007, *92*, 1873–1882.
52. "Organic-Inorganic Hybrids as Transparent Coatings for UV and X-ray Shielding" Mazzocchetti, L.; Cortecchia, E.; Scandola, M. *ACS Applied Materials & Interfaces* 2009, *1*, 726–734.
53. "Self-Passivation of Polymer-Layered Silicate Nanocomposites" Fong, H.; Vaia, R. A.; Sanders, J. H.; Lincoln, D.; Vreugdenhil, A. J.; Liu, W.; Bultman, J.; Chen, C. *Chem. Mater.* 2001, *13*, 4123–4129.
54. "Surface Controlled Fire Retardancy of Polymers Using Plasma Polymerisation" Schartel, B.; Kuhn, G.; Mix, R.; Friedrich, J. *Macromol. Mater. Eng.* 2002, *287*, 579–582.
55. "New approach to flame retardancy using plasma assisted surface polymerisation techniques" Bourbigot, S.; Jama, C.; Le Bras, M.; Delobel, R.; Dessaux, O.; Goudmand, P. *Polym. Degrad. Stab.* 1999, *66*, 153–155.
56. "Evaluation of Intumescent Coatings for Shipboard Fire Protection" Sorathia, U.; Gracik, T.; Ness, J.; Durkin, A.; Williams, F.; Hunstad, M.; Berry, F. *J. Fire Sci.* 2003, *21*, 423–450.

57. "Thermoplastic resins for thin film intumescent coatings – towards a better understanding of their effect on intumescence efficiency" Duquesne, S.; Magnet, S.; Jama, C.; Delobel, R. *Polym. Degrad. Stab.* 2005, *88*, 63–69.
58. "Thermal Barrier Effect of Intumescent Coatings and Mats on Fibre-Reinforced Polymeric Composites" Kandola, B. K.; Kandare, E.; Myler, P.; Chukwudolue, C.; Bhatti, W. *Proceedings of Fire and Materials 2009*, San Francisco, CA. pp 33–45.
59. "Fire retardancy of a buckypaper membrane" Wu, Q.; Zhang, C.; Liang, R.; Wang, B. *Carbon* 2008, *46*, 1159–1174.
60. "Ultrasound assisted twin screw extrusion of polymer-nanocomposites containing carbon nanotubes" Isayev, A. I.; Kumar, R.; Lewis, T. M. *Polymer* 2009, *50*, 250–260.
61. "CH-π Interactions as the Driving Force for Silicone-Based Nanocomposites with Exceptional Properties" Beigbender, A.; Linares, M.; Devalckenaere, M.; Degee, P.; Claes, M.; Deljonne, D.; Lazzaroni, R.; Dubois, P. *Adv. Mater.* 2008, *20*, 1003–1007.
62. "Noncovalent Functionalization as an Alternative to Oxidative Acid Treatment of Single Wall Carbon Nanotubes with Applications for Polymer Composites" Simmons, T. J.; Bult, J.; Hashim, D. P.; Linhardt, R. J.; Ajayan, P. M. *ACS Nano* 2009, *3*, 865–870.
63. "Thermal Degradation Chemistry of Alkyl Quaternary Ammonium Montmorillonite" Xie, W.; Gao, Z.; Pan, W-P.; Hunter, D.; Singh, A.; Vaia, R. *Chem. Mater.* 2001, *13*, 2979–2990.
64. "Thermal decomposition of alkyl ammonium ions and its effect on surface polarity of organically treated nanoclay" Dharaiya, D.; Jana, S. C. *Polymer* 2005, *46*, 10139–10147.
65. "Thermal degradation of commercially available organoclays studied by TGA-FTIR" Cervantes-Uc, J. M.; Cauich-Rodriguies, J. W.; Vasquez-Torres, H.; Garfias-Mesias, L. F.; Paul, D. R. *Thermochimica Acta* 2007, *457*, 92–102.
66. "Organoclay degradation in melt processed polyethylene nanocomposites" Shah, R. K.; Paul, D. R. *Polymer* 2006, *47*, 4084.
67. "Influence of compatibilizer degradation on formation and properties of PA6/organoclay nanocomposites" Monticelli, O.; Musina, Z.; Frache, A.; Bullucci, F.; Camino, G.; Russo, S. *Polym. Degrad. Stab.* 2007, *92*, 370–378.
68. "Effect of organoclay purity and degradation on nanocomposite performance, Part 1: Surfactant degradation" Cui, L.; Khramov, D. M.; Dielawski, C. W.; Hunter, D. L.; Yoon, P. J.; Paul, D. R. *Polymer* 2008, *49*, 3751–3761.
69. "Effect of organoclay purity and degradation on nanocomposite performance, Part 2: Morphology and properties of nanocomposites" Cui, L.; Hunter, D. L.; Yoon, P. J.; Paul, D. R. *Polymer* 2008, *49*, 3762–3769.
70. "Polymer/Layered Silicate Nanocomposites from Thermally Stable Trialkylimidazolium-Treated Montmorillonite" Gilman, J. W.; Awad, W. H.; Davis, R. D.; Shields, J.; Harris, R. H. Jr.; Davis, C.; Morgan, A. B.; Sutto, T. E.; Callahan, J.; Trulove, P. C.; DeLong, H. C. *Chem. Mater.* 2002, *14*, 3776.
71. "Melt-Processable Syndiotactic Polystyrene/Montmorillonite Nanocomposites" Wang, Z. M.; Chung, T. C.; Gilman, J. W. Manias, E. *J. Polym. Sci. Part B.* 2003, *41*, 3173–3187.
72. "Synthesis of imidazolium salts and their application in epoxy montmorillonite nanocomposites" Langat, J.; Bellayer, S.; Hudrlik, P.; Hudrlik, A.; Maupin, P. H.; Gilman, J. W.; Raghavan, D. *Polymer* 2006, *47*, 6698–6709.
73. "Benzimidazolium surfactations for modifications of clays for use with styrenic polymers" Costache, M. C.; Heidecker, M. J.; Manias, E.; Gupta, R. K.; Wilkie, C. A. *Polym. Degrad. Stab.* 2007, *92*, 1753–1762.
74. "Thermal Stability of Quaternary Phosphonium Modified Montmorillonites" Xie, W.; Xie, R.; Pan, W-P.; Hunter, D.; Koene, B.; Tan, L-S.; Vaia, R. *Chem. Mater.* 2002, *14*, 4837–4845.
75. "Poly(propylene)/organoclay nanocomposite formation: Influence of compatibilizer functionality and organoclay modification" Reichert, P.; Nitz, H.; Klinke, S.; Brandsch, R.; Thomann, R.; Mulhaupt, R. *Macromol. Mater. Eng.* 2000, *275*, 8–17.

76. "Swelling behavior of montmorillonite cation exchanged for ω-amino acids by ε-caprolactam" Usuki, A.; Kawasumi, M.; Kojima, Y.; Okada, A.; Kurauchi, T.; Kamigaito, O. *J. Mater. Res.* 1993, *8*, 1174–1178.
77. "New Polylactide/Layered Silicate Nanocomposites: Role of Organoclays" Maiti, P.; Yamada, K.; Okamoto, M.; Ueda, K.; Okamoto, K. *Chem. Mater.* 2002, *14*, 4654–4661.
78. Self-Assembly of Alkylammonium Chains on Montmorillonite: Effect of Chain Length, Head Group Structure, and Cation Exchange Capacity" Heinz, H.; Vaia, R. A.; Krishnamoorti, R.; Farmer, B. L. *Chem. Mater.* 2007, *19*, 59–68.
79. "Polyamide- and polycarbonate-based nanocomposites prepared from thermally stable imidazolium organoclay" Cui, L.; Bara, J. E.; Brun, Y.; Yoo, Y.; Yoon, P. J.; Paul, D. R. *Polymer* 2009, *50*, 2492–2502.
80. "Ionic Liquid Modification of Layered Silicates for Enhanced Thermal Stability" Byrne, C.; McNally, T. *Macromol. Rapid Commun.* 2007, *28*, 780–794.
81. "Phosphonium-based layered silicate—Poly(ethylene terephthalate) nanocomposites: Stability, thermal and mechanical properties" Patro, T. Umasankar; Khakhar, Devang V.; Misra, Ashok *J. App. Polym. Sci.* 2009, *113*, 1720 – 1732.
82. "A flammability performance comparison between synthetic and natural clays in polystyrene nanocomposites" Morgan, A. B.; Chu, L-L.; Harris, J. D. *Fire Mater.* 2005, *29*, 213–229.
83. "A study of the flammability reduction mechanism of polystyrene-layered silicate nanocomposite: layered silicate reinforced carbonaceous char" Gilman, J. W.; Harris, R. H.; Shields, J. R.; Kashiwagi, T.; Morgan, A. B. *Polym. Adv. Technol.* 2006, *17*, 263–271.
84. "Maleated polypropylene OMMT nanocomposite: Annealing, structural changes, exfoliated and migration" Tang, Y.; Lewin, M. *Polym. Degrad. Stab.* 2006, *92*, 53–60.
85. "Nanocomposites at elevated temperatures: migration and structural changes" Lewin, M.; Pearce, E. M.; Levon, K.; Mey-Marom, A.; Zammarano, M.; Wilkie, C. A.; Jang, B. N. *Polym. Adv. Technol.* 2006, *17*, 226–234.
86. "Flammability Properties of Polymer-Layered Silicate Nanocomposites. Polypropylene and Polystyrene Nanocomposites" Gilman, J. W.; Jackson, C. L.; Morgan, A. B.; Harris, R.; Manias, E.; Giannelis, E. P.; Wuthenow, M.; Hilton, D.; Phillips, S. H. *Chem. Mater.* 2000, *12*, 1866–1873.
87. "LDPE/Mg-Al layered double hydroxide nanocomposite: Thermal and flammability properties" Costa, R. R.; Wagenknecht, U.; Heinrich, G. *Polym. Degrad. Stab.* 2007, *92*, 1813–1823.
88. "Comparative study on the flammability of polyethylene modified with commercial fire retardants and a zinc aluminum oleate layered double hydroxide" Manzi-Nshuti, C.; Hossenlopp, J. M.; Wilkie, C. A. *Polym. Degrad. Stab.* 2009, *94*, 782–788.
89. "Flame-retarded polystyrene: Investigating chemical interactions between ammonium polyphosphate and MgAl layered double hydroxide" Nyambo, C.; Kandare, E.; Wang, D.; Wilkie, C. A. *Polym. Degrad. Stab.* 2008, *93*, 1656–1663.
90. "Epoxy nanocomposites based on the synthetic alpha-zirconium phosphate layer structure" Sue, H. J.; Gam, K. T.; Bestaoui, N.; Spurr, N.; Clearfield, A. *Chem. Mater.* 2004, *16*, 242–249.
91. "Effect of nanoplatelet aspect ratio on mechanical properties of epoxy nanocomposites" Boo, W-J.; Sun, L.; Warren, G. L.; Moghbelli, E.; Pham, H.; Clearfield, A.; Sue, H-J. *Polymer* 2007, *48*, 1075–1082.
92. "Mechanical Reinforcement of Polymers Using Carbon Nanotubes" Coleman, J. N.; Khan, U.; Gun'ko, Y. K. *Advanced Materials* 2006, *18*, 689–706.
93. "The incorporation of carbon nanofibres to enhance the properties of self-reinforced, single polymer composites" Hine, P.; Broome, V.; Ward, I. *Polymer* 2005, *46*, 10936–10944.
94. "Control of Carbon Nanotube-Surface Interactions: The Role of Grafted Polymers" Nap, R.; Szleifer, I. *Langmuir* 2005, *21*, 12072–12075.

95. "Toolbox for Dispersing Carbon Nanotubes into Polymers to Get Conductive Nanocomposites" Grossiord, N.; Loos, J.; Regev, O.; Koning, C. E. *Chem. Mater.* 2006, *18*, 1089–1099.
96. "A new approach to functionalize multi-wall carbon nanotubes by the use of functional polymers" Hong, C-Y.; You, Y-Z.; Pan, C-Y. *Polymer* 2006, *47*, 4300–4309.
97. "Covalent Surface Chemistry of Single-Walled Carbon Nanotubes" Bannerjee, S.; Hemraj-Benny, T.; Wong, S. S. *Advanced Materials* 2006, *17*, 17–29.
98. "Effects of Surface Modification, Carbon Nanofiber Concentration, and Dispersion Time on the Mechanical Properties of Carbon-Nanofiber-Polycarbonate Composites" Gao, Y.; He, P.; Lian, J.; Schulz, M. J.; Zhao, J.; Wang, W.; Wang, X.; Zhang, J.; Zhou, X.; Shi, D. *J. App. Polym. Sci.* 2007, *103*, 3792–3797.
99. "Photophysical properties of noncovalently functionalized multi-walled carbon nanotubes with poly-para-hydroxystyrene" Park, Sungjin; Huh, Jung Oh; Kim, Nam Gwang; et al. *Carbon* 2008, *46*, 714 –716.
100. "Noncovalent Functionalization as an Alternative to Oxidative Acid Treatment of Single Wall Carbon Nanotubes with Applications for Polymer Composites" Simmons, T. J.; Bult, J.; Hashim, D. P.; Linhardt, R. J.; Ajayan, P. M. *ACS Nano* 2009, *3*, 865–870.
101. 'Polyhedral oligomeric silsesquioxane(POSS)-based polymers', J. J. Schwab, J. D. Lichtenhan 12(10–11):707–713 (1998).
102. "Special Control of Chemistry on the Inside and Outside of Inorganic Nanocrystals" Murphy, C. J. *ACS Nano* 2009, *3*, 770–774.
103. "Applications of hybrid organic-inorganic nanocomposites" Sanchez, C.; Julian, B.; Belleville, P.; Popall, M. *J. Mater. Chem.* 2005, *15*, 3559–3592.
104. "Nanostructured Organic-Inorganic Composite Materials by Twin Polymerization of Hybrid Monomers" Spange, S.; Grund, S. *Adv. Mater.* 2009, *21*, 2111–2116.

Chapter 3
Interphases in Composites

Ranga Pitchumani

Abstract During the cure of thermosetting polymer–matrix composites, the presence of reinforcing fibers significantly alters the resin composition in the vicinity of the fiber surface via several microscale processes, forming an interphase region with different chemical and physical properties from the bulk resin. The interphase composition is an important parameter that determines the composite micromechanical properties and the durability of the products in service. Historically, the description of the interphase has been rather empirical and because of the complexities of the molecular level mechanisms near the fiber surface, few studies have been carried out on the prediction of the interphase evolution as function of the process parameters. This chapter provides an overview of the existing thermodynamic and kinetics-based models for describing the interphase in composites and related experimental studies. The models are used to develop the processing–interphase and interphase–property linkages, which form the basis for guiding the process design for tailored composite properties and performance.

3.1 Introduction

Fabrication of fiber-reinforced thermosetting matrix composite parts consists of exposing a net-shaped mixture of fibers and catalyzed resin to a prescribed temperature variation with time – referred to as the cure cycle. The elevated temperatures initiate and sustain cross-linking polymerization reactions – called the cure – in the resin whereby the starting materials are irreversibly transformed to a structurally hard composite product.

R. Pitchumani (✉)
Advanced Materials and Technologies Laboratory, Department of Mechanical Engineering, Virginia Tech, Blacksburg, VA 24061-0238, USA
e-mail: pitchu@vt.edu

K.V. Pochiraju et al. (eds.), *Long-Term Durability of Polymeric Matrix Composites*,
DOI 10.1007/978-1-4419-9308-3_3,

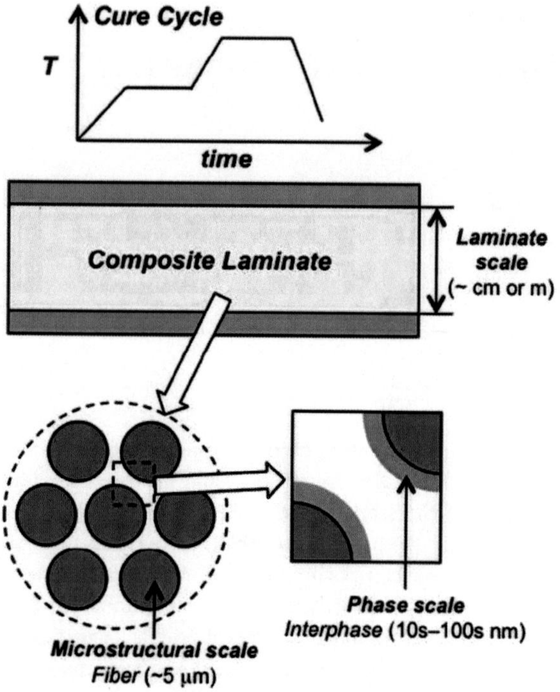

Fig. 3.1 Schematic of the range of length scales involved in the curing of thermosetting matrix composites

The properties and performance of the composite depend on the properties of the constituent matrix and reinforcement phases, and additionally on the physical and chemical interactions between them during processing which lead to formation of an interphase region adjacent to the reinforcement surfaces. The interphase has different composition and mechanical properties from the bulk matrix and fiber, and serves as a buffer between the original components. Formation of the interphase is governed by the transport processes of heat, mass, and momentum transfer at multiple levels ranging from the macroscale (at the laminate level, on the order of the structural dimensions) to the microscale (encompassing the microstructural and phase level length scales, on the order of microstructural dimensions or less). The transport processes at the microscale are strongly interlinked to the macroscale transport phenomena, which determine the local temperature and extent of cure at the microscale. A schematic of the range of scales involved in the problem is depicted in Fig. 3.1.

While the description of the macroscale phenomena provides the link to the process parameters, it is the microscale phenomena that determine the evolution of the microstructure, quality, and properties in the final products. Although significant advances have been made over the years in the understanding of the transport processes at the macro level, descriptions of the microscale phenomena have been steeped in considerable empiricism owing to the fundamental challenges posed by (a) the mechanistic and geometric complexities inherent at the microscale, and (b) the coupling of the microscale and the macroscale phenomena. This chapter provides a summary of the studies that have sought to address these critical challenges and establish a fundamental description of the phenomena at the phase and microstructural length scales integrated with the macroscale processes for a comprehensive processing–property relationship.

As mentioned above, the interphase possesses neither the properties of the reinforcement nor those of the homogeneous neat resin [4]. The size and chemical and structural characteristics of the interphase are, therefore, particularly important in determining the overall thermal, mechanical, and thermomechanical properties of the composite [5, 16, 34, 35, 38]. The interphase formation itself is related to the complex interplay among the processing and material variables including the cure temperature cycle, resin kinetics, choice of fibers, surface treatments, and surface coatings (called sizings). From a design standpoint, the interphase region can potentially be tailored, through appropriate manipulation of the processing and material variables, derived from fundamental understanding and modeling, to achieve the desired properties and performance of the composite product. Accurate prediction of the service life of a composite structure relies on a detailed understanding of the processing–property relationships and the effects of the service environment on the constituent, interface, and interphase properties.

Prediction of the overall composite properties in the presence of the interphase region involves the following steps:

1. First, the manufacturing parameters must be linked with the interphase structures. For example, given a cure temperature and pressure cycle, the chemical composition of the interphase should be determined. In this step, physical and chemical mechanisms must be identified and modeled to predict the interphase structure. This step forms the basis of the other two steps below.
2. In the second step, the known interphase structure is related to the interphase material properties such as glass-transition temperature (T_g), flexural modulus, or thermal expansion coefficient. This step is primarily based on the experimental correlation of the interphase chemical composition to the measured interphase material properties, and forms the basis of the third step below.
3. The last step is to link the interphase material properties to overall composite properties such as the strength, fracture, and environmental resistance.

A majority of the earlier studies in the literature have focused on experimental determination of the influence of interphase layers on the behavior of composite materials, i.e., steps 2 and 3 [4, 5, 7, 8, 20, 28, 29, 34, 36]. Fiber surface modifications such as high temperature treatment and sizing are commonly implemented to tailor the structure of the interphase regions and to investigate their effects on the mechanical and other properties of the composite materials. The effects of interphase property gradients on the overall composite properties are extensively investigated in the literature [1, 3, 6, 14, 19, 24, 33, 37, 39, 41].

However, the studies are commonly based on *assumed or empirical* interphase thickness and modulus. Tsai et al. [37] used an axisymmetric finite element model to study the interface stress, displacement, and fracture toughness. An elastic shear-lag analysis was developed and correlated with the micro-debonding test data to determine the thickness and material properties of the interphase. Boundary element method was adopted by Liu et al. [19] to predict transverse moduli of fibrous composites. Tsui et al. [39] studied the effects of different interphase properties on Youngs modulus, maximum stress concentration factor, and stress distribution in particle-filled polymer composites. Transverse Youngs storage/loss moduli and physical aging of a viscoelastic composite with fiber reinforcement were investigated by Fisher and Brinson [6].

Some empirical relations have also been proposed for the interphase property gradients. By assuming that the rate of change of a property is proportional to the value of the property, the interphase modulus variation with radius, $E_i(r)$, was given as [1, 24]

$$E_i(r) = E_m\left[1 + \left(p\frac{E_f}{E_m} - 1\right)\left(\frac{1 - re^{1-r/r_i}/r_i}{1 - r_f e^{1-r_f/r_i}/r_i}\right)\right], \qquad (3.1)$$

where $E_i(r)$ is the interphase modulus, r is the radius, the subscripts i, m, and f refer to interphase, matrix, and fiber, respectively, and r_f is the radius of the fiber. The parameter p is called an adhesion factor, which is the ratio between the interphase modulus at $r = r_f$ and the fiber modulus: $p = E_i(r_f)/E_f$.

Similar expressions were applied to other properties, such as the Poisson's ratio. Thermal stresses due to the interphase property gradients are predicted by Sottos et al. [33], where the elastic constants are assumed to vary linearly within the interphase. A power law relation was used by Wacker et al. [41] to calculate transverse Youngs modulus of composites:

$$E_i(r) = \left(\alpha E_f - E_m\right)\left[\frac{r_i - r}{r_i - r_f}\right]^n + E_m, \qquad (3.2)$$

where $0 \leq \alpha \leq 1$, and $n = 2,3, \ldots$. A review on the empirical interphase models and their features was given by Jayaraman et al. [14].

Due to complexities of the molecular-level mechanisms that occur in the vicinity of the fibers during the process, few investigations have been conducted in this area. The overhead comes from the fact that a dozen physical and chemical mechanisms contribute simultaneously to the formation of the interphase region, and only few of these mechanisms can be rigorously described in mathematical models. Garton et al. [8] showed that the carbon surfaces influenced the cross-linking reaction in an anhydride–epoxy system by adsorbing the tertiary amine catalyst and forming amine-rich interphase regions near the carbon surfaces. Similarly, Hirschfield [12] and Sellitti et al. [31] used Fourier transform IR attenuated total reflection spectroscopy to characterize the interphase phenomena in an epoxy–anhydride– catalyst system and showed that the surface species introduced on graphitized carbon fibers could promote or inhibit the cross-linking process by the preferential adsorption of the catalyst. Other possible interphase mechanisms were proposed by Drzal [4] including: the skin area of the fiber might have morphological deviation from the bulk fiber; the surface proximity of the fiber changes the structure of the resin in the interphase; surface treatments give rise to chemically and structurally different regions near the fiber surface; exposure to air before composite processing results in the adsorption of impurities that are desorbed at elevated temperatures; and the presence of a thin monomer coating on the surface of the fiber. Experimental studies by Sergeyeva et al. [32], Ko et al. [17], and Arayasantiparb et al. [2] have shown that the adsorption of an epoxy system can form an interphase layer 100–500 nm in thickness, which indicates that adsorption effects penetrate beyond one molecular layer.

The first work on modeling interphase formation in thermosetting composites was presented by Palmese [23]. The model predicts the interphase composition under thermodynamic equilibrium conditions of an epoxy–amine resin mixture near a fiber surface. The Gibbs phase rule was used to set up the equilibrium state, accounting for the enthalpy interaction between fiber surface and resin components, and the calculation of Gibbs free energy was based on a Flory–Huggins type lattice structure. The model does not take the chemical reactions into account, and furthermore, because of the assumption of thermodynamic equilibrium, it cannot predict the interphase evolution with time during processing. Hrivnak [13] extended Palmese's model to a reacting system by using renewal theory models to construct the assembly Gibbs free energy and the associated chemical potential. Adsorption from a mixture of polymer chains and solvent molecules near a surface was presented by Scheutjens and Fleer [30] with a statistical approach. The partition function for the mixture was evaluated with a quasicrystalline lattice model, which in turn, gave the number of chains in each conformation in equilibrium. The focus of the work was on polymer adsorption, whereas the interphase formation in thermosetting materials is based on monomer transport.

Yang and Pitchumani [42–44] presented a kinetics-based description of the governing phenomena to predict the interphase development during thermosetting composite processing. In this method, mass conservation principle was employed to describe the transport processes of multilayer adsorption, desorption, and diffusion near a fiber surface, which are accompanied by simultaneous cure reaction between the resin components. The time evolution of interphase concentration profile gradients before the gelation of the thermosetting system was predicted as function of material and process parameters. The interphase composition and thickness obtained from the models were used in composite property prediction without any assumed variations [45] and in providing comprehensive processing–interphase–property relationships [46].

The focus of the present chapter is on illustrating the first principles approach to prediction of the interphase evolution as function of the processing and material parameters, and the use of the physics-based description to predicting the composite properties as well as in establishing the complete processing–property relationships including the effects of the interphase. The interphase formation models are reviewed in Sect. 3.2 followed by a description of an approach to relating the interphase parameters to the composite properties in Sect. 3.3. Section 3.4 discusses coupling of the molecular-scale interphase evolution modeling with the macroscopic thermochemical description of the composite curing process toward establishing processing–property relationships including the effects of the interphase. The chapter concludes with a summary of the state-of-the-art and some avenues for future work on the predictive description of the effects of interphases in composites in Sect. 3.5.

3.2 Interphase Formation Models

Two interphase formation models are considered in the subsections below: the first is based on the kinetics of the interphase formation mechanisms as presented in Yang and Pitchumani [42–44], and the second is based on thermodynamic considerations at the fiber–resin interface as presented originally by Palmese [23] for a non-reacting system under thermodynamic equilibrium and later extended in Hrivnak [13] for reacting systems.

3.2.1 Kinetics Model

The cure of thermosetting resin systems is characterized by the reaction between prepolymer (or monomer) molecules and a curing agent to form a cross-linked network that cannot flow on vitrification. The reinforcing fibers alter the cure characteristics by selective adsorption of resin components, which changes the concentration of the reacting species in the vicinity of the fiber surfaces. The goal of the modeling is to predict the concentration profiles of the constituent species near the fiber surface by consideration of the processes that occur in the cure reaction. An inorganic fiber/epoxy–amine thermosetting system is considered in the following discussion; however, the formulation and results are applicable to a general two-component thermosetting system.

The geometry considered in the cure kinetics model is the interfiber space in a composite having a typical staggered fiber arrangement (Fig. 3.2a). The domain in the model development is idealized as the region between two identical infinite planes representing the fiber surfaces (Fig. 3.2b), which may be justified by the fact that the interphase thickness is often small in comparison to the fiber diameter. The fiber is considered to have an epoxy sizing layer applied to its surface. The sized fiber is exposed to an epoxy–amine resin mixture, and the epoxy and amine species begin reacting with each other at $t = 0^+$. Accompanying the chemical

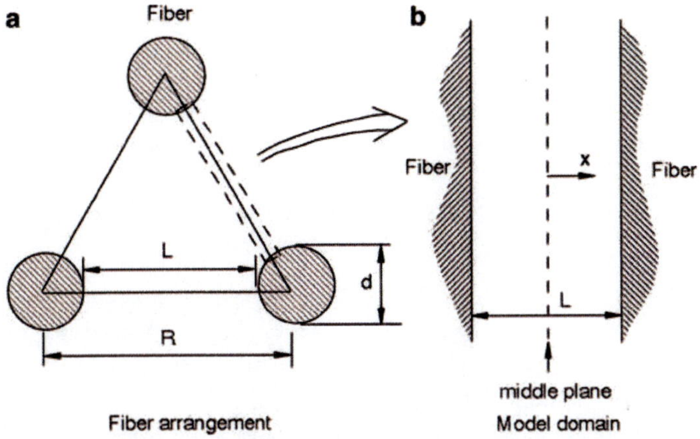

Fig. 3.2 (**a**) A representative fiber arrangement in the composite and (**b**) the one-dimensional domain between two fiber surfaces considered in the modeling

reaction, epoxy molecules diffuse away from the sizing layer adjacent to the fiber. In a reverse movement, the amine molecules diffuse into the sizing layer due to the relatively higher amine concentration in the bulk resin region. The diffusion process has a tendency to eliminate the concentration gradients. In addition, the "force field" of the fiber surfaces causes the epoxy and amine molecules to migrate in the direction toward the surfaces, a process referred to as adsorption. Also, the adsorbed molecules may be desorbed to the bulk resin in a desorption process. Unlike the diffusion process, the net effect of the adsorption and desorption processes is the buildup of concentration gradients. All the processes mentioned above take place simultaneously, resulting in a continuously evolving concentration profile that is "frozen" in space upon gelation of the thermosetting system.

The adsorption phenomenon at the equilibrium state near solid surfaces was described by Brunauer, Emmett, and Teller (referred to as the BET theory) [11, 21, 25]. In this development, the concepts in BET theory were extended to the transient processes in the interphase evolution. Following the approach of the BET theory, the domain between the two fiber surfaces is divided into molecular layers, such that there are N_L layers in the half domain. Each layer is further discretized along the fiber direction, as shown in Fig. 3.3, and a molecule of epoxy or amine can occupy only one of these discretized cells. Due to the interaction between resin molecules and the fiber surface, as well as those among resin molecules themselves, epoxy and amine molecules can move from layer to layer. The solid surface can adsorb molecules from a "bulk" state into an "adsorbed" state, and conversely, molecules in the "adsorbed" state may be desorbed into the "bulk" state. Molecules in the "adsorbed" state are treated to be fixed in the space and are not permitted to diffuse, while molecules in the "bulk" state may diffuse within the resin mixture. Figure 3.3 shows the molecular layers in the model domain, where the shaded cells

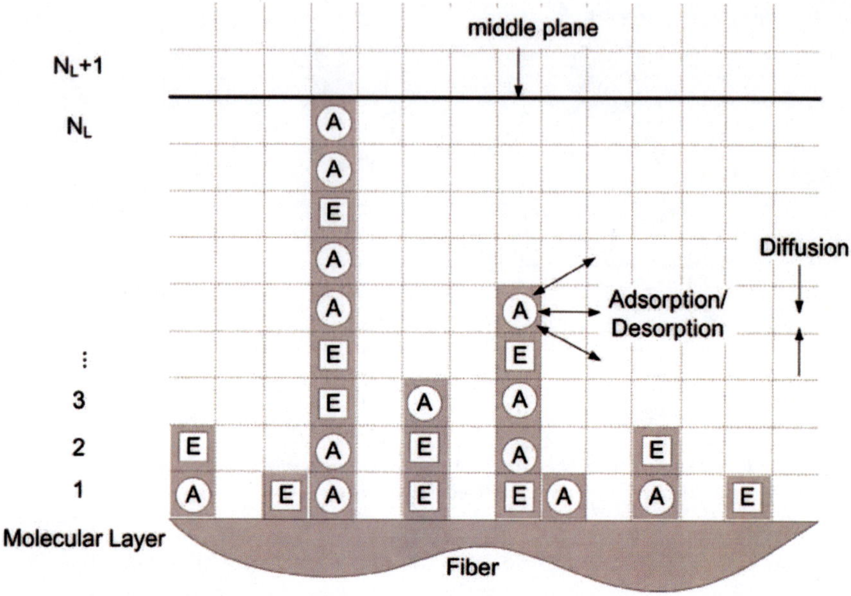

Fig. 3.3 Schematic of the adsorption, desorption, and diffusion processes near a fiber surface in a fiber–epoxy–amine system

contain molecules in the "adsorbed" state, while the blank cells hold molecules in the "bulk" state. Chemical reaction between epoxy (E) and amine (A) happens simultaneously during the adsorption, desorption, and diffusion processes, resulting in a continuously evolving concentration profile that is "frozen" in space upon gelation of the thermosetting system. The reaction equation may be written as $n_1E + n_2A \rightarrow P$ where P denotes the product and n_1 and n_2 are the molar numbers of the reactants needed to produce 1 mol of product.

The mass balance for the epoxy molecules in the "adsorbed" state in any (i)th molecular layer yields [44].

$$\underbrace{\frac{dN_{E,i}}{dt}}_{\text{storage}} = \underbrace{R_{a,E}(i-1,i) + R_{a,E}(i,i) + R_{a,E}(i+1,i)}_{\text{adsorption, } \Re_{a,E}}$$

$$- \underbrace{R_{d,E}(i-1,i) + R_{d,E}(i,i) + R_{d,E}(i+1,i)}_{\text{desorption, } \Re_{d,E}} - \underbrace{\Re_{r,E}(i)}_{\text{depletion(reaction)}} \quad . \tag{3.3}$$

In Eq. 3.3, $dN_{E,i}/dt$ is the rate of change of total number of adsorbed epoxy molecules in the (i)th molecular layer, and the subscripts E and i denote epoxy and (i)th layer, respectively. Epoxy molecules in the bulk state in the ($i-1$)th, (i)th, and

$(i+1)$th layers may be adsorbed into the (i)th molecular layer, denoted by the rate terms $R_{a,E}(i-1,i)$, $R_{a,E}(i,i)$, and $R_{a,E}(i+1,i)$, respectively; in a reverse process, the adsorbed epoxy molecules in the (i)th layer can also be desorbed into the bulk in the $(i-1)$th, (i)th, and $(i+1)$th layers through the rate terms $R_{d,E}(i-1,i)$, $R_{d,E}(i,i)$, and $R_{d,E}(i+1,i)$. Further, the depletion of adsorbed epoxy in the (i)th molecular layer through chemical reaction is determined by the rate term $\Re_{r,E}(i)$.

The adsorption of epoxy molecules from a (j)th layer of the bulk state, where $j=i-1$, i or $i+1$, to the (i)th layer of the adsorbed state is given as

$$R_{a,E}(j,i) = k_{a,E}(N_j - N_i)\exp\left(\frac{-E_{a,E}}{RT}\right)\frac{N_{E\infty,j}}{N_{\infty,i-1}+N_{\infty,i}+N_{\infty,i+1}}. \quad (3.4)$$

The parameter $k_{a,E}$ in the above equation is the frequency factor in the adsorption rate of epoxy molecules, and N_i is the total number of epoxy and amine molecules adsorbed in the (i)th layer, i.e., $N_i = N_{E,i} + N_{A,i}$. Because a molecule adsorbed into the (i)th layer must adjoin an adsorbed epoxy or amine molecule in the $(i-1)$th layer, the term $N_{i-1} - N_i$ yields the number of available sites in the $(i-1)$th layer which are open for adsorption. Note that for $i=1$, the term $N_0 - N_1$ denotes the available sites in the first adsorption layer, where N_0 is the number of adsorption sites on the bare fiber surface. The activation energy of adsorption for the epoxy molecules is denoted as $E_{a,E}$, which defines the energy barrier to be crossed for an epoxy molecule to be adsorbed. The parameters $N_{E\infty,i}$ and $N_{A\infty,i}$ are the number of epoxy and amine molecules in the bulk state in the (i)th molecular layer, and $N_{\infty,i} = N_{E\infty,i}\ N_{A\infty,i}$. Since the epoxy molecules adsorbed into the (i)th layer come from the $(i-1)$th, (i)th, and $(i+1)$th layers, the fraction $N_{E\infty,i-1}/N_{\infty,i-1}+N_{\infty,i}+N_{\infty,i+1}$ in the above equation denotes the probability that a site can capture an epoxy molecule from the $(i-1)$th layer of the bulk state.

The desorption rate terms in Eq. 3.4 are expressed as follows:

$$R_{d,E}(j,i) = \frac{1}{3}k_{d,E}(N_i - N_{i+1})\exp\left(\frac{-E_{d,E}}{RT}\right)\frac{N_{E,i}}{N_i}; \quad j = i-1, i, i+1, \quad (3.5)$$

where the fraction 1/3 arises from the assumption that the probabilities of desorption from the (i)th adsorbed layer to each of the three neighborhood bulk layers are identical. The parameter $k_{d,E}$ is the frequency factor in the desorption rate of epoxy molecules. Because a molecule in the adsorbed layer (i) can be desorbed into the bulk state only if the molecule has no molecule adjoining it in the $(i+1)$th adsorbed layer, the term $(N_i - N_{i+1})$ yields the number of molecules in the (i)th adsorbed layer that may be desorbed into the bulk state. The activation energy of desorption $E_{d,E}$ for the epoxy molecules defines the energy barrier to be crossed for an epoxy molecule to be desorbed. The fraction $N_{E,i}/N_i$ arises from the fact that among all the molecules desorbed from the (i)th adsorbed layer, the probability of finding an epoxy molecule is its molar fraction in the adsorbed layer.

The "depletion" term, $\Re_{r,E}(i)$, is determined by the cross-linking chemical reaction between epoxy and amine, and may be defined as: $\Re_{r,E}(i) = n_1 k_r N_{E,i}$, where k_r is the reaction rate. The other rate terms and the mass conservation analysis to the amine and product in the adsorbed state (i.e., $N_{A,i}$ and $N_{P,i}$) may be discussed in a similar way, and the relevant equations are summarized below from Yang and Pitchumani [44]:

$$\underbrace{\frac{dN_{A,i}}{dt}}_{\text{storage}} = \underbrace{R_{a,A}(i-1,i) + R_{a,A}(i,i) + R_{a,A}(i+1,i)}_{\text{adsorption, } \Re_{a,A}}$$
$$- \underbrace{R_{d,A}(i-1,i) + R_{d,A}(i,i) + R_{d,A}(i+1,i)}_{\text{desorption, } \Re_{d,A}} - \underbrace{\Re_{r,A}(i)}_{\text{depletion (reaction)}} , \qquad (3.6)$$

$$\frac{dN_{P,i}}{dt} = k_r N_{E,i}. \qquad (3.7)$$

The adsorbed state exchanges mass with the bulk state, in which the molecules undergo diffusion in addition to the adsorption, desorption, and reaction processes. Evidently, the following processes are responsible for the change of number of the epoxy molecules in the bulk state in the (i)th layer: (1) diffusion of epoxy molecules in the "bulk" state from (i + 1)th and (i – 1)th layers to (i)th layer, which increases $N_{E\infty,i}$; (2) desorption of epoxy molecules in the "adsorbed" state in the (i – 1)th, (i)th, and (i + 1)th layers to the (i)th layer of the "bulk" state, which increases $N_{E\infty,i}$; (3) epoxy molecules in the "bulk" state in the (i)th layer being adsorbed to the (i – 1)th, (i)th, and (i + 1)th layers of the "adsorbed" state, which reduces $N_{E\infty,i}$; and (4) chemical reaction in the bulk in (i)th layer, which depletes $N_{E\infty,i}$. Furthermore, it is assumed that any discrete cell within a molecular layer is filled with one and only one resin molecule at any time instant, which assures constant number density of resin molecules near the surface. Consequently, the transfer (caused by either adsorption or desorption) of one molecule (epoxy or amine) from the (i)th layer to a neighborhood layer j (where $j = i+1$, i, or $i+1$) creates a vacant square box in (i)th layer, which is assumed to be filled by a molecules in the bulk state from the (j)th layer. The probability that the vacant site is filled by an epoxy molecule in the bulk state from the (j)th is $N_{E\infty,i}/N_{\infty,j}$. Conversely, the transfer (caused by either adsorption or desorption) of a molecule (epoxy or amine) from a neighborhood (j)th layer into the (i)th molecular layer repeals one of the molecules in the bulk state in the (i)th layer to the (j)th layer, and the epoxy molecules are reduced with the probability $N_{E\infty,i}/N_{\infty,i}$.

The mass balance for the epoxy molecules in the "bulk" state in any (i)th molecular layer yields:

$$\frac{dN_{E\infty,i}}{dt} = \frac{D_{EA}}{\Delta L^2}\left(N_{E\infty,i+1}\frac{N_0 - N(i)}{N_0} - N_{E\infty,i}\frac{N_0 - N(i+1)}{N_0}\right)$$
$$+\frac{D_{EA}}{\Delta L^2}\left(N_{E\infty,i-1}\frac{N_0 - N(i)}{N_0} - N_{E\infty,i}\frac{N_0 - N(i-1)}{N_0}\right)$$
$$+R_{d,E}(i,i-1) + R_{d,E}(i,i) + R_{d,E}(i,i+1)$$
$$-R_{a,E}(i,i-1) - R_{a,E}(i,i) - R_{a,E}(i,i+1) - n_1 k_r N_{E\infty,i}$$
$$+\sum_{j=i-1,i+1}\left[\frac{N_{E\infty,j}}{N_{\infty,j}}\left(R_{d,E}(j,i) + R_{d,A}(j,i) + R_{a,E}(i,j) + R_{a,A}(i,j)\right)\right]$$
$$-\frac{N_{E\infty,i}}{N_{\infty,i}}\sum_{j=i-1,i+1}\left[R_{a,E}(j,i) + R_{a,A}(j,i) + R_{d,E}(i,j) + R_{d,A}(i,j)\right], \tag{3.8}$$

where the mutual diffusion coefficient D_{EA} in the binary epoxy–amine mixture is a function of temperature and degree of cure [44] and ΔL corresponds to the physical size of a molecular layer. Since molecules in the adsorbed state are assumed to be fixed in the space, a molecule can diffuse into the (i)th layer only into an available space that is unoccupied by an adsorbed molecule. Consequently, the diffusion rate of epoxy molecules from ($i+1$)th layer to (i)th layer is $\frac{D_{EA}}{\Delta L^2}N_{E\infty,i+1}\frac{N_0-N(i)}{N_0}$, where $\frac{N_0-N(i)}{N_0}$ denotes the probability of the above-mentioned diffusion. Similar considerations are incorporated in Eq. 3.8 for the other diffusion terms, i.e., from (i)th to ($i+1$)th layer, and those between (i)th and ($i-1$)th layers. The terms in the last two lines of Eq. 3.8 ensure a constant number density in the mixture.

Similarly, the rate equation for $N_{A\infty}$ is obtained from a mass balance for the amine species in the "bulk" state in any (i)th molecular layer:

$$\frac{dN_{A\infty,i}}{dt} = \frac{D_{EA}}{\Delta L^2}\left(N_{A\infty,i+1}\frac{N_0 - N(i)}{N_0} - N_{A\infty,i}\frac{N_0 - N(i+1)}{N_0}\right)$$
$$+\frac{D_{EA}}{\Delta L^2}\left(N_{A\infty,i-1}\frac{N_0 - N(i)}{N_0} - N_{A\infty,i}\frac{N_0 - N(i-1)}{N_0}\right)$$
$$+R_{d,A}(i,i-1) + R_{d,A}(i,i) + R_{d,A}(i,i+1)$$
$$-R_{a,A}(i,i-1) - R_{a,A}(i,i) - R_{a,A}(i,i+1) - n_2 k_r N_{E\infty,i}$$
$$+\sum_{j=i-1,i+1}\left[\frac{N_{A\infty,j}}{N_{\infty,j}}\left(R_{d,E}(j,i) + R_{d,A}(j,i) + R_{a,E}(i,j) + R_{a,A}(i,j)\right)\right]$$
$$-\frac{N_{A\infty,i}}{N_{\infty,i}}\sum_{j=i-1,i+1}\left[R_{a,E}(j,i) + R_{a,A}(j,i) + R_{d,E}(i,j) + R_{d,A}(i,j)\right], \tag{3.9}$$

and the equation for the product species, $N_{P\infty}$, following the form in Eq. 3.7, is expressed as

$$\frac{dN_{P\infty,i}}{dt} = k_r N_{E\infty,i}. \qquad (3.10)$$

The diffusivity, D_{EA}, and reaction rate, k_r, appearing in the above equations are functions of the degree of cure, $\varepsilon = (N_{E,0} - N_E)/N_{E,0}$. The diffusivity is described by free-volume theory as [29]

$$D_{EA} = D_0 \exp\left(\frac{-E_D}{RT}\right) \exp\left[\frac{-b_D}{f_g + \alpha_f (T - T_g(\varepsilon))}\right], \qquad (3.11)$$

where D_0, f_g, and α_f are constants, E_D is the activation energy, b_D is an empirical constant, and $T_g(\varepsilon)$ defines the available free volume and degree of rotational restriction, which in turn are functions of the degree of cure, ε. The parameters D_0, b_D, f_g, α_f, and E_D in the model depend on the type of thermosetting system and can be determined by the approach described in Sanford [29]. The DiBenedetto equation relates T_g to the degree of cure as [29]: $\frac{T_g(\varepsilon)-T_g^0}{T_g^0} = \frac{(E_x/E_m - F_x/F_m)\cdot\varepsilon}{1-(1-F_x/F_m)\cdot\varepsilon}$, where the constants T_g^0, E_x/E_m, and F_x/F_m can be obtained from experimental data of T_g versus ε.

The reaction rate k_r is controlled by the retarded diffusion process at later stages of cure and is given as

$$k_r = \frac{k_{r0} \exp(-E_a/RT)}{1 + (\delta_0/D_{EA}) \exp(-E_a/RT)}, \qquad (3.12)$$

in which k_{r0} is the Arrhenius pre-exponential constant, δ_0 is the coordination sphere reaction parameter, and E_a is the reaction rate activation energy [29].

The rate equation for species in the bulk state, Eqs. 3.8, 3.9, and 3.10, must be solved simultaneously with those corresponding to the adsorbed state, Eqs. 3.3, 3.6, and 3.7, for the six unknowns: $N_{E,i}$, $N_{A,i}$, $N_{P,i}$, $N_{E\infty,i}$, $N_{A\infty,i}$, and $N_{P\infty,i}$ in each of the molecular layers, $i = 1, 2, \ldots, N_L$. The symmetry of the geometry in Fig. 3.2b suggests the following conditions:

$$N_{X,N_L} = N_{X,N_L+1}; \quad X = \begin{cases} E, A, P \text{ [adsorbed state]}, \\ E_\infty, A_\infty, P_\infty \text{ [bulk state]}. \end{cases} \qquad (3.13)$$

For a thermosetting system with fiber sizing thickness of N_S molecular layers, the initial conditions associated with the species rate equations are: (1) the number of molecules for each species in the adsorbed state and the number of product in the bulk state are zero, (2) within the epoxy sizing layer, the number of epoxy is a constant, $N_{E,1}$, while the number of amine is zero, and (3) beyond the sizing layer, the numbers of epoxy and amine species are constants, N_{E0} and N_{A0}, respectively. The mathematical expressions for the initial conditions may be written as

Table 3.1 Summary of nondimensional parameters representing the kinetics model [44]

Parameter	Definition	Expression
t'	Dimensionless time	$k_{a,E}\exp\left(-E_{a,E}/RT_0\right)\cdot t$
β_E	Epoxy desorption ratio	$\left(k_{d,E}/k_{a,E}\right)e^{-\left(E_{d,E}-E_{a,E}\right)/RT_0}$
β_A	Amine desorption ratio	$\left(k_{d,A}/k_{a,E}\right)e^{-\left(E_{d,A}-E_{a,E}\right)/RT_0}$
α_A	Amine adsorption ratio	$\left(k_{a,A}/k_{a,E}\right)e^{-\left(E_{a,A}-E_{a,E}\right)/RT_0}$
γ	Adsorption Damköhler number	$\left(k_{r_0}/k_{a,E}\right)e^{-\left(E_a-E_{a,E}\right)/RT_0}$
ϕ_{EA}	Diffusion ratio	$\left(D_0/\Delta L^2 k_{a,E}\right)e^{-\left(E_D-E_{a,E}\right)/RT_0}$

$$N_{E,i}=N_{A,i}=N_{P,i}=N_{P\infty,i}=0;\quad i=1,2,\ldots,N_L,$$

$$N_{E\infty,i}=\begin{cases}N_{E,1} & i=1,2,\ldots,N_S;\\ N_{E0} & i=N_S+1,\ldots,N_L;\end{cases}\qquad N_{A\infty,i}=\begin{cases}0 & i=1,2,\ldots\;,N_S,\\ N_{A,0} & i=N_S+1,\ldots\;,N_L.\end{cases} \tag{3.14}$$

For a given temperature history, the microscale cure kinetics model (Eqs. 3.3–3.14) may be used to predict the evolution of the concentration profiles – $N_{E,i}$, $N_{A,i}$, $N_{P,i}$, $N_{E\infty,i}$, $N_{A\infty,i}$, and $N_{P\infty,i}$ – with time near a fiber surface. The governing equations may be cast in a dimensionless form to obtain the principal nondimensional parameters summarized in Table 3.1 [44]. In addition, all the number of molecules of each species ($N_{E,i}$, $N_{A,i}$, N_0) is divided by the initial number of epoxy molecules in the far region layers (N_{E0}) to obtain a nondimensional species concentration; for example: $N'_{E,i}=N_{E,i}/N_{E0}$.

Arayasantiparb et al. [2] presented an experimental study on the measurement of the composition of an epoxy–aluminum interphase with spatially resolved electron energy-loss spectroscopy (EELS) in a scanning transmission electron microscope. The material system consisted of an aliphatic bis(p-aminocyclohexyl)methane (PACM20) curing agent of density 960 kg/m^3 and molecular weight 198.3 g/mol, an aromatic diglycidyl ether of bisphenol-A (DGEBA) epoxy resin of density 1,170 kg/m^3 and molecular weight 382.4 g/mol, and a single aluminum fiber with a diameter of 125 μm. Energy-loss spectra were collected for the DGEBA/PACM20 system at different locations of the sample and were used to determine the local volume fraction of PACM20.

The kinetics model presented in this section involves a number of parameters, some of which are available in the literature, and others need to be determined for the specific resin/catalyst/fiber system. Yang and Pitchumani [44] correlated the kinetics model presented in this section to the experimental data of Arayasantiparb et al. [2] to determine the parameters of the model. The parameters from the literature and those obtained from the correlation are summarized in Table 3.2. The following observations are noteworthy among the parameters in Table 3.2: (1) the adsorption rate of DGEBA, $k_{a,E}\exp\left(-E_{a,E}/RT\right)$, is three orders of magnitude smaller than its desorption rate, $k_{d,E}\exp\left(-E_{d,E}/RT\right)$, indicating that the aluminum

Table 3.2 Kinetics model parameters for a DGEBA/PACM20/aluminum system [29, 44]

Parameter	Value	Parameter	Value
E_x/E_m	0.337	$D_0 \exp(-E_D/RT)/\Delta L^2$ [1/s]	6.05×10^4
T_g (K)	254	$k_{r_0} \exp(-E_a/RT)$ [1/s]	1.08×10^{-4}
α_f (K^{-1})	$5 \times 10^{-4}\,(T>T_g)$ $5 \times 10^{-5}\,(T<T_g)$	$k_{a,E} \exp(-E_{a,E}/RT)$ [1/s]	2.04×10^{-5}
F_x/F_m	0.194	$k_{a,A} \exp(-E_{a,A}/RT)$ [1/s]	0.85
f_g	0.025	$k_{d,E} \exp(-E_{d,E}/RT)$ [1/s]	1.58×10^{-2}
δ_0 (cm^2/s)	0.220	$k_{d,A} \exp(-E_{d,A}/RT)$ [1/s]	1.03×10^{-2}
b_D	5.48×10^{-2}	N_0/N_{E0}	0.65

surface does not have affinity with the DGEBA molecules; (2) the adsorption rate of PACM20, $k_{a,A} \exp(-E_{a,A}/RT)$, is two orders of magnitude larger than its desorption rate, $k_{d,A} \exp(-E_{d,A}/RT)$, denoting a preferential adsorption on the PACM20 species as reported by Arayasantiparb et al. [2]; (3) the value of reaction rate, $k_{r0} \exp(-E_a/RT)$, is similar to that reported by Sanford [29] and (4) the value of diffusivity determined by the parameters $D_0 \exp(-E_D/RT)/\Delta L^2$ and b_D is roughly an order of magnitude larger than that reported by Sanford [29], and (5) the number of adsorption sites on the aluminum surface (N_0) is about 65% of the initial number of DGEBA molecules in the bulk (N_{E0}). In most cases, the fitting results were consistent with the data in the literature.

Figure 3.4a shows the correlation results of Yang and Pitchumani [44] in terms of the PACM20 volume fraction as a function of distance from the aluminum wire surface. The dashed line denotes the experimental data, and the solid line corresponds to model prediction. The prediction closely followed the data over the entire range. The concentration of PACM20 in terms of percentage volume was a large value, 80%, at the aluminum surface, indicating a preferential adsorption on the species, and decreased sharply away from the fiber surface. In the region between 100 and 1,500 nm, experimental measurements were not reported, and Arayasantiparb et al. [2] stated that the concentration was a constant bulk value of 25%. As shown in Fig. 3.4a, the model predicted a decrease in the concentration to 22% before the recovery of the bulk concentration at 25%. The model prediction pointed to the fact that the mass aggregation in the region 0–100 nm had to be compensated by the mass deficit beyond the region. However, the deficit may be too small to have been discerned in the experimental measurement technique. Figure 3.4b shows a close-up view of Fig. 3.4a in the region 0–100 nm. The error bars of the experimental data as provided in Ref. 14 are included for comparison. Overall, Fig. 3.4 demonstrates the ability of the model to represent the physical trends.

Figure 3.5a–f shows an example distribution of the number of epoxy and amine molecules from the fiber surface (layer 1) to the far region (layer 100) at different nondimensional times during the process. Figure 3.5a, b presents the concentration profiles for the molecules in the adsorbed state, Fig. 3.5c, d corresponds to bulk state concentration profiles, and Fig. 3.5e, f shows the total concentrations of epoxy

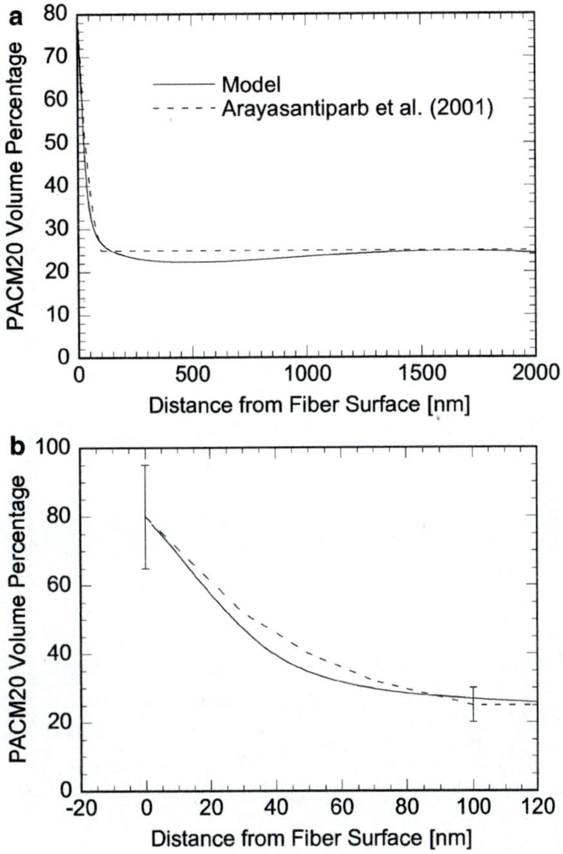

Fig. 3.4 Comparison of the interphase composition profile predicted by the model with experimental data over (**a**) a relatively large region from the fiber surface and (**b**) the interphase region (after [44])

and amine, $N'_{E,tot}$ and $N'_{A,tot}$, respectively. The results correspond to the parameter combination of $N_S = 0$ (i.e., without a sizing layer), $\gamma = 0$ (corresponding to an absence of chemical reactions), $\beta_E = \beta_A = 0.5$, $\alpha_A = 1.5$, $N'_{E0} = N'_{A0} = 1$, $N'_{E,1} = 2$, $N'_0 (= N_0/N_{E0}) = 1$, and $\phi_{EA} = 60$. The desorption ratios β_E and β_A were relatively small, which indicated that the resin molecules were easily adsorbed onto the fiber surface.

In the adsorbed state profiles, Fig. 3.5a, b, the number of molecules for both of the species increases due to adsorption onto the surface, resulting in a high concentration region near the fiber surface. The concentration profiles propagated from a small region near the fiber surface at $t' = 0.2$ to the far region at the final time $t' = 102$, which is identified as a strong adsorption effect. The number of epoxy molecules near the fiber surface is smaller than that of the amine molecules because of the larger amine adsorption rate ($\alpha_A = 1.5$). The adsorption/desorption processes causes the deficit of species in the bulk state near the fiber surface, as shown in Fig. 3.5c, d from $t' = 0.2$ to $t' = 25$. However, the diffusion process compensates for the deficit at the final time, $t' = 102$, when the concentration gradients in the bulk

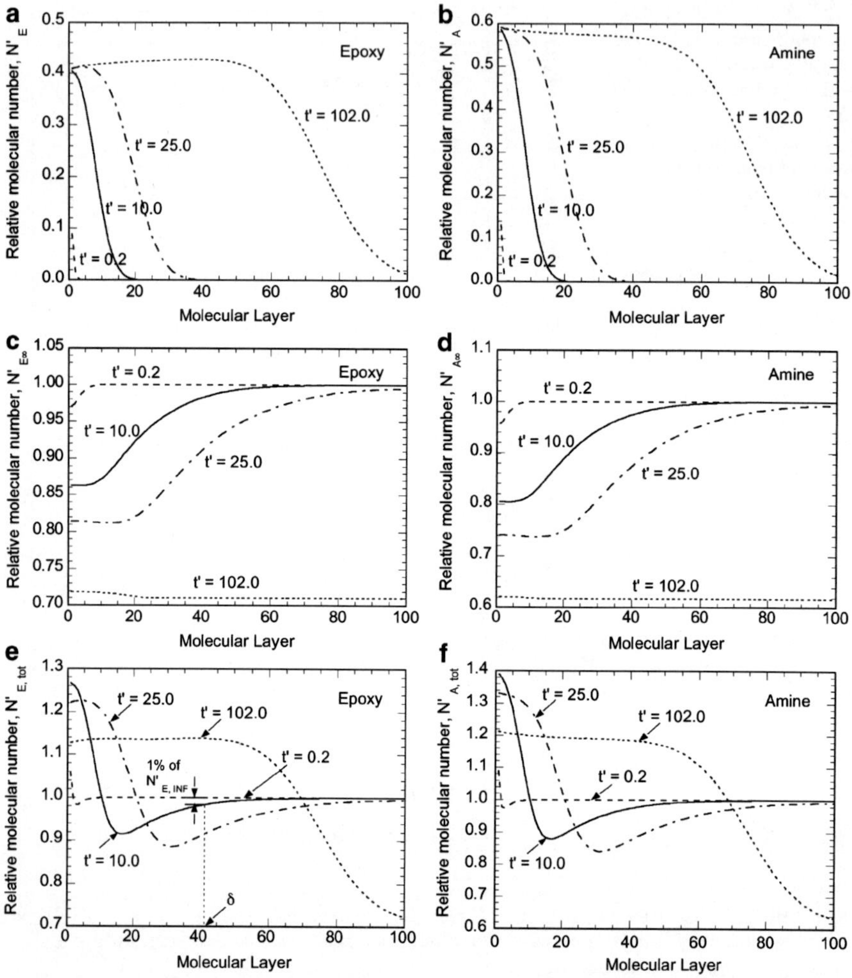

Fig. 3.5 Interphase concentration profiles in terms of the relative number of (**a**) epoxy and (**b**) amine molecules in the adsorbed state; (**c**) epoxy and (**d**) amine molecules in the bulk state; and total (**e**) epoxy and (**f**) amine molecules in both states (after [44])

state approaches zero. The total concentration profiles of each species (Fig. 3.5e, f) shows minima at $t' = 10$ and $t' = 25$, which could be explained by the coupled influence of the adsorption/desorption and the diffusion processes as discussed.

The interphase thickness is a critical factor influencing the overall composite performance and properties, as noted in the literature [19, 28]. For example, Rydin et al. [28] reported that a thin ductile interphase increases the interlaminar toughness through crack blunting, enhanced frictional sliding, and greater part deflections prior to fracture. However, a further increase in the thickness was shown to facilitate debonding and delamination because the strong adhesion between fiber and matrix was replaced by weak dipole–dipole interactions between the interphase and the matrix. In another study, Liu et al. [19] found that when the modulus of the interphase was greater than that of the matrix, an increase in the interphase thickness (δ) led to an increase in the overall composite modulus.

Interphase thickness may be defined as the number of layers from the fiber surface beyond which the epoxy concentration is within 1% of the epoxy concentration in the far region layers (Fig. 3.5e). A similar thickness, δ_A, may be defined on the basis of the amine concentration profile. With the thickness based on the epoxy concentration profile denoted as δ_E, an overall δ may be determined as the larger of the two values obtained from the epoxy and amine profiles, i.e., $\delta = \max(\delta_\varsigma, \delta_E)$. As shown in Fig. 3.5e, f, the interphase is thin at an early time ($t' = 0.2$) and grows as the process progresses ($t' = 102$); the profiles at $t' = 102$ are the equilibrium profiles representing the balance among the adsorption, desorption, and diffusion processes. For the combination of parameters in Fig. 3.5, because of the absence of the reaction and the relatively small desorption, the influence of the fiber surface propagates all the way to the far region layer, leading to a very thick or no distinct interphase formation at the final time.

Figure 3.6 shows the concentration profiles evolution of a system with sizing corresponding to the parameter combination of $N_S = 5$, $\gamma = 5$, $\beta_E = \beta_A = 0.5$, $\alpha_A = 1.5$, and all other parameters retaining the same values as previously stated for Fig. 3.5. Recall that the mass transfer through the adsorption, desorption, and diffusion processes is dramatically slowed when the reacting resin system reaches the gelation point, and the final concentration profiles can be approximated by the profiles at the gelation point. Because γ determines the gelation time and, in turn, the available time for the mass transfer processes, as γ increases, the available time decreases. In Fig. 3.6, γ is set to be relatively large, providing only a limited time for the mass transfer processes to develop.

The concentration of epoxy in the adsorbed state increases with time, and the growth is stopped by the reaction at the gelation time $t' = 51.1$, as shown in Fig. 3.6a. The amine concentration profiles, shown in Fig. 3.6b, have maxima near the fiber surface, which may be explained as follows: because the epoxy sizing directly contacts the fiber surface, epoxy molecules occupy most of the adsorption sites near the fiber surface, and most of the amine molecules could only be adsorbed on top of the epoxy molecules. Therefore, the amine concentration is small at the fiber surface, followed by an increase within a few molecular layers around the fiber due to adsorption and then a decrease as the net adsorption diminishes in the region away from the fiber. As shown in Fig. 3.6c, d, the initially large concentration of epoxy and zero concentration of amine (at $t' = 0.2$) near the fiber surface correspond to the epoxy sizing layer applied on the fiber. At the gelation time ($t' = 51.1$), the concentration gradients in the bulk are small but greater than zero because the gelation time is not long enough for the diffusion process to reach equilibrium.

At time $t' = 10$, the total epoxy profile (Fig. 3.6e) has two distinct regions: a large gradient region near the fiber surface, followed by a small gradient region. An examination of the concentration profiles of the epoxy species in the adsorbed state (Fig. 3.6a) and bulk state (Fig. 3.6c) separately reveals that the large gradient region

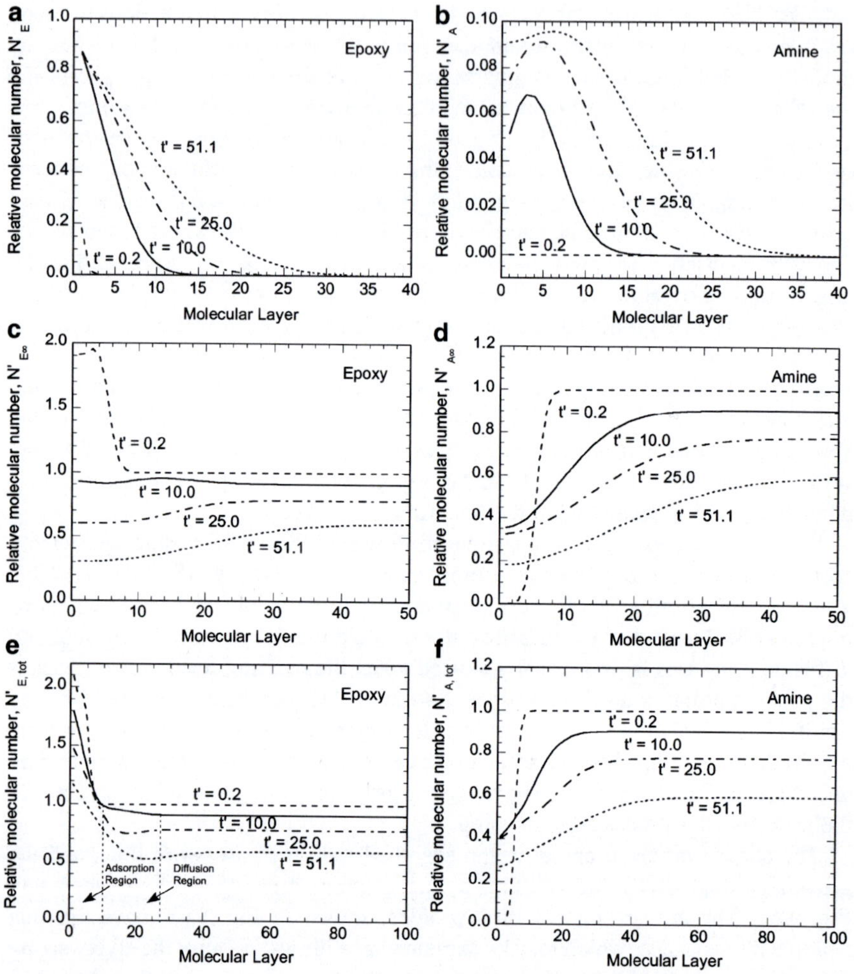

Fig. 3.6 Interphase concentration profiles of (**a**) epoxy and (**b**) amine molecules in the adsorbed state; (**c**) epoxy and (**d**) amine molecules in the bulk state; and total (**e**) epoxy and (**f**) amine molecules in both states (after [44])

comes from the profile in the adsorbed state, whereas the small gradient region is determined by the profile in the bulk state. These are shown in Fig. 3.6e as the adsorption region and the diffusion region, respectively, for the particular time instant $t' = 10$. These regions grow away from the fiber with time, with the gradients in the diffusion region approaching zero because the tendency of the diffusion to equilibrate the concentration, whereas the gradient in the adsorption region approached an equilibrium value, corresponding to the net balance of the adsorption and desorption effects. Two minima of the epoxy concentration are shown in Fig. 3.6e, at $t' = 25$ and $t' = 51.1$, respectively, which can be explained

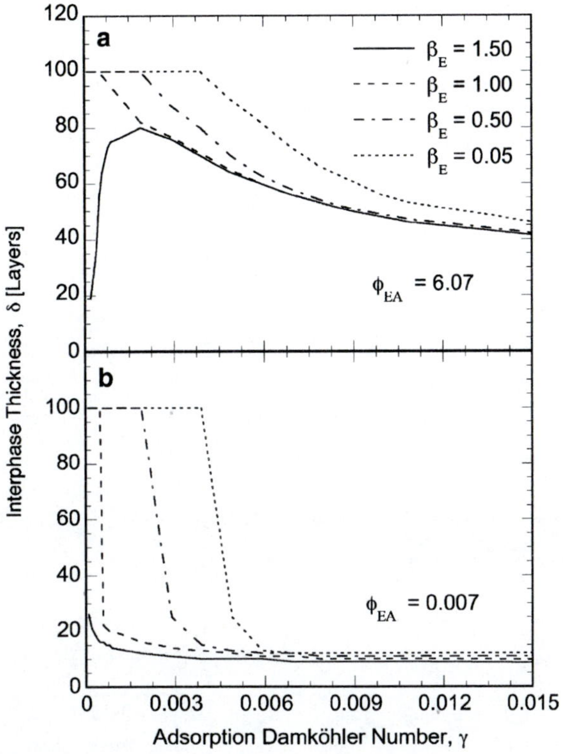

Fig. 3.7 Interphase thickness as a function of adsorption Damköhler number at various epoxy desorption ratios for (**a**) $\phi_{EA} = 6.07$ and (**b**) $\phi_{EA} = 0.007$ (after [44])

by the fact that epoxy molecules are adsorbed onto the fiber surface from the neighborhood, resulting in the deficiency of the epoxy species, which is not sufficiently replenished through the diffusion process. Figure 3.6f presents the amine concentration evolution with time. In this case, the diffusion and adsorption mass transfer are in the same direction, toward the fiber surface, as opposed to that in Fig. 3.6e, where the adsorption causes epoxy migration toward the fiber, and diffusion tends to move the epoxy molecules away from the fiber.

An interphase thickness could be identified on the basis of the total concentration profiles, as shown in Fig. 3.5e, and forms a concise representation of the interphase concentration profiles. As discussed previously, it critically influences the composite properties and constitutes an important input in the micromechanical models [23, 28]. It is, therefore, instructive to examine the influence of each mechanism studied in the model on the interphase thickness, as presented in Figs. 3.7–3.9. In Fig. 3.7a, the interphase thickness at the gelation time is plotted as a function of γ for different values of β_E for the parameter combination of $N_S = 5$, $\alpha_A = \beta_A = 1.0$, $\phi_{EA} = 6.07$ and the other parameter values as in Fig. 3.5. In the parametric studies of the epoxy–amine system, the maximum thickness value was taken to be $N_L = 100$ layers. For the case of no reaction (i.e., $\gamma = 0$), which corresponds to an infinitely

Fig. 3.8 Interphase thickness as a function of adsorption Damköhler number at various amine adsorption ratios for (**a**) $\phi_{EA} = 6.07$ and (**b**) $\phi_{EA} = 0.007$ (after [44])

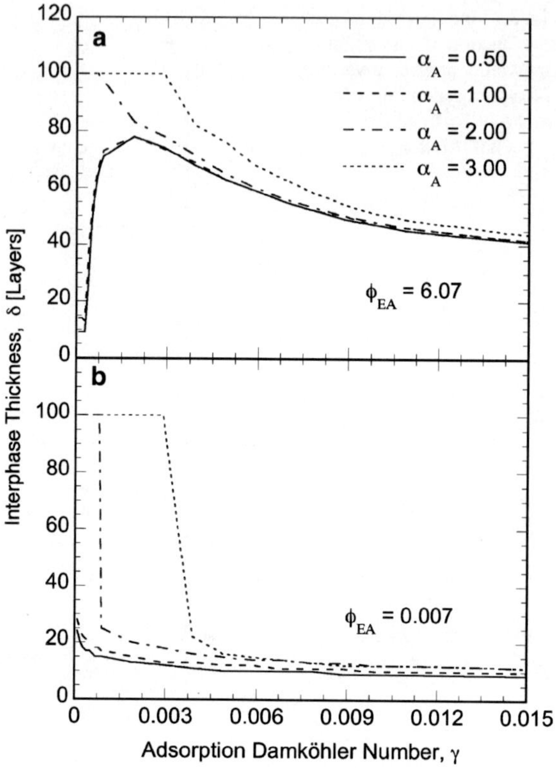

long interphase growth time, the diffusion process is fully developed, and the interphase concentration gradient and thickness are only determined by the net adsorption. For the case of $\beta_E = 0.05$, the small value indicates that the fiber surface has a strong net adsorption, which may penetrate to the entire resin domain given enough time. Therefore, when $\gamma = 0$, the curve starts at the maximum thickness of 100 layers. As γ increases, less time was available for the interphase development, and the thickness decreases monotonically. The same trend is observed for $\beta_E = 0.50$ and 1.00, which also has relatively strong net adsorptions.

With the increase in γ from zero, the increased reaction rate corresponds to less time available for the transport processes prior to gelation. The diffusion profile is, therefore, arrested before completion, which leads to a thick diffusion. Because the adsorption/desorption concentration gradients are confined to a region near the fibers, the overall interphase thickness was governed by the diffusion profile, which leads to an increased interphase thickness, as shown in Fig. 3.7a for $\beta_E = 1.50$. The increase in the interphase thickness from $\gamma = 0$ to a nonzero value is primarily due to the shift in the contribution to the interphase gradient from that of adsorption/desorption only (for $\gamma = 0$) to that of diffusion (for a small γ not equal to zero). With a further increase in the reaction rate (i.e., an increase in γ), the progressive decrease in the available time for the transport processes leads to

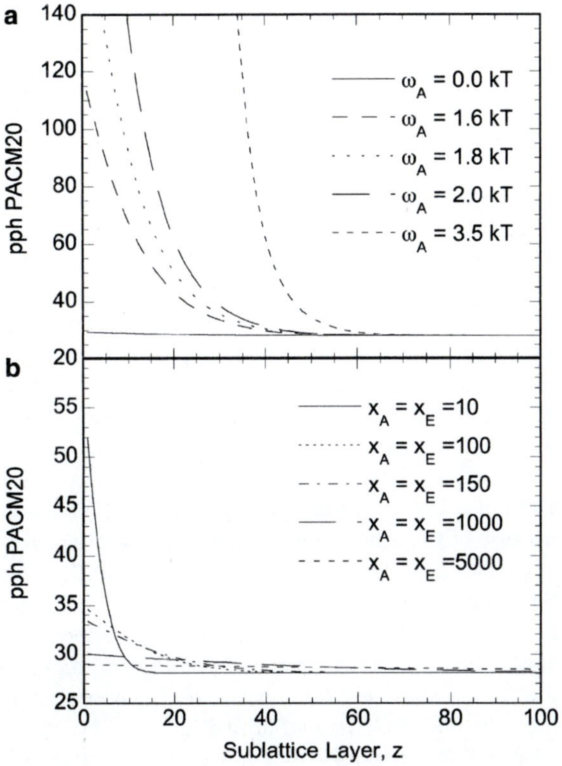

Fig. 3.9 Interphase concentration profiles predicted by the thermodynamic model for different: (**a**) surface potentials and (**b**) chain lengths (after [45])

a monotonic decrease in the interphase thickness. In fact, as γ approaches infinity (the case of an infinitely fast reaction), the interphase composition should be that of the initial condition, and the interphase thickness should be the sizing thickness, N_S. As shown in the trend in Fig. 3.7a, all of the curves asymptotically approach this value (beyond the range of the plot). Furthermore, for a fixed γ, the interphase thickness decreases monotonically with increasing β_E as more molecules are desorbed from the interphase region.

Figure 3.7b has the same parameter combination as Fig. 3.7a except that the diffusion ratio is set to $\phi_{EA} = 0.007$ to show the effect of sluggish diffusion. The interphase thicknesses predicted in Fig. 3.7b reflect the negligible contribution of the diffusion process and are always smaller than or equal to the corresponding values shown in Fig. 3.7a. Also, for the weak net adsorption cases, Fig. 3.7b does not exhibit the peaks shown in Fig. 3.7a, again due to the small diffusion.

From Eqs. 3.3 and 3.6, it may be reasoned that β_A has the same parametric effect on the interphase thickness variation as β_E in that large values of β_A and β_E corresponds to weak net adsorption effects. It must be noted, however, that although the interphase thickness variations with corresponding desorption rates β_A and β_E are similar, the interphase compositions in the two cases are different: a large β_E

value would correspond to an epoxy-deficient interphase, whereas a large β_A value would correspond to an amine-deficient interphase. These two types of interphase would, in turn, lead to completely different properties of the overall composite.

Figure 3.8a presents the interphase thickness as a function of γ and α_A for $N_S = 5$, $\beta_A = 2.0$, $\beta_E = 1.0$, $\phi_{EA} = 6.07$ and the other parameter values as in Fig. 3.5. The diffusion ratio ϕ_{EA} of 6.07 denotes a relatively active diffusion process. The parameter α_A reflects the relative attraction strength of the fiber surface to the epoxy and amine molecules, with $\alpha_A > 1$ denoting a preferential adsorption of the amine molecules. Through an increase in α_A, surface attraction to amine molecules is strengthened, which leads to a thicker, amine-rich interphase. The influence of γ can be discussed by similar considerations to those for the β_E effect in Fig. 3.7. For strong net adsorption (corresponding to $\alpha_A = 2.0$ and 3.0), the thickness decreases monotonically from the maximum thickness of 100 layers with the increase of γ, whereas in the cases of weak net adsorption (corresponding to $\alpha_A = 0.5$ and 1.0), the thickness increases first due to the contribution of the diffusion and then decreases gradually. All the curves tend to the initial sizing thickness for $\gamma \to \infty$. At a constant value of γ, the interphase thickness increases monotonically with the increase in α_A because of the enhancement of the net adsorption. Figure 3.8b illustrates the effect of a weak diffusion represented by $\phi_{EA} = 0.007$ on the interphase thickness variation. Similar to the discussion of Fig. 3.7b, the interphase thicknesses predicted in Fig. 3.8b do not have a significant contribution of the diffusion process and are always smaller than or equal to the corresponding values shown in Fig. 3.8a. Again, for the weak net adsorption cases, Fig. 3.8b does not exhibit the peaks shown in Fig. 3.8a due to the lack of diffusion.

While the kinetics-based model presented in this subsection represents one approach to predicting the interphase in composite materials, an alternative approach based on thermodynamic considerations is briefly reviewed in the following subsection.

3.2.2 Thermodynamic Model

The interphase formation model summarized in this subsection is that of Hrivnak [13], which builds upon an earlier model presented by Palmese [23]. The model examines the chain/surface and chain/chain interactions of an epoxy/amine binary thermosetting resin mixture in the vicinity of a fiber surface. The interaction of a single polymer chain with a surface is described by an analytical molecular partition function, which is incorporated into a sublattice model to derive the Gibbs free energy of the chain assembly. By minimizing the assembly Gibbs free energy, the volume fraction of the amine species at the (z)th lattice layer, $\nu_A(z)$, is obtained as follows.

$$v_A(z) = \kappa v_A(\infty) \underbrace{q(x_A, z) \exp[S_A(z)]}_{\text{chain/surfaceinteraction}} \underbrace{\exp\left[\chi\left(v_E^2(\infty) - v_E^2(z)\right)\right]}_{\text{chain/chaininteraction}}$$
$$\underbrace{\times \exp\left[\frac{v_A(z) - v_A(\infty)}{N_L}\left(1 - \frac{x_A}{x_E}\right)\right]}_{\text{excess mixing}}, \tag{3.15}$$

where

$$q(x_A, z) = 1 - \sqrt{\frac{2}{\pi x_A}}\left(1 - e^{\omega_A/kT}\right) erfc\left(\frac{z}{\sigma_A}\right), \tag{3.16}$$

$$S_A(z) = \frac{\omega_A}{kT} \exp\left(\frac{\omega_A}{kT}\right) \sqrt{\frac{2}{\pi x_A}} erfc\left(\frac{z}{\sigma_A}\right), \tag{3.17}$$

$$\sigma_A^2 = c\frac{x_A l^2}{3}. \tag{3.18}$$

Similar expressions for the epoxy species may be obtained by switching the subscripts A (amine) and E (epoxy) in the above equations.

The parameter κ is introduced as a normalization constant to ensure that volume fractions $v_A(z)$ and $v_E(z)$ sum up to unity, and $v_A(\infty)$ is the volume fraction of amine in the far region. Equations 3.16 and 3.17 and similar equations for the epoxy species show that chain/surface interaction depends on the surface potentials, ω_A and ω_E, chain lengths, x_A and x_E, number of the lattice layer from the surface, z, Boltzmann constant, k, and the temperature, T. Positive, negative, and zero values of ω_A (and ω_E) correspond to attractive, repulsive, and neutral surfaces to the amine (and epoxy) species, respectively. The second exponential term in Eq. 3.15 accounts for the chain/chain interaction between the two species in the binary resin mixture, where χ is the interaction parameter. For $\chi > 0$, the two species are repulsive to each other. If the surface prefers to adsorb amine, a positive chain/chain interaction pushes the epoxy molecules further away from the surface, leading to enhanced preferential adsorption of amine. The last exponential term in Eq. 3.3 represents the effect of excess mixing caused by different chain lengths (x_A and x_E) of the two species. The quantity N_L is the number of lattice layers, each of thickness l^0, contained within a sublattice, and may be treated as a weighting factor of the excess mixing. Note that large values of N_L reduce the effect of excess mixing. The parameters c and l in Eq. 3.18 are the characteristic constant and the C–C bond length of the polymer chain, respectively.

Figure 3.9 presents two sample concentration profiles predicted by the above model for adsorption from a thermosetting mixture consists of PACM20 curing agent and a DGEBA epoxy resin. In the following discussion, the subscripts A and E denote the resin components PACM20 and DGEBA, respectively. Note that the volume fraction of PACM20 is converted to parts per hundred (i.e., *pph*) concentration, defined as the weight percentage of PACM20 to DGEBA in the mixture as *pph* PACM20 $= 100 \cdot \rho_A v_A / \rho_E(1 - v_A)$ where ρ_A and ρ_E are density of PACM20 and

DGEBA, respectively. Figure 3.9a shows the distribution of *pph* PACM20 from the fiber surface (Layer 0) to the far region (Layer 100) for different amine surface potentials ω_A. The results correspond to the parameter combination of $\omega_E = -0.28kT$, $x_E = x_A = 150$, $\chi = 1.0$, $N_L = 2.5$ as determined by Hrivnak [13] for the DGEBA/PACM20 system. For a neutral surface, the *pph* PACM20 values near the fiber surface are close to the bulk value of 28. The surface concentrations increase as the surface becomes more attractive to PACM20, and sharp increases are observed for $\omega_A > 1.6kT$. As the distance increases, the *pph* PACM20 decreases asymptotically to the bulk value (28) for all the values of ω_A. Figure 3.9b presents the effect of chain lengths, x_A and x_E, following the presentation format of Fig. 3.9a, for $\omega_E = 0.43$ kT and all other parameters retaining their values as in Fig. 3.9a. Figure 3.9b shows that the surface may easily adsorb shorter chains, and the *pph* concentration at the surface is almost doubled as compared to the bulk value for the case of $x_E = x_A = 10$. However, the magnitudes of the surface concentration increases are not as significant as those in Fig. 3.9a. It is also observed that the region perturbed by the surface becomes smaller as the chain length decreases.

The discussion on the interphase formation model in this subsection only provides sufficient information that will be used in the construction of the interphase–property relationships in the next section. The readers are referred to Palmese [23] and Hrivnak [13] for additional details on the thermodynamic modeling.

3.3 Interphase–Property Relationships

The interphase composition profiles predicted by either of the two models discussed in the preceding section can be used to predict the interphase properties, and in turn, the overall composite properties without the need for assumed interphase thickness and properties in the micromechanical analyses. In this section, the interphase–property relationship is illustrated by considering the elastic moduli and stress concentration factors in fiber-reinforced composites. Since matrix modulus is determined by the resin component compositions, the concentration profiles such as those in Figs. 3.5, 3.6, or 3.9 need to be mapped to corresponding interphase modulus profiles. Figure 3.10 presents experimental data on matrix modulus as a function of the parameter $X = pph - 1/pph$ for the DGEBA/PACM20 system at 30°C as reported by VanLandingham et al. [40]. Two peaks in the modulus variation are observed: one around 18 *pph* PACM20 with a value of 3.2 GPa, and the other between 48 and 56 *pph* PACM20 having a value of 2.4 GPa. Similar twin peak profiles were reported in VanLandingham et al. [40] for a variety of other epoxy–amine systems and was explained in terms of the relationship of the matrix modulus to an effective chain aspect ratio, defined as the ratio of the length to the effective diameter of the polymer chains, and the volume fraction of the microgel structures in the cured matrix. Larger values of the effective aspect ratio and the microgel volume fraction result in higher matrix modulus. At the stoichiometric point of the thermosetting system, the effective aspect ratio and the microgel volume

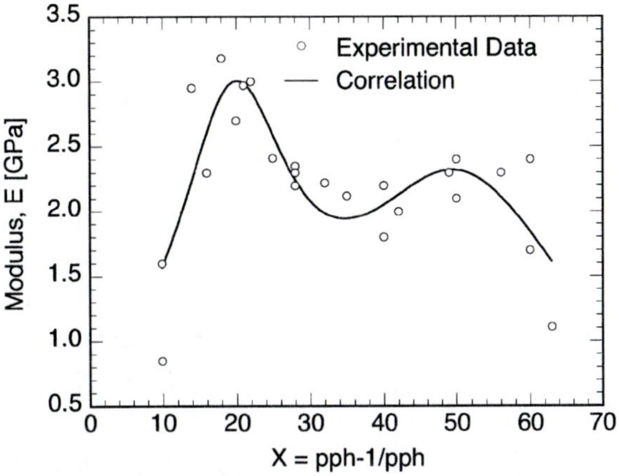

Fig. 3.10 Variation of Young's modulus with stoichiometry for a DGEBA/PACM20 system (Data from [40]; correlation after [45])

fraction were argued to exhibit minimum and maximum, respectively. Consequently, the modulus is not monotonic with respect to the *pph* values near the stoichiometric ratio. The interested reader is referred to VanLandingham et al. [40] for more details. For concentrations between 18 and 56 *pph*, the value of modulus varies about 60%. The modulus drops sharply on both ends of the concentration axis due to the fact that the matrix is in extreme deficit of either resin component and cannot be cured sufficiently. Based on these observations, a twin peak function was derived by Yang and Pitchumani [45] to fit the data as follows:

$$E = \frac{1}{(X - 19.57)^2/214.42 + 0.42} + \frac{1}{(X - 50.59)^2/803.09 + 0.47}. \tag{3.19}$$

Equation 3.19, which has an asymptotic value of 0 as *pph* approaches 0 and infinity, is used in this study to map the concentration profiles, $v_A(z)$ and $v_E(z)$, to modulus profile, $E(r)$. Note that the radial distance from a fiber center, r, may be obtained from the fiber radius, r_f, and the lattice layer, r, as $r = r_f + zl^0$.

Since the interphase region is typically thin (around 10–500 nm), most micromechanical analyses consider only a single uniform interphase layer of thickness δ_i and effective modulus E_i. From the species concentration profiles, $v_A(z)$ and $v_E(z)$, the interphase thickness may be defined as before, and determined as $\delta_i = \Delta z \cdot l^0$, where Δz is the number of lattice layers within the interphase region. The modulus profile, $E(r)$, may be used to determine the effective interphase modulus, E_i, by considering a series connection of infinite springs as follows [41]:

$$E_i = \left(r_i - r_f\right) \Big/ \int_{r_f}^{r_i} \frac{dr}{E(r)}. \tag{3.20}$$

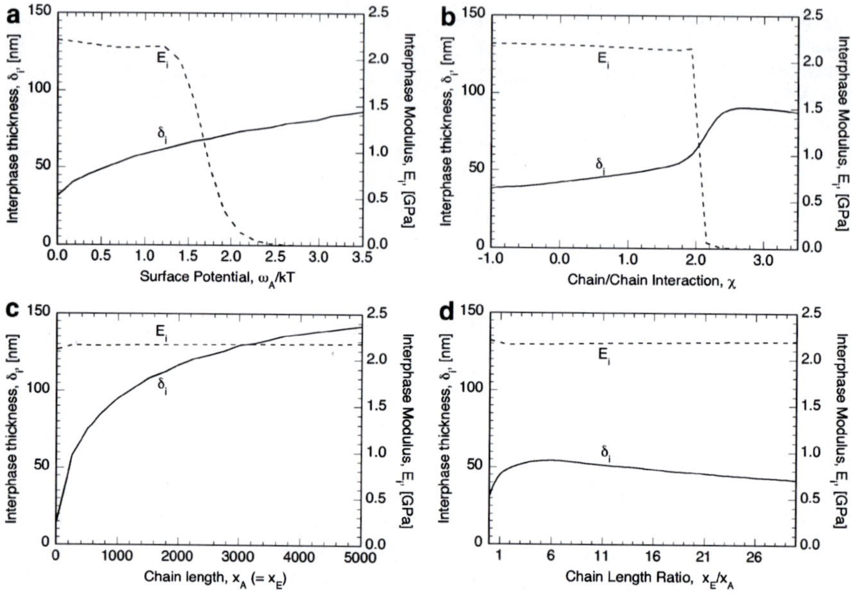

Fig. 3.11 Interphase thickness and modulus as functions of (**a**) surface potential, (**b**) chain/chain interaction, (**c**) chain length of amine, and (**d**) chain length ratio (after [45])

The interphase thickness and the effective interphase modulus were found to remain invariant with the number of lattice layers in a sublattice, N_L, and the lattice layer thickness, l^0 by Yang and Pitchumani [45], and the influence of the other parameters is considered here. Figure 3.11a shows the interphase thickness, δ_i, and the effective interphase modulus, E_i, as functions of the surface potential, ω_A, for the parameter combination in Fig. 3.9a. As the surface potential, ω_A, increases up to around 1.2*kT*, the maximum *pph* PACM20 value in the interphase region changes from 28 to about 64. The corresponding interphase modulus profiles have relatively small variations from 1.6 to 2.3 GPa. Consequently, the effective interphase modulus evaluated from Eq. 3.20 has negligible changes as $\omega_A < 1.2kT$. Further increase in ω_A results in the formation of interphase region with very high PACM20 concentration, as shown by Fig. 3.9a. The effective interphase modulus starts to drop and asymptotically approaches zero, which corresponds to the modulus of pure PACM20. Arayasantiparb et al. [2] reported a DGEBA/PACM20 interphase concentration profile with the maximum *pph* value as high as 1,560, for which E_i may be determined to be 0.0022 GPa using Eqs. 3.19 and 3.20. The interphase thickness increases from 32 to 86 nm as ω_A increases from 0 to 3.5*kT*, since more PACM20 molecules are adsorbed into the interphase region.

The influence of the chain/chain interaction parameter, χ, on the interphase properties is shown in Fig. 3.11b. The results correspond to $\omega_A = 0.43kT$, and all other parameters retain their values as in Fig. 3.9a. Recall that the positive chain/chain interaction enhances the preferential adsorption of PACM20; therefore, an increase in χ has the same effect as an increase in ω_A. Consequently, the overall trends in Fig. 3.11b are similar to those in Fig. 3.11a. For $\chi < 2.0$, the interphase modulus profiles are characterized by relatively small variations and the effective interphase modulus exhibits little change with respect to χ. Further increase in χ causes significant adsorption of PACM20, and E_i decreases asymptotically to zero as in Fig. 3.11a. A sharp increase in the interphase thickness is observed for the interaction parameter χ in the interval 1.8–2.6.

Figure 3.11c illustrates the effect of the chain lengths, x_A and x_E, on E_i and δ_i, and the results correspond to the same parameter combination as in Fig. 3.9b. The interphase thickness increases monotonically with increasing x_A and x_E, which is consistent with the concentration profiles shown in Fig. 3.9b. It must be pointed out that the interphase modulus is almost independent of the chain lengths, owing to the small variation of the modulus in the *pph* range 28–55, as noted in Fig. 3.10. Figure 3.11d presents the effect of increasing chain length x_E for a fixed chain length $x_A = 150$, and other parameters have the same values as in Fig. 3.11c. Again, the effective interphase modulus is insensitive to x_E due to the small variation of the interphase modulus profiles. A non-monotonic trend is observed for the interphase thickness, and may be explained as follows. The increase in x_E results in an increase in the excess mixing term and a decrease in the chain/chain interaction term in Eq. 3.15. As x_E increases from 1 to around 900, the increase in the excess mixing term leads to more adsorption of PACM20 and increased interphase thickness. The decrease in the chain/chain interaction term overcomes the increase in the excess mixing term as x_E increases further, yielding less PACM20 adsorption and thinner interphase.

The effective interphase modulus, the interphase thickness, and the properties of the fiber and the matrix influence the effective composite properties. In the thermodynamic model, the interphase formation is governed by the principal parameters: (1) ω_A and ω_E, (2) χ, (3) l^0, (4) N_L, and (5) x_A and x_E, of which ω_A, ω_E, and χ have been reported to be the most significant [45]. In addition, the effective properties also depend on the fiber microarchitectural parameters. In order to investigate the effect of the fiber arrangement on the composite properties, two types of representative volume elements (RVEs) are considered in the present discussion, corresponding to a staggered and rectangular arrangement of fibers as depicted in Figs. 3.12a, b respectively. The cross sections of the RVE in the *x*–*y* plane are rectangular with length *a*, height *b*, and array angle γ_o. The composite lamina is treated to be an orthotropic material with two independent transverse Young's moduli, E_2 and E_3, in the *x*- and *y*- directions in Fig. 3.12, respectively.

Yang and Pitchumani [45] presented a systematic finite element analysis of the effects of the two classes of parameters – interphase and fiber architectural – on the transverse Young's moduli, E_2 and E_3, and the stress concentration factor, K_S. For each combination of model parameters, Eq. 3.15 is used to determine the interphase composition profile, which was transformed to the interphase modulus profile $E(r)$ by Eq. 3.19, and the effective interphase modulus, E_i, was determined by Eq. 3.20. Stress fields of the three deformation states were calculated with $E(r)$ or E_i as input information. The stress field analyses provided the values of strain energies, which, in turn, were used to determine the composite moduli, E_2 and E_3. The stress concentration factor, K_S, was defined from the stress field as the ratio

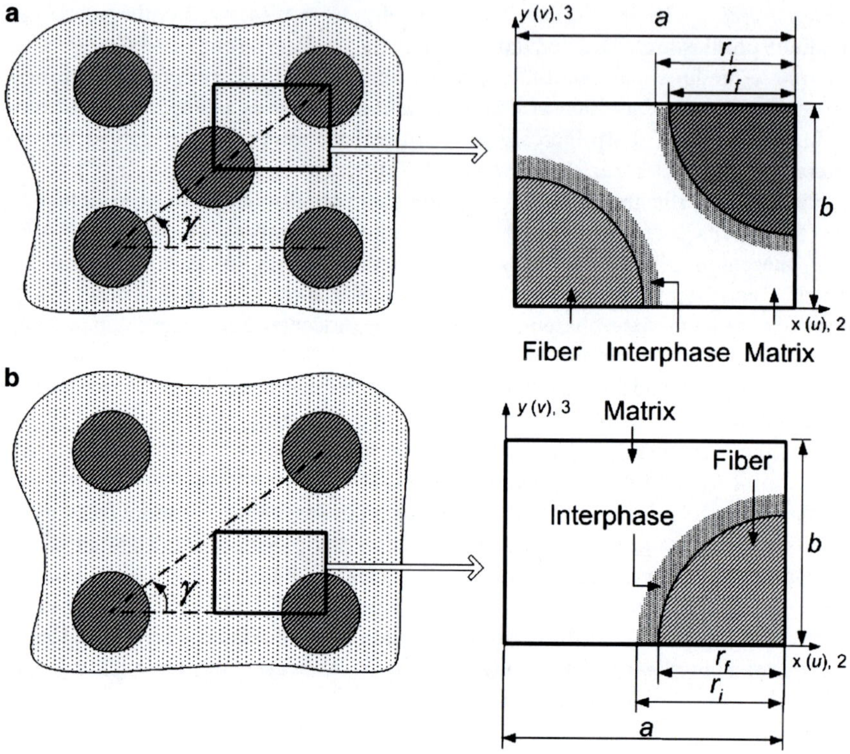

Fig. 3.12 (**a**) Staggered and (**b**) rectangular unit cells considered for the finite element analysis

between the maximum stress in the domain to the average applied stress. Illustrative results from Yang and Pitchumani [45] on the variation of E_2, E_3, and K_S with the interphase formation parameters and the geometric parameters are discussed below.

Figure 3.13a shows a dimensionless composite transverse Young's modulus, E_2/E_m, as a function of fiber modulus ratio, E_f/E_m at different surface potentials ω_A. The results correspond to a rectangular array with a packing angle, $\gamma_o = 45°$, for which the fiber arrangement is identical in the 2- and the 3-directions, and consequently, $E_2/E_m = E_3/E_m$. For the case of $\omega_A = 0$, the fiber surface is neutral to PACM20, and its concentration near the surface is almost identical to that of the far region (see Fig. 3.9a). Consequently, the composite material is similar to a two-component (fiber and matrix) system without distinct interphase regions. When $E_f/E_m = 1$, the composite modulus ratio $E_2/E_m(= E_3/E_m) = 1$ since the material is essentially homogeneous. The composite modulus is seen to increase with increasing fiber modulus, as physically expected. However, the increase rate of $E_2(= E_3)$ slows down at larger E_f/E_m values; as $E_f/E_m \to \infty$, $E_2/E_m(= E_3/E_m)$ approaches an asymptotic value corresponding to an ideal rigid body reinforcement.

Interphases richer in PACM20 are developed as ω_A increases. When ω_A increases from 0 to 1.6*kT*, the *pph* PACM20 at the fiber surface increases from 28 to 115, while the interphase modulus E_i decreases 38% from 2.30 to 1.43 GPa (Fig. 3.10). Since the

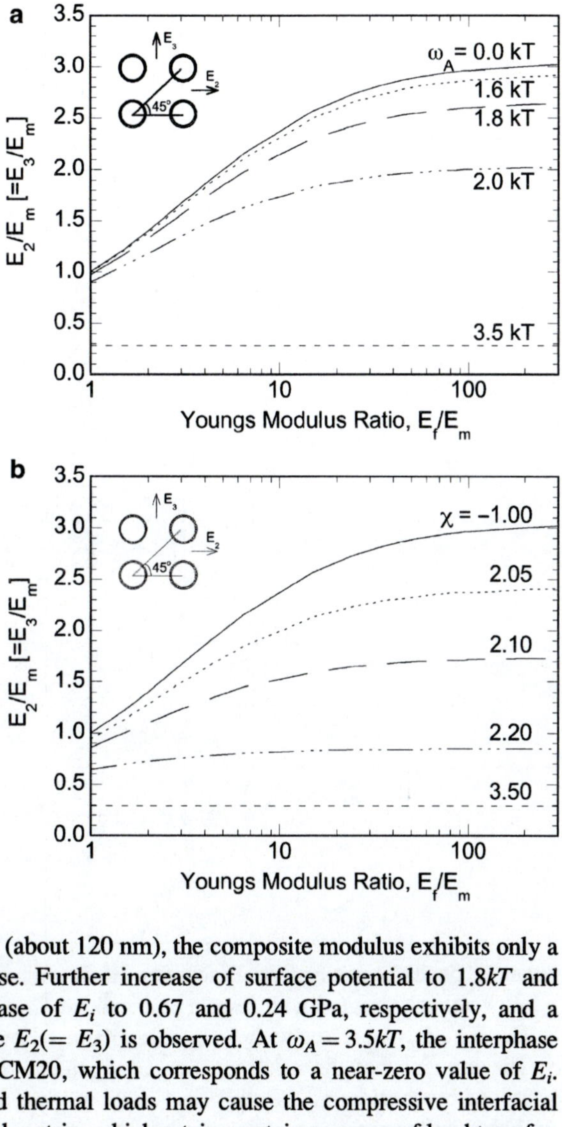

Fig. 3.13 Transverse Young's modulus of a composite with *rectangular* fiber arrangement, as a function of the fiber–matrix modulus ratio for different (**a**) surface potentials and (**b**) chain/chain interaction parameters (after [45])

interphase thickness is small (about 120 nm), the composite modulus exhibits only a slight corresponding decrease. Further increase of surface potential to 1.8kT and 2.0kT yields a sharp decrease of E_i to 0.67 and 0.24 GPa, respectively, and a significant drop in the value $E_2(= E_3)$ is observed. At $\omega_A = 3.5kT$, the interphase consists of almost pure PACM20, which corresponds to a near-zero value of E_i. In this case, mechanical and thermal loads may cause the compressive interfacial contact between the fiber and matrix, which retains certain measure of load transfer.

In Fig. 3.13b, the influence of the chain/chain interaction, χ, on the composite modulus is demonstrated. The result corresponds to $\omega_A = 0.43kT$, with all the other parameters retaining their values as in Fig. 3.13a. Once again, by virtue of isotropy in the 2- and 3-directions corresponding to $\gamma_o = 45°$, $E_2/E_m = E_3/E_m$, as in Fig. 3.13a. Recall that positive chain/chain interaction enhances the preferential adsorption of PACM20; therefore, the increase of χ has the same effect as the increase of surface potential ω_A. Since the fiber surface typically exhibits preferential adsorption of the PACM20 species [4, 8, 23, 28, 31, 36], the values of ω_A are taken to be positive. However, both repulsive and attractive chain/chain interactions are considered, and

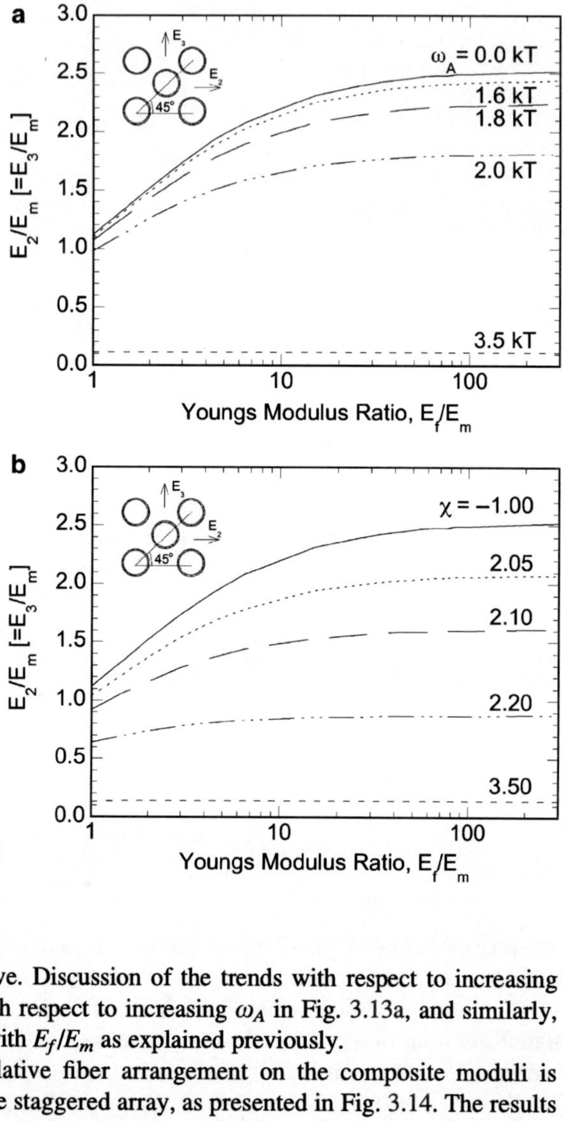

Fig. 3.14 Transverse Young's modulus of a composite with *staggered* fiber arrangement, as a function of the fiber–matrix modulus ratio for different (**a**) surface potentials and (**b**) chain/chain interaction parameters (after [45])

χ can be positive or negative. Discussion of the trends with respect to increasing χ follows that presented with respect to increasing ω_A in Fig. 3.13a, and similarly, $E_2/E_m(= E_3/E_m)$ increases with E_f/E_m as explained previously.

The influence of the relative fiber arrangement on the composite moduli is examined by considering the staggered array, as presented in Fig. 3.14. The results are based on the same parameter combinations as in Fig. 3.13, except for the fiber microarchitecture. The trends in Fig. 3.14a, b follow those in Fig. 3.13a, b respectively, namely, that $E_2/E_m(= E_3/E_m)$ increases with increasing ω_A and χ, and with increasing E_f/E_m. However, it is observed that the values of $E_2/E_m(= E_3/E_m)$ for the staggered arrays are, in general, smaller than those for the corresponding rectangular arrays (cf. Fig. 3.13).

From the stress fields solved using the finite element analysis of the RVEs in Fig. 3.12, the stress concentration factors (K_S) can be obtained for different combination of interphase and geometric parameters. Figure 3.15a–c presents the

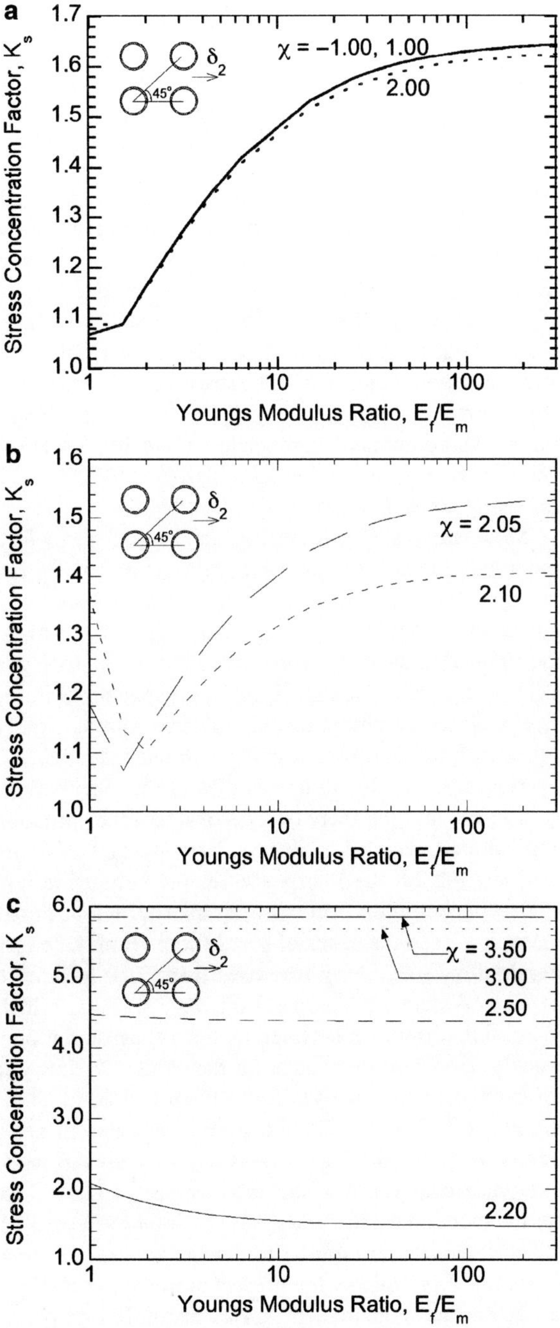

Fig. 3.15 Stress concentration factor in a composite with *rectangular* fiber arrangement as a function of E_f/E_m for **(a)** $\chi = -1.00$, 1.00 and 2.00, **(b)** $\chi = 2.05$ and 2.10, and **(c)** $\chi = 2.20$, 2.50, 3.00 and 3.50 (after [45])

Table 3.3 Interphase moduli at different values of χ and ω_A

χ	−1.00	1.00	2.00	2.05	2.10	2.20	2.50	3.00	3.50
E_i (GPa)	2.19	2.16	1.44	0.43	0.16	0.04	0.0039	0.00062	0.00032
ω_A	0.0	0.5	1.4	1.6	1.8	2.0	2.8	3.5	
E_i (GPa)	2.21	2.15	2.00	1.03	0.68	0.24	0.00114	0.00001	

stress concentration curves corresponding to nine values of χ in the range -1.00 to 3.50, which covers the range of interphase concentration values reported by Arayasantiparb et al. [2]. The interphase modulus values corresponding to the different values of χ are tabulated in the top two rows of Table 3.3. For the cases of $\chi = -1.00, 1.00, 2.00$ (Fig. 3.15a), it is noted from Table 3.3 that the values of E_i are close to that of E_m (= 2.30 GPa); therefore, the composite is similar to a two-component material. The stress concentration factor increases monotonically with increasing E_f/E_m in Fig. 3.15a, which is consistent with the trend reported for a two-component composite in the literature [4]. At fixed E_f/E_m, a slight decrease of K_S is observed as χ increases to 2.00, which may be attributed to the relatively softer interphase.

Moderately stiff interphases are formed for χ in the range 2.05–2.10. Non-monotonic trends are observed in the stress concentration factor, K_S, for the cases in Fig. 3.15b, which may be explained as follows. As E_f/E_m increases from 1, the composite modulus, E_2, experiences a rapid initial increase (Fig. 3.13b), and the applied load being directly proportional to E_2, increases at the same rate. On the other hand, the maximum principal stress in the fibers exhibits a relatively smaller increase since the load transfer through the interphase is not sufficient, causing an initial decrease in K_S with increasing E_f/E_m.

Figure 3.13b also indicates that further increase in E_f/E_m yields only limited gain in E_2, and the increase rate of maximum principal stress exceeds that of the applied load, which explains the later increase of K_S in Fig. 3.15b.

For $\chi \geq 2.20$, the interphases in the composite are soft, where the maximum stress concentration occurs in the matrix region. For the case of $\chi = 2.20$, similar reasoning as in the cases of $\chi = 2{:}05$, 2.10 may be used to explain the decreasing trend with E_f/E_m. As χ increases from 2.20 to 3.50, the value of K_S at the ratio $E_f/E_m = 1$ increases sharply from 2.07 to 5.85, which may be explained by the increasing difference between the interphase and the matrix moduli. The fibers are weakly bonded to the matrix for the cases $\chi \geq 2.50$, and K_S is seen to be relatively insensitive to the change in the fiber–matrix modulus ratio (Fig. 3.15c). It was reported by Yang and Pitchumani [45] that the surface potential, ω_A, shows similar trends as χ in Fig. 3.15a–c. As ω_A increases from 0 to 3.5kT, the type of the interphase changes from stiff to moderate, and finally to soft; non-monotonic trends for the moderately stiff interphases (similar to Fig. 3.15b) and higher K_S values for the soft interphases (similar to Fig. 3.15c) are seen again.

Figure 3.16 follows the presentation format of Fig. 3.15, and depicts the effect of the surface potential, ω_A as an example interphase formation parameter. The interphase modulus values corresponding to the different values of ω_A are tabulated in the bottom two rows of Table 3.3. Figure 3.16a–c corresponds to the results of

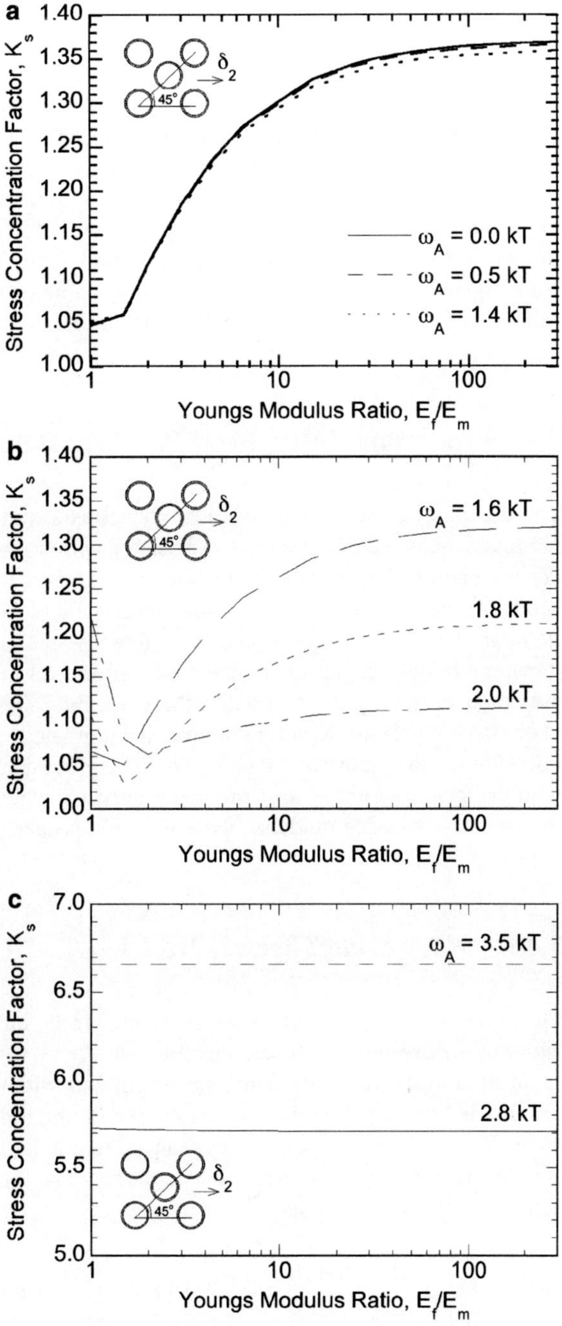

Fig. 3.16 Stress concentration factor in a composite with *staggered* fiber arrangement as a function of E_f/E_m and ω_A (after [45])

stiff, moderate, and soft interphase regions, respectively. As mentioned before, increases in ω_A and χ both lead to interphase regions richer in PACM20; therefore, the observations and explanations of trends in Fig. 3.16 are similar to those in Fig. 3.15. For $E_f/E_m > 10$ (typical range of practical interest), the stress concentration factor initially decreases with the increase of ω_A due to softer interphase; however, K_S increases as ω_A increases further to form weak interphases.

The example results presented in this section illustrated the methodology for an explicit relationship between the interphase formation parameters to the composite properties, without the need for an assumed interphase modulus or thickness. Building upon this, relating the processing parameters to the interphase and, in turn, the composite properties, is discussed in the next section.

3.4 Processing–Interphase–Property Relationships

The dominant physical and chemical phenomena during the curing of composite laminates are: (1) the heat transfer associated with the heating of the composite, and (2) the chemical reaction leading the cure process [9, 10, 15, 18, 22, 26, 47]. Modeling of the heat transfer in the laminate has been well established in the literature (see, e.g., [27, 47], and the references therein). In this section, the heat transfer model at the laminate scale is coupled with the interphase kinetics model presented in Sect. 3.2.1 to predict the cure, interphase, and composite property evolution at various locations across the thickness of the laminate under the influence of non-isothermal cure cycles. The processing conditions are linked first with the local interphase structure and properties, which in turn are used to predict the overall composite modulus, as an example property under consideration.

3.4.1 *Macroscale Thermal Model*

The heat transfer in the autoclave is assumed to be uniform across the length and width of the assembly (y- and z-directions in Fig. 3.17), and the layup thickness is small in comparison to the laminate length and width. It is further assumed that symmetric boundary conditions are imposed on the top and bottom surfaces of the layup. The problem domain is thus reduced to one half of the assembly thickness, and the one-dimensional heat transfer equations for the laminate and tooling materials may be written as

$$
\begin{aligned}
\frac{\partial[(\rho c_p T)_L]}{\partial t} &= \frac{\partial}{\partial x}\left(k_L \frac{\partial T_L}{\partial x}\right) + C_{E0}\Delta H_r(1-v_f)\frac{d\varepsilon}{dt} \quad 0 \le x \le L(\text{laminate}),\\
\frac{\partial[(\rho c_p T)_T]}{\partial t} &= \frac{\partial}{\partial x}\left(k_T \frac{\partial T_T}{\partial x}\right) \quad L < x \le L + L_T(\text{tooling}), \qquad (3.21)
\end{aligned}
$$

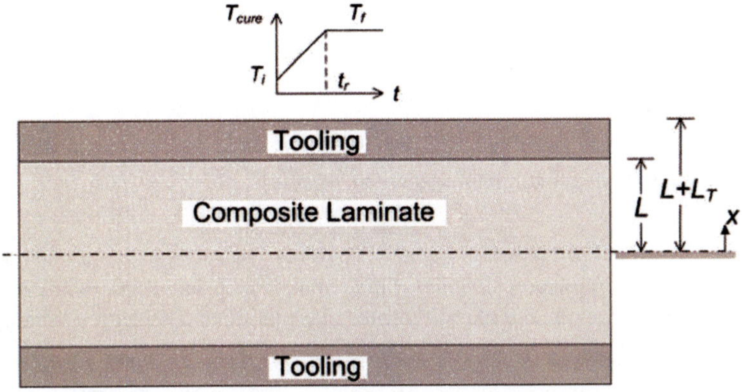

Fig. 3.17 Schematic of a laminate cross section sandwiched between tooling layers in an autoclave process for manufacturing thermosetting matrix composites

where the variables t and x are the time and location coordinates, respectively, ΔH_r is the heat of the cure reaction, C_{E0} is the initial concentration of the epoxy resin, v_f is the fiber volume fraction, and T is the local temperature of the layup assembly. The subscripts L and T refer to the laminate and the tooling, respectively. Note that a heat source term is present in the laminate equation to account for the heat of the exothermic cure reaction in the composite. The three material properties of the laminate, namely, the thermal conductivity, k_L, the density, ρ_L, and the specific heat, c_{pL}, are evaluated based on the weight fraction of the fiber in the fiber–resin mixture [22].

The initial temperature field in the laminate and tooling for the governing equation, Eq. 3.21, is given by $T_L(x,0) = T_T(x,0) = T_i$, and the boundary conditions associated with Eq. 3.21 are convective heating of the top tooling surface by the cure cycle $T_{cure}(t)$ in the autoclave ($x = L + L_T$ in Fig. 3.17) and an insulated condition at the symmetry line of the layup ($x = 0$ in Fig. 3.17). Furthermore, temperature and heat flux must be continuous at the laminate-tooling interface ($x = L$). The mathematical expressions for these conditions may be written as follows:

$$
\begin{aligned}
&-k_T\frac{\partial T_T}{\partial x}(L+L_T,t) = h[T_T - T_{cure}(t)]; \quad \frac{\partial T_L}{\partial x}(0,t) = 0,\\
&T_T(L,t) = T_L(L,t); \quad k_T\frac{\partial T_T}{\partial x}(L,t) = k_L\frac{\partial T_L}{\partial x}(L,t),
\end{aligned}
\tag{3.22}
$$

where h is the heat transfer coefficient and all the other terms in the above equations are defined previously. The cure cycle $T_{cure}(t)$ considered begins with a linear ramp from T_i to T_f within a ramp time t_r, followed by a hold stage with a constant temperature T_f, as shown schematically in Fig. 3.17.

The term $d\varepsilon/dt$ is the local reaction rate, and its determination is the focus of the discussion in the next subsection.

3.4.2 Multiscale Thermochemical Model

The kinetics model discussed in Sect. 3.2.1 paves the way for the assembling of the multiscale thermochemical model, which involves the determination of the term $d\varepsilon/dt$ in Eq. 3.21 from the microscale species rate equations. The degree of cure, ε, is defined as

$$\varepsilon = \frac{N_{E0} - N_E}{N_{E0}}, \tag{3.23}$$

where N_{E0} is the initial number of epoxy molecules in the half domain between two fiber surfaces (Fig. 3.3), and N_E is the instantaneous number of epoxy molecules in the half domain:

$$N_E = \sum_{i=1}^{N_L} \left(N_{E,i} + N_{E\infty,i}\right). \tag{3.24}$$

Differentiating Eqs. 3.23 and 3.24 with respect to time yields:

$$\frac{d\varepsilon}{dt} = -\frac{1}{N_{E0}}\frac{dN_E}{dt} = -\frac{1}{N_{E0}}\sum_{i=1}^{N_L}\left(\frac{dN_{E,i}}{dt} + \frac{dN_{E\infty,i}}{dt}\right), \tag{3.25}$$

where the rates of change $dN_{E,i}/dt$ and $dN_{E\infty,i}/dt$ are given by Eqs. 3.3 and 3.8, respectively. The energy equation, Eq. 3.21, for the macroscale composite domain in Fig. 3.17 and the reaction rate equation, Eq. 3.25, for the microscale model domain in Fig. 3.3 together constitute a multiscale thermochemical model. Note that Eqs. 3.21 and 3.25 are coupled through the temperature dependence of the rate terms in Eqs. 3.3 and 3.8.

The governing equations, Eqs. 3.21, 3.25, and the rate equations of the cure kinetics model, along with the initial and boundary conditions in Eqs. 3.13, 3.14, and 3.22 can be solved numerically [46]. Depending on the fiber volume fraction, each computational grid may consist of a large number of fiber surfaces with the adsorption–desorption–diffusion–reaction processes such as that illustrated in Fig. 3.3 occurring near each surface. All the fibers within a computational grid experience the same temperature history, and, hence, the same concentration profile evolution history. Within each time step, the temperature field obtained from Eq. 3.21 is used to calculate the cure reaction rate $d\varepsilon/dt$ using Eq. 3.25, which in turn is substituted into Eq. 3.21 to update the temperature field; and the procedure is performed until the temperature field is converged. The stopping criterion for the numerical simulation is that all the sections of the composite are completely cured, which corresponds to $\varepsilon = 0.95$. At convergence, the concentration profiles predicted by the multiscale thermochemical model are used to determine the interphase and composite material properties, following the process in Sect. 3.3, and the effects of the cure cycle on the interphase and the resulting composite properties are examined in the discussion below.

A nondimensional form of the governing thermochemical equation (Eq. 3.21) and the associated boundary and initial conditions is obtained by introducing the dimensionless temperature, location, and time as [46]: $\theta = (T - T_0)/T_0$, $x' = x/L$, $t' = k_{d,E}\exp\left(-E_{d,E}/RT_0\right) \cdot t$, where T_0 is a reference temperature; while the dimensionless form of the reaction rate equation, Eq. 3.25, and its initial conditions is derived by introducing the dimensionless time t' and dividing all the number of molecules (e.g., $N_{E,i}$, $N_{A,i}$) by the initial number of epoxy molecules in the layer next to the middle plane N_{E0} (e.g., $N'_{E,i} = N_{E,i}/N_{E0}$). The principal dimensionless groups that govern the interphase concentration evolution at various locations in a composite were identified using the epoxy desorption rate as the scaling factor instead of the epoxy adsorption rate in Table 3.1, and may be summarized as follows:

1. Adiabatic reaction temperature $\theta_{ad} = \Delta H_r C_{E0}\left(1 - v_f\right)/\rho c_p T_0$;
2. Epoxy adsorption ratio $\alpha_E = \left(k_{a,E}/k_{d,E}\right)e^{-\left(E_{a,E}-E_{d,E}\right)/RT_0}$;
3. Mutual diffusion ratio $\phi_{EA} = \left(D_0/\Delta L^2 k_{d,E}\right)e^{-\left(E_D-E_{d,E}\right)/RT_0}$;
4. Amine adsorption ratio $\alpha_A = \left(k_{a,A}/k_{d,E}\right)e^{-\left(E_{a,A}-E_{d,E}\right)/RT_0}$; and
5. Desorption Damköhler number $\gamma = \left(k_{r0}/k_{d,E}\right)e^{-\left(E_a-E_{d,E}\right)/RT_0}$.

In the following results, the number of molecular layers in the model domain is kept fixed at $N_L = 200$, the reference temperature is kept at $T_0 = 80°\mathrm{C}$, the half thickness of the laminate $L = 0.00635$ m, the fiber volume fraction $v_f = 0.5$, the tooling thickness $L_T = 0.00318$ m, the number of sizing layers $N_L = 5$, and the physical size of the molecular layer $\Delta L = 1$ nm. Furthermore, the cure cycles begin with a linear ramp from $\theta_i = -0.33$ to θ_f within a ramp time t'_r, followed by a hold stage with a constant temperature θ_f until cure was completed. The effects of the processing parameters on the interphase thickness δ_i and interphase modulus E_i (determined from the interphase composition and modulus profiles), and the transverse composite modulus E_2 are examined in terms of the various nondimensional groups.

Figure 3.18a presents the interphase thickness as a function of epoxy adsorption ratio, α_E, at three locations in the composite. For the locations $x/L = 0.0$ and 0.5, δ_i increases monotonically with the increase in α_E since more epoxy molecules are adsorbed into the interphase region. At the outer surface $x/L = 1.0$, a relatively large value of δ_i is observed for $\alpha_E = 0.001$, which is attributed to the pronounced adsorption of amine molecules, and the additional adsorption of epoxy molecules does not have significant influence on δ_i as α_E increases. For a fixed epoxy adsorption ratio, δ_i increases with x/L due to enhanced adsorption processes at the higher temperatures at the outer locations.

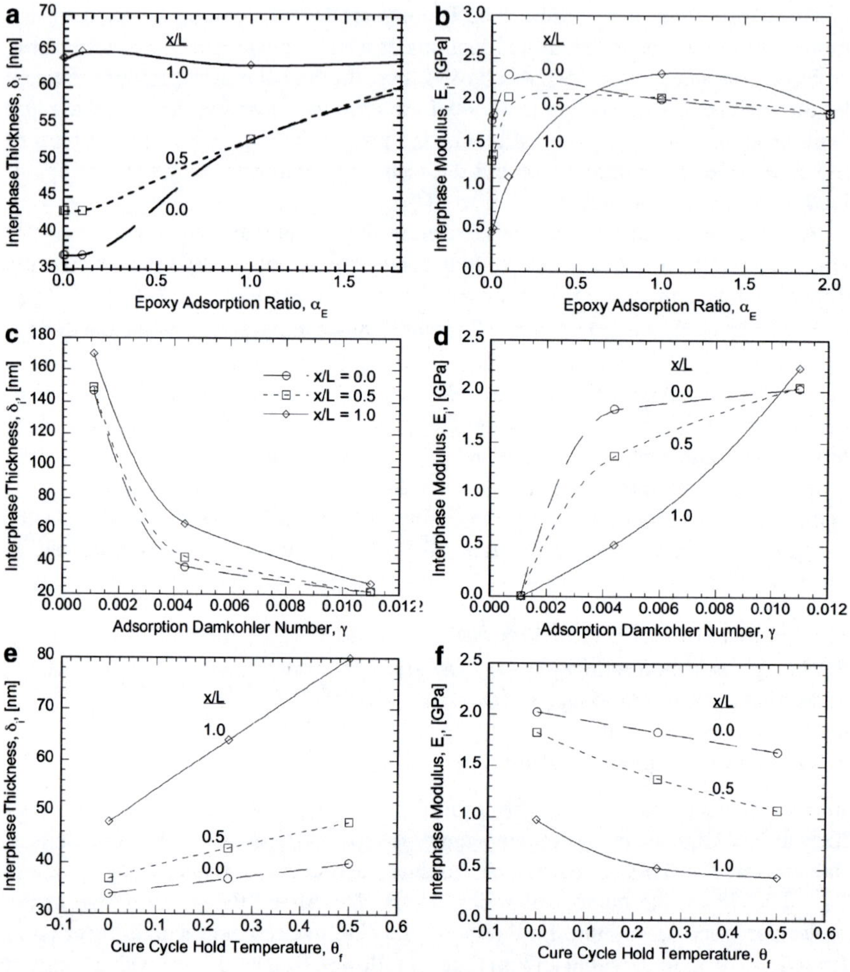

Fig. 3.18 Interphase thickness and modulus as functions of the interphase formation parameters: **(a, b)** α_E and **(c, d)** γ; and **(e, f)** cure cycle hold temperature, θ_f (after [46])

Figure 3.18b shows the interphase modulus E_i as a function of α_E at different values of *x/L*. The average *pph* value decreases monotonically with increasing α_E, which indicates that E_i should vary non-monotonically with α_E similar to that of Fig. 3.10. For the location $x/L = 0.0$, the average pph in the interphase region decreases from 60 to 28 when α_E increases from 0.001 to 0.1, which corresponds to the initial increase of interphase modulus from $E_i = 1.77$ to 2.32 GPa. When α_E increases to 1.0 and 2.0, the average *pph* value decreases to 12.8 and 12.0, and E_i is seen to decrease to 2.02 and 1.86 GPa, respectively. Note that the twin-peak shape in Fig. 3.10 is not exactly duplicated since relatively small number of α_E values are investigated; however, the range of α_E is sufficiently large to cover a comprehensive range of interphase modulus values. At the other two locations, $x/L = 0.5$ and 1.0, it is also seen that E_i reaches maxima as α_E increases. As $\alpha_E \rightarrow \infty$, all the curves in Fig. 3.18b are expected to approach zero modulus corresponding to a pure epoxy interphase. It is observed that the curve at $x/L = 1.0$ intersects with the other two curves, which is again caused by the twin-peak nature of the experimental modulus data. The interphase modulus at the outer surface may be either larger (e.g., $\alpha_E = 1.0$) or smaller (e.g., $\alpha_E = 0.1$) than that at the centerline location.

The influence of desorption Damköhler number, γ, and cure cycle hold temperature, θ_f, on the interphase thickness, δ_i, and interphase modulus, E_i, is illustrated in Fig. 3.18c–f. Since γ determines the speed of the cure process, as γ increases, less time is available for the interphase development, and the thickness decreases monotonically at each location of the composite (Fig. 3.18c). Significant drops of δ_i are observed as γ increases from 0.0011 to 0.011 (e.g., from 170 to 27 nm at the outer surface). When $\gamma = 0.0011$, the interphase regions consist of nearly pure amine due to excessive adsorption of the amine species, and E_i is close to zero at all the locations (Fig. 3.18d). The interphase modulus increases with increasing γ due to reduced amine adsorption. Note that the curve at $x/L = 1.0$ crosses the other two curves, which may be explained by similar discussions in Fig. 3.18b.

The effects of the cure cycle hold temperature θ_f on the interphase thickness and modulus are presented in Figs. 3.18e, f which directly link the cure cycle parameters to the interphase properties in a predictive manner. Figure 3.18e shows the interphase thickness as a function of the cure cycle hold temperature at three locations of the composite. The interphase thickness, δ_i, increases monotonically with increasing θ_f for a fixed x/L and with increasing x/L for a fixed θ_f, owing to enhanced adsorption of amine molecules at higher temperatures. A large increase in δ_i from 48 to 80 nm is observed at the outer surface location as θ_f increases from 0 to 0.5, while the increases in δ_i are moderate at the other two locations. The interphase modulus decreases with θ_f at a fixed location due to more adsorption of amine molecules into aminerich interphase regions (Fig. 3.18f). For the parameter combination, the maximum interphase modulus at the outer surface location equals to 0.98 GPa as $\theta_f = 0$, which is only 44% of the bulk matrix modulus. Thus, a lower hold temperature ($\theta_f < 0$) is needed in the cure cycle to improve the modulus at the outer surface location. A large difference of modulus is observed at different locations due to the cure temperature gradient in the composite, which may be made uniform through uniformity of the temperature across the laminate.

The effects of the cure cycle parameters, t_r' and θ_f, on the transverse composite modulus are presented in Figs. 3.19a, b, respectively. In Fig. 3.19a, E_2 increases monotonically with increasing t_r' at all the locations, owing to increasing interphase modulus. For the case $t_r' = 20$, the composite transverse modulus ($E_2 = 4.4$ GPa) is relatively small at the outer surface location; however, when t_r' is increased to 160, a composite with uniform transverse modulus (about 5.5 GPa) across the thickness is obtained. Figure 3.19b shows that the composite transverse modulus decreases monotonically with increasing θ_f, which is in agreement with the trend in Fig. 3.18f. For the two locations, $x/L = 0.0$ and 0.5, E_2 has limited variation from 5.56 to 5.42 GPa, while a significant drop to 4.44 GPa is observed at the outer surface, $x/L = 1.0$.

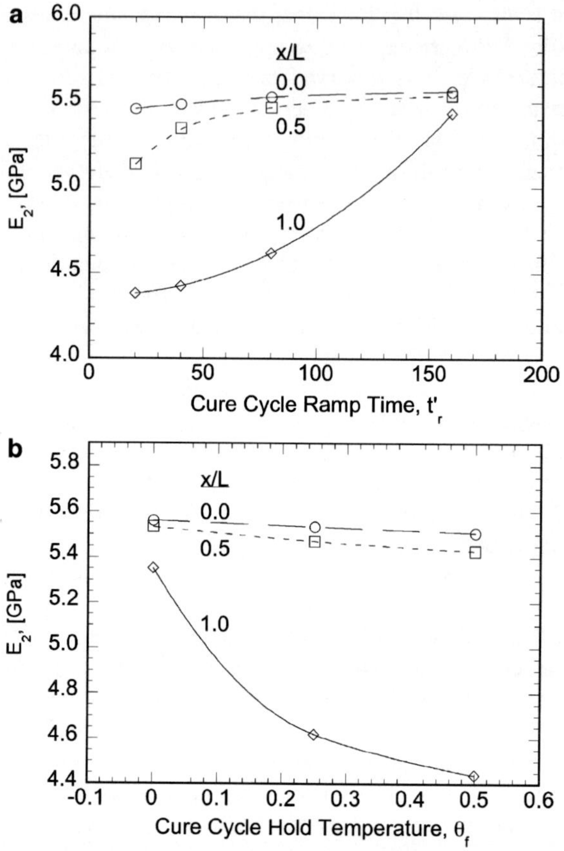

Fig. 3.19 Composite transverse modulus as a function of (**a**) cure cycle ramp time and (**b**) cure cycle hold temperature (after [46])

The results provide a direct linkage between the cure cycle and composite properties and demonstrate the feasibility of tailoring the composite property through cure cycle optimization (such as in [22]) using the predictive capability provided by the model.

3.5 Concluding Remarks

Two models for interphase composition prediction in fiber-reinforced thermosetting matrix composites were reviewed in this chapter. The thermodynamic or statistical model is based on detailed consideration of the driving forces and the complex motion and configuration of polymer molecules under these forces. The kinetics model, on the other hand, is based on a phenomenological description of the governing transport processes. The two models are equivalent if the relationship between the driving forces and the kinetic parameters can be established, which offers an area of future study. In any case, the models provide for predicting the interphase composition as function of the processing conditions, and in turn, provide processing–interphase–property relationships without the need for an *assumed* interphase thickness and properties. This predictive approach was illustrated in this chapter by considering the Young's modulus and the stress concentration factor; however, the methodology applies to prediction of other properties including thermo-oxidative stability and durability of composites. The predictive relationship will provide for tailoring the cure cycles with consideration of the properties, performance, and life of composites in use. The effects of the processing conditions on the Young's modulus were illustrated in this chapter, which may be extended to other properties. Moreover, the cure pressure cycle in an autoclave processing may also influence the interphase evolution, which needs to be studied. Flow processes in composite materials processing may also play a role in governing the fiber–matrix interphase which need to be accounted for in an interphase prediction and tailoring endeavor.

References

1. Anifantis, NK 2000, 'Micromechanical stress analysis of closely packed fibrous composites', *Composites Science and Technology*, vol. 60, pp. 1241–1248.
2. Arayasantiparb, D, McKnight, S and Libera, M 2001, 'Compositional variation within the epoxy/adherend interphase', *Journal of Adhesion Science and Technology*, vol. 15, pp. 1463–1484.
3. Dong, Z and Wu, Y 1996, 'Micromechanics analysis of particulate-reinforced composites and their failure mechanisms', *Journal of Materials Science*, vol. 31, no. 16, pp. 4401–4405.
4. Drzal, LT 1986, 'The Interphase in Epoxy Composites', in K Dusek (ed.), *Advances in Polymer Science*, Springer-Verlag, Berlin.
5. Drzal, LT 1990, 'Fiber-Matrix Interphase Structure and its Effect on Adhesion and Composite Mechanical Properties', in H Ishida (ed.), *Controlled Interphases in Composite Materials*, Elsevier Applied Science, New York.
6. Fisher, FT and Brinson, LC 2001, 'Viscoelastic interphases in polymer–matrix composites: theoretical models and finite-element analysis', *Composites Science and Technology*, vol. 61, pp. 731–748.
7. Garton, A and Daly, JH 1985, 'Characterization of the Aramid:Epoxy and Carbon:Epoxy Interphases', *Polymer Composites*, vol. 6, no. 4, pp. 195–200.
8. Garton, A, Stevenson, WTK and Wang, S 1987, 'Interfacial reactions in carbon-epoxy composites', *British Polymer Journal*, vol. 19, no. 5, pp. 459–467.
9. Han, SM and Aydil, ES 1997, 'Silanol concentration depth profiling during plasma deposition of SiO_2 using multiple internal reflection infrared spectroscopy', *Journal of the Electrochemical Society*, vol. 144, pp. 3963–3967.
10. Han, CD, Lee, DS and Chin, HB 1986, 'Development of a mathematical model for the pultrusion process, *Polymer Engineering and Science*, vol. 26, no. 6, pp. 393–404.
11. Hill, TL 1946, 'Theory of multimolecular adsorption from a mixture of gases', *Journal of Chemical Physics*, vol. 14, pp. 268–275.
12. Hirschfield, T 1977, 'Subsurface layer studies by attenuated total reflection Fourier transform spectroscopy', *Applied Spectroscopy*, vol. 31, pp. 289–292.
13. Hrivnak, J 1997, 'Interphase Formation in Reacting Systems', Technical Report Number 97-05, University of Delaware Center for Composite Materials, Newark, DE.

14. Jayaraman, K, Reifsnider, KL and Swain, RE 1993, 'Elastic and thermal effects in the interphase: part II. Comments on modeling studies', *Journal of Composites Technology and Research*, vol. 15, pp. 14–22.
15. Kamal, MR and Sourour, S 1973, 'Kinetics and thermal characterization of thermoset cure', *Polymer Engineering and Science*, vol. 13, pp. 59–64.
16. Kardos, JL 1983, 'The Role of the Interface in Polymer Composites – Some Myths, Mechanisms, and Modifications', in Ishida, H and Kumar, G (eds.), *Symposium on Polymer Composites and Interfaces*, Plenum, New York.
17. Ko, YS, Forsman, WC and Dziemanowicz, TS 1982, 'Carbon fiber-reinforced composites: Effect of fiber surface on polymer properties', *Polymer Engineering and Science*, vol. 22, no. 13, pp. 805–814.
18. Lem, KW and Han, CD 1983, 'Chemorheology of Thermosetting Resin', *Journal of Applied Polymer Science*, vol. 28, pp. 3155–3183.
19. Liu, YJ, Xu, N and Luo, JF 2000, 'Modeling of interphases in fiber-reinforced composites under transverse loading using the boundary element method', *ASME Journal of Applied Mechanics*, vol. 67, pp. 41–49.
20. Madhukar, MS and Drzal, LT 1991, 'Fiber-matrix adhesion and its effect on composite mechanical properties: II Longitudinal (0°) and transverse (90°) tensile and flexure behavior of graphite/epoxy composites', *Journal of Composite Materials*, vol. 25, pp. 958–991.
21. Masel, RI 1996, *Principles of Adsorption and Reaction on Solid Surfaces*, Wiley, New York.
22. Mawardi, A and Pitchumani, R 2003, 'Optimal temperature and current cycles for curing of composites using internal resistive heating', *ASME Journal of Heat Transfer*, vol. 125, no. 1, pp. 126–136.
23. Palmese, GR 1992, 'Origin and Influence of Interphase Material Property Gradients in Thermosetting Composites', Report CCM 92-25, Center for Composite Materials, University of Delaware, Newark, DE.
24. Papanocolaou, GC, Michalopoulou, MV and Anifantis, NK 2002, 'Thermal stresses in fibrous composites incorporating hybrid interphase regions', *Composites Science and Technology*, vol. 62, pp. 1881–1894.
25. Ponec, V, Knor, Z and Cerny, S 1974, *Adsorption on Solids*, Butterworth, London.
26. Rai, N and Pitchumani, R 1997, 'Optimal cure cycles for the fabrication of thermosetting-matrix composites,' *Polymer Composites*, vol. 18, no. 4, pp. 566–581.
27. Ramakrishnan, B, Zhu, L and Pitchumani, R 2000, 'Curing of composites using internal resistive heating', *ASME Journal of Manufacturing Science and Engineering*, vol. 122, no. 1, pp. 124–131.
28. Rydin, RW, Varelidis, PC, Papaspyrides, CD and Karbhari, VM 1997, 'Glass fabric vinyl-ester composites: Tailoring the fiber bundle/matrix interphase with nylon coating to modify energy absorption behavior', *Journal of Composite Materials*, vol. 31, pp. 182–209.
29. Sanford, WM 1987, 'Cure Behavior of Thermosetting Resin Composites', PhD. Dissertation, University of Delaware, Newark, DE.
30. Scheutjens, JMHM and Fleer, GJ 1979, 'Statistical theory of the adsorption of interacting chain molecules. 1. Partition function, segment density distribution, and adsorption isotherms', *Journal of Physical Chemistry*, vol. 83, no. 12, pp. 1619–1635.
31. Sellitti, C, Koenig, JL and Ishida, H 1990, 'Surface characterization of carbon fibers and interphase phenomena in epoxy-reinforced composites', *Materials Science and Engineering*, vol. A126, pp. 235–244.
32. Sergeyeva, LM, Todosiychuk, TT and Fabulyak, FG 1974, Getero-gennye, Polim Mater, p. 79.
33. Sottos, NR, McCullough, RL and Guçeri, SI 1989, 'Thermal stresses due to property gradients at the fiber/matrix interface', in: Reddy, JN and Teply, JL (eds.), *Mechanics of composites materials and structures*, American Society of Mechanical Engineers, New York, pp. 11–20.
34. Sottos, NR 1990, 'The Influence of the Interphase on Local Thermal Stresses and Deformations in Composites', Ph.D. Dissertation, University of Delaware, Newark, DE.

35. Sottos, NR, McCullough, RL and White, SR 1992, 'The influence of interphase regions on local thermal displacements in composites', *Composites Science and Technology*, vol. 44, no. 4, pp. 319–332.
36. Subramanian, S, Lesko, JJ, Reifsnider, KL and Stinchcomb, WW 1996, 'Characterization of the fiber-matrix interphase and its influence on mechanical properties of unidirectional composites', *Journal of Composites Materials*, vol. 30, pp. 309–332.
37. Tsai, HC, Arocho, AM and Gause, LW 1990, 'Prediction of fiber–matrix interphase properties and their influence on interface stress, displacement and fracture toughness of composite material', *Materials Science and Engineering: A*, vol. 126, pp. 295–304.
38. Tsai, SW and Hahn, HT 1979, 'Role of interface in polymer composites', in Lee, LH (ed.), *International Conference on Adhesion and Adsorption of Polymers*, Plenum, New York.
39. Tsui, CP, Tang, CY and Lee, TC 2001, 'Finite element analysis of polymer composites filled by interphase coated particles', *Journal of Materials Processing Technology*, vol. 117, pp. 105–110.
40. VanLandingham MR, Eduljee RF and Gillespie Jr, JW 1999, 'Relationships between stoichiometry, microstructure, and properties for aminecured epoxies', *Journal of Applied Polymer Science*, vol. 71, pp. 699–712.
41. Wacker, G, Bledzk, AK and Chate, A 1998, 'Effect of interphase on the transverse Young's modulus of glass/epoxy composites', *Composites Part A*, vol. 29, pp. 619–626.
42. Yang, F and Pitchumani, R 2001, 'Kinetics of interphase formation in thermosetting composites', Paper Number 078 in Proceedings (CD-ROM) of 16th Technical Conference of the American Society for Composites, Blacksburg, VA.
43. Yang, F and Pitchumani, R 2002, 'Studies on Fiber/Matrix Interphase Development in Thermosetting Matrix Composites', in Proceedings of the 12th International Heat Transfer Conference, Grenoble, France, vol. 3, pp. 153–158.
44. Yang, F and Pitchumani, R 2003, 'A kinetics model for interphase formation in thermosetting matrix composites', *Journal of Applied Polymer Science*, vol. 89, pp. 3220–3236.
45. Yang, F and Pitchumani, R 2004, 'Effects of interphase formation on the modulus and stress concentration factor of fiber-reinforced thermosetting-matrix composites', *Composites Science and Technology*, vol. 64, pp. 1437–1452.
46. Yang, F and Pitchumani, R 2005, 'Processing-interphase-property relationship in fiber-reinforced thermosetting-matrix composites', *Polymer Composites*, vol. 26, no. 2, pp. 193–208.
47. Zhu, L and Pitchumani, R 2000, 'Analysis of a process for curing composites by the use of embedded resistive heating elements', *Composites Science and Technology*, vol. 60, no. 14, pp. 2699–2712.

Chapter 4
Modeling of Response of Composite Materials with Damage

Ramesh Talreja

Abstract Damage treated here is a multitude of surfaces formed within a composite that permanently change its response to external impulses. Examples of such damage are matrix cracking at different scales, fiber breakage, fiber/matrix debonding, and interply cracking (delamination). The response affected could be mechanical (stiffness properties), thermal (expansion and conductivity), time-dependent (viscoelastic), and in general any that is sensitive to the presence of internal surfaces. The representation of damage is by internal variables and the response functions are formulated in a thermodynamics framework. Although specific cases considered here are composite laminates with multiple sets of intralaminar cracks, the formulation has sufficient generality to treat other composite configurations and other energy-dissipating mechanisms such as thermal oxidation and radiation-induced morphological changes in polymers.

4.1 Introduction

Composite materials have gained a wide range of applications in recent years and the accompanying advances in manufacturing have given rise to a rich variety of fiber architectures. Apart from the defects that may be present, the volume of a composite material usually contains numerous bonded interfaces between constituents and between layers that under loading can disbond and create internal surfaces. These surfaces can then produce larger and more surfaces by processes of merger and multiplication. The remarkable aspect of composite materials is that in spite of the large number of the internal surfaces, these materials can continue to perform satisfactorily until certain critical condition develops. With proper design,

R. Talreja (✉)
Department of Aerospace Engineering, Texas A&M University,
736A H.R. Bright Building, 3141 TAMU, College Station, TX 77843-3141, USA
e-mail: talreja@aero.tamu.edu

K.V. Pochiraju et al. (eds.), *Long-Term Durability of Polymeric Matrix Composites*,
DOI 10.1007/978-1-4419-9308-3_4, © Springer Science+Business Media, LLC 2012

the critical condition can be kept away, allowing safe operation in a substantial subcritical range. This requires, however, that the response of composite materials to external impulses in the presence of internal surfaces can be reliably determined. This chapter is aimed at an approach suited for this purpose.

The term damage is used as a collective reference to the internal surfaces and the field dealing with the characterization of damage and its effects on averaged (global) material response characteristics is called continuum damage mechanics (CDM). In recent years, this author and his associates have expanded the scope of CDM by combining it with micromechanics treatments of damage, referred to as micro-damage mechanics (MDM). The combined field thus developed has been named synergistic damage mechanics (SDM). This chapter will give an exposition of the fundamentals of CDM and present the rationale for the development of SDM. Finally, in view of the computational implications of the SDM approach, the multiscale aspects of damage mechanics will be discussed.

4.2 Mechanisms of Damage

Before discussing mechanics of damage, it is useful to review the nature of mechanisms underlying formation and progression of damage. Since this book volume addresses polymeric matrix composites, the examples taken to illustrate damage are for this material system. We begin first by considering a carbon/epoxy cross-ply laminate subjected to axial tension–tension fatigue. The x-ray radiograph shown in Fig. 4.1 illustrates a number of characteristic features of damage in composite laminates. The damage modes seen in the x-ray radiograph are schematically illustrated in the accompanying figure that shows the details through the thickness of the laminate. The main mode of damage of interest is the transverse cracking. The characteristic feature here is the multiple cracking in this mode. While the formation of individual cracks is rich in mechanisms such as fiber/matrix debonding, matrix cracking and crack growth through the ply thickness and along the fibers in the ply, this stage of damage has less influence on the composite deformational response than in the later stage when the fully grown transverse cracks multiply and form a parallel array of cracks. Other modes of damage indicated in Fig. 4.1 are axial splitting and interior delamination. These mechanisms are usually not pronounced under monotonic loading because of their slow growth, but under cyclic loading they can be observed after a sufficient number of cycles. The effect of these damage modes on in-plane laminate properties is small compared to the changes induced by multiple transverse cracking.

The transverse cracking is further illustrated in Fig. 4.2 where a view of a free edge of a cross-ply laminate is shown (upper figure). When the cracks have grown in the fiber direction in the transverse plies, a longitudinal cut displays a similar image. A higher magnification detail of one crack from the array of the transverse cracks is also shown in the figure. As seen, the path of the crack follows fiber surfaces for the most part. Detailed studies indicate that at the fiber diameter scale,

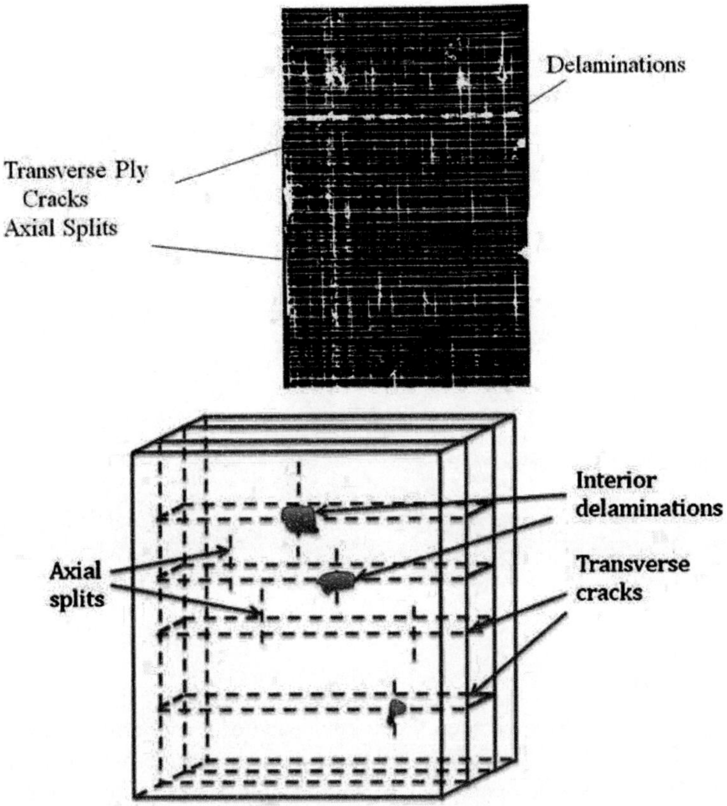

Fig. 4.1 X-ray radiograph of a cross-ply laminate showing transverse ply cracks, axial splits, and internal delaminations. The schematic illustrates these details through the laminate thickness

the microcracks formed at or close to the fiber surfaces as well as in the matrix-rich regions are involved in the coalescence process, leading to a continuous crack. The multiple cracking nature of the transverse cracks is attributed to the so-called "shear-lag" process, which essentially involves addition of a shear stress on the 90/0 ply interfaces that lasts over a distance ("lag") from the transverse crack edge. Over this distance, the longitudinal normal stress in the transverse plies is reduced, protecting the plies from formation of another crack at the load level that formed the previous crack. In the early stages of transverse cracking, individual cracks form in random sites, not influenced by other cracks. As more cracks form, the probability of overlapping of their shear-lag distances increases, leading to interaction between the adjacent cracks. In the interactive mode of transverse cracking, crack multiplication occurs under increasing load by new cracks forming in regions between existing cracks. The rate of crack multiplication per unit of applied load decreases as more cracks per unit laminate length form. The cracking process eventually tends to a saturation state. If cyclic loading is applied, the associated fatigue process can enhance the crack multiplication rate.

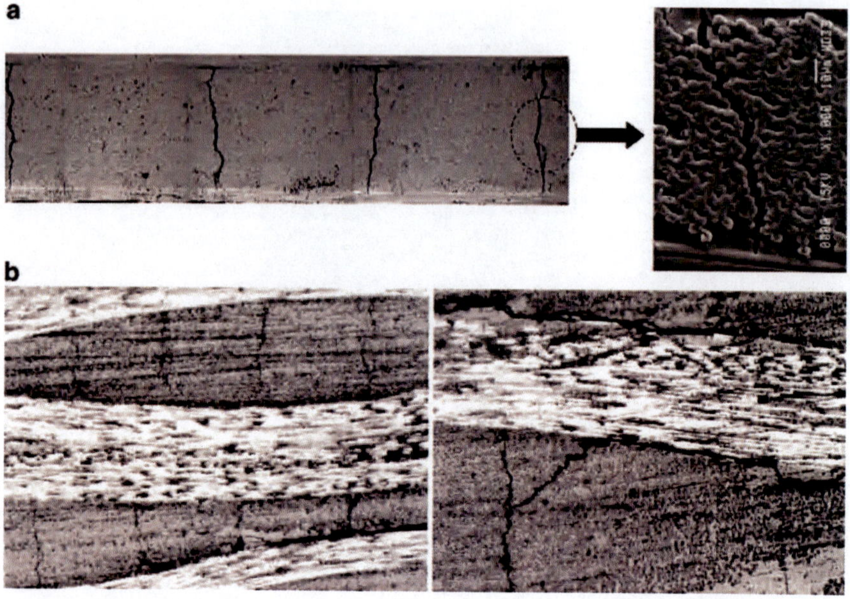

Fig. 4.2 Transverse cracks in a cross-ply laminate (**a**) and in a woven fabric laminate (**b**)

Figure 4.2 also shows a woven fabric composite of eight-harness satin weave with arrays of cracks that resemble the transverse cracks in cross-ply laminates. These cracks are within fiber bundles running transverse to the applied load. The close-up picture in the lower right shows a crack edge that has caused cracking of the interface between the longitudinal and transverse fiber bundles. This inter-bundle cracking resembles delamination in cross-ply laminates.

The transverse crack multiplication process requires load transfer from a cracking element (straight-fiber ply, fiber bundle, etc.) to a neighboring (constraining) element. In some woven fiber architectures, this is possible, e.g. in a eight-harness satin weave, and to a lesser degree in a four-harness satin weave, while in a plain weave composite the interfacial distance normal to the transverse crack plane may not be sufficient to complete the shear-lag process. Other mechanisms of progressive failure, such as inter-bundle cracks and fiber failures, are still likely if local stress conditions caused by fiber waviness reach critical levels under applied loading.

The evolution of the transverse crack multiplication process is illustrated in Fig. 4.3 (upper part) where cracks observed in x-ray radiographs of the type shown in Fig. 4.1 are drawn in the specimen-width direction. The pictures from left to right in the figure are at increasing levels of the tensile load in the longitudinal direction. The feature of interest to note is the noninteractive early stage of the cracking process which manifests in random locations of cracks, while the cracks become increasingly equally spaced as interaction described above increases. If an average value of cracks per unit of specimen length is plotted against the applied tensile load, an S-shaped curve results, as illustrated in Fig. 4.3 (lower part). For further treatment of transverse cracks, the reader is referred to a thorough exposition of the subject in Nairn [5].

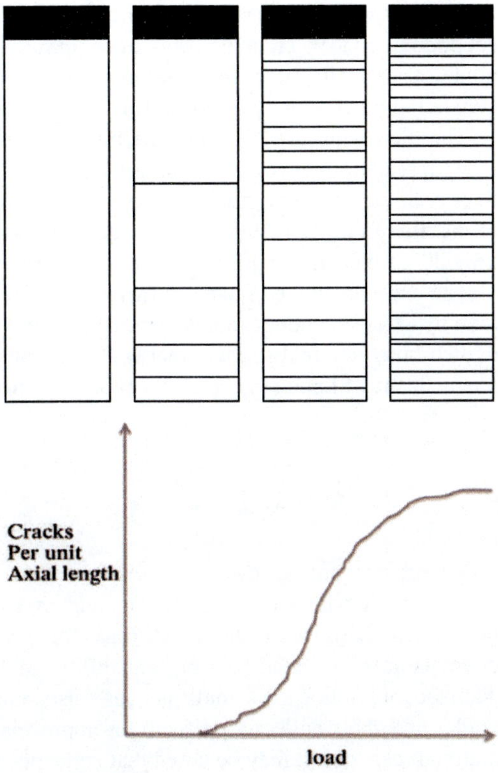

Fig. 4.3 Multiple transverse cracking. Cracking evolution, left to right (*upper figure*). S-shaped curve of crack density vs. load (*lower figure*)

When plies in a laminate are placed in oblique orientations, such as in $[0/\theta]$, $\theta \neq 90$, the cracks form along fibers, i.e. in θ-orientation, and crack multiplication process occurs by a shear-lag process similar to that in cross-ply laminates. Oblique cracks (also known as off-axis cracks) have been treated in Varna et al. [16, 17]; Singh and Talreja [7]) for $[0/\pm\theta_4/0_{1/2}]_s$ laminates, and in Singh and Talreja [9] for $[0/90_m/\pm\theta_n]_s$ laminates. The evolution of crack multiplication has been reported in Singh and Talreja [8]. These recent treatments of multiple intralaminar cracks in multiple orientations illustrate combining the micro-damage mechanics (MDM) and continuum damage mechanics (CDM) into a synergistic damage mechanics (SDM), as stated in Introduction.

The multiple intralaminar cracking is the most extensively studied damage in straight-fiber laminates. The distributed internal surfaces formed in the interlaminar planes, also called internal delaminations, illustrated in Fig. 4.1, are less common and are found to develop mostly under repeated loading. The internal delaminations can however form in woven fabric composites in fiber bundle crossover regions under the induced shear stresses and if the applied loading generates through-thickness tensile stresses.

Distributed internal surfaces form in short fiber composites by fiber/matrix debonding, matrix cracking, and fiber-tip/matrix separation. All these cracking modes are usually subcritical and can lead to the critical condition of failure by generating a sufficiently large continuous crack, which eventually grows unstably.

The increasing use of composite materials in aerospace industry, as well as in other applications such as automotive and wind turbines has led to a wide range of fiber architectures that induce complex local stress fields. This leads to formation of distributed internal surfaces that can remain restricted to local regions and thus provide a subcritical regime in which safe performance can be possible.

The main objective of damage mechanics is to describe the mechanical response (and by extension nonmechanical response) of composite materials in the presence of distributed internal surfaces that when activated by external mechanical loads (or nonmechanical impulses) cause local perturbations of relevant fields.

4.2.1 Mechanics of Damage

The field of damage mechanics can in some ways be better appreciated by contrasting it with the more known field of fracture mechanics. While fracture mechanics deals with the conditions for growth of a single crack and its consequences on load-bearing capability of a structure, damage mechanics addresses evolution of a multitude of (distributed) internal surfaces (cracks and voids) and their collective effects on appropriate volume-averaged measures of material response. It may be noted that concepts and methods developed in fracture mechanics are often useful in damage mechanics, but in a general case, other concepts and methods are necessary to meet the objectives of damage mechanics. In the following, we shall discuss the fundamental definitions and concepts underlying damage mechanics before treating specifics of approaches described as continuum damage mechanics (CDM) and synergistic damage mechanics (SDM).

4.2.1.1 Homogenization and Representative Volume Element (RVE)

Any continuum description of a solid entails homogenization since materials are inherently heterogeneous. For polycrystalline metals, for instance, the scale of heterogeneity (e.g., grain size) is often small compared to the scale at which material response characteristics (e.g., the elastic constants) are measured, allowing the stress and strain states to be defined as continuous fields. For commonly used fiber-reinforced solids, such as glass/epoxy and carbon/epoxy, the fiber diameter of approximately 10 μm allows treating these materials as a homogeneous continuum with good accuracy. When internal surfaces in composite materials form, their characteristic dimensions and mutual distances between them are orders of magnitude larger than the scale of heterogeneities underlying the homogenized pristine composite. Furthermore, on application of external loads, the internal surfaces are

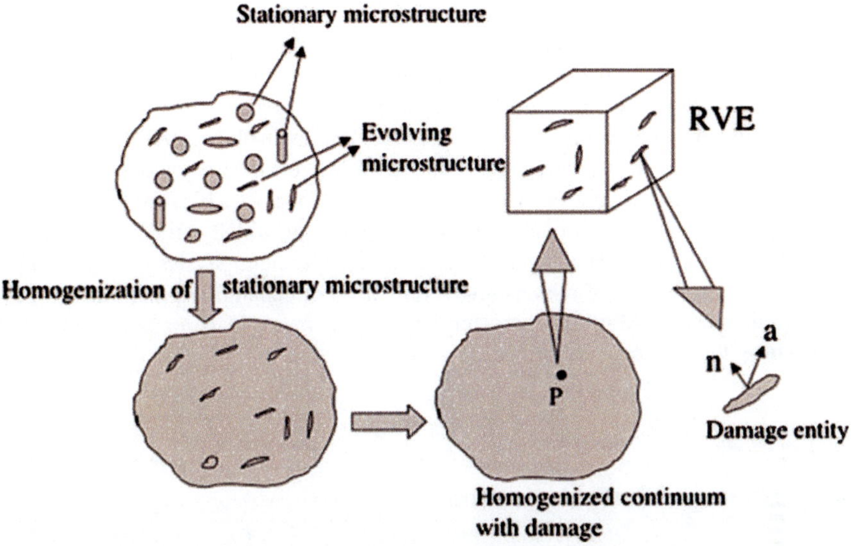

Fig. 4.4 Two-step homogenization of a heterogeneous solid with damage

subject to evolution (enlargement and multiplication), as discussed above. This warrants a separate (and different) homogenization of the composite with internal surfaces (collectively called damage). Such a procedure is discussed next.

Figure 4.4 depicts a homogenization procedure for a composite solid containing damage. The heterogeneities in the pristine (undamaged) composite are referred to as "stationary microstructure" and are homogenized first. This may also be called "classical" homogenization. Textbooks on mechanics of composite materials usually begin with this homogenization. In fact, the classical laminate theory goes one step further by developing homogeneous constitutive relations for laminates consisting of stacked layers of homogenized unidirectionally reinforced composite (ply or lamina). Returning to Fig. 4.4, the second homogenization pertains to the internal surfaces, collectively named as damage or "evolving microstructure" to highlight their ability to permanently change by processes of energy dissipation. Homogenization of the evolving microstructure necessitates employing the notion of a representative volume element (RVE), which will be discussed next.

A general and thorough exposition of the RVE notion in the context of micromechanics is given in Nemat-Nasser and Hori [6]. Here, we shall apply this notion to the particular case of composite materials with damage. With reference to Fig. 4.4 again, a generic point P in the homogenized composite with damage has associated with it a *damage state* (in addition to stress and strain states), which is given by an appropriate volume-averaged measure of the presence of internal surfaces that affect the constitutive behavior (stress–strain relations) at the point P. The volume over which the averaging is performed must be representative of the neighborhood of point P that can be associated with P. This neighborhood is the RVE, whose volume is not fixed but depends on the geometrical configuration (size, spacing, etc.) of the internal surfaces around P. As this configuration changes under applied loading, the RVE size changes.

With the notion of RVE at hand, the damage state at P can be defined by a set of variables obtained by averaging over the RVE. The choice of the variables is guided by the type of internal surfaces formed, and in this respect the knowledge of damage mechanisms discussed above is useful. In general, the variables can be scalars, vectors, or tensors of second order or higher. Settling on which variables to employ is a matter of finding a balance between capturing sufficient physics of the damage process and usefulness of the ensuing formulation of constitutive relations. In the following, the second-order characterization of damage in composite materials adopted by this author is described.

4.3 Damage Characterization

Characterization of damage in composite materials can be approached in different ways. A formal approach would be to view damage in terms of transformation of the elastic stiffness tensor (which is of fourth order) from pristine (undamaged) material state to the damage state. Although correct, this approach will not be feasible in a general case of anisotropic composites with multiple modes of damage.

Beginning with his first contribution in the damage mechanics area Talreja [11], this author has pursued a physically based damage characterization that does not leave behind essential features of damage in the context of the material response of interest while keeping in sight practical implementation of the resulting methodology. As argued in Talreja [11] and elaborated further in Talreja [13,15], a scalar-valued characterization would be inadequate in view of the oriented nature of the internal surfaces formed. Although a vector-valued characterization was employed at first, the issue of ambiguity of the sense of a vector could be addressed more elegantly by using a second-order tensor instead. Referring to the previous works cited above, for more details a brief description of the damage as second-order tensors is given below.

As shown in Fig. 4.4, a single internal surface within a RVE, called a damage entity from now on, can be characterized by two vectors: a unit outward normal $\boldsymbol{n}$ at a point on the surface, and an "influence" vector $\boldsymbol{a}$ at the same point. A dyadic product of the two vectors, integrated over the surface S, is denoted damage entity tensor, and given by

$$d_{ij} = \int_S a_i n_j dS \tag{4.1}$$

where the components of the vectors are with reference to a Cartesian coordinate system. The dyadic product assures consistency of the signs of the two vectors.

The physical significance of this characterization is that it represents the oriented nature of the presence of internal surfaces. As illustrated by the examples of damage discussed above, common internal surfaces are cracks (flat or curved) generated by interface debonding and matrix failure. The unit normal vector at a point on the damage entity carries the information on orientation of the surface (with respect to the frame of reference), while the other vector represents an appropriate influence induced by activation of the considered point on the surface. This influence is generally also directed in nature. For the case of mechanical response, the appropriate influence would be the displacement of the activated point on the damage entity surface. For a nonmechanical response, such as thermal or electrical conductivity, the perturbation induced by an internal surface can also be cast as a vector-valued quantity.

Integrating the dyadic product in Eq. 4.1 over the damage entity surface provides the total net effect of the entity. For example, if the entity is a flat crack, then taking $\boldsymbol{a}$ as the displacement vector in the integral gives the crack surface separation times the crack surface area. This product may be viewed as an affected volume associated with the crack. For a penny-shaped crack with the two surfaces separating symmetrically about the initial crack plane, the sole surviving term of the damage entity tensor represents an ellipsoidal-shaped volume.

Referring once again to Fig. 4.4, the RVE associated with a generic point P carries a sufficiently large number of the discrete damage entities to represent the collective effect on the homogenized constitutive response at the point. The number of damage entities needed for this representation, and the consequent RVE size, depend on the distribution of the entities. For instance, if the entities are sparsely distributed, then the RVE size would be large, while for densely distributed case a small RVE would suffice. Furthermore, for uniformly distributed entities of the same geometry, a repeating unit cell containing a single entity can replace the RVE, while for the cases of nonuniform distribution of unequal entities, the RVE size will increase until a statistically homogeneous representation is attained. This implies that further increasing the RVE size will have no impact on the averages of the selected characteristics. As an example, if the selected characteristic is the affected matrix volume by a damage entity, as mentioned above, then the average value of this quantity will vary as the RVE size increases and will approach a constant value at a certain RVE size. The minimum RVE size beyond which no appreciable change in the considered average is found may be taken as the needed RVE. It is apparent that the RVE is not unique but is subject to the choice made for the particular formulation of the constitutive response of a continuum with damage. Consequently, there is no unique constitutive theory of a continuum with damage; however, the use of the concept of an internal state in a given theory requires specifying RVE in a consistent manner and assuring that the conditions for its existence are present.

From the cases of damage mechanisms reviewed above, it can be noted that in composite laminates, the damage tends to occur as sets of parallel cracks within the plies, each oriented along fibers in the given ply. It is therefore convenient to separate each set of ply cracks according to its orientation, referred to a fixed frame of reference, and assign it a damage mode number. Denoting damage mode by $\alpha = 1,2, \ldots n$, a damage mode tensor can be defined as

$$D_{ij}^{(\alpha)} = \frac{1}{V} \sum_{k_\alpha} \left(d_{ij}^{(\alpha)}\right)_{k_\alpha} \tag{4.2}$$

where k_α is the number of damage entities in the αth mode, and V is the RVE volume. As noted above, if the ply cracks of a given orientation are uniformly spaced, then the RVE will reduce to the unit cell containing one crack. For nonuniform distribution of ply cracks, V must be large enough to provide a steady average of the damage mode tensor components.

As defined by Eq. 4.2, the damage mode tensor will in general be asymmetrical. Decomposing the influence vector $\boldsymbol{a}$ along directions normal and tangential to the damage entity surface S gives

$$a_i = an_i + bm_i \tag{4.3}$$

where n and m are unit normal and tangential vectors on S.

Using Eq. 4.3 in Eq. 4.2, the damage entity tensor can be written in two parts as

$$d_{ij} = d_{ij}^1 + d_{ij}^2 = \int_S an_i n_j dS + \int_S bm_i n_j dS \tag{4.4}$$

The damage mode tensor for a given mode can now be written as (dropping mode number α for convenience)

$$D_{ij} = D_{ij}^1 + D_{ij}^2 = \frac{1}{V} \sum_k \left(d_{ij}^1\right) + \frac{1}{V} \sum_k \left(d_{ij}^2\right) \tag{4.5}$$

This separation of the damage mode tensor in two parts allows simplifying the analysis so that dealing with asymmetric tensors is avoided. For instance, for damage entities consisting of flat cracks, the two parts of the damage mode tensor represent the two crack surface separation modes. If an assumption can be made that only the symmetric crack surface separation (known as mode I or crack opening mode in fracture mechanics) is significant, then the second term in Eq. 4.5 can be neglected. This will render the damage mode tensor symmetrical and it can then be written as

$$D_{ij} = D_{ij}^1 = \frac{1}{V} \sum_k \left[\int_S an_i n_j dS \right]_k \tag{4.6}$$

The consequence of this assumption was examined by Varna [20] for one class of laminates and it was found that not including the crack sliding displacement (CSD) for ply cracks inclined to the laminate symmetry directions results in errors in estimating degradation of average elastic properties of laminates. However, these errors were found to be small in absolute values while being significant in percentages. In fact, for those ply crack orientations where CSD dominates, the cracks are difficult to initiate until high loads close to failure load are applied.

For cases where the damage entity surfaces conduct tangential displacements only (e.g., CSD by flat cracks), it is possible to formulate the damage mode tensor as a symmetric tensor. One example of this is sliding of fiber/matrix interface in ceramic matrix composites [14].

With stress, strain, and damage, all expressed as symmetric second-order tensors, a constitutive theory can now be formulated to have a convenient, usable form. Such a formulation is described next.

4.4 Constitutive Framework for Material Response with Damage

Referring once again to Fig. 4.4, a formulation of the constitutive response of a homogenized continuum with damage will now be discussed. In view of the observed behavior of common composite materials such as glass/epoxy and carbon/epoxy, only elastic response will be considered. Theoretical treatment of elastic response of solids is classical and can be found in textbooks. Incorporating damage is, however, not a simple extension of the classical theory of elasticity. So far, the efforts made in this respect can be categorized as micro-damage mechanics (MIDM) and macro-damage mechanics (MADM), the latter often described as continuum damage mechanics (CDM). A framework for CDM to be described here is based on thermodynamics and is naturally suited for thermo-mechanical response. It can be extended to incorporate nonmechanical effects, such as electrical and magnetic, as well as chemical. Every extension, however, comes with the price of having to determine associated response coefficients (material constants) by certain identification procedure. In the treatment presented here, the task of determining material constants is reduced by use of selected micromechanics. This way of combining micromechanics with CDM generates useful synergism, justifying the characterization of the combined approach as synergistic damage mechanics (SDM). We begin with the conventional CDM framework first.

At the foundation of CDM are the first and second laws of thermodynamics. Additionally, use is made of the concept of an internal state, which is identified here as the evolving microstructure depicted in Fig. 4.4. As discussed before, this microstructure is homogenized into a damage field characterized by the set of damage mode tensors $D_{ij}^{(\alpha)}$. The collection of all variables resulting from thermodynamics with internal state can now be placed in two categories: state variables and response functions. The former for the case of small deformation is given by the strain tensor $\varepsilon_{ij} = 1/2\,(u_{i,j} + u_{j,i})$, where u_i are the displacement components, the absolute temperature T, the temperature gradient $g_i = T_i$, and damage mode tensors $D_{ij}^{(\alpha)}$. The response functions are the Cauchy stress tensor σ_{ij}, the specific Helmholz free energy ψ, the specific entropy η, the heat flux vector q_i, and the damage rate tensors $D_{ij}^{(\alpha)}$.

Following Truesdell's principle of equipresence, which states that all state variables should be present in all response functions unless thermodynamics or other relevant considerations preclude their dependency, we write

$$\sigma_{ij} = \sigma_{ij}(\varepsilon_{kl}, T, g_k, D_{kl}^{(\alpha)}) \tag{4.7}$$

$$\psi = \psi(\varepsilon_{kl}, T, g_k, D_{kl}^{(\alpha)}) \tag{4.8}$$

$$\eta = \eta(\varepsilon_{kl}, T, g_k, D_{kl}^{(\alpha)}) \tag{4.9}$$

$$q_i = q_i(\varepsilon_{kl}, T, g_k, D_{kl}^{(\alpha)}) \tag{4.10}$$

$$D_{ij}^{(\alpha)} = D_{ij}^{(\alpha)}(\varepsilon_{kl}, T, g_k, D_{kl}^{(\beta)}) \tag{4.11}$$

The dependency of the response functions listed above on the temperature gradient is examined by applying the second law of thermodynamics, which expressed in the form of the Clausius–Duhem inequality, is

$$\sigma_{ij}\varepsilon_{ij} - p\psi - \rho T\eta - \frac{q_i g_i}{T} \geq 0 \tag{4.12}$$

where ρ is the mass density. Time differentiation of the Helmholz free-energy function gives

$$\psi = \frac{\partial \psi}{\partial \varepsilon_{ij}}\varepsilon_{ij} + \frac{\partial \psi}{\partial T}T + \frac{\partial \psi}{\partial g_i}g + \sum_{\alpha}\frac{\partial \psi}{\partial D_{ij}^{(\alpha)}}D_{ij}^{(\alpha)} \tag{4.13}$$

Substitution of Eq. 4.13 in Eq. 4.12 yields

$$\left(\sigma_{ij} - \rho\frac{\partial \psi}{\partial \varepsilon_{ij}}\right)\varepsilon_{ij} - \rho\left(\eta + \frac{\partial \psi}{\partial T}\right)T - \rho\frac{\partial \psi}{\partial g_i}g_i - \rho\sum_{\alpha}\frac{\partial \psi}{\partial D_{ij}^{(\alpha)}}D_{ij}^{(\alpha)} - \frac{q_i g_i}{T} \geq 0 \tag{4.14}$$

Requiring Eq. 4.14 to hold for all rates of the independently varying strain, temperature and temperature gradient gives the following results.

$$\sigma_{ij} = \rho\frac{\partial \psi}{\partial \varepsilon_{ij}} \tag{4.15}$$

$$\eta = -\frac{\partial \psi}{\partial T} \tag{4.16}$$

$$\frac{\partial \psi}{\partial g_i} = 0 \tag{4.17}$$

Equation 4.17 states that the Helmholz free-energy function does not depend on the temperature gradient, and consequently, Eqs. 4.15 and 4.16 eliminate this dependency from stress and entropy. Equations 4.10 and 4.11 in the response function set remain unaffected. From Eq. 4.14, the following restriction (also known as internal dissipation inequality) results.

$$\sum_{\alpha} R_{ij}^{(\alpha)} \dot{D}_{ij}^{(\alpha)} - \frac{q_i g_i}{T} \geq 0 \tag{4.18}$$

where $R_{ij}^{(\alpha)}$ are the thermodynamic forces conjugate to the damage mode tensors $D_{ij}^{(\alpha)}$, and are given by

$$R_{ij}^{(\alpha)} = -\rho \frac{\partial y}{\partial D_{ij}^{(\alpha)}} \tag{4.19}$$

Each of these forces is analogous to the crack extension force (i.e., energy release rate) for a single crack. As an example, for a damage mode component, say D_{11} of mode $\alpha = 1$, the quantity $R_{11}^{(1)}$ can be interpreted as the "force" causing an infinitesimal change in the internal state represented by $D_{11}^{(1)}$. Equation 4.18 expresses the condition these forces must satisfy as damage evolves under thermo-mechanical impulses.

The collection of response functions and the associated restrictions on the interrelations among these, discussed above, form the framework for a rational description of constitutive behavior of heterogeneous solids containing evolving internal surfaces (damage). This framework is based on the homogenization depicted in Fig. 4.4, the concept of RVE for characterization of the evolving internal state, and the two laws of thermodynamics. It is possible to extend the framework by including other energy-dissipative entities than the internal surfaces formed by atomic bond breakages, e.g. metal crystalline slip and polymer morphological changes or rearrangements such as viscoelasticity and aging. The kinetics of dissipative mechanisms is treated in the response function, exemplified by Eq. 4.11, and the associated restriction, Eq. 4.18.

In view of the experimental data, which are mostly available for polymer matrix composites at room temperature, the thermo-mechanical framework will be developed further for the mechanical response. Thus, for isothermal conditions ($\dot{T} = 0$, $g_i = 0$), the set of response functions is reduced to the following.

$$\psi = \psi(\varepsilon_{ij}, D_{ij}^{(\alpha)}) \tag{4.20}$$

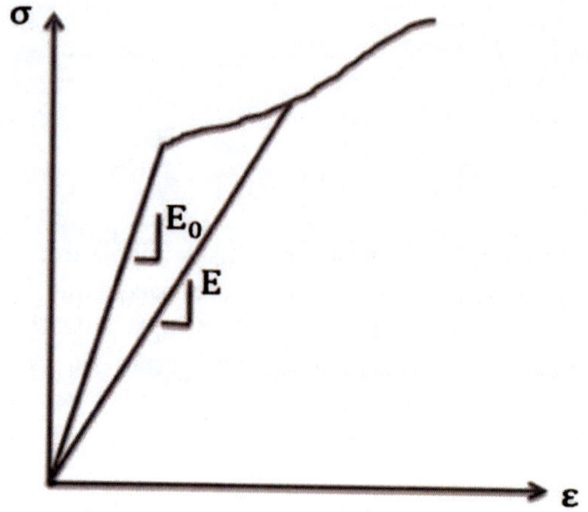

Fig. 4.5 Stress–strain curve of a composite with damage. Secant modulus E varies with the state of damage

and

$$D_{ij}^{(\alpha)} = D_{ij}^{(\alpha)}(\varepsilon_{kl}, D_{kl}^{(\alpha)}) \tag{4.21}$$

Note that stress is derivable from the Helmholz free-energy function according to Eq. 4.15. Thus, dealing with this scalar-valued function as the sole response function for a given internal state of damage provides a favorable situation for further development of the theory. The form of the Helmholz free-energy function can be chosen in different ways. A powerful way is possible by use of the theory of invariants for polynomial functions [10]. In the following, it is illustrated for one case of orthotropic composites containing one damage mode.

Consider an initially orthotropic composite laminate with one array of intralaminar cracks as illustrated by Fig. 4.5. Denoting this damage mode by $\alpha = 1$, the damage mode tensor components can be written as

$$D_{ij}^{(1)} = \frac{\kappa t_c^2}{st\cos\theta} n_i n_j \tag{4.22}$$

where Eq. 4.6 is used in which $V = LWt$, where L, W, and t are the length, width, and thickness of the representative volume V of the laminate, respectively, $S = t_c(W/\cos\,\theta)$, and $a = \kappa t_c$, where κ is an assumed constant of proportionality between the influence vector magnitude (here, crack opening displacement) and crack dimension t_c (same as the thickness of cracked ply). Finally, $n_i = (\cos\,\theta, \sin\,\theta, 0)$.

Choosing the function in Eq. 4.20 as a polynomial function of strain tensor components and damage tensor components, both symmetrical second-order tensors, gives

$$\psi = P(\varepsilon_1, \varepsilon_2, \ldots, \varepsilon_6, D_1, D_2, \ldots, D_6) \tag{4.23}$$

where P stands for a polynomial function and the tensor components are expressed in the Voigt notation, $\varepsilon_1 = \varepsilon_{11}$, $\varepsilon_2 = \varepsilon_{22}$, $\varepsilon_3 = \varepsilon_{33}$, $\varepsilon_4 = \varepsilon_{23}$, $\varepsilon_5 = \varepsilon_{13}$, $\varepsilon_6 = \varepsilon_{12}$, and similarly for the D-components. The superscript on D-components indicating the damage mode has been dropped for convenience.

Expansion of the polynomial function (23) can in general have infinite terms, which will obviously present an impractical situation. One way to restrict the functional form is by expanding the polynomial in terms that account for the initial material symmetry. This is done in the polynomial invariant theory by using the so-called integrity bases [10]. Such bases have been developed for scalar functions of various vector and tensor variables. For the case of two symmetric second-order tensors, such as in Eq. 4.23, the integrity bases for orthotropic symmetry are given by Adkins [1]

$$
\begin{aligned}
&\varepsilon_1, \varepsilon_2, \varepsilon_3, \varepsilon_4^2, \varepsilon_5^2, \varepsilon_6^2, \varepsilon_4\varepsilon_5\varepsilon_6, \\
&D_1, D_2, D_3, D_4^2, D_5^2, D_6^2, D_4D_5D_6, \\
&\varepsilon_4D_4, \varepsilon_5D_5, \varepsilon_6D_6, \\
&D_4\varepsilon_5\varepsilon_6, D_5\varepsilon_6\varepsilon_4, D_6\varepsilon_4\varepsilon_5, \\
&\varepsilon_4D_5D_6, \varepsilon_5D_6D_4, \varepsilon_6D_4D_5.
\end{aligned} \tag{4.24}
$$

For the sake of applying the constitutive theory to thin laminates where only in-plane strains are of interest, and for small strains, the expansion of the function (23) can be restricted to no more than quadratic terms in strain components ε_1, ε_2, and ε_6. To what extent the damage tensor components are to be taken in the expansion depends on the nature and amount of information that can be acquired for evaluation of the material constants that will appear in the polynomial function. This issue will be discussed later. To begin with the simplest possible case, we will include only linear terms in D_1, D_2, and D_6, which are the nonzero components for intralaminar cracks. Thus,

$$
\begin{aligned}
\rho\psi = {} & c_1\varepsilon_1^2 + c_2\varepsilon_1\varepsilon_2 + c_3\varepsilon_1^2D_1 + c_4\varepsilon_1^2D_2 + c_5\varepsilon_1\varepsilon_6D_6 + c_6\varepsilon_2^2 + c_7\varepsilon_2^2D_1 \\
& + c_8c_2^2D_2 + c_9\varepsilon_2\varepsilon_6D_6 + c_{10}\varepsilon_6^2 + c_{11}\varepsilon_6^2D_1 + c_{12}\varepsilon_6^2D_2 \\
& + c_{13}\varepsilon_1\varepsilon_2D_1 + c_{14}\varepsilon_1\varepsilon_2D_2 + P_0 + P_1(\varepsilon_p, D_q) + P_2(D_q)
\end{aligned} \tag{4.25}
$$

where c_i, $i = 1,2, \ldots, 14$ are material constants, P_0 is a constant, P_1 is a linear function of strain and damage tensor components, and P_2 is a linear function of the damage tensor components. It can be seen that $P_0 = 0$, if the free-energy value in the undeformed and undamaged state is set to zero. Letting the unstrained material of any damage state to be stress-free, one gets $P_1 = 0$, on using Eq. 4.15.

Equation 4.15 expressed in the Voigt notation is given by

$$
\sigma_p = \rho \frac{\partial\psi(\varepsilon_q, D_r)}{\partial\varepsilon_p} \tag{4.26}
$$

A differential in stress can now be written as

$$d\sigma_p = \rho \frac{\partial \psi}{\partial \varepsilon_p \partial \varepsilon_q} d\varepsilon_q + \rho \frac{\partial \psi}{\partial \varepsilon_p \partial D_r} dD_r = C_{pq} d\varepsilon_q + K_{pr} dD_r \tag{4.27}$$

where

$$C_{pq} = \rho \frac{\partial \psi}{\partial \varepsilon_p \partial \varepsilon_q} \tag{4.28}$$

is the stiffness matrix when $dD_r = 0$, i.e. at constant damage. This is illustrated for uniaxial stress–strain response in Fig. 4.5. As seen there, the elastic modulus at any point on the stress–strain curve is the secant modulus, not the tangent modulus.

Using Eqs. 4.25 and 4.28, one obtains

$$C_{pq} = C^0_{pq} + C^1_{pq} \tag{4.29}$$

where

$$C^0_{pq} = \begin{bmatrix} 2c_1 & c_2 & 0 \\ c_2 & 2c_6 & 0 \\ 0 & 0 & 2c_{10} \end{bmatrix} \tag{4.30}$$

and

$$\begin{aligned} C^1_{11} &= 2c_3 D_1 + 2c_4 D_2 \\ C^1_{12} &= C^1_{21} = c_{13} D_1 + c_{14} D_2 \\ C^1_{16} &= C^1_{61} = c_5 D_6 \\ C^1_{22} &= 2c_7 D_1 + 2c_8 D_2 \\ C^1_{26} &= C^1_{62} = c_9 D_6 \\ C^1_{66} &= 2c_{11} D_1 + 2c_{12} D_2 \end{aligned} \tag{4.31}$$

It can be noted here that Eqs. 4.29–4.31 show linear dependence of the stiffness properties on damage tensor components. This is the consequence of including only linear terms in these components in the polynomial expansion of the free-energy function (Eq. 4.25). Including higher order terms will add additional constants c_i, which will need to be evaluated. The evaluation procedure is described below, but it is remarked here that the formulation of constitutive response is in no way restricted only to linear dependence on the chosen damage measure.

The evaluation of the material constants c_i ($i = 1, \ldots, 14$ for one intralaminar cracking mode) will first be illustrated by the case of transverse cracks, i.e., $\theta = 0$ in Fig. 4.5. For this case, $D_2 = D_6 = 0$, and D_1 is given by

$$D_1 = \frac{\kappa t_c^2}{st} \tag{4.32}$$

and the C^1_{pq} matrix takes the form

$$C^1_{pq} = \frac{\kappa t_c^2}{st}\begin{bmatrix} 2c_3 & c_{13} & 0 \\ c_{13} & 2c_7 & 0 \\ 0 & 0 & 2c_{11} \end{bmatrix} \tag{4.33}$$

Thus, with the transverse cracking mode, the elastic response of the initially orthotropic laminate retains this symmetry. The number of constants that need evaluation is now eight, four of which (c_1, c_2, c_6, and c_{10}) correspond to the initial elastic response for which the constants can either be measured experimentally or calculated by the laminate plate theory using ply properties. The remaining four constants can be deduced from measurements of the four independent elastic constants, the Young's moduli E_1 and E_2, in the axial and transverse directions, respectively, the major Poisson's ratio ν_{12} and the shear modulus G_{12}, all at a given value of D_1. Expressions for these constants linearized in D_1 are as follows.

$$\begin{aligned} E_1 &= E_1^0 + 2\frac{\kappa t_c^2}{st}\left[c_3 + c_7(\nu_{12}^0)^2 - c_{13}\nu_{12}^0\right] \\ E_2 &= E_2^0 + 2\frac{\kappa t_c^2}{st}\left[c_7 + c_3(\nu_{12}^2)^2 - c_{13}\nu_{12}^0\right] \\ \nu_{12} &= \nu_{12}^0 + \frac{\kappa t_c^2}{st}\left[\frac{1 - \nu_{12}^0\nu_{21}^0}{E_2^0}\right](c_{13} - 2c_7\nu_{12}^0) \\ G_{12} &= G_{12}^0 + 2\frac{\kappa t_c^2}{st}c_{11} \end{aligned} \tag{4.34}$$

where quantities with superscript 0 correspond to initial (undamaged) state, and $\nu_{21}^0 = \nu_{12}^0(E_2^0/E_1^0)$.

The unknown constants in Eq. 4.34 can now be expressed as

$$\begin{aligned} \kappa c_3 &= A_1(1 - 2\nu_{12}^0\nu_{21}^0) + A_2(\nu_{12}^0)^2 + 2A_3E_2^0\nu_{12}^0 \\ \kappa c_7 &= A_1(\nu_{21}^0)^2 + A_2 + 2A_3E_2^0\nu_{21}^0 \\ \kappa c_{11} &= A_4 \\ \kappa c_{13} &= -2A_1\nu_{12}^0(\nu_{21}^0)^2 + 2A_2\nu_{12}^0 + 2A_3E_2^0(1 + \nu_{12}^0\nu_{21}^0) \end{aligned} \tag{4.35}$$

where

$$\begin{aligned} A_1 &= Q(E_1 - E_1^0) \\ A_2 &= Q(E_2 - E_2^0) \\ A_3 &= Q(\nu_{12} - \nu_{12}^0) \\ A_4 &= \frac{Q}{(1 - \nu_{12}^0\nu_{21}^0)^2}(G_{12} - G_{12}^0) \end{aligned} \tag{4.36}$$

with

$$Q = \frac{ts}{2t_c^2\left(1 - \nu_{12}^0\nu_{21}^0\right)^2} \tag{4.37}$$

From Eqs. 4.35–4.37, it can be seen that in the case of elastic response linearized in damage components, the material constants c_i are not evaluated but instead their products with the crack opening parameter κ are determined. While the values of c_i are fixed for a given composite laminate (that has been homogenized), the parameter κ depends on the ability of the cracks to perform surface displacements under applied mechanical impulse. Thus this parameter may be viewed as a measure of the constraint to the crack surface separation imposed by the material surrounding the crack. One way to view this is by considering a crack of a given size embedded in an infinite isotropic material, in which case the crack surface separation is unconstrained and can be calculated by fracture mechanics methods. When the laminate geometry is finite and its symmetry is different from isotropic, the κ parameter will take a value less than that for the infinite isotropic medium. This consideration allows us to assign κ an undermined value, say κ_0, for a reference laminate under reference loading conditions, and evaluate a change from this value for another crack orientation. This approach is discussed further below.

4.5 Synergistic Damage Mechanics (SDM)

The observation that the κ-parameter (hitherto referred to as constraint parameter) may be viewed as a carrier of the local effects on damage entities within a RVE, while the c_i-constants are material constants, led to a number of studies to explore prediction of elastic property changes due to damage in different modes. To be sure, the elastic properties are the averages over appropriate RVEs.

At first, it was found that from changes in E_1 and ν_{12} due to transverse cracking in $[0/90_3]_s$ glass/epoxy laminates reported in Highsmith and Reifsnider [3] and assuming no changes in E_2, the constants calculated by the procedure described above, Eqs. 4.35–4.37, gave the values $\kappa c_3 = -6.712$ GPa, $\kappa c_7 = -0.770$ GPa, and $\kappa c_{13} = -4.455$ GPa. Using these values, changes in E_1 for the same glass/epoxy of $[0/90]_s$ configuration could be predicted with good accuracy. Also, in $[0/\pm45]_s$ laminate of the same glass/epoxy, the change in E_1 could be predicted by setting $D_1 = D_2$ (a good approximation, supported by crack density data). These results have been reported in Talreja [12, 13].

Later, a systematic study of the effect of constraint on the κ-parameter was done by experimentally measuring the crack opening displacement (COD) in $[\pm\theta/90_2]_s$ laminates [16, 17] for different θ-values. By relating these values to the COD at $\theta = 0$ normalized and by a unit applied strain, the predictions of E_1 and ν_{12} for different θ could be made. Another study of the constraint effects was made by examining $[0/\pm\theta_4/0_{1/2}]_s$ laminates, where the ply orientation θ was varied.

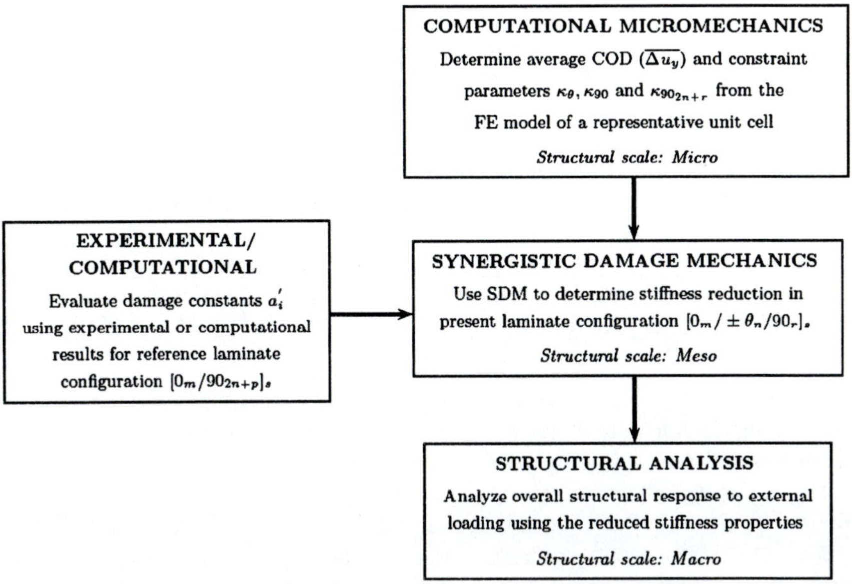

Fig. 4.6 Multiscale synergistic methodology for analyzing damage behavior in a general symmetric laminate ($[0_m/\pm\theta_n/90_r]_s$) with matrix cracks in $+\theta$, $-\theta$, and 90° layers

Once again, using experimentally measured COD for $\theta = 0$ as the reference, the κ-parameter for other ply orientations was evaluated from the COD values and E_1 and ν_{12} for different θ were predicted [16, 17].

While the experimental studies supported the idea of using the κ-parameter as a carrier of local constraints, the scatter in test data and the cost of testing do not make the experimental approach attractive. Therefore, another systematic study of $[0_m/\pm\theta_n/0_{m/2}]_s$ laminates was undertaken [7] where computational micromechanics was employed instead of physical testing. An elaborate parametric study of the κ-parameter allowed developing a master curve for elastic property predictions.

The most recent study [9] examines damage modes consisting of transverse ply cracks as well as inclined cracks of different orientations in $[0_m/\pm\theta_n/90_r]_s$ and $[0_m/90_r/\pm\theta_n]_s$ laminates [18, 19]. The SDM approach is developed and its predictions are compared with available experimental data for $[0/90/-45/+45]_s$ laminate.

At the current state of its development, the SDM methodology has the ingredients of a multiscale damage mechanics depicted in Fig. 4.6. As illustrated there, structural (macroscale) analysis of a part containing damage is accomplished by a mesoscale SDM analysis of appropriate RVE in which microscale input from MIDM (computed CODs) as well as materials constants from a reference laminate configuration are entered. What remains is to codify this approach in software to be used by design engineers.

The treatment above used only elastic response to illustrate CDM/SDM formulation. Polymer matrix composites are found to have time-dependent (viscoelastic) response at elevated temperatures. For the linear viscoelastic case, Kumar and

Talreja [4] developed a methodology for predicting the material constants at fixed damage using the correspondence principle. If the viscoelastic deformation becomes nonlinear, the correspondence principle does not apply in which case the internal state can be characterized by variables that represent the molecular morphology changes in addition to the damage variables. An approach along these lines was developed by Ahci and Talreja [2] where a mixed experimental and computational procedure for identification of the material constants was proposed.

4.6 Concluding Remarks

This chapter has provided essential fundamentals of a CDM approach giving specific details for a baseline case of purely mechanical response of composite laminates with one array of cracks. While references to works dealing with more complex damage modes are given, the versatility of the approach in dealing with more general cases of damage and associated thermo-mechanical response has been discussed. The conventional reservations in the use of CDM rooted in the difficulties of identifying material response characteristics have been dealt with by outlining the so-called synergistic damage mechanics (SDM) where this problem is alleviated by a judicial use of micromechanics of damage. A recently developed multiscale computational framework further enhances the practical use of the approach. To keep the discussions focused here on these aspects of CDM, the treatment of damage evolution is not presented. Integrating SDM and damage evolution in a computational code remain tasks for future activities.

References

1. Adkins, J.E., *Symmetry relations for orthotropic and transversely isotropic materials*, Arch. Rat. Mech. An. 1960. **4**: pp. 193–213.
2. Ahci, E. and Talreja, R., *Characterization of viscoelasticity and damage in high temperature polymer matrix composites*. Composites Science and Technology, 2006. **66:** p. 2506–2519.
3. Highsmith, A.L. and K.L. Reifsnider, *Stiffness-reduction mechanisms in composite laminates*, in *Damage in Composite Materials*, K.L. Reifsnider, Editor. 1982, ASTM: Philadelphia. p. 103–117.
4. Kumar, R. and Talreja, R., *Linear viscoelastic behavior of matrix cracked cross ply laminates*. Mechanics of Materials, 2001. **33:** p. 139–154.
5. Nairn, J.A., *Matrix Microcracking in Composites*, in *Polymer Matrix Composites*, R. Talreja and J.A.E. Manson, Editors. 2000, Elsevier Science. p. 403–432.
6. Nemat-Nasser, S. and M. Hori, *Micromechanics: Overall Properties of Heterogeneous Materials*. 2 ed. 1999, Amsterdam: North Holland.
7. Singh, C.V. and R. Talreja, *Analysis of multiple off-axis ply cracks in composite laminates*. International Journal of Solids and Structures, 2008. **45**(16): p. 4574–4589.
8. Singh, C.V. and R. Talreja, *Evolution of ply cracks in multidirectional composite laminates*. International Journal of Solids and Structures, 2010. **47**: p. 1338–1349.

9. Singh, C.V. and R. Talreja, *A synergistic damage mechanics approach for composite laminates with matrix cracks in multiple orientations*. Mechanics of Materials, 2009. **41**(8): p. 954–968.
10. Spencer, A.J.M., *Theory of invariants*. In: C. Eringen, Editor, Continuum Physics, Academic Press, 1971. pp. 239–353.
11. Talreja, R., *A Continuum-Mechanics Characterization of Damage in Composite-Materials*. Proceedings of the Royal Society of London Series a-Mathematical Physical and Engineering Sciences, 1985. **399**(1817): p. 195–216.
12. Talreja, R., *Transverse Cracking and Stiffness Reduction in Composite Laminates*. Journal of Composite Materials, 1985. **19**(4): p. 355–375.
13. Talreja, R., *Internal Variable Damage Mechanics of Composite Materials*, in *Yielding, Damage, and Failure of Anisotropic Solids*, J.P. Boehler, Editor. 1990, Mechanical Engineering Publications: London. p. 509–533.
14. Talreja, R., *Continuum Modeling of Damage in Ceramic Matrix Composites*. Mechanics of Materials, 1991. **12**(2): p. 165–180.
15. Talreja, R., *Damage Characterization by Internal Variables*, in *Damage Mechanics of Composite Materials*, R. Talreja, Editor. 1994, Elsevier: Amsterdam. p. 53–78.
16. Varna, J., Joffe, R., Akshantala and Talreja, R., *Damage in composite laminates with off-axis plies*. Composites Science and Technology, 1999. **59**(14): p. 2139–2147.
17. Varna, J., N.V. Akshantala, and R. Talreja, *Crack opening displacement and the associated response of laminates with varying constraints*. International Journal of Damage Mechanics, **1999**(8): p. 174–193.
18. Varna, J., R. Joffe, and R. Talreja, *A synergistic damage-mechanics analysis of transverse cracking in [+/−theta/90(4)](s) laminates*. Composites Science and Technology, 2001. **61**(5): p. 657–665.
19. Varna, J., R. Joffe, and R. Talreja, *Mixed micromechanics and continuum damage mechanics approach to transverse cracking in [S,90(n)](s) laminates*. Mechanics of Composite Materials, 2001. **37**(2): p. 115–126.
20. Varna, J., *Physical interpretation of parameters in synergistic continuum damage mechanics model for laminates*, Composites Science and Technology, 2008. **68**(13): pp. 2592–2600.

Chapter 5
Impact Damage of Composite Structures

Hyonny Kim, John C. Halpin, and Gabriela K. DeFrancisci

Abstract The damage tolerance of composite structures is intimately linked to the morphology and extent of flaws or damage. Impacts are a commonly occurring source of threat to composites that can produce a "seed" damage state which inherently controls the subsequent durability and damage tolerance of the affected structure. Transverse impact to composites is of particular concern due to the possibility of exciting damage modes that are difficult, or even not possible, to visually detect from the exterior (impact-side) surface. Some examples of such damage are delamination, backside-only fiber failure, debonding of internal substructure (e.g., stringers and stiffeners, doublers, joints), and crushing and separation of sandwich panel core. Impact damage is highly dependent upon the nature of the threat and conditions associated with the impact event. This chapter will provide an overview of impact damage threats that are common to composite aircraft structures and describe the relationship of these threat sources to the damage that is seeded. An historical overview of the impact damage tolerance methodologies developed by military aircraft programs, and subsequently widely adopted across the composite structures community, is also provided.

5.1 Introduction

The current body of knowledge on impact damage to composites exists mostly as a result of studies focused on aviation. Therefore, this chapter on impact damage to composites primarily draws upon aircraft-related applications. However, the phenomena described are also applicable to other high-performance structures

H. Kim (✉)
University of California, San Diego, 9500 Gilman Drive MC 0085,
La Jolla, CA 92093-0085, USA
e-mail: hyonny@ucsd.edu

K.V. Pochiraju et al. (eds.), *Long-Term Durability of Polymeric Matrix Composites*,
DOI 10.1007/978-1-4419-9308-3_5,

such as vehicles (e.g., cars, ships) and wind turbine blades, etc., where damage tolerance is an important safety issue.

Composite materials are being used extensively to build lightweight and highly loaded structures. In particular, the aerospace industry has deployed composites in both primary and secondary structures, the former type of structure being one carrying significant flight, ground, or pressure loads, as defined in Federal Aviation Administration Advisory Circular 25.1529 [22]. Economical factors realized in the form of fuel savings in lighter-weight vehicles have directed the commercial aircraft industry to aggressively pursue "all" composite structures (both wing and fuselage) for a wide range of vehicles, from the four-seat general aviation aircraft (examples are Cirrus SR20 and Lancair Columbia 300) to the business and regional jets (e.g., Raytheon Premier I) to the larger transport aircraft such as the Boeing 787 and the emerging Airbus A350 and Bombardier 300. With significantly increased acreage of primary structural area exposed to external impact "threats," in combination with the inherent resiliency characteristic of composites to not show dents as easily as metals, impact damage remains a topic of even more heightened concern as these aircraft are deployed and used throughout their design life spans (and often beyond).

Technology advancements in engineering structures and airframes generate a need for additional understanding of the new vulnerabilities that are introduced as the response of new materials change. In the age of fabric and wood airframes, structural life was limited by the rotting of the materials. With the introduction of aluminum alloys, bacterial degradation was replaced by corrosion and fatigue cracking. Metallic airframe construction permitted higher aircraft velocities. As aviation evolved in the 1950s with the introduction of the jet engines, a variety of impact threats received attention: in-flight bird and hail strike, foreign object damage, and tire rupture. Examples of early work addressing hail impact threats for metallic structures are by Souter and Emerson [68], McNaughton and Chisman [69], and Thomson and Hayduk [70].

The transition from homogeneous sheet metal construction to layered (laminated) material systems again changed the sensitivities of the air vehicle. Layered materials, plywood, composites, exhibit increased sensitivity to the transverse impact threats already of concern to metallic construction. While the introduction of graphite fiber-reinforced composites has reduced the fatigue cracking and corrosion mechanisms typical of metallic construction, the airframe parts, regardless of being metallic or composite construction, are still subject to impact threats from a variety of sources such as birds, hailstones, tire and engine fragments, tools dropped by maintenance personnel, access panels lost during flight, collisions by service vehicles, or foreign objects thrown up from the runway. It is important to know first of all how to detect the occurrence of such events (not all being self-evident), the discrimination of whether the event is prospectively damaging or not, determining where and how to inspect for the damage, and finally how such damage affects a structure's near- and long-term performance as the existence of damage will generally result in some amount of strength and stiffness degradation [36, 37].

Aside from the immediate degradation in load-bearing capability, impact damage can affect the long-term durability of composite structures. Damage "seeded" by the impact event can grow under repeated load cycles, particularly if the damage remains undetected. When impact damage occurs, it generally exists in the form of cracks (micro- to mesosized matrix cracks, delamination, fiber failure) and therefore can serve to increase absorption of fluid/gas [55] which can have especially damaging effects to lightweight sandwich structures [58]. The damage tolerance and long-term durability of a composite structure is therefore dependent on the exact morphology and extent of the seed damage caused by an impact event.

Of particular concern during the transverse impact of laminated composite structures is the development of interlaminar failures and subsurface damage which are likely to occur in modes that are difficult to visually detect from the exterior surface, i.e., the impact-side. Some of these damage threats are well known and "mature" in the methods by which they are characterized (e.g., "tool drop" and bird strike). Other threats, such as ground vehicle impacts (large-mass, wide-contact area) and even hail ice impacts are not mature in the methods by which these impact threats are characterized, and thus future (ongoing) developments are needed for establishing appropriate test methods.

The methods one establishes for characterizing prospective impact damage are critical to the development of the "seed" damage state. In turn, this strongly affects the subsequent durability and damage tolerance of the impacted structure. The following guidelines are recommended for the development of an *impact testing methodology*:

1. An impact threat should be defined in an extrinsic manner that is independent of structural platform. While the structure of interest inherently defines what impact threats the structure is exposed to, the description of the impact threat itself is independent of what structure it might hit – i.e., the threat is defined by the physical characteristics of the impactor (e.g., mass, geometry, hard or soft body) and its velocity as related to operational conditions. For example, the threat potential of hailstones failing at terminal velocity can be defined independently of what it will ultimately strike, be it a direct (normal) impact onto horizontal structure, glancing strike on a near-vertical surface, or even hitting the ground.
2. The damage state developed by an impact is critically dependent upon the physical attributes of the impactor and object being impacted, as well as the velocity regime at which the impact event occurs. Thus characterization methodologies must be tailored for each impact threat type in order to correctly capture the important physics and phenomena that properly represent the actual projectile. For example, conducting impact studies for hail ice using a metallic drop-weight test is a significant departure from both the actual physical attributes of ice and velocity regime of the real impact event itself. Since the damage morphology seeded by the impact event is controlled by the nature of the impactor, a metal projectile is not able to, in general, produce the same damage state as an ice projectile.

3. The damage threshold concept should be used as a part of preliminary design for damage resistance and tolerance assessment of composite structures. i.e., damage threshold should be the basis for the thickness sizing of laminates exposed to impact threats. The damage thresholds will inherently depend on the nature of the threat (e.g., tool drop, hail ice, bird strike, rubber puck for tire damage are all associated with different damage thresholds).

5.2 Historical Overview of Damage Tolerance Methodology

The historical approach to impact damage resistance has evolved from the early experiences addressing tool drop and "runway" foreign object damage. This basis evolved in the 1970s as technology preparations were being made for the B-2 bomber. Characterization of material resistance to transverse impact utilized a quasi-static test concept, namely, falling weight impact testing. Typically, this test method utilizes a hemispherical metallic impactor. The selection of the falling weight method with metallic impactors was judged to be reasonable at that time as the impact threats of tool drop and foreign object damage (FOD) were judged to involve "low-impact velocities," involving "hard" objects, and localized impact conditions.

Very early (1975–1985 timeframe) in the characterization of composite material impact damage resistance, data appeared suggesting that there was a level of impact kinetic energies below which a specific laminated composite plate would not experience delamination, resin cracking, or fiber failure. At higher impacting energy levels, damage occurred, increasing in severity with increasing impacting energies until penetration. By the late 1980s, it was apparent that the observation of a defined *damage threshold* was consistent among a variety of industrial teams [47] and in open referred journal papers [17, 29, 54].

It was also apparent that this damage threshold was strongly dependent on laminate thickness, local bending stiffness, and the fracture toughness of the matrix. This is generically illustrated in Fig. 5.1 as a conceptual plot of damage size vs., impact energy developed during the impact event. Note that a distinct damage threshold is typically observed in composites, where damage size follows a steep relationship relative to energy levels beyond the damage threshold. A leveling out of damage size often occurs, with even a reduction due to penetration. Damage modes could be changing along the progression of this curve, often starting as delamination, followed by some amount of back-side fiber cracking, then finally penetration.

Damage thresholds are either referred to as the Failure Threshold Energy (FTE), Failure Threshold Force (FTF), or the Damage Threshold Load (DTL). The work of Davies and coworkers [13–15] articulated the dependence of the threshold conditions in terms of the laminate thickness and interlaminar

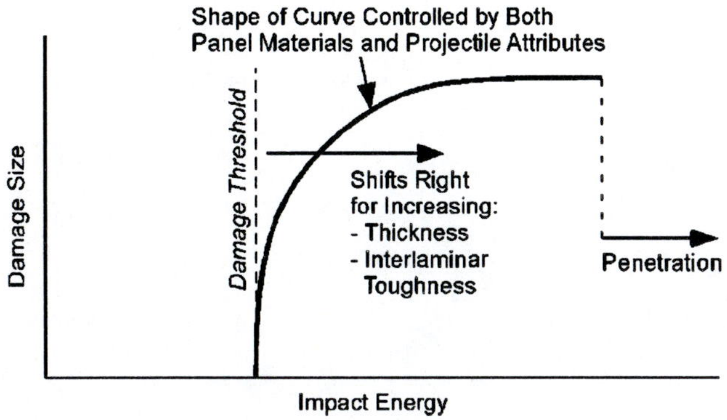

Fig. 5.1 Conceptual damage size vs. KE/contact force

mode 2 fracture toughness. Davies and Zhang [15] suggest a relationship for predicting the DTL:

$$DTL = \left[\frac{8\pi^2 E t^3 G_{IIc}}{9(1-\nu^2)}\right]^{1/2} \rightarrow Ct^{3/2} \tag{5.1}$$

where G_{IIc} is the mode 2 interlaminar critical strain energy release rate, E is the panel in-plane modulus, and C is a constant indicating the DTL to scale with the panel thickness t by the 3/2 power. Schoeppner and Abrate [57], and many others have confirmed the dependence of damage threshold on laminate thickness for quasi-isotropic laminates. This historical testing has provided data in terms of critical incident-impacting energies or the critical contact force. Transformations between force and energy are complex as they involve the local dynamic stiffness of the target which is dependent upon the composite structure as well as the boundary conditions. The concept of impacting energy and the contact force developed during the impact was very clearly articulated by Jackson and Poe [33]. *The energy format is useful for developing requirements and comparing materials, while the force format is better suited for structural analysis. Both are necessary and useful concepts.*

A major consequence of the impact-induced damage is a loss of compression strength capacity. The loss in residual strength follows similar strength reductions to that observed in "notched" laminates. This strength penalty is typically a function of induced damage state. Being "fiber dominated," it shows tendency to be insensitive to matrix toughness, as shown in Fig. 5.2 [17, 47].

In summary, the past 30 years have provided a useful understanding of the impact threat environments and the characterization of small, hard object, low-velocity impact conditions. The current recognition of this threat to airframe structures is discussed in detail in the Composite Materials Handbook 17 (vol. 3, Chap. 12 of [47]).

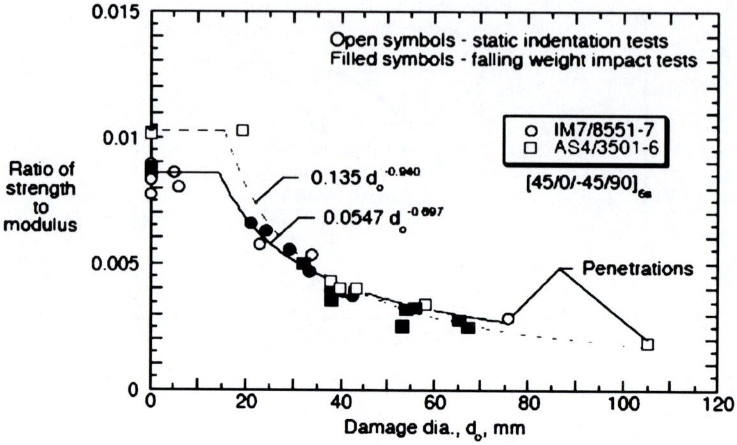

Fig. 5.2 Compression after impact (CAI) test results for untoughened and toughened carbon/epoxy laminates [17, 47]

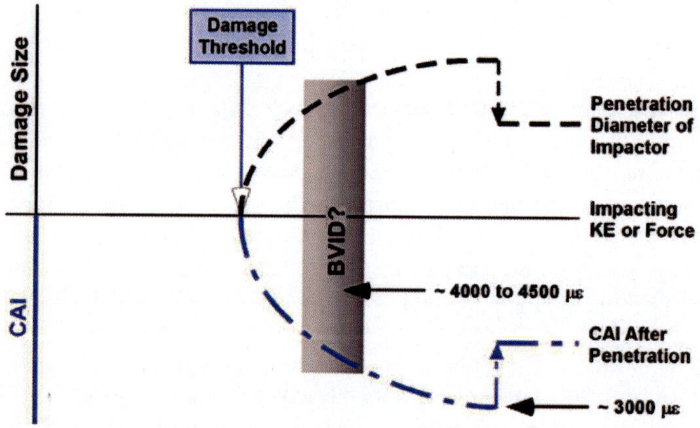

Fig. 5.3 Impact damage size and corresponding reduction in compression after impact (CAI) strength [31]

The combination of Figs. 5.1 and 5.2 is useful to understand the interaction of material characteristics, the damage threshold, damage size, and the resulting reduction of compression strength (see Fig. 5.3) for a flat plate. Impact events at relatively low energy levels do not induce resin cracking, local delamination, or fiber breakage. At a critical impacting energy level, local matrix cracking and delamination occurs with some visible surface roughness. With increasing impacting energies, the damage area increases in size. Typically, the impacted surface returns back to its initial contour until local fiber breakage occurs, indicated as the gray shaded area labeled as BVID (barely visible impact damage) in Fig. 5.3. At this point, a visible dent is typically observed. Additional strength reduction occurs in a proportionate manner to the increase in the damage area (see Fig. 5.2). As the intensity of the impact event increases, the damage area stabilizes, as does the compression strength reduction. The compression strength after impact, at this plateau is typical of a plate with a open hole stress concentration, corresponding to about 3,200 microstrain.

This generic relationship between the impact-induced damage state and the residual compression strength is independent of the variables cited in Fig. 5.1. The absolute robustness of a structural panel, however, is a direct function of the variables cited in this figure. The concept of the Failure Threshold Energy is introduced to provide a measure of the interaction of material toughness, panel thickness, and the specific material characteristics of the impacting object. These interactions will be developed in the following sections. Other chapters in this book will address specific material topics influencing the fracture toughness of resins and the interlaminar strength of laminated composite materials.

The threshold conditions, DTL and/or FTE cited earlier, were reported in research papers but their use in the engineering design process did not occur. Work in the 1980s and 1990s implemented a design requirement for tool drop and foreign object damage (FOD) by bounding the impacting kinetic energy, an energy cutoff [47]. Structural damage tolerance management focused on Compression after Impact (CAI) and visual inspection. Discrete source impact conditions for both bird strike and tire rupture were also implemented in this time frame.

5.3 Damage Threat Sources

Composite structures, in particular aircraft airframes, can operate in a variety of discrete threat environments. Airframes fabricated with composite materials have different capabilities than metallic airframes and they also have increased sensitivity in particular to impact threats.

In-service damage can come from a variety of sources, each having a different area of the structure that it may affect. The most common in-service damage is due to an impact event. Sources of in-service damage threats and the areas of aircraft structure they typically affect are described in Table 5.1. Lightning strike has been included in this listing of impact threats due to it being similar in the time-scale over which it occurs and the localization of damage produced by the strike event.

Impact events can generally cause combinations of damage. For example, high-energy impacts by large objects (e.g., liberated turbine blades) may lead to penetration of skins, broken elements, and failed attachments. The resulting damage may include significant fiber failure, matrix cracking, delamination, broken fasteners, and disbonded elements. Damage caused by low-energy impact is more contained, but may also include a combination of broken fibers, matrix cracks, debonding, and multiple delaminations.

Table 5.1 Impact damage threat sources and locations affected

Damage threat sources	Locations affected
Tool drops	Horizontal upper surfaces
Runway debris	Lower surfaces around and behind landing gear, empennage
Ground vehicles and equipment	Almost any exterior surface
Bird strike	Nose and leading edges, windshield, engines, all forward-facing surfaces
Hailstones	Upper surfaces, leading edges, and side surfaces
Maintenance actions	Removable panels and structures
Turbine engine blade separation	Structure near engine
Ballistic damage	Military aircraft
Rain erosion	Nose and leading edges
Lightning strike	Nose and tail, leading and trailing edges

Damage from any of these sources may be exacerbated by subsequent repeated loads and environmental exposure. Growth of cracks from damage locations must be evaluated to ensure that either growth will not occur during the life of the structure or will not become critical prior to being found during periodic inspections, depending on the damage tolerance approach one follows. Two special areas for concern are: the possible growth of delaminations due to repeated compressive or shear loads, and the possible growth of delaminations or disbonds at stress concentrations such as ply drops and bondlines. The type and extent of damage will determine its effect on structural performance and its detectability.

Due to the concerns regarding the potential for defects and damage to go undetected, it is crucial for composite aircraft structures to be damage tolerant, i.e., to continue to perform their intended functions with specific levels of damage present. To accomplish this, it is important to understand the physics and phenomena governing different types of defects and damage that can occur in composite structures.

A status matrix for aeronautical service-induced impact threat environments ranging from low-velocity to relatively high-velocity impact conditions is shown in Table 5.2. This matrix summarizes the nature of the threat, test method for characterizing the response of a material or structure to the impact, status of simulation models of the structural response to the impact event, the probability for the severity of the threat to be addressed in the structural design and substantiation of the structure, whether this is a self evident event or incident that equipment operators and maintenance staff will be aware of, and if expected locations where the impact events would typically occur are known (i.e., zoned).

The following subsections describe each of these impact threats, with the exception of the panels lost in flight, providing a general description about the phenomena, as well as information about how the threat is handled experimentally and what types of damage are produced.

Table 5.2 Status matrix for aeronautical service-induced impact threat environment

Threat	Test protocol	Simulation models	Threat allowable	Self-evident event	Impact location(s) zoned
Bird strike	Gel pack, real bird	Yes	"B-basis" FAR's (defines weight and velocity)	Yes	Yes
Hail ice	Simulated hail ice (cast ice)	Yes/maturing	"B-basis" Update MIL-HDBK-310 Company specific	Yes	Yes
Runway debris	Metal ball, drop tower	Maturing	"B-basis" Update JSSG-2006	Sometimes	Usually
Tire rupture	Rubber puck	Yes/maturing	AC25.963-1 [20]	Yes	Sometimes
Panels lost in-flight	Not defined	Not defined	Not defined	Yes	Sometimes
Tool drop	Steel or aluminum hemisphere, drop weight or pendulum	Yes/maturing	JSSG-2006 Structures	Sometimes	Yes
Ground equipment contact	Not defined	Not defined	Not defined	Sometimes	Yes

Definitions
"*B-basis*", threat definition based on 90th percentile of cumulative probability
FAR, Federal Air Regulations issued by the FAA
Mil-Hdbk-310, DoD Global Climatic Data for Developing Military Products
JSSG-2006, DoD Joint Service Specification Guide for Aircraft Structures

5.3.1 Tool Drop and Foreign Object Damage (FOD)

A major threat to composites is caused by dropped tools or impact by foreign objects. Damage from these sources can occur during routine operation and while the vehicle is receiving maintenance. Foreign object damage (FOD) is defined as damage produced by any foreign object or debris, such as rocks, runway/pavement fragments, loose tools and fasteners, parts, etc. FOD can lift up off the ground and impact an aircraft while taking off or landing. Inadvertent impact damage that can occur during maintenance consists of tools dropped, personnel stepping on delicate locations, part handling, and storage accidents [57]. These impact sources can lead to significant damage to the composite structure, including the internal structure, that is prospectively difficult to detect [1].

Typically, tool drop and FOD are experimentally handled by drop weight or swinging pendulum impact tests. These test methods have traditionally been used to produce damage in the low-velocity range, using a heavy mass to produce a kinetic energy level of interest. The basic principle of these low-velocity test setups is to apply an impact threat at a known energy level and observe what damage, if any, develops. The impact energy level can be determined by the known mass and drop height, based on energy balance (i.e., initial potential energy = kinetic energy at moment of impact) or can more accurately be known by measuring the velocity of the impactor just prior to the impact event, from which the projectile kinetic energy can be calculated. The drop weight and pendulum impact setups are also amenable to the measurement of contact force history that develops between the impactor and target, by use of a dynamic load cell. The load cell must be dynamically rated and calibrated since a statically calibrated load cell can give a high percent error in the actual dynamic load measurements [60].

A pendulum impact system typically consists of a rotating rigid arm with an impactor head, or an impactor mass swinging on cables. The impactor swings into a test specimen that is typically mounted in the vertical plane. The impactor is released from a predetermined height corresponding to a desired impact energy level [35]. Similarly, a drop test system consists of vertically dropping impactors from a predetermined height onto a panel that is typically mounted in the horizontal plane [10]. The falling weight is usually guided by a tube or rail system so as to achieve better targeting accuracy. The energy of the event is determined by the mass of the impactor and the height from which the impactor is dropped.

Some advantages of the drop weight and pendulum impactors are: fairly simple setup of test apparatus, impact energy levels can be easily calculated by energy balance, impactor tip can be instrumented allowing the energy level to be correlated to the developing impact forces, and the ability to easily test a variety of tip geometries. For drop weight and pendulum impact systems, the impactor should be caught on the first rebound, eliminating multiple impacts [35, 60].

The impactor geometry plays a significant role in the damage produced. Poe [53] found that thick composites impacted with blunt impactors reduced the tensile strength of the laminate without producing visible damage, while sharp impactors caused similar reductions in postimpact tensile strength but with visible damage. Elber [18] used a 25-mm diameter steel ball with a pendulum impactor to classify tool drop. Typically, metal (often hardened steel), with hemispherical tips, is used as the impactor and is essentially nondeforming relative to the target. However, the pendulum impactor and drop test allow for a variety of tip geometries, ranging in bluntness, to be tested [10, 16].

Low-velocity impacts representing tool drop can be characterized as exciting deformation modes in the target that are primarily described by the first mode of vibration, or the "static mode" [18]. Higher modes can be neglected because the contact force duration is much longer than the time required for the impact wave to reach the boundaries and return. The phenomenon can be approximated by an energy balance model where the total energy of the system is conserved, neglecting energy of higher vibration modes, friction, and other losses. While energy-balance models are capable of estimating the peak contact force and are simple to implement, they cannot accurately generate force, velocity, or displacement histories during the impact [59].

Kinetic energy, peak force, and force history are common parameters used to classify impact. Other considerations are the impactor geometry, material, and velocity of the impactor just prior to impact, and the shape and boundary conditions of the impacted structure. A critical aspect of any impact event is its velocity regime (high vs. low) which controls the type of target deformation state that develops. The velocity will determine the amount of global deformation that will occur and the extent that the boundary conditions affect the deformation state [33, 59]. For a high-velocity event, the deformation of the plate is still low at the time the peak contact forces are developed and therefore the boundary conditions become less important. This is due to the fact that, as the duration of contact force history decreases, the plate cannot respond quickly enough to develop a global response [9]. Thus for increasingly higher projectile velocity, elastic wave propagation effects replaces boundary conditions as the most important parameter. Alternatively, for low-velocity tests (typical of drop weight and pendulum), a static-like structural response develops and therefore boundary conditions become crucial when attempting to accurately correlate the response of test specimens to the actual response in the structure of interest. Damage created from a dynamic impact event is often more strongly influenced by the boundary conditions and the test fixture than the material properties [61].

For low-velocity events, dynamic impact scenarios can usually be experimentally represented using equivalent quasi-static tests. Many researchers show equivalence between quasi-static and low-velocity impact tests, in particular Maikuma et al. [45], Murthy and Chamis [49], Cheverton et al. [11], and Crosley and Ripling [12]. Equivalence for flat composite plates is shown by Elber [18], Kwon and Sankar [43], Lee and Zahuta [44], Sjoblom et al. [61], Jackson and Poe [33], Tan and Sun [62], and Wu and Shyu [67]. Equivalence for shells has been shown by Meyer [46] and Wardle and Lagace [64]. For example, Tan and Sun [62] have shown correlation between the strain response predicted from finite element models with quasi-static indentation and low-velocity impact tests.

Wardle and Lagace [64] found that composite structures impacted at 3 m/s resulted in the same damage behavior as when they were quasi-statically loaded. Vibrations that occur in the dynamic tests were found to have negligible effects on the damage produced when a composite shell is impacted with a low-velocity high mass object. Damage states created for quasi-static tests and dynamic tests are equivalent if the peak loads are the same [53, 64, 67]. Due to their more controlled nature, quasi-static indentation tests can provide more consistent insight to the damage progression and interaction of damage modes than impact tests, allowing damage mechanisms to be compared between specimens [64]. Figure 5.4 shows the static and dynamic equivalence for a 16 ply composite plate in the form of measured contact force vs. displacement at the impact/indentation point. The vibrations in the dynamic test are a result of vibration of the impactor [61].

Sjoblom [60] found that the lower energy levels created minor matrix damage, followed by delamination, and back face damage. Figure 5.5 shows a comparison of static and dynamic impact damage in 48 ply Graphite/PEEK composite plates. The static test showed unstable delamination growth and matrix cracking with no

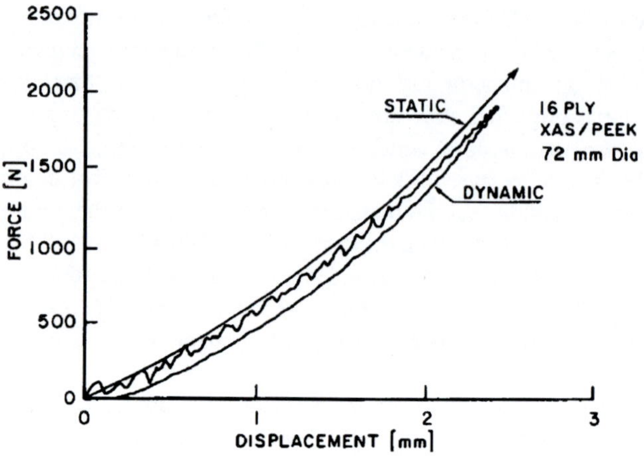

Fig. 5.4 Comparison between static and dynamic response: 16 ply Gr/PEEK [61]

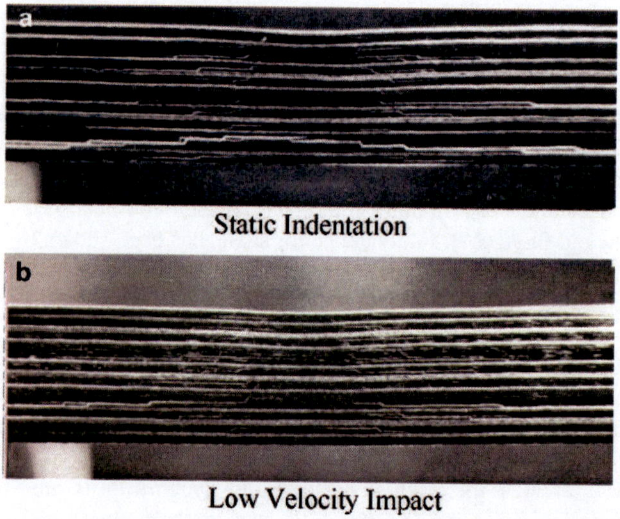

Fig. 5.5 Damage of 48 ply graphite/PEEK plates subjected to static and dynamic loading [61]: (**a**) static indentation and (**b**) low velocity impact

visible surface damage, with the final damage state being backside transverse matrix cracking and fiber breakage [61]. Elber [18] found matrix damage in the form of delamination and matrix cracking along the fiber direction for both static and impact tests for quasi-isotropic eight ply plates. For tool drop, typical failure mechanisms are a combination of delamination, matrix failure, or back-face tension failure, with back-face fiber damage being the first external visible damage [15]. Wardle and

Lagace [63] found there is no correlation between dent depth, peak force of impact, and development of significant internal damage for thin composite plate and shell structures.

5.3.2 Ground Service Equipment (GSE)

In addition to tool drop and FOD, damage to aircraft often occurs during normal aircraft ground operations (e.g., cargo and passenger loading, refueling, etc.) due to accidental contact between the aircraft and surrounding equipment, which is generally referred to as ground service equipment (GSE). These include (but are not limited to): ground vehicles (cargo loaders, catering and fueling trucks, other vehicles), ground stairs or passenger loading bridges, various ground service equipment such as ladders and work stands that can be accidentally rolled or blown into the aircraft. Some equipment can positioned such that contact can be made due to the settling of the aircraft during the fueling and passenger/cargo loading process.

Ground service vehicles account for a significant percentage of damage occurring to commercial transport aircraft. 50% of *major* damage has been recorded to be caused by baggage vehicles and 60% of all *minor* damage caused by collision with ground vehicles/equipment [32], costing the industry US$4 billion per year. This occurs during cargo movement while loading the aircraft, or docking of GSE around the aircraft doors. The areas in close vicinity of the doorway openings, while reinforced with metallic "scuff plates" to protect against accidental damage, are not sufficient as damage is often incurred beyond the coverage of the reinforcement.

Different aircraft geometries and sizes influence the possible impact sources and their frequency of occurrence. Smaller aircraft models, such as the Boeing 737, typically have a shorter turnaround time between flights compared to larger aircraft, and therefore operations around these aircraft can be more rushed. Additionally, there tends to be less space around the smaller aircraft gates, leading to more concentrated GSE traffic levels in tight proximity of each other. As shown in Fig. 5.6, the Boeing 737 fuselage is at a high risk of direct contact with the GSE due to its low height from the ground. Indirect impact between the aircraft and cargo loading equipment can also occur when, e.g., a luggage-towing vehicle approaches too fast and accidentally collides with a belt loader, which then moves and impacts the aircraft.

There is a different set of threats associated with larger aircraft models, such as the Boeing 747 and 767. The angle of approach for docking is more difficult toward the tail of the aircraft than for the smaller aircrafts, due to higher degree of curvature of the large aircraft toward the tail, as shown in Fig. 5.7. The bottom of the fuselage of the large aircraft is much higher, thereby eliminating contact from sources such as vehicles and luggage carts driving by or under (which officially is not permitted, but still does occur).

Fig. 5.6 Boeing 737 with belt loader and truck

Fig. 5.7 Difficult approach of belt loader and boeing 747

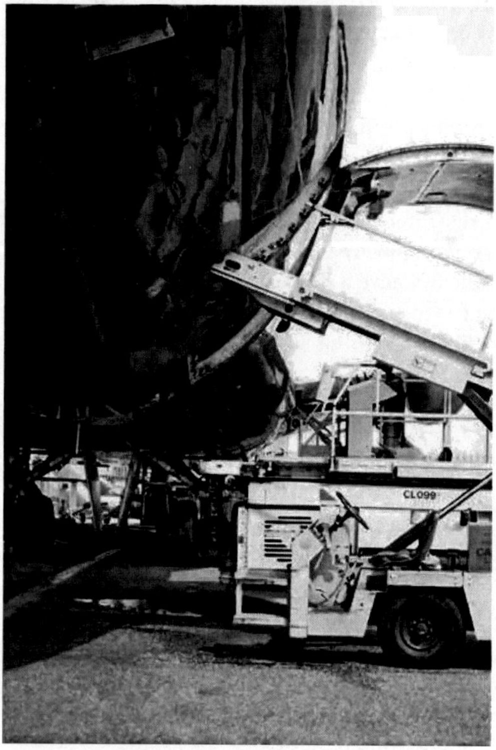

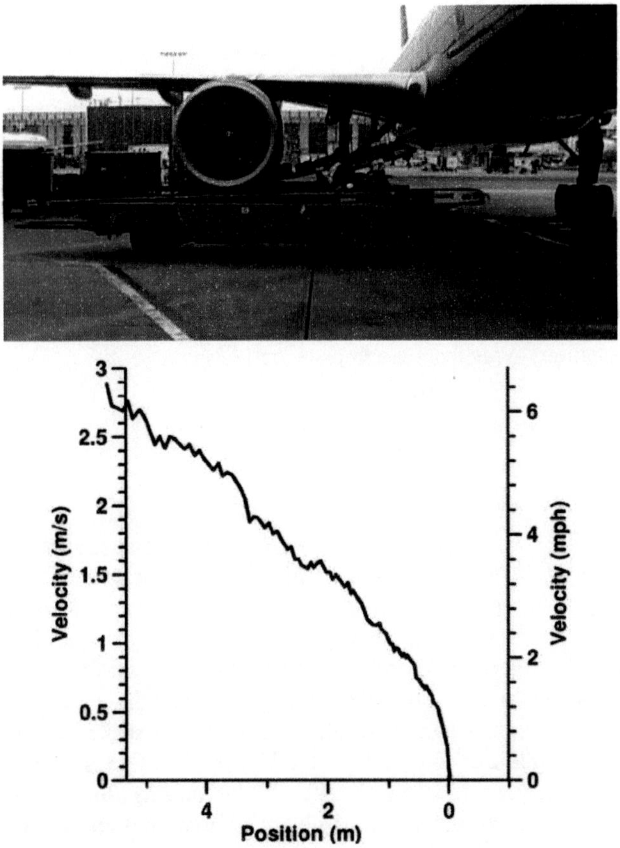

Fig. 5.8 Velocity of TUG 660E Belt loader approaching boeing 757 aircraft

Of particular concern is the large mass of GSE, especially ground vehicles, in combination with the practice of using deformable bumpers to "protect" the aircraft in the event of inadvertent contact. The bumpers, typically made from dense rubber or foam, do serve this function for light contacts where the bumper is not significantly deformed, and therefore high contact forces are not developed. However, for a high-energy impact event, where the GSE interacts with the aircraft in a manner generating significant force, the bumpers deform significantly and effectively spread the contact over a wide area ($>10^5$ mm^2 order of magnitude compared with $<10^2$ mm^2 for tool drop and FOD), thus decreasing the detectability of damage.

The impact energy possible for ground vehicle impacts has been quantified recently (DeFrancisci and Kim, unpublished work) as part of a research program funded by the FAA. Figure 5.8 shows the velocity of a belt loader (model TUG 660E) approaching an aircraft together with a plot of vehicle velocity as a function of relative position from stop. As shown in Fig. 5.8, speeds up to 2 mph are realistic within close proximity of the aircraft. For a TUG 660E Belt Loader with a mass of 3,030 kg (6,680 lbs), the kinetic energy associated with 1 m/s (2.24 mph) is 1,515 J (1,117 ft lb). This is an extremely high energy level, not typically tested in a laboratory environment which usually considers only relatively small panel specimens mounted in a drop tower.

The physics that describe the low-velocity impact tests, such as tool drop and FOD, also apply to impact due to ground service equipment, i.e., can be represented by quasi-static indentation. The fundamental differences are the potentially widespread nature of the impact contact area and the much larger magnitude of energy of impact, thereby generating considerable contact forces. For these reasons, widespread impact from GSE must also take into consideration the potential of creating damage at locations away from the impact site.

5.3.3 *Bird Strike*

Birds present a significant threat to fast-moving vehicle structures, particularly the forward-facing surfaces of in-flight aircraft, and rotating fan and propeller blades. Bird strikes have been an issue since the earliest days of manned flight. The first bird strike on a powered aircraft occurred over Huffman Prairie, Dayton, Ohio, on September 7, 1905 as the Wright brothers were in flight development testing. The first recorded human fatality resulting from a bird strike occurred in 1912. Cal Rodgers, the first man to fly across the United States, crashed into the ocean after a gull became jammed in his aircraft's flight controls. Since then, birdstrikes have become an increasingly serious problem in both civil and military aviation, with many thousands of strikes occurring every year. Since the 1950s, there have been extensive efforts to characterize bird-strike events, providing a technical basis for understanding the bird-strike threat environment, and the development of procedures to mitigate the threat environment and provide a level of damage tolerance to bird strikes through regulations and design practices.

There has been a coordinated international effort to accumulate bird-strike data for both civilian and military aviation. Through its Bird/Wildlife Aircraft Strike Hazard (BASH) team (US [73]), the USAF initiated and maintains a comprehensive database. Currently, the United States, Canadian, and European Union civil authorities have organized data collection systems. Recently, the FAA has provided access to their bird-strike data since January 1990 [23]. Figure 5.9 shows a cumulative distribution of bird weights based on a subset of the FAA database from January 1990 to August 2005 (Halpin, unpublished). The data labeled 1977 is from an internal USAF study by Halpin using both USAF BASH and FAA data. The bird-strike database identifies the bird species. An average bird weight for the individual species was used to develop the weight distributions. Similar results in Fig. 5.10 show the cumulative distribution of bird weights [24] based on a detailed study conducted by the Federal Aviation Administration (FAA) between 1981 to 1983.

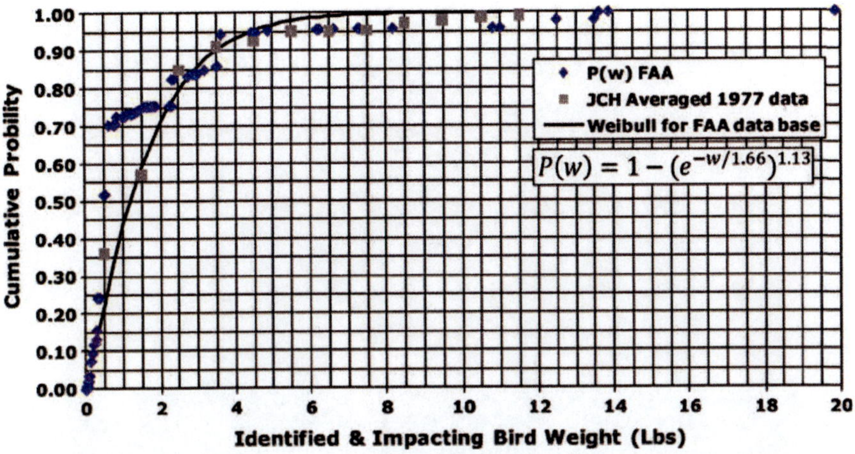

Fig. 5.9 Cumulative distributions of U.S. bird weights (unpublished database composed by J.C. Halpin)

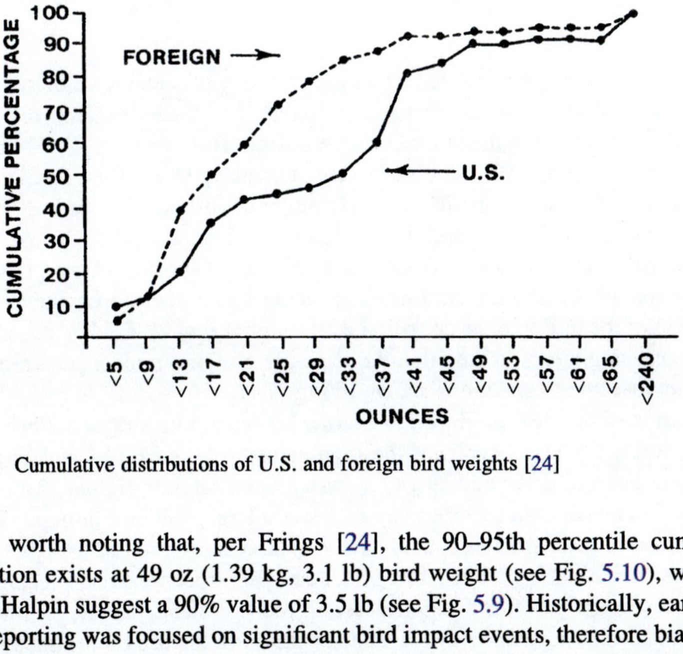

Fig. 5.10 Cumulative distributions of U.S. and foreign bird weights [24]

It is worth noting that, per Frings [24], the 90–95th percentile cumulative distribution exists at 49 oz (1.39 kg, 3.1 lb) bird weight (see Fig. 5.10), while the data by Halpin suggest a 90% value of 3.5 lb (see Fig. 5.9). Historically, early bird-strike reporting was focused on significant bird impact events, therefore biasing the available database. In the 1950–1970 timeframe, the 4-lb bird was selected based on the USAF damage tolerance philosophy of defining a threat at the 90% cumulative probability level. In fact, the FAA specifies the requirement in Advisory Circular 25-21, Section 25.571 [21] that structures be tolerant to damage that may be caused by in-flight impact with a 4-lb (64 oz, or 1.81 kg) bird. It was understood at that time

Fig. 5.11 Damage to nose cone of C-130 due to bird strike (USAF public domain photo)

that the velocity of impact is defined based on the relative closing velocity between the moving bird and moving structure (aircraft, fan blade, other vehicle). The velocity chosen for civil applications was the design cruise velocity, V_c. For military applications, historical 95% cumulative velocity data up to 10,000 ft (above ground level) was used for fighter applications (Halpin unpublished F-16 studies).

Significant damage does occur, particularly as the velocity and mass of the bird increases with respect to the reference regulations. Of special concern is whether the ingestion of birds into aircraft engines will halt their operation, and the penetration of the bird into the cockpit, i.e., through the windshield or surroundings structure, causing injury to the pilot. An example of a bird impact penetration to a composite nose cone structure is shown in Fig. 5.11.

Evaluation of damage developed in a structure due to bird strikes has historically been dominated by tests involving the high-velocity projection of real birds onto structures of interest. In the mid 1950s, at de Havilland Aircraft, Hatfield, UK, freshly killed chickens were used to accurately simulate a large, live bird in flight. The test target is fixed in place on a test stand, and a cannon is used to fire the chicken into the engine, windshield, or other test structure. High-speed cameras recorded the complete test activity. After firing, the test objects were taken away, examined for damage, and design adjustments were made to provide tolerance for the impact event. Other early use of a gas gun was by the Royal Aeronautical (Aircraft) Establishment (RAE) in Farnborough, UK in 1961 [50]. The gun was driven by a compressed air tank. In the 1970s, Goodyear Aerospace in Litchfield Park, Arizona, used a gun with a ceramic diaphragm to seal the compressed air in the tank from the gun's barrel. To fire the gun, a solenoid-driven needle struck and ruptured the diaphragm, allowing the compressed air to drive the chicken (in its container – a cylindrical cardboard ice cream carton) down the barrel. At the muzzle, a metal ring stopped the carton, but allowed the chicken to pass through. High-speed cameras photographed the chicken impacting a fighter windshield in the test bed. These cameras were started in time with the breaking of the diaphragm.

The forces generated by a bird strike have been found to be dominated by a fluidlike behavior [66] of the projectile and thus a gelatin substitute having the same total mass, density, and compressibility characteristics was developed as an equivalent replacement [7, 65]. However, Airoldi and Cacchione [2] suggest that fluid-only behavior can be nonconservative in its representation of a bird, with significant shear strength existing due to the tissue material playing an important role in the development of damage to the impacted structure.

The numerical simulation of bird-strike events, and the use of the simulation to evaluate structural response and damage accumulation due to bird impacts have become regularly practiced. Recent examples of works include Georgiadis et al. [27], Mao et al. [46], Goncalves et al. [28], Airoldi and Cacchione [2], and Anghileri et al. [3]. Certification of composite flight hardware for the Boeing 787 has been supported using numerical simulations of bird impacts. An example shown in Fig. 5.12 of a gelatin bird impact onto a composite moveable wing trailing edge [27] demonstrates the capability of modern computational models to replicate the physical characteristics of the soft, deformable bird projectile, as well as the damage incurred by the impacted composite structure (see Fig. 5.13).

Goncalves et al. [28] summarize the use of numerical simulation to support the development of the Embraer ERJ-170 family of aircraft. They define a methodology by which accurate simulations are developed on a *representative* structure in conjunction with experiments used for verification and calibration. The validated numerical simulation models can then be used to predict bird impact damage to the *real* structure of interest, supporting the structure's development program and ultimately certification of the structure.

Despite the advances in simulation, system-level testing will continue to be a requirement. Figures 5.14 and 5.15 are illustrations of bird strikes in which the bird struck a windshield, was compressed on impact, but retained a structural mass while it slid across the windshield, over the window frame, and then penetrated into the cockpit. Figure 5.14 shows the aftermath of a B737 incident over Canada [71]. The bird penetrated into the cockpit above the window injuring the pilot. Figure 5.15 documents a bird-strike test on a helicopter cockpit using a 2-lb bird at 83.8 m/s [2].

In another incident, an American Airlines aircraft, Boeing 767-323ER aircraft experienced a bird-strike incident while climbing out of Charles de Gaulle Airport on April 2, 2001 at 12,000–14,000 ft. altitude and a speed of 330 KIAS (knots indicated air speed) when it hit a flock of large birds causing multiple penetrations, with a bird entering the cockpit and depressurizing the cabin (reported by Aviation Safety Network [6] and Bird Strike Committee USA [8]). The windshield was covered with blood and bird debris and insulation was splattered on captain and copilot. Seventeen strikes were found with 11 punctures. The birds were identified by the Smithsonian Institution to be the Northern Shoveler, weighing ~1.8 lbs.

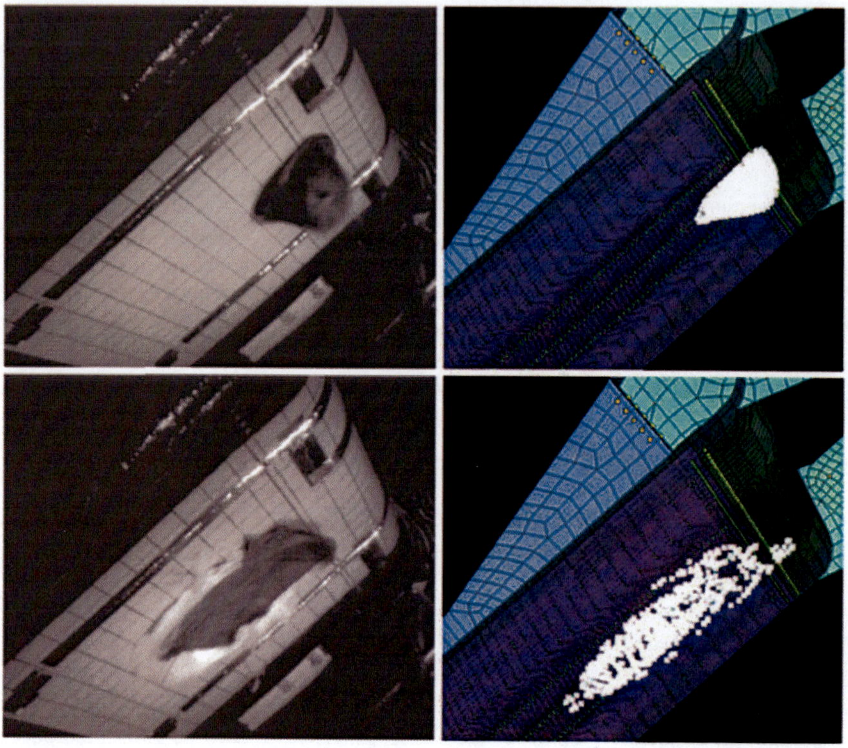

Fig. 5.12 Correlation between experiment (*left*) and simulation (*right*) of bird projectile deformation during impact [27]

Areas dented include above the windshield, fuselage, left wing, horizontal stabilizer, edge of a hatch door. Areas penetrated include radome, wing leading edges, lower fuselage, lower wing surface, and nose. At the time of impact, the copilot was flying, the captain took over after putting on an oxygen mask and recovered the aircraft, landing successfully.

These are only three incidents, but they illustrate the challenging conditions bird-strike threats pose for an integrated air vehicle system. These three illustrations involved the integrated response of the windscreens and surrounding structure, which is a system-level challenge. In the past, the development of air regulations has been reactive: most aviation industry rules have been drafted in response to accidents. The reader is referred to the specific FAA regulations for the various classes of air vehicles. A good summary of the requirements can be found in the document "Sharing the Skies" [71].

It has long been recognized that bird-strike incidents can occur at weights greater than cited regulation weights and velocities. An additional precaution has been introduced to moderate transport aircraft velocities at altitudes below 10,000 ft. In a Notice of Proposed Amendment (NPA) 2002-022 to *Canadian*

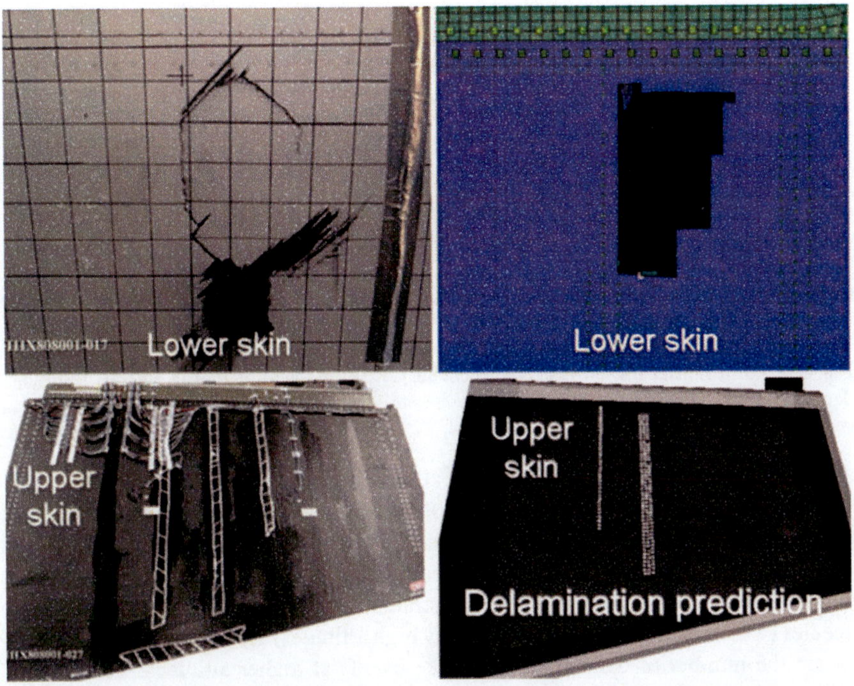

Fig. 5.13 Correlation between experiment (*left*) and simulation (*right*) of damage produced in composite structure by bird impact [27]

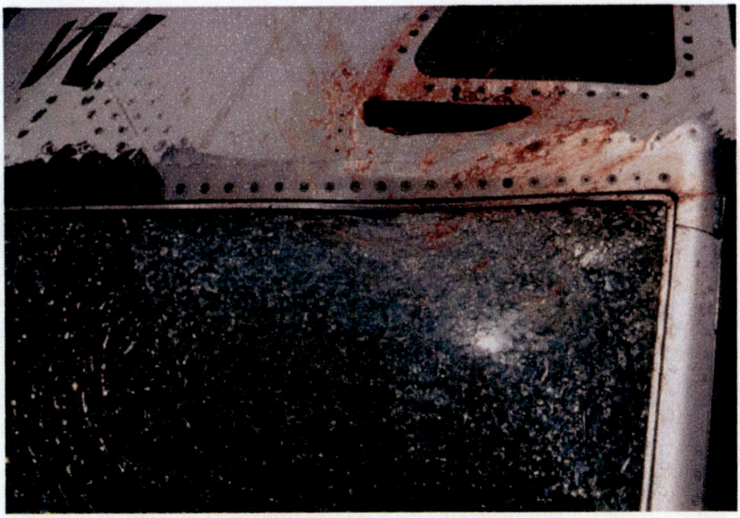

Fig. 5.14 Penetration damage to boeing 737 by bird impact at 10,000 ft altitude and 250 Kts causing injury to pilot (colour plate 11 [71])

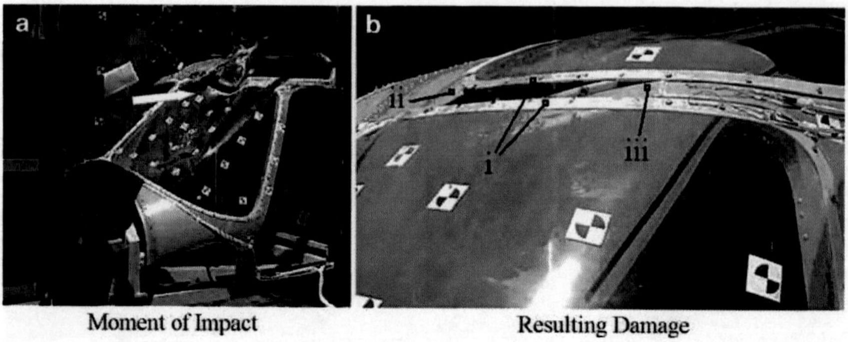

Fig. 5.15 Bird impact test on helicopter cockpit [2]: (**a**) moment of impact and (**b**) resulting damage

Aviation Regulation 602.32 Transport Canada defined a maximum departing velocity. On February 26, 2002, the NPA proposed to eliminate the air traffic control practice of allowing aircraft departing from Canadian airports to exceed 250 KIAS (knots indicated air speed) below 10,000 ft. MSL (mean sea level). The NPA responded to data that demonstrated populations of large flocking-bird species (which tend to migrate at relatively high altitudes) are increasing in Canada, as are the number of damaging bird-strike events at higher altitudes. To provide additional data in support of the amendment, Transport Canada conducted a risk analysis of the air traffic control practice. The resulting study, *Risk Analysis of High Speed Aircraft Departures* [72], was a system-safety examination of risks associated with high-speed departures. Many international airports, operating in areas of high bird congestion, have adopted this protocol in recognition of the limitations of precaution design and the practical weight implications.

5.3.4 Tire Rupture

A tire rupture event can produce a discrete impact threat risk in the form of projected tire fragments impacting the underside of the aircraft. This phenomenon is infamously known to ultimately have caused the loss of a Concorde aircraft shortly after takeoff at Charles de Gaulle airport in Paris, France, on July 25, 2000 (Davidson and Labib [13]). Rupture and disintegration of an aircraft tire can project tire fragments at high velocities, particularly at takeoff speeds, due to the tangential wheel surface speed of the rolling tire.

The FAA provides a clear, well-defined requirement regarding resistance to tire rupture damage in Advisory Circular 25.963-1 [20]. Here, it is specified that fuel access covers be as impact resistant as the contiguous fuel tank structure (but does not need to be more resistant) and that the resistance be proven via tests to show that no "hazardous quantity" of fuel leakage occurs as a result of the impact.

AC 25.963-1 further recommends that impact tests be conducted using a tire tread segment equal to 1% of the tire mass, travelling at a speed corresponding to the airplane rotational (takeoff) speed (V_R), and distributed over an impact area equal to 1.5% of the total tread area. The possible locations of impact that must be considered are from 30° inboard to 30° outboard of the tire's plane of rotation. Note that AC 25.963-1 [20] is a good example of a requirement which clearly defines the threat based on a realistic physical representation of a ruptured tire fragment associated with the most severe velocity (V_R), based on operational conditions.

A thorough characterization of aircraft rubber tire fragments has been conducted recently by Mines and coauthors in a two-part series of papers on this topic [38, 48]. Mines et al. [48] methodically measured the basic mechanical properties of tire material in both perpendicular and along the tread directions. They then fired cuboid- and ribbon-shaped tire fragments at up to 135 m/s onto aluminum alloy plates at incidence angles ranging from 30° to 90°. Visual observations and strain gage measurements were recorded to be used in the validation of their simulation models [38]. The behavior of the tire fragment projectiles during the impact event was that of a rubber-elastic solid, maintaining physical integrity while undergoing large deformations and not shattering or flowing fluidly as bird projectile material does [66].

The experiments reported by Mines et al. [48] directly supported the development of a numerical simulation model [38]. In these models, clear correlation between experimentally measured/observed physical response was achieved by the finite element-based models which successfully represented the tire fragments using the Mooney–Rivlin rubber material model. The authors extend their simulation methodology, developed via smaller tire fragment test correlations, to large-sized tire fragment impacts onto aircraft structure (as shown in Fig. 5.16), and convincingly argue the widespread applicability of the models to simulate impacts on a general host of structures of interest.

5.3.5 *Hail Ice*

With the recent widespread deployment of all-composite commercial aircraft, including wing and fuselage primary structures (e.g., Boeing 787 and Airbus A350), a significant factor driving the sizing of skin thickness is hail ice, or hailstone, impact, particularly for smaller-size aircraft. Mike Bair, former head of the Boeing 787 program, had stated that the next-generation 737 model, in consideration of using composites, must have upper fuselage thickness capable of withstanding heavy hail, thereby adding proportionally more weight to a smaller jet [26].

Hail, both on-ground and in-flight, is a commonly enough occurring impact threat that it needs to be considered in laminate sizing, particularly for the thin facesheets of sandwich structures. The distribution of hail ice diameters falling to the ground has recently been characterized for 11 major airports across the U.S.

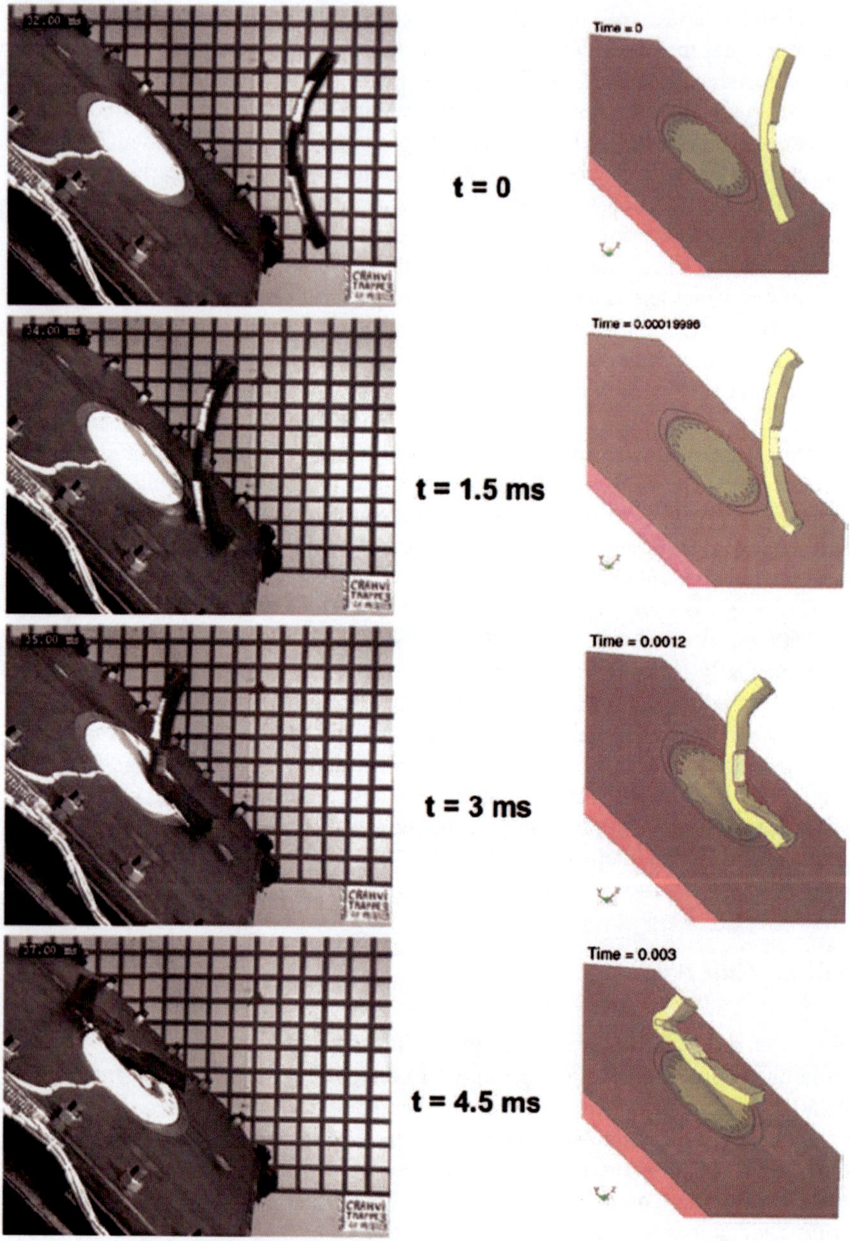

Fig. 5.16 Comparison of tire fragment impact experiment and simulation for wing access panel test at 110 m/s at 30° angle [38]

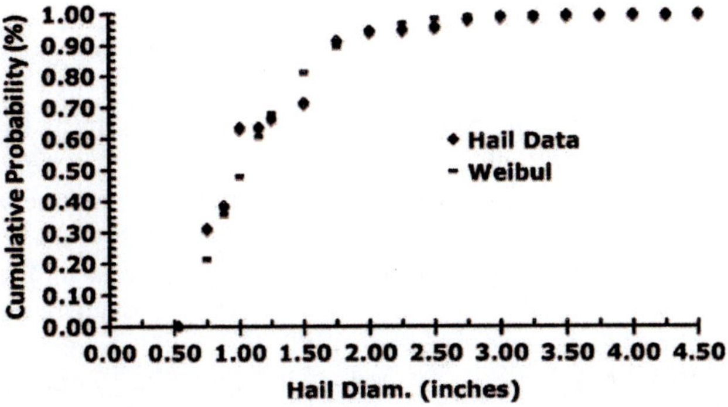

Fig. 5.17 Cumulative probability for hail diameter [30]

using 2,829 severe hail reports made between 1955 and 2006 [30]. These data are summarized in Fig. 5.17 as a cumulative probability of hail size, with 90th, 95th, and 99th percentile hail sizes being 1.76, 2.04, and 2.63 in., respectively.

Impacts with hail ice can occur both as in-flight, and as "ground hail" events. Ground hail has the potential to damage a large number of aircraft, as had occurred in 1995 in a severe Dallas, TX area storm where grapefruit-sized ice was reported, damaging over 100 aircraft on the ground [56]. For ground hail, all outer-exposed upper and side facing surfaces of the aircraft are subject to impact by the ice falling at terminal velocity. Additionally, wind gusts can give the hail a significant horizontal velocity component. In-flight hail presents an even more severe threat than ground hail due to the relatively high closing velocity between the fast-moving aircraft and hail ice. All forward-exposed surfaces, including engines and windshields, are subject to the in-flight hail threat. Rotating engine fan and propeller blades are subject to the greatest hail impact velocities due to the combination of forward motion and high rotational speeds.

Most large, naturally occurring hailstones have been found to be triaxial ellipsoids with less than 10% difference between the two longest axes, thereby allowing them to be treated as oblate spheroids [25]. Typically, hail has a spherically layered or onionskin-like construction resulting from two growth mechanisms: "dry growth mode" and "wet growth mode" [19]. Low water content and/or cold temperatures characterize the environment related to the dry growth mode. This environment allows the supercooled water droplets to freeze upon contact, resulting in air bubbles being trapped in the ice. These air bubbles give this ice, referred to as rime, an opaque appearance and lower density. The wet growth environment has high water content and/or low supercooling. Not all of the water is frozen and the ice, known as glaze, assumes a clear appearance and higher density.

The density of these layers can vary between 170 (lowest density found for rime ice [19]) and 920 kg/m^3 (the density of polycrystalline ice at -16°C [52]). Ground observations of hailstones have found overall density values between 870 and 920 kg/m^3. These seemingly high values, as compared to the range of possible densities, can be explained by the fact that some of the excess water can be incorporated into the less dense layers of rime growth during wet growth [34].

For purposes of conducting impact tests onto vehicle structures of interest, it is impractical to collect natural hail or make accurate artificial hail due to a variety of factors, including the complex physical construction of hail and lack of mechanical property data for properly composing each layer of the hail. Furthermore, if one collects naturally occurring hail and stores it in a freezer prior to testing, the act of normalizing the entire volume to one temperature would fully freeze all the layers and thereby inherently and irreversibly change the hail. Therefore, polycrystalline simulated hail ice (SHI) is usually used for experimental investigations. SHI differs from natural hail in that it is usually created in only a spherical shape and it is made either monolithic (i.e., mold-casted in single pour) or layered (flatwise or spherically layered, see [41]) or with cotton filler [5]. The density of SHI doesn't vary as greatly, having values closer to that of pure glaze ice (920 kg/m^3).

Unlike the bird impact projectile, ice is a relatively hard material, possessing an elastic modulus of 9.3 GPa [52]. During impact, the ice behaves elastically up until compressive stress exceeds its crush strength. Kim and Keune [40] have measured the strain rate sensitive compressive strength of ice at impact strain rates (10^3 s^{-1}) to be 19.7 MPa (3.9 MPa standard deviation). Once ice crushes, the material locally breaks down and flows away from the impact site, as shown in the sequence of photos in Fig. 5.18.

Note that the ice sphere material away from the crushing impact front still maintains its shape long after initial contact, indicating that the sphere's entire volume does not shatter immediately upon contact with the target. By the time the ice breaks into larger fragments, the major portion of the impact event has been completed, as observed by Kim and Kedward [39] via contact force vs. time measurements compared with high-speed film stills. In Fig. 5.18, it can be seen that at $t = 91$ μs after impact, corresponding to peak contact force ($t = 97$ μs), only a small portion of the ice sphere volume has crushed down. By the next frame ($t = 182$ μs), the impact force pulse is mostly over, while the blue dye colored ice is showing signs of whitening, indicating development of fractures through the ice's still-spherical volume.

The damage produced to thin-gage (0.8–2.6 mm thickness) composite panels by ice impact has been investigated by Kim and coauthors [39, 41, 42, 51] and is found to produce a progression of different damage modes as the severity of the impact increases. This damage mode progression is summarized in Fig. 5.19, with each damage type (I–V) described in Table 5.3. The visibility of the resulting damage, described in the table, is a key attribute of interest as this is related to the threshold of visual detectability of any seed damage for which a durability and damage tolerance investigation would be based. Additionally, the transition of damage modes from, e.g., delamination (type I) to penetration with large cracking (type IV), means that the sensitivity of the structure to in-plane compression or tension loading after impact will be directly affected by the mode of damage that was excited.

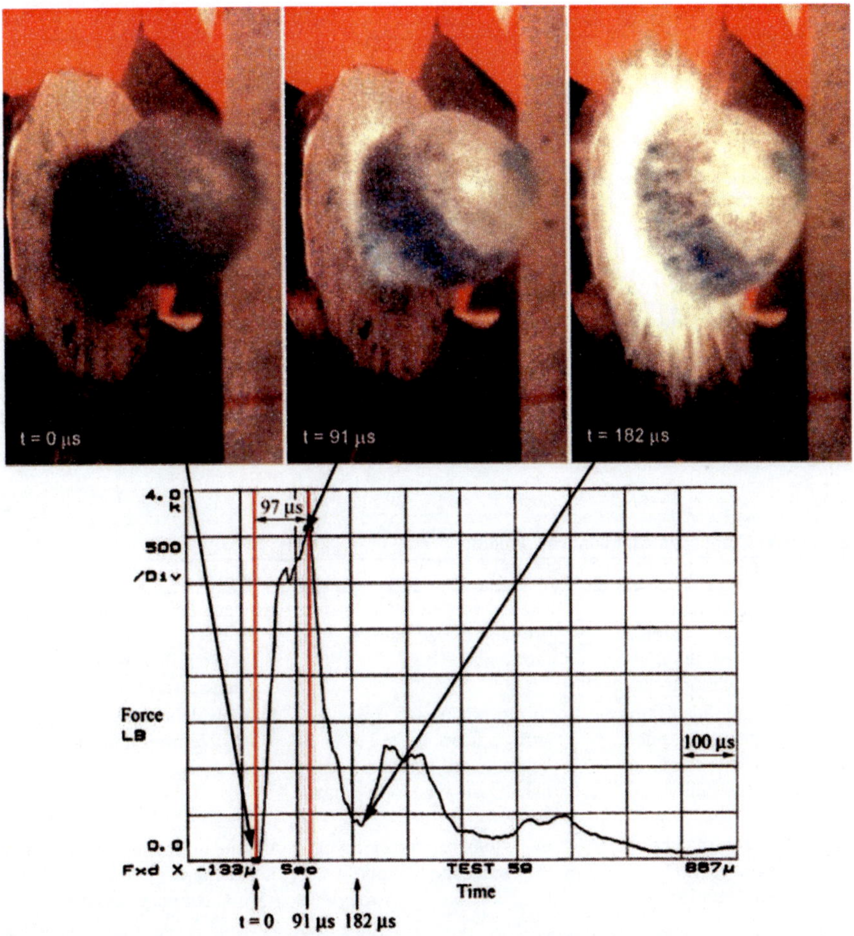

Fig. 5.18 Still images from high-speed film of 42.7 mm ice sphere impacting rigid target at 73.5 m/s (*above*); corresponding force time history (*below*)

The clean hole (type V) will have significantly less postimpact strength reduction than the large cracks (type IV), despite the higher energy of impact needed to produce type V damage. This is reflected by the damage modes and normalized damage size vs. projectile kinetic energy plot shown in Fig. 5.20, as well as the comparison of damage size and reduction of CAI strength illustrated in Fig. 5.3. Note that a failure threshold energy (FTE) can be clearly defined for ice impacts, where impacts at energy levels below the FTE do not produce any detectable damage to the composite.

As shown by the two trends in Fig. 5.20 for SHI made by both unfilled and cotton-filled (per ASTM F320) ice, the resulting initial damage modes are sensitive to the composition of the SHI projectile. Either delamination (type I) or backside fiber

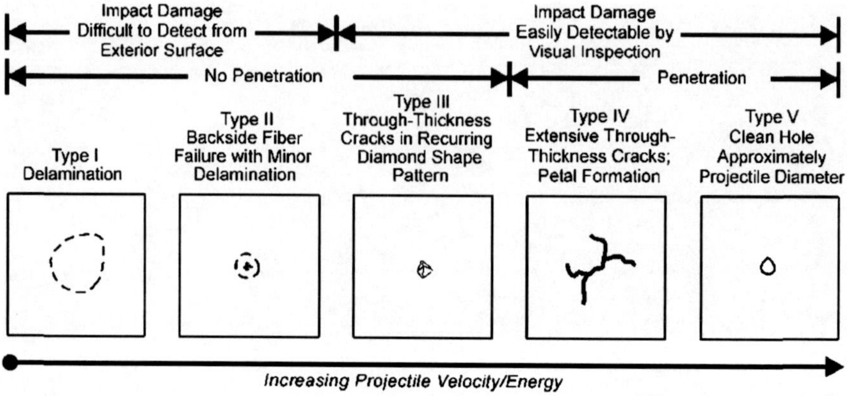

Fig. 5.19 Progression of damage modes for ice sphere impacts onto woven carbon epoxy composite panels

Table 5.3 Composite panel damage modes

Type	Damage mode	Description	Visibility
I	Delamination; no surface-visible fiber failure; no penetration	Ellipsoidal-shaped delamination, growing in size with increasing energy	No dent. Possibly slight surface aberration. Nearly no visibility from impact-side surface
II	Backside fiber failure; no impact-side fiber failure; no penetration	Localized failure, often just surface fibers broken at backside and small amount of delamination near the impact zone	Shallow dent nearly non-visible from impact-side surface
III	Through-thickness cracks; no penetration	Through-thickness cracking of the composite without projectile passing through the panel	Visible indentation with fiber failure at both impact-side and backside surfaces
IV	Extensive through-thickness cracks; penetration	Large characteristic cracking patterns; petals	Clearly visible large cracks
V	Clean hole; penetration	Nearly round hole roughly same size as projectile diameter	Clearly visible hole

failure (type II) is the initial mode of damage. Increasing projectile energy levels results in increasing damage area. For the unfilled ice, the delamination increases in size, until fiber failure occurs, at which point the delamination area drops dramatically. This marks the transition to fiber failure damage modes, with increasing total through-thickness crack length (types III and IV) and ultimately a clean hole is produced (type V). The total damage length (i.e., perimeter) of the clean hole (type V) is actually significantly lower than the crack length of the type IV damage mode and therefore type V damage would have less affect on postimpact strength reduction. Impacts at

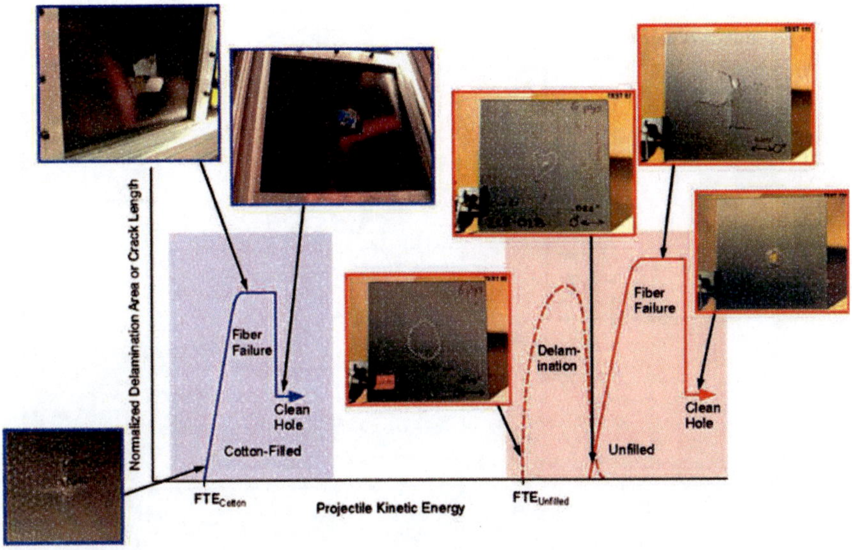

Fig. 5.20 Damage size vs. projectile kinetic energy for unfilled and cotton-filled ice impact

energy levels higher than the clean hole level will not produce different-sized damage. The cotton-filled ice was found to not excite delamination, and directly produce fiber failure modes (type II initial) of damage, as indicated in Fig. 5.20.

The FTE is a useful design-orientated data point, unique for each combination of ice sphere diameter and panel thickness. The summary of FTE data [41, 42], plotted in Fig. 5.21, defines the damage resistance of a monolithic (i.e., non-sandwich) composite panel to the ice impact threat. The vertical axis of the plot is a normalized nondimensional FTE, defined in the figure, and the horizontal axis is the ratio of panel thickness (H) over ice ball diameter (D). Note that the data in Fig. 5.21 include two trends: one for unfilled ice, the other for cotton-filled (per ASTM F320). The comparison of trends shows that the cotton-filled ice spheres are significantly more lethal than unfilled ice. The lethality difference is more pronounced for thinner panels impacted by larger diameter ice (i.e., small values of H/D). This difference in lethality is due to the strengthening reinforcement that the cotton fibers provide to the ice projectile, making the projectile itself a composite of cotton fibers reinforcing an ice matrix.

While the unfilled ice produced delamination (type I) as the initial damage mode, the cotton-filled ice produced backside fiber failure (type II). This initial impact damage mode transition is due to the cotton-filled ice not breaking up during the impact event, as the unfilled ice does (see inset images in Fig. 5.21). The lack of breakup results in contact with the target panel occurring over a much more concentrated area, thereby forcing a higher local curvature response at the location of impact.

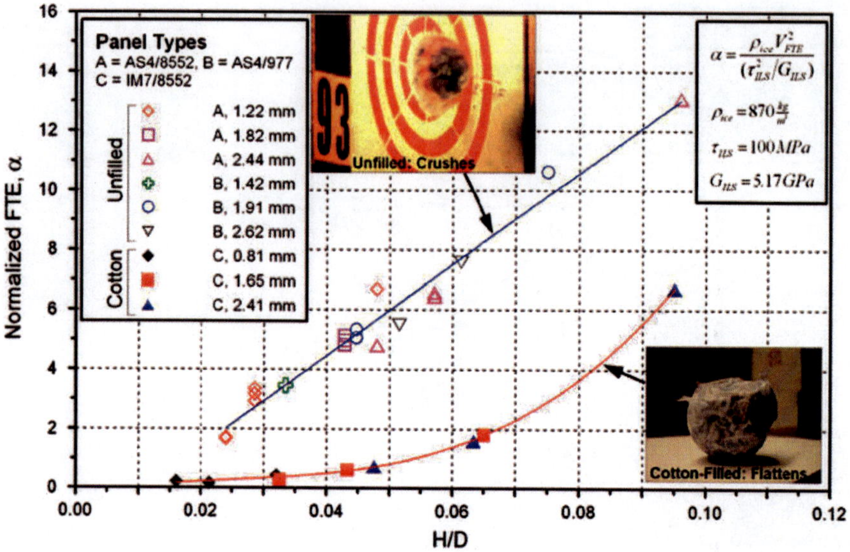

Fig. 5.21 Normalized failure threshold energy for unfilled and cotton-filled ice

Impact response of sandwich panels differs from monolithic (single skin) panels. Sandwich construction exists to provide stiffness in bending in the same manner as an I-beam, where the facesheets are analogous to the I-beam flanges. The relatively thin facesheets are also exposed to the surface-impact threat environment. Several phenomena govern the formation of damage to thin facesheet sandwich panels. While the sandwich construction greatly increases the panel global bending stiffness relative to monolithic panels having the same total thickness as the facesheets (see Fig. 5.22), the sandwich core acts as an elastic foundation to the relatively thin impact-side facesheet. The interaction between the facesheet and core, behaving like a beam on elastic foundation (BOEF), localizes the deflection to primarily be on the thin impact-side facesheet during the early timeframe of the impact event, as described in Fig. 5.22, and as a consequence concentrates the incident energy within close proximity of the impact contact area. Additionally, with increased global bending stiffness, higher contact forces are developed between the impactor and panel. Since the thin facesheet is more locally compliant than a thicker monolithic panel, localized damage occurs as a "shear plug" and core crushing under the contact, with a significant increase in the interlaminar shear stresses. Penetration of the impact-side facesheet is a common response to the concentration of the impacting energy, as shown by the photos in Figs. 5.23 and 5.24 of damage produced by 61-mm diameter ice impacting onto a composite sandwich panel at 35 m/s. Non-penetrating damage modes are of perhaps more concern since extensive separation of the adhesive-to-facesheet bond can occur while the impact-side facesheet returns to its original contour soon after the impact event. Subsequent

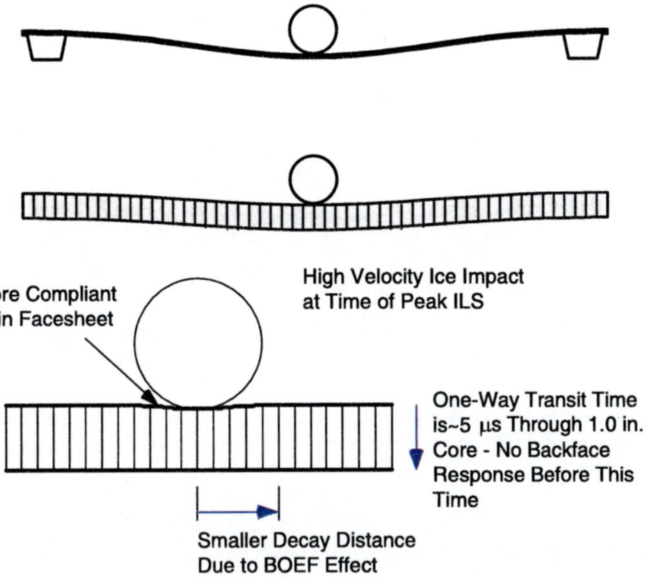

Fig. 5.22 Global stiffening and local response phenomena leading to concentration of impacting energy in sandwich panels

Fig. 5.23 High-speed video stills showing 61 mm ice impacting carbon/epoxy sandwich panel at 35 m/s producing localized damage to impact-side facesheet and core crushing; Note: ball stopped and rebounded by back-side facesheet

post-impact structural evaluations must then account for damage states growing as interlaminar separation within the facesheets, as well as disbonding between the facesheet and core and fracture of the core material itself. Additionally, since sandwich construction is often used in flight control surfaces, the potentially negative aeroelastic effects of local changes in compliance should also be considered.

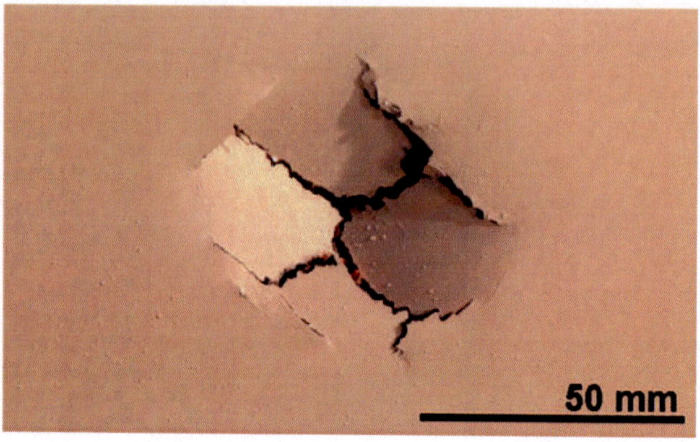

Fig. 5.24 Damage to impact-side facesheet of carbon/epoxy sandwich panel by 61 mm ice at 35 m/s; Note: overall dimension of damage is less than ice diameter due to stoppage of ball by back-side facesheet

Due to the inherent localizing response of sandwich panels, the formation of damage is relatively easier to achieve than for monolithic composite panels, and thus the resulting damage state becomes less dependent on the exact details of how the impact threat is applied, i.e., equivalent damage state can be produced onto sandwich panels by dropping metal balls (high mass, slow speed) as with actually shooting the panel with ice at operational velocity (low mass, high speed). Therefore, the sandwich panel can lead to some confusion regarding the generalization of a test method in its applicability to other types of structures, namely, monolithic panels and the internal stiffening elements attached to them.

5.4 Influence of Impact Damage in Composite Structure Design

The design of composite structures requires many simultaneous considerations to be weighed in the final determination of the structure's overall geometry and laminate configuration. As an example, aircraft structures have significant operational-related loading from flight maneuvers, gusts, landing, etc., which provide the basis for laminate thickness sizing and structural configuration (including the internal stiffening elements such as stringers, spars, etc). Wing and empennage experience bending and torsional loading while the fuselage is more complex, experiencing tension loads at the crown, compression loads in the lower keel areas and biaxial loading in the mid-side area (window and door areas), in addition to internal pressure of the cabin.

The initial design of the structure addresses the multiple design needs for structural performance. In the case of wing and empennage structure, the torque boxes reacting the bending and torsional loading tend to be about 80% stiffness-based

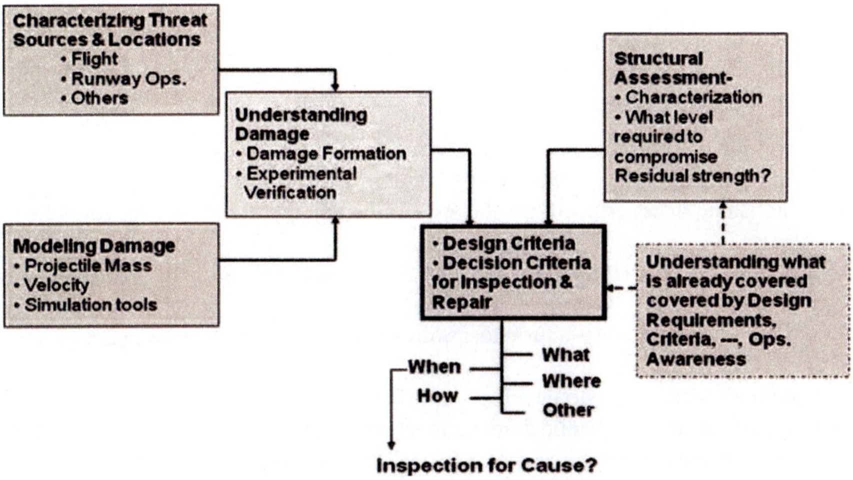

Fig. 5.25 Impact damage assessment

with local points sized based on strength requirements, usually for load transfer reasons. Material selection typically focuses on *material damage resistance* employing the concepts illustrated in Fig. 5.3. In many cases, operating strain levels through this phase of the design cycle are determined by the notch sensitivity of the material system in response to stress concentrations for bolted joints and structural openings.

As the preliminary design evolves, "Build Block" testing is conducted at the element and subcomponent levels. Table 5.2 and the discussions in Sect. 5.3 govern this activity. The data collected is used to assess the local *damage tolerance*. Figure 5.3 illustrates one aspect of damage tolerance in terms of laminate damage. An equally important aspect is the assessment of hidden structural damage to the stiffening elements behind the laminate surface skins. This database is incorporated into the structural models for the simulation of structural response to various damage scenarios. Local adjustments to the structure, e.g. thicknesses, are typically made at this point. Structural validation by test and simulation follows in support of the certification process. The reader is referred to Mil-Hdbk-17 [47] vol. 3, Chap. 12 for a summary of these activities and the applicable regulations.

The flow chart in Fig. 5.25 provides a logic diagram for assessing a structure's capability to withstand impact damage. On the left, the focus is on the characterization of the impact sources and understanding what damage modes and sizes will result. Impact simulation and modeling tools that can be developed will aid in the understanding of the relationship between the threats and the damage produced. On the right of the diagram, the tolerance of the structure to the impact damage is understood – e.g., the residual strength characterized as a reduction in the compression after impact (CAI) strength shown in Figs. 5.2 and 5.3. These collectively combine to aid in the definition of design criteria for sizing the structure, as well as for establishing criteria for conducting maintenance inspections and determining when repairs are necessary.

In summary, the management of various discrete impact threats outlined in this chapter are:

1. Definition of the individual discrete threats
2. Quantifying the level of the threat for assuring damage tolerance
3. Establishing a test protocol for the test simulation appropriate for the individual threats
4. Development of simulation models for design and structural assessment purposes
5. Zoning of vulnerable structural locations for the purposes of focusing design and in-service inspections
6. Building block testing to develop a certification database
7. Full-scale structural demonstrations to show damage tolerance
8. In-service inspection requirements to assure continuing airworthiness
9. Operational guidelines to mitigate the potential severity or frequency of an in-service impact event

Mitigation actions include FOD control on runways, establishing training and operational practices that prevent ground equipment contact, limiting aircraft velocities below 10,000 ft to moderate bird-strike kinetic energy, flying 25 miles away from severe hail storm fronts. The discussion presented in this chapter summarizes the evolution of a multiple layered proactive approach for the mitigation of in-service discrete impact threats.

References

1. Abrate, S, 1991, "Impact on Composite Materials," *Applied Mechanics Review,* vol. 44, no. 4, pp. 155–190.
2. Airoldi, A, Cacchione B, 2005, "Numerical Analysis of Bird Impact on Aircraft Structures Undergoing Large Deformations and Localized Failures," *Impact Loading of Lightweight Structures*, M. Alves and N. Jones (Eds), WIT Transactions on Engineering Sciences, Ashurst Lodge, Ashurst, Southampton SO40 7AA, UK, vol. 49, WIT Press, pp. 1–18.
3. Anghileri, M, Castelletti, LML, and Mazza, V 2005, "Birdstrike: Approaches to the Analysis of Impacts with Penetration," *Impact Loading of Lightweight Structures*, M. Alves and N. Jones (Eds), WIT Transactions on Engineering Sciences, Ashurst Lodge, Ashurst, Southampton SO40 7AA, UK, vol. 49, pp. 63–74.
5. ASTM International 1994, Standard Test Method for Hail Impact Resistance of Aerospace Transparent Enclosures, ASTM F320-94, ASTM Intl, West Conshohocken, PA.
6. Aviation Safety Network 2010, ASN Aviation Safety Database 2001, viewed 20 January, 2010, <http://aviation-safety.net/database/record.php?id=20010402-1>.
7. Barber, J.P., Taylor, H.R., Wilbeck, J.S.: Bird impact forces and pressures on rigid and compliant targets. Technical Report AFFDL-TR-77–60. University of Dayton Research Institute, Dayton, OH, USA (1978).
8. Bird Strike Committee USA 2010, Significant Bird Strikes, viewed 20 January 2010, <http://www.birdstrike.org/events/signif.htm>.

9. Cairns, D, Lagace, PA 1989, "Transient Response of Graphite/Epoxy and Kevlar/Epoxy Laminates Subjected to Impact," *AIAA Journal,* vol. 27, no. 11, pp. 1590–1596.
10. Cantwell, WJ, Morton, J 1991, "The impact resistance of composite materials – a review" *Composites*, vol 22. no.5, pp. 347–362.
11. Cheverton, RD, Gejlen, PC, Haun, GT. Iskander, SK, 1980, "Application of Crack Arrest Theory to a Thermal Shock Experiment," *Crack Arrest Methondology and Applications, ASTM STP 711,* pp. 392–421.
12. Crosley, PB, Ripling, EJ, 1980, "Significance of Crack Arrest Toughness(Kla) Testing," *Crack Arrest Methodology and Applications, ASTM STP 711,* pp 301–337.
13. Davidson, GG, Labib, AW, 2003, "Learning from failures: design improvements using a multiple criteria decision-making process," *Proceedings of the Institution of Mechanical Engineers; Part G; Journal of Aerospace Engineering*, vol. 217, no. 4, pp. 207–216.
13. Davies, GAO, Robinson, P, 1992, "Predicting Failure by Debonding/Delamination," AGARD: 7th Structures and Materials Meeting, Debonding/Delamination of Composites.
14. Davies, GAO, Zhang, X, Zhou, G, Watson, S, 1994, "Numerical Modeling of Impact Damage," *Composites*, Vol. 25, No. 5, pp. 342–350.
15. Davies, GAO, Zhang, X 1995, "Impact Damage Prediction in Carbon Composite Structures," *Journal of Impact Engineering,* vol. 16, no. 1, pp. 149–170.
16. Dost, EF, Avery, WB, Ilcewicz, LB, Grande, DH, Coxon, BR 1992, "Impact Damage Resistance of Composite Fuselage Structure, Part 1," FAA, *Ninth DOD(NASA)FAA Conference on Fibrous Composites in Structural Design*, vol. 2, pp 1037–1069.
17. Dost, EF, Ilcewicz, LB, Avery, WB, Coxon, BR, 1991, "Effects of Stacking Sequence on Impact Damage Resistance and Residual Strength for Quasi-Isotropic Laminates," *Composite Materials: Fatigue and Fracture (3rd Volume)*, ASTM STP 1110, TK O'Brien, Ed. pp. 476–500.
18. Elber, W 1983. "Failure Mechanics in Low-Velocity Impacts on Thin Composite Plates," NASA TP-2152.
19. Farley, Richard D., "Numerical Modeling of Hailstorms and Hailstone Growth. Part II: The Role of Low-Density Riming Growth in Hail Production," *Journal of Climate and Applied Meteorology*, Vol. 26, 1987, pp. 234–254.
20. Federal Aviation Administration 1992, Fuel Tank Access Covers, Advisory Circular 25. 963–1.
21. Federal Aviation Administration 1999, Certification of Transport Aircraft Structure, Advisory Circular 25–21.
22. Federal Aviation Administration 2007, Instructions for Continued Airworthiness of Structural Repairs on Transport Airplanes, Advisory Circular 25.1529-1A.
23. Federal Aviation Administration 2010, FAA Wildlife Strike Database, viewed 19 January 2010, <http://wildlife-mitigation.tc.faa.gov/wildlife>.
24. Frings, G 1984. "A Study of Bird Ingestions Into Large High Bypass Ratio Turbine Aircraft Engines," Federal Aviation Administration Final Report DOT/FAA/CT-84/13, September.
25. Garcia-Garcia, F., and List, R., "Laboratory measurements and Parameterizations of Supercooled Water Skin Temperatures and Bulk Properties of Gyrating Hailstones," *Journal of the Atmospheric Sciences*, Vol. 49, No. 22, 1992, pp. 2058–2073.
26. Gates, D 2008 'Bair pilots Boeing effort to replace best-selling 737' *The Seattle Times*, 17 July, viewed 8 October 2009, retrieved from <http://seattletimes.nwsource.com>.
27. Georgiadis, S, Gunnion, AJ, Thomson, RS & Cartwright, BK 2008, 'Bird-strike simulation for certification of the Boeing 787 composite moveable trailing edge', *Composite Structures,* vol. 86, no. 1–3, pp. 258–268.
28. Goncalves, W, Olmi, F, Chaves, CE, and Feitoza, AP 2005, "High Velocity Impact Against Aeronautical Structures," *Impact Loading of Lightweight Structures*, M. Alves and N. Jones (Eds), WIT Transactions on Engineering Sciences, Ashurst Lodge, Ashurst, Southampton SO40 7AA, UK, vol. 49, pp. 231–240.
29. Halpin, JC 1980, "Practical Applications of Composite Materials," UCLA Short Course Lecture Notes.

30. Halpin, JC and Kim, H 2007, 'Managing Impact Risk for Composite Structures: Unifying Durability and Damage Tolerance Perspective', presentation given at FAA/EASA/Industry Composite Damage Tolerance and Maintenance Workshop, Amsterdam, 7–11 May.
31. Halpin, JC and Kim, H 2009, "Managing Damage Threats for Composite Structures: Unifying Durability and Damage Tolerance Perspective," presentation given at 3rd FAA/EASA/Industry Composite Damage Tolerance and Maintenance Workshop, Tokyo, 1–5 June.
32. International Air Transportation Association 2005, "Ground Damage Prevention Programme Targets 10% Cost Reduction," Industry Times, Edition 7, September, Article 4.
33. Jackson, WC and Poe, CC, 1992, "The Use of Impact Force as a Scale Parameter for the Impact Response of Composite Laminates," NASA Technical Memorandum 104189.
34. Johnson, D. B., and Rasmussen, R. M., "Hail Growth Hysteresis," *Journal of the Atmospheric Sciences*, Vol. 49, No. 24, 1992, pp. 2525–2532.
35. Kakarala, S, Roche, JL 1987, "Experimental Comparison of Several Impact Test Methods" *Instrumented Impact Testing of Plastics and Composite Materials, ASTM STP 936*, pp. 144–162.
36. Kan, HP, 1998, "Enhanced Reliability Prediction Methodology for Impact Damaged Composite Structures," FAA Final Report DOT/FAA/AR-97/79.
37. Kan, HP, Cordero, R and Whitehead, RS, 1997, "Advanced Certification Methodology for Composite Structures," FAA Final Report DOT/FAA/AR-96/111.
38. Karagiozova, D and Mines, RAW 2007, 'Impact of aircraft rubber tyre fragments onto aluminum alloy plates: II – Numerical simulations using LS-DYNA', *International Journal of Impact Engineering*, vol. 34, no. 4, pp. 647–667.
39. Kim, H. and Kedward, K. T. 2000, "Modeling Hail Ice Impacts and Predicting Impact Damage Initiation in Composite Structures," AIAA Journal, Vol. 38, No. 7, 2000, pp. 1278–1288.
40. Kim, H. and Keune, J. N. 2007, "Compressive Strength of Ice Measured at High Strain Rate," *Journal of Materials Science*, Vol. 42, No. 8, 2007, pp. 2802–2806.
41. Kim, H., Kedward, K.T., and Welch, D.A. 2003, "Experimental Investigation of High Velocity Ice Impacts on Woven Carbon/Epoxy Composite Panels," Composites Part A, Vol. 34, No. 1, 2003, pp. 25–41.
42. Kim, H., Nightingale, J., and Park, H. 2007, "Impact Damage Resistance of Composite Panels Impacted by Cotton-Filled and Unfilled Ice," Proceedings of 16th International Conference on Composite Materials, July 9–13, 2007, Kyoto, Japan.
43. Kwon, YS, Sankar, BV 1993, "Indentation-Flexure and Low-Velocity Impact Damage in Graphite Epoxy Laminates," *Journal of Composite Technology and Research*, vol. 15, no. 2, pp. 101–111.
44. Lee, SM, Zahuta, P 1991, "Instrumented Impact and Static Indentation of Composites," *Journal of Composite Materials*, vol. 25, no. 2, pp. 204–222.
45. Maikuma, H, Gillespie, JW Jr., Wilkins, DJ 1990, "Mode II Interlaminar Fracture of the Center Notch Flexural Specimen Under Impact Loading," *Journal of Composite Materials*, vol. 24, no. 2, pp 124–149.
46. Mao, RH, Meguid, SA, Ng, TY 2008, "Transient three dimensional finite element analysis of a bird striking a fan blade," *International Journal of Mechanics and Materials in Design*, vol. 4, no. 1, pp. 79–96.
46. Meyer, PI 1988, "Low-Velocity Hard-Object Impact of a Filament-Wound Kevlar/Epoxy Composite," *Composite Science and Technology*, vol. 33, no. 4, pp. 279–293.
47. Military Handbook Mil-Hdbk-17 Rev. F, 2002. Composite Materials Handbook, June 17, 2002.
48. Mines, RAW, McKwon, S, and Birch, RS 2007, 'Impact of aircraft rubber tyre fragments onto aluminum alloy plates: I – Experimental', *International Journal of Impact Engineering*, vol. 34, no. 4, pp. 627–646.
49. Murthy, PLN, Chamis, CC 1986, "Dynamic Stress Analysis of Smooth and Notched Fiber Composite Flexural Specimens," *Composite Materials: Testing and Design (Seventh Conference), ASTM STP 893*, pp 368–391.

50. National Research Council Canada 2010, "Highlights – It's a Bird, It's a Plane... It's a Bird Striking a Plane," Viewed 19 January 2010, <http://www.nrc-cnrc.gc.ca/eng/news/nrc/2007/01/07/bird-plane.html>.
51. Park, H., and Kim, H., "Damage Resistance Of Single Lap Adhesive Composite Joints By Transverse Ice Impact" International Journal of Impact Engineering, in press – submitted Aug. 10, 2008, accepted Aug. 17, 2009.
52. Petrenko, V. F., and Whitworth, R. W., *Physics of Ice*, New York: Oxford University Press, 1999.
53. Poe, CC, Jr. 1988, "Simulated Impact Damage in a Thick Graphite/Epoxy Laminate Using Spherical Indenters," NASA TM-100539.
54. Poe, CC, Jr., 1992, "Impact Damage and Residual Tension Strength of a Thick Graphite/Epoxy Rocket Motor Case," *Journal of Spacecraft and Rockets*, Vol. 29, No. 3, pp. 394–404.
55. Sala, G, 2000, "Composite Degradation Due to Fluid Absorption," *Composites Part B*, Vol. 31, No. 5, pp. 357–373.
56. Saul, M., Lunford, J.L., "Damage From Hail Grounds Nearly 100 D/FW Planes Passengers Scramble as Scores of Flights Cancelled," *Dallas Morning News*, May 1, 1995, pp. 1A.
57. Schoeppner, GA, Abrate, S 2000, "Delamination threshold loads for low velocity impact on composite laminates," *Composites: Part A*, vol. 31, no. 9, pp. 903–915.
58. Shafizadeh, JE, Seferis, JC, Chesmar, EF, and Geyer, R, 1999, "Evaluation of the In-Service Performance Behavior of Honeycomb Composite Sandwich Structures," *Journal of Materials Engineering and Performance*, Vol. 8, No. 6, pp. 661–668.
59. Shivakumar, KN, Elber, W, Illg, W 1985. "Prediction of Impact Force and Duration Due to Low-Velocity Impact on Circular Composite Laminates" *Journal of Applied Mechanics*, Transactions of the ASME, vol. 52, no. 3, pp. 674–680.
60. Sjoblom, P 1987, "Simple Design Approach Against Low-Velocity Impact Damage" *32nd International SAMPE Symposium*, April 6–9, pp. 529–539.
61. Sjoblom, PO, Hartness, TJ, Cordell, TM 1988. "On Low-Velocity Impact Testing of Composite Materials," *Journal of Composite Materials*, vol. 22, no. 30, pp. 30–52.
62. Tan, TM, Sun, CT 1985, "Use of Statical Indentation Laws in the Impact Analysis of Laminated Composite Plates," *Journal of Applied Mechanics*, Transactions of the ASME, vol. 52, no. 7, pp. 6–12.
63. Wardle, BL, Lagace, PA 1997, "On the Use of Dent Depth as an Impact Damage Metric for Thin Composite Structures," *Journal of Reinforced Plastics and Composites*, vol. 16, no. 12, pp. 1093–1110.
64. Wardle, BL, Lagace, PA 1998, "On the Use of Quasi-Static Testing to Assess Impact Damage Resistance of Composite Shell Structures," *Mech. Of Advanced Materials and Structures*, vol. 5, no. 1, pp 103–121.
65. Wilbeck, J.S., Rand, J.L.: The development of a substitute bird model. ASME Paper ASME 81-GT-23, Gas turbine conference and products show, Houston, TX, USA (1981).
66. Wilbeck, J.S.: Impact behavior of low strength projectiles. Report No. AFML-TR-77-134, Air Force Materials Lab., Air Force Wright Aeronautical Lab's, Wright-Patterson Air Force base, OH (1977).
67. Wu, E, Shyu, K 1993, "Response of Composite Laminates to Contact Loads and Relationship to Low-Velocity Impact," *Journal of Composite Materials*, vol. 27, no. 15, pp. 1443–1464.
68. Souter, RK and Emerson, JB, 1952, "Summary of Available Hail Literature and the Effect of Hail on Aircraft in Flight," National Advisory Committee for Aeronautics Technical Note, NACA TN 2734, September.
69. McNaughton, II and Chisman SW, 1969, "A Study of Hail Impact at High Speed on Light Alloy Plates," Proc. 9th Annual National Conference on Environmental Effects on Aircraft and Propulsion Systems, Nav. Air Propulsion Test Center, pp. 16–1 to 16–7.
70. Thomson, RG and Hayduk, RJ, 1971, "An Improved Analytical Treatment of the Denting of Thin Sheets by Hail," National Aeronautics and Space Administration Technical Note, NASA TN D-6102, January.

71. Transport Canada 2004, "Sharing the Skies, An Aviation Industry Guide to the Management of Wildlife Hazards, 2nd Ed," Transport Canada Publication TP13549E.
72. Transport Canada 2010, "Airport Wildlife Management Bulletin No. 33,".
73. US Air Force 2010, Air Force Safety Center BASH, viewed 19 January 2010, <http://www.afsc.af.mil/organizations/bash>.

Chapter 6
Moisture-Induced Degradation

Samit Roy

Abstract Composites provide advantages over conventional structural upgrade systems by offering up to 50% first-cost savings and lower life-cycle costs, often with additional benefits such as easier installation and improved safety. Fiber-reinforced polymer (FRP) composites are finding increasing applications as primary structural components in aerospace and automotive applications, bridges, building repair, and the oil and gas pipeline industry. These composites are typically exposed to a variety of aggressive environments, such as extreme temperature cycles, ultraviolet (UV) radiation, moisture, alkaline/salt environments, etc. However, no capability currently exists for reliably projecting the future state and conditions of composites used in various environments. The accurate determination of diffusivity and moisture uptake in a polymer composite is a key step in the accurate prediction of moisture-induced degradation. With this in mind, the chapter is subdivided into three sections: (1) the combined influence of damage and stress on moisture diffusion within the (bulk) polymer matrix in a polymer composite, (2) the combined influence of strain gradient, relative humidity, and temperature on moisture diffusion at the fiber–matrix and/or interlaminar interface, and (3) a simple mechanism-based model to predict strength degradation in a composite due to moisture ingress. The discussions presented in this chapter are primarily directed toward thermoset resins, such as epoxy.

6.1 Introduction and Background

It is now widely recognized that moisture plays a significant role in influencing the mechanical behavior, and therefore, long-term durability of polymers and polymer matrix composites (PMC). Numerous diffusion models have been proposed over

S. Roy (✉)
Department of Aerospace Engineering and Mechanics, University of Alabama,
Tuscaloosa, AL 35487, USA
e-mail: sroy@eng.ua.edu

K.V. Pochiraju et al. (eds.), *Long-Term Durability of Polymeric Matrix Composites*,
DOI 10.1007/978-1-4419-9308-3_6, © Springer Science+Business Media, LLC 2012

the years for modeling hygrothermal effects in PMC. The one most frequently used by researchers is the one-dimensional Fickian model due to its simplicity and mathematical tractability. Unfortunately, this model tends to overestimate the moisture absorption in panels for short diffusion times, even when it is modified to take into account edge effects [1]. Some researchers have suggested that the deviation can be explained by a two-stage Fickian process [2, 3]. Others claim that the diffusion process in a PMC is really non-Fickian [4, 5]. The applicability of Fick's law for a given material system under a specified loading cannot be guessed "a priori" but must be determined from moisture absorption/desorption test data.

Frisch [6] and Crank [7] were among the first researchers to recognize and attribute non-Fickian moisture transport in resins to time-dependent molecular mechanisms within a polymer. To model this phenomenon, Weitsman [8] applied basic principles of continuum mechanics and irreversible thermodynamics to derive governing equations and boundary conditions for coupled stress-assisted diffusion in elastic and viscoelastic materials. Following an approach originally proposed by Biot [9] and Schapery [10], viscoelasticity was introduced by means of scalar-valued internal state variables, also referred to as hidden coordinates. The results of the analysis allowed an insight into the mechanism that causes a time drift toward equilibrium at the boundary of a viscoelastic material subjected to a constant chemical potential of the ambient vapor. The governing equations also indicated that the saturation levels vary quadratically with stress and that they can be expressed in terms of the dilatational and the deviatoric stress invariant. Subsequently, Weitsman [11] extended the rigorous phenomenological model to incorporate polymer "free volume" as a thermodynamic internal state variable in order to include the effect of physical aging on moisture diffusion. The resulting governing equations were not amenable to closed-form solution due to mathematical complexity although it was observed that the diffusion equation followed a time-retardation process analogous to mechanical viscoelastic response, and that it exhibited an aging behavior characteristic of glassy polymers. Employing a similar phenomenological approach, Weitsman [12] developed a model for coupled damage and moisture transport in an elastic, transversely isotropic, fiber-reinforced polymer composite. The damage entity was represented as a skew-symmetric tensor and was included in the model as an internal state variable.

Moisture diffusing into the epoxy matrix in a PMC generally leads to (1) plasticization of the polymer matrix, and (2) hydrolysis [13]. While plasticization is a physical change that can be reversed through desorption of moisture, hydrolysis leads to a permanent scission of chemical bonds in the resin that can cause irreversible damage to the PMC. Both these processes can result in a significant reduction in the modulus, strength, and glass transition temperature (T_g) of the PMC. Concurrently, rapid moisture ingress along the fiber–matrix interface through capillary action can lead to chemical degradation at the fiber–matrix interface.

As mentioned previously, the accurate determination of diffusivity and moisture uptake in a polymer composite is a key step in the accurate prediction of moisture-induced degradation. Moisture absorption characteristics comprise an important part of moisture/durability studies, as is evident from the schematic moisture

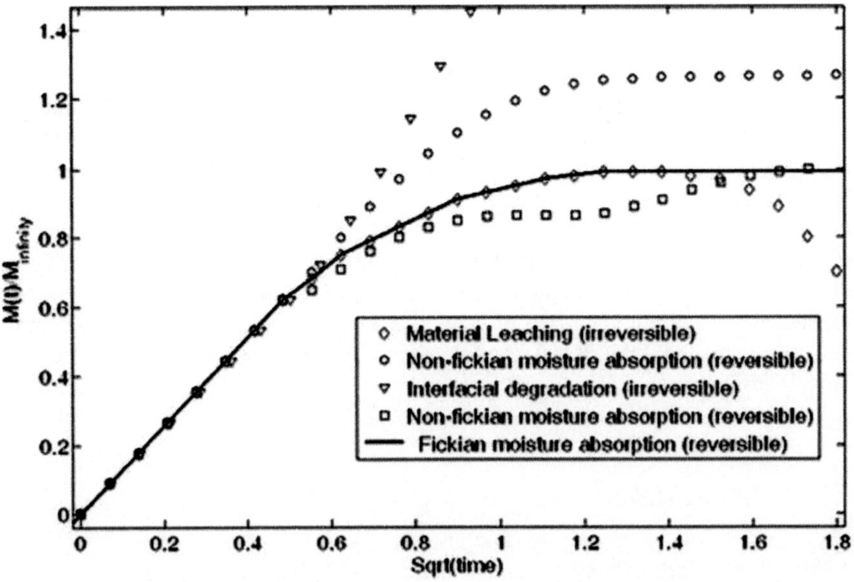

Fig. 6.1 Different types of moisture absorption behavior in polymer matrix composites [14]

absorption curves shown in Fig. 6.1. The solid line indicates classical Fickian moisture absorption, which is typically reversible. The curve indicated by the triangular symbols represents a material that has undergone significant interfacial degradation, while the curve represented by the diamond symbols is indicative of significant material loss through leaching. Both these types of absorption behavior are deemed irreversible. The remaining curves represent reversible moisture absorption behavior [14]. Consequently, accurate prediction of moisture absorption and desorption characteristics in a polymer or polymer composite can provide qualitative as well as quantitative insight into the degradation mechanism present in the material.

6.2 Modeling of Moisture Diffusion in the Presence of Biaxial Damage in Polymer Matrix Composite Laminates

It is now well known that Fick's Law is frequently inadequate for describing moisture diffusion in polymers and polymer composites. Non-Fickian or anomalous diffusion is likely to occur when a polymer composite laminate is subjected to external stresses that could give rise to internal damage in the form of matrix cracks. As a result, it is necessary to take into account the combined effects of temperature, stress (or strain), and damage in the construction of such a model. In this section, a modeling methodology based on irreversible thermodynamics applied within the framework of composite macro-mechanics is extended to the case of a biaxially damaged laminate. The model allows characterization of non-Fickian diffusion coefficients as well as moisture saturation level from moisture weight gain data for laminates with preexisting and non-evolving damage. A symmetric damage tensor based on continuum damage mechanics is incorporated in this model by invoking the principle of invariance with respect to coordinate transformations. For tractability, the diffusion-governing equations are simplified for the special case of a laminate with biaxially oriented matrix cracks that is subjected to a uniaxial tensile stress. The final equations obtained from this derivation indicate that both effective diffusivity and maximum saturation level for this particular case can be expressed as quadratic functions of crack density. Comparisons with test data for a biaxially damaged AS4/PR500 graphite/epoxy woven composite are provided for model verifications.

6.2.1 Introduction

Although thousands of polymer matrix composite (PMC) components are currently in service in the aerospace industry as well as in civil infrastructure repair and rehabilitation, barriers still exist to further implementation in more structurally critical and extreme temperature applications. Most of these barriers are associated with the inability to accurately predict component lives, and therefore, component life cycle costs. A fiber-reinforced composite material with a polymer matrix will typically absorb moisture in a humid environment and at elevated temperatures. Combined exposure to heat and moisture affects a PMC in a variety of ways. First, the hygrothermal swelling causes a change in the residual stresses within the composite that could lead to microcrack formation. These microcracks in turn provide fast diffusion paths and thus alter the moisture absorption characteristics of the laminate. Second, heat and humidity may cause the matrix to become plasticized, thus causing an increase in the strain to failure of the matrix. Third, the chemical bond at the interface between fiber and matrix may be affected which in turn would influence strength and toughness. Finally, in the event of cyclic heating and cooling with a sustained use-temperature above the boiling point of water, vaporization and out-gassing of absorbed moisture may take place, leading to physical damage and chemical changes within the PMC, especially at temperatures greater than the glass transition temperature (T_g) of the matrix. Continuous exposure to high moisture concentrations at the exposed surfaces of a PMC component could also lead to damage in the form of matrix cracking, matrix dissolution, and peeling.

In reality, the nature of the diffusion process depends on the material and on the environmental conditions that the material is exposed to. For example, if the rate of viscoelastic relaxation in a polymer is comparable to the rate of moisture diffusion, then the diffusion is likely to be non-Fickian. In addition, the presence of strong temperature and stress and/or strain gradients has been known to engender non-Fickian driving forces. The presence of damage in the form of matrix cracks is very likely to result in anomalous diffusion. In this context, Roy and Xu [15] developed a model employing theory of irreversible thermodynamics within the framework of macro (continuum) mechanics to derive governing equations for diffusion in a PMC from first principles. A special form for Gibbs potential was formulated for an orthotropic material using stress, temperature, damage, and moisture concentration as independent state variables. A symmetric damage tensor originally proposed by Talreja [16] was used for this purpose. The resulting governing equations incorporated coupled interactions between stress, temperature, damage, and moisture concentration on the diffusion process within an orthotropic material. For model validation, the model predictions were compared with experimental data for the special case of isothermal diffusion in an unstressed graphite/epoxy five-harness satin woven laminate with preexisting uniaxial damage.

In addition to predicting changes in laminate diffusion coefficient with the degree of damage, the proposed model also allows characterization and prediction of moisture saturation levels from moisture weight gain data for laminates with preexisting damage. Damage evolution is not included in the present analysis in the interest of tractability. This approach provides a significant improvement over methods reported in the literature for problems of this nature. For model validation, the model predictions are compared with experimental data for the special case of isothermal diffusion in an unstressed five-harness satin woven laminate with preexisting biaxial damage states.

6.2.2 Model Development

The Gibbs potential for an orthotropic material subjected to applied stress and internal damage must be expressed in a polynomial form that remains invariant to coordinate transformations. Such a polynomial can be mathematically expressed as a combination of invariant terms obtained from the so-called irreducible integrity bases [16, 17]. The irreducible integrity bases for orthotropic materials are

$$\begin{aligned}&\overline{\sigma}_{11},\overline{\sigma}_{22},\overline{\sigma}_{33},d_{11},d_{22},d_{33},\overline{\sigma}_{23}^2,\overline{\sigma}_{13}^2,\overline{\sigma}_{12}^2,\overline{\sigma}_{23}\overline{\sigma}_{13}\overline{\sigma}_{12},d_{23}^2,d_{13}^2,d_{12}^2,d_{23}d_{13}d_{12},\\&\overline{\sigma}_{23}d_{23},\overline{\sigma}_{13}d_{13},\overline{\sigma}_{12}d_{12},d_{23}\overline{\sigma}_{13}\overline{\sigma}_{12},d_{13}\overline{\sigma}_{12}\overline{\sigma}_{23},d_{12}\overline{\sigma}_{23}\overline{\sigma}_{13},\overline{\sigma}_{23}d_{13}d_{12},\overline{\sigma}_{13}d_{12}d_{23},\overline{\sigma}_{12}d_{23}d_{13}\end{aligned} \tag{6.1}$$

where the normalized stress

$$\overline{\sigma}_{ij} = \frac{\sigma_{ij}}{\sigma_f} \tag{6.2}$$

where σ_f is the ultimate stress in a material principal direction, and the damage tensor d_{ij} is a symmetric tensor of the second rank [15, 16]. One advantage of selecting this damage description is that the magnitude of the damage entity is typically less than one.

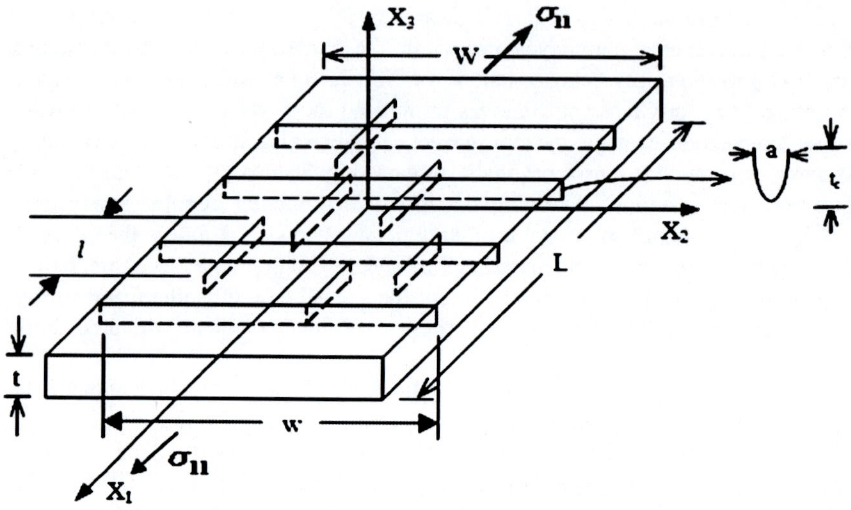

Fig. 6.2 Idealized schematic of a biaxially micro-cracked laminate under uniaxial stress-state

6.2.2.1 Special Case

Consider a laminate with biaxial microcracks oriented perpendicular to the X_1 and X_2 axis, subjected to in-plane uniaxial loading in the X_1 direction under isothermal conditions as shown schematically in Fig. 6.2. Note that the axial cracks (i.e., cracks parallel to the X_1 axis) are assumed discontinuous in order to reflect realistic damage states in woven laminates subjected to uniaxial loading. For the special case of uniaxial loading, the states of stress and damage reduce to

$$\overline{\sigma}_{22} = \overline{\sigma}_{33} = \overline{\sigma}_{13} = \overline{\sigma}_{23} = \overline{\sigma}_{12} = 0$$

and

$$d_{33} = d_{13} = d_{23} = d_{12} = 0$$

Using the irreducible integrity bases from Eq. 6.1 and retaining terms up to second order in the normalized stress component $\overline{\sigma}_{11}$, and the two nonzero damage components d_{11} and d_{22}, the Gibbs potential becomes,

$$\rho_s\phi = \hat{C}_0 + \hat{C}_1(d_{11} + d_{22}) + \hat{C}_2 d_{11} d_{22} + \hat{C}_3(d_{11}^2 + d_{22}^2) + \hat{C}_4\overline{\sigma}_{11} + \hat{C}_5\overline{\sigma}_{11} d_{11} + \hat{C}_6\overline{\sigma}_{11} d_{22} + \hat{C}_7\overline{\sigma}_{11}^2 \tag{6.3}$$

where ρ_s is the mass density of the laminate, and $\hat{C}_i$ are material coefficients independent of stress and damage, but may depend on concentration (m) and temperature (T). Equation 6.3 was derived specifically for the case of a woven fabric laminate, and not for a monotape cross-ply laminate. It should be noted that Eq. 6.3 incorporates the special material symmetry condition in the plane of the laminate (X_1–X_2 plane) that is characteristic of a $[0/90/0/90]_s$ weave architecture, i.e., two aligned woven laminas stacked on two identical woven laminas that have been flipped over about the central plane of the laminate. As a direct consequence of the weave architecture, orthogonal material symmetry exists for the in-plane directions, that is, e.g., $E_{11} = E_{22}$ (refer to Fig. 6.2). Analogously, from a moisture diffusion standpoint, it can be argued that the through-thickness diffusivity (D_3) is identically affected by moisture diffusing through cracks that are either parallel or perpendicular to the X_1 direction, in the absence of any applied stress. Therefore, the d_{11} and d_{22} terms should share the same coefficient in the Gibbs potential, as should the d_{11}^2 and d_{22}^2 terms. However, when applied stress (σ_{11}) is introduced, then this biaxial symmetry is modified and the Gibbs potential must therefore include separate coefficients for the $\sigma_{11}d_{11}$ and the $\sigma_{11}d_{22}$ terms, as reflected in the use of C_5 and C_6.

The chemical potential of moisture in the polymer composite for this special case is defined as

$$\begin{aligned}\mu = \rho_s\frac{\partial\phi}{\partial m} = {} & \frac{\partial\hat{C}_0}{\partial m} + \frac{\partial\hat{C}_1}{\partial m}(d_{11}+d_{22}) + \hat{C}_1\left(\frac{\partial d_{11}}{\partial m}+\frac{\partial d_{22}}{\partial m}\right) + \frac{\partial\hat{C}_2}{\partial m}d_{11}d_{22} \\ & + \hat{C}_2\frac{\partial d_{11}}{\partial m}d_{22} + \hat{C}_2\frac{\partial d_{22}}{\partial m}d_{11} + \frac{\partial\hat{C}_3}{\partial m}(d_{11}^2+d_{22}^2) + \hat{C}_3\left(2d_{11}\frac{\partial d_{11}}{\partial m}+2d_{22}\frac{\partial d_{22}}{\partial m}\right) \\ & + \frac{\partial\hat{C}_4}{\partial m}\overline{\sigma}_{11} + \hat{C}_4\frac{\partial\overline{\sigma}_{11}}{\partial m} + \frac{\partial\hat{C}_5}{\partial m}\overline{\sigma}_{11}d_{11} + \hat{C}_5\overline{\sigma}_{11}\frac{\partial d_{11}}{\partial m} + \hat{C}_5 d_{11}\frac{\partial\overline{\sigma}_{11}}{\partial m} \\ & + \frac{\partial\hat{C}_6}{\partial m}\overline{\sigma}_{11}d_{22} + \hat{C}_6\overline{\sigma}_{11}\frac{\partial d_{22}}{\partial m} + \hat{C}_6 d_{22}\frac{\partial\overline{\sigma}_{11}}{\partial m} + \frac{\partial\hat{C}_7}{\partial m}\overline{\sigma}_{11}{}^2 + 2\hat{C}_7\overline{\sigma}_{11}\frac{\partial\overline{\sigma}_{11}}{\partial m}\end{aligned} \tag{6.4}$$

Because chemical potential is, in this case, a function of σ_{11}, d_{11}, d_{22}, T, and m, hence,

$$\frac{\partial\mu}{\partial X_i} = \frac{\partial\mu}{\partial m}\cdot\frac{\partial m}{\partial X_i} + \frac{\partial\mu}{\partial T}\cdot\frac{\partial T}{\partial X_i} + \frac{\partial\mu}{\partial\overline{\sigma}_{11}}\cdot\frac{\partial\overline{\sigma}_{11}}{\partial X_i} + \frac{\partial\mu}{\partial d_{11}}\cdot\frac{\partial d_{11}}{\partial X_i} + \frac{\partial\mu}{\partial d_{22}}\cdot\frac{\partial d_{22}}{\partial X_i}$$

For the special case of uniform stress, damage, and temperature distributions, only the moisture concentration gradient term is dominant, giving the moisture flux term as

$$D_i\frac{\partial\mu}{\partial X_i} = D_i\left(\frac{\partial\mu}{\partial m}\right)\left(\frac{\partial m}{\partial X_i}\right) = \overline{D}_i\left(\frac{\partial m}{\partial X_i}\right) \tag{6.5}$$

where

$$\overline{D}_i = D_i\left(\frac{\partial\mu}{\partial m}\right)$$

In Eq. 6.5, D_i are the components of diffusivity in the material principal directions, and D_i are the components of effective diffusivity in the material principal directions. Note that repeated indices do not imply summation in Eq. 6.5. Substitute Eq. 6.4 into Eq. 6.5 gives

$$
\begin{aligned}
D_i\frac{\partial \mu}{\partial X_i} &= D_i\left(\frac{\partial \mu}{\partial m}\right)\left(\frac{\partial m}{\partial X_i}\right)\\
&= D_i\left[\begin{array}{c}
\frac{\partial^2 \hat{C}_0}{\partial m^2}+\frac{\partial^2 \hat{C}_1}{\partial m^2}(d_{11}+d_{22})+2\frac{\partial \hat{C}_1}{\partial m}\left(\frac{\partial d_{11}}{\partial m}+\frac{\partial d_{22}}{\partial m}\right)+\hat{C}_1\left(\frac{\partial^2 d_{11}}{\partial m^2}+\frac{\partial^2 d_{22}}{\partial m^2}\right)\\
+\frac{\partial^2 \hat{C}_2}{\partial m^2}d_{11}d_{22}+2\frac{\partial \hat{C}_2}{\partial m}\frac{\partial d_{11}}{\partial m}d_{22}+2\frac{\partial \hat{C}_2}{\partial m}\frac{\partial d_{22}}{\partial m}d_{11}+\hat{C}_2\frac{\partial^2 d_{11}}{\partial m^2}d_{22}+2\hat{C}_2\frac{\partial d_{11}}{\partial m}\frac{\partial d_{22}}{\partial m}\\
+\hat{C}_2\frac{\partial^2 d_{22}}{\partial m^2}d_{11}+\frac{\partial^2 \hat{C}_3}{\partial m^2}(d_{11}^2+d_{22}^2)+4\frac{\partial \hat{C}_3}{\partial m}\left(d_{11}\frac{\partial d_{11}}{\partial m}+d_{22}\frac{\partial d_{22}}{\partial m}\right)\\
+2\hat{C}_3\left(\frac{\partial d_{11}}{\partial m}\frac{\partial d_{11}}{\partial m}+d_{11}\frac{\partial^2 d_{11}}{\partial m^2}+\frac{\partial d_{22}}{\partial m}\frac{\partial d_{22}}{\partial m}+d_{22}\frac{\partial^2 d_{22}}{\partial m^2}\right)+\frac{\partial^2 \hat{C}_4}{\partial m^2}\overline{\sigma}_{11}+2\frac{\partial \hat{C}_4}{\partial m}\frac{\partial \overline{\sigma}_{11}}{\partial m}\\
+\hat{C}_4\frac{\partial^2 \overline{\sigma}_{11}}{\partial m^2}+\frac{\partial^2 \hat{C}_5}{\partial m^2}\overline{\sigma}_{11}d_{11}+2\frac{\partial \hat{C}_5}{\partial m}\frac{\partial \overline{\sigma}_{11}}{\partial m}d_{11}+2\frac{\partial \hat{C}_5}{\partial m}\frac{\partial d_{11}}{\partial m}\overline{\sigma}_{11}+2\hat{C}_5\frac{\partial \overline{\sigma}_{11}}{\partial m}\frac{\partial d_{11}}{\partial m}\\
+\hat{C}_5\overline{\sigma}_{11}\frac{\partial^2 d_{11}}{\partial m^2}+\hat{C}_5 d_{11}\frac{\partial^2 \overline{\sigma}_{11}}{\partial m^2}+\frac{\partial^2 \hat{C}_6}{\partial m^2}\overline{\sigma}_{11}d_{22}+2\frac{\partial \hat{C}_6}{\partial m}\frac{\partial \overline{\sigma}_{11}}{\partial m}d_{22}+2\frac{\partial \hat{C}_6}{\partial m}\frac{\partial d_{22}}{\partial m}\overline{\sigma}_{11}\\
+2\hat{C}_6\frac{\partial \overline{\sigma}_{11}}{\partial m}\frac{\partial d_{22}}{\partial m}+\hat{C}_6\overline{\sigma}_{11}\frac{\partial^2 d_{22}}{\partial m^2}+\hat{C}_6 d_{22}\frac{\partial^2 \overline{\sigma}_{11}}{\partial m^2}+\frac{\partial^2 \hat{C}_7}{\partial m^2}\overline{\sigma}_{11}{}^2+4\frac{\partial \hat{C}_7}{\partial m}\frac{\partial \overline{\sigma}_{11}}{\partial m}\overline{\sigma}_{11}\\
+2\hat{C}_7\frac{\partial \overline{\sigma}_{11}}{\partial m}\frac{\partial \overline{\sigma}_{11}}{\partial m}+2\hat{C}_7\overline{\sigma}_{11}\frac{\partial^2 \overline{\sigma}_{11}}{\partial m^2}
\end{array}\right]\left(\frac{\partial m}{\partial X_i}\right)
\end{aligned}
\tag{6.6}
$$

For the special external stress-free case where $\overline{\sigma}_{11} = 0$, Eq. 6.6 reduces to

$$
D_i\frac{\partial \mu}{\partial X_i} = D_i\left[\begin{array}{c}
\frac{\partial^2 \hat{C}_0}{\partial m^2}+\frac{\partial^2 \hat{C}_1}{\partial m^2}(d_{11}+d_{22})+2\frac{\partial \hat{C}_1}{\partial m}\left(\frac{\partial d_{11}}{\partial m}+\frac{\partial d_{22}}{\partial m}\right)+\hat{C}_1\left(\frac{\partial^2 d_{11}}{\partial m^2}+\frac{\partial^2 d_{22}}{\partial m^2}\right)\\
+\frac{\partial^2 \hat{C}_2}{\partial m^2}d_{11}d_{22}+2\frac{\partial \hat{C}_2}{\partial m}\frac{\partial d_{11}}{\partial m}d_{22}+2\frac{\partial \hat{C}_2}{\partial m}\frac{\partial d_{22}}{\partial m}d_{11}+\hat{C}_2\frac{\partial^2 d_{11}}{\partial m^2}d_{22}\\
+2\hat{C}_2\frac{\partial d_{11}}{\partial m}\frac{\partial d_{22}}{\partial m}+\hat{C}_2\frac{\partial^2 d_{22}}{\partial m^2}d_{11}+\frac{\partial^2 \hat{C}_3}{\partial m^2}(d_{11}^2+d_{22}^2)\\
+4\frac{\partial \hat{C}_3}{\partial m}\left(d_{11}\frac{\partial d_{11}}{\partial m}+d_{22}\frac{\partial d_{22}}{\partial m}\right)+2\hat{C}_3\left(\begin{array}{c}\frac{\partial d_{11}}{\partial m}\frac{\partial d_{11}}{\partial m}+d_{11}\frac{\partial^2 d_{11}}{\partial m^2}\\ +\frac{\partial d_{22}}{\partial m}\frac{\partial d_{22}}{\partial m}+d_{22}\frac{\partial^2 d_{22}}{\partial m^2}\end{array}\right)
\end{array}\right]\left(\frac{\partial m}{\partial X_i}\right)
\tag{6.7}
$$

Based on a definition of a damage entity originally proposed by Talreja [16], it can be shown that the *preexisting* nonzero damage components d_{11}, d_{22} are given by

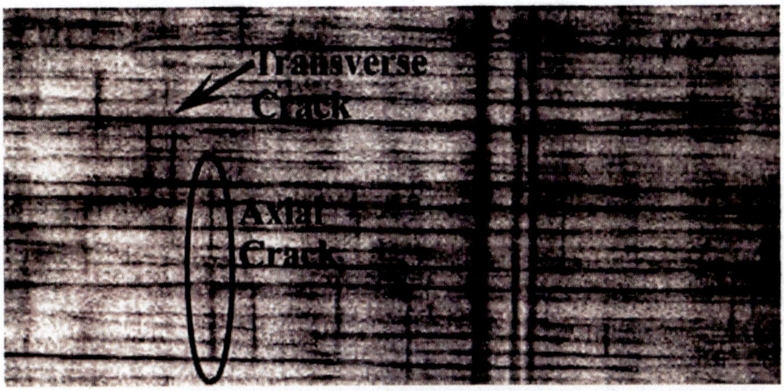

Fig. 6.3 Radiograph from a specimen containing fatigue induced biaxial cracks

$$d_{11} = \frac{\kappa(m,T)t_c^2\delta_1}{t}$$
$$d_{22} = \frac{\kappa(m,T)t_c^2\delta_2}{t} \tag{6.8}$$

where $\kappa(m,T)$ is an experimentally determined influence parameter on crack opening displacement, t_c is the crack size as shown in Fig. 6.2, t is the total thickness of the laminate, and δ_1, δ_2 are crack densities in laminate longitudinal (X_1) and transverse (X_2) directions, respectively. In the definition of the Gibbs potential, it is assumed that damage and stress are the two independent state variables in the formulation. In addition, the focus of this chapter is on the influence of preexisting damage on the diffusion characteristics of the laminate, i.e., the microcracks have been generated by load application prior to the diffusion experiments. When the applied load is removed, it is assumed that the microcracks still remain sufficiently open, as quantified by $\kappa(m,T)t_c$, to allow ingress of water vapor into the cracks and perhaps even liquid water through capillary action. It should be noted that $\kappa(m,T)$ is defined to depend only on the moisture concentration and the ambient temperature but not on stress. The diffusion-enhancing effect of applied tensile stress in the presence of preexisting damage is included in the model through the damage–stress interaction terms in the Gibbs potential given by $C_5\sigma_{11}d_{11}$ and $C_5\sigma_{11}d_{22}$.

Subsequent to the moisture uptake experiments, penetrant-enhanced radiography was employed to assess the damage state in the composite laminates. An example of a section taken from such a radiograph is shown in Fig. 6.3 for a specimen that was subjected to 75,000 cycles of uniaxial loading ($R = 0.1$, $f = 10$ Hz) at 521 MPa (75.6 ksi, 70% UTS). We note that while some inter-bundle separation (commonly referred to as a meta-delamination) occurs, delamination damage mode is not incorporated into the present diffusion model because the influence of meta-delaminations on laminate through-thickness diffusion is deemed insignificant. As may be observed from the radiograph shown in Fig. 6.3, the longitudinal cracks do not extend the full length of the laminate in uniaxial fatigue of AS4/PR500 woven composite, where as the transverse cracks span the entire width. Consequently, for the longitudinal cracks, the crack density is calculated as

$$\delta_1 = \frac{\sum_{i=1}^{M} l_i}{WL}$$

where M is the number of longitudinal cracks each having a length of l_i within a specimen volume of WLt.

The transverse crack density is calculated as

$$\delta_2 = \frac{\sum_{i=1}^{N} w_i}{WL}$$

where N is the number of transverse cracks, and w_i is the length of the ith transverse crack. For transverse cracks, the crack length is equal to the specimen width W, so that we have

$$\delta_2 = \frac{N}{L}$$

If it is assumed that swelling due to moisture absorption influences only the opening displacement without significantly affecting either the crack size, t_c, or the crack density, δ_1, δ_2, then,

$$\begin{aligned}
\frac{\partial d_{11}}{\partial m} &= \frac{t_c^2\delta_1}{t}\frac{\partial \kappa}{\partial m} \\
\frac{\partial^2 d_{11}}{\partial m^2} &= \frac{t_c^2\delta_1}{t}\frac{\partial^2 \kappa}{\partial m^2} \\
\frac{\partial d_{22}}{\partial m} &= \frac{t_c^2\delta_2}{t}\frac{\partial \kappa}{\partial m} \\
\frac{\partial^2 d_{22}}{\partial m^2} &= \frac{t_c^2\delta_2}{t}\frac{\partial^2 \kappa}{\partial m^2}
\end{aligned} \tag{6.9}$$

Taking the definitions of damage and its derivative given, respectively, by Eqs. 6.8 and 6.9 and substituting in Eq. 6.7 results in an expression for moisture flux in terms of an effective diffusivity and concentration gradient as indicated on the right-hand side of Eq. 6.10,

$$D_i\frac{\partial \mu}{\partial X_i} = \overline{D}_i\frac{\partial m}{\partial X_i} \tag{6.10}$$

where each orthotropic component of effective diffusivity ($\overline{D}_i$) can be obtained as a quadratic function of the crack density given by

$$\overline{D}_i = D_i \frac{\partial \mu}{\partial m}$$

$$= D_i \left\{ \begin{aligned} & \frac{\partial^2 \hat{C}_0}{\partial m^2} + \left(\kappa \frac{\partial^2 \hat{C}_1}{\partial m^2} + 2 \frac{\partial \hat{C}_1}{\partial m} \frac{\partial \kappa}{\partial m} + \hat{C}_1 \frac{\partial^2 \kappa}{\partial m^2} \right) \frac{t_c^2}{t} (\delta_1 + \delta_2) \\ & + \left[\frac{\partial^2 \hat{C}_2}{\partial m^2} \kappa^2 + 4\kappa \frac{\partial \hat{C}_2}{\partial m} \frac{\partial \kappa}{\partial m} + 2\kappa \hat{C}_2 \frac{\partial^2 \kappa}{\partial m^2} + 2\hat{C}_2 \left(\frac{\partial \kappa}{\partial m} \right)^2 \right] \left(\frac{t_c^2}{t} \right)^2 \delta_1 \delta_2 \\ & + \left[\frac{\partial^2 \hat{C}_3}{\partial m^2} \kappa^2 + 4\kappa \frac{\partial \hat{C}_3}{\partial m} \frac{\partial \kappa}{\partial m} + 2\kappa \hat{C}_3 \frac{\partial^2 \kappa}{\partial m^2} + 2\hat{C}_3 \left(\frac{\partial \kappa}{\partial m} \right)^2 \right] \left(\frac{t_c^2}{t} \right)^2 (\delta_1^2 + \delta_2^2) \end{aligned} \right\} \tag{6.11}$$

Arranging terms in Eq. 6.11 yields,

$$\overline{D}_i = C_0 + C_1(\delta_1 + \delta_2) + C_2 \delta_1 \delta_2 + C_3(\delta_1^2 + {\delta_1}^2) \tag{6.12}$$

where

$$C_0(m, T) = D_i \frac{\partial^2 \hat{C}_0}{\partial m^2} \tag{6.13}$$

$$C_1(m, T) = D_i \left(\kappa \frac{\partial^2 \hat{C}_1}{\partial m^2} + 2 \frac{\partial \hat{C}_1}{\partial m} \frac{\partial \kappa}{\partial m} + \hat{C}_1 \frac{\partial^2 \kappa}{\partial m^2} \right) \frac{t_c^2}{t} \tag{6.14}$$

$$C_2(m, T) = D_i \left[\frac{\partial^2 \hat{C}_2}{\partial m^2} \kappa^2 + 4\kappa \frac{\partial \hat{C}_2}{\partial m} \frac{\partial \kappa}{\partial m} + 2\kappa \hat{C}_2 \frac{\partial^2 \kappa}{\partial m^2} + 2\hat{C}_2 \left(\frac{\partial \kappa}{\partial m} \right)^2 \right] \left(\frac{t_c^2}{t} \right)^2 \tag{6.15}$$

$$C_3(m, T) = D_i \left[\frac{\partial^2 \hat{C}_3}{\partial m^2} \kappa^2 + 4\kappa \frac{\partial \hat{C}_3}{\partial m} \frac{\partial \kappa}{\partial m} + 2\kappa \hat{C}_3 \frac{\partial^2 \kappa}{\partial m^2} + 2\hat{C}_3 \left(\frac{\partial \kappa}{\partial m} \right)^2 \right] \left(\frac{t_c^2}{t} \right)^2 \tag{6.16}$$

Conservation of diffusing mass of penetrant within a unit volume of the polymer requires

$$\frac{\partial m}{\partial \hat{t}} = -\frac{\partial f_i}{\partial X_i} \tag{6.17}$$

where $\hat{t}$ is time, and in the absence of temperature, stress, and damage gradients, moisture flux f_i for orthotropic symmetry is assumed to be of the form

$$f_i = -D_i \frac{\partial \mu}{\partial X_i} \tag{6.18}$$

where D_i are the diffusion coefficients in material principal directions. Note that while repeated indices imply summation in Eq. 6.18, summation is not implied in Eq. 6.17. Combining Eqs. 6.17 and 6.18, gives the governing equation for diffusion in an orthotropic medium,

$$\frac{\partial m}{\partial t} = \frac{\partial}{\partial X_i}\left(D_i \frac{\partial \mu}{\partial X_i}\right) \tag{6.19}$$

Repeated indices imply summation in Eq. 6.19, with $i = 1,2,3$ for three-dimensional diffusion. Substituting Eqs. 6.10 and 6.12 in Eq. 6.19 yields the governing equation for moisture diffusion in an unloaded orthotropic laminate with uniform damage in a volume-average sense,

$$\frac{\partial m}{\partial \hat{t}} = \frac{\partial}{\partial X_i}\left\{\left[C_0 + C_1(\delta_1 + \delta_2) + C_2\delta_1\delta_2 + C_3\left(\delta_1^2 + \delta_2^2\right)\right]\frac{\partial m}{\partial X_i}\right\} \tag{6.20}$$

For modeling absorption and/or desorption in a damaged laminate in the through-thickness, i.e., in the X_3 direction, Eq. 6.20 reduces to

$$\frac{\partial m}{\partial \hat{t}} = \frac{\partial}{\partial X_3}\left\{\left[C_0 + C_1(\delta_1 + \delta_2) + C_2\delta_1\delta_2 + C_3\left(\delta_1^2 + \delta_2^2\right)\right]\frac{\partial m}{\partial X_3}\right\} \tag{6.21}$$

6.2.2.2 Boundary Condition and Maximum Saturation Level

For a linear elastic composite laminate, the chemical potential of the ambient vapor μ_b on the boundary is a constant with respect to time, i.e.,

$$\mu\left(m, T, d_{ij}\right) = \mu_b = \text{constant on boundary} \tag{6.22}$$

Assuming negligible change in stress and damage with moisture concentration, Eq. 6.4 reduces to

$$\begin{aligned}\mu\left(m, T, d_{ij}\right) &= \frac{\partial \hat{C}_0}{\partial m} + \frac{\partial \hat{C}_1}{\partial m}(d_{11} + d_{22}) + \frac{\partial \hat{C}_2}{\partial m} d_{11}d_{22} + \frac{\partial \hat{C}_3}{\partial m}(d_{11}^2 + d_{22}^2)\\ &= A_0(m, T) + A_1(m, T)(d_{11} + d_{22})\\ &\quad + A_2(m, T)d_{11}d_{22} + A_3(m, T)(d_{11}^2 + d_{22}^2)\end{aligned} \tag{6.23}$$

Assuming that moisture concentration $m \ll 1$, as is typically the case for a PMC, and expanding the coefficients in Eq. 6.23 in powers of m and retaining only first-order terms in m yields

$$\begin{aligned}\mu\left(m, T, d_{ij}\right) = {} & \left[\bar{A}_0(T) + \hat{A}_0(T)m\right] + \left[\bar{A}_1(T) + \hat{A}_1(T)m\right](d_{11} + d_{22})\\ & + \left[\bar{A}_2(T) + \hat{A}_2(T)m\right]d_{11}d_{22} + \left[\bar{A}_3(T) + \hat{A}_3(T)m\right](d_{11}^2 + d_{22}^2)\end{aligned} \tag{6.24}$$

Combining Eqs. 6.22 and 6.24 gives concentration on the boundary (m_b) as a function of damage

$$m_b = \frac{\mu_b - \overline{A}_0(T) - \overline{A}_1(T)(d_{11} + d_{22}) - \overline{A}_2(T)d_{11}d_{22} - \overline{A}_3(T)(d_{11}^2 + d_{22}^2)}{\hat{A}_0(T) + \hat{A}_1(T)(d_{11} + d_{22}) + \hat{A}_2(T)d_{11}d_{22} + \hat{A}_3(T)(d_{11}^2 + d_{22}^2)} \quad (6.25)$$

Dividing numerator and denominator in Eq. 6.25 by $\hat{A}_0(T)$,

$$m_b = \frac{\hat{a}_0(T) + \hat{a}_1(T)\,(d_{11} + d_{22}) + \hat{a}_2(T)d_{11}d_{22} + \hat{a}_3(T)\,(d_{11}^2 + d_{22}^2)}{1 + \hat{a}_4(T)\,(d_{11} + d_{22}) + \hat{a}_5(T)d_{11}d_{22} + \hat{a}_6(T)\,(d_{11}^2 + d_{22}^2)} \quad (6.26)$$

where

$$\hat{a}_0(T) = \frac{\mu_b - \overline{A}_0(T)}{\hat{A}_0(T)}$$

$$\hat{a}_1(T) = \frac{\overline{A}_1(T)}{\hat{A}_0(T)}$$

and so on.

In the interest of solution tractability, if it is assumed that $d_{11}, d_{22} \ll 1$, then we may ignore the damage terms beyond first order in the numerator and denominator of Eq. 6.26, giving,

$$m_b = \frac{\hat{a}_0(T) + \hat{a}_1(T)(d_{11} + d_{22})}{1 + \hat{a}_4(T)(d_{11} + d_{22})} \quad (6.27)$$

Expanding Eq. 6.27 in terms of (d_{11} + d_{22}) using Taylor series and retaining terms up to second order provides an approximate expression for the concentration on the boundary in terms of biaxial damage state,

$$m_b = \overline{a}_0(T) + \overline{a}_1(T)(d_{11} + d_{22}) + \overline{a}_2(T)(d_{11} + d_{22})^2 \quad (6.28)$$

where

$$\overline{a}_0(T) = \hat{a}_0(T)$$

$$\overline{a}_1(T) = \hat{a}_1(T) - \hat{a}_0(T)\hat{a}_4(T)$$

$$\overline{a}_2(T) = \hat{a}_4(T)[\hat{a}_0(T)\hat{a}_4(T) - \hat{a}_1(T)]$$

Recalling from Eq. 6.8,

$$d_{11} = \frac{\kappa(m,T)t_c^2\delta_1}{t}, \quad d_{22} = \frac{\kappa(m,T)t_c^2\delta_2}{t}$$

and substituting in Eq. 6.28, the boundary moisture concentration given in Eq. 6.28 can now be expressed in terms of biaxial crack densities as

$$m_b = a_0(T) + a_1(m,T)(\delta_1 + \delta_2) + a_2(m,T)(\delta_1 + \delta_2)^2 \tag{6.29}$$

where a_0, a_1, and a_2 are constants associated with the moisture concentration level and temperature, and are given by

$$a_0(T) = \bar{a}_0(T)$$

$$a_1(m,T) = \bar{a}_1(T)\left\{\frac{\kappa(m,T)t_c^2}{t}\right\}$$

$$a_2(m,T) = \bar{a}_2(T)\left\{\frac{\kappa(m,T)t_c^2}{t}\right\}^2$$

Finally, the maximum moisture content in the specimen at saturation can then be obtained as

$$M_m = m_b \times (\text{volume of specimen}) \tag{6.30}$$

6.2.3 *Closed-Form Solution for Diffusion Through the Laminate Thickness*

If the crack densities are uniform in a volume-averaged sense, the diffusion-governing equation given by Eq. 6.21 is amenable to a closed-form solution subject to the boundary condition defined by Eq. 6.29. The resulting through-thickness, time-varying moisture concentration profile in terms of biaxial crack densities is given by

$$m_t = \left[a_0(T) + a_1(m,T)(\delta_1 + \delta_2) + a_2(m,T)(\delta_1 + \delta_2)^2\right] \left\{ \begin{array}{l} 1 - \frac{4}{\pi}\sum_{n=0}^{\infty}\frac{(-1)^n}{(2n+1)} \\ \exp\left[-(2n+1)^2\left(\frac{\pi}{h}\right)^2\left[C_0 + C_1(\delta_1+\delta_2) + C_2\delta_1\delta_2 + C_3({\delta_1}^2 + {\delta_2}^2)\right]\hat{t}\right] \\ \cos\frac{(2n+1)\pi x}{h} \end{array} \right\} \tag{6.31}$$

where h is laminate thickness and t is time. Integrating Eq. 6.31 over the laminate thickness results in an expression for the total moisture uptake per unit area of the laminate surface as a function of biaxial crack densities,

$$M_t = \left[a_0(T) + a_1(m,T)(\delta_1+\delta_2) + a_2(m,T)(\delta_1+\delta_2)^2\right] \left\{1 - \frac{8}{\pi^2}\sum_{n=0}^{\infty}\frac{1}{(2n+1)^2}\exp\left[-(2n+1)^2\left(\frac{\pi}{h}\right)^2 \left[C_0 + C_1(\delta_1+\delta_2) + C_2\delta_1\delta_2 + C_3({\delta_1}^2+{\delta_2}^2)\right]\hat{t}\right]\right\} \tag{6.32}$$

6.2.4 *Hygrothermal Testing and Diffusion Data Characterization*

In order to characterize the damage–diffusivity influence coefficients C_0, C_1, C_2, C_3 and the damage–saturation influence coefficients a_0, a_1, a_2, moisture weight gain data for graphite/epoxy five-harness satin $[0/90/0/90]_s$ laminate with different biaxial microcrack densities were obtained from hygrothermal tests performed. Specifically, test specimens were mechanically pre-cracked by uniaxial fatigue or by static loading and then fully immersed in water at 65°C. The biaxial microcrack densities in terms of cracks/cm for the specimens are listed in Table 6.1 together with the type of loading that was used for pre-cracking the specimen. A set of uncracked specimens was also included in the test matrix as a control set. The specimens were not subjected to any applied mechanical stress during moisture absorption. The last two columns of Table 6.1 list the diffusivity and maximum moisture weight gain (%) measured for each specimen type from test data. Figure 6.4 shows the moisture weight gain (%) vs. square root of time data for the five test specimens. The damage–diffusivity influence coefficients C_0, C_1, C_2, C_3 and the damage–saturation influence coefficients, a_0, a_1, a_2 for each test specimen were extracted from weight gain data using a simple quadratic least-squares curve-fit procedure outlined below. It should be noted that test data from only the first four specimens (i.e., specimens 1 through 4) were employed to characterize the influence coefficients. The fifth data set (specimen 5) was held in reserve for verification of model prediction.

6.2.4.1 Least-Squares Curve-Fit Procedure

Let

$$D(\delta_1^k, \delta_2^k) = C_0 + C_1(\delta_1^k + \delta_2^k) + C_2\delta_1^k\delta_2^k + C_3(\delta_1^{k2} + \delta_2^{k2}) \tag{6.33}$$

Table 6.1 Diffusivity and maximum moisture uptake obtained from hygrothermal tests performed at different damage levels

Specimen	Load type	δ_1 (cm^{-1})	δ_2 (cm^{-1})	D (cm^2/s)	M_m (%)
1	No damage	0	0	8.621×10^{-9}	0.40808
2	21.4 KN (static)	0	7.35	1.265×10^{-8}	0.45255
3	100,000 cycles	1.44	24.7	5.341×10^{-8}	0.50715
4	200,000 cycles	4.95	25.2	1.936×10^{-7}	0.60336
5	300,000 cycles	1.71	25.2	9.634×10^{-8}	0.55616

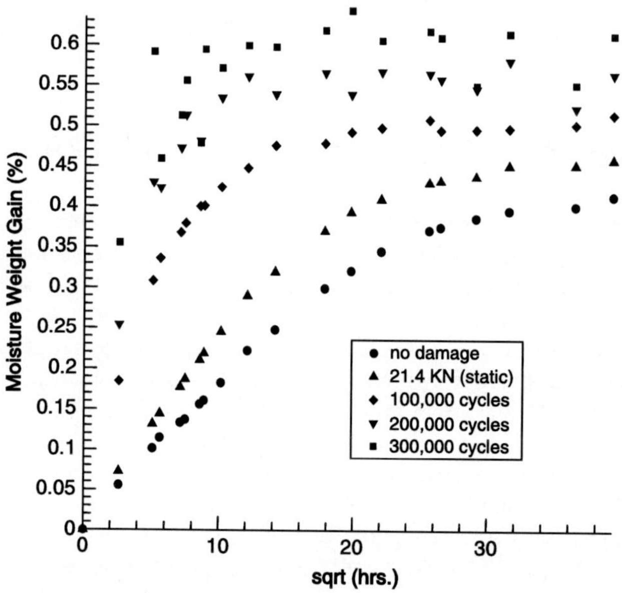

Fig. 6.4 Moisture uptake test data for specimens 1 through 5

and

$$L_E = \sum_{k=1}^{N} \left[D_k - D(\delta_1^k, \delta_2^k)\right]^2 \tag{6.34}$$

where

L_E is the least square error
N is the number of test data points
D_k is the kth test data point
δ_1^k is the crack density in laminate 1-direction corresponding to kth data point
δ_2^k is the crack density in laminate 2-direction corresponding to kth data point

The variation of the least square error should be zero, i.e.,

$$\delta L_E = \sum_{k=1}^{N} (D_k - D)\delta D = 0 \tag{6.35}$$

Substituting Eq. 6.12 into Eq. 6.35 gives

$$\sum_{k=1}^{N} (D_k - D)\left[\delta C_0 + \delta C_1(\delta_1^k + \delta_2^k) + \delta C_2 \delta_1^k \delta_2^k + \delta C_3(\delta_1^{k^2} + \delta_2^{k^2})\right] = 0 \tag{6.36}$$

Since all δC_i are arbitrary variations, the coefficients of δC_i must be zero, giving four Eqs. 6.37–6.40 that can be solved for the four influence coefficients C_0, C_1, C_2, C_3,

$$NC_0 + \left[\sum_{k=1}^{N} (\delta_1^k + \delta_2^k)\right] C_1 + \left[\sum_{k=1}^{N} (\delta_1^k \delta_2^k)\right] C_2 + \left[\sum_{k=1}^{N} (\delta_1^{k^2} + \delta_2^{k^2})\right] C_3 = \sum_{k=1}^{N} D_k \tag{6.37}$$

$$\left[\sum_{k=1}^{N} (\delta_1^k + \delta_2^k)\right] C_0 + \left[\sum_{k=1}^{N} (\delta_1^k + \delta_2^k)^2\right] C_1 + \left[\sum_{k=1}^{N} (\delta_1^k + \delta_2^k)(\delta_1^k \delta_2^k)\right] C_2$$
$$+ \left[\sum_{k=1}^{N} (\delta_1^k + \delta_2^k)(\delta_1^{k^2} + \delta_2^{k^2})\right] C_3 = \sum_{k=1}^{N} D_k(\delta_1^k + \delta_2^k) \tag{6.38}$$

$$\left[\sum_{k=1}^{N} (\delta_1^k \delta_2^k)\right] C_0 + \left[\sum_{k=1}^{N} (\delta_1^k + \delta_2^k)(\delta_1^k \delta_2^k)\right] C_1 + \left[\sum_{k=1}^{N} (\delta_1^k \delta_2^k)^2\right] C_2$$
$$+ \left[\sum_{k=1}^{N} (\delta_1^{k^2} + \delta_2^{k^2})(\delta_1^k \delta_2^k)\right] C_3 = \sum_{k=1}^{N} D_k(\delta_1^k \delta_2^k) \tag{6.39}$$

$$\left[\sum_{k=1}^{N} (\delta_1^{k^2} + \delta_2^{k^2})\right] C_0 + \left[\sum_{k=1}^{N} (\delta_1^k + \delta_2^k)(\delta_1^{k^2} + \delta_2^{k^2})\right] C_1 + \left[\sum_{k=1}^{N} (\delta_1^k \delta_2^k)(\delta_1^{k^2} + \delta_2^{k^2})\right] C_2$$
$$+ \left[\sum_{k=1}^{N} (\delta_1^{k^2} + \delta_2^{k^2})^2\right] C_3 = \sum_{k=1}^{N} D_k(\delta_1^{k^2} + \delta_2^{k^2}) \tag{6.40}$$

A similar procedure was employed to obtain the three damage–saturation influence coefficients a_0, a_1, a_2 and is not presented separately here. The characterized values of the influence coefficients are listed in Tables 6.2 and 6.3, respectively. Upon characterization, these coefficients were then used to calculate the effective diffusivity given by Eq. 6.12, and the maximum saturation level given by Eqs. 6.29 and 6.30. The values of diffusivity and maximum saturation level for the first four

Table 6.2 Damage-diffusivity influence coefficients characterized using specimens 1 through 4

C_0 (cm^2/s)	0.86210×10^{-08}
C_1 (cm^3/s)	0.98573×10^{-09}
C_2 (cm^4/s)	0.15594×10^{-08}
C_3 (cm^4/s)	-0.59533×10^{-10}

Table 6.3 Damage-saturation influence coefficients characterized using specimens 1 through 4

a_0 (gm)	$0.41959 \times 10^{+00}$
a_1 (gm cm)	0.41541×10^{-03}
a_2 (gm cm^2)	0.16178×10^{-03}

Table 6.4 Diffusivity and maximum moisture uptake obtained by using the proposed model at various damage levels for specimens 1 through 4

Specimen	δ_1 (cm^{-1})	δ_2 (cm^{-1})	D (cm^2/s)	M_m (%)
1	0	0	8.6210×10^{-09}	0.41959
2	0	7.35	1.2653×10^{-08}	0.43138
3	1.44	24.7	5.3409×10^{-08}	0.50499
4	4.95	25.2	1.9306×10^{-07}	0.57918

test specimens were computed by employing the influence coefficients from Tables 6.2 and 6.3 in the proposed model and are listed in Table 6.4. As can be expected, a comparison of the calculated results presented in Table 6.4 with the test data in Table 6.1 reveals that the values calculated using the model agree closely with the original test data. While this constitutes a necessary check, it does not conclusively validate the model because the same set of test data was used to obtain the influence coefficients. The four solid lines in Fig. 6.5 are model results for moisture uptake obtained using Eq. 6.32 and they merely serve to verify the accuracy of the influence coefficient characterization procedure for specimens 1 through 4.

6.2.4.2 Model Verification

For actual model verification, Eq. 6.32 was applied to simulate the moisture uptake in test specimen 5 that was not included in the influence coefficient characterization data set. The influence coefficients listed in Tables 6.2 and 6.3 were used for this purpose, in conjunction with crack densities of $\delta_1 = 1.71$ and $\delta_2 = 25.2$ cracks/cm listed in Table 6.1 for specimen 5. The solid line in Fig. 6.6 shows predicted moisture weight gain as a function of time. The solid circles depict test data. Despite the scatter in the test data, the agreement between the test data and model prediction is deemed quite reasonable. The laminate diffusivity and maximum saturation level predicted by the model for this case are listed in Table 6.5 together

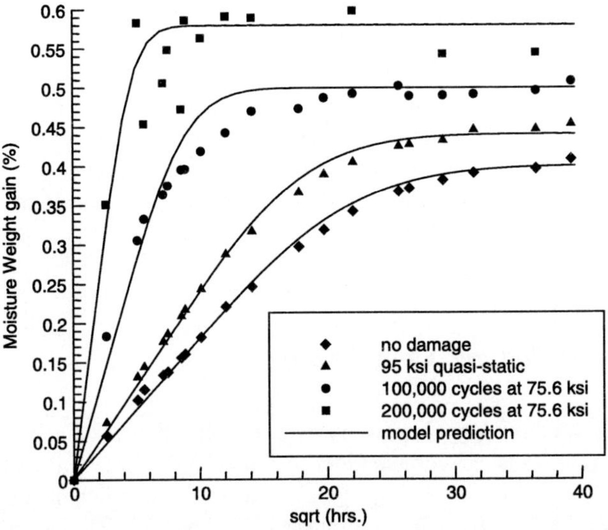

Fig. 6.5 Moisture uptake test data for specimens 1, 2, 3, and 4 for comparison with model prediction

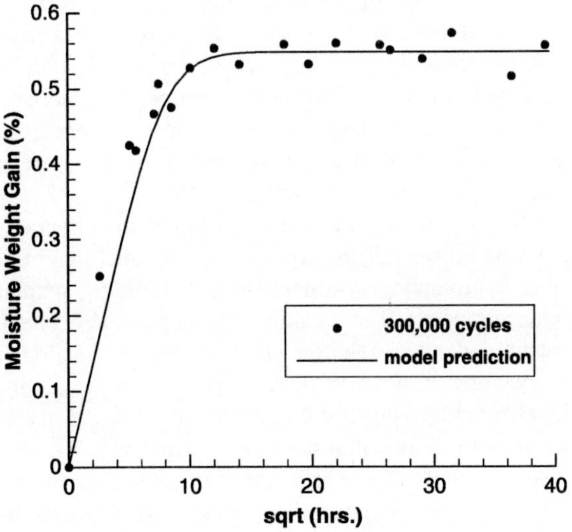

Fig. 6.6 Moisture uptake test data for specimen 5 and comparison with model prediction

Table 6.5 Prediction of diffusivity and maximum saturation level for crack density $\delta_1 = 1.71$ and $\delta_2 = 25.2$ cracks/cm using proposed model

	Diffusivity (cm^2/s)	% Error in D	M_m (%)	% Error in M_m
Model prediction	6.437×10^{-08}	33.2	0.54792	1.5
Test data	9.634×10^{-08}	–	0.55616	–

with test data for comparison. While there is a 33% error in predicted diffusivity, the error in the predicted maximum saturation level is less than 2%. It is likely that the accuracy of the diffusivity prediction can be greatly improved through the collection of more data points during the early stages of each experiment.

These results seem to confirm that retaining only up to the quadratic terms in damage in the expansion of the Gibbs potential is a valid modeling approximation for the material system under consideration. More model verification tests will be necessary to fully validate the hypothesis.

6.2.5 Discussion and Conclusions

A modeling methodology based on irreversible thermodynamics developed within the framework of composite macro-mechanics was presented to allow characterization of anomalous (non-Fickian) diffusion coefficients from moisture weight gain data for laminated composites with biaxial damage. A symmetric damage tensor based on continuum damage mechanics was incorporated in this model by invoking the principle of invariance with respect to coordinate transformations. To maintain tractability, the diffusion governing equations were simplified for the special case of a laminate with uniform matrix cracks that is subjected to a uniaxial tensile stress state. Because of the macro-mechanics formulation employed in developing this model, the model is currently restricted to the study of global laminate absorption/desorption characteristics in the presence of non-evolving damage states. An alternative mesomechanics approach that considers the more detailed physical aspects of moisture ingress into a micro-cracked laminate is presented elsewhere [18].

The material influence coefficients needed to model the effect of biaxial matrix micro-cracking on laminate diffusivity and maximum saturation level were evaluated by using hygrothermal test data for a $[0/90/0/90]_s$ graphite/epoxy five-harness satin textile composite. The moisture weight gain curves predicted by using a quadratic diffusivity–damage relation yielded good correlation with test data. While these observations do not conclusively prove that the laminate diffusivity and saturation level are indeed quadratic functions of crack density, it does indicate that retaining only up to the quadratic terms in damage in the expansion of the Gibbs potential is an acceptable modeling approximation for the material system under consideration. The primary purpose of this section is to establish a theoretical framework for using the proposed modeling approach to characterize, and eventually, to predict absorption and desorption in biaxially micro-cracked laminates. The formulation presented in this chapter is deemed to provide a small yet positive step toward that goal.

6.3 Characterization of Strain-Assisted Diffusion in an Epoxy-Based Cohesive Layer Material

The overall objective of this section is to develop a model for the synergistic bond degradation mechanism that may occur at the bonded interface between a fiber-reinforced polymer (FRP) and a substrate, which could be another FRP laminate or even concrete for civil infrastructure repair/retrofit. For this purpose, a two-dimensional cohesive layer constitutive model with a prescribed traction–separation (stress–strain) law can be constructed from basic principles of continuum mechanics and thermodynamics, taking into account non-Fickian hygrothermal effects, such as strain gradients, that are likely to occur within the cohesive layer. It is a well-established fact that Fick's law is frequently inadequate for describing moisture diffusion in polymers and polymer composites. Non-Fickian or anomalous diffusion is likely to occur when a polymer is subjected to external stresses and strains, as well as elevated temperature and humidity. As a result, it is necessary to take into account the combined effects of temperature, humidity, and stress in the construction of such a model for which the characterization of the material under these conditions is required. In this section, a modeling methodology based on the basic principles of continuum mechanics and thermodynamics is developed which allows characterization of non-Fickian diffusion coefficients as well as moisture saturation level from moisture weight gain data. For tractability, the diffusion-governing equations are simplified for the special case of 1D diffusion subjected to uniaxial strain and a uniform strain gradient. Comparisons with test data for an epoxy based primer adhesive are provided for model verifications.

6.3.1 Introduction

Fiber-reinforced polymer (FRP) composites have been extensively used as light-weight, performance-enhancing materials in the aerospace and defense industries for quite some time. However, the actual application of FRP in the civil engineering sector has been slow. One of the chief reasons for their slow acceptance is because of a lack of reliable predictive models and sound design guidelines for their use in civil infrastructure applications. One promising prospect of FRP application in civil engineering is infrastructure repair and retrofit. A major concern for such retrofitting is the debonding of polymeric adhesive that could compromise the reinforcing effect of the FRP. When exposed to harsh environment, degradation of the adhesive bond could lead to delamination of the FRP reinforcement that could ultimately lead to catastrophic failure. Combined exposure to heat and moisture affects a polymer in several ways. First, the hygrothermal swelling causes a change in the residual stresses within the polymer that could lead to degradation. Second, heat and humidity may cause the matrix to become plasticized, thus causing an increase in the strain to failure of the polymer. Finally, in the event of cyclic heating and cooling with a sustained use-temperature above the boiling point of water, vaporization and out-gassing of absorbed moisture may take place, leading to physical damage and chemical changes within the polymer, especially at temperatures greater than glass transition temperature of the polymer matrix. Continuous exposure to high moisture concentrations at the exposed surfaces of the polymer could also lead to damage in the form of polymer cracking, dissolution, and peeling.

The processes of sorption in polymeric materials were described in detail by Crank [7]. The influence of moisture diffusion on crack growth along an interface is not yet fully understood. Environmental cracking in a polymer typically occurs in the presence of a penetrant, such as moisture, and stress (or strain). It has been postulated that the mechanism involved in environmental crack growth in a polymer involves a small zone of craze formation and/or plasticization at the crack tip. However, for most thermoset resins such as epoxy, energy absorption at the crack tip is primarily by a shear yielding process and not by crazing. Consequently, for a thermoset epoxy, the zone of plasticization ahead of the crack tip must be determined using a diffusion law for nonporous media, such as Fick's law. However, quite frequently, polymer composites exhibit deviations from the classical Fickian treatment, termed as anomalous or non-Fickian diffusion, especially at elevated temperatures and stress levels, and at high relative humidity. For stretched polymer sheets where the diffusion-governing equations are coupled with mechanical response through volumetric strain, Roy et al. [19] presented a numerical procedure for solving coupled strain-assisted diffusion equations using an approach based on free-volume theory. Sancaktar and Baechtle [20] showed that there is a substantial change in the free-volume ratio in a polymer as a result of stress whitening, which, in turn, results in an increase in moisture uptake in the stress-whitened region. A multi-valued diffusion coefficient, based on an earlier model proposed by Wong and Broutman [21, 22], was employed to model this effect. More recently, Roy [23] derived governing equations for history-dependent diffusion using irreversible thermodynamics, and developed a novel numerical framework for solving the complex non-Fickian governing equations using the finite element method.

Stress-assisted diffusion in polymers was observed by Fahmy and Hurt [24] who used a four-point bend specimen to study the effect of bending stress on diffusion in a polymer. They observed that more water uptake occurred on the tensile side than on the compressive side for NARMCO 5208 Epoxy. Subsequently, governing equations for stress-assisted diffusion were derived by Weitsman [8] using principles of continuum thermodynamics. It was reported that both polymer diffusivity and saturation level depend upon stress and that these dependencies can stem from separate aspects of material response, i.e., elastic or viscoelastic material behavior. Various hygrothermal models have been developed and verified by the authors of this chapter to account for various types of anomalous diffusion (Roy et al. [15, 18, 19, 23–26]).

In this section, a coupled model for strain-assisted diffusion in a "cohesive layer" between two substrates is derived from the basic principles of continuum

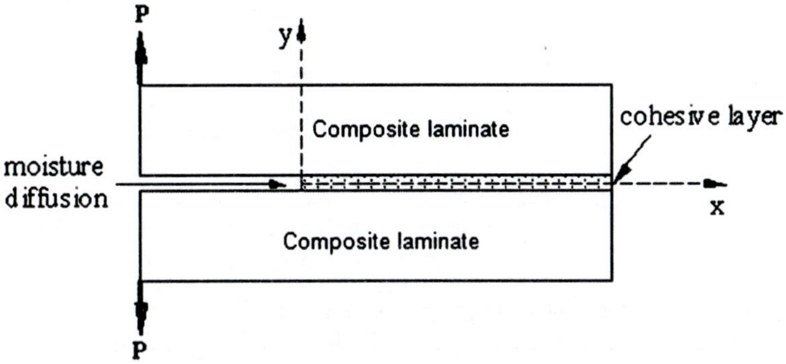

Fig. 6.7 A cohesive layer depicting adhesive bonded laminates with moisture diffusion

mechanics and thermodynamics, which also provides a consistent framework for the derivation of a coupled cohesive-layer constitutive law [27]. A series of experiments were conducted to collect moisture weight gain data in an epoxy specimen under different conditions of temperature, humidity, applied transverse strain, and applied strain gradient. The transverse strain as well as transverse strain gradient are likely to exist within a cohesive layer undergoing Mode I type debond growth during delamination as depicted in Fig. 6.7. For this purpose, a novel diffusion test specimen was developed for the application of transverse strain as well as strain gradient during moisture absorption, and this is described in a subsequent section. The effect of strain history in the epoxy layer on moisture diffusion is not included in the current model in the interest of tractability.

6.3.2 *Basic Equations*

6.3.2.1 Two-Dimensional Governing Equation

For a two-dimensional cohesive layer of finite thickness h, under plane-strain conditions as shown in Fig. 6.7, the Helmholtz free energy per unit volume is given by

$$\begin{aligned}\rho\psi = {} & C_0(m,T) + C_1(m,T)\varepsilon_{11} + C_2(m,T)\varepsilon_{22} + C_3(m,T)\varepsilon_{12} + C_4(m,T)\varepsilon_{11}^2 \\ & + C_5(m,T)\varepsilon_{22}^2 + C_6(m,T)\varepsilon_{12}^2 + C_7(m,T)\varepsilon_{11}\varepsilon_{12} + C_8(m,T)\varepsilon_{11}\varepsilon_{22} \\ & + C_9(m,T)\varepsilon_{12}\varepsilon_{22} + C_{10}(m,T)\varepsilon_{22}^3 + C_{11}(m,T)\varepsilon_{22}^2\varepsilon_{12} + C_{12}(m,T)\varepsilon_{22}\varepsilon_{12}^2 \\ & + C_{13}(m,T)\varepsilon_{12}^3 + C_{14}(m,T)\varepsilon_{22}^4 + C_{15}(m,T)\varepsilon_{22}^3\varepsilon_{12} \\ & + C_{16}(m,T)\varepsilon_{22}^2\varepsilon_{12}^2 + C_{17}(m,T)\varepsilon_{22}\varepsilon_{12}^3 + C_{18}(m,T)\varepsilon_{12}^4 \end{aligned} \tag{6.41}$$

where the mechanical strain components in two dimensions are defined as

$$\varepsilon_{11} = E_{11} - \alpha(T - T_{REF}) - \beta(m - m_{REF})$$

$$\varepsilon_{22} = E_{22} - \alpha(T - T_{REF}) - \beta(m - m_{REF})$$

$$\varepsilon_{12} = E_{12}$$

and

ρ is the mass density of epoxy in the cohesive layer
ε_{11} is the mechanical strain component in X_1 direction (*x*-direction)
ε_{22} is the mechanical strain component normal to crack face in X_2 direction (*y*-direction)
ε_{12} is the in-plane shear strain component tangential to crack face
E_{ij} is the total (kinematic) strain components
*m*is the moisture concentration in the cohesive layer at time *t*
m_{REF} is the reference moisture concentration
T is the temperature in the cohesive layer at time *t*
T_{REF} is the reference temperature
$\alpha(T)$ is the isotropic linear coefficient of thermal expansion of polymer
$\beta(T)$ is the isotropic linear coefficient of moisture expansion of polymer

Chemical potential of the diffusing vapor in a polymer can be defined as [5]

$$\mu = \rho \frac{\partial \psi}{\partial m}$$

or

$$\mu = \frac{\partial C_0}{\partial m} + \left[\frac{\partial C_4}{\partial m}\varepsilon_{11}^2 - 2C_4(m,T)\beta(T)\varepsilon_{11}\right] + \left[\frac{\partial C_5}{\partial m}\varepsilon_{22}^2 - 2C_5(m,T)\beta(T)\varepsilon_{22}\right] + \left[\frac{\partial C_{10}}{\partial m}\varepsilon_{22}^3 - 3C_{10}(m,T)\beta(T)\varepsilon_{22}^2\right] + \left[\frac{\partial C_{14}}{\partial m}\varepsilon_{22}^4 - 4C_{14}(m,T)\beta(T)\varepsilon_{22}^3\right] \quad (6.42)$$

From conservation of mass, the governing equation for two-dimensional moisture diffusion is

$$\frac{\partial m}{\partial t} = -\left(\frac{\partial f_x}{\partial x} + \frac{\partial f_y}{\partial y}\right) \quad (6.43)$$

where *t* is time and the moisture flux, $\vec{f} = f_x\hat{n}_x + f_y\hat{n}_y$, in the absence of temperature gradients is given by

$$f_x = -\hat{D}\frac{\partial \mu}{\partial x}$$
$$f_y = -\hat{D}\frac{\partial \mu}{\partial y} \quad (6.44)$$

Assuming isotropic material behavior and using the chain rule,

$$\begin{aligned} f_x &= -\hat{D}\left(\frac{\partial\mu}{\partial m}\right)\frac{\partial m}{\partial x} - \hat{D}\left(\frac{\partial\mu}{\partial T}\right)\frac{\partial T}{\partial x} - \hat{D}\left(\frac{\partial\mu}{\partial\varepsilon_{kk}^m}\right)\frac{\partial\varepsilon_{kk}^m}{\partial x} \\ &= -\hat{D}\left(\frac{\partial\mu}{\partial m}\right)\frac{\partial m}{\partial x} - \hat{D}\left(\frac{\partial\mu}{\partial T}\right)\frac{\partial T}{\partial x} - \hat{D}\left(\frac{\partial\mu}{\partial\varepsilon_{11}}\right)\frac{\partial\varepsilon_{11}}{\partial x} - \hat{D}\left(\frac{\partial\mu}{\partial\varepsilon_{22}}\right)\frac{\partial\varepsilon_{22}}{\partial x} \end{aligned} \tag{6.45}$$

$$f_y = -\hat{D}\left(\frac{\partial\mu}{\partial m}\right)\frac{\partial m}{\partial y} - \hat{D}\left(\frac{\partial\mu}{\partial T}\right)\frac{\partial T}{\partial y} - \hat{D}\left(\frac{\partial\mu}{\partial\varepsilon_{11}}\right)\frac{\partial\varepsilon_{11}}{\partial y} - \hat{D}\left(\frac{\partial\mu}{\partial\varepsilon_{22}}\right)\frac{\partial\varepsilon_{22}}{\partial y}$$

Assuming isothermal condition and substituting Eq. 6.45 in Eq. 6.43, gives

$$\frac{\partial m}{\partial t} = \frac{\partial}{\partial x}\left(D_m\frac{\partial m}{\partial x} + D_{\varepsilon 1}\frac{\partial\varepsilon_{11}}{\partial x} + D_{\varepsilon 2}\frac{\partial\varepsilon_{22}}{\partial x}\right) + \frac{\partial}{\partial y}\left(D_m\frac{\partial m}{\partial y} + D_{\varepsilon 1}\frac{\partial\varepsilon_{11}}{\partial y} + D_{\varepsilon 2}\frac{\partial\varepsilon_{22}}{\partial y}\right) \tag{6.46}$$

where the diffusion coefficients are given by

$$D_m = \hat{D}\left(\frac{\partial\mu}{\partial m}\right), \quad D_{\varepsilon 1} = \hat{D}\left(\frac{\partial\mu}{\partial\varepsilon_{11}}\right), \quad D_{\varepsilon 2} = \hat{D}\left(\frac{\partial\mu}{\partial\varepsilon_{22}}\right)$$

with

$$\begin{aligned} D_m = \hat{D}\Bigg[&\frac{\partial^2 C_0}{\partial m^2} + \frac{\partial^2 C_4}{\partial m^2}\varepsilon_{11}^2 - 4\frac{\partial C_4}{\partial m}\beta(T)\varepsilon_{11} + 2C_4(m,T)\beta(T)^2 \\ &+ \frac{\partial^2 C_5}{\partial m^2}\varepsilon_{22}^2 - 4\frac{\partial C_5}{\partial m}\beta(T)\varepsilon_{22} + 2C_5(m,T)\beta(T)^2 \\ &+ \frac{\partial^2 C_{10}}{\partial m^2}\varepsilon_{22}^3 - 6\frac{\partial C_{10}}{\partial m}\beta(T)\varepsilon_{22}^2 + 6C_{10}(m,T)\beta(T)^2\varepsilon_{22} \\ &+ \frac{\partial^2 C_{14}}{\partial m^2}\varepsilon_{22}^4 - 8\frac{\partial C_{14}}{\partial m}\beta(T)\varepsilon_{22}^3 + 12C_{14}(m,T)\beta(T)^2\varepsilon_{22}^2\Bigg] \end{aligned} \tag{6.47a}$$

$$D_{\varepsilon 1} = \hat{D}\left[2\frac{\partial C_4}{\partial m}\varepsilon_{11} - 2\beta(T)C_4(m,T)\right] \tag{6.47b}$$

$$\begin{aligned} D_{\varepsilon 2} = \hat{D}\Bigg[&2\frac{\partial C_5}{\partial m}\varepsilon_{22} - 2\beta(T)C_5(m,T) \\ &+ 3\frac{\partial C_{10}}{\partial m}\varepsilon_{22}^2 - 6\beta(T)C_{10}(m,T)\varepsilon_{22} \\ &+ 4\frac{\partial C_{14}}{\partial m}\varepsilon_{22}^3 - 12\beta(T)C_{14}(m,T)\varepsilon_{22}^2\Bigg] \end{aligned} \tag{6.47c}$$

It should be noted that in Eq. 6.46, the diffusion coefficient D_m is associated with the moisture flux term involving concentration gradient, where as the diffusion coefficients $D_{\varepsilon 1}$ and $D_{\varepsilon 2}$ are associated with flux terms involving strain gradients.

6.3.2.2 Characterization of Boundary Condition and Saturation Concentration

Assuming that the chemical potential of the ambient vapor on the exposed boundary of the cohesive zone remains constant with respect to time [5], the resulting concentration at the boundary of the cohesive zone can be derived as

$$\mu(T, m, \varepsilon_{ij})\big|_{BOUNDARY} = \mu_b(RH, T, \varepsilon_b) \tag{6.48}$$

where ε_b is the value of the transverse mechanical strain ε_{22} at the boundary, and RH is the environmental relative humidity. Assuming that in the expression for chemical potential defined in Eq. 6.42, axial strain $\varepsilon_{11} = 0$, and retaining terms up to second order in ε_{22} (for $\varepsilon_{22} <<1$), then Eq. 6.48 reduces to [27]

$$\left\{\hat{C}_{10}.(\overline{C}_0 + 2\tilde{C}_0\Delta m) + \frac{27}{8}\frac{\sigma_{MAX}(T)}{\varepsilon_{MAX}}\left[(\overline{C}_5 + 2\tilde{C}_5\Delta m)\varepsilon_{22}^2 - 2\beta(1 + \overline{C}_5\Delta m)\varepsilon_{22}\right]\right.$$
$$\left. + \frac{27}{2}\frac{\sigma_{MAX}(T)}{\varepsilon_{MAX}^2}\left[\beta(1 + \overline{C}_{10}\Delta m)\varepsilon_{22}^2\right]\right\}_{BOUNDARY} = \mu_b(RH, T, \varepsilon_b) \tag{6.49}$$

Substituting for the change in concentration $\Delta m = m_b - m_{REF}$ in Eq. 6.49 and solving for m_b gives

$$m_b = B_0(RH, T) + B_1(RH, T)\varepsilon_b + B_2(RH, T)\varepsilon_b^2 \tag{6.50}$$

where B_0, B_1, B_2 are material constants that need to be characterized.

Assuming maximum moisture concentration in the specimen at saturation can be expressed as

$$m_{MAX} = m_b$$
$$m_{MAX} = B_0(RH, T) + B_1(RH, T)\varepsilon_b + B_2(RH, T)\varepsilon_b^2 \tag{6.51}$$

6.3.2.3 Design of Experiment

The governing equation for diffusion given by Eq. 6.46 is a strain-based model. Hence for this study, the design of experiment should satisfy the condition that the diffusion is strain assisted rather than stress assisted. The epoxy material used in this study was found to experience large increase in tensile strain due to viscoelastic creep, and so, introduction of constant strain by standard method of loading a specimen by deadweight becomes highly inaccurate. This was the motivation to develop a novel constant-strain test specimen based on the theory of beam bending that achieves the design of experiment objective and yet remains practical.

The experiment is designed with two objectives: (a) being able to introduce a predetermined strain without any change in strain due to creep, and, (b) being able to measure the sample in a laboratory accuracy-balance without much loss of weight gain measurement sensitivity. For this purpose, a simple but novel method is developed. The epoxy primer is molded onto an aluminum sheet base, which is then bent to calculated radius of curvature to introduce a predetermined strain in the primer as depicted in Fig. 6.8. An epoxy-based primer commercially available under the name Waco Mbrace primer is used for this study. This primer is mixed with the hardener and degassed in a degassing chamber to remove any air bubbles present. The surface of the aluminum is treated with etching agent to ensure proper adhesiveness between the aluminum and primer. Then the primer is molded onto the aluminum sheet in the shape of a 25.4 mm × 25.4 mm coupon with an average thickness of 0.74 mm. Then these specimens are allowed to cure at room temperature for 72 h and strains are introduced by bending the specimens to the required radius, which is calculated as defined in the following paragraph. These specimens are then preconditioned in an electric oven at 93.3°C for 36 h for drying as well as post-curing. After recording the dry weight, the specimens are put in an environmental chamber under various conditions of relative humidity and temperature as listed in Table 6.3. Knowing the weight of the aluminum sheet base, the weight of the primer alone can be calculated, and the weight gain due to moisture can also be calculated. A schematic diagram of the specimens used for strain-assisted diffusion test is shown in Fig. 6.8a and a photograph of the actual specimen is shown in Fig. 6.8b.

As the modulus of aluminum sheet is much greater than that of the epoxy primer, when the aluminum sheet is bent, we can assume that the aluminum and primer are still in linear elastic state and the bonded primers have little effect on the curvature of the aluminum sheet or the composite beam of aluminum and epoxy primer. Therefore, Kirchhoff's hypothesis of "plane sections remain plane" is valid in this case.

From the force equilibrium on the beam cross section, the x coordinate of the neutral line for the composite beam is given as

$$x_0 = \frac{E_2 h_2 (h_1 + h_2)}{2E_1 h_1 + 2E_2 h_2} = \frac{h_1 + h_2}{2\left(1 + \frac{E_1 h_1}{E_2 h_2}\right)} \tag{6.52}$$

where

h_1 is the thickness of aluminum sheet
h_2 is the thickness of epoxy primer
E_1 is the Young's modulus of aluminum sheet

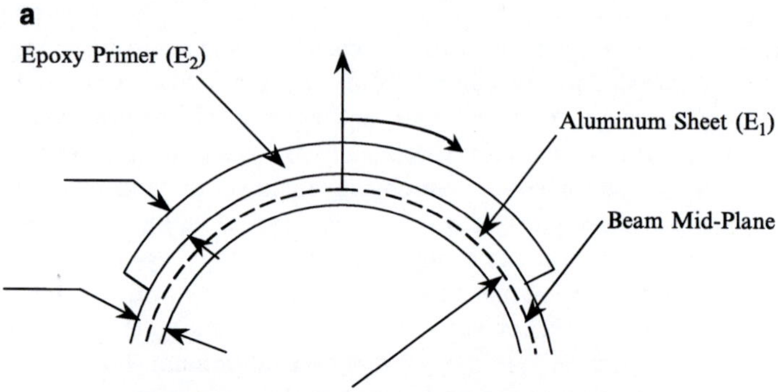

Fig. 6.8 (**a**) Schematic diagram of specimen and (**b**) photograph of the actual specimen

E_2 is the Young's modulus of epoxy primer
R is the radius of curvature

The total longitudinal strain at the inner surface of the primer is given by Gere and Timoshenko [28]

$$\varepsilon_{in} = \frac{\frac{1}{2}h_1 - x_0}{R + x_0} \tag{6.53}$$

Strain at the outer surface,

$$\varepsilon_{out} = \frac{\frac{1}{2}h_1 + h_2 - x_0}{R + x_0} \tag{6.54}$$

Then, the average tensile strain at the mid-plane of the primer is

$$\varepsilon_{avg} = \frac{h_1 + \frac{1}{2}h_2 - x_0}{R + x_0} \tag{6.55}$$

To obtain a specified average total tensile strain ε_{avg} in primer, the radius of the cylinder needs to be

$$R = \frac{h_1 + \frac{1}{2}h_2 - x_0}{\varepsilon_{avg}} - x_0 \tag{6.56}$$

For this present study, the Young's modulus of aluminum sheet $E_1 = 75$ GPa, thickness $h_1 = 0.51$ mm and for epoxy primer, $E_2 = 3.85$ GPa, $h_2 = 0.74$ mm. For the specimen with prescribed 5% average mechanical strain, the radius of curvature is calculated to be $R = 16.69$ mm and for 10% average mechanical strain, the radius is calculated to be $R = 8.35$ mm. Strain gauge tests were carried out to calibrate the radius of the specimen to get the required strain.

6.3.2.4 Effect of Temperature on the Radius of the Specimen Ring

As the primer specimen is molded onto the aluminum base at room temperature, and the specimen ring as a whole is subjected to elevated temperatures, the resulting change in radius of curvature of the aluminum ring needs to be evaluated. Linear expansion of the aluminum ring by itself is calculated to study its effect on the specimen as a whole. The specimen is cured at room temperature (T_{REF}) of 25.5°C and then kept in a environmental chamber at temperatures (T) of 32.2°C, 40.5°C, and 48.9°C for the diffusion tests. The change in length in aluminum ring due to the change in temperature is given by

$$\Delta l = l_0 \alpha_{Al} \Delta T$$

where

l_0 is the initial circumference length at room temperature
α_{Al} is the coefficient of linear expansion for aluminum (23×10^{-6}/°C)
ΔT is the difference in temperature ($T - T_{REF}$)

For a 10% strain specimen at 48.9°C, the radius is 8.350 mm, so the initial circumferential length l_0 is 52.46 mm. Therefore, the change in length at 48.9°C is given as 0.03 mm. The final circumference length is given as $l = 52.46 + 0.03 = 52.49$ mm. The radius of the ring after thermal expansion is given as

$$r = \frac{l}{2\pi}$$

Table 6.6 Mechanical strain in each specimen at various environmental conditions

Specimen type	Mech. strain location	Relative humidity (%) 75, Temperature (°C) 32.2	75, 40.5	75, 48.9	85, 32.2	85, 40.5	85, 48.9	95, 32.2	95, 40.5	95, 48.9
		Mechanical strain (%)			Mechanical strain (%)			Mechanical strain (%)		
Flat (radius = ∞)		0	0	0	0	0	0	0	0	0
Circular (radius = 16.69 mm)	ε_{max}	5.15	5.06	4.99	5.05	4.96	4.89	4.95	4.86	4.79
	ε_{min}	0.72	0.63	0.56	0.62	0.53	0.46	0.52	0.43	0.36
	ε_{avg}	4.46	4.37	4.30	4.36	4.27	4.20	4.26	4.17	4.10
Circular (radius = 8.35 mm)	ε_{max}	11.09	11.01	10.94	10.99	10.91	10.84	10.89	10.81	10.74
	ε_{min}	2.23	2.15	2.08	2.14	2.05	1.98	2.04	1.95	1.88
	ε_{avg}	9.72	9.63	9.56	9.62	9.53	9.46	9.52	9.43	9.36

which is calculated to be 8.354 mm. Therefore, the change in ring radius is only 0.048%, which can be neglected without adversely affecting the accuracy of the calculation of strain in epoxy.

6.3.2.5 Calculation of Mechanical Strain in the Primer

In the formulation used for calculating Helmholtz free energy per unit volume, Eq. 6.41, the strain parameters involved are mechanical strains whereas the strains calculated in Eqs. 6.53–6.55 are total strains. The relationship between total strain and mechanical strain is given as

$$\varepsilon_{Mechanical} = \varepsilon_{Total} - \varepsilon_{Hygral} - \varepsilon_{Thermal} \tag{6.57}$$

where

$$\varepsilon_{Hygral} = \beta \Delta m$$

$$\varepsilon_{Thermal} = \alpha \Delta T$$

Notice that the relationship between hygral strain and moisture concentration is assumed to be linear even though in some cases, especially for composites, a slight nonlinearity has been found to exist [29]. From our experiments on epoxy primer, we determined the value of $\beta = 100.18 \times 10^{-6}$ mm/mm/RH%, and $\alpha = 97 \times 10^{-6}$ mm/mm/°C, respectively. Variations in these coefficients as functions of temperature are ignored because of the small temperature range over which tests were conducted. From these data, the maximum, minimum, and average mechanical strains and the mechanical strain gradients are calculated and tabulated in Tables 6.6

Table 6.7 Through-thickness strain gradient in epoxy specimen at various environmental conditions

		Relative humidity (%)								
		75			85			95		
		Temperature (°C)								
		32.2	40.5	48.9	32.2	40.5	48.9	32.2	40.5	8.9
Specimen type	Radius (mm)	Mechanical strain gradient (% strain/mm)								
Flat	∞	0	0	0	0	0	0	0	0	0
Ring	16.69	5.98	5.98	5.98	5.98	5.98	5.98	5.98	5.98	5.98
Ring	8.35	11.97	11.97	11.97	11.97	11.97	11.97	11.97	11.97	11.97

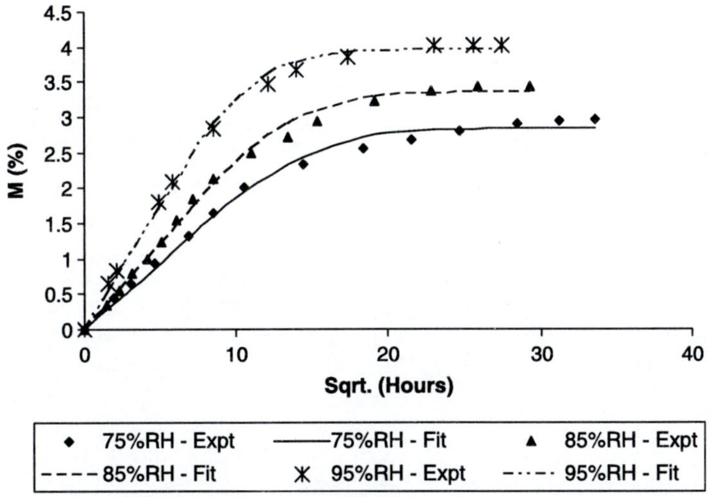

Fig. 6.9 Moisture uptake vs. sqrt. (time) (0% strain, 32.2°C)

and 6.7, respectively. Since the variation of moisture concentration through thickness of the specimen cannot be measured exactly, therefore the relative humidity (RH) at the specimen boundaries is used in Eq. 6.57 to compute mechanical strain.

The moisture weight gain for each of the specimens is recorded with respect to time, until saturation occurs. Tests are carried out under three different temperatures of 32.2°C, 40.5°C, and 48.9°C, and three different environmental moisture concentrations of 75% RH, 85% RH, and 95% RH. For each environmental condition (specified temperature and relative humidity), three total strain levels of 0%, 5%, and 10% are tested. The corresponding mechanical strain for each case computed using Eq. 6.57 is listed in Table 6.6. Each test data point in Figs. 6.9–6.17 represents the average of four test specimens. From these test data, diffusivity and saturation moisture concentration could be derived through a nonlinear least-squares technique that is described in the following section.

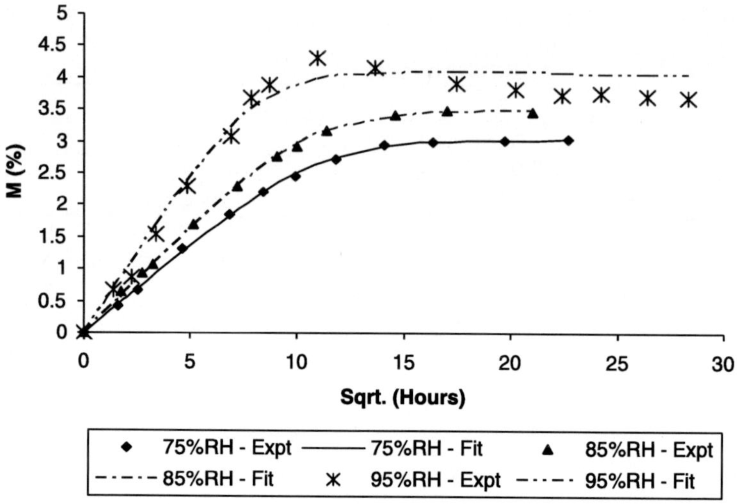

Fig. 6.10 Moisture uptake vs. sqrt. (time) (0% strain, 40.5°C)

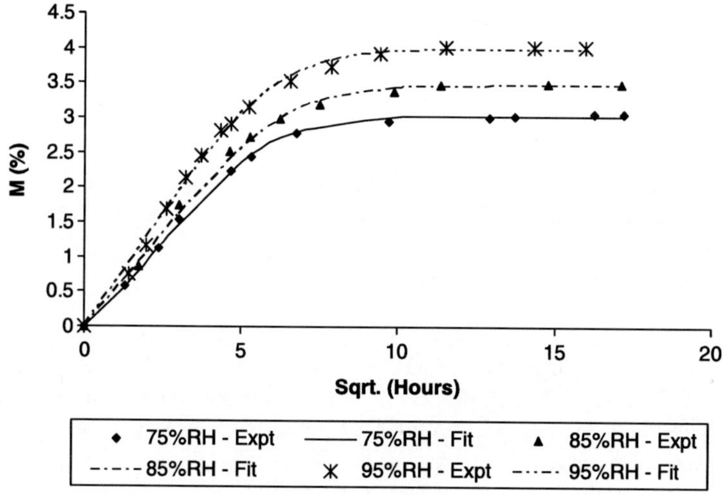

Fig. 6.11 Moisture uptake vs. sqrt. (time) (0% strain, 48.9°C)

The bonded electrical resistance strain gage is used to verify that the theoretically computed strain is indeed the strain that is actually introduced in the specimen. A quarter bridge three-wire system with a large strain measurement gage is used. A 10% strain specimen is used to verify the actual strain introduced and thereby to calculate the strain correction factor.

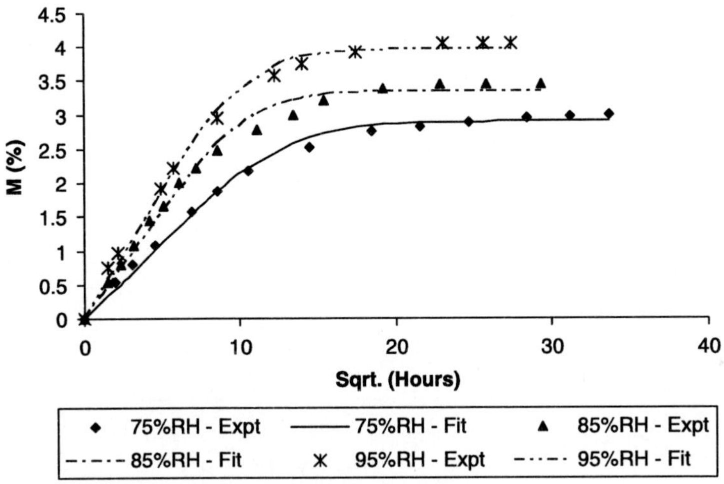

Fig. 6.12 Moisture uptake vs. sqrt. (time) (5% strain, 32.2°C)

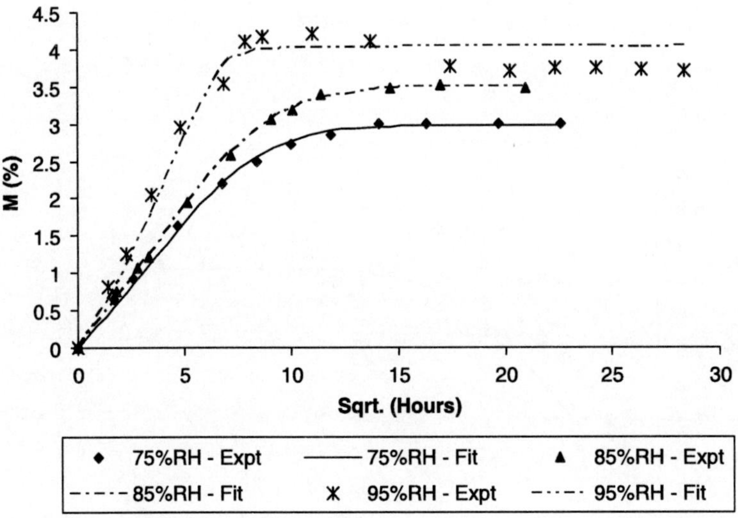

Fig. 6.13 Moisture uptake vs. sqrt. (time) (5% strain, 40.5°C)

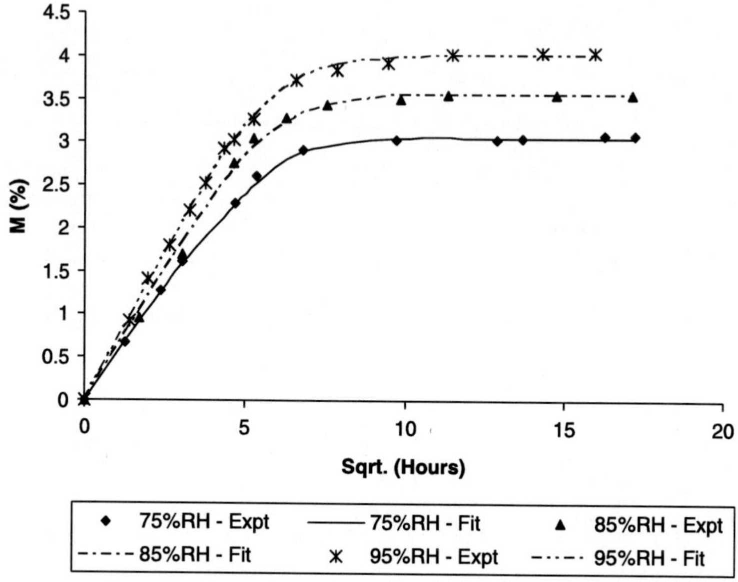

Fig. 6.14 Moisture uptake vs. sqrt. (time) (5% strain, 48.9°C)

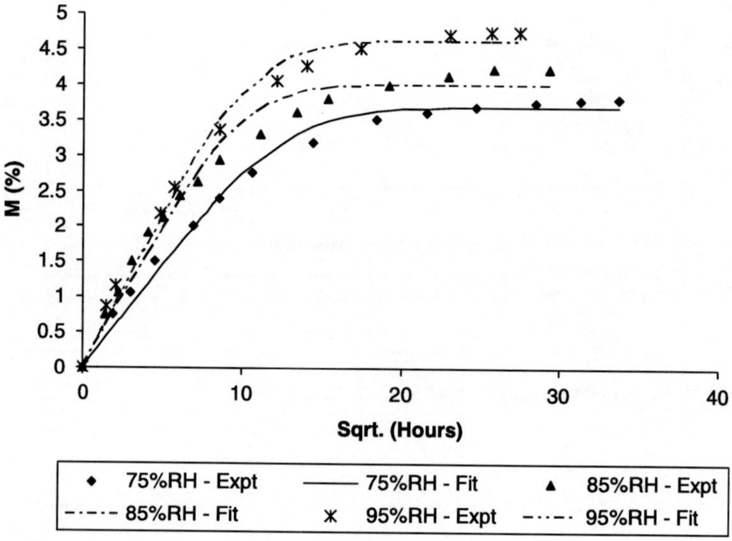

Fig. 6.15 Moisture uptake vs. sqrt. (time) (10% strain, 32.2°C)

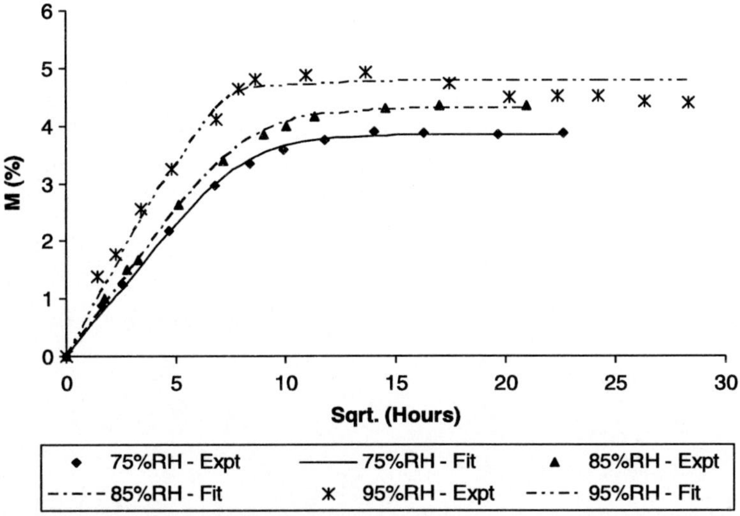

Fig. 6.16 Moisture uptake vs. sqrt. (time) (10% strain, 40.5°C)

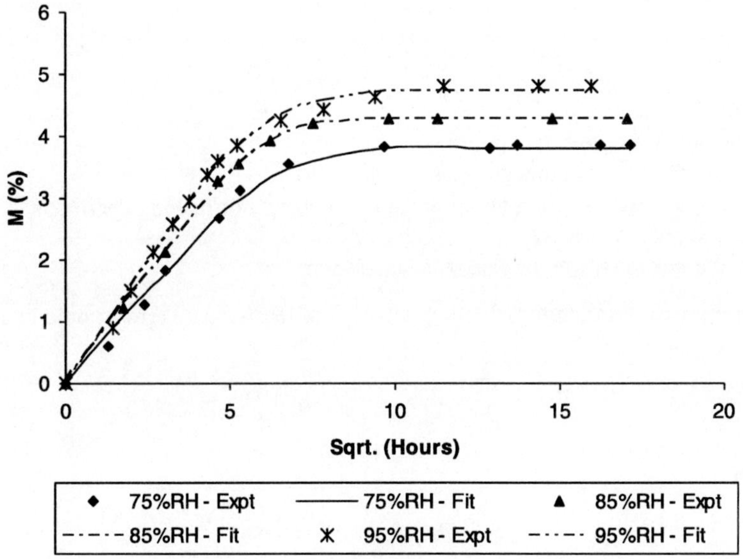

Fig. 6.17 Moisture uptake vs. sqrt. (time) (10% strain, 48.9°C)

$$\text{Strain correction factor } (\lambda) = \frac{\varepsilon_{actual}}{\varepsilon_{theory}}$$

From the strain gage test, the strain correction factor λ is found to be 0.87, presumably due to the fact that that the ring is not perfectly circular. This correction factor is applied to the calculated radii of all specimens to introduce the correct mid-plane strain.

6.3.2.6 Calculation of Diffusion Coefficient and Saturation Mass Uptake

Let

$$M(t_k) = M_\infty \left\{ 1 - \frac{8}{\pi^2} \sum_{n=0}^{\infty} \frac{1}{(2n+1)^2} e^{-(2n+1)^2 \left(\frac{\pi}{h}\right)^2 D t_k} \right\} \tag{6.58}$$

and

$$L_E = \sum_{k=1}^{N} [M_k - M(t_k)]^2 \tag{6.59}$$

where

L_E is the least square error
N is the number of test data points
M_k is the kth test data point for mass uptake
t_k is the time corresponding to kth data point during a diffusion experiment
D is the unknown diffusivity
M_∞ is the unknown weight gain % at saturation

For error to be minimum, the first variation of least-square error should be zero.

$$\delta L_E = \sum_{k=1}^{N} -2[M_k - M(t_k)]\delta M(t_k) = 0 \tag{6.60}$$

But

$$\delta M(t_k) = \frac{\partial M(t_k)}{\partial M_\infty}\delta M_\infty + \frac{\partial M(t_k)}{\partial D}\delta D \tag{6.61}$$

Substituting Eq. 6.21 in Eq. 6.20

$$\delta L_E = \left\{ \sum_{k=1}^{N} [M_k - M(t_k)] \frac{\partial M(t_k)}{\partial M_\infty} \right\} \delta M_\infty + \left\{ \sum_{k=1}^{N} [M_k - M(t_k)] \frac{\partial M(t_k)}{\partial D} \right\} \delta D = 0 \tag{6.62}$$

Since δM_∞ and δD are arbitrary variations, each of the terms in parenthesis must independently go to zero, i.e.,

$$\sum_{k=1}^{N}[M_k - M(t_k)]\frac{\partial M(t_k)}{\partial M_\infty} = 0 \tag{6.63}$$

$$\sum_{k=1}^{N}[M_k - M(t_k)]\frac{\partial M(t_k)}{\partial D} = 0 \tag{6.64}$$

Expanding Eq. 6.63

$$\sum_{k=1}^{N}\left[M_k - M_\infty\left\{1-\frac{8}{\pi^2}\sum_{n=0}^{\infty}\frac{1}{(2n+1)^2}e^{-(2n+1)^2\left(\frac{\pi}{h}\right)^2 Dt_k}\right\}\right]$$
$$-\left\{1-\frac{8}{\pi^2}\sum_{n=0}^{\infty}\frac{1}{(2n+1)^2}e^{-(2n+1)^2\left(\frac{\pi}{h}\right)^2 Dt_k}\right\} = 0$$

$$M_\infty = \frac{\sum_{k=1}^{N}\left[M_k\left\{1-\frac{8}{\pi^2}\sum_{n=0}^{\infty}\frac{1}{(2n+1)^2}e^{-(2n+1)^2\left(\frac{\pi}{h}\right)^2 Dt_k}\right\}\right]}{\sum_{k=1}^{N}\left[1-\frac{8}{\pi^2}\sum_{n=0}^{\infty}\frac{1}{(2n+1)^2}e^{-(2n+1)^2\left(\frac{\pi}{h}\right)^2 Dt_k}\right]^2} \tag{6.65}$$

Expanding Eq. 6.64,

$$\sum_{k=1}^{N}\left[M_k - M_\infty\left\{1-\frac{8}{\pi^2}\sum_{n=0}^{\infty}\frac{1}{(2n+1)^2}e^{-(2n+1)^2\left(\frac{\pi}{h}\right)^2 Dt_k}\right\}\right]$$
$$-\left\{(M_\infty)\frac{8}{\pi^2}\sum_{n=0}^{\infty}-\frac{1}{(2n+1)^2}(2n+1)^2\left(\frac{\pi}{h}\right)^2 t_k e^{-(2n+1)^2\left(\frac{\pi}{h}\right)^2 Dt_k}\right\} = 0$$

$$\sum_{k=1}^{N}\left[\begin{array}{l}(M_k)(M_\infty)\frac{8}{\pi^2}t_k\sum_{n=0}^{\infty}\frac{1}{(2n+1)^2}e^{-(2n+1)^2\left(\frac{\pi}{h}\right)^2 Dt_k} - \\ (M_\infty)^2\left\{1-\frac{8}{\pi^2}\sum_{n=0}^{\infty}-\frac{1}{(2n+1)^2}e^{-(2n+1)^2\left(\frac{\pi}{h}\right)^2 Dt_k}\right\} \\ \left\{\frac{8}{\pi^2}t_k\sum_{n=0}^{\infty}\frac{1}{(2n+1)^2}e^{-(2n+1)^2\left(\frac{\pi}{h}\right)^2 Dt_k}\right\}\end{array}\right] = 0 \tag{6.66}$$

Substituting Eq. 6.65 in Eq. 6.66 results in a highly nonlinear equation in diffusivity (D) that must be solved using an iterative procedure. A software code in C language

Table 6.8 Diffusivity and m_{sat} for different conditions of temperature, relative humidity, and strain

Temperature (°C)	Relative humidity (%)	Total strain (%)	No. of replicate specimens	Diffusivity (cm^2/s)× 10^{-9}	m_{sat} (%)
32.2	75	0	4	5.8220	2.9486
		5	4	5.9532	2.9848
		10	4	6.2772	3.7839
	85	0	4	7.7720	3.3402
		5	4	8.4432	3.3398
		10	4	10.3720	4.1039
	95	0	4	12.0880	3.9786
		5	4	13.2530	3.9848
		10	4	19.1720	4.7139
40.5	75	0	4	10.1770	3.0302
		5	4	10.4780	3.0219
		10	4	10.6510	3.8484
	85	0	4	12.6470	3.4749
		5	4	13.8740	3.5028
		10	4	16.7710	4.2985
	95	0	4	18.7780	4.0400
		5	4	20.7380	4.0360
		10	4	25.0860	4.7730
48.9	75	0	4	11.5210	3.0150
		5	4	12.0070	3.0457
		10	4	13.1110	3.8064
	85	0	4	17.4410	3.4602
		5	4	19.5980	3.5298
		10	4	24.6970	4.2795
	95	0	4	22.7000	4.0244
		5	4	24.6090	4.0312
		10	4	30.0080	4.7529

was written to solve this numerically. Once D is obtained from Eq. 6.66, it can be substituted in Eq. 6.65 to evaluate M_∞. The computed diffusivity and saturation concentration for all test conditions are listed in Table 6.8.

6.3.2.7 Characterization of Nonlinear Diffusion Coefficients

Revisiting the diffusion-governing equation, Eq. 6.6,

$$\frac{\partial m}{\partial t} = \frac{\partial}{\partial x}\left(D_m \frac{\partial m}{\partial x} + D_{\varepsilon 1}\frac{\partial \varepsilon_{11}}{\partial x} + D_{\varepsilon 2}\frac{\partial \varepsilon_{22}}{\partial x}\right) + \frac{\partial}{\partial y}\left(D_m \frac{\partial m}{\partial y} + D_{\varepsilon 1}\frac{\partial \varepsilon_{11}}{\partial y} + D_{\varepsilon 2}\frac{\partial \varepsilon_{22}}{\partial y}\right) \tag{6.67}$$

and imposing the condition that the diffusion experiments conducted in this study are a special case of one-dimensional diffusion in the x-direction, with an uniform transverse tensile strain in the y-direction having a uniform tensile strain gradient through the thickness (refer to 6.8(b)), Eq. 6.67 reduces to

$$\frac{\partial m}{\partial t} = \frac{\partial}{\partial x}\left(D_m \frac{\partial m}{\partial x} + D_{\varepsilon 1}\frac{\partial \varepsilon_{11}}{\partial x} + D_{\varepsilon 2}\frac{\partial \varepsilon_{22}}{\partial x}\right) \tag{6.68}$$

In the absence of applied strain in the x-direction, $\varepsilon_{11} = -v\varepsilon_{22}$, and $\frac{\partial \varepsilon_{11}}{\partial x} = -v\frac{\partial \varepsilon_{22}}{\partial x}$, where v is the Poisson's ratio of the epoxy. Hence, the governing equation for this particular study reduces to

$$\frac{\partial m}{\partial t} = \frac{\partial}{\partial x}\left(D_m \frac{\partial m}{\partial x} + (D_{\varepsilon 2} - vD_{\varepsilon 1})\frac{\partial \varepsilon_{22}}{\partial x}\right) \tag{6.69}$$

From the data presented in Table 6.7, it becomes quickly evident that the through-thickness gradient of the transverse strain ($\frac{\partial \varepsilon_{22}}{\partial x}$) is sufficiently large that it cannot be neglected. Hence the expected diffusion behavior in the epoxy specimens should be non-Fickian. However, the actual diffusion experimental data (refer to Figs. 6.9–6.17) indicate that the moisture uptake in the epoxy primer generally obeys Fick's law at least for the range of temperatures, relative humidity, and strains considered in this study, although the magnitude of the diffusivity and saturation levels for each case depends on the imposed boundary conditions. It can therefore be concluded without loss of generality that this particular epoxy material behaves in a nonlinear Fickian manner rather than in a non-Fickian manner, because the non-Fickian term comprising of $(D_{\varepsilon 2} - vD_{\varepsilon 1})\frac{\partial \varepsilon_{22}}{\partial x}$ in Eq. 6.69 is presumably weakly coupled. Consequently, the non-Fickian diffusion coefficient $(D_{\varepsilon 2} - vD_{\varepsilon 1})$ can be characterized as zero for this material and the final governing equation reduces to

$$\frac{\partial m}{\partial t} = \frac{\partial}{\partial x}\left(D_m \frac{\partial m}{\partial x}\right) \tag{6.70}$$

which is essentially Fick's law, with a diffusion coefficient D_m that depends on temperature, moisture concentration, and applied strains as defined in Eq. 6.47a. Assuming $\beta(T) << 1$, we may ignore all the terms beyond first order, while for the transverse strain ε_{22}, we retain only up to quadratic term in the expression with $\varepsilon_{11} = -v\varepsilon_{22}$ under uniaxial tension. Consequently, Eq. 6.47a can be expressed as

$$D_m = D_0(m, T) + D_1(m, T)\varepsilon_{22} + D_2(m, T)\varepsilon_{22}^2 \tag{6.71}$$

Introducing normalized temperature

$$\overline{T} = \frac{T}{T_{REF}} \tag{6.72}$$

where $T_{REF} = 40.5°\text{C}$ and normalized moisture concentration per unit volume,

$$\Delta\overline{m} = \frac{\Delta m}{\rho_{epoxy}} = M(\%)$$

$$M(\%) = \left(\frac{M_k - M_{initial}}{M_{initial}}\right) \times 100 \tag{6.73}$$

where

M_k is the mass of the polymer at kth data point
$M_{initial}$ is the initial mass of the polymer (dry weight)

Because there is a significant variation in the mechanical strain through the thickness of the epoxy layer (see Table 6.6), the average mechanical strain ($\overline{\varepsilon}_{22}$) at the mid-plane of the epoxy layer was used to compute diffusivity. Using separation of variables, Eq. 6.71 can be written as

$$\begin{aligned} D(\Delta\overline{m}, \overline{T}, \overline{\varepsilon}_{22}) &= F_1(\Delta\overline{m})F_2(\overline{T})F_3(\overline{\varepsilon}_{22}) \\ &= (a_{11} + a_{12}\Delta\overline{m} + a_{13}\Delta\overline{m}^2)(a_{21} + a_{22}\overline{T} + a_{23}\overline{T}^2) \\ &\quad (a_{31} + a_{32}\overline{\varepsilon}_{22} + a_{33}\overline{\varepsilon}_{22}^2) \end{aligned} \tag{6.74}$$

Expanding the above equation, we have

$$\begin{aligned} D(\Delta\overline{m}, \overline{T}, \varepsilon_{22}) = {} & (D_1 + D_2\Delta\overline{m} + D_3\overline{T} + D_4\Delta\overline{m}^2 + D_5\overline{T}^2 + D_6\Delta\overline{m}\overline{T} \\ & + D_7\Delta\overline{m}\overline{T}^2 + D_8\Delta\overline{m}^2\overline{T} + D_9\Delta\overline{m}^2\overline{T}^2) \\ & + (D_{10} + D_{11}\Delta\overline{m} + D_{12}\overline{T} + D_{13}\Delta\overline{m}^2 + D_{14}\overline{T}^2 + D_{15}\Delta\overline{m}\hat{T} \\ & + D_{16}\Delta\overline{m}\overline{T}^2 + D_{17}\Delta\overline{m}^2\overline{T} + D_{18}\Delta\overline{m}^2\overline{T}^2)\overline{\varepsilon}_{22} \\ & + (D_{19} + D_{20}\Delta\overline{m} + D_{21}\overline{T} + D_{22}\Delta\overline{m}^2 + D_{23}\overline{T}^2 + D_{24}\Delta\overline{m}\overline{T} \\ & + D_{25}\Delta\overline{m}\overline{T}^2 + D_{26}\Delta\overline{m}^2\overline{T} + D_{27}\Delta\overline{m}^2\overline{T}^2)\overline{\varepsilon}_{22}^2 \end{aligned} \tag{6.75}$$

$$D_M(\Delta\overline{m}, \overline{T}, \varepsilon_{22}) = D_0(\Delta\overline{m}, \overline{T}) + D_1(\Delta\overline{m}, \overline{T})\varepsilon_{22} + D_2(\Delta\overline{m}, \overline{T})\varepsilon_{22}^2$$

where

$$\begin{aligned} D_0(\Delta\overline{m}, \overline{T}) = {} & D_1 + D_2\Delta\overline{m} + D_3\overline{T} + D_4\Delta\overline{m}^2 + D_5\overline{T}^2 + D_6\Delta\overline{m}\overline{T} + D_7\Delta\overline{m}\overline{T}^2 \\ & + D_8\Delta\overline{m}^2\overline{T} + D_9\Delta\overline{m}^2\overline{T}^2 \end{aligned}$$

$$D_1(\Delta\overline{m},\overline{T}) = D_{10} + D_{11}\Delta\overline{m} + D_{12}\overline{T} + D_{13}\Delta\overline{m}^2 + D_{14}\overline{T}^2 + D_{15}\Delta\overline{m}\hat{T} + D_{16}\Delta\overline{m}\overline{T}^2 + D_{17}\Delta\overline{m}^2\overline{T} + D_{18}\Delta\overline{m}^2\overline{T}^2 \tag{6.76}$$

$$D_2(\Delta\overline{m},\overline{T}) = D_{19} + D_{20}\Delta\overline{m} + D_{21}\overline{T} + D_{22}\Delta\overline{m}^2 + D_{23}\overline{T}^2 + D_{24}\Delta\overline{m}\overline{T} + D_{25}\Delta\overline{m}\overline{T}^2 + D_{26}\Delta\overline{m}^2\overline{T} + D_{27}\Delta\overline{m}^2\overline{T}^2$$

There are a total of 27 unknowns which can be computed from the 27 data points for diffusion coefficient listed in Table 6.8 by means of a least-squares technique as described below.

Equation 6.75 can be expressed in the following general form

$$D(\Delta\overline{m},\overline{T},\overline{\varepsilon}_{22}) = \sum_{i=1}^{N} D_i f_i(\Delta\overline{m},\overline{T},\overline{\varepsilon}_{22}) \tag{6.77}$$

where N is the number of unknowns ($N = 27$ in this study), D_i are the unknown coefficients, $f_i(k) \equiv f_i(\Delta m_k, \overline{T}_k, \overline{\varepsilon}_k)$ are the known functions of $\Delta\overline{m}, \overline{T}$, and $\overline{\varepsilon}_{22}$. These 27 known functions are defined in Appendix A. From Eq. 6.77, summation of square error L_E can be formed as

$$L_E = \sum_{k=1}^{M} (\tilde{D}_k - D(\Delta\overline{m}_k, \overline{T}_k, \overline{\varepsilon}_k))^2 = \sum_{k=1}^{M} \left[\tilde{D}_k - \sum_{i=1}^{N} D_i f_i(\Delta\overline{m}_k, \overline{T}_k, \overline{\varepsilon}_k)\right]^2 = \sum_{k=1}^{M} \left[\tilde{D}_k - \sum_{i=1}^{N} D_i f_i(k)\right]^2 \tag{6.78}$$

where M is the total number of test data points ($M = N = 27$ in this study, but in general M and N need not be equal), $f_i(k) \equiv f_i(\Delta\overline{m}_k, \overline{T}_k, \overline{\varepsilon}_k)$, $\tilde{D}_k$ is the value of diffusivity for the kth test, $\hat{D}_k$ is the derived diffusivity from curve fit, and $\Delta\overline{m}_k$, $\overline{T}_k$, and $\overline{\varepsilon}_k$ are environmental humidity, temperature, and the average transverse mechanical strain, respectively, corresponding to the kth test.To minimize the least-square error,

$$\frac{\partial L_E}{\partial D_i} = 0 \quad (i = 1, 2, 3, \ldots, N)$$

Therefore,

$$\begin{cases} D_1 \sum_{k=1}^{M} f_1(k)f_1(k) + D_2 \sum_{k=1}^{M} f_1(k)f_2(k) + D_3 \sum_{k=1}^{M} f_1(k)f_3(k) + \cdots + \\ D_N \sum_{k=1}^{M} f_1(k)f_N(k) = \sum_{k=1}^{M} \widetilde{D}_k f_1(k) \\ D_1 \sum_{k=1}^{M} f_2(k)f_1(k) + D_2 \sum_{k=1}^{M} f_2(k)f_2(k) + D_3 \sum_{k=1}^{M} f_2(k)f_3(k) + \cdots + \\ D_N \sum_{k=1}^{M} f_2(k)f_N(k) = \sum_{k=1}^{M} \widetilde{D}_k f_2(k) \\ \cdots\cdots \\ D_1 \sum_{k=1}^{M} f_N(k)f_1(k) + D_2 \sum_{k=1}^{M} f_N(k)f_2(k) + D_3 \sum_{k=1}^{M} f_N(k)f_3(k) + \cdots + \\ D_N \sum_{k=1}^{M} f_N(k)f_N(k) = \sum_{k=1}^{M} \widetilde{D}_k f_N(k) \end{cases} \tag{6.79}$$

The unknown diffusivity coefficients D_3, D_2, …, D_{27} are obtained by solving Eq. 6.79 simultaneously.

6.3.2.8 Characterization of Nonlinear Moisture Saturation Coefficients

Revisiting Eq. 6.51 and using separation of variables,

$$\begin{aligned} m_{MAX}(RH,\overline{T},\varepsilon_{22}) &= B_0(RH,\overline{T}) + B_1(RH,\overline{T})\varepsilon_{22} + B_2(RH,\overline{T})\varepsilon_{22}^2 \\ &= F_1(RH)F_2(\overline{T})F_3(\varepsilon_{22}) \\ &= (c_{11} + c_{12}RH + c_{13}RH^2)(c_{21} + c_{22}\overline{T} + c_{23}\overline{T}^2) \\ &\quad (c_{31} + c_{32}\varepsilon_{22} + c_{33}\varepsilon_{22}{}^2) \end{aligned} \tag{6.80}$$

Expanding the above equation, we have

$$\begin{aligned} m_{MAX}(RH,\overline{T},\varepsilon_{22}) = {} & (B_1 + B_2RH + B_3\overline{T} + B_4RH^2 + B_5\overline{T}^2 + B_6RH\overline{T} + B_7RH\overline{T}^2 \\ & + B_8RH^2\overline{T} + B_9RH^2\overline{T}^2) \\ & + (B_{10} + B_{11}RH + B_{12}\overline{T} + B_{13}RH^2 + B_{14}\overline{T}^2 + B_{15}RH\overline{T} \\ & + B_{16}RH\overline{T}^2 + B_{17}RH^2\overline{T} + B_{18}RH^2\overline{T}^2)\varepsilon_{22} \\ & + (B_{19} + B_{20}RH + B_{21}\overline{T} + B_{22}RH^2 + B_{23}\overline{T}^2 + B_{24}RH\overline{T} \\ & + B_{25}RH\overline{T}^2 + B_{26}RH^2\overline{T} + B_{27}RH^2\overline{T}^2)\varepsilon_{22}^2 \end{aligned} \tag{6.81}$$

where

$$
\begin{aligned}
B_0(RH,\overline{T}) &= B_1 + B_2RH + B_3\overline{T} + B_4RH^2 + B_5\overline{T}^2 + B_6RH\overline{T} + B_7RH\overline{T}^2 \\
&\quad + B_8RH^2\overline{T} + B_9RH^2\overline{T}^2 \\
B_1(RH,\overline{T}) &= B_{10} + B_{11}RH + B_{12}\overline{T} + B_{13}RH^2 + B_{14}\overline{T}^2 + B_{15}RH\overline{T} \\
&\quad + B_{16}RH\overline{T}^2 + B_{17}RH^2\overline{T} + B_{18}RH^2\overline{T}^2 \\
B_2(RH,\overline{T}) &= B_{19} + B_{20}RH + B_{21}\overline{T} + B_{22}RH^2 + B_{23}\overline{T}^2 + B_{24}RH\overline{T} \\
&\quad + B_{25}RH\overline{T}^2 + B_{26}RH^2\overline{T} + B_{27}RH^2\overline{T}^2
\end{aligned}
\tag{6.82}
$$

There are again a total of 27 unknowns which are computed from the 27 test diffusion coefficient data by curve-fit technique as described below.

Equation 6.81 can be expressed in the following general form

$$
m_{MAX}(RH,\overline{T},\varepsilon_{22}) = \sum_{i=1}^{27} B_i f_i(RH,\overline{T},\varepsilon_{22}) \tag{6.83}
$$

where B_i are the unknown coefficients, $f_i(k) \equiv f_i(RH_k, \overline{T}_k, \varepsilon_k)$ $(i = 1, 2, 3, \ldots, 27)$ are known functions of RH, $\overline{T}$, ε_{22}, as discussed in the previous section. These 27 known functions are defined in Appendix B.

From Eq. 6.83, summation of squares of error L_E can be formed as

$$
\begin{aligned}
L_E &= \sum_{k=1}^{N} (M_k - m_{MAX})^2 = \sum_{k=1}^{N} \left[M_k - \sum_{i=1}^{27} B_i f_i(RH_k, \overline{T}_k, \varepsilon_k) \right]^2 \\
&= \sum_{k=1}^{N} \left[M_k - \sum_{i=1}^{27} \hat{B}_i f_i(k) \right]^2
\end{aligned}
\tag{6.84}
$$

where $N = 27$ is the number of test data points, $f_i(k) \equiv f_i(RH_k, \overline{T}_k, \varepsilon_k)$, M_k the saturation concentration for the kth test data point, and RH_k, $\overline{T}_k$, and ε_k are environmental humidity, temperature, and transverse strain, respectively, corresponding to the kth test.

To minimize the least-squares error,

$$
\frac{\partial L_E}{\partial B_i} = 0 \quad (i = 1, 2, 3, \ldots, 27)
$$

Therefore,

$$
\begin{aligned}
&B_1\sum_{k=1}^{N} f_1(k)f_1(k) + B_2\sum_{k=1}^{N} f_1(k)f_2(k) + B_3\sum_{k=1}^{N} f_1(k)f_3(k) + \cdots \\
&\quad + B_N\sum_{k=1}^{N} f_1(k)f_N(k) = \sum_{k=1}^{N} M_k f_1(k)
\end{aligned}
$$

$$
\begin{aligned}
&B_1\sum_{k=1}^{N} f_2(k)f_1(k) + B_2\sum_{k=1}^{N} f_2(k)f_2(k) + B_3\sum_{k=1}^{N} f_2(k)f_3(k) + \cdots \\
&\quad + B_N\sum_{k=1}^{N} f_2(k)f_N(k) = \sum_{k=1}^{N} M_k f_2(k) \qquad (6.85)
\end{aligned}
$$

$$\cdots$$

$$
\begin{aligned}
&B_1\sum_{k=1}^{N} f_N(k)f_1(k) + B_2\sum_{k=1}^{N} f_N(k)f_2(k) + B_3\sum_{k=1}^{N} f_N(k)f_3(k) + \cdots \\
&\quad + B_N\sum_{k=1}^{N} f_N(k)f_N(k) = \sum_{k=1}^{N} M_k f_N(k)
\end{aligned}
$$

The unknown saturation coefficients B_1, B_2, …, B_{27} are obtained by solving Eq. 6.85 simultaneously.

6.3.3 Results and Discussions

The diffusivity and mass saturation for various conditions of temperature, humidity, and strain are tabulated in Table 6.8. As is evident from the table, the influence of temperature, relative humidity, and strain on the diffusivity values is very significant. In the case of maximum saturation, although there is significant influence of the above parameters, the effect of strain is clearly evident only for total strains greater than 5%. The mass saturation values of both 0% and 5% strain for a given test condition of temperature and humidity remain approximately constant. As presented in Figs. 6.9–6.17, the moisture uptake data indicates that the moisture diffusion in this material occurs in a nonlinear Fickian manner rather than non-Fickian. This nonlinearity is due to the presence of strain and environmental conditions of elevated temperature and relative humidity.

The diffusivity (D_0, D_1, D_2) and moisture coefficients (B_0, B_1, B_2) are tabulated in Tables 6.9 and 6.10, respectively. The behavior of diffusivity and mass saturation under various conditions of temperature and relative humidity, with respect to transverse strain, is shown in Figs. 6.18 and 6.19, respectively.

Table 6.9 Diffusivity coefficients D_0, D_1, D_2

	Environmental relative humidity (%)								
	75%			85%			95%		
Temp. (°C)	D_0	D_1	D_2	D_0	D_1	D_2	D_0	D_1	D_2
32.2	0.5822	0.0690	3.860	0.7772	1.6840	9.16	1.2088	2.424	95.08
40.5	1.018	0.730	2.60	1.265	0.760	33.60	1.878	1.530	47.80
48.9	1.152	0.370	12.20	1.744	1.380	58.80	2.270	0.290	70.20

Table 6.10 Moisture Saturation coefficients B_0, B_1, B_2

	Environmental relative humidity (%)								
	75%			85%			95%		
Temp. (°C)	B_0	B_1	B_2	B_0	B_1	B_2	B_0	B_1	B_2
32.2	2.9486	7.187	155.23	3.3402	8.215	121.09	3.9786	5.939	114.50
40.5	3.030	9.120	169.63	3.475	7.638	121.66	4.040	6.261	117.37
48.9	3.015	7.154	148.28	3.460	5.797	107.71	4.024	5.852	113.26

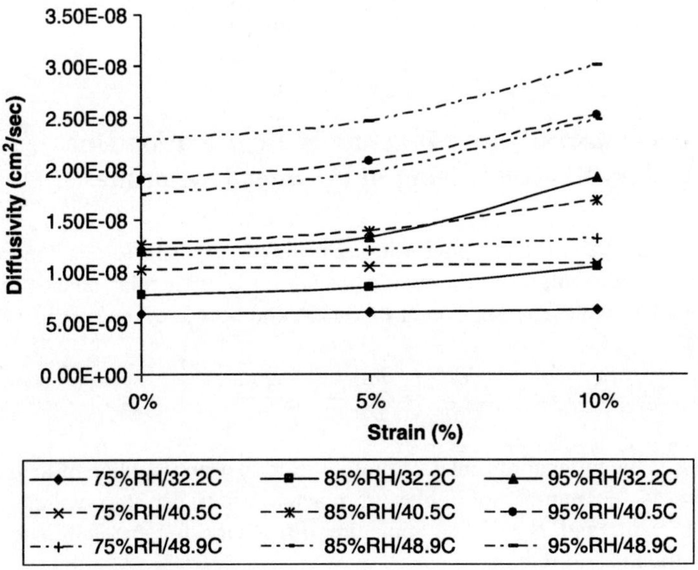

Fig. 6.18 Influence of total strain on diffusivity

From the figures and the tables, it is evident that strain has a significant influence over diffusivity and moisture concentration coefficients, whereas the effect of temperature on diffusion coefficients and relative humidity on moisture concentration coefficients, respectively, is not very significant.

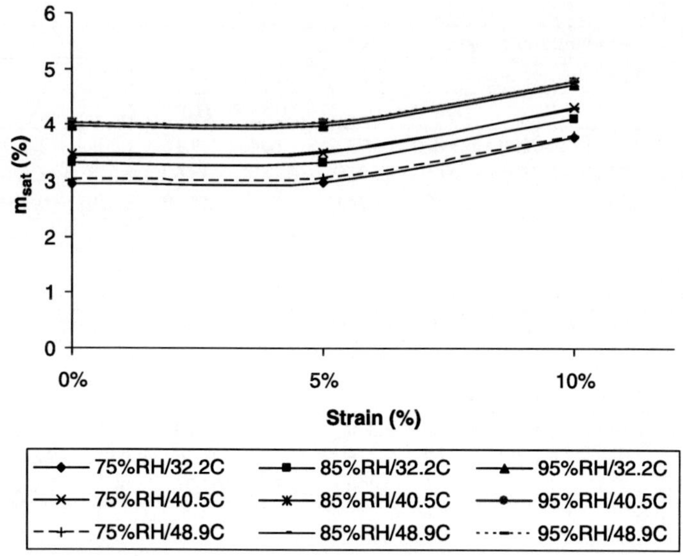

Fig. 6.19 Influence of total strain on saturation concentration

6.4 Mechanism-Based Modeling of Degradation Due to Hygrothermal Aging in Polymer Composites

Assuming that one-dimensional (through-thickness) diffusion is the primary mode of moisture ingress in a composite laminate, molecular concentration of water in a laminate as a function of thickness and time is given by [30]

$$\frac{\partial m_1}{\partial t} + \frac{\partial m_2}{\partial t} = D_m \frac{\partial^2 m_1}{\partial x^2} \tag{6.86}$$

where m_1 is the number of moles of mobile water per unit volume of the polymer resin, m_2 is the number of moles of reacted water per unit volume of the polymer resin, and D_m is the moisture diffusivity for the laminate in the through-thickness direction.

It is assumed that a relationship exists between the reacted and the mobile moles of water, given by

$$m_2 = r m_1 \tag{6.87}$$

where the moisture degradation parameter, r, is assumed to be a material constant.

6.4.1 Calculation of Moisture Degradation Parameter 'r'

Assuming $m_{1\infty}$ to be the saturation mass of water uptake at the end of the sorption phase, and $m_{2\infty}$ to be the mass of reacted water remaining in the laminate at the end of desorption phase, from moisture absorption and desorption experiments on a unidirectional carbon/epoxy laminate ($V_f = 0.5$), we get

$$m_{1\infty} + m_{2\infty} = 1.33403\%$$

$$m_{2\infty} = 0.15760\%$$

$$\frac{m_{2\infty}}{m_{1\infty} + m_{2\infty}} = \frac{0.1576}{1.33403} = 0.11814 = \gamma$$

But from $m_{2\infty} = r(m_{1\infty})$ it can be shown that,

$$\frac{r}{(1+r)} = \gamma$$

Hence, for a carbon/epoxy laminate,

$$r = 0.13396$$

6.4.2 Derivation of Internal State Variable (ISV) for Moisture-Induced Degradation

Defining,

$$D'_m = \frac{D_m}{(1+r)}$$

$$m_{1\infty} = \frac{M_\infty \rho_{polymer}}{(1+r)m_w}$$

where $\rho_{polymer}$ is the density of the polymer resin. The solution to Eq. 6.86 for total moisture uptake in terms of moles per unit surface area of the laminate is given by

For

$$t \leq \frac{0.05d^2}{D_m{}'}$$

$$M_2(t) = \frac{4rm_{1\infty}}{d}\sqrt{\frac{D_m{}'t}{\pi}}$$

For

$$t > \frac{0.05d^2}{D_m{}'}$$

$$M_2(t) = rm_{1\infty}\left[1 - \frac{8}{\pi^2}e^{-\left[\frac{D_m{}'\pi^2 t}{d^2}\right]}\right]$$

where d is the transverse thickness of the laminate.

If $M_2(t)$ moles of water reacted with the polymer at a given time t, then there are $M_2(t)N_A$ bonds broken, where N_A is Avogadro's number. The probability of a chain segment being cut Y times is given by [30]

$$P\{M_2(t)N_A, Y\} = \left(\frac{1}{N_A N_0}\right)^Y \left(1 - \frac{1}{N_A N_0}\right)^{(M_2(t)N_A - Y)} \left(\frac{(M_2(t)N_A)!}{Y!(M_2(t)N_A - Y)!}\right) \quad (6.88)$$

Since $\frac{1}{N_A N_0}$ is very small compared with $M_2(t)N_A$, Eq. 6.88 can be written as

$$P\{M_2(t)N_A, Y\} = \frac{\left(\frac{M_2(t)}{N_0}\right)^Y \exp\left(-\frac{M_2(t)}{N_0}\right)}{Y!}$$

If $P(0)$ is the probability of no bond breakage, then the probability of at least one bond breakage is given by

$$P(Y \geq 1) = 1 - P(0)$$

$$P(Y \geq 1) = 1 - \exp\left(-\frac{M_2(t)}{N_0}\right)$$

Therefore, the number of moles of intercrosslink chains having been cut per unit volume at time t is

$$R = N_0\left[1 - \exp\left(-\frac{M_2(t)}{N_0}\right)\right] \quad (6.89)$$

Hence, the remaining number of moles of intercrosslink chains will be

$$N(t) = N_0 - ZR \quad (6.90)$$

where Z is constant and has a value of $1 < Z < 3$, depending on the crosslink morphology.

Substituting Eq. 6.89 in Eq. 6.90,

$$N(t) = N_0\left[Z\exp\left(-\frac{M_2(t)}{N_0}\right) - (Z-1)\right] \tag{6.91}$$

$$S_H(t) = \frac{N(t)}{N_0} = Ze^{-\left(\frac{M_2(t)}{N_0}\right)} - (Z-1) \tag{6.92}$$

where

N_0 is the initial number of network crosslinks per mole at time, $t = 0$
$N(t)$ is the number of remaining network crosslinks per mole at any time, t
Z is a parameter determining the type of crosslink ($1 < Z < 3$)
$S_H(t)$ is a mechanism-based internal state variable (ISV) quantifying change in crosslink density due to hygrothermal aging

6.4.3 Strength Degradation Modeling

For the case of hygrothermal degradation in a polymer composite, it may be conjectured that, with time, the number of intercrosslink chains is reduced as given by Eq. 6.91, but the strength of each of the remaining bonds remains relatively unchanged with time. Therefore, defining the individual bond strength as λ, and calculating the critical load carrying capacity at failure initiation in a mole of the polymer resin before aging (i.e., at time $t = 0$) gives,

$$P_{cr}(0) = \lambda N_0 \tag{6.93}$$

Similarly, the critical load carrying capacity at failure initiation in a mole of the polymer after aging (i.e., at time t) is

$$P_{cr}(t) = \lambda N(t) \tag{6.94}$$

Combining (93) and (94), gives,

$$\frac{P(t)}{P(0)} = \frac{N(t)}{N_0} = S_H(t) \tag{6.95}$$

where $S_H(t)$ is an internal state variable (ISV) defined earlier.

6.4.3.1 Case Study: Delamination Failure at the Interface Between Adjacent Lamina in a Unidirectional Carbon/Epoxy Laminate

In a short-beam shear test, delamination typically initiates at an interlaminar interface close to the loading point, and then propagates outward to the free

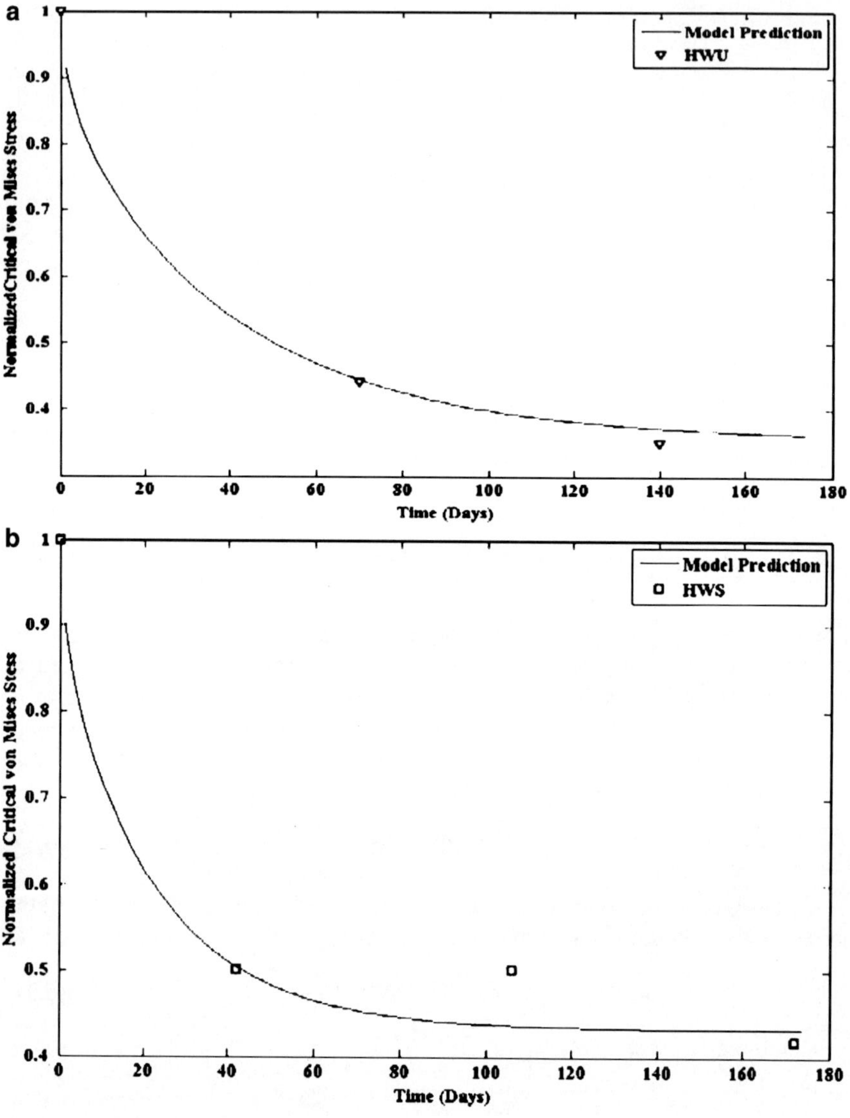

Fig. 6.20 Comparison of model prediction of degradation in ILSS in carbon/epoxy composite with test results for (**a**) HWU and (**b**) HWS

edge of the test specimen. In this case, failure initiation in the interlaminar region is dictated by the transverse shear stress at the interfaces in a laminate.

Extending the strength degradation concept of the previous section to the critical interlaminar transverse shear stress in a composite laminate,

$$\sigma_{VM}(t) = S_H(t)\sigma_{VM0} \tag{6.96}$$

where $\sigma_{VM}(t)$ is the critical von Mises stress at time t after environmental aging, σ_{VM0} is the critical von Mises stress before environmental degradation.

6.4.3.2 Comparison of Model Prediction with Test Data for Interlaminar Shear Strength

For model verification, hygrothermal aging was carried out on unidirectional carbon/epoxy laminates for two separate conditions: hot, wet, unstressed (HWU), and hot, wet, stressed (HWS). In both cases, test panels were immersed in a bath of deionized water in an environmental chamber and held at a uniform temperature of 70°C for up to 180 days. For the HWS case, during hygrothermal aging, an applied strain equivalent to 5% ultimate tensile strength (UTS) of the epoxy resin was applied (transverse to the fibers) to the test panel using a unique constant-strain, aging test fixture described elsewhere [31]. Subsequently, the hygrothermally aged laminates were removed from the environmental chamber, machined into 1 in. × 0.25 in. × 0.2 in. test specimens for short-beam shear testing using ASTM D2344-84 standard test protocol, and tested for interlaminar shear strength at aging time intervals of approximately 60 days.

The experimentally observed degradation in interlaminar shear strength (ILSS) in carbon/epoxy laminate is plotted as a function of time in Fig. 6.20, and compared with the degradation in ILSS predicted by the proposed mechanism-based ISV model. Excellent agreement between model prediction and test data is observed for both the HWU and the HWS cases. The synergistic effect of applied strain on the initial rate of degradation of ILSS for HWU and HWS can be readily observed by comparing Figs. 6.20a, b. This difference in the rate of degradation is captured by the proposed model through the definition of moisture diffusivity as a linear function of applied strain (refer to Eq. 6.71), thereby allowing accelerated moisture diffusion due to elevated strain levels in the polymer resin. The effect of temperature on ILSS was found to be negligible for this case, and hence was not included in the model.

6.5 Conclusions

A strain-assisted diffusion model was developed and experiments were conducted to characterize the material coefficients. It was shown that both the diffusivity and mass saturation are dependent on strain. The present experiment indicates that the presence of strain in an epoxy-based polymer influences the diffusion of water molecules. It is evident from the results that there is a consistent increase in diffusivity with strain whereas the mass saturation is affected only beyond 5% strain level. The mass saturation for 0% and 5% strain level for a given environmental temperature and relative humidity remains approximately the same. This may be due to the reason that the fractional free volume is not affected till 5% strain, for this particular material. As polymers tend to exhibit more non-Fickian behavior at high strain, temperature, and humidity, the experimental data (refer Fig. 6.19) for high strain of 10%, temperature of 48.9°C, and relative humidity of 95% concludes that the material behaves in nonlinear Fickian diffusion rather than non-Fickian due to the weakly coupled non-Fickian term. This proves that in general, except for very few anomalies, the material behaves in a nonlinear Fickian manner. If the material were to behave in a fully non-Fickian manner, then the whole set of experiments would have to be repeated with very small strain, say 0.5%, so that the strain gradient is negligible and $D_{\varepsilon 2}$ is characterized by numerical iterative procedure.

As a case study, the experimentally observed degradation in interlaminar shear strength (ILSS) in carbon/epoxy laminate was compared with the degradation in ILSS predicted by the proposed mechanism-based ISV model. Excellent agreement was observed for both the HWU and the HWS cases. The synergistic effect of applied strain on the initial rate of degradation of ILSS for HWU and HWS was captured in the model by defining moisture diffusivity as a linear function of applied strain, thereby allowing accelerated diffusion due to elevated strain levels in the polymer resin.

Acknowledgment The author would like to acknowledge the support of the National Science Foundation, Grant Number CMS-0296167. The author would also like to thank Mr. Avinash Reddy Akepati for his help in preparing the manuscript.

Appendix A

$$f_1(\Delta\overline{m}, \overline{T}, \varepsilon_{22}) = 1$$

$$f_2(\Delta\overline{m}, \overline{T}, \varepsilon_{22}) = \Delta\overline{m}$$

$$f_3(\Delta\overline{m}, \overline{T}, \varepsilon_{22}) = \overline{T}$$

$$f_4(\Delta\overline{m}, \overline{T}, \varepsilon_{22}) = \Delta\overline{m}^2$$

$$f_5(\Delta\overline{m}, \overline{T}, \varepsilon_{22}) = \overline{T}^2$$

$$f_6(\Delta\overline{m}, \overline{T}, \varepsilon_{22}) = \Delta\overline{m}\overline{T}$$

$$f_7(\Delta\overline{m}, \overline{T}, \varepsilon_{22}) = \Delta\overline{m}\overline{T}^2$$

$$f_8(\Delta\overline{m}, \overline{T}, \varepsilon_{22}) = \Delta\overline{m}^2\overline{T}$$

$$f_9(\Delta\overline{m}, \overline{T}, \varepsilon_{22}) = \Delta\overline{m}^2\overline{T}^2$$

$$f_{10}(\Delta\overline{m}, \overline{T}, \varepsilon_{22}) = \varepsilon_{22}$$

$$f_{11}(\Delta\overline{m}, \overline{T}, \varepsilon_{22}) = \Delta\overline{m}\varepsilon_{22}$$

$$f_{12}(\Delta\overline{m}, \overline{T}, \varepsilon_{22}) = \overline{T}\varepsilon_{22}$$

$$f_{13}(\Delta\overline{m}, \overline{T}, \varepsilon_{22}) = \Delta\overline{m}^2\varepsilon_{22}$$

$$f_{14}(\Delta\overline{m}, \overline{T}, \varepsilon_{22}) = \overline{T}^2\varepsilon_{22}$$

$$f_{15}(\Delta\overline{m}, \overline{T}, \varepsilon_{22}) = \Delta\overline{m}\overline{T}\varepsilon_{22}$$

$$f_{16}(\Delta\overline{m}, \overline{T}, \varepsilon_{22}) = \Delta\overline{m}\overline{T}^2\varepsilon_{22}$$

$$f_{17}(\Delta\overline{m}, \overline{T}, \varepsilon_{22}) = \Delta\overline{m}^2\overline{T}\varepsilon_{22}$$

$$f_{18}(\Delta\overline{m}, \overline{T}, \varepsilon_{22}) = \Delta\overline{m}^2\overline{T}^2\varepsilon_{22}$$

$$f_{19}(\Delta\overline{m}, \overline{T}, \varepsilon_{22}) = {\varepsilon_{22}}^2$$

$$f_{20}(\Delta\overline{m}, \overline{T}, \varepsilon_{22}) = \Delta\overline{m}{\varepsilon_{22}}^2$$

$$f_{21}(\Delta\overline{m}, \overline{T}, \varepsilon_{22}) = \overline{T}{\varepsilon_{22}}^2$$

$$f_{22}(\Delta\overline{m}, \overline{T}, \varepsilon_{22}) = \Delta\overline{m}^2{\varepsilon_{22}}^2$$

$$f_{23}(\Delta\overline{m}, \overline{T}, \varepsilon_{22}) = \overline{T}^2{\varepsilon_{22}}^2$$

$$f_{24}(\Delta\overline{m}, \overline{T}, \varepsilon_{22}) = \Delta\overline{m}\overline{T}{\varepsilon_{22}}^2$$

$$f_{25}(\Delta\overline{m}, \overline{T}, \varepsilon_{22}) = \Delta\overline{m}\overline{T}^2{\varepsilon_{22}}^2$$

$$f_{26}(\Delta\overline{m}, \overline{T}, \varepsilon_{22}) = \Delta\overline{m}^2\overline{T}{\varepsilon_{22}}^2$$

$$f_{27}(\Delta\overline{m}, \overline{T}, \varepsilon_{22}) = \Delta\overline{m}^2\overline{T}^2\varepsilon_{22}^2$$

Appendix B

$$f_1(RH,\overline{T},\varepsilon_{22}) = 1$$

$$f_2(RH,\overline{T},\varepsilon_{22}) = RH$$

$$f_3(RH,\overline{T},\varepsilon_{22}) = \overline{T}$$

$$f_4(RH,\overline{T},\varepsilon_{22}) = RH^2$$

$$f_5(RH,\overline{T},\varepsilon_{22}) = \overline{T}^2$$

$$f_6(RH,\overline{T},\varepsilon_{22}) = RH\overline{T}$$

$$f_7(RH,\overline{T},\varepsilon_{22}) = RH\overline{T}^2$$

$$f_8(RH,\overline{T},\varepsilon_{22}) = RH^2\overline{T}$$

$$f_9(RH,\overline{T},\varepsilon_{22}) = RH^2\overline{T}^2$$

$$f_{10}(RH,\overline{T},\varepsilon_{22}) = \varepsilon_{22}$$

$$f_{11}(RH,\overline{T},\varepsilon_{22}) = RH\varepsilon_{22}$$

$$f_{12}(RH,\overline{T},\varepsilon_{22}) = \overline{T}\varepsilon_{22}$$

$$f_{13}(RH,\overline{T},\varepsilon_{22}) = RH^2\varepsilon_{22}$$

$$f_{14}(RH,\overline{T},\varepsilon_{22}) = \overline{T}^2\varepsilon_{22}$$

$$f_{15}(RH,\overline{T},\varepsilon_{22}) = RH\overline{T}\varepsilon_{22}$$

$$f_{16}(RH,\overline{T},\varepsilon_{22}) = RH\overline{T}^2\varepsilon_{22}$$

$$f_{17}(RH,\overline{T},\varepsilon_{22}) = RH^2\overline{T}\varepsilon_{22}$$

$$f_{18}(RH,\overline{T},\varepsilon_{22}) = RH^2\overline{T}^2\varepsilon_{22}$$

$$f_{19}(RH,\overline{T},\varepsilon_{22}) = \varepsilon_{22}^2$$

$$f_{20}(RH,\overline{T},\varepsilon_{22}) = RH\varepsilon_{22}^2$$

$$f_{21}(RH,\overline{T},\varepsilon_{22}) = \overline{T}\varepsilon_{22}^2$$

$$f_{22}(RH,\overline{T},\varepsilon_{22}) = RH^2\varepsilon_{22}^2$$

$$f_{23}(RH,\overline{T},\varepsilon_{22}) = \overline{T}^2\varepsilon_{22}^2$$

$$f_{24}(RH,\overline{T},\varepsilon_{22}) = RH\overline{T}\varepsilon_{22}^2$$

$$f_{25}(RH,\overline{T},\varepsilon_{22}) = RH\overline{T}^2\varepsilon_{22}^2$$

$$f_{26}(RH,\overline{T},\varepsilon_{22}) = RH^2\overline{T}\varepsilon_{22}^2$$

$$f_{27}(RH,\overline{T},\varepsilon_{22}) = RH^2\overline{T}^2\varepsilon_{22}^2$$

References

1. Shen, C. H. and G.S. Springer (1981) "Effects of Moisture and Temperature on the Tensile Strength of Composite Materials," *Environmental Effects on Composite Materials*, G. S. Springer, ed., Lancaster, PA: Technomic Publishing Co., Inc., pp. 79–93.
2. Gurtin, M.E., and Yatomi, C. (1979) "On a model for Two Phase Diffusion in Composite Materials," *Journal of Composite Materials*, Vol. 13, pp. 126–130.
3. Carter, H.G., and Kibler, K.G. (1978) "Langmuir-Type Model for Anomalous Diffusion in Composite Resins," *Journal of Composite Materials*, Vol. 12, pp. 118–130.
4. Shirrell, C. D., Leisler, W. H. and Sandow, F. A. (1979) "Moisture-Induced Surface Damage in T300/5208 Graphite/Epoxy Laminates," *Nondestructive Evaluation and Flaw Criticality for Composite Materials*, ASTM STP 696, R. B. Pipes, ed., American Society for Testing and Materials, pp. 209–222.
5. Weitsman, Y. (1991) "Moisture in Composites: Sorption and Damage,", *Fatigue of Composite Materials*, K. L. Reifsnider, ed., Elsevier Science Publishers B.V., pp. 385–429.
6. Frisch, H.J. (1966) "Irreversible Thermodynamics of Internally relaxing Systems in the Vicinity of the Glass Transition," In "*Non-Equilibrium Thermodynamics, Variational Techniques, and Stability*", Edited by R.J. Dennelly, R. Herman, and I. Prigogine. University of Chicago Press, pp. 277–280.
7. Crank, J. (1975) *The Mathematics of Diffusion*, Oxford University Press.
8. Weitsman, Y. (1987) "Stress Assisted Diffusion in Elastic and Viscoelastic Materials," *Journal of Mechanics and Physics of Solids*, 35(1):73–93.
9. Biot, M.A. (1956) "Thermoelasticity and Irreversible Thermodynamics," *Journal of Applied Physics*, 27(3): 240–253.
10. Schapery, R.A. (1969) "Further Development of a Thermodynamic Constitutive Theory: Stress Formulation," *A&S Report* No. 69–2, Purdue University, West Lafayette.
11. Weitsman, Y. (1990) "A Continuum Diffusion Model for Viscoelastic Materials," *Journal of Physical Chemistry*, 94(2): 961–968.
12. Weitsman, Y. (1987) "Coupled Damage and Moisture Transport in Fiber- Reinforced, Polymeric Composites," *International Journal of Solids and Structures*, 23(7):1003–1025.

13. Kumar, Bhavesh G., Singh, Raman P., and Nakamura, Toshio. (2002) “Degradation of Carbon Fiber-reinforced Epoxy Composites by Ultraviolet Radiation and Condensation”, *Journal of Composite Materials,* Vol. 36, No. 24.
14. Weitsman,, Y.J. and Elahi, M., (2000) *Mechanics of Time dependent Materials,* 3, 107–126.
15. Roy S., and Xu W., (2001). "Modeling of Diffusion in the Presence of Damage in Polymer Matrix Composites," *International Journal of Solids and Structures,* 38, pp. 115–125.
16. Talreja, R., (1994) “Damage Characterization by Internal Variables,” *Damage Mechanics of Composite Materials,* Edited by R. Talreja, Elsevier Science, pp. 53–78.
17. Adkins, J.E., (1959) “Symmetry Relations for Orthotropic and Transversely Isotropic Materials,” Arch. Rational Mech. Anal., Vol. 4, pp. 193–213.
18. Roy, S. and Bandorawalla T., (1999) " Modeling of Diffusion in a Micro-cracked Composite Laminate using Approximate Solutions,” *Journal of Composite Materials,* Vol. 33, No.10, pp. 872–905.
19. Roy, S., Lefebvre, D. R., Dillard, D. A. and Reddy, J. N. (1989) "A Model for the Diffusion of Moisture in Adhesive Joints. Part III: Numerical Simulations," *Journal of Adhesion,* Vol. 27, pp. 41–62.
20. Sancaktar, E., and Baechtle, D., (1993)” The Effect of Stress Whitening on Moisture Diffusion in Thermosetting Polymers,” *J. Adhesion,* Vol. 42, pp. 65–85.
21. Wong, T., and Broutman, L., (1985a.) “Moisture Diffusion in Epoxy Resins Part I: Non-Fickian Sorption Processes,” *Polymer Eng. & Sci.,* Vol. 25, No. 9, pp. 521–528.
22. Wong, T., and Broutman, L., (1985b.) “Water in Epoxy Resins Part II: Diffusion Mechanisms,” *Polymer Eng. & Sci.,* Vol. 25, No. 9, pp. 529–534.
23. Roy, S., (1999) “Modeling of Anomalous Diffusion in Polymer Composites: A Finite Element Approach,” *Journal of Composite Materials,* Vol. 33, No. 14, pp. 1318–1343.
24. Fahmy, A. A., and Hurt, J. C. (1980) “Stress dependence of water diffusion in epoxy resin,” *Polymer Composites,* 1, pp. 77–80.
25. Roy, S., W. Xu, S.J. Park, and K.M. Liechti (2000) "Anomalous Moisture Diffusion in Viscoelastic Polymers: Modeling and Testing”, *Journal of Applied Mechanics,* Vol. 67, pp. 391–396.
26. Roy, S., W. Xu, S. Patel, and S. Case, (2001) "Modeling of Moisture Diffusion in the Presence of Bi-axial Damage in Polymer Matrix Composite Laminates,” *International Journal of Solids and Structures,* Vol. 38, issue 42–43, pp. 7627 – 7641.
27. Roy S. and Shiue, F.W., (2003) “A Coupled Hygrothermal Cohesive-Layer Constitutive Model for Simulating Debond Growth”, *Polymer and Polymer Composites,* Vol. 11, No. 8, pp. 633–647.
28. Gere, J. and Timoshenko, S. (1984) “Mechanics of Materials,” *PWS-Kent Publishing Company.*
29. Tsai, C. Cheng, M., Hwang, S., Tsai, Y., (2004) “Characterizing the Hygric Behavior of Composites by Suspending Method,” *Journal of Composite Materials,* Vol. 38, No. 9.
30. Xiao, G.Z., and M.E.R. Shanahan, (1997) "Water Absorption and Desorption in an Epoxy Resin with Degradation," *J. Polym. Sci. B: Polym. Phys.,* 35, pp. 2659–2670.
31. Goruganthu, S., Jason Elwell, Arun Ramasetty, Abilash R. Nair, Samit Roy, Anwarul Haque, Piyush K. Dutta and Ashok Kumar, (2008) “Characterization and Modeling of the Effect of Environmental Degradation on Interlaminar Shear Strength of Carbon/Epoxy Composites,” *Polymer and Polymer Composites,* Vol. 16, No. 3, pp. 165–179.

Chapter 7
Physical Aging in Glasses and Composites

Gregory B. McKenna

Abstract Physical aging is observed in all glassy materials because of the fact that they are out of equilibrium. The ways in which aging manifests itself are the results of the thermal history of the materials, the environment, and even the constraint of, e.g., fibers or particles. In the present chapter, the fundamentals of aging of glasses are summarized by considering first structural recovery, which is the kinetics of the thermodynamic-type variables such as volume or enthalpy, and its impact on the mechanical response, which is the physical aging. Linear viscoelastic and nonlinear viscoelastic properties as well as yield behaviors will be considered. Furthermore, we will consider environmental effects on physical aging behaviors. The work will end with a perspective on aging in composites and where further research is needed.

7.1 Introduction

A unique feature of glasses, be they polymeric, organic, inorganic, or metallic, is that they are used in a nonequilibrium condition [1–20]. In the case of these materials being used in "high performance" applications, this fact leads to their properties constantly evolving because they are generally used, then, at relatively high fractions of their glass transition temperatures T_g. In the case of polymeric materials, this is especially evident in even relatively benign conditions because these materials have T_g values that materials scientists work very hard to raise above 300°C (573 K). A simple example is in a relatively high-performance thermoplastic resin, polycarbonate. This material has a T_g value reported at between

G.B. McKenna (✉)
Department of Chemical Engineering, Texas Tech University,
Lubbock, TX 79409-3121, USA

Procédés et Ingénierie en Mécanique et Matériaux (UMR CNRS 8006),
Arts et Métiers ParisTech, 151 Boulevard de l'Hôpital, 75013 Paris, France
e-mail: greg.mckenna@ttu.edu

K.V. Pochiraju et al. (eds.), *Long-Term Durability of Polymeric Matrix Composites*,
DOI 10.1007/978-1-4419-9308-3_7,

140°C and nearly 150°C (413 and 423 K) [21, 22]. Even at ambient conditions of 25°C (298 K), this leads to the material being used at $T > 0.70\ T_g$. Hence, a full understanding of the nature of the glass transition and the surrounding kinetics is essential to understanding how these materials will perform in general applications. When glassy polymers are used as the matrices in composite materials, the issues surrounding the glassy behavior of the polymers is highly important in determining the ultimate performance of the composite material essentially because the polymer matrix is frequently the "weakest link" in the composite [23–28]. In addition, once a polymeric material is reinforced with fibers or fillers, the interactions of the reinforcement with the matrix can alter the behavior of the resin, and this can have consequences for the expected behavior of the composite material. The purpose of the present article is to provide a fundamental knowledge of the physical aging process and how it applies to glasses in general and, then, in a more specific manner to the long-term performance of composite materials that are based on polymeric matrixes.

The chapter is organized as follows. First, a thorough review of structural recovery and physical aging of unreinforced materials is provided. This includes an overview of the phenomenology of the glass transition event to include kinetics of structural recovery, physical aging, and how these impact engineering properties. Some time is also spent on the influence of environmental factors such as water on the performance of the resins. A discussion is then provided for the behavior of materials confined at the nanoscale as these effects are directly relevant to composite materials, and especially the growing field of nanocomposites. Emphasis is made on purely physical processes and chemical processes are assumed to be negligible. We then turn to the behavior of composite materials and how one expects their properties to evolve with time based on the fundamentals of the response of the glassy matrix and we then examine the experimental state of knowledge as it applies to composite materials. We conclude with a section on perspectives for the future use of polymeric resin-based composite materials as well as the potential of present models to predict long-term behavior or durability.

7.2 The Glass Transition and Structural Recovery

7.2.1 General Phenomenology

The fundamental nature of the glassy material is that it conserves a state of liquid-like disorder but with solid-like properties. This is the result of the vitrification process occurring from either a supercooled liquid state in materials that can crystallize but do not in the specific conditions of interest or from an equilibrium liquid state in the case of materials that cannot crystallize such as is the case for many polymers. There are many books and reviews that discuss this process and the reader is referred to them for detail [1–20]. Essentially, however, as the liquid is

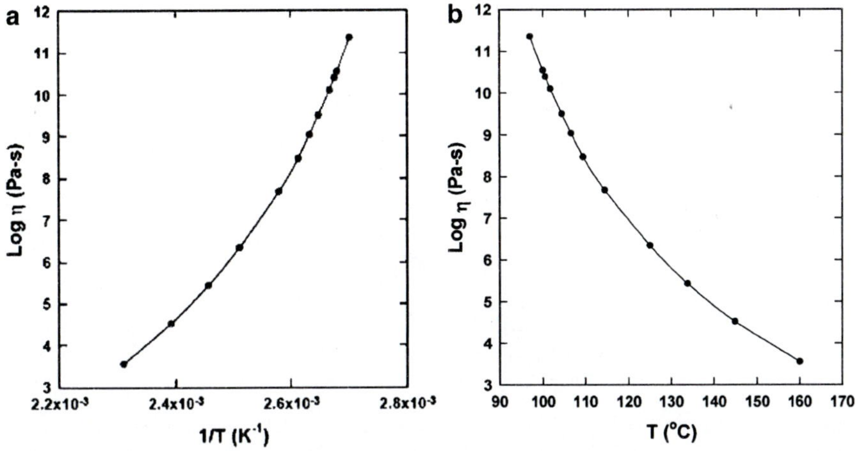

Fig. 7.1 Logarithm of viscosity for a narrow molecular weight distribution polystyrene material having a molecular weight of 46,900 g/mol. versus (**a**) $1/T$ or (**b**) T. *Lines* represent VFT/WLF fits to the data (Data replotted from Plazek [29])

cooled, the molecular mobility decreases in a way that is termed super-Arrhenius in either relaxation time or viscosity. Figure 7.1a, b show [29] how the logarithm of the viscosity of a typical polymer changes as a function of T and $1/T$ where the strong curvature of the response represents the non-Arrhenius nature of the behavior that is often characterized by either a Vogel–Fulcher–Tammann [30–32] (VFT) equation

$$\eta = A_0 e^{\frac{B}{(T-T_\infty)}} \tag{7.1}$$

where η is the viscosity, A_0 is a material constant (prefactor), B is a material parameter akin to an activation energy, T is temperature, and T_∞ is the so-called Vogel temperature or VFT temperature and is the point at which the viscosity apparently goes to infinity.

In the polymer community, it is more common to use the Williams–Landel–Ferry [33, 34] (WLF) equation written as

$$\log a_T = \log\left(\frac{\eta}{\eta_{ref}}\right) = \frac{C_1(T - T_{ref})}{C_2 + T - T_{ref}} \tag{7.2}$$

where a_T is a shift factor, the symbol η is again the viscosity, and η_{ref} is a reference viscosity at a reference temperature T_{ref}, C_1 and C_2 are material constants. The two equations are equivalent and when the reference temperature is taken as $T_{ref} = T_g$, we obtain [34].

$$B = 2.303 C_1 C_2; \quad T_\infty = T_g - C_2 \tag{7.3}$$

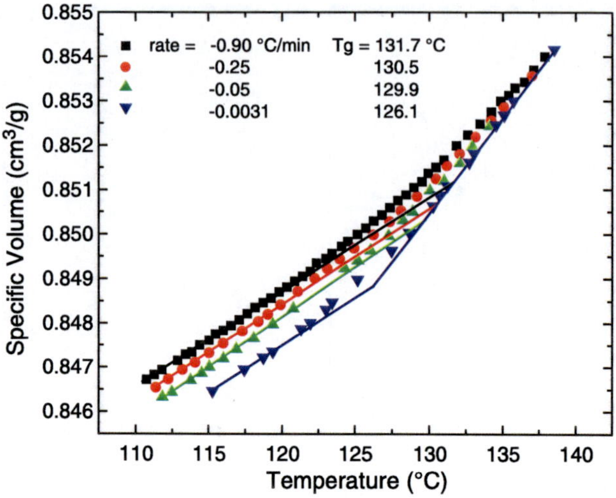

Fig. 7.2 Specific volume in dilatometric experiments for an epoxy thermoset showing the effects of cooling rate on response and on the apparent glass transition temperature (Data replotted from Bero and Plazek [37])

Importantly, both the WLF and VFT expressions predict that the relaxation time or the viscosity becomes infinite at a temperature below the nominal glass transition temperature but above absolute zero. The consequence of this rapid slowing of molecular mobility is that over a span of some 20–30°C, the material goes from a liquid-like mobility to a nearly solid behavior. However, it is important to realize that this nearly solid behavior does in fact have a molecular mobility associated with it and this aspect makes polymers viscoelastic and it provides a mechanism for their glassy states to change or age with the progress of time. It is also worth noting that the divergence of viscosity or time scale at the VFT temperature T_∞ has recently been questioned [35, 36] and the problem of time scale or viscosity divergence remains a problem of interest as it is fundamental to a full understanding of the behavior of glass-forming materials.

Another manifestation of the vitrification process can be seen in a cooling experiment in which the volume is measured. The result of such an experiment is depicted in Fig. 7.2 for an amorphous polymer in which several different speeds of cooling were used [37]. As can be seen in the figure, as the system is cooled, the volume follows a straight line in temperature until there is a region where the molecular mobility has declined sufficiently that the volume begins to deviate from the equilibrium state. Eventually, the volume finds a second linear dependence on temperature and this is the glassy state. Between the two is the glass transition region and for convenience one determines the glass transition temperature T_g from the intersection of the two straight-line regimes. Importantly, one can see from the figure that the glassy state is clearly a nonequilibrium state. The second aspect of Fig. 7.2 is that the actual value of T_g is different depending on the time scale of the experiment. When one cools over long periods of time, the glass transition

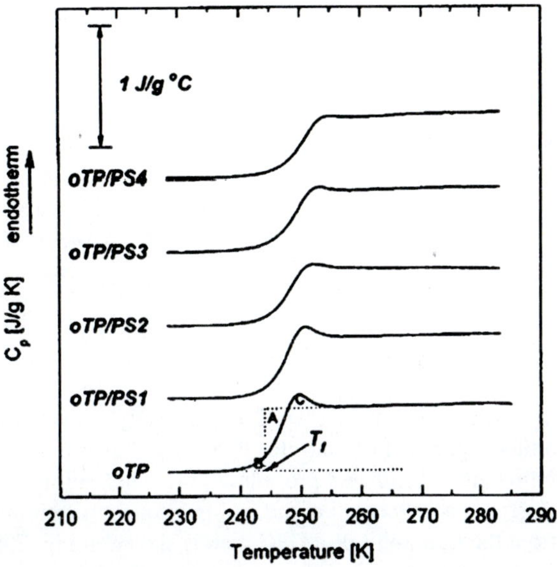

Fig. 7.3 Heat capacity measurements after cooling samples of polystyrene mixed in ortho-terphenyl at different mass fractions polystyrene: oTP/PS1 (0.0468); oTP/PS2 (0.0923); oTP/PS3 (0.1282); oTP/PS4 (0.1640). Lower trace for the pure OTP shows construction needed to determine the limiting fictive temperature (After Park and McKenna [40])

temperature itself decreases and this is important because the implication is that a rapid determination of the glass transition temperature results in an overestimate of the value of T_g and a consequent overestimate of the long-term useful temperature of a material. In polymers, this overestimate can be approximately 3–5°C per logarithmic decade of time. Thus, a 100 s determination of T_g in the laboratory can overestimate the effective T_g in a material's use condition of, e.g., 10 years (3.15×10^8 s) by approximately 20–32°C or more.

In addition to volume measurements of the glass transition, it is also common practice to make measurements using a differential scanning calorimeter. In this case, one measures the heat flow response which is related to the heat capacity C_p which is a derivative of the heat flow with changing temperature [14].

$$HF = C_{p,app} \frac{dT}{dt} \tag{7.4}$$

And we use the term $C_{p,app}$ here to bring attention to the fact that the heat flow also can contain contributions due to structural recovery, chemical reaction, and so forth [14].

A typical experiment is to cool the sample and then reheat at the same rate to obtain T'_F the limiting fictive temperature [14, 38, 39]. The fictive temperature T_F itself is defined as the intersection of the extrapolation of the glassy enthalpy and the liquid enthalpy and the limiting fictive temperature is that obtained from an experiment upon cooling at a constant rate followed by an immediate reheating of the material (typically, DSC uses heating experiments because the transition is sharper than upon cooling) at the same rate. Figure 7.3 shows the heat capacity

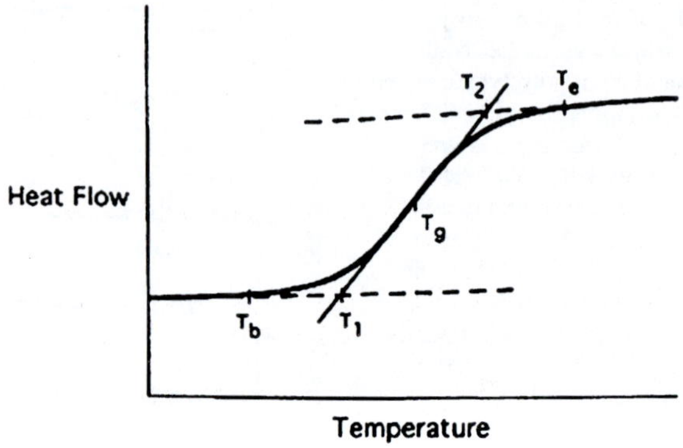

Fig. 7.4 Schematic of the determination of the glass transition from the step change in heat capacity (heat flow) in a typical DSC experiment (After Simon and McKenna [14])

(heat flow) in typical DSC experiments for several polystyrene solutions in ortho-terphenyl [40] and illustrates also how to determine the fictive temperature following Moynihan's procedure [38]. The limiting fictive temperature in unaged glasses is nearly equal to the mid-point glass transition temperature. T_g determined from the mid-point of a DSC scan is shown in Fig. 7.4 [14].

It is important here to remark that the DSC measurements are also kinetic. Hence, the T_g (or T_F') depends on cooling rate. As with the volume measurements, the value of T_g decreases as cooling rate decreases. Full understanding of these kinetics is important to developing the ability to make predictions of long-term performance. We now turn to a discussion of the glass transition kinetics and the models that are currently used to describe the observed kinetic behavior.

7.2.2 *Structural Recovery: Experiments*

7.2.2.1 Intrinsic Isotherms

If we return to Fig. 7.2 and the fact that the polymer glass is not in equilibrium, it becomes important to establish what happens as a consequence. Unlike window glass at room temperature where the molecular mobility is so slow that the evolution of the properties takes "geological" ages,[1] [41, 42], the polymer glass still evolves spontaneously and when temperatures get to a higher fraction of the T_g, the evolution occurs at a higher rate. The volume of the material changes with time

[1] Silica glass at 1,000°C has a viscosity of $10^{15.82}$ Pa s[41] and at room temperature, from the reported activation energy, one would estimate the value to be >> 10^{30} Pa s. In this case, since the Maxwell model [34] estimation of the relaxation time is $\tau = \eta_0/G$ and $G \approx 28$ GPa for a window glass [41], then $t \approx 10^{30}/28 \times 10^9$ s $\approx 10^{12}$ years which is much longer than the times available for European cathedral windows to have flowed.

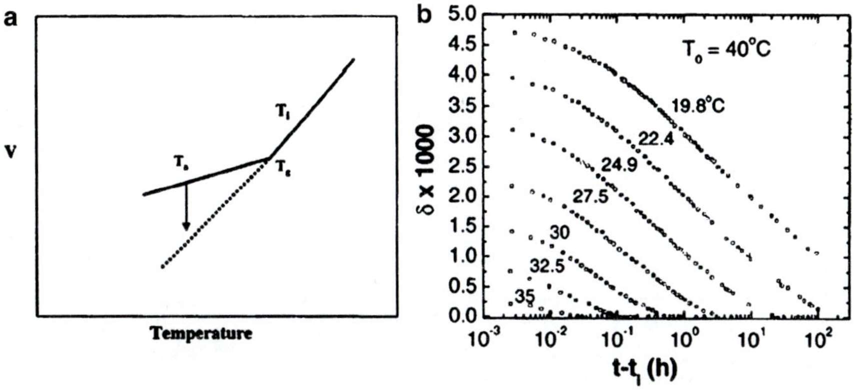

Fig. 7.5 (**a**) Schematic of thermal history used in obtaining a single intrinsic isotherm curve in a down-jump experiment. (**b**) Intrinsic isotherms for glucose glass (Figure from Zheng and McKenna [43], data from Kovacs [1])

and this is referred to as volume recovery or volume relaxation. A similar evolution of other state-like variables such as enthalpy will also take place. An important experiment to demonstrate this structural recovery (the structure of interest is the state-like variable or liquid-like structure frozen into the glass) is the down-jump experiment from close to but above the glass transition temperature. A series of such experiments is referred to as the *intrinsic isotherms* [1] of the material and one such set of curves is shown in Fig. 7.5 [43] as a plot of the volume departure ($\delta = (v{-}v_0/v_0)$) from equilibrium where v is the instantaneous volume and v_0 is the volume in equilibrium. δ itself defines the glassy structure and is nondimensional much as a volumetric strain is. It can be seen in the figure that the further one is below the starting temperature T_0 (chosen to be close to the nominal glass transition temperature), the longer the time required for the structure to recover or relax into equilibrium. In fact, once about 20°C below T_0, the times required to reach equilibrium became longer than the experimental times of 100 h in the measurement. However, the question mentioned above about modeling the behavior cannot be accomplished by the determination of the intrinsic isotherms alone. Kovacs [1] in early studies used two other experiments which form the minimal or essential behaviors that need to be captured in models of structural recovery. These were the *asymmetry of approach* experiment that demonstrated the nonlinearity of the glass transition kinetics and the *memory experiment* that demonstrated the relaxation process is non-exponential in nature.

7.2.2.2 Asymmetry of Approach

The asymmetry of approach [1, 2] experiment demonstrates a fundamental nonlinearity in the structural recovery process. This experiment provides a comparison of behaviors in an up-jump condition and in a down-jump condition. In both cases, the experiment

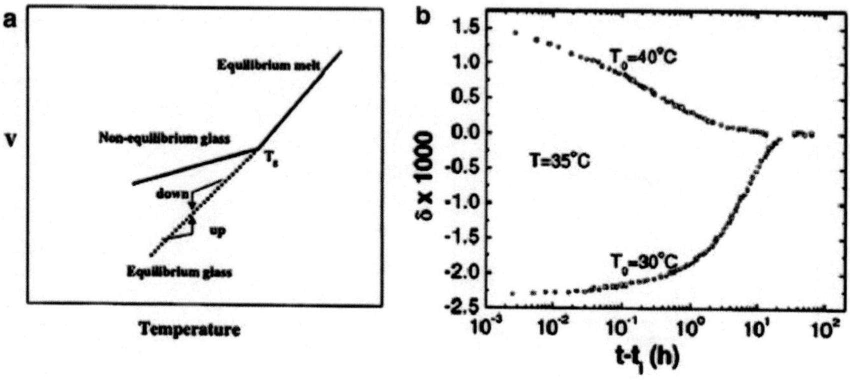

Fig. 7.6 (**a**) Schematic of temperature histories for down- and up-jumps for the asymmetry of approach experiment. (**b**) Asymmetry of approach results for a poly(vinyl acetate) polymer glass (Figures from Zheng and McKenna [43], data from Kovacs [1])

ends at the same temperature, e.g. T_1 but begins at a temperature either ΔT below T_1 or ΔT above T_1. In this case, if the response were linear the dependence of δ on the time after the jump would be the same in both cases just with opposite signs. As shown in Fig. 7.6, this is not the case when the magnitude of ΔT is large enough. The current explanation of this phenomenon is that in the down-jump there is an excess in δ and this results in higher molecular mobility and a rapid recovery of the structure toward equilibrium. As time goes on, the departure from equilibrium decreases in magnitude and the molecular mobility similarly decreases. This leads to what appears to be an auto-retarded process. In the case of the up-jump, one begins with a deficit of δ (it is negative) and this means that the molecular mobility is reduced and the sample is very slow to recover toward equilibrium. As time progresses, the deficit in δ decreases and the molecular mobility increases progressively, leading to what appears to be an auto-catalytic behavior. This asymmetry shows the nonlinearity of the structural recovery process and it is interpreted to be related to the fact that the molecular mobility depends on the instantaneous state of the glass as well as the temperature. In the terminology being used here, the relaxation time for structural recovery depends on the departure from equilibrium δ.

7.2.2.3 Memory

The third experiment that needs to be understood for any successful model of the kinetics of structural recovery is that of the memory or cross-over effect. In this case, one subjects the sample to a two-step temperature history in which the first step is to a temperature lower than the second step and the sample is allowed to partially recover before the temperature is increased in the second step. This is shown schematically in Fig. 7.7a. The results from such an experiment are shown in Fig. 7.7b. Prior to

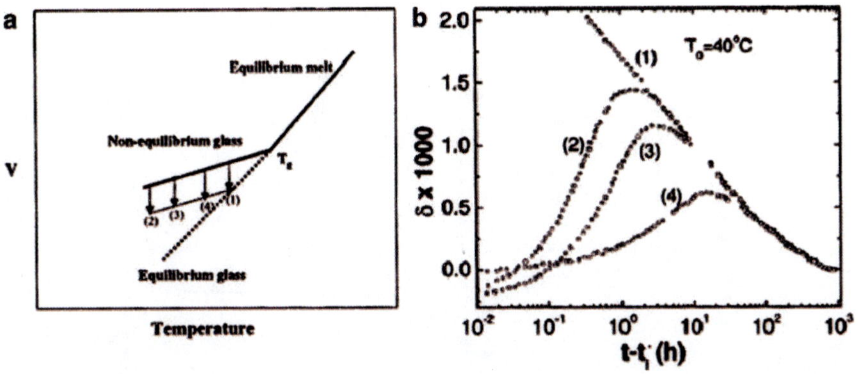

Fig. 7.7 (a) Schematic of thermal histories used to show memory effect. (b) Volume response in memory experiments for a poly(vinyl acetate) polymer glass. History 1: direct jump from 40°C to 30°C. History 2: jump from 40°C to 10°C and anneal for 160 h before making second jump to $T = 30°C$. History 3: jump from 40°C to 15°C and anneal for 140 h before making second jump to $T = 30°C$. History 4: jump from 40°C to 25°C and anneal for 90 h before making second jump to $T = 30°C$ (Figures from Zheng and McKenna [43], data from Kovacs [1])

Kovacs' work [1] on the memory effect in polymers, it had been seen in inorganic glasses and referred to as a cross-over effect [5, 6]. The important thing that Kovacs brought to the understanding was that he performed experiments in which the partial recovery and the final temperature were chosen such that the initial condition for the second step was that the departure from equilibrium would be very close to zero (ideally zero as depicted in Fig. 7.7a). The result that the departure from equilibrium evolves, and does so non-monotonically, can only be explained by a non-exponential relaxation process in the isothermal conditions of the second step. The results from the Kovacs memory experiments are shown in Fig. 7.7b.

7.2.2.4 Enthalpy Measurements

While the essentials of structural recovery are contained in the instrinsic isotherms, the asymmetry of approach, and the memory effect, such results are much easier to obtain dilatometrically than they are to obtain using a calorimeter, especially a differential scanning calorimeter (DSC), such as is a common practice in the study of the structural recovery of polymer glasses. For completeness, then, we consider typical DSC measurements of glasses subjected to thermal annealing treatments that lead to enthalpy overshoots typical of structural recovery. The experimental procedure used is to cool a material through the glass transition and to then either reheat immediately at the same or a higher rate as the cooling rate or to let the sample sit isothermally (age or anneal) so that the structure (now enthalpy instead of volume) can recover. After a given annealing time, the sample is then reheated and the enthalpy measured. Figure 7.8 shows the heat flow response of a polyetherimide material which has been annealed for different times and then reheated. As can be

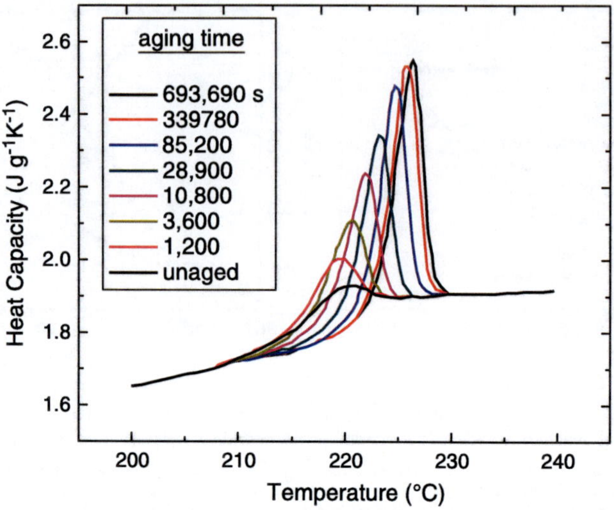

Fig. 7.8 Heat capacity from scanning calorimetry experiments as a function of temperature for a poly(ether imide) after aging, for the times indicated, at 201.3°C. Figure shows progression of the excess enthalpy peak as a function of aging time (Data replotted from Echeverria et al. [44])

Fig. 7.9 Enthalpy departure from equilibrium $\delta = \Delta H_\infty - \Delta H_a$ versus aging time for a poly(ether imide) material at temperatures of (●) 211.3°C; (○) 207.5°C; (■) 206.3°C; (□) 202.5°C; (▲) 201.3°C (Data replotted from Simon [8])

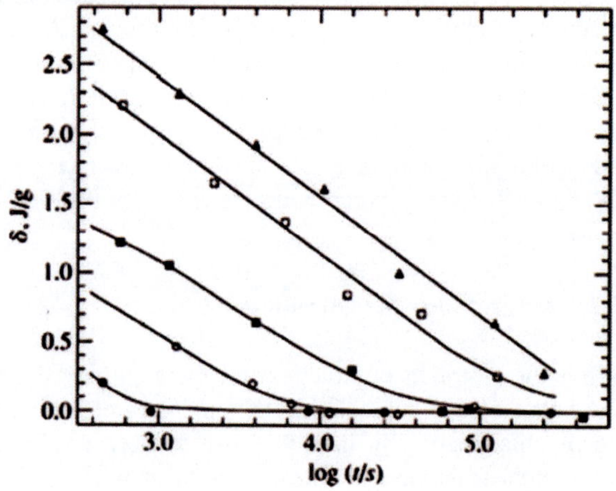

seen, with no annealing there is a very small peak in the curve and the peak grows with annealing time as the structure (enthalpy) recovers toward equilibrium. The data can be analyzed either as the enthalpy departure from equilibrium ΔH or through the fictive temperature T_F. The enthalpy departure equivalent to the intrinsic isotherms shown in Fig. 7.5b for the same polyetherimide is shown [8, 44], in Fig. 7.9.

7.2.3 Structural Recovery: Models

7.2.3.1 The Kovacs, Aklonis, Hutchinson, Ramos (KAHR) and Tool–Narayanaswamy–Moynihan (TNM) Models

The most commonly used models for describing the structural recovery of glassy materials are the Kovacs, Aklonis, Hutchinson, Ramos [12] (KAHR) model and the Tool [2], Narayanaswamy [3], Moynihan [4] (TNM) model. These models are essentially the same, though the physical meaning of the parameters in them do differ somewhat. The important aspect of the models is that they provide a zeroth-order model to structural recovery that captures much of the phenomenology just described. In particular, the models capture the nonlinearity embodied in the asymmetry of approach experiment and the non-exponentiality of the behavior seen in the memory experiment. Both models do this by building a set of response equations that are based on a Boltzmann [45] –superposition formalism that is linear in the reduced time framework. The response is the structural recovery parameter of interest (KAHR did it for volume departure from equilibrium; TNM did it for the enthalpy.) The stimulus is the thermal history dT/dz where z is the reduced time. The KAHR expressions are [12, 14]

$$\delta = -\Delta\alpha \int_0^z R(z - z') \frac{dT}{dz'} dz' \tag{7.5}$$

where δ is the departure from equilibrium, described above, and is the Kovacs measure of the glassy structure. $R(z)$ is the response function, which KAHR consider as a sum of exponentials, and $\Delta\alpha$ is the change in the coefficient of thermal expansion (or the change in heat capacity ΔC_p) at the glass transition temperature. We write the reduced time z as

$$z = \int_0^t \frac{d\xi}{a_\delta a_T} \tag{7.6}$$

the response function as

$$R(t) = \sum_{i=1}^{N} \delta_i e^{-t/\tau_i} \tag{7.7}$$

where the δ_i are prefactors (local departures from equilibrium) and the τ_i are the relaxation times, thought to be related to local molecular dynamics. The temperature dependence is expressed in terms of the shift factors: a_T for temperature and a_δ for the structure and are defined as

$$\frac{\tau_i(T,\delta)}{\tau_{i,ref}} = a_T a_\delta = e^{-\Theta(T-T_{ref})} e^{\frac{-(1-\Theta\delta)}{\Delta\alpha}} \tag{7.8}$$

And the τ_i refer either to the value at the current T and δ or to the value in the reference condition of T_{ref} and $\delta = 0$. Θ is a temperature parameter and KAHR used $\theta \approx \frac{E_a}{RT_g^2}$ where E_a is an apparent activation energy (at T_g).

And the TNM equations are written in terms of fictive temperature, and generally in differential form [2–4, 13, 14].

$$\frac{dT_F}{dt} = 1 - e^{-\int_0^t \left(\frac{dt}{\tau_0}\right)^{\beta}} \tag{7.9}$$

The parameter β is the stretching exponent in the response function, which is generally taken to be of the Kohlrausch–Williams-Watts [46, 47] (KWW) form

$$R(z) = e^{-\left(\left(\frac{t}{\tau_0}\right)^{\beta}\right)} \tag{7.10}$$

And the reduced time is written in terms of the temperature shift factor a_T and the structure shift factor a_{T_F} (remark the difference with the KAHR model where structure is measured by δ rather than by T_F) as

$$z = \int_0^t \frac{d\xi}{a_T a_{T_F}} \tag{7.11}$$

The temperature and structure dependences are expressed in terms of the structure and temperature shift factors as

$$\frac{\tau_0(T, T_F)}{\tau_{0,ref}} = a_T a_{T_F} = e^{-\frac{x\Delta h}{R}\left(\frac{1}{T} - \frac{1}{T_{ref}}\right)} e^{\frac{(1-x)\Delta h}{R}\left(\frac{1}{T} - \frac{1}{T_F}\right)} \tag{7.12}$$

The equations generally capture the qualitative features of the intrinsic isotherms, asymmetry of approach, and memory effect as well as the enthalpy overshoot seen in typical DSC experiments. Comparisons of the model with actual data are shown [48–51] in Figs. 7.10–7.13. The interesting aspect here is that the models do well, but not perfectly. What is less well seen in the figures is that if a set of data over a temperature range of, e.g., T_g−10 K to T_g is used to determine material parameters, when the temperature range is changed to T_g−20 K to T_g−10 K the parameters will change. This is obvious in Fig. 7.12 where the data from asymmetry of approach experiments was used to obtain KAHR model parameters and then the memory effect for the same polymer was calculated. Another aspect of the models is that the parameters are not independent, though in principle they should be. This is seen in the statistical correlation between the stretching parameter β in the TNM model and the nonlinearity parameter x as shown in Fig. 7.13. Such discrepancies indicate that there is something missing from the models and, at this point, one can only say that the missing “ingredients”

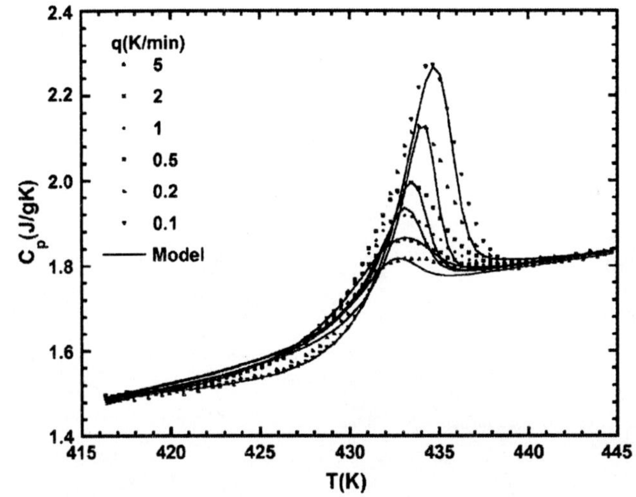

Fig. 7.10 Comparison between enthalpy recovery after quenches at different cooling rates and TNM model calculations for a polycarbonate copolymer (After Badrinarayanan et al. [48]. Reprinted with permission from Elsevier, Copyright 2008)

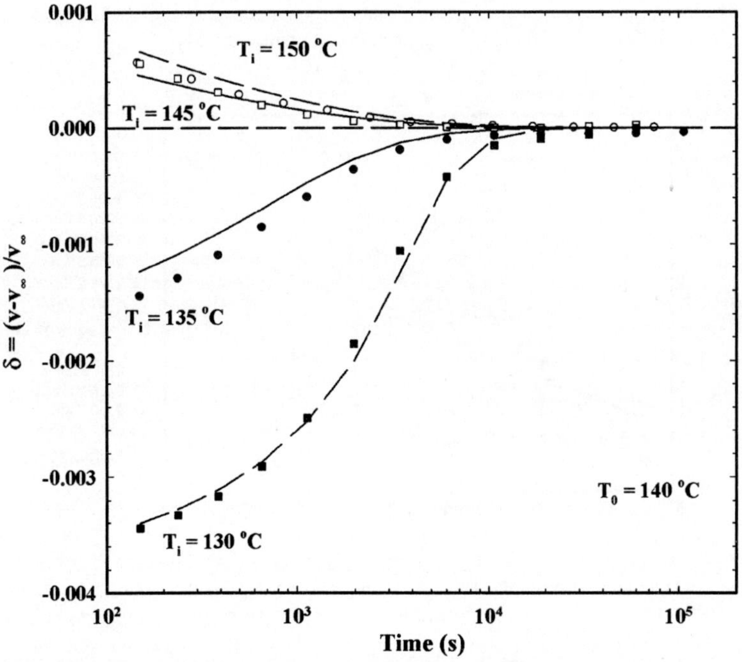

Fig. 7.11 Comparison between volume recovery in polycarbonate in asymmetry of approach conditions. Final temperature is $T_0 = 140$ C and initial equilibrium conditions are T_1 values indicated in figure (Figure from McKenna and Simon [49] and data from Schultheisz and McKenna [50])

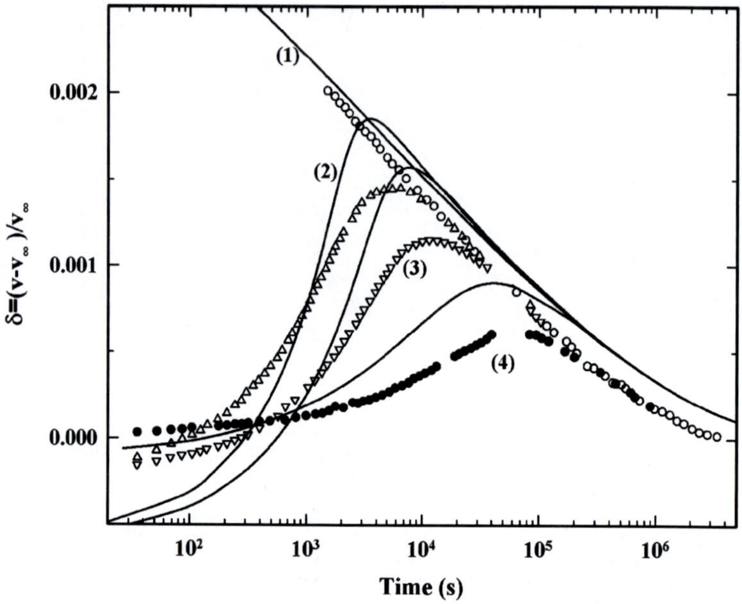

Fig. 7.12 Memory effect results of Kovacs [1] for poly(vinyl acetate) modeled using data from asymmetry of approach experiments. Points are data and lines are KAHR model results (Figure from McKenna and Simon [49] and data from Schultheisz [51])

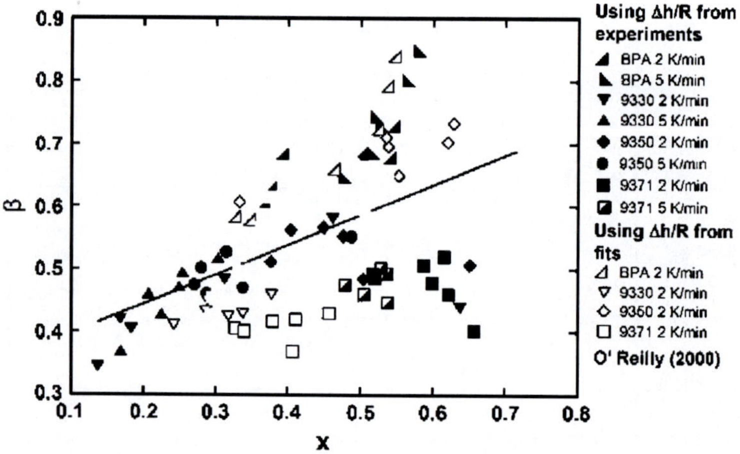

Fig. 7.13 Correlations between the TNM stretching exponent β and the nonlinearity parameter x (After Badrinarayanan et al. [48]. Reprinted with permission from Elsevier, Copyright 2008)

for the models are not well known. This has been the case for some time [52] and there has been relatively little progress in solving the dilemma; though there are some attempts [53, 54] to modify the TNM/KAHR approaches, there has been no generally recognized solution to these difficulties. Perhaps the most promising of new approaches comes from the thermoviscoelastic model, described subsequently, that provides improvement but at the expense of a heavy computational requirement. From the experimental view, there has also been some work, especially by Simon and coworkers [55, 56], that suggests that any full description of the structural recovery of glassy materials will need an approach that includes a dependence of the material response not only on the instantaneous state of the glass (through the reduced time) but also on how the glass got to any specific instantaneous state (the response depends on the full history). The reader is referred to those works for further discussion of the experiments.

Another possibility is that the general continuum-type of approach embodied in the KAHR and TNM models (and the thermoviscoelastic model described next) may be doomed to failure because of the inherent heterogeneity of the response of glassy materials. Heterogeneity in glasses has been well known for some time [57–64], but the full appreciation of its importance and intense investigation into the heterogeneity of glass-forming materials have only occurred in the past two decades [65–74]. Again, the interested reader is referred to the relevant references.

7.2.3.2 The Thermoviscoelastic [75–84] Model

While the TNM-KAHR models are currently widely used as the general paradigm for understanding the kinetics of the glass transition event, and may well remain so for some time to come because of the relative simplicity of the models, there has been some recent progress from the J.M. Caruthers group [75–81] at Purdue University, to extend the framework of Rational Mechanics or Rational Thermodynamics [85] to the problem of, not only structural recovery and aging phenomena, but also to the nonlinear mechanics of glassy polymers. In this section, we look at the developments of the model for structural recovery and, in later sections, will examine its usefulness in describing physical aging and other mechanical properties.

From this author's perspective [14], a conceptual difficulty with the TNM-KAHR models is that they are empirical and not truly fundamental either from a microscopic/molecular view or from a nonequilibrium thermodynamics perspective. Rather, they are built on an edifice in which it is assumed that the lower derivatives of the thermodynamic properties are dependent on the higher level ones. Hence, we write, as in Eq. 7.5, a relationship in which the volume depends on the coefficient of thermal expansion difference at the T_g rather than beginning with, e.g., the free energy function and determining the other functions from its derivatives, as is done in classical thermodynamics [86, 87]. This creates an operationally useful approach, but it seems less than satisfying from a fundamental perspective as it does not say directly that what is important in the system is the (nonequilibrium) free energy function. This is where the Caruthers model makes a major step forward. By beginning with a nonequilibrium thermodynamics framework, it is rigorously correct within that framework. In this case, the choice is to use Rational Mechanics [85] as the starting point and make approximations within that framework. It is important to remark that there are other nonequilibrium frameworks that might be invoked [88–92], but a discussion of these is far beyond the scope of the present work and these other frameworks are not as well developed for the engineering world as the Caruthers' model. Hence, we, here, use the approach developed and elaborated by Caruthers and coworkers [75–84] and that we refer to as the "thermoviscoelastic model."

7.2.3.3 General Approach

The thermoviscoelastic model is based on the ideas of Coleman [93, 94] and Noll [95] that builds a nonequilibrium thermodynamics framework referred to in the literature [85] as Rational Mechanics or Rational Thermodynamics and that has been a bulwark of general understanding of the nonlinear deformation and flow behavior of, e.g., polymeric materials. The first major article describing the model can be attributed to Lustig, Shay, and Caruthers [79] but we follow the more recent approach of Adolf, Chambers, and Caruthers [80] that has been extensively used by the group at Sandia National Laboratories to model behavior of polymers and polymer composites.

Importantly, the model begins with a free energy functional that is 'approximated by a Frechet expansion in temperature and strain histories about the equilibrated state at the current temperature and strain [80]. Then one finds that the stress (Cauchy) is given by:

$$\overline{\overline{\sigma}} = \frac{\rho}{\rho_{ref}} \overline{\overline{F}} \left[\overline{\overline{S}}_H : \frac{d\overline{\overline{H}}}{d\overline{\overline{E}}} \right] F^T \tag{7.13}$$

and where

$$\begin{aligned} \overline{\overline{S}}_H = {} & \overline{\overline{S}}_{H_\infty} + \rho_{ref}\,\Psi_1 \int_0^t ds f_1(t^* - s^*, 0) \frac{dI_H}{ds}(s)\overline{\overline{I}} \\ & + 2\rho_{ref} \left[\Psi_2^{ref} + \left(\frac{\partial \Psi_2}{\partial T} \right)_{II_H}^{ref} \Delta T \right] \int_0^t ds f_2(t^* - s^*, 0) \frac{d\overline{\overline{H}}}{ds} \\ & + \rho_{ref} \left[\Psi_3^{ref} + \left(\frac{\partial \Psi_3}{\partial T} \right)_{I_H}^{ref} \Delta T + \left(\frac{\partial \Psi_2}{\partial I_H} \right)_T^{ref} I_H \right] \\ & \times \int_0^t ds f_3(t^* - s^*, 0) \frac{dT}{ds}(S)\overline{\overline{I}} + \rho_{ref} \left(\frac{\partial \Psi_3}{\partial I_H} \right)_T^{ref} \\ & \times \left[\int_0^t ds \int_0^t du f_3(t^* - s^*,\ t^* - u^*) \frac{dI_H}{ds}(s) \frac{dT}{du}(u) \right] \overline{\overline{I}} \end{aligned} \tag{7.14}$$

And $\overline{\overline{F}}$ is the deformation gradient tensor, $\overline{\overline{H}} = \left(\frac{1}{2}\right) ln \left(\overline{\overline{F}}^T \overline{\overline{F}}\right)$ is the Hencky strain measure, I_H is the first invariant of $\overline{\overline{H}}$ and is a function only of volume, $\overline{\overline{S}}_H$ is the stress that is conjugant to the Hencky strain, $\overline{\overline{E}} = (\overline{\overline{F}}^T\overline{\overline{F}} - 1)/2$ is the Green-Lagrange strain measure, T is the current temperature, ρ is the current density, and ρ_{ref} is the density in the reference state.

The prefactors Ψ_i are temperature and volume dependent in order to account for the experimentally observed dependence of the moduli and coefficient of thermal expansion on temperature and volume. Adolf et al. [80] used expression (7.14) to perform stress calculations and we return to this subsequently. However, the point of interest now is the determination of the entropy in the system in order that the heat capacity in different thermal histories can be determined and we then compare the predictions of the thermoviscoelastic model with experimental results.

The entropy symbol used by Adolf et al. [80] is the Greek symbol η and we follow this convention in what follows.

$$\begin{aligned}
\eta = \eta_\infty &- \left[\Psi_4^{ref} + \left(\frac{\partial\Psi_4}{\partial T}\right)_{I_H}^{ref}\Delta T + \frac{1}{2}\left(\frac{\partial^2\Psi_4}{\partial T^2}\right)_{I_H}^{ref}\Delta T^2\right] \\
&\times \int_0^t dsf_4(t^* - s^*, 0)\frac{dT}{ds}(s) - \left[\Psi_3^{ref} + \left(\frac{\partial\Psi_3}{\partial T}\right)_{I_H}^{ref}\Delta T + \left(\frac{\partial\Psi_3}{\partial I_H}\right)_{T}^{ref} I_H\right] \\
&\times \int_0^t dsf_3(t^* - s^*, 0)\frac{dI_H}{ds}(s) - \frac{1}{2}\left[\left(\frac{\partial\Psi_4}{\partial T}\right)_{I_H}^{ref} + \left(\frac{\partial^2\Psi_4}{\partial T^2}\right)_{I_H}^{ref}\Delta T\right] \\
&\times \left[\int_0^t ds \int_0^t duf_4(t^* - s^*, t^* - u^*)\frac{dT}{ds}(s)\frac{dT}{du}(u)\right] \\
&- \left(\frac{\partial\Psi_3}{\partial T}\right)_{I_H}^{ref}\left[\int_0^t ds \int_0^t duf_3(t^* - s^*, t^* - u^*)\frac{dI_H}{ds}(s)\frac{dT}{du}(u)\right] \\
&- \left(\frac{\partial\Psi_2}{\partial T}\right)_{I_H}^{ref}\left[\int_0^t ds \int_0^t duf_2(t^* - s^*, t^* - u^*)\frac{d\overline{\overline{H}}}{ds}(s)\frac{d\overline{\overline{H}}}{du}(u)\right]
\end{aligned} \tag{7.15}$$

where

$$\begin{aligned}
\eta_\infty = \eta_\infty(\rho_{ref}, T_{ref}) &- \Psi_{TT}I_H - \frac{1}{2}\Psi_{TI}I_H^2 - \Psi_{TTT}\Delta T I_H - \Psi_{TT}\Delta T \\
&- \frac{1}{2}\Psi_{TTT}\Delta T^2 - \frac{1}{6}\Psi_{TTTT}\Delta T^3
\end{aligned} \tag{7.16}$$

When the prefactors to the convolution integrals in the free energy are constant, the entropy becomes more familiar looking:

$$\eta = \eta_\infty + \frac{\Delta C_v}{T}\int_0^t dsf_4(t^* - s^*)\frac{dT}{ds}(s) + \frac{\Delta(K\alpha)}{\rho}\int_0^t dsdf_3(t^* - s^*)\frac{dI_H}{ds}(s) \tag{7.17}$$

where C_v is the constant volume heat capacity. However, the simple form of Eq. 7.17 cannot be used due to the temperature dependence of the constant pressure heat capacity C_p [80, 82].

Adolf et al. [80] define the shift factor a_T for the time-dependent functions (a viscoelastic shift factor) as $t^* - s^* = \int_s^t \frac{dx}{a_T(x)}$ where the logarithm of the shift factor is determined in terms of the potential or "configurational entropy" U_c of the system. They write

$$\log(a_T) = B\left(\frac{1}{U_c} - \frac{1}{U_c^{ref}}\right) = C_1\left(\frac{U_c^{ref}}{U_c} - 1\right) \tag{7.18}$$

And B is a material constant, C_1 is the first WLF [33] constant, B is a material constant, and $U_c^{ref} = U_{\infty_{pot}}(\rho_{ref}, T_{ref}) = C_2$, i.e., the second WLF [33] constant [80]. The configurational entropy is given by

$$U_c = U_{\infty_{pot}} + (\Psi - \Psi_\infty) + T(\eta - \eta_\infty) \tag{7.19}$$

And $U_{\infty_{pot}}$ lies between the glassy and the equilibrium limits of the configurational entropy as expressed in Eqs. 66 and 67 of Caruthers et al. [81]. The second term in Eq. 7.19 can be written as

$$\begin{aligned}
\Psi - \Psi_\infty = \frac{\Psi_1}{2} \times & \left[\int_0^t ds \int_0^t du f_1(t^* - s^*, t^* - u^*) \frac{dI_H}{ds}(s) \frac{dI_H}{du}(u)\right] \\
+\Psi_2 & \left[\int_0^t ds \int_0^t du f_2(t^* - s^*, t^* - u^*) \frac{d\overline{\overline{H}}}{ds}(s) \frac{d\overline{\overline{H}}}{du}(u)\right] \\
+\Psi_3 & \left[\int_0^t ds \int_0^t du f_3(t^* - s^*, t^* - u^*) \frac{dI_H}{ds}(s) \frac{dT}{du}(u)\right] \\
+\frac{\Psi_4}{2} & \left[\int_0^t ds \int_0^t du f_4(t^* - s^*, t^* - u^*) \frac{dT}{ds}(s) \frac{dT}{du}(u)\right]
\end{aligned} \tag{7.20}$$

The four functions f_1, f_2, f_3, f_4 are linear viscoelastic relaxation functions. Adolf et al. [80] use a simple stretched exponential type of function (see Eq. 7.10) to represent these behaviors. We remark that the machinery here is fairly heavy computationally, but in principle the material parameters needed to obtain the full nonlinear response (structural recovery, mechanical nonlinearity) are obtained from linear thermal and linear viscoelastic measurements. In addition to the four linear relaxation functions, there are an "additional eight constants for parameterization." While this seems extensive, these are related to physical parameters and are obtained by conventional measurement methods such as dynamic mechanical, dilatometric, and calorimetric experiments [80]. We also note here that the nonlinearity of the thermal history response (e.g., the asymmetry of approach) arises due to both the four functions just commented upon and due to the introduction of the reduced time that depends on configurational entropy (Eqs. 7.18–7.20). Furthermore, more recently Adolf et al. [96] have recently introduced a reduced time that depends on the potential energy of the system. Hence, it is clear that the model itself is evolving and the future may hold further successes.

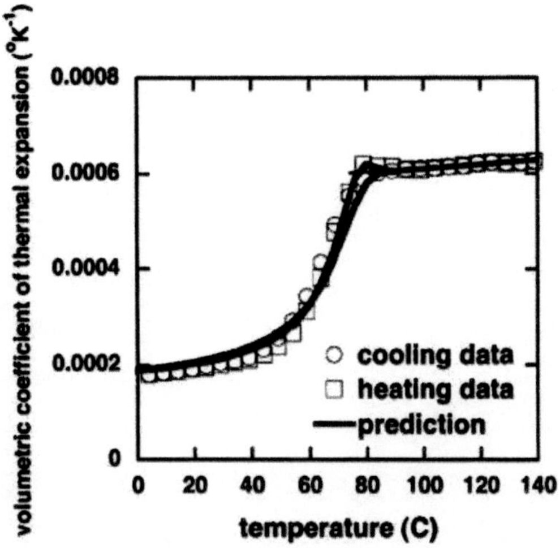

Fig. 7.14 Thermoviscoelastic model predictions compared with cooling and heating (dilatometric) data for a model epoxy system (After Adolf et al. [80] Reprinted with permission from Elsevier, Copyright 2004)

The purpose here is not to fully develop the model but to give the reader a sense of the model. We now turn to the question of how well does the model predict the structural recovery behavior. We will see that for this aspect of the physical aging and structural recovery response, one may do as well by simply using the standard methods described above. However, we will see later that within the same formalism, expanding now on Eq. 7.11, the mechanical behavior can be predicted and this is something that cannot be done in the TNM and KAHR frameworks, except in the linear viscoelastic regime, i.e., physical aging behavior can be accounted for.

7.2.3.4 The Thermoviscoelastic Model: Comparison with Calorimetric Experiments

For simple cooling and heating at constant rate, the thermoviscoelastic model provides very good predictions of the apparent coefficient of thermal expansion or the apparent heat capacity for the material. This is shown in Figs. 7.14 and 7.15 for a cross-linked epoxy material that has been cooled from above the T_g to below it at 5 K/min and then reheated at the same rate. The comparison between model and experiments is quite encouraging for these simple histories, where the structural recovery is somewhat minimized due to the correspondence between cooling and heating rates.

For more complex thermal histories, the agreement is not as good as in the simple history and more like that obtained with the TNM and KAHR models discussed above. This is shown in Fig. 7.16 for the same epoxy system, but now for different aging temperatures and aging times of 120 min.

Next we look at the physical aging response of materials in both linear and nonlinear regimes and we will return to the thermoviscoelastic model and its predictions of mechanical behavior in the appropriate sections.

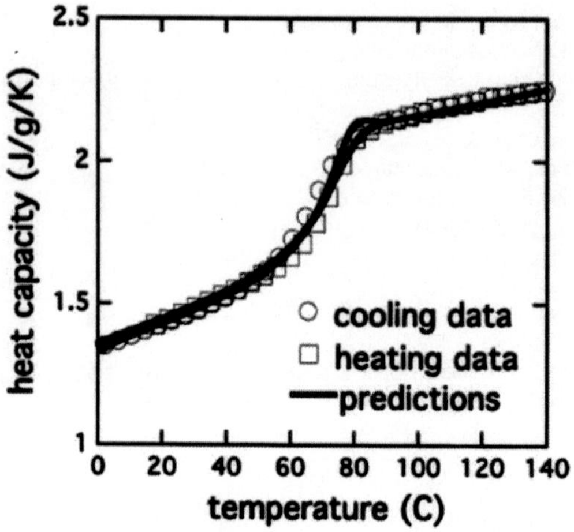

Fig. 7.15 Comparison of the thermoviscoelastic model predictions with cooling and heating data for calorimetric measurements on a model epoxy system (After Adolf et al. [80] Reprinted with permission from Elsevier, Copyright 2004)

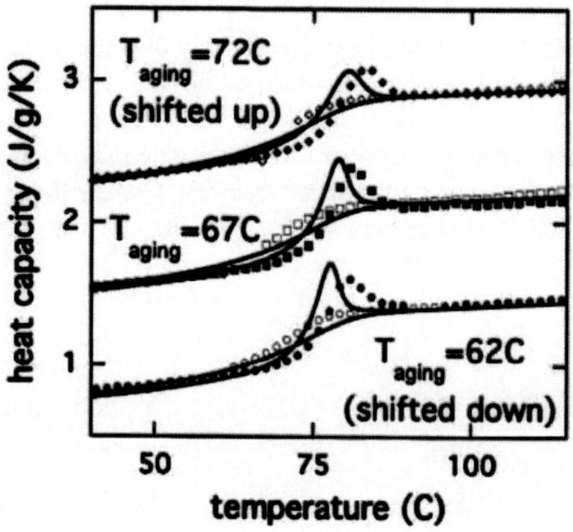

Fig. 7.16 Comparison of the thermoviscoelastic model predictions with heating data for calorimetric measurements on a model epoxy system that has been aged at different temperatures for 120 min prior to the heating (After Adolf et al. [80] Reprinted with permission from Elsevier, Copyright 2004)

7.2.4 Physical Aging of Glassy Polymers

7.2.4.1 Viscoelastic Properties

Just as is assumed in the structural recovery models, physical aging is an outcome of the viscoelastic response function or spectrum varying as a function of the changing thermodynamic state of the glass, either volume or enthalpy in the usual case. If, e.g., the material is subjected to a thermal history such that the temperature is changed rapidly from above to below the glass transition temperature, one finds that the sample is out of equilibrium after this down-jump and the properties evolve with aging or elapsed time after the temperature jump. Results from the pioneering

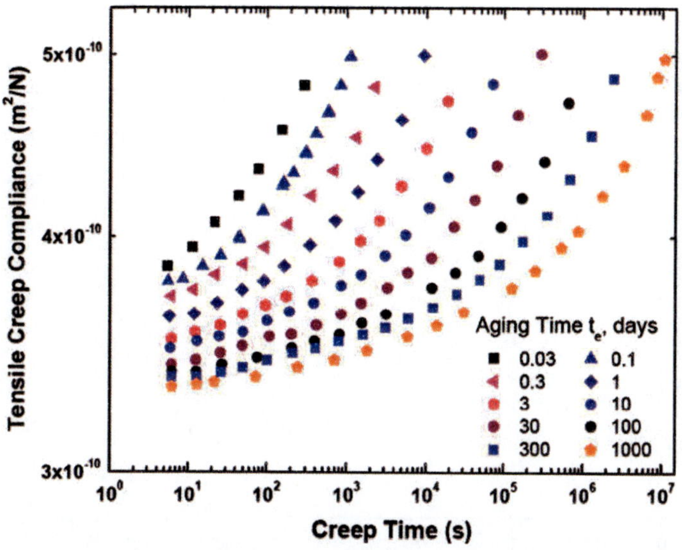

Fig. 7.17 Creep compliance for a PVC quenched from 90°C to 40°C and aged for 1,000 days. In the original work, Struik [9] showed that the data superimposed onto the longest aging time with minor vertical shifts (Data replotted from Struik [9])

work of Struik are shown for a series of creep experiments for a poly(vinyl chloride) (PVC) glass in Fig. 7.17 and for a series of relaxation experiments for a polycarbonate from the present authors' own work [97] in Fig. 7.18. As can be seen in these figures, as aging time progresses, the curves shift to longer times. If one has a spectrum of relaxation times τ_i then the shift is represented by $a_{te} = \frac{\tau_i}{\tau_{i,ref}}$ where the $\tau_{i,ref}$ are the relaxation times at the relevant reference condition which would be the reference temperature T_{ref} and the reference departure from equilibrium, which would normally be taken as equilibrium, i.e., $\delta = 0$. A common representation of the relaxation function is the so-called Kohlrausch-Williams-Watts function (KWW) [46, 47] or stretched exponential function discussed in the context of the structural recovery models above. In that case, the shear modulus can be represented as $G(t) = G_0 e^{-(t/\tau_0)^\beta}$ where the τ_0 is the characteristic relaxation time and the stretching exponent β characterizes the width of the relaxation spectrum or response function. $0 < \beta < 1$ and when $\beta = 1$, the relaxation is a single exponential function and the spectrum broadens progressively as β decreases toward 0. Then the aging time shift factor is represented as $a_{te} = \frac{\tau_0}{\tau_{0,ref}}$ and for time-structure superposition to hold, the value of β would be constant as the glassy structure evolves.

Figure 7.19 shows the time-aging time shift factors for down-jump experiments performed on an epoxy glass [98]. The results are for experiments performed near to the glass transition regime and one sees that at the higher temperatures, the polymer nears equilibrium as would be expected from the volume recovery or enthalpy recovery plots shown previously. In addition, one sees at the lower temperatures that the slope of the aging line reaches a constancy that gives a

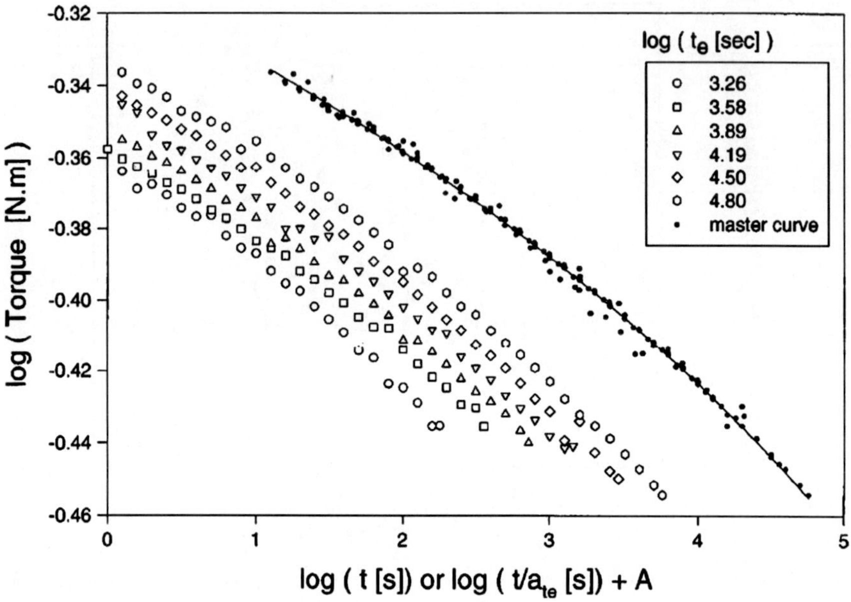

Fig. 7.18 Stress (as torque in a torsional experiment) relaxation response at different aging times for a polycarbonate polymer quenched from above T_g (≈140°C) to 70°C (After O'Connell and McKenna [97])

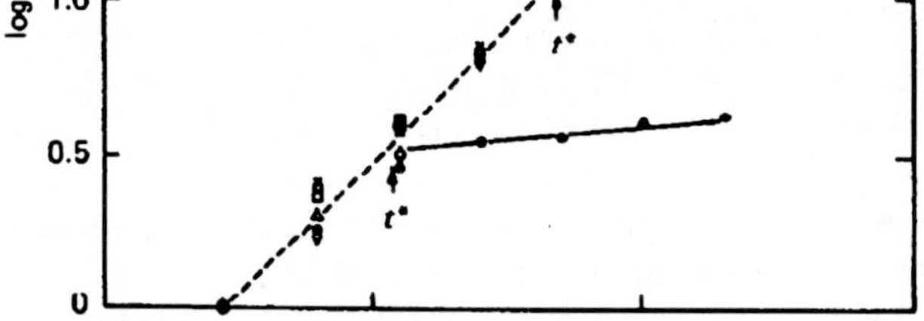

Fig. 7.19 Logarithm of aging time shift factor versus logarithm of aging time for an epoxy glass at different temperatures below the T_g. T_g–T: (○) 30.1 K, (×) 24 K, (□) 20.8 K, (◇) 10.3 K, (●) 6.3 K (After Lee and McKenna [98])

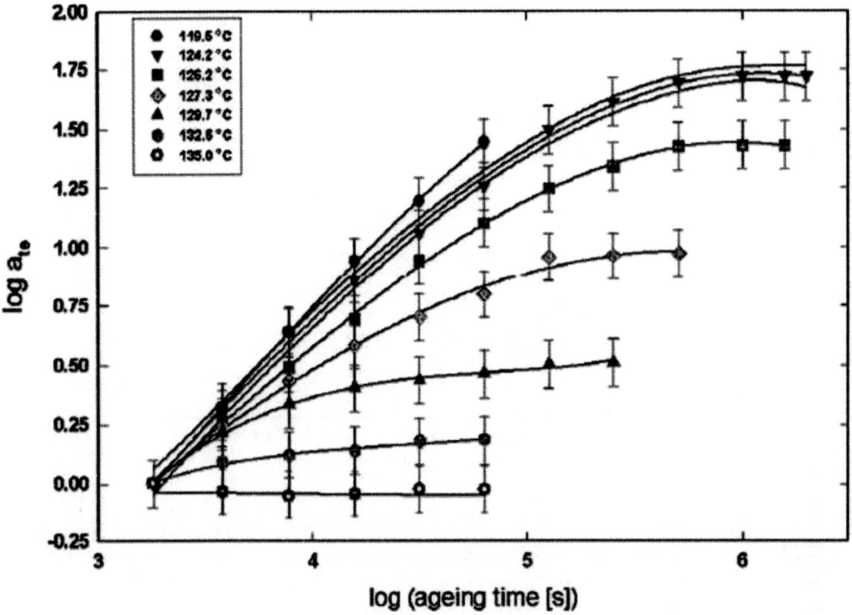

Fig. 7.20 Aging time shift factors as a function of aging time at different temperatures, as indicated, for a polycarbonate glass (After O'Connell and McKenna [99])

power-law behavior typical of the aging response. The power-law behavior is referred to as the shift rate $\mu = \frac{d \log a_{te}}{d \log t_e}$ and it implies that, at least over a limited time regime, one can express the relaxation time for the material as $\tau_0 = \tau_{0,ref} t_e^{\mu}$ from the expression above for the aging time shift factor [9]. A caution needs to be noted here. First, as seen in Fig. 7.19, the shift factor is not a power law at all aging times [99] and, as can be seen in Fig. 7.20, it is not necessarily independent of temperature. Thus, the idea of a power-law shifting of the relaxation times or the relaxation/creep curves is at best an approximation. In reality, the power-law regime itself can be quite large and is an approximation to a sigmoidal-shaped curve as indicated in Fig. 7.21 and discussed in more detail in reference [100].

An important aspect of time-structure (or time-aging time) superposition is that it is valid for other histories than the down-jump type of result shown in Figs. 7.17–7.21. Time-aging time superposition holds for all thermal histories, but can appear to be non-monotonic with aging time when the history is complicated, as in a memory experiment where the volume is non-monotonic (see discussion of memory effect in structural recovery above). This is shown in Fig. 7.22a, b where the volume response and the negative of the logarithm of the aging time shift factor are shown against aging time. However, as shown in Fig. 7.23, when the data are plotted against the volume (similar to departure from equilibrium), one finds that the data fall on a single curve of shift factor versus volume.

These results imply that the long-term prediction of material behavior for complicated thermal histories should, in reality, be based on measurements that

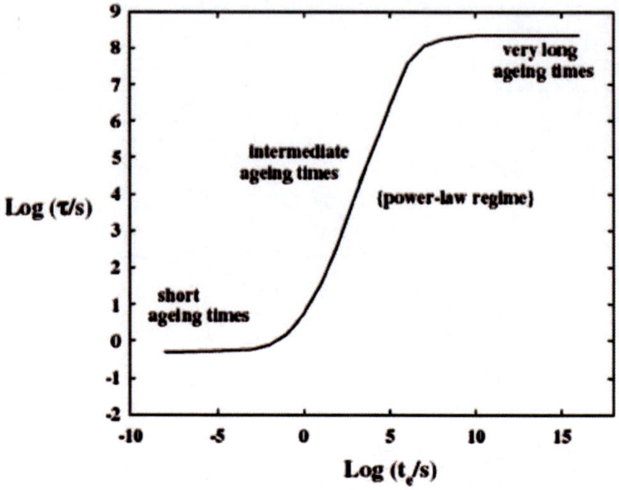

Fig. 7.21 Semiquantitative schematic of the shift factor versus aging or elapsed time for a material in a down-jump experiment showing the sigmoidal shape that the curve must have due to physical limitations at the short times and the fact that the material reaches equilibrium at long times (After McKenna [100])

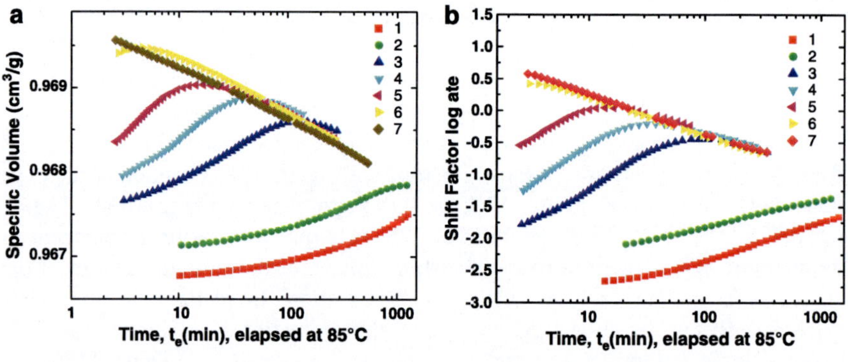

Fig. 7.22 (**a**) Volume as a function of elapsed time after the second step of two-step temperature histories. (**b**) Aging time shift factors corresponding to the changing specific volume for histories as indicated. Numbers correspond to the same histories for both graphs. All tests are for final jump to 85°C. Prior histories before the jump to 85°C are: Anneal at 100°C and: (1) Age to equilibrium at 80°C; (2) Age for 144 h at 80°C. For the following histories, the sample was annealed at 100°C and aged at each temperature for 96 h prior to the jump to 85°C: (3) $T_a = 70°C$; (4) $T_a = 65°C$; (5) $T_a = 60°C$; (6) $T_a = 50°C$. History 7 is a direct quench from 100°C to 85°C (intrinsic isotherm) (Data replotted from Struik [9])

include both the viscoelastic response and the volumetric response. Because the viscoelastic response depends directly on the volume itself and its evolution and not, in reality, on the aging time. This, of course, makes the set of measurements required for long-term predictions more cumbersome and few laboratories actually

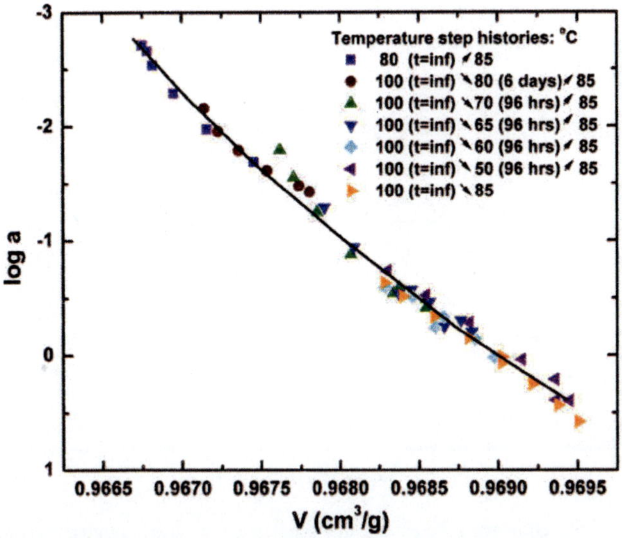

Fig. 7.23 Shift factors versus specific volume for a polymer glass in different temperature histories, as indicated. Because the viscoelastic response during aging depends on the specific volume or other state variable and not on the elapsed time per se, the shift factors are functions of volume and temperature, but not the specific history (Data replotted from Struik [9])

perform dilatometric (or enthalpic) aging experiments. Yet the glassy structure is of primordial importance when it comes to understanding and quantitatively predicting behavior. Though there has been considerable effort to use empirical approaches to non-isothermal aging time shift factors and with some success, it is this author's view that progress will only come with further exploitation of the volume (or enthalpy) data in conjunction with the viscoelastic data.

Finally, from a fundamental view it is worth noting that the finding that the viscoelastic (aging) response depends on volume or enthalpy has a strong logic to it as one might expect the dynamics to vary with these state-like parameters. However, it is worth noting that in recent work deviations from this picture seem to occur and, while volume or enthalpy might be the major considerations in many simple thermal histories, it appears that the full history of the structural recovery may need to be considered. Further discussion of this has been developed by Bernazzani and Simon [55, 56].

7.2.4.2 Nonlinear Response and Engineering Properties

Nonlinear Viscoelastic Response

One of the important findings according to the work of Struik [9] was that the aging response of polymeric glasses in the nonlinear viscoelastic regime is very different from that in the linear regime and that was discussed just above. The single most

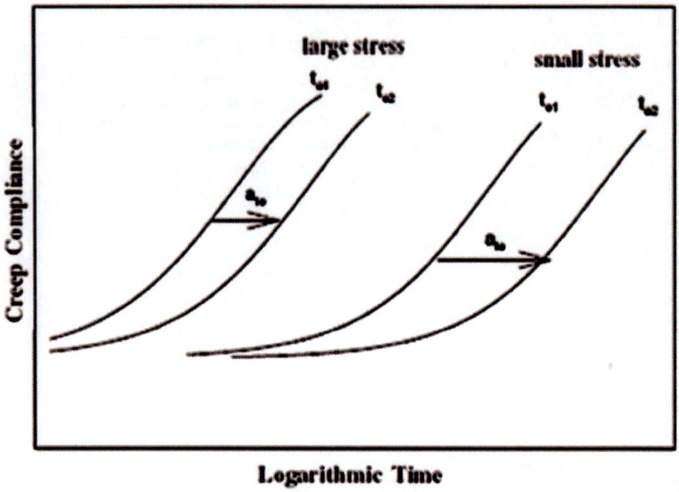

Fig. 7.24 Schematic comparing the change in viscoelastic response in large stress (nonlinear) and small stress aging experiments (After McKenna and Simon [14])

important finding is shown schematically in Fig. 7.24 where the creep response in the small deformation regime and the creep response in the nonlinear regime are compared.

The observations of importance are that as the stress increases in magnitude, there are two effects. First, the creep curve at a given short aging time shifts to shorter times at the high stresses [9]. This has been referred to as time–stress superposition and has been known for some time [101–105] though its quantitative validity remains to be fully proven [106, 107]. The second point of importance is that in going from the short aging time to the long aging time, one finds that the shift factor for the large deformations is smaller than for the small deformations. Hence, it is clear that the impact of the structural recovery on the viscoelastic properties is less for large stresses (or strains) than it is for the linear viscoelastic response. Figure 7.25 [108] shows the shift factor versus aging time for different applied stresses where it is readily seen that it decreases as applied stress for the robe stress is increased. The full development of the topic of the nonlinear viscoelastic response and physical aging is beyond the scope of the present chapter, but it is important to be aware of it because the thermal strains in composites can be large enough to move the resin into the nonlinear viscoelastic regime and, should this be the case, the expected correlation between the aging response of the resin and that of the composite may breakdown. This will be discussed in the section on aging in composites.

Yield Response

Related to the aging in the nonlinear regime is the observation of the aging of the yield stress of glassy polymers [9, 11, 23, 100, 109–125]. The yield response, to the 7extent it is related to the viscoelasticity of the polymer, will also be affected by

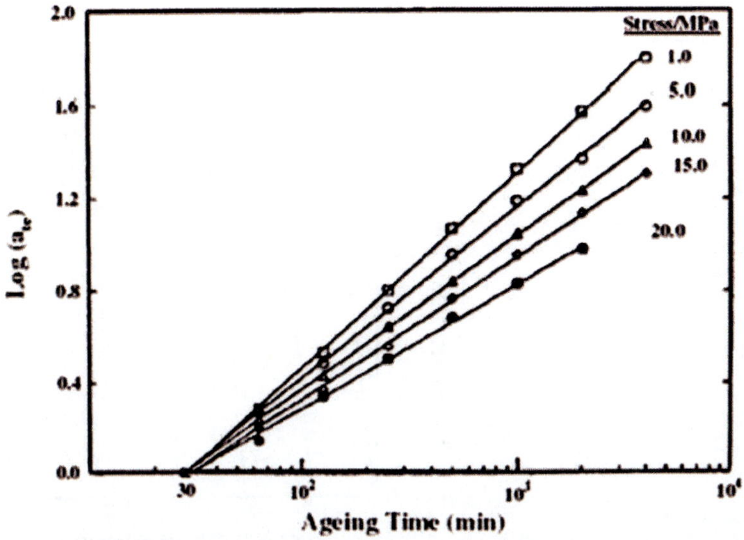

Fig. 7.25 Aging time shift factor versus aging time in double logarithmic representation for an epoxy glass in which the aging is "probed" at different stress levels (After Lee and McKenna [108])

the structural recovery. However, yield is somewhat more complicated than is the viscoelastic response and therefore the simple linear aging events are not adequate to fully describe the changing yield behavior upon aging. It is worth noting as well that there are many fewer aging and yield studies than aging and viscoelasticity studies. Figure 7.26 shows the impact of aging on the compressive yield response of an epoxy resin [109]. Near to the glass transition ($T_g - 15$ K in this case), we see a significant increase in the yield stress for the epoxy as aging times go from 0.17 h to over 1,100 h (10 min to 46 days). If one plots yield stress versus logarithm of the strain rate for different aging times, master curves of yield versus reduced strain rate can be created. The major features of yield and aging can then be treated empirically in a similar fashion to the viscoelastic response. Figure 7.27 shows one such construction for an epoxy material for aging experiments performed at two different temperatures. Within the set of data, one can see that as the compression rate (shifted rate) becomes larger, the yield strength increases at a diminishing rate, indicating an approach to an asymtotic value. Similarly, for any given aging time, one finds that the yield stress is sigmoidal in temperature, as shown in Fig. 7.28 where yield stress is plotted for several aging times as a function of "distance" below the glass transition temperature. At each temperature, one anticipates that there is an asymptotic yield stress with aging time as shown in Fig. 7.29. Other data show strong effects of aging on yield and that the impact depends on the cross-linking agent used to cure the system. Figure 7.30 presents data for yield stress as a function of aging time for a set of DGEBA epoxies cured with various cross-linking

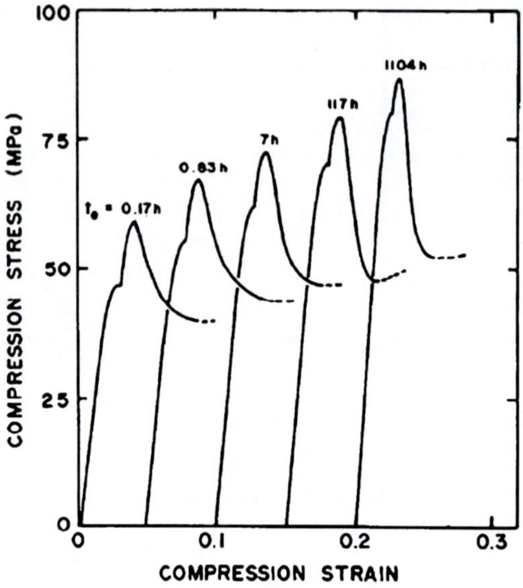

Fig. 7.26 Compressive stress–strain curves at different aging times. Note that the method of test was to begin the test at a strain rate $\dot{\varepsilon}_1$ and when the yield began to increase the strain rate by an order of magnitude, thus obtaining two yield peak data in a single test (After G'Sell and McKenna [109])

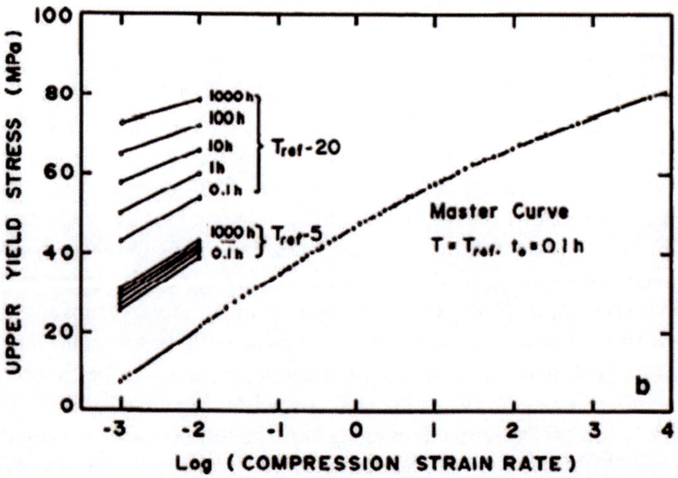

Fig. 7.27 Yield stress versus logarithm of the strain rate for different temperatures and aging times, as indicated. This shows the master curve construction (After G'Sell and McKenna [109])

agents. The plateau with aging, on this linear time scale, is clear and it depends on aging temperature and type of cross-linking agent [126].

To the extent that yield of the resin impacts the response of composite materials, the effect of aging on the yield of polymeric materials can be expected to have major impact on, e.g., compressive failure in composites, interlaminar fracture in composites, and so forth.

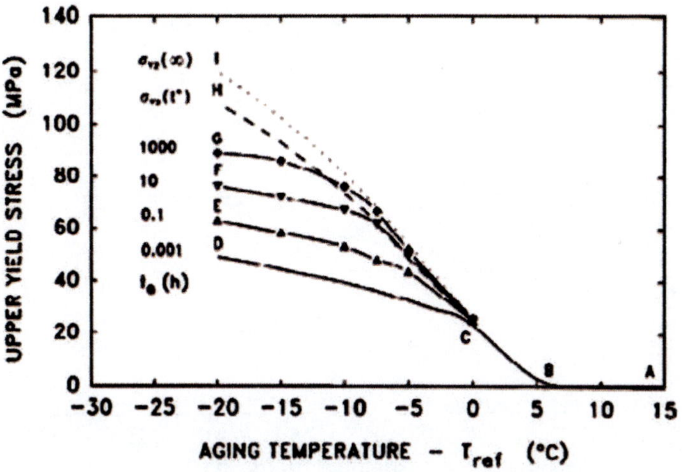

Fig. 7.28 Yield stress of an epoxy glass as a function of temperature below the glass transition temperature (T_{ref}) for different aging times (After G'Sell and McKenna [109])

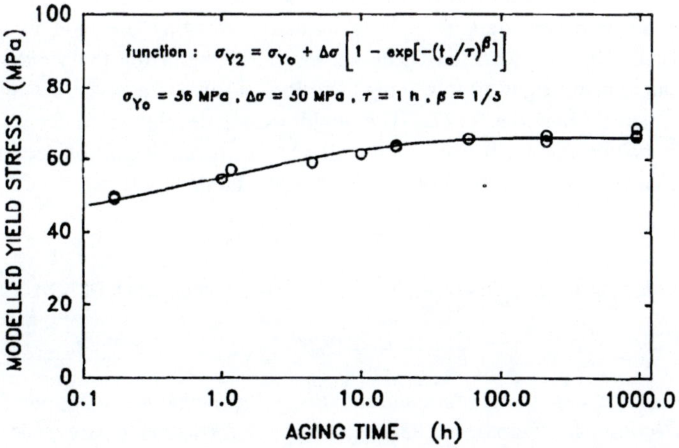

Fig. 7.29 Yield stress versus aging time for an epoxy glass aged near to the T_g and showing the asymptotic approach to a constant yield stress at long aging times (After G'Sell and McKenna [109])

Semiempirical Model of Yield-Failure and Aging: Background

The above information gives a sense that the yield response of polymeric glasses follows very similar behavior to the viscoelastic aging response and, in fact, this is probably fairly widely recognized. However, the first real attempts to put the information into a methodology that has predictive power comes from the group at the Technical University Eindhoven and began with the early development by

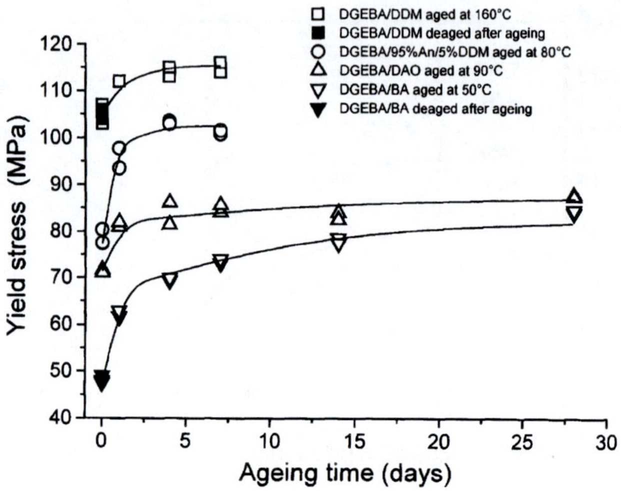

Fig. 7.30 Variation of yield stress with aging time for differently cross-linked epoxy systems (After Cook et al. [126] Reprinted with permission from Elsevier, Copyright 1999)

Tervoort, Klompen, and Govaert [105] of a nonlinear viscoelastic model and the work evolved into consideration of the influence of aging on yield in different conditions. Briefly, the work has been successful in predicting the yield stress in simple loading histories, in simple aging histories, and also after so-called "mechanical rejuvenation" histories [127]. The yield model developed in Ref. [105] is adapted to include aging and rejuvenation effects. The model begins by decomposing the total Cauchy stress into a driving stress σ_s and a hardening stress σ_r as

$$\sigma = \sigma_s + \sigma_r \tag{7.21}$$

The hardening response is treated using a neo-Hookean model [128]

$$\sigma_r = G_r \tilde{B}^d \tag{7.22}$$

where G_r is the hardening modulus and $\tilde{B}^d$ is the deviatoric component to the isochoric (constant volume) left Cauchy-Green deformation tensor. The driving stress is decomposed into a deviatoric stress σ_d^s and a hydrostatic component σ_s^h [127]

$$\sigma_s^d = G\tilde{B}_e^d \quad \text{and} \quad \sigma_s^h = \kappa(J-1)I \tag{7.23}$$

where G is the shear modulus, $\tilde{B}_B^d$ is the deviatoric part of the isochoric left Cauchy-Green deformation tensor, κ is the bulk modulus, J is the volume change factor, and I is the identity tensor. The evolution of the volume change and the deviatoric part of the left Cauchy-Green deformation tensor is given as

$$J = Jtr(D) \tag{7.24}$$

$$\dot{\tilde{B}}_e = (D^d - D_p) \cdot \tilde{B}_e \cdot (D^d - D_p) \tag{7.25}$$

and $\dot{\tilde{B}}_e$ is the Jaumann rate of $\tilde{B}_e$, D^d the deviatoric part of the rate of deformation tensor, and D_p the plastic part of the rate of deformation tensor.

The model uses a non-Newtonian flow rule and a stress-dependent Eyring viscosity to relate the plastic deformation rate tensor to the deviatoric driving stress [129]

$$D_p = \frac{\sigma_s^d}{2\eta(T,\rho,\overline{\tau},D)} \tag{7.26}$$

where the viscosity depends strongly on the equivalent stress $\overline{\tau}$ and the Eyring viscosity also is assumed to depend on the pressure through a coefficient μ, and on the intrinsic strain softening D [129, 130]

$$\eta(T,p,\overline{\tau},D) = A_0(T)e^{\mu p/\tau_0}\frac{\overline{\tau}/\tau_0}{\sin h(\overline{\tau}/\tau_0)}e^{-D} \tag{7.27}$$

and the prefactor $A_0(T)$ is written as

$$A_0(T) = A_0 e^{\Delta U/RT} \tag{7.28}$$

with A_0 a constant, ΔU the activation energy, R the gas constant, and T the absolute temperature. And the characteristic stress and the equivalent stress are defined as

$$\tau_0 = \frac{kT}{V*}; \quad \overline{\tau} = \sqrt{\frac{1}{2}tr(\sigma_s^d\sigma_s^d)} \tag{7.29}$$

and V^* is the activation volume and k is the Boltzmann constant.

The intrinsic strain softening is postulated to be represented by the parameter D which evolves from some initial value D_0 to an equilibrium value $D_\infty > D_0$ as the equivalent plastic strain $\overline{\gamma}_p$ increases, hence causing a large reduction in the viscosity of the polymer (this is the viscosity relevant to the glassy mobility and not the molecular weight entanglement viscosity). The evolution of D is then written as [127]

$$\dot{D} = h\left(1 - \frac{D}{D_\infty}\right)\dot{\overline{\gamma}}_p \tag{7.30}$$

h is the relative softening rate and the equivalent plastic strain rate $\dot{\overline{\gamma}}_p$ is given by [127]

$$\dot{\overline{\gamma}}_p = \sqrt{2tr(D_p \cdot D_p)} \tag{7.31}$$

The model has had considerable success in capturing the yield response of materials, but is too simple to capture the onset of yield, giving too linear a response of the stress–strain curve up to yield. It does capture, e.g., the strain-rate dependence of the yield stress itself and the rate dependence of the post-yield behavior.

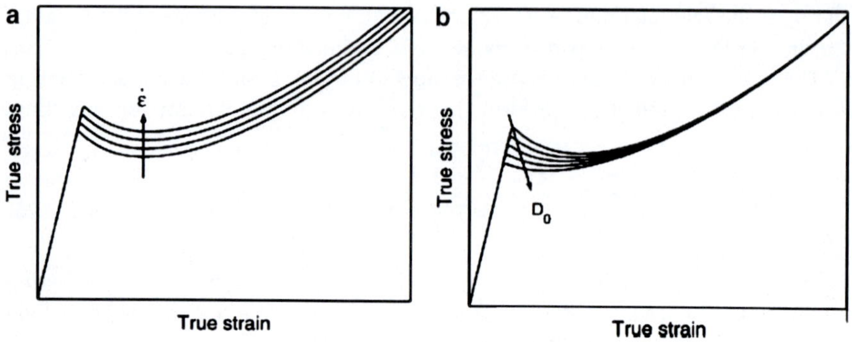

Fig. 7.31 Schematics of stress–strain response of glassy polymer predicted by the multimode elasto-viscoplastic model for (**a**) different strain rates and (**b**) different initial states of the glass (After Klompen et al. [127] Reprinted with permission from the American Chemical Society, Copyright 2005)

Furthermore, if the state of the glass, e.g., the structure, changes, one might see a change in the yield stress. Schematics of this behavior are shown in Fig. 7.31a, b. The reader is referred to Fig. 7.26 for actual stress–strain behavior of a glassy polymer that shows the "nonlinear" approach to yield that is not captured by the multimode elasto-viscoplastic model as implemented here.

The interesting aspect of the work from the Eindhoven group is that they were able to take the simple model just developed and incorporate ideas of aging kinetics into it and use these ideas to provide quantitative predictions of the yield behavior of polymers as a function of thermal history. We turn to their development of the aging kinetics [127].

The first point is that the initial condition D_0 can be taken to evolve with aging time since this impacts the response at $t = 0$. It is worth noting that Hasan et al. [131] had already recognized that the formalism of Eq. 7.15, that the state would conceptually vary with temperature, pressure, and structural state, though they did not provide an explicit form for this [127]. Klompen et al. [127] provide a combined description of the aging and the softening behavior by decomposing the flow stress into three components:

$$\sigma(\dot{\varepsilon}, S, \varepsilon) = \sigma_{rej}(\dot{\varepsilon}) + \Delta\sigma_y(S) + \sigma_r(\varepsilon) \tag{7.32}$$

where σ_{rej} flow stress of the "fully rejuvenated state,"[2] [132–134] and σ_r represents the strain-hardening stress. The non-steady state or transient contribution to the behavior is the yield drop $\Delta\sigma_y$ and this depends on the thermomechanical history

[2] We remark that the concept of rejuvenation is not completely clear. The idea of a freshly quenched state is often used when one talks of thermal rejuvenation. However, other workers sometimes refer to mechanical rejuvenation and this is more controversial. The reader is advised to see the paper by the current author on mechanical rejuvenation [100] and also the more recent work of Isner and Lacks [132] that investigate the issue using molecular simulations.

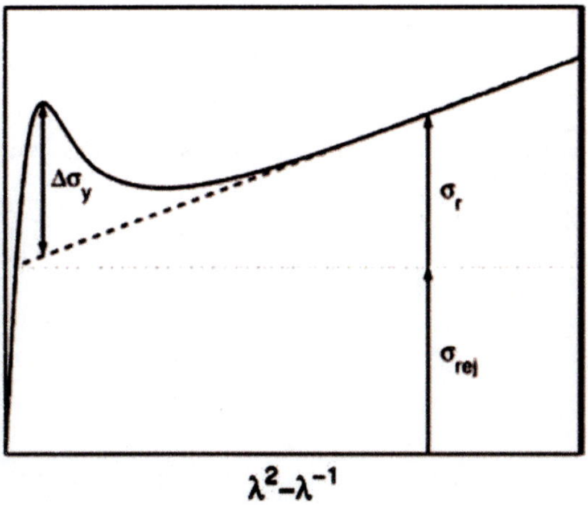

Fig. 7.32 Schematic representation of the true stress versus the deformation measure $\lambda^2 - \lambda^{-1}$ showing the decomposition of the total stress into the three components of Eq. 7.32 (After Klompen et al. [127] Reprinted with permission from the American Chemical Society, Copyright 2005)

of the material which determines the state of the glass, i.e., the parameter S. This decomposition is represented in Fig. 7.32 [127] and where the operational definitions of the parameters in Eq. 7.32 is clear.

Now the multimode elasto-viscoplastic model can be adapted to this framework by rewriting Eq. 7.27 for the viscosity as

$$\eta(T, p, \bar{\tau}) = \eta_{0,ref}(T) e^{\mu p/\tau_0} \frac{\bar{\bar{\tau}}/\tau_0}{\sin h(\bar{\tau}/\tau_0)} \tag{7.33}$$

and the value of the parameters are as defined previously except for $\eta_{0,r}(T)$ which is written as

$$\eta_{0,r}(T) = A_{rej}(T) \cdot \tau_0 \tag{7.34}$$

where $A_{rej}(T)$ is the value of $A_0(T)$ used to describe the "rejuvenated" material.

At this point, the model gives a flow behavior that depends on temperature, pressure, and rate, but not on history, viz., aging. The history dependence is put into the model by putting the state parameter S in Eq. 7.33 [127]

$$\eta(T, p, \bar{\tau}, S) = \eta_{0,r}(T) e^{\mu p/\tau_0} \frac{\bar{\bar{\tau}}/\tau_0}{\sin h(\bar{\tau}/\tau_0)} e^S \tag{7.35}$$

Klompen et al. [127] use the following expression to describe the evolution of the state parameter S:

$$S(t, T, \bar{\gamma}_p) = S_a(t, T) R_\gamma(\bar{\gamma}_p) \tag{7.36}$$

and $S_a(t,T)$ includes the aging kinetics and the fact that they depend on both time and temperature. The softening kinetics, $R_\gamma(\bar{\gamma}_p)$ are determined by the equivalent

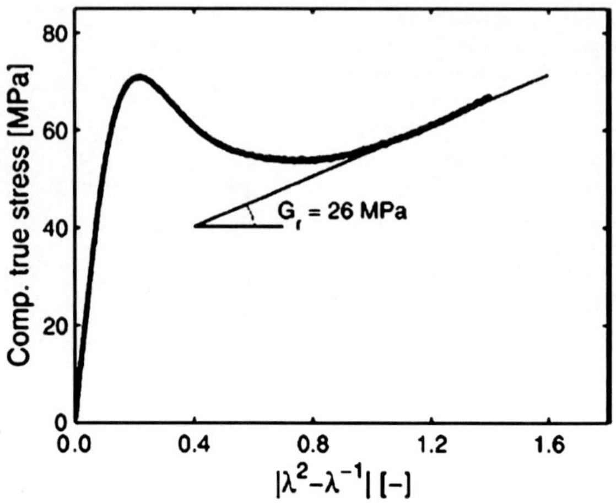

Fig. 7.33 Intrinsic deformation behavior of a commercial polycarbonate measured at a true compressive strain rate of $-10^{-3}\,s^{-1}$ (After Klompen et al. [127] Reprinted with permission from the American Chemical Society, Copyright 2005)

plastic strain. They also make the assumption that the aging kinetics are determined fully by the $S_a(t,T)$ term while the softening kinetics are normalized to unity initially and decrease to zero with increasing equivalent plastic strain. S, while conceptually similar to, e.g., volume departure from equilibrium or fictive temperature, is here used as a fitting parameter rather than as a physical parameter [127].

Semiempirical Model of Yield-Failure and Aging: Comparison with Data

Though it is semiempirical, the model proposed by Klompen et al. [127] is very useful as evidenced by applied works in which property distributions and failure due to yield have been successfully predicted [135–137]. In order to apply the model, it is necessary to first carry out experiments to obtain the "intrinsic behavior" of the material. For a material that is aged sufficiently that the test time is short relative to the total aging time, i.e., the structure of the glass is not changing during the aging experiment (see the discussion of Struik's [9] aging protocol above), then S_a in Eq. 7.36 is constant. Klompen et al. [127] use compression at a true strain rate in order to avoid necking or shear banding, i.e., to assure homogeneous deformations. Figure 7.33 shows the behavior for a commercial polycarbonate. We see immediately the similarity to Fig. 7.32 (and the difference from Fig. 7.31 in the sub-yield regime) and the definition of the strain hardening modulus G_r.

From the definition of the total stress in Eq. 7.32, we can determine the driving stress during plastic deformation (assuming incompressibility during the flow):

$$\sigma_s = \sigma - \sigma_r = \sigma - G_r(\lambda^2 - \lambda^{-1}) \tag{7.37}$$

where σ is the experimental true stress, and the hardening stress determined from the hardening modulus G_r and the draw or stretch λ.

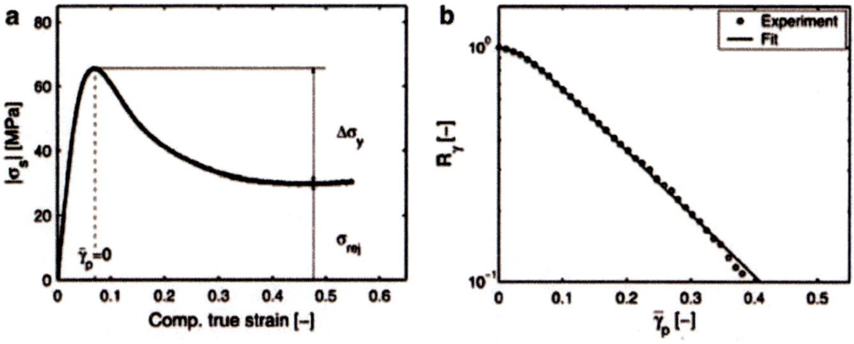

Fig. 7.34 (**a**) Driving stress and (**b**) softening characteristics for the commercial polycarbonate of Figure 7.33 (After Klompen et al. [127] Reprinted with permission from the American Chemical Society, Copyright 2005)

The driving stress, shown in Fig. 7.34a, can be split into two contributions, the rejuvenated stress (which is history independent) and the yield drop (which depends on history):

$$\sigma_s(\dot{\varepsilon}, S) = \sigma_{rej}(\dot{\varepsilon}) + \Delta\sigma_y \tag{7.38}$$

Hence, the model has a separable strain rate and history dependence.

Because, as noted before, R_γ is equal to unity at the yield point, the softening characteristic is obtained from:

$$R_\gamma(\bar{\gamma}_p) = \frac{\sigma_s - \sigma_{rej}}{|\Delta\sigma_y|} \tag{7.39}$$

Integration of Eq. 7.31 gives the value of the equivalent plastic strain [127] when we assume that plastic deformation begins at yield:

$$\bar{\gamma}_p = \sqrt{3}|\dot{\varepsilon}|(t - t_y) \tag{7.40}$$

And Fig. 7.34b shows the experimentally determined softening characteristic for the same polycarbonate as for Fig. 7.33. The problem now is to input the aging kinetics, and Klompen et al. [127] do this using an empirical equation for the yield stress as a function of aging time:

$$\sigma_y(t) = \sigma_{y,0} + c\log\left(\frac{t + t_a}{t_0}\right) \tag{7.41}$$

where t_a is an initial aging time parameter required to account for unknown initial state of the glass, and it and c and $\sigma_{y,0}$ are fitting parameters and $t_0 = 1$ s. In the case of the fit shown in Fig. 7.35 to data for a mechanically "rejuvenated" polycarbonate,

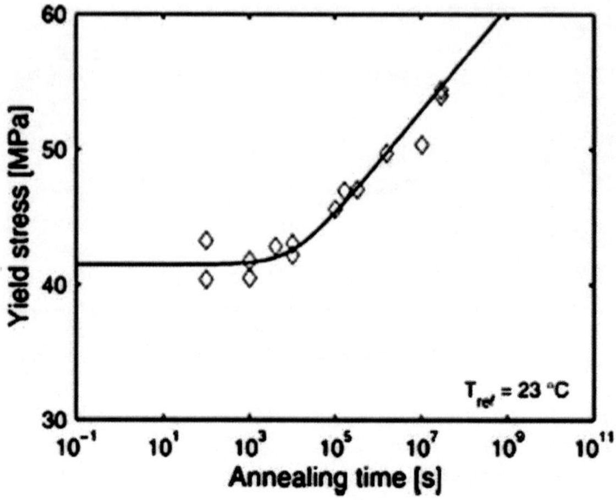

Fig. 7.35 Evolution of the yield stress for polycarbonate at a strain rate of $-10^{-2}\ s^{-1}$ as a function of aging time after "mechanical rejuvenation" showing data (*points*) and fit to Eq. 7.41 (*solid line*) (After Klompen et al. [127] Reprinted with permission from the American Chemical Society, Copyright 2005)

the fit parameters are $\sigma_{y,0} = 26.1$ MPa, the slope $c = 3.82$ MPa, and the initial "age" $t_a = 10{,}715$ s. Remark that the lack of aging or evolution of the yield stress at early times is a result of the rejuvenation being performed in torsion, which leaves the center of the sample "unrejuvenated."

The full description of the behavior of the material also requires determination of the annealing time–temperature superposition parameters and a stress shift factor. For the polycarbonate, Klompen et al. [127] used an equation for the state variable evolution similar to Eq. 7.41 for the yield stress, but including aging time, temperature and equivalent stress $\bar{\tau}$ and formulated in terms of the effective or reduced time similar to the KAHR and TNM models described above.

$$S_a(t_{eff}) = c_0 + c_1 \log\left(\frac{t_{eff}(t,T,\bar{\tau}) + t_a}{t_0}\right) \tag{7.42}$$

where the effective time is written as

$$t_{eff}(t,T,\bar{\tau}) = \int_0^t \frac{d\xi}{a_T(T(\xi))a_\sigma(\bar{\tau}(\xi))} \tag{7.43}$$

And Eq. 7.43 has the same form for reduced time as Eq. 7.11, but now the shifts are for temperature and stress. The equations for a_T and a_σ are

$$a_T(T) = e^{\left(\frac{\Delta U_a}{R}\left(\frac{1}{T} - \frac{1}{T_{ref}}\right)\right)} \tag{7.44}$$

$$a_\sigma(\bar{\tau}) = \frac{\bar{\tau}/\tau_a}{\sin h(\bar{\tau}/\tau_a)} \quad \text{and with} \quad \tau_a = \frac{RT}{\bar{v}_a} \tag{7.45}$$

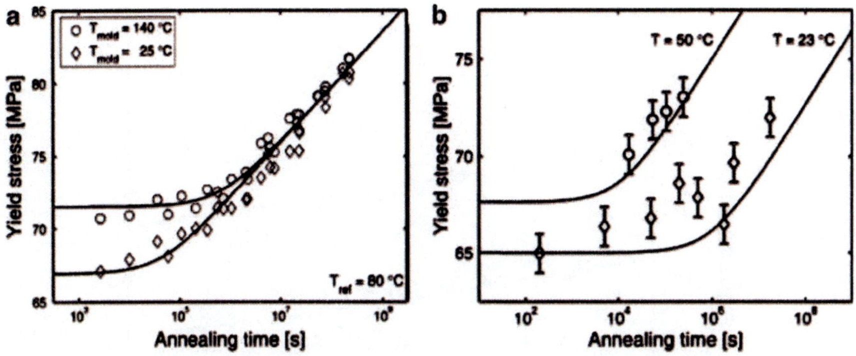

Fig. 7.36 Evolution of yield stress as a function of aging time for (**a**) a polycarbonate with two different initial ages determined by the mold temperature (indicated) and (**b**) a polycarbonate aged under 40 MPa of stress at two different temperatures, as indicated (After Klompen et al. [127] Reprinted with permission from the American Chemical Society, Copyright 2005)

And we note that the temperature shift factor is an Arrhenius form and the stress shift factor is in the form of an activated process with the aging activation volume found from fitting the data [127] to have a value $\bar{v}_a = 1.33 \times 10^{-3}$ m^3/mol and $\bar{\tau}$ is the equivalent applied stress. Furthermore, the other parameters in Eq. 7.42 are:

$$c_0 = \frac{\sqrt{3}+\mu}{3\tau_0}\left(\sigma_{y,0}(\dot{\varepsilon}_0) - \sigma_{rej}(\dot{\varepsilon}_0) - \frac{\sqrt{3}}{\sqrt{3}+\mu}\sigma_r(\lambda_y)\right) \tag{7.46}$$

$$c_1 = \frac{\sqrt{3}+\mu}{3\tau_0}c \tag{7.47}$$

Figure 7.36a, b show the success of the model in predicting the yield stress as a function of aging time for two samples with different ages and for samples loaded at 40 MPa at two different temperatures for different annealing times. We see from the figures that the semiempirical model just discussed provides reasonable predictions of the aging/annealing effects on yield stress for the polycarbonate studied.

Finally, the semiempirical model has been extended to investigate cyclic failure in which the failure mode is yield rather than fracture [137]. In this instance, there is again good agreement between the model and the experimental results, as depicted in Fig. 7.37. Interestingly, there is a better prediction in the case of the quenched material when the actual aging kinetics of the fatigue experiment are included in the model. This is, possibly, related to the observation of Yee et al. [138] that cyclic loading seems to change the structure of amorphous polymers.

In sum, the work from the Eindhoven group that provides a semiempirical, elasto-viscoplastic model of yielding in amorphous glassy polymers provides an excellent design tool and, because of the extensive experimental work that has gone into the validation of the approach, there is also a good database to evaluate other models, should one so desire.

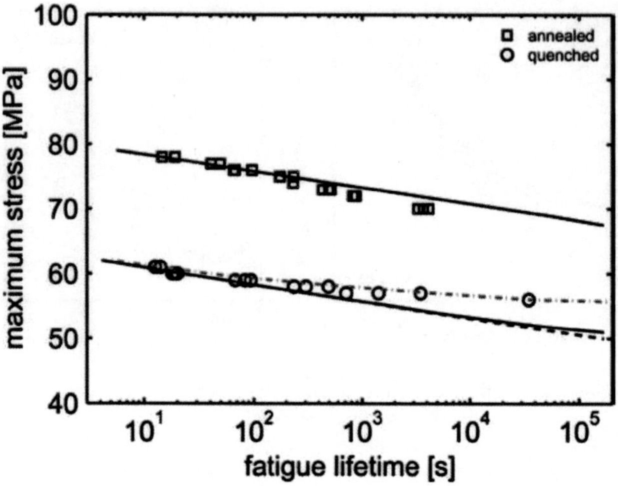

Fig. 7.37 Fatigue lifetime (in yield) for two different thermal histories and for a commercial polycarbonate. Points are data, solid lines are model predictions without aging, and dashed lines are model predictions including aging kinetics. For annealed system, the two solutions are identical. The quenched sample, the lighter dash-dot line uses the aging kinetics observed in cyclic loading tests (After Janssen et al. [137] Reprinted with permission from the American Chemical Society, Copyright 2008)

We next turn to examine the thermoviscoelastic model discussed previously in the context of structural recovery and how it describes yield and aging.

The Thermoviscoelastic Model: Yield and Aging

As indicated earlier, the thermoviscoelastic model has the intrinsic advantage over the TNM and KAHR models because it begins from a more fundamental basis, i.e., the time-dependent free energy is appropriately differentiated into the time and history dependent material properties. Above, we looked at results for the nonequilibrium apparent coefficient of thermal expansion and the apparent heat capacity. The generality of the thermoviscoelastic model is evidenced in the next paragraphs by examining the stress–strain history constitutive model that arises upon solution of Eq. 7.14. The details are beyond the present article, but briefly, the nonlinearity of the behavior arises from both the large strain deformation tensor formalism of Rational Mechanics that is used in the model and by the introduction of the configurational entropy clock term that has the time scale dependent on the entropy, which depends on the thermal history and the deformation history. Typical results for the stress–strain behavior at two different aging times are shown in Fig. 7.38 for an epoxy glass. While the results are not in perfect agreement, they do point to the power of the thermoviscoelastic model to make predictions of nonlinear response

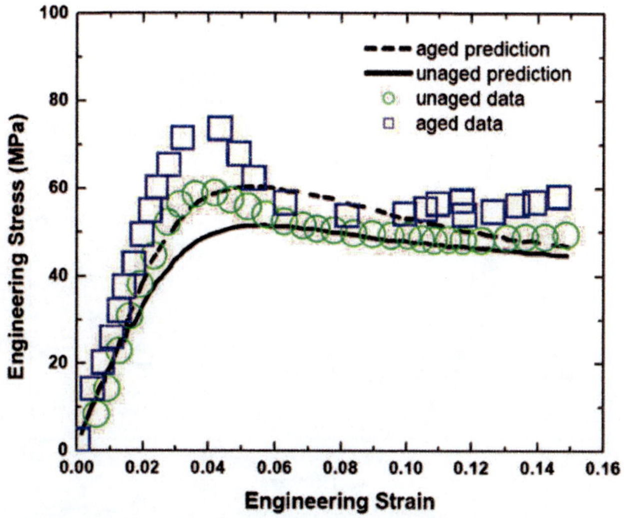

Fig. 7.38 Variation of stress–strain response, including yield stress, with aging time for a model epoxy system compared with the predictions from the thermoviscoelastic model (Replotted from Adolf et al. [80])

from linear viscoelastic parameters. Unlike the semiempirical elasto-viscoplastic model just discussed, the thermoviscoelastic model captures the curvature in the stress–strain response on the way to yield due to the fact that it is a fully nonlinear viscoelastic constitutive law, hence, naturally captures some of the time and/or rate response typical of the glassy regime viscoelasticity seen amorphous polymers. In addition to the simple yield experiment, even more complicated deformation histories can be modeled fairly successfully with the thermoviscoelastic model. This is shown in Fig. 7.39 where the epoxy sample is subjected to a three-step torsional history where the yield is reached in both the first and third steps.

Before going on to discuss failure-related properties (other than yield), it is important to mention that the thermoviscoelastic model in its most recent implementation by Adolf et al. [96] has been able to predict, at least quantitatively, the effects of aging on yield of the epoxy glass discussed above (Figs. 7.26–7.29), as well as observations such as the "implosion" event originally reported by Colucci et al. [139] in which glassy polycarbonate deformed into the nonlinear, but sub-yield, regime far below the glass transition temperature can densify in stress relaxation conditions. Hence this model, in spite of the reservation about heterogeneities in the material mentioned above, seems to have a great potential for describing general nonlinear and thermal and viscoelastic (or "viscoplastic") behaviors in polymer glasses. Further work undoubtedly should be carried out and codes made available so that groups outside of the large National labs can perform experiments and modeling in a way that fully tests the model and shows its range of strengths and limitations.

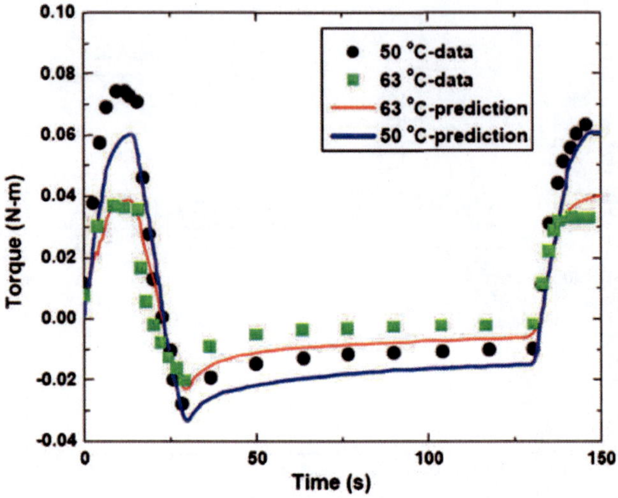

Fig. 7.39 Comparison of three-step torsional responses for a model epoxy at different temperatures with the predictions from the thermoviscoelastic model (Replotted from Adolf et al. [80])

Failure-Related Properties

The long-term performance of polymer matrix composites is affected by the ability of the resins to resist the formation of cracks and to resist fatigue or static fatigue-related processes. In addition, embrittlement due to aging would be an unacceptable outcome. Hence, it is important to understand that all of the processes described above and related to structural recovery and physical aging impact the failure response of neat polymers and, therefore, can be expected to be active in polymer matrix composites. For the most part, there are relatively few studies of the influence of structural recovery or physical aging on the failure behavior of polymers [23, 140–149]. The most probable reason for this is the large amount of time needed for the specific experiments for a single polymer and the fact that detailed failure mechanisms can vary from polymer to polymer, hence making generalizations difficult. Here, we look at several aspects of the aging and failure of glassy polymers in an attempt to illustrate that aging effects can be important but recognizing that the specific observations provided may not be universal. One important type of polymer rupture is that which occurs under a constant load and is referred to as creep rupture or static fatigue. As shown in Fig. 7.40, the creep rupture of a poly(methyl methacrylate) polymer (PMMA) is not only a very strong power of the applied stress, but it also changes with aging time. In the case of the PMMA, it appears that structural recovery (aging) increases the lifetime of the material. However, if one looks at a different polymer and at the critical strain for craze initiation, we see that the strain at crazing decreases as aging time increases. This is shown in Fig. 7.41.

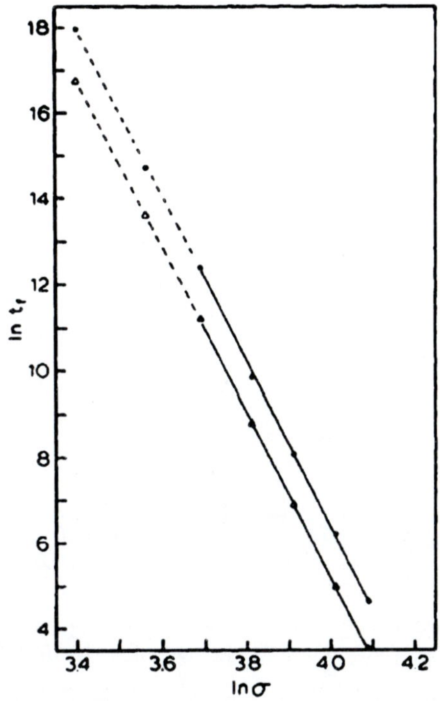

Fig. 7.40 Time to failure in creep rupture experiments for a PMMA at room temperature. (▲,Δ) Freshly quenched from above T_g, (●,○) aged for 5 years at room temperature (After Crissman and McKenna [140])

The full understanding of the relationships between structural recovery (aging) and failure is not very well developed and much further work is required even for the neat, unreinforced polymers. We now turn to the influence of the environment, specifically moisture, on the structural recovery behavior of an epoxy glass.

7.2.4.3 Environmental Effects and Structural Recovery and Aging

While it is commonly known and understood that small molecules plasticize polymers, hence reduce the glass transition temperature, there is relatively little work done to investigate how the structural recovery or aging response of polymers might be affected by changes in moisture content. In our group at Texas Tech University, we performed some experiments in which thin films of epoxy were subjected to rapid changes in moisture content and we followed either the volume recovery or the physical aging response [43, 150]. Similar works were performed using carbon dioxide as the plasticizer [151, 152] but we limit ourselves here to the findings in the "humidity-jump" experiments as moisture is the most common plasticizing agent in service conditions.

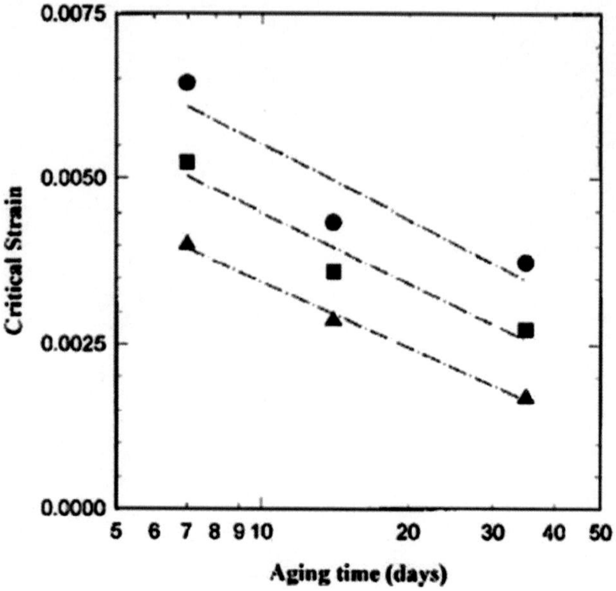

Fig. 7.41 Critical strain at which crazing initiates as a function of aging time and at different times after load application: (▲) 10^5 s, (■) 10^4 s, (●) 10^3 s (After Gusler and McKenna [142])

The Basic Hypothesis

From the above discussions of structural recovery and physical aging, one can understand the possibility that if one has a glass transition temperature that depends on the concentration of the small molecule plasticizer, then one should have an isothermal glass transition concentration. Then, similar to Figs. 7.2 or 7.5a, one can imagine that there is a nonequilibrium glassy state that arises upon "drying" the sample at a rate that is too rapid for the molecules to relax into equilibrium, but we remark, is slow enough that the moisture can diffuse out of the sample. In fact, when we first undertook the work, we thought that the "thermodynamic surface" for volume–temperature–moisture would look like that shown in Fig. 7.42. Furthermore, the surface could, in principle, be described in terms of a volume departure from equilibrium for any given temperature, humidity path. As it turned out, while the experiments in humidity-jump conditions could reproduce results similar to the intrinsic isotherms (intrinsic isopiestics for the humidity-jump experiments), asymmetry of approach and memory effect, it was found that quantitatively the glass formed by the humidity-jump experiment had different structural recovery properties than that formed by temperature-jump experiments even though the response was observed for the same final conditions of temperature and humidity. We next show the three classic signatures of structural recovery and how the relaxation response during aging varies with the departure from equilibrium, just as we saw above for the temperature-jump experiments. We then show how the

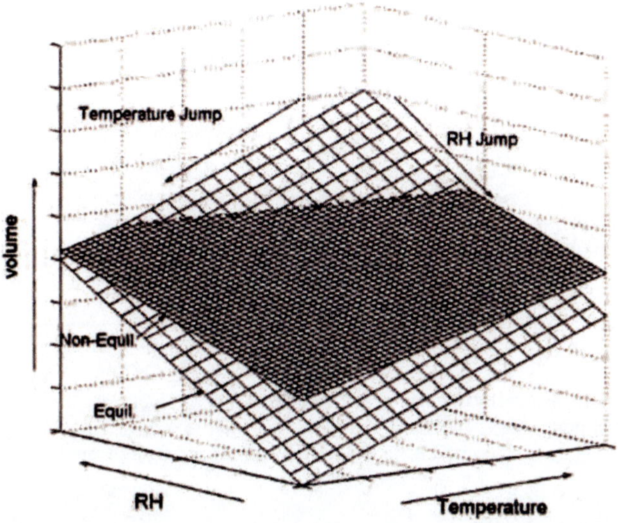

Fig. 7.42 Schematic of volume, relative humidity (moisture content), temperature surfaces for equilibrium liquid (rubbery state of polymer) (*large squares*), and for the glassy state (*small squares*). It was anticipated that the volume departure from equilibrium would determine the relaxation behavior (After Zheng and McKenna [43])

responses for temperature-jump paths through the glass transition create a different glass from the humidity jumps through the glass transition.

The Three Classic Structural Recovery Signatures

The first of the classic signatures of structural recovery is the intrinsic isotherm when one performs a temperature jump from above the glass transition temperature to below it (See Fig. 7.5). In the case of the relative humidity-jump (RH-jump), one takes a sample that is above the glass transition RH (which corresponds to a concentration) at the temperature of interest and then performs RH-jumps to lower values of RH and monitors the changing volume. Figure 7.43 shows the structural recovery in a set of RH-jump experiments and we refer to the family of curves as the "intrinsic isopiestics" [43, 153].

Figure 7.44 shows the asymmetry of approach for RH-jump experiments and we see that the figure looks similar to Fig. 7.6b, although the strength of the asymmetry is not as strong for the RH-jump as for the temperature-jump conditions. Figure 7.45 shows the memory response for the same epoxy in two-step RH-jump conditions [154]. Again, we see that the RH-jump mimics the response in temperature-jump conditions.

The above type of result was expected and supported the basic hypothesis outlined above that the plasticizer-jump should be equivalent to a temperature jump.

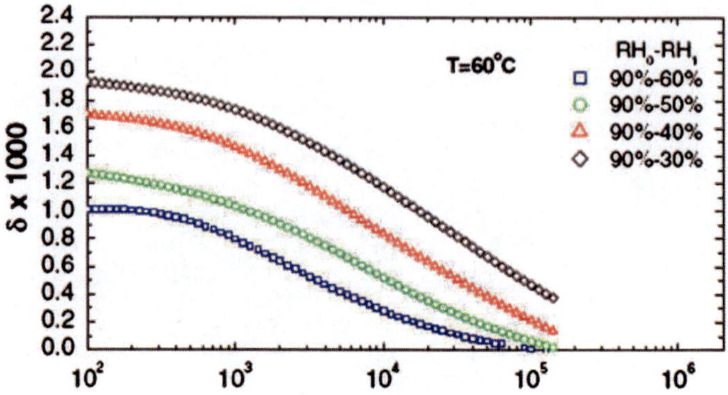

Fig. 7.43 Intrinsic isopiestics showing structural recovery (volume departure) for an epoxy subjected to RH-jumps from above the RH-glass transition to below it at the final RH values indicated in the figure (Data from Zheng and McKenna [43] and figure from McKenna [153])

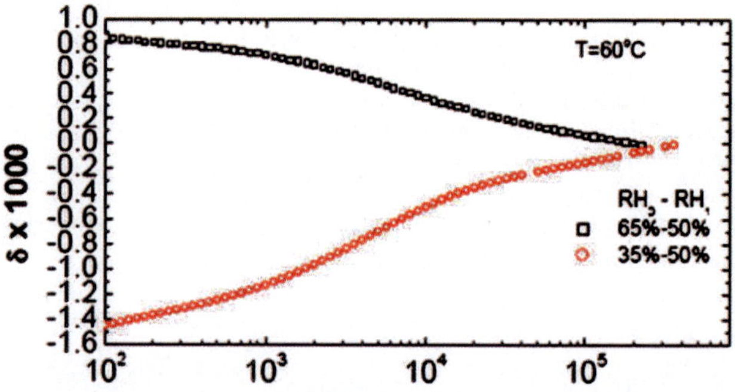

Fig. 7.44 Asymmetry of approach results for structural recovery (volume departure) for an epoxy subjected to RH-jumps from above the RH-glass transition and from below it (Data from Zheng and McKenna [43] and figure from McKenna [153])

In fact, Zheng and McKenna [43] showed that an extended form of the TNM/KAHR formalism could describe the data. However, the surprise came when physical aging experiments were performed and compared with the volume recovery data [150]. Figure 7.46 shows the comparison of the relaxation times determined during physical aging in both temperature jumps and RH jumps to the same final condition of 50% relative humidity and 60°C. As can be seen in the figure, for the same departure from equilibrium, the creep retardation time for the samples subjected to the T-jump conditions is shorter than that of the sample subjected to the RH-jump condition. This suggests that the glassy state that results from the RH jump is different from that which results from the T-jump [43, 150–154]. This is not per se surprising, but is not generally recognized. Furthermore, it implies that description of the nonequilibrium state of the glass is more difficult than would have been necessary if the glassy state did not depend on both the departure from equilibrium and the path to that departure from

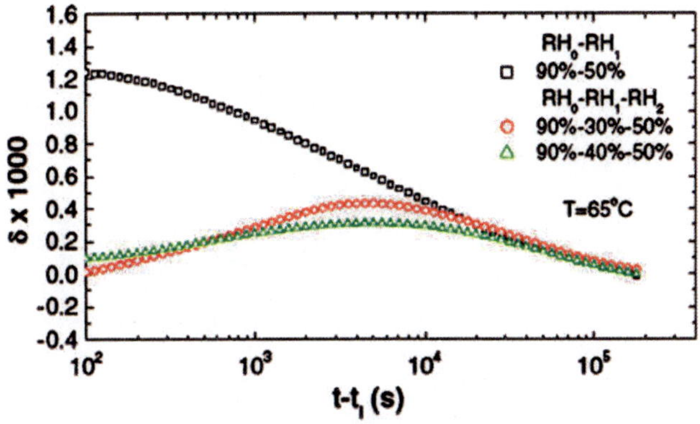

Fig. 7.45 Results from two-step RH-experiments showing memory effect for structural recovery (volume departure) for an epoxy subjected to RH-jumps as indicated in figure (Data from Zheng and McKenna [43] and figure from Zheng [154])

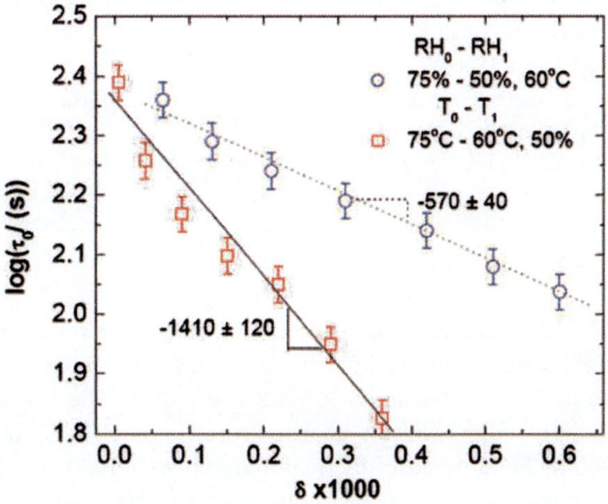

Fig. 7.46 The creep retardation time in physical aging experiments versus the volume departure from equilibrium showing that the viscoelastic response depends on whether the sample achieves the structural state (δ) via a temperature jump (*squares*) or a relative humidity jump (*circles*) (Data from Zheng and McKenna [43] and figure from Zheng [154])

equilibrium. A question to ask is whether this sort of "stable" glass for the down-jump from moisture is at all related to the ultra stable glasses formed by vapor deposition in the work of Swalen et al. [155] in spite of the differences in density, viz., that work has a much denser glass where as the present work has a less dense glass.

A final point is important here. The observation of different time scales for the two different glassy states seems to hold. However, within a given route to the glass formation (i.e., temperature jump or RH-jump), the departure from equilibrium determines the characteristic viscoelastic time of the material. This was seen in Fig. 7.23 for different temperature histories. We see it is also true for the humidity jumps in Fig. 7.47 where the aging time shift factors are seen to depend only

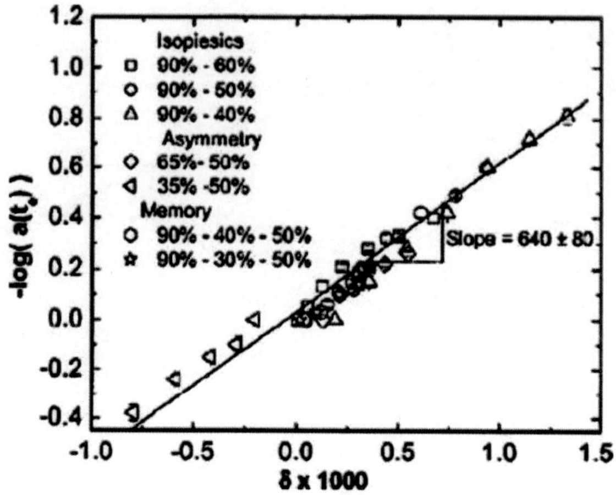

Fig. 7.47 Aging time shift factors versus volume departure from equilibrium for an epoxy glass subjected to different relative humidity-jump histories as indicated in the figure. $T = 60°C$ (After Zheng and McKenna [43])

on the volume departure from equilibrium for the different relative humidity jump conditions of intrinsic isopiestic (down-jump), asymmetry of approach (up-jump), and memory experiment (two-step RH history).

In the next section, we discuss structural recovery in materials subjected to confinement at the nanometer size scale.

7.2.4.4 Behavior at the Nanoscale

The final section on aging of unreinforced materials is related to the behavior at the nanoscale. While ordinarily, this might not seem to be a topic of relevance to composites, in the author's mind it should be. The reason for this is twofold. Even in "micro" composites or fiber-reinforced composites, there is a strong chance that a significant amount of the material between fibers or particles will be within a few nanometers (nanometers implies less than 100 nm), hence this length scale becomes important. Furthermore, there has recently been significant interest in nanocomposite materials and, in this instance, the glassy matrix material between reinforcements is going to be of nanometer dimensions.

There are only a handful of investigations of aging of materials at the nanometer size scale. The earliest was done by Mckenna et al. [156] in a structural recovery investigation of ortho-terphenyl (o-TP) confined in controlled pore glasses having different pore sizes. This work is of interest because of the errors of interpretation made in that study. The work treated the structural recovery by examining the enthalpy overshoot response in differential scanning calorimetry experiments after different aging times and within the framework of the Tool Narayanaswamy-Moynihan (TNM) model of structural recovery discussed previously. Figure 7.48 shows the results of the enthalpy recovery of o-TP in 11.2-nm diameter pores,

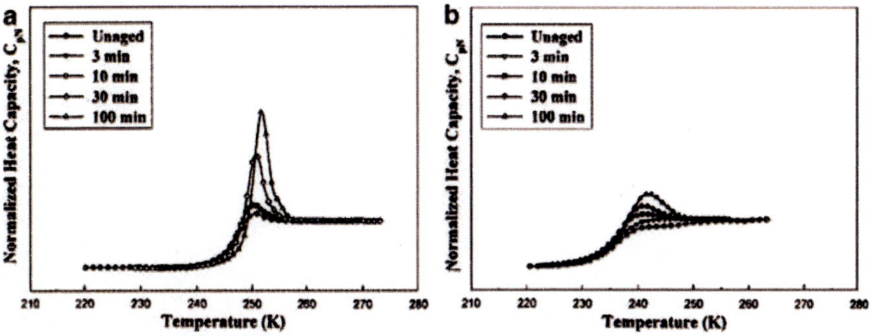

Fig. 7.48 Enthalpy recovery results for ortho-terphenyl aged at T_g–11°C in (**a**) bulk state and (**b**) confined in an 11.6 nm pore diameter controlled pore glass material showing much smaller buildup of enthalpy overshoot upon aging of the confined material. Such data were originally interpreted [156] to imply reduced aging in confined systems (Data after Simon, Park and McKenna [157])

performed in a subsequent study [157] compared to that of the bulk, unconfined material. The important feature observed is that the enthalpy overshoot in the material confined to the nanoscale is greatly reduced. In the original work, the TNM model fits to the data were interpreted to mean that the reduced enthalpy overshoot and its build-up with aging time corresponded to increased relaxation times, i.e., in spite of a reduced T_g at the nanoscale, the structural recovery at a constant temperature below the reduced T_g seemed to imply that the material was less mobile than at the macroscale. While possible, this did not seem very satisfying and that work only appeared in an abstract publication. The later work from Park and McKenna [158] and from Simon, Park, and McKenna [157] showed the correct analysis to be related to the constraints introduced in the liquid as it vitrified. The *o-TP* sticks to the walls of the nanopores and undergoes an isochoric (constant volume) glass transition. The key additional feature that they recognized in the enthalpy recovery of the o-TP was that the fictive temperature does not age to the annealing temperature for the material in the pores, as it should (see above). This is shown in Fig. 7.49 where the difference between the fictive temperature (a measure of the structure of the glass) and the aging temperature is shown as a function of aging time. As seen in the figure, the difference $T_f - T_a$ does not go to zero but remains finite. Simon, Park, and McKenna had the insight to see that the fact that $T_f - T_a$ does not go to zero implies that the confined material is in a different structural state than is the bulk material. The kinetics of the structural recovery, in this case, were successfully modeled using an extension of the TNM and KAHR models to the case in which the material in the nanopores is aged isochorically (at constant volume) and with a T_g that is reduced from that of the bulk due to the nanoconfinement. Importantly, the modeling gave a quantitative explanation for the reduced enthalpy overshoots seen in Fig. 7.48b and also provided a clear view that the behavior in the confining pores seemed to age in a way that was only related to the reduced glass transition and the increased glassy enthalpy caused by the pore

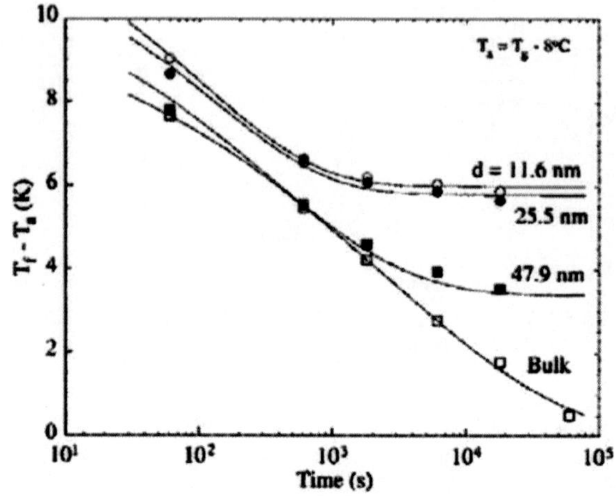

Fig. 7.49 Enthalpy recovery as $T_f - T_a$ for ortho-terphenyl in bulk and confined in nanopores as indicated. Test temperatures are at $T_g - 8°C$. Importantly, $T_f - T_a$ does not go to zero for the confined materials. See text (After Simon, Park and McKenna [157])

confinement of the *o-TP* to an isochoric state. The interpretation of a retarded process made originally was incorrect and, from Fig. 7.49, one could even argue that the sample equilibrates somewhat faster in the pores simply because the material evolves to a different state than the bulk state. In the confined system, it in fact is a state of higher enthalpy and higher free volume. Of course, if this were the final figure, we could imagine that the behavior of aging at the nanoscale is resolved. Unfortunately, for ultrathin polymer films, and even just for very thin polymer films, the situation is not yet fully understood with different results being obtained from different methods and, possibly for different materials. This is described next.

The earliest aging experiments in ultrathin polymer films were performed by Kawana and Jones [159] on polystyrene on a substrate. Thickness measurements by ellipsometry were used as an "elongational dilatometer" and volume overshoots akin to the enthalpy overshoots in Fig. 7.35 were observed. Similar to the results from the enthalpy measurements, the overshoots were observed to decrease with decreasing film thickness, though the results were interpreted to be due to a gradient of properties from bulk to liquid surface layer rather than as being due to an isochoric transition as was the case for the *o-TP* confined to nanopores. Also, the work by Kawana and Jones did not follow the kinetics as they were more interested in the film thickness effect and, in fact, observed no aging or glass transition in 10-nm thick films. Of, perhaps, greater interest is a recent series of papers from J.M. Torkelson's group in which fluorescent probes have been used to probe not only glass transition gradients in ultrathin films supported on substrates, but also to investigate the physical aging response in these thin films [160–163]. The work has also been complemented with some dielectric spectroscopy [164, 165]. The first thing to comment upon is that in the bulk, the fluorescent probe intensity is a function of density of the host matrix at a constant temperature. Therefore, making measurements of intensity versus time is nominally equivalent to making measurements of specific volume or density versus time, i.e. these are structural

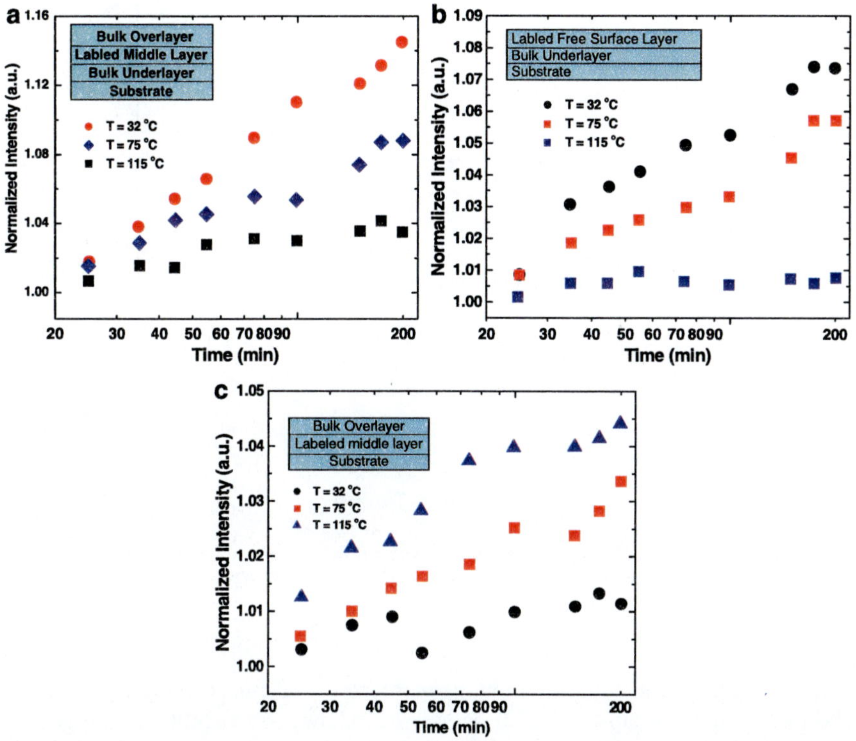

Fig. 7.50 "Structural recovery" at different temperatures as measured by fluorescence intensity of probes in PMMA layers located at different distances from the substrate or free surface showing that the aging depends on position. (**a**) middle layer exhibiting bulk-like behavior. (**b**) surface layer showing reduced T_g as evidenced by no aging at a temperature below the bulk T_g but still exhibiting reduced aging at the lower temperatures. (**c**) substrate layer showing impact of constraining substrate to reduce aging at all temperatures. (○) $T = 32°C$; (◊)$T = 75°C$; (□) $T = 115°C$. $T_{g,bulk} = 120°C$ (Data replotted from Priestley, et al. [160])

recovery measurements. One of the major findings of the work is that confinement of the ultrathin film on a substrate leads to reduced structural recovery in the ultrathin films and the amount of recovery decreases as film thickness decreases. Furthermore, the works from the Torkelson group suggest that the aging depends on where in the thin film one places the probes [160] and this is interpreted to be consistent with a gradient in T_gs reported previously by the group. Typical results are shown in Fig. 7.50 where we see the effects of confinement and position on the evolution of the normalized fluorescence intensity for poly(methyl methacrylate) layers at different locations relative to the supporting substrate.

An interesting problem arises in considering the results from work initiated in the studies of Pfromm and Koros [166] on polymer membranes in which the film dimensions are not as small as those in the aging studies above. This work has been

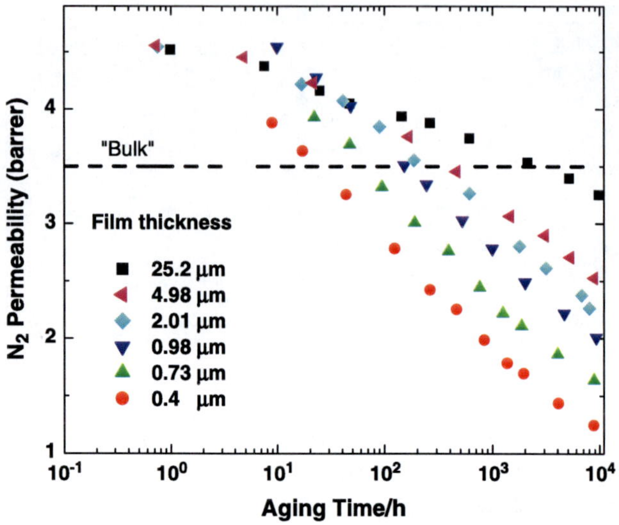

Fig. 7.51 Evolution of the nitrogen permeability with aging time for a poly(phenylene oxide) film at 35°C as a function of film thickness. The T_g of the material is 210°C (Data replotted from Huang and Paul [167])

greatly enlarged upon both in continuing work from Paul and coworkers [167–170] and by Pfromm's group [171, 172]. Figure 7.51 shows permeability of nitrogen in poly(phenylene oxide) films having thicknesses of between 400 nm and 25 μm. It can be seen that the thinnest films show more rapid aging than the thickest films and all films exhibit decreases in permeability below that for the bulk at longer aging times. Importantly, unlike the experimental results on confined films as reported above, aging is accelerated rather than retarded. Such results are not totally incongruous, but it does surprise that the aging of the free surfaces in the Torkelson experiments (Fig. 7.50b) does not show the accelerated aging observed in the permeability of the thin membranes. In fact, the results from Huang and Paul [167] are reminiscent of results from Swallen et al. [155] in which vapor-deposited samples are found to stabilize because of enhanced mobility of surfaces.

Interestingly, in earlier work, McCaig, Paul and Barlow [168] had been able to describe the aging of such films nearer to the T_g using a combined free volume recovery model and a free volume diffusion model. Although such modeling is promising, there is also evidence from dilatometry of micron-sized particles from Braun and Kovacs [173] that suggests that free volume does not diffuse and this has been an issue over the years with free volume descriptions of the glass transition. In any event, the results of Fig. 7.51 were not so-modeled undoubtedly because the aging here takes place far below the glass transition temperature and the sort of models that form the basis of our understanding of aging breakdown in such conditions. Further work, as suggested by these authors, is clearly required. One

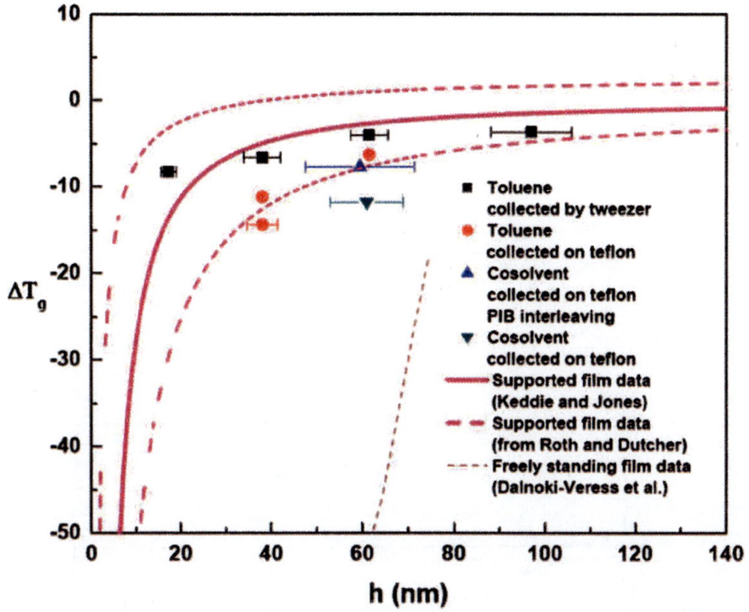

Fig. 7.52 Reduction in the glass transition in stacked, ultrathin polystyrene films compared with thin layers either constrained on a substrate or freely standing (Data replotted from Koh and Simon [174])

item of interest here is also that the observed reduction of the permeability to below that of the bulk is reminiscent of the so-called implosion or densification event observed in mechanical straining of glassy polymers far below their T_gs [139]. The similarities may suggest that there are unaccounted residual stresses in the thin films or that the diffusion/permeability process itself induces stresses sufficient to cause implosion.

One final work that needs to be commented upon here is that of Koh and Simon [174] where they took ultrathin films of polystyrene, stacked them up to 200 films in thickness, and measured the enthalpy recovery response. The response was compared with that of the bulk and it was analyzed using the TNM model discussed above. There are two important findings from that work. First, the T_g is depressed, though less than in freely standing films even though the molecular weight investigated was approximately 2.0×10^6 g/mol. Figure 7.52 shows the depression of the T_g for these samples compared with reported results for supported and freely standing polystyrene films from the literature [175]. The second important finding is that the rate of structural recovery seems to be not significantly different from the macroscopic rate of structural recovery when measured at the same distance from the glass transition temperature. Figure 7.53 shows the changing fictive temperature (minus the aging temperature) versus aging time for the macroscopic material and the stacked films. Note that here the differences between T_f and T_a go to zero at long times, as expected, unlike the case of the *o-TP* confined in nanopores discussed

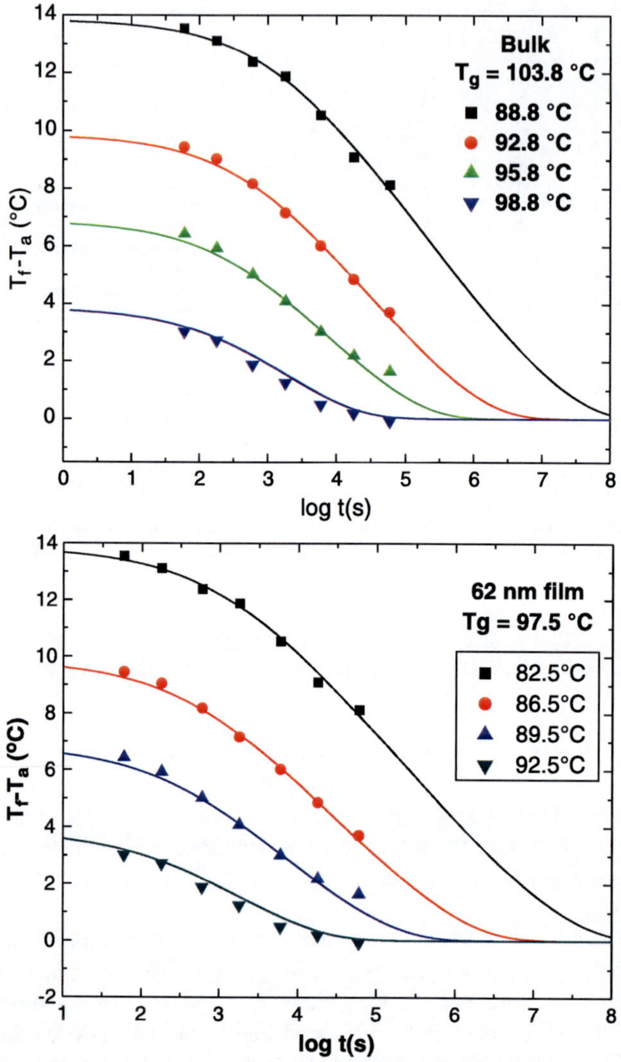

Fig. 7.53 Evolution of fictive temperature during isothermal aging for the bulk (*upper figure*) and for the stacked 62 nm ultrathin film sample (*lower figure*) at the same distance from T_g: (●) $T_a = T_g - 5$ K, (▲) $T_a = T_g - 8$ K, (■) $T_a = T_g - 11$ K, (◆) $T_a = T_g - 15$ K. The *dotted lines* are the results of TNM model fitting (Data replotted from Koh and Simon [174])

previously. Figure 7.54 shows the aging rate ($R = dT_f/d\log t_a$) as a function of the distance from T_g and as a function of temperature (insert) for both stacked 62-nm films and the bulk. Another finding in the work was that the breadth of the glass transition widened in the thin films. This latter result is consistent with either a

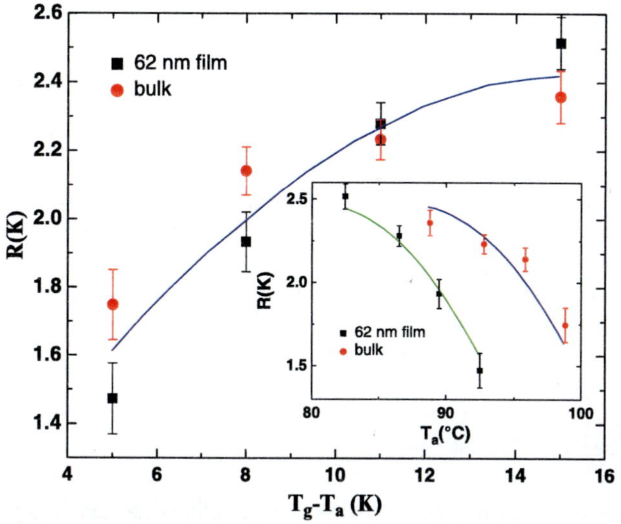

Fig. 7.54 The aging rate ($R = dT_f/d\log t_a$) versus the size of the temperature down-jump for bulk and 62-nm ultrathin film samples (Data replotted from Koh and Simon [174])

broadening of the relaxation time distribution or with a gradient of behaviors through the ultrathin film layers.

7.2.5 Physical Aging of Composite Materials

7.2.5.1 General

One important point to keep in mind as one discusses the physical aging behavior of composite materials is that, unless the fibers or particles somehow interact with the polymer resin to change the aging response of the resin, the composite aging will be the same as that of the resin. This is most easily seen in the case of viscoelastic properties where the aging is determined by shifts in the viscoelastic response of the resin and the composite properties are governed by viscoelastic mixture rules that are similar to those determined for elastic behavior, but now describing the response in complex modulus parameters. For the engineering properties of the resins, the same holds true but the models of composite failure are probably less well developed than are viscoelastic models and, therefore, specific behaviors such as interlaminar failures or compressive failures may not be as easily mapped to, e.g., effect of aging on the yield response or fracture response of the resin matrix.

Of course, if there are complex interactions between the fibers or particles and the reinforced resin, the description of the aging response can be dramatically changed. One example might be observations of reduced or increased glass transitions as reported for polymers at interfaces or at the nanoscale. Another possible effect could be macroscopic constraints that lead to, e.g., negative hydrostatic tension on the composite that could affect aging. Thermal residual stresses can also cause viscoelastic effects in these systems that have a behavior that looks like aging, but is in fact not the aging induced by structural recovery. Hence, the apparent effects of structural recovery on aging behavior of composite materials can be the same or more complicated than that of the neat polymer resin.

Finally, it is important to remark that, generally, physical aging effects are less severe than are chemical effects. This needs to be kept in mind when considering how important physical aging of the composite might be. In moderate temperature applications where chemistry is not important (or may not be important), one may still find important physical aging effects. On the other hand, if chemical degradation is easily identified as a mechanism of aging in the specific application, physical aging will be a less important, perhaps, second-order effect.

7.2.5.2 Fiber-Reinforced Composites

There are several extremely good works [176–186] that describe the aging of fiber-reinforced composites within the general framework of effective time theory as originally put forward by Struik [9]. In this author's view, however, these works suffer from the same conceptual weaknesses as the original development by Struik. It is not that the development is, per se, wrong, rather it is limited especially in the usual implementation that uses only down-jump experiments. In the case of a simple and isotropic material, as originally studied by Struik, the down-jump experiment can be readily modeled simply by noting that the curves at increasing aging times can be shifted by an amount a_{te} and that far from the glass transition temperature, this shift factor follows a power law in the aging time such that $a_{te} = a_0 {t_e}^{\mu(te)}$ and where $\mu\ (t_e)$ is the shift rate. More generally, $\mu(t)$ is not a constant as discussed above, but has a shape that results from a_{te} having a sigmoidal shape over the entire range of possible aging times. Furthermore, if the thermal history is more complicated than just a simple down-jump, such as in a memory-type [1] of experiment, one anticipates a non-monotonic behavior of a_{te} and this is a result of the shift factor depending on the structure and not on the aging time. If the measure of structure is the volume departure from equilibrium, then the response of the aging time shift factor is actually a more complicated function of time because the volume departure from equilibrium depends on the entire thermal history as shown in Eqs.7.5–7.10 if one considers the KAHR [12] or TNM [2–4] type of models. Within such a framework, observations of differing aging time shift rates, as reported by the several authors just referred to, is simply not possible unless there is something more complex happening in the system. Of course, complexity is readily envisionable. For example, a tension experiment can add hydrostatic tensions to the changing volume which would reduce the shift rate. Hydrostatic compression will do the opposite. In addition, it is now known that shearing experiments, even in the not strongly nonlinear regime, also contribute an "internal" clock to the material response [187]. In composite materials, one anticipates that there can be complicated residual stresses and these interact with the ongoing structural recovery in ways that have not been fully explored and, to the extent explored, may even be considered controversial. Saying this, then, it is worth the effort to examine some of the works on aging in fiber-reinforced composites so that the actual aging behaviors can be seen and some of the observations of differences in aging or shift rates on geometry of deformation can be considered.

There was a significant amount of work done on aging in composites, perhaps beginning with the paper by Sullivan [176] in 1990 in which it was essentially found that the aging of a composite is dominated by the resin aging properties, as indicated above should be the case. One issue in the description of the aging of a composite is how to treat the material. Sullivan, and subsequently several other workers, made the choice to treat the systems as anisotropic viscoelastic materials such that one could write the Boltzmann superposition integral as

$$\varepsilon_{ij}(t) = \int_{-\infty}^{t} \frac{S_{ijkl}(t-t')(d\sigma_{kl})}{dt'} dt' \tag{7.48}$$

and for the special case of an orthotropic material that is transversely isotropic, the compliance tensor S_{ijkl} reduces to six components S_{11}, S_{12}, S_{23}, S_{22}, S_{66}, and S_{21}. The key question arises vis-à-vis the time dependence of these functions, which ought to be the same because the resin that determines the time dependence is the same with the anisotropy coming from the fibers and their alignments. If this is so, then the equations for the time dependence should look much like those for the structural recovery using a material time that depends on the structure, e.g. departure from equilibrium δ, enthalpy or T_f. Then the equation for an aging material would become

$$\varepsilon_{ij}(t) = \int_{-\infty}^{z} \frac{S_{ijkl}(z-z')(d\sigma_{kl})}{dz'} dz' \tag{7.49}$$

and where the reduced time is defined in terms of the relevant structural measure. Following the KAHR [12] approach, one would use Eq. 7.6 and write

$$z = \int_{0}^{t} \frac{d\xi}{a_T a_\delta} \tag{7.6}$$

and the meaning of a_T and a_δ are the same as in the KAHR model. However, the need in this formulation to know how the volume of the resin changes as a function of the aging history has been little, if at all, investigated for composites. The result is the normal approach to empirically determine the material time by shifting the response curves through time-aging time superposition (and time temperature) superposition. We now examine this approach.

7.2.5.3 Time-Aging Time Superposition

There is significant work to investigate the time-aging time behavior of fiber-reinforced composites. As just indicated, most is related to the empirical

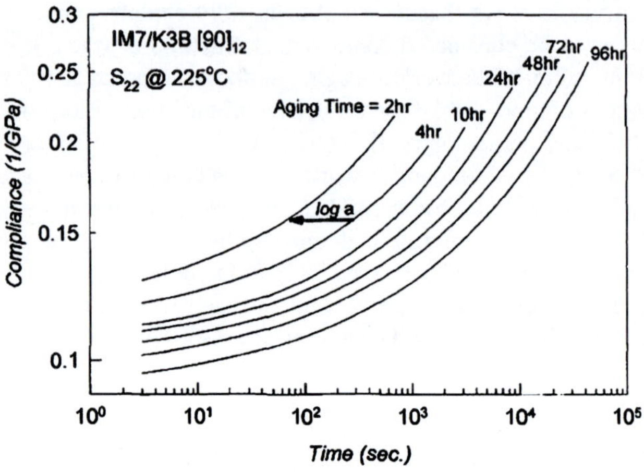

Fig. 7.55 Physical aging response of an IM7/K3B fiber-reinforced composite. Stress probes were applied transverse to the fibers (After Gates et al. [181] Reprinted with permission from Sage Publications, Copyright 1997)

time-aging time shifting and the so-called equivalent time approach to analyzing the data. In particular, rather than writing the equivalent of Eq. 7.49 with Eq. 7.6 as the definition of the reduced time, it is common to write the relevant aging time shift factor a_{te} as

$$\log(a_{te}) = \log(\alpha t_e^{\mu}) \tag{7.50}$$

where α is a number, t_e is the aging time, μ is the shift rate defined previously. Equation 7.6 would become in this instance

$$z = \int_0^t \frac{d\xi}{a_T a_{te}} \tag{7.51}$$

and the structure shift factor is replaced by the aging time shift factor. Here, one loses some of the physics underlying the process of physical aging, but the gain is that the structure itself need not be measured.

Figure 7.55 shows aging results from the work of Gates, et al. [181] for the transverse response of an IM7/K3B composite system and Fig. 7.56 shows, "momentary master curves" (MMCs) obtained from the longest aging times and in tension for the same material for several temperatures. For comparison, the neat (unreinforced) resin data are also shown. Figure 7.57 shows predictions for the long-term creep response for the same material based on using the effective time theory just discussed. Several things are to be noted from Figs. 7.55–7.57. First, aging in composites occurs in a fashion similar to that of the neat resin, at least qualitatively. Also, both time-aging time and time–temperature superposition seem to hold for the materials investigated. Finally, at least for the tension data,

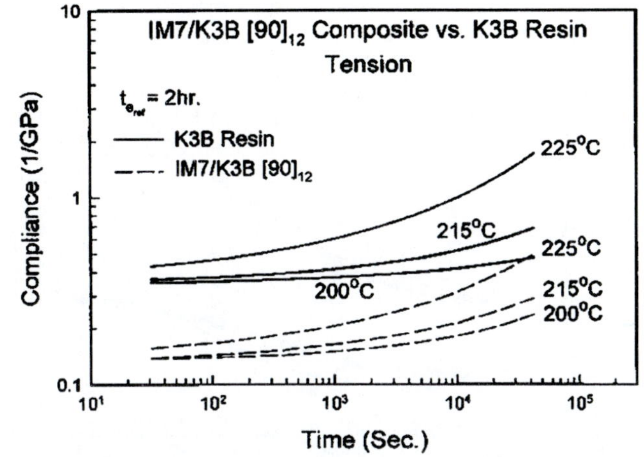

Fig. 7.56 Comparison of aging of neat resin and IM7/K3B composite in tension experiments. (After Gates et al. [181] Reprinted with permission from Sage Publications, Copyright 1997)

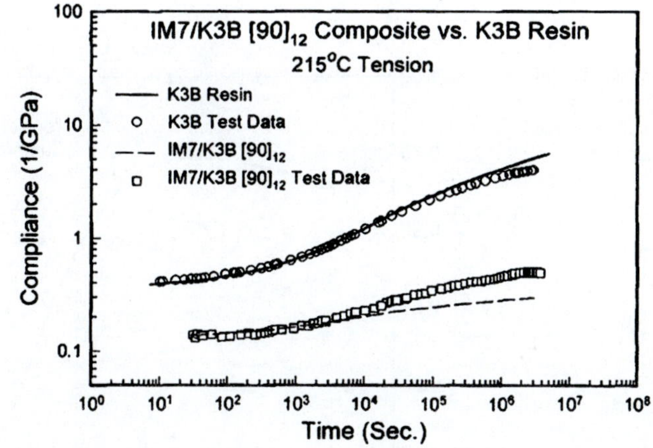

Fig. 7.57 Comparison of effective time theory predictions with the creep response for the K3B resin and its composite with IM7 fiber. Experiments in tension (After Gates et al. [181] Reprinted with permission from Sage Publications, Copyright 1997)

the effective time theory provides a reasonable prediction of the long-term creep behavior of the composite.

Several things do, however, lead to difficulties in the scheme presented here. First, as shown in Fig. 7.58, the predictions of long-term creep are not always as good as presented in Fig. 7.57 and this result is the same material, but now in compression rather than tension. Of particular interest here is the observations by Sullivan et al. [176, 177] in which there was less of a problem in the long time predictions that, in fact, the aging itself can be affected by the magnitude of,

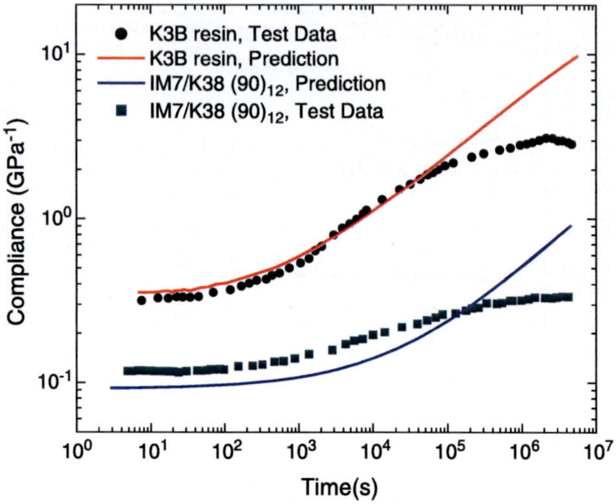

Fig. 7.58 Same as Fig. 7.57, but now for compression rather than tension (Data replotted from Gates et al. [181])

Fig. 7.59 Shift rate for the neat K3B resin in tension and compression (Data replotted from Gates et al. [181])

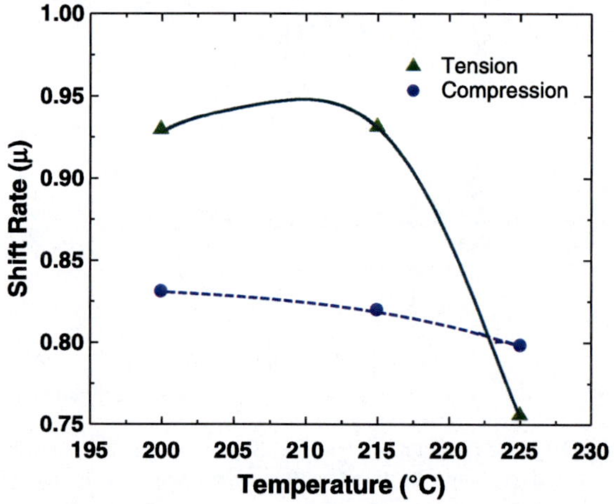

e.g., the applied stresses as originally found by Struik [9] and subsequently became an important area of study. What is true is that it seems that internal stresses in a composite material can lead to different aging rates. The data of Fig. 7.58, however, do not necessarily justify this interpretation because the lack of long-time agreement of the creep curves is equally evident in the neat resin as it is in the composite.

Another aspect of the aging of fiber-reinforced composites is the observation that aging in different geometries of deformation can be different. This is shown in Figs. 7.59 and 7.60 for the neat resin and the composite from the data of Gates,

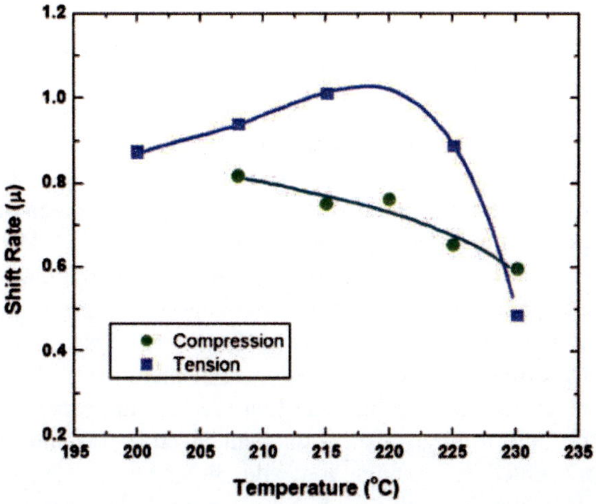

Fig. 7.60 Shift rates in temperature and compression as functions of temperature for the IM7/K3B composite material (Data replotted from Gates et al. [181])

et al. [181] just discussed. The surprising thing about the data in Figs. 7.59 and 7.60 is the observation of different shift rates for the different modes of deformation, not only in the composite, but also in the neat resin. In the linear regime, it is difficult to imagine such behavior. A partial explanation is found in the fact that one often needs vertical shifts to superimpose the data, either in aging experiments or temperature-shifting experiment and a large sensitivity to the vertical shifting of the horizontal shift factors may play an important role in the application of time-aging time superposition principles and especially for applications where one is further below the T_g than most laboratory experiments are performed. This is an important point because the implication is that more and longer tests need to be performed to evaluate a material's long-term performance than would otherwise be the case.

A set of comments is needed here. There is little reason to think that the principles of time- aging time (time-structure) superposition should not be applicable in a straightforward manner to the viscoelastic behavior of composite materials. Yet, the above discussion certainly suggests that this is not the case as long-term predictions are not as good as one might expect and the physical parameters are often not consistent, though they should be, from one set of measurements to another even for the neat resins. Hence, it is clear that there is a need for both further investigation of the origins of the breakdown of the general laws governing the time-structure superposition in the linear regime (if it is not due to error of test or analysis), as well as a need to find other than the empirical laws to relate aging time to the viscoelastic response, viz., using a true structure shift factor a_δ or a_{Tf}. Increasing use of resin-based materials in engineering applications demands this.

To provide a specific suggestion, while much of this author's own work has either been in the empirical vein or looking at the volume as a measure of structure, this may not be possible in the case of composites. Yet, there is the possibility of using the enthalpy recovery as providing the structural parameter in fiber-reinforced composites. This has been done to study aging in composites, hence indicating the possibility of using calorimetric methods to determine the structure of the composite resin. The missing link, to this author's knowledge, has been tying the structure as determined by enthalpy recovery to the viscoelastic response in the composites. In that case, the TNM version of the structural recovery would be applicable to the reduced time equations. It is worth noting that Simon and coworkers have examined the relationships among volume, enthalpy, and creep of unreinforced polymers and they find that these are not uniquely related [188–190].

Finally, in the present discussion, it is important to remark that if the presence of the reinforcing fibers causes a change in the properties of the resin, the problem becomes significantly more difficult. The first possible cause of different resin properties is chemical. In the case of thermosets, this could be extremely difficult to control and it could lead to a gradient in properties from the fiber surface into the bulk through, e.g., a gradient in cross-link density. If known, this could be accounted for, though it would require a detailed model of the T_g gradient in the system as well as a full description of the thermal histories in the sample. To the extent that time-crosslink superposition can be expected to work [98, 191–193] one might be able to take this into account. While property differences have been reported in fiber-reinforced systems, there is little done to examine systematically the potential impact on modeling of the long-term aging of composites of such a property gradient. The other aspect of the fibers having a large impact on the resin properties is more subtle. Due to coefficient of thermal expansion differences between the resin and the fiber, there can be reasonably large residual stresses in the composite; the important ones for the present discussion are those in the resin. Another point is that the impact of the changing structure of an aging glass is lessened for large stresses as compared to small stresses. The origin of this difference in aging behaviors is not completely clear and is a subject of current research. What is clear is that the description of aging in a system that contains stress gradients can be different from one without such gradients. Since residual stresses are tensorial in nature, this may also partially explain some of the differences in aging observed for different geometries of deformation, at least in composite materials. It is clear that further research is required to understand the relevant phenomena.

7.2.5.4 Particle-Reinforced Polymers

The general rules governing aging and structural recovery have been discussed and these would be expected to apply to particle-reinforced polymers. However, the same sort of issue discussed for the fiber-reinforced composites may prove to give rise to similar apparently inconsistent results for the particle-reinforced materials. That is, such aspects as residual stress may be important. Furthermore, if one considers microparticle versus nanoparticle reinforced systems, there may be a difference in behaviors if the properties of the resins in the nanocomposites show

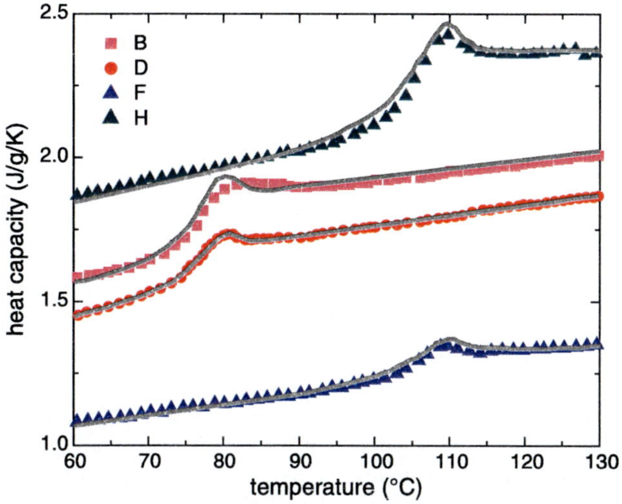

Fig. 7.61 Comparison of predictions from the thermoviscoelastic model of the enthalpy recovery response for several microparticle-reinforced polymer composites that have been aged for 120 min close to the T_g of the system before reheating. *Solid lines* are predictions, *symbols* are data (Data replotted from Adolf and Chambers [82])

the same sort of variation as one sees in confined glass-forming liquids or ultrathin polymer films [194–200].

In the case of aging of microparticle-reinforced composites, there may also be manifestations of property changes such as found by Fraga et al. [201] where different activation energies for aging in filled and unfilled epoxy were observed. In addition, phenomena related to residual stresses (perhaps similar to fiber-reinforced systems) can be seen. Yilmazer and Farris [202] attribute enhanced ductility in filled systems to relaxation of residual stresses surrounding the glass sphere filler particles during aging. They also attributed some differences in behavior to dewetting of the reinforcing particles due to the hydrostatic tension surrounding spherical particles in the tensile deformation of the composite. One promising avenue of research is the use by Adolf and Chambers [82] of the thermoviscoelastic model described above for particle-filled composites. They were able to predict the enthalpy response after 120 min of aging near to the glass transition temperature. The results are shown in Fig. 7.61. They also showed some success in describing the yield response of these same materials. As long as the resin behaves as it did in the unreinforced state, simple rules of mixtures should apply, but nonlinear events may make superposition rules difficult to formulate.

In the case of nanocomposites, perhaps it is first useful to discuss briefly the behavior of semicrystalline polymers, which have an amorphous phase of resin reinforced by nanometers thick lamellar crystals. The nano-confined amorphous regions between crystalline lamellae can be either "rigid amorphous" [203] or simply confined or constrained [204–207] depending on the view that one takes. Aging of such confined systems has been little studied, though some thoughtful results are found in early measurements of aging of the rigid amorphous phase [208–210] in semicrystalline polymers and simply the aging of semicrystalline polymers in the broadened glass transition regime of the constrained amorphous phase [204–207, 211]. More recent work on nanocomposites has suggested that aging can be suppressed by making the nanoparticles attractive [165], which can also increase the glass transition temperature. But there are few studies of nanocomposite aging [212–214] and this is a field that is ripe for systematic and extensive investigation.

One final point of interest concerning the nanocomposite materials is their behavior in the rubbery state. The oldest example of such composites, of course, is carbon black-filled rubber. In that example, it is well known that there is a potential for so-called bound rubber [215] to exist because the polymer chains are attracted to the high-energy carbon black surface. In such an instance, one sees similar behavior to the semicrystalline polymers and the bound rubber can cause the filled rubber to exhibit a broadened, toward high temperatures, glass transition [216] just as is seen in some of the semicrystalline polymers [204–207, 211, 217]. However, it is also true that the impact of the interface in the nanocomposites seems to depend on the strength of the interactions between the particles and the polymer with the result that, even above the glass transition temperature, the behavior remains controversial [216, 218–223]. Finally, it is well worth noting that in his seminal work on physical aging, Struik already had observed that aging could occur above the unfilled polymer glass transition due to the adherence of the polymer chains to the filler particles [9].

7.3 Perspectives

The continued increase of polymer resin-based composite materials in applications that demand long-term performance, e.g. aerospace structures, biomedical implants, automotive applications, and civil engineering structures, demands improved methods of estimating long-term performance. Although chemical stability is often the point of first concern, it is clear that the closeness of the use temperature of most polymer-based composites to the glass transition temperatures of the "weakest link" polymer will mean that models that can incorporate structural recovery effects on material behavior will become increasingly important in material and structural design. Today, even with the unreinforced resins, it is difficult to make long-term predictions of lifetime. This is partially because the fundamentals of the failure mechanisms of the individual materials is often poorly understood and, even more relevant in the present context, the impact of the structural recovery or physical aging process on failure processes is not well understood and very little studied. This implies that the long-term use of the reinforced materials demands further investigation not only of the durability of the reinforced materials but also of the neat resins. This needs to be done in a fundamental context, such as that currently being attempted by the Caruthers group and Sandia National Laboratory collaborations. If polymers and nanocomposites are to be more widely used, a similar set of comments is relevant to the understanding of their behaviors. There is relatively little work on durability of nanocomposites in the context of physical aging or structural recovery and the lack of full understanding of, e.g., the impact of nanoconfinement on even the glass transition of materials at the nanoscale implies that much more fundamental work is required. If this author has a recommendation to make, though, it is not simply for further research, but it is for research that couples modeling approaches that span the range of molecular to continuum and that are strongly linked to experimental investigations that are relevant to each scale. Much progress continues to be made by individual teams. More could be made with coordination among the teams. The future development of the composite materials technologies of the twenty-first century demands it.

Acknowledgments The author is grateful for the generous support of the National Science Foundation under grant DMR-0804438, the Office of Naval Research under project number N00014-06-1-0922, and the John R. Bradford Endowment at Texas Tech University for partial support of this work. He is also thankful to Arts et Métiers ParisTech in Paris, France for partial support of the work during the summer of 2009. The author is also grateful to Ben Xu for help in creating figures.

References

1. A.J. Kovacs, "Transition Vitreuse dans les Polyméres Amorphes. Etude Phénoménologique," Fortschritte der Hochpolymeren-Forschung, **3**, 394–507 (1963).
2. A.Q. Tool, "Relation Between Inelastic Deformability and Thermal Expansion of Glass in Its Annealing Range," J. Amer. Ceram. Soc., **29**, 240–253 (1946); A.Q. Tool, "Viscosity and the Extraordinary Heat Effects in Glass," J. Research National Bureau of Standards (USA), **37**, 73–90 (1946).
3. O.S. Narayanaswamy, "A Model of Structural Relaxation in Glass, "J. Am. Ceram. Soc., **54**, 491–498 (1971).
4. C.T. Moynihan, P.B. Macedo, C.J. Montrose, P.K. Gupta, M.A. DeBolt, J.F. Dill, B.E. Dom, P.W. Drake, A.J. Esteal, P.B. Elterman, R.P. Moeller, H. Sasabe and J.A. Wilder, "Structural Relaxation in Vitreous Materials," Ann. N.Y. Acad. Sci., **279**, 15–35 (1976).
5. G.W. Scherer, *Relaxation in Glass and Composites*, Krieger Publishing Co., Malabar, Florida (1992).
6. G.B. McKenna, "Glass Formation and Glassy Behavior," in *Comprehensive Polymer Science, Vol. 2: Polymer Properties*, ed. by C. Booth and C. Price, Pergamon Press, Oxford, (1989) pp. 311–362.
7. J.M. Hutchinson, "Physical Aging in Polymers," Progress in Polymer Science, **20**, 703–760 (1995).
8. S.L. Simon, "Physical Aging," Chapter in Encyclopedia of Polymer Science, this edition (2001).
9. L.C.E. Struik, *Physical Aging in Polymers and Other Amorphous Materials*, Elsevier, Amsterdam (1976).
10. C.A. Angell, K.L. Ngai, G.B. McKenna, P.F. McMillan and S.W. Martin, "Relaxation in Glassforming Liquids and Amorphous Solids," J. Appl. Phys., **88**, 3113–3157 (2000).
11. S.E.B. Petrie, "Thermal behavior of annealed organic glasses," J. Polym. Sci. A-2, **10**, 1255–1272 (1972).

12. A.J. Kovacs, J.J. Aklonis, J.M. Hutchinson and A.R. Ramos, "Isobaric Volume and Enthalpy Recovery of Glasses. II. A Transparent Multiparameter Model," J. Polym. Sci., Polym. Phys. Ed., **17**, 1097–1162 (1979).
13. I.M. Hodge, "Enthalpy Relaxation and Recovery in Amorphous Materials," J. Non-Crystalline Solids, **169**, 211–266 (1984).
14. G.B. McKenna and S.L. Simon, "The glass transition: Its measurement and underlying physics," *in Handbook of Thermal Analysis and Calorimetry, 3rd. Ed.*, S.Z.D. Cheng, editor. Elsevier, Amsterdam (2002).
15. J.M. O'Reilly, "Review of structure and mobility in amorphous polymers," CRC Critical Reviews in Solid State and Materials Sciences, **13**, 259–277 (1987).
16. R.N. Haward, *The Physics of Glassy Polymers*, Applied Science, London (1973).
17. *Assignment of the Glass Transition*, R.J. Seiler, ed., ASTM STP 1249, American Society for Testing and Materials, Philadelphia, PA (1994).
18. S.V. Nemilov, *Thermodynamic and Kinetic Aspects of the Vitreous State,* CRC Press, Boca Raton, FL (1995).
19. E.J. Donth, *The glass transition: relaxation dynamics in liquids and disordered materials,* Springer-Verlag, New York (2001).
20. M.R. Tant and A.J. Hill, *Structure and Properties of Glassy Polymers,* ACS Symposium Series 710, American Chemical Society, Washington, DC (1998).
21. P.A. O'Connell and G.B. McKenna, "Large Deformation Response of Polycarbonate: Time-temperature and Time-Aging Time Superposition," in *Handbook of Polycarbonate Science and Technology*, ed. by D.G. LeGrand and J.T. Bendler, Marcel Dekker, New York, Chapter 10, pp 225–254 (1999).
22. S. Arrese-Igor, O.Mitxelena, A. Arbe, A. Alegría, J. Colmenero and B. Frick, "Effect of stretching on the sub-T_g phenylene-ring dynamics of polycarbonate by neutron scattering," Phys. Rev. E., **78**, 021801–1 - 021801–8 (2008).
23. G.B. McKenna, "On the Physics Required for the Prediction of Long Term Performance of Polymers and Their Composites," J. Res. NIST, **99**, 169–189 (1994).
24. G.B. McKenna, "Interlaminar Effects in Fiber Reinforced Plastics–A Review," Polymer-Plast. Technol. Eng., **5**, 23–53 (1975).
25. *Composite Materials: Testing and Design, ASTM STP 460,* American Society for Testing and Materials, Philadelphia, PA (1969).
26. *Composite Materials: Testing and Design (Third Conference), ASTM STP 546,* American Society for Testing and Materials, Philadelphia, PA (1973).
27. J.R. Vinson and T.W. Chou, *Composite Materials and Their Use in Structures,* Applied Science Publishers, London (1975).
28. *Composite Materials, ed. by L.J. Broutman and R.H. Krock, Vol. 2. Mechanics of Composite Materials, ed. by G.P. Sendeckyj,* Academic Press, New York (1971).
29. D.J. Plazek,"Temperature Dependence of the Viscoelastic Behavior of Polystyrene," J. Phys. Chem., **69**, 3480–3487 (1965).
30. H. Vogel, "Das Temperaaturabhängigkeitsgesetz der Viskosität Flüssigkeiten," Phys. Z., **22**, 645–646 (1921).
31. G. S. Fulcher, "Analysis of Recent Measurements of the Viscosity of Glasses," J. Am. Ceram. Soc., **8**, 339–355 (1925).
32. G. Tammann, "Glasses as supercooled liquids," J. Soc. Glass Technol. **9**, 166–185 (1925).
33. M.L. Williams, R.F. Landel and J.D. Ferry, "The Temperature Dependence of Relaxation Mechanisms in Amorphous Polymers and Other Glass-forming Liquids," Journal of the American Chemical Society, **77**, 3701–3707 (1955).
34. J.D. Ferry,*Viscoelastic Properties of Polymers, 3rd. Edition*, J. Wiley and Sons, New York (1980).
35. T. Hecksher, A.I. Nielsen, N.B. Olsen and J.C. Dyre, "Little evidence for dynamic divergences in ultraviscous molecular liquids," Nature Physics, **4**, 737–741 (2008).

36. G.B. McKenna, "Glass dynamics - Diverging views on glass transition, Nature Physics, **4**, 673–674 (2008).
37. C.A. Bero and D.J. Plazek, "Volume-dependent rate processes in an epoxy resin," J. Polym. Sci. Part B: Polymer Physics, **29**, 39–47 (1991).
38. C.T. Moynihan, A.J. Easteal, M.A. DeBolt and J. Tucker, "Dependence of the fictive temperature of glass on cooling rate," J. Am. Ceramic Soc., **59**(1–2), 12–16 (1976).
39. P. Badrinarayanan, W. Zheng, Q.X. Li and S.L. Simon, "The glass transition temperature versus the fictive temperature," J. Non-Crystalline Solids, **353**, 2603–2612 (2007).
40. J.Y. Park and G.B. McKenna, "Size and confinement effects on the glass transition behavior of polystyrene/o-terphenyl polymer solutions, Phys. Rev. B., **61**, 6667–6676 (2000).
41. W.D. Callister, Jr. and D.G. Rethwisch, *Fundamentals of Materials Science and Engineering. An Integrated Approach, 3rd. Ed.*, J. Wiley and Sons, Hoboken, NJ, USA (2008).
42. R.H. Doremus, "Viscosity of Silica," J. Appl. Physics, **92**, 7619–7629 (2002).
43. Y. Zheng and G.B. McKenna, "Structural Recovery in a Model Epoxy: Comparison of Responses after Temperature and Relative Humidity Jumps," Macromolecules, **36**, 2387–2396 (2003).
44. I. Echeverria, P.-C. Su, S.L. Simon and D.J. Plazek, "Physical aging of a polyetherimide: Creep and DSC measurements," J. Polymer Science. Part B: Polymer Physics, **33**, 2457–2468 (1995).
45. L. Boltzmann, "Zur Theorie der Elastischen Nachwirkung," Sitzungsber. Akad. Wiss. Wien. Mathem.-Naturwiss. Kl., **70**, 2. Abt. 275–300 (1874).
46. R. Kohlrausch, "Theorie des Elektrischen Rückstandes in der Leidener Flasche," Annalen der Physik und Chemie von J.C. Poggendorff, **91**, 179–214 (1854).
47. G. Williams and D.C. Watts, "Non-symmetrical dielectric relaxation behaviour arising from a simple empirical decay function," Trans. Faraday Soc. **66**, 80–85 (1970).
48. P. Badrinarayanan, S.L. Simon, R.J. Lyn and J.M. O'Reilly, "Effect of structure on enthalpy relaxation of polycarbonate: Experiments and modeling," Polymer, **49**, 3554–3560 (2008).
49. G.B. McKenna and S.L. Simon, "Time-dependent volume and enthalpy responses in polymers," in *Time dependent and nonlinear effects in polymers and composites*, ASTM STP 1357, R.A. Schapery and C.T. Sun, eds., West Conshohocken, PA, pp 18–46 (2000).
50. C.R. Schultheisz and G.B. McKenna, "Volume recovery, physical aging and the tau-effective paradox in glassy polycarbonate following temperature jumps," Proceedings of the 25th Annual Conference of the North American Thermal Analysis Society, September 7–9, McClean, VA, pp 366–373 (1997).
51. C.R. Schultheisz, NIST, unpublished.
52. G.B. McKenna, C.A. Angell, R.W. Rendell, C.T. Moynihan, A.J. Kovacs,J.M. Hutchinson, M. Oguni, J. O'Reilly, L. Struik, I.M. Hodge, J.C. Bauwens, E. Oleynick, S. Rekhson, G. Williams and S. Matsuoka, "The phenomenology and models of the kinetics of volume and enthalpy in the glass transition range," J. Non-Crystalline Solids, **131**, 528–536 (1991).
53. L. Andreozzia, M. Faetti, F. Zulli, and M. Giordano, "Enthalpy relaxation of polymers: comparing the predictive power of two configurational entropy models extending the AGV approach," Eur. Phys. J. B., **41**, 383–393 (2004).
54. J.M. Hutchinson, S. Montserrat, Y. Calventus and P. Cortes, "Application of the Adam-Gibbs equation to the non-equilibrium glassy state," Macromolecules, **33**, 5252–5262 (2000).
55. P. Bernazanni and S.L. Simon," Volume Recovery of Polystyrene: Evolution of the Characteristic Relaxation Time," J. Non-Crystalline Solids, **307**, 470–480 (2002).
56. S.L. Simon and P. Bernazanni, "Structural Relaxation in the Glass: Evidence for a Path Dependence of the Relaxation Time," J. Non-Crystalline Solids, **352**, 4763–4768 (2006).
57. It is important to note that ideas such as free volume and its distribution are inherently heterogeneous. For instance, in the KAHR model[12], it was recognized that the local relaxation time could depend on the local free volume or local departure from equilibrium δ_I but the choice was made to simply use the global departure from equilibrium because of limited computational capacity available at that time.

58. R.E. Robertson, R. Simha and J.G. Curro, "Free-volume and the kinetics of aging of polymer glasses," Macromolecules, **17**, 911–919 (1984).
59. R. Simha and T. Somcynsky, "Statistical thermodynamics of spherical and chain molecule fluid," Macromolecules, **2**, 342–350 (1969).
60. G. Adam and J.H. Gibbs, "The temperature dependence of cooperative relaxation properties in glass-forming liquids," J. Chem. Phys., **43**, 139–146 (1965).
61. T.S. Chow, "Free volume distributions in amorphous polymers," Macromolecular Theory and Simulations," **4**, 397-4-4 (1995).
62. J. Liu, Q. Deng and Y.C. Jean, "Free-Volume Distributions of Polystyrene Probed by Positron-annihilation–Comparison with Free-volume Theories," Macromolecules, **26**, 7149–7155 (1993).
63. M. Schmidt and F.H.J. Maurer, "Isotropic pressure-densified atactic poly(methyl methacrylate) glasses: Free-volume properties from equation-of-state data and positron annihilation lifetime spectroscopy," Macromolecules, **33**, 3879–3891 (2000).
64. Y.C. Jean, "Can positron annihilation lifetime spectroscopy measure the free-volume hole size distribution in amorphous polymers? Comment," Macromolecules, **29**, 5756–5757 (1996).
65. R. Richert, "Heterogeneous Dynamics in Liquids: Fluctuations in Time and Space," J. Phys.: Condens. Matter, **14**, R703 - R738 (2002).
66. M. Ediger, "Spatially Heterogeneous Dynamics in Supercooled Liquids," Annu. Rev. Phys. Chem. **51**, 99–128 (2000).
67. S. C. Glotzer, "Spatially Heterogeneous Dynamics in Liquids: Insights from Simulation," J. Non-Cryst. Solids, **274**, 342–355 (2000).
68. R. Bohmer, "Non-exponential Relaxation in Disordered Materials: Phenomenological Correlations and Spectrally Selective Experiments," Phase Transitions, **65**, 211 (1998).
69. R. Bohmer, "Nanoscale Heterogeneity of Glass-Forming Liquids: Experimental Advances," Curr. Opin. Solid State Mater. Sci., **3**, 378–385 (1998).
70. C. T. Thurau and M. D. Ediger, "Influence of Spatially Heterogeneous Dynamics on Physical Aging of Polystyrene," J. Chem. Phys., **116**, 9089–9099 (2002).
71. P. Chaudhuri, L. Berthier, S. Sastry and W. Kob, "On the relaxation dynamics of glass-forming systems: Insights from computer simulations," *Modeling and Simulation of New Materials,* AIP Conference Proceedings, Volume 1091, 95–108 (2009).
72. T. R. Bohme and J. J. dePablo, "Evidence for Size-Dependent Mechanical Properties from Simulations of Nanoscopic Polymeric Structures," J. Chem. Phys., 116 (22) 9939–9951 (2002).
73. J. Dudowicz, K.F. Freed KF and J.F. Douglas, "Generalized entropy theory of polymer glass formation," *Advances in Chemical Physics*, **137**, 125–222 (2008).
74. K. Chen, E.J. Saltzman and K.S. Schweizer, "Segmental dynamics in polymers: from cold melts to ageing and stressed glasses," J. Phys.: Cond. Matter, **21**, 503101–1 – 503101–20 (2009).
75. J.M. Caruthers, School of Chemical Engineering, Purdue University, Lafayette, IN. The work is originally published in Ph.D. theses referenced below. The first paper by Lustig, Shay and Caruthers that put the model into the literature is also cited below. The more recent developments by Caruthers in collaboration with Adolf and Chambers and then extended by Adolf and others are broad and well implemented attempts to make the model more useful and to develop fuller understanding of the strengths and limitations of the model from an engineering perspective.
76. S.R. Lustig (1989) "A Continuum Thermodynamics Theory for Transport in Polymer/Fluid Systems," Ph.D. Thesis, Purdue University, Lafayette, IN.
77. Colucci, D.M., (1995) "The Effect of Temperature and Deformation on the Relaxation Behavior in the Glass Transition Region," Ph.D. Thesis, School of Chemical Engineering, Purdue University, Lafayette, IN.
78. D.S. McWilliams, (1996) "Study of the Effect of Thermal History on the Structural Relaxation and Thermoviscoelasticity of Amorphous Polymers," Ph.D. Thesis, Purdue University, Lafayette, IN.

79. S.R. Lustig, R.M. Shay and J.M. Caruthers, "Thermodynamic Constitutive Equations for Materials with Memory on a Material Time Scale," Journal of Rheology, **40**, 69–106 (1996).
80. D.B. Adolf, R.S. Chambers and J.M. Caruthers, "Extensive Validation of a Thermodynamically Consistent, Nonlinear Viscoelastic Model for Glassy Polymers," Polymer, **45**, 4599–4621 (2004).
81. J.M. Caruthers, D.B. Adolf, R.S. Chambers, P. Shrikhande, "A Thermodynamically Consistent, Nonlinear Viscoelastic Approach for Modeling Glassy Polymers," Polymer, **45**, 4577–4597 (2004).
82. D.B. Adolf and R.S. Chambers, "Application of a Nonlinear Viscoelastic Model to Glassy, Particulate-Filled Polymers," Journal of Polymer Science: Part B. Polymer Physics, **43**, 3135–3150 (2005).
83. D.B. Adolf and R.S. Chambers, "A Thermodynamically Consistent, Nonlinear Viscoelastic Approach for Modeling Thermosets during Cure," J. Rheol., **51**, 23–50 (2007).
84. D.B. Adolf, R.S. Chambers, J. Flemming, J. Budzien and J. McCoy, "Potential Energy Clock Model: justification and Challenging Predictions," J. Rheology, **51**, 517–540 (2007).
85. C. Truesdell, *Rational Thermodynamics*, 2nd ed., Springer-Verlag, New York (1984).
86. Typically one uses Maxwell relations to relate the different thermodynamic variables (specific volume, enthalpy, coefficient of thermal expansion, heat capacity for example) to the free energy of the system[6,87].
87. Typically any general thermodynamics text will have the appropriate relations. See, e.g., R.A. Swalin, *Thermodynamics of Solids, second edition*, Wiley-Interscience, 1972.
88. S.R. de Groot and P. Mazur, *Nonequilibrium Thermodynamics*, Dover Publications, Mineola, NY (1984).
89. S. Lengyel, "On Classical Nonequililbrium Thermodynamics and Its Extensions," Lecture Notes in Physics, **199**, 398–406 (1984).
90. F. Bampi and A. Morro, "Nonequilibrium thermodynamics—A hidden variable approach," Lecture Notes in Physics, **199**, 211–232 (1984).
91. H.C. Ottinger, *Beyond Equilibrium Thermodynamics*, Wiley, Hoboken, NJ, USA (2005).
92. I. Prigogine, *Introduction to Thermodynamics of Irreversible Processes*. 3 rd edition, Wiley Interscience, New York (1967).
93. B.D. Coleman, "Thermodynamics of Materials with Memory," Archives of Rational Mechanics and Analysis, **17**, 1–46 (1964).
94. B.D. Coleman, "On Thermodynamics, Strain Impulses, and Viscoelasticity," Archives of Rational Mechanics and Analysis, **17**, 230–254 (1964).
95. W.Noll, "A Mathematical Theory of the Mechanical Behavior of Continuous Media," Archives of Rational Mechanics and Analysis, **2**,197–226 (1958).
96. D.B. Adolf, R.S. Chambers, J. Flemming, J. Budzien and J.McCoy, "Potential Energy Clock Model: Justification and Challenging Predictions," J. Rheol., **51**, 517–540 (2007).
97. P.A. O'Connell and G.B. McKenna, "Large Deformation Response of Polycarbonate: Time-Temperature, Time-Aging Time, and Time-Strain Superposition," Polymer Engineering and Science, **37**, 1485–1495 (1997).
98. A.Lee and G.B. McKenna, "Effect of Crosslink Density on Physical Aging of Epoxy Networks," Polymer, **29**, 1812–1817 (1988).
99. P.A. O'Connell and G.B. McKenna, "Arrhenius like Temperature Dependence of the Segmental Relaxation below Tg," Journal of Chemical Physics, **110**, 11054–11060 (1999).
100. G.B. McKenna, "Mechanical Rejuvenation in Polymer Glasses: Fact or Fallacy?," J. Phys.: Condens. Matter, **15**, S737–S763 (2003).
101. S. Matsuoka, S.J. Aloisio and H.E. Bair, "Interpretation of shift of relaxation time with deformation in glassy polymers in terms of excess enthalpy," J. Appl. Phys., **44**, 4265–4268 (1973).
102. R.A. Schapery, Polym. Eng. Sci.,"Characterization of nonlinear viscoelastic materials," **9**, 295–310 (1969).

103. Y.C. Lou and R.A. Schapery, J. Comp. Matls., "Viscoelastic characterization of a nonlinear fiber-reinforced plastic," **5**, 208–234 (1971).
104. B. Bernstein and A. Shokooh, "The Stress Clock Function in Viscoelasticity," J. Rheol., **24**, 189–211 (1980).
105. T.A. Tervoort, E.T. J. Klompen and L.E. Govaert, "A multi-mode approach to finite, three-dimensional, nonlinear viscoelastic behavior of polymer glasses," J. Rheol., **40**, 779–797 (1996).
106. J.-J. Pesce and G. B. McKenna, "Prediction of the Sub-Yield Extension and Compression Responses of Glassy Polycarbonate from Torsional Measurements," J. Rheology, **41**, 929–942 (1997).
107. S. Jazouli, W.B. Luo, F. Bremand and T. Vu-Khanh," Application of time-stress equivalence to nonlinear creep of polycarbonate," Polymer Testing, **24**, 463–467 (2005).
108. A. Lee and G.B. McKenna,, "The Physical Aging Response of an Epoxy Glass Subjected to Large Stresses," Polymer, **31**, 423–430 (1990).
109. C.G'Sell and G.B. McKenna,, "Influence of Physical Aging on the Yield Behavior of Model DGEBA/Poly(propylene oxide) Epoxy Glasses," Polymer, **33**, 2103–2113 (1992).
110. M. Aboulfaraj, C. G'Sell, D. Mangelinck, and G.B. McKenna, "Physical Aging of Epoxy Networks after Quenching and/or Plastic Cycling," J. Non-Crystalline Solids, **172–174**, 615–621 (1994).
111. C.H. Huu and T. Vu-Khanh, "Effects of physical aging on yielding kinetics of polycarbonate," Theoretical and Applied Fracture Mechanics, **40**, 75–83 (2003).
112. J.M. Hutchinson, S. Smith, B. Horne and G.M. Gourlay, "Physical aging of polycarbonate: Enthalpy relaxation, creep response, and yielding behavior," Macromolecules, **32**, 5046–5061 (1999).
113. M. Aboulfaraj, C. G'Sell, D. Mangelinck, and G.B. McKenna, "Physical Aging of Epoxy Networks after Quenching and/or Plastic Cycling," J. Non-Crystalline Solids, **172–174**, 615–621 (1994).
114. C.G. Robertson, J.E. Monat and G.L. Wilkes, "Physical aging of an amorphous polyimide: Enthalpy relaxation and mechanical property changes," J. Polym. Sci. Part B: Polymer Physics, **37**, 1931–1946 (1999).
115. R. Song R, J. Chen, J.G. Gao, S. Lin and Q.R. Fan, "The effect of physical aging on the properties of atactic polystyrene," Acta Polymerica Sinica, **1**, 61–66 (1998).
116. L. Teze, J.L. Halary, L. Monnerie and L. Canova, "On the viscoelastic and plastic behaviour of methylmethacrylate-co-N-methylglutarimide copolymers," Polymer, **40**, 971–981 (1999).
117. J.P. Armistead and A.W. Snow, "Influence of matrix properties on fragmentation test," J. Adhesion, **1–4**, 209–222 (1995).
118. T.S. Chow, "Stress-Strain Behavior of Physically Aging Polymers," Polymer, **34**, 541–545 (1993).
119. B. Haidar and T.L. Smith, "History-dependent and temperature-dependent yield phenomena of polycarbonate related to its rate of physical aging," Polymer, **32**, 2594–2600 (1991).
120. W.H. Jo and K.J. Ko, "The effects of physical aging on the thermal and mechanical-properties of an epoxy polymer," Polym. Eng. Sci., **31**, 239–244 (1991).
121. O.A. Hasan, M.C. Boyce, X.S. Li and S. Berko, "An investigation of the yield and post-yield behavior and corresponding structure of poly(methyl methacrylate)," J. Polym. Sci. Part B: Polymer Physics, **31**, 186–197 (1993).
122. R.A. Bubeck, S.E. Bales and H.D. Lee, "Changes in yield and deformation of polycarbonates caused by physical aging," Polym. Eng. Sci., **24**, 1142–1148 (1984).
123. C. Bauwens-Crowet and J.C. Bauwens, "Annealing of polycarbonate below the glass transition-Quantitative interpretation of the effect on yield stress and differential scanning calorimetry measurements," Polymer, **23**, 1599–1604 (1982).
124. C. Bauwens-Crowet and J.P. Bauwens, "Effect of thermal history on the tensile yield stress of polycarbonate in the beta-transition range," Polymer, **24**, 921–924 (1983).
125. J.J. Martinez-Vega, H. Trumel and J. L. Gacougnolle, "Plastic deformation and physical ageing in PMMA," Polymer, **43**, 4979–4987 (2002).

126. W.D. Cook, M. Mehrabi and G.H. Edward, "Ageing and Yielding in Model Epoxy Thermosets," Polymer, **40**, 1209–1218 (1999).
127. E.T.J. Klompen, T.A.P. Engels, L.E. Govaert and H.E.H. Meijer, "Modeling of the postyield response of glassy polymers: Influence of thermomechanical history," Macromolecules, **38**, 6997–7008 (2005).
128. T.A. Tervoort and L.E. Govaert, "Strain-hardening behavior of polycarbonate in the glassy state", J. Rheol., **44**, 1263–1277 (2000).
129. T.A. Tervoort, R.J.M. Smit, W.A.M. Brekelmans and L.E. Govaert, "A constitutive equation for the elasto-viscoplastic deformation of glassy polymers," Mech. Time-Dep. Materials, **1**, 269–291 (1998).
130. L.E. Govaert, P.H.M. Timmermans, W.A.M. Brekelmans, "The influence of intrinsic strain softening on strain localization in polycarbonate: Modeling and experimental validation" J. Eng. Mater. Techn., **122**, 177–185 (2000).
131. O.A. Hasan, M.C. Boyce, X.S. Li and S. Berko, "An investigation of the yield and post-yield behavior and corresponding structure of poly(methyl methacrylate)," J. Polym. Sci. Part B: Polymer Physics, **31**, 185–197 (1993).
132. B.A. Isner and D.J. Lacks, "Generic Rugged Landscapes under Strain and the Possibility of Rejuvenation in Glasses," Phys. Rev. Lett., **96**, 025506–1 – 025506–4 (2006).
133. G.B. McKenna and A.J.Kovacs, "Physical Ageing of Poly(methyl methacrylate) in the Nonlinear Range: Torque and Normal Force Measurements," Polym. Eng. and Sci., **24**, 1138-1141 (1984).
134. H.G.H. van Melick, L.E. Govaert, B. Raas, W.J. Nauta and H.E.H. Meijer, "Kinetics of ageing and re-embrittlement of mechanically rejuvenated polystyrene," Polymer, **44**, 1171–1179 (2003).
135. H.E.H. Meijer and L.E. Govaert, "Mechanical performance of polymer systems: The relation between structure and properties," Prog. Polym. Sci., **30**, 915–938 (2005).
136. T.A.P. Engels, L.C.A. van Breemen, L.E. Govaert and H.E.H. Meijer, "Predicting the long-term mechanical performance of polycarbonate from thermal history during injection molding," Marcomolecular Materials and Engineering, **294**, 829–838 (2009).
137. R. P.M. Janssen, D. de Kanter, L.E. Govaert and H.E.H. Meijer, "Fatigue life predictions for glassy polymers: A constitutive approach," Macromolecules, **41**, 2520–2530 (2008).
138. L.B. Liu, D. Gidley and A.F. Yee, "Effect of cyclic stress on structural changes in polycarbonate as probed by positron-annihilation lifetime spectroscopy," J. Polym. Sci. Part B: Polymer Physics, **30**, 230–238 (1992).
139. D.M. Colucci, P.A O'Connell and G.B. McKenna, "Stress Relaxation Experiments in Polycarbonate: A Comparison of Volume Changes for Two Commercial Grades," Polymer Engineering and Science, **37**, 1469–1474 (1997).
140. J.M. Crissman and G.B. McKenna,"Relating Creep and Creep Rupture in PMMA Using a Reduced Variables Approach," J. Polym. Sci., Phys. Ed., **25**, 1667–1677 (1987).
141. J.M. Crissman and G.B. McKenna, "Physical and Chemical Aging in PMMA and Their Effects on Creep and Creep Rupture Behavior," J. Polymer Science, Phys. Ed., **28**, 1463–1473 (1990).
142. G.M. Gusler and G.B. McKenna, "The Craze Initiation Response of A Polystyrene and a Styrene-Acrylonitrile Copolymer During Physical Aging," Polymer Engineering and Science, **37**, 1442–1448 (1997).
143. M. Delin and G.B. McKenna, "The Craze Growth Response in Stress Relaxation Conditions for a Styrene Acrylonitrile Copolymer During Physical Aging," Mechanics of Time Dependent Materials, **4**, 231–255 (2000).
144. J.E. Lincoln, R.J. Morgan and E.E. Shin, "Effect of thermal history on the deformation and failure of polyimides," J. Polym. Sci. Part B: Polymer Physics, **39**, 2947–2959 (2001).
145. S. Sacks and W. S. Johnson, "Effects of thermal aging on the mechanical behavior of K3B matrix material," J. Thermoplastic Composites, **11**, 429–442 (1998).

146. H. Parvatareddy, J.Z. Wang, D.A. Dillard, T.C. Ward and M.E. Rogalski, "Environmental aging of high performance polymeric composites-Effects on durability," Composites Science and Technology, **53**, 399-409 (1995).
147. J.C. Arnold, "The effects of physical aging on the brittle-fracture behavior of polymers," Polym. Eng. Sci., **35**, 165–169 (1995).
148. J.C. Arnold, "The influence of physical aging on the creep-rupture behavior of polystyrene," J. Polym. Sci. Part B: Polymer Physics, **31**, 1451–1458 (1993).
149. V.T. Truong and B.C. Ennis, "Effect of physical aging on the fracture-behavior of cross-linked epoxies," Polym. Eng. Sci., **31**, 548–557 (1991).
150. Y. Zheng, R.D. Priestley and G.B. McKenna, "Physical Aging of an Epoxy Subsequent to Relative Humidity Jumps through the Glass Concentration," J. Polym. Sci., B: Polym. Phys.,**42**, 2107–2121 (2004).
151. M. Alcoutlabi, F. Briatico-Vangosa and G.B. McKenna, "Effect of Chemical Activity Jumps on the Viscoelastic Behavior of an Epoxy Resin: The Physical Aging Response in Carbon Dioxide Pressure-Jumps," J. Polym. Sci., Part B: Polymer Physics, **40**, 2050–2064 (2002).
152. M. Alcoutlabi, L. Banda, G. B. McKenna, "A Comparison of Concentration-Glasses and Temperature-Hyperquenched Glasses: CO2-Formed Glass vs. Temperature-Formed Glass," Polymer, **45**, 5629–5634 (2004).
153. G.B. McKenna, "Glassy States: Concentration Glasses and Temperature Glasses Compared," J. Non-Crystalline Solids, **353**, 3820–3828 (2007).
154. Y. Zheng, *Effects of Moisture on the Dimensional and Viscoelastic Properties of Glassy Polymers*, Ph.D. Thesis, Department of Chemical Engineering, Texas Tech University, Lubbock, TX (2003).
155. S.F. Swallen, K.L. Kearns, M.K. Mapes, Y.S. Kim, R.J. McMahon, M.D. Ediger, T. Wu, L. Yu and S. Satija, "Organic glasses with exceptional thermodynamic and kinetic stability," Science, **315**, 353–356 (2007).
156. G.B. McKenna, C.L. Jackson, J.M. O'Reilly, and J. S. Sedita "Kinetics of Enthalpy Recovery near the Glass Transition of Small Molecule Glasses at Nanometer Size Scales,"Polymer Preprints, **33**(1),118–119 (1992).
157. S. L. Simon, J.-Y. Park and G. B. McKenna, "Enthalpy Recovery of a Glass-Forming Liquid Constrained in a Nanoporous Matrix: Negative Pressure Effects," European Physical Journal E: Soft Matter, **8**, 209–216 (2002).
158. J.Y. Park and G.B. McKenna, "Size and Confinement Effects on the Glass Transition Behavior of oTP/PS Polymer Solutions," Physical Review B, **61**, 6667–6676 (2000).
159. S. Kawana and R.A.L. Jones, "Effect of physical aging in thin glassy polymer films," Eur. Phys. J. E, **10**, 223–230 (2003).
160. R.D. Priestley, C.J. Ellison, L.J. Broadbelt and J.M. Torkelson, "Structural relaxation of polymer glasses at surfaces, interfaces and in between," Science, **309**, 456–459 (2005).
161. R.D. Priestley, L.J. Broadbelt and J.M. Torkelson, "Physical aging of ultrathin polymer films above and below the bulk glass transition temperature: Effects of attractive vs neutral polymer-substrate interactions measured by fluorescence," Macromolecules, **38**, 654–657 (2005).
162. C.J. Ellison and J.M. Torkelson, "The distribution of glass-transition temperatures in nanoscopically confined glass formers," Nature Materials, **2**, 695–700 (2003).
163. M.K. Mundra, C.J. Ellison, R.E. Behling and J.M. Torkelson, "Confinement, composition and spin coating effects on the glass transition and stress relaxation of thin films of polystyrene and styrene containing random copolymers: Sensing by intrinsic fluorescence," Polymer, **47**, 7747–7759 (2006).
164. R.D. Priestley, L.J. Broadbelt, J. M. Torkelson and K. Fukao, "Glass transition and β-relaxation of labeled polystyrene," Phys. Rev. E., **75**, 061806–1 – 061806–10 (2007).
165. R.D. Priestley, P. Rittigstein, L.J. Broadbelt, K. Fukao and J.M. Torkelson, "Evidence for the molecular-scale origin of the suppression of physical ageing in confined polymer:

Fluorescence and dielectric spectroscopy studies of polymer-silica nanocomposites," J. Phys.: Condens. Matter, **19**, 205120–1 – 205120–12 (2007).
166. P.H. Pfromm and W.J. Koros, "Accelerated physical aging of thin glassy polymer-films-Evidence from gas- transport measurements," Polymer, **36**, 2379–2387 (1995).
167. Y. Huang and D.R. Paul, "Effect of film thickness on the gas-permeation characteristics of glassy polymer membranes," Ind. Eng. Chem. Res., **46**, 2342–2347 (2007).
168. M. S. McCaig, D.R. Paul and J.W. Barlow, "Effect of film thickness on the changes in gas permeability of a glassy polyarylate due to physical aging. Part II. Mathematical model," Polymer, **41**, 639–648 (2000).
169. Y. Huang and D.R. Paul, Effect of temperature on physical aging of thin glassy polymer films," Macromolecules, **38**, 10148–10154 (2005).
170. Y. Huang and D.R. Paul, "Effect of molecular weight and temperature on physical aging of thin glassy poly(2,6-dimethyl-1,4-phenylene oxide) films," J. Polym. Sci. Part B: Polym. Physics, **45**, 1390–1398 (2007).
171. K.D. Dorkenoo and P.H. Pfromm, "Experimental evidence and theoretical analysis of physical aging in thin and thick amorphous glassy polymer films," J. Polym. Sci. Part B: Polym. Physics, **37**, 2239–2251 (1999).
172. K.D. Dorkenoo and P.H. Pfromm, "Accelerated Physical Aging of Thin Poly[1-(trimethylsilyl)-1-propyne] Films," Macromolecules, **33**, 3747–3751 (2000).
173. G. Braun and A.J. Kovacs, "Glass transition in powdered polystyrene," Phys. Chem. Glasses, **4**,1152–160 (1963).
174. Y.P. Koh and S.L. Simon, Structural Relaxation of Stacked Ultrathin Polystyrene Films," Journal of Polymer Science: Part B: Polymer Physics, **46**, 2741–2753 (2008).
175. K. Dalnoki-Veress, J.A. Forrest, C. Murray, C. Gigault and J.R. Dutcher, "Molecular weight dependence of the reduction in the glass transiiton temperature of thin, freely-standing polymer films," Phys. Rev. E., **63**, 031801–1 -13801-10 (2001).
176. J.L. Sullivan, E.J. Blais and D. Houston, "Physical aging and the creep behavior of thermosetting and thermoplastic composites," Composites Science and Technology, **47**, 389–403 (1993).
177. J.L. Sullivan, "Creep and physical aging of composites," Composites Science and Technology, **39**, 207–232 (1990).
178. A. d'Amore, F. Cocchini, A. Pompo, A. Apicella and L. Nicolais, "The effects of physical aging on long- term properties of poly-ether-keton (PEEK) and PEEK-based composites," J. Appl. Polym. Sci., **39**, 1163–1174 (1990)
179. T.S. Gates, D.R. Veazie and L.C. Brinson, "Comparison of physical aging effects on the tension and compression creep of the IM7/K3B composite," Proc. ASME Aerospace Division, **AD-52**, American Society of Mechanical Engineers, New York, pp. 361–365 (1996).
180. M.E. Nichols, S.S. Wang and P.H. Geil, "Creep and physical aging in a polyamideimide carbon fiber composite," J. Macromol. Sci.-Physics, **B29**, 303–336 (1990).
181. T.S. Gates, D.R. Veazie and L.C. Brinson, "Creep and Physical Aging in a Polymeric Composite: Comparison of Tension and Compresssion," J. Comp. Matls., **31**, 2478–2505 (1997).
182. R.D. Bradshaw and L.C. Brinson, "Physical Aging in Polymers and Polymer Composites: An Analysis and Method for Time-Aging Time Superposition," Polym. Eng. Sci., **37**, 31–44 (1997).
183. J.Z. Wang, H. Pavatareddy, D.A. Dillard and G.L. Wilkes, "Studies on the physical aging behavior of cyanate ester resin and its graphite fiber composites," Proc. ASME Symposium on Reliability, Stress Analysis, and Failure Prevention Aspects of Composite and Active Materials, **DE-79**, American Society of Mechanical Engineers, New York, pp.15-29 (1994).
184. E.S. Kong, S.M. Lee and H.G. Nelson, "Physical Aging in Graphite/Epoxy Composites," Polymer Composites, **3**, 29–33 (1982).
185. H.W. Hu, "Physical aging in long term creep of polymeric composite laminates," J. Mechanics, **23**, 245–252 (2007).

186. H. Hu and C.T. Sun, "The characterization of physical aging in polymeric composites," Composites Science and Technology," **60**, 2693–2698 (2000).
187. W.G. Knauss and W. Zhu, "Nonlinearly viscoelastic behavior of polycarbonate. II. The role of volumetric strain," Mech. Time Dependent Matls., **6**, 301–322 (2002).
188. I. Echeverria, P.L. Kolek, D.J. Plazek and S.L. Simon, "Enthalpy recovery, creep and creep-recovery measurements during physical aging of amorphous selenium," J. Non-Crystalline Solids, **324**, 242–255 (2003).
189. P. Badrinarayanan and S.L. Simon, "Origin of the divergence of the timescales for volume and enthalpy recovery," Polymer, **48**, 1464–1470 (2007).
190. S.L. Simon and G.B. McKenna, "Experimental evidence against the existence of an ideal glass transition," J. Non-Crystalline Solids, **355**, 672–675 (2009).
191. P. Prasatya, G.B. McKenna and S.L. Simon," A viscoelastic model for predicting isotropic residual stresses in thermosetting materials: Effects of processing parameters," J. Composite Materials, **35**, 826–848 (2001).
192. Y.K. Kim and S.R. White, "Stress relaxation behavior of 3501–6 epoxy resin during cure," Polym. Eng. Sci., **36**, 2852–2862 (1996).
193. D.J. Plazek and I.C. Chay, "The evolution of the viscoelastic retardation spectrum during the development of an epoxy-resin network," J. Polym. Sci. Part B: Polymer Physics, **29**, 17–29 (1991).
194. M. Alcoutlabi and G.B. McKenna, "Effects of Confinement on Material Behaviour at the Nanometre Size Scale," J. Phys.: Condensed Matter, **17**, R461-R524 (2005).
195. G.B. McKenna, "Confit III. Summary and perspectives on dynamics in confinement," European Physical Journal Special Topics, **141**, 291–300 (2007).
196. C.B. Roth and J.R. Dutcher, "Glass transition and chain mobility in thin polymer films," J. Electroanalytical Chemistry, **584**, 13–22 (2005).
197. J.A. Forrest and K. Dalnoki-Veress, "The glass transition in thin polymer films," Adv. in Coll. Interface. Sci., **94**, 167–196 (2001).
198. F. Varnik and J. Baschnagel, "Computer simulations of supercooled polymer melts in the bulk and in-confined geometry," J. Phys.: Cond. Matter, **17**, R851-R953 (2005).
199. D.R. Paul and L.M. Robeson, "Polymer nanotechnology: Nanocomposites," Polymer, **49**, 3187–3204 (2008).
200. A.J. Crosby and J.Y. Lee, "Polymer Nanocomposites: The "Nano" Effect on Mechanical Properties," Polymer Reviews, **47**, 217–229 (2007).
201. F. Fraga, M. Lopez, V.H. Soto Tellini, E. Rodriguez-Nunez, J.M. Martinez-Ageitos and J. Mirayaga, "Study of the physical aging of the epoxy system BADGE n = 0/m-XDA/ $CaCO_3$,"J. Appl. Polym. Sci., **113**, 2456–2461 (2009).
202. U. Yilmazer and R.J. Farris, "Physical aging in particulate-filled composites with an amorphous glassy matrix," J. Appl. Polym. Sci., **28**, 3269–3280 (1983).
203. J. Menczel and B. Wunderlich, "Heat capacity hysteresis of semicrystalline macromolecular glasses," J. Polym. Sci. Polym. Lett., **19**, 261–264 (1981).
204. L.C.E. Struik, "The mechanical and physical aging of semicrystalline polymers.1.," Polymer, **28**,1521–1533 (1987).
205. L.C.E. Struik, "The mechanical and physical aging of semicrystalline polymers.2.," Polymer, **28**,1534–1542 (1987).
206. L.C.E. Struik, "Mechanical behavior and physical aging of semicrystalline polymers. 3. Prediction of long-term creep from short-time tests," Polymer, **30**, 799–814 (1989).
207. L.C.E. Struik, "Mechanical behavior and physical aging of semicrystalline polymers: 4," Polymer, **30**, 815–8830 (1989).
208. B. Wunderlich, "Glass transition of partially ordered macromolecules," Prog. Coll. Polym. Sci., **96**, 22–28 (1994).
209. R.K. Krishnaswamy, J.F. Geibel and B.J.Lewis, "Influence of Semicrystalline Morphology on the Physical Aging Characteristics of Poly(phenylene sulfide)," Macromolecules, **36**, 2907–2914 (2003).

210. P. Huo and P. Cebe, "Effects of thermal history on the rigid amorphous phase in poly (phenylene sulfide), "Coll. Polym. Science, **270**, 840–852 (1992).
211. J. Beckmann, G.B. McKenna, B.G. Landes, D.H. Bank, and R.A. Bubeck, "Physical Aging Kinetics of Syndiotactic Polystyrene as Determined from Creep Behavior," Polymer Engineering and Science, **37**, 1459–1468 (1997).
212. A.Y.H. Liu and J. Rottler, "Physical aging and structural recovery in nanocomposites," J. Poly. Sci. Part B: Polymer Physics, **47**,1789–1798 (2009).
213. C. Thenau, M. Salmeron Sanchez, J.C. Rodriguez Hernandez, M. Monleon Pradas, J.M. Saiter and J.L. Gomez Ribelles, "The kinetics of the structural relaxation process in PHEMA-silica nanocomposites based on an equation for the configurational entropy," Eur. Phys. J. E., **24**, 69–77 (2007).
214. H. Lu and S. Nutt, "Enthalpy relaxation of layered silicate-epoxy nanocomposites," Macromol. Chem. and Physics, **204**, 1832–1841 (2003).
215. J.B. Donnet and A. Voet, *Carbon Black: Physics, Chemistry and Elastomer Reinforcement,"* Marcel Dekker, New York (1976).
216. C.G. Robertson and C.M. Roland, "Glass transition and interfacial segmental dynamics in polymer-particle composites," Rubber Chemistry and Technology, **81**, 506–522 (2008).
217. S.M. Aharoni, "Increased glass transition temperature in motionally constrained semicrystalline polymers," Polym. Adv. Tech., **9**, 169–201, (1998).
218. S. Ammanuel, A.M. Gaudette and S.S. Sternstein, "Enthalpic relaxation of silica-polyvinylacetate nanocomposites," J. Polym. Sci. Part B: Polymer Physics, **46**, 2733–2740 (2008).
219. B. Haidar, H.Salah Deradji, A. Vidal and E. Papirer, "Physical aging phenomena in silica and glass beads filled elastomers (EPDM)," Macromol. Symp., **108**, 147–161 (1996).
220. S.S. Sternstein and A.J. Zhu, "Reinforcement Mechanism of Nanofilled Polymer Melts As Elucidated by Nonlinear Viscoelastic Behavior," Macromolecules, **35**, 7262–7273 (2002).
221. H. Montes, F. Lequeux and J. Berriot, "Influence of the glass transition temperature gradient on the nonlinear viscoelastic behavior in reinforced elastomers," Macromolecules, **36**, 8107–8118 (2003).
222. J. Jancar, "The Thickness Dependence of Elastic Modulus of Organosilane Interphases," Polym. Comp., **29**, 1372–1377 (2008).
223. S. Merabia, P. Sotto and D.R. Long, "A Microscopic Model for the Reinforcement and the Nonlinear Behavior of Filled Elastomers and Thermoplastic Elastomers (Payne and Mullins Effects)," Macromolecules, **41**, 8252–8266 (2008).

Chapter 8
Mechanisms and Kinetics of Organic Matrix Thermal Oxidation

Xavier Colin and Jacques Verdu

Abstract It is now well recognized that during thermal aging at moderate temperatures, for example, typically below the glass transition temperature, organic matrix composites perish mainly by matrix embrittlement resulting from its thermo-oxidation. The present chapter aims to briefly introduce this domain. The chapter consists of a brief history of polymer oxidation and description of mechanisms and kinetics. The radical character of oxidation processes; the main elementary steps: propagation, termination, initiation processes, and initial steps; structure–property relationships; the nature of oxidation products; and experimental methods for the study of oxidation mechanisms are also discussed. The standard kinetic scheme, case of oxygen excess and general shape of oxidation kinetic curves, the induction period, departure from Arrhenius law, and case of oxygen lack are described. Consequences of oxidation on matrix thermomechanical properties including chain scission and cross-linking physical approaches are presented.

8.1 Introduction

One can situate the birth date of research in composite thermal aging in the 1960s, as soon as these materials appeared in the market. In the following decades, practically with the beginning of the 1990s, lifetime prediction studies were based on only two experimental tools: gravimetry and mechanical testing, and one theoretical tool: Arrhenius law to extrapolate results obtained during accelerated aging (at high temperature) to use temperature. A first problem, systematically found in the case of thermoset matrices and sometimes in thermoplastic ones, is illustrated by the example of polyester (25%)–glass fiber (75%) composites in the 180–260°C temperature range [48]. These authors tried to

X. Colin (✉)
ARTS ET METIERS ParisTech, PIMM, 151 boulevard de l'Hôpital, 75013 Paris, France
e-mail: Xavier.COLIN@ensam.eu

K.V. Pochiraju et al. (eds.), *Long-Term Durability of Polymeric Matrix Composites*, DOI 10.1007/978-1-4419-9308-3_8, © Springer Science+Business Media, LLC 2012

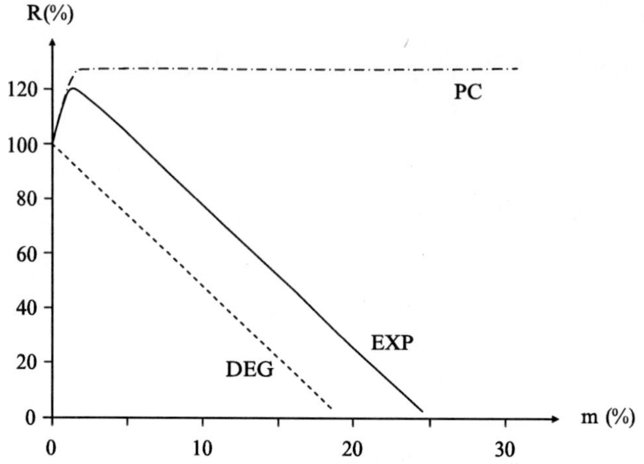

Fig. 8.1 Schematic shape of residual strength-mass loss curve (*EXP*) and its deconvolution into two components: post-cure (*PC*) and degradation (*DEG*)

correlate the residual flexural strength R, expressed in percents of the initial value, with the relative weight loss: $m = \Delta w/w_0$ (Fig. 8.1).

A linear fit of experimental points gave, approximately,

$$R \approx 130 - 5m \tag{8.1}$$

This result calls for the following comments:

- The residual strength varies non-monotonically with mass loss (and time) because its variations result from two opposite processes: post-cure, which tends to increase the mechanical strength, and degradation, which leads always to a strength decrease. Here, post-cure is faster than degradation so that it predominates in a short initial period, after which it stops because all the reactive functions available for cross-linking have been consumed. Then degradation predominates. It is noteworthy that since post-cure and degradation have different activation energies, except for a coincidence, this peculiar behavior can exist only in a more or less restricted temperature interval and disappears far above and far below the temperature range of accelerated aging. Indeed, the amplitude of post-cure effects is expected to sharply depend on cure conditions, especially cure temperature (Fig. 8.2).

  T_g increases continuously with the cure conversion in thermosets [35]. Its (real or virtual) maximum value $T_{g\infty}$ corresponds to a full conversion ($y = 1$). Let us first consider a cure performed at a temperature $T_{c1} < T_{g\infty}$. It will be stopped at the sample vitrification ($y = y_1$). Indeed, in further aging at temperature $T_a > T_{c1}$, the polymer will undergo a post-cure of amplitude ($y_a - y_1$). For a cure at $T_a > T_{g\infty}$, the conversion of cure reactions is expected to be complete; no post-cure effect will be observed during further thermal aging. In the case where $T_{g\infty}$ is very high, typically above 450–550 K depending

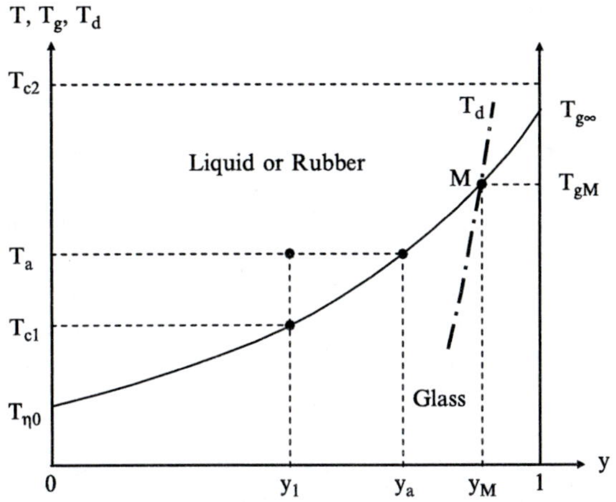

Fig. 8.2 Shape of the change of glass transition temperature T_g with cure conversion ratio y. *Full line*: Variation of T_g. *Dashed line*: Degradation temperature (T_d)

on polymer structure, degradation and/or side reactions occurring in the timescale of cure can limit cross-link density. Using an arbitrary end-life criterion, one can define a "degradation ceiling" surface of which the projection in the plane of Fig. 8.2 crosses the curve $T_g = f(y)$ at the point M (y_M, T_{gM}). At $T > T_{gM}$, degradation predominates over cure and T_g decreases. There is no way to obtain T_g values higher than T_{gM}. The problem of degradation during processing has raised a great amount of literature in the field of thermostable polymers (polyimides, for instance) and high T_g epoxies (tetraglycidyl derivative of diaminodiphenylmethane systems, for instance). In all these cases, the presence of oxygen aggravates the problem.

- The strength vanishes only when the matrix is totally volatilized, which indicates a relatively low sensitivity of composite mechanical properties to structural changes occurring in the matrix. There are two possible ways to explain such a behavior: (1) Degradation is homogeneously distributed into the sample volume, the matrix is progressively converted in more and more small (more and more volatile) molecular fragments. In this case, however, mass loss is expected to autoaccelerate catastrophically beyond a certain time, which is not observed here. (2) Degradation occurs in a limited superficial layer which is progressively eroded. Schematically, the residual strength would decrease in proportion to the thickness of the (sound) core layer.

The second "scenario" seems to be more plausible than the first one, but it is only realistic if a diffusion-controlled process explains why the superficial layer degrades preferentially. In 1972, there was sufficient scientific knowledge to identify diffusion-controlled thermal oxidation as the cause of this behavior. Polymer oxidation mechanisms and kinetics were abundantly studied in the preceding quarter of a century, but essentially in the rubber community, and scarce publications were available on the reaction–diffusion coupling [30], but only in the case of hydrolysis.

Unfortunately, these results remained ignored by the composite community until the turn of the century. Starting from the assumption that composite aging results from the diffusion-controlled oxidation of the matrix, one can hold serious reservations about the use of Arrhenius law to extrapolate accelerated aging data. As a matter of fact, rigorously speaking, Arrhenius law applies only to an elementary process. It can apply also to a complex process if the global rate r of this latter can be expressed as a product of elementary rate constants:

$$r = A\,k_1^{a_1}k_2^{a_2}...k_n^{a_n} \tag{8.2}$$

In this case, the apparent activation energy of r is

$$H = a_1H_1 + a_2H_2 + \cdots + a_nH_n \tag{8.3}$$

where $H_1, H_2, \ldots H_n$ are the activation energies of $k_1, k_2, \ldots k_n$.

Arrhenius law is no longer valid if the global rate is a sum of rate constants:

$$r = A_1k_1^{a_1} + A_2k_2^{a_2} + \cdots + A_nk_n^{a_n} \tag{8.4}$$

In the case under study, oxidation is a very complex process involving, at least, six elementary reactions. As will be shown in the following, the global oxidation rate does not obey Arrhenius law. However, it is diffusion controlled, oxidation occurs in a superficial layer, and the thickness distribution of oxidation products is temperature dependent. In such a case, the fact that the global change of composite properties obeys Arrhenius law would result from a very surprising coincidence. When such a problem reaches such a degree of complexity, it is tempting to search for an empirical solution. However, empirical approaches, which can be efficient in interpolations, are inadequate for extrapolations as in the case of aging. The only reasonable way is, therefore, the scientific way in which the kinetic scheme is derived from a mechanistic scheme, the latter being established from analytical investigations.

For composites, the first attempts at kinetic modeling (considering diffusion-controlled oxidation) were made in the 1980s–1990s [9, 10, 20, 22, 33, 34, 43, 44, 47]. Their common characteristic is that they ignored the work published in the 1940s–1980s on the oxidation of hydrocarbon polymers, including diffusion-controlled oxidation [21, 27, 45, 49, 51], and they used fully empirical kinetic models. A detailed analysis of these works would be out of the scope of this chapter. The model of McManus and colleagues [22] will be quoted just to illustrate this empirical character and the difference with further approaches. These authors proposed the following kinetic expression:

$$\frac{\partial a}{\partial t} = k\,(1 - a)^n C^p \tag{8.5}$$

where a is the conversion degree of the whole oxidation process, C is the oxygen concentration, k is a rate constant obeying Arrhenius law, n and p are apparent reaction orders.

Here, oxidation is practically considered as a single polymer–oxygen reaction but, indeed, the partial order (n and p) values would be difficult to justify from mechanistic considerations. They are, in fact, adjustable parameters. At the turn of the century, our research group at Arts et Metiers ParisTech, proposed to build the lifetime prediction approach, around a mechanistic scheme profiting from the advances made by the hydrocarbon polymer oxidation community in the previous half century, trying to consider composite (or matrix) aging as a multiscale problem, and giving to the model an open structure to be progressively complexified, as well in terms of mechanisms as in terms of structure–property relationships.

This chapter is aimed at describing the recent advances in matrix oxidation mechanisms and kinetics. It will be organized into three sections:

1. Mechanisms, in which the main elementary steps will be described, structure–property relationships will be considered when they are available, and a brief description of most common experimental methods will be made
2. Kinetics, in which standard kinetic schemes, distinguishing the cases of oxygen excess and oxygen lack, will be examined, and where the problem of reaction–diffusion coupling will be briefly treated
3. Consequences of oxidation on matrix physical properties, including mass variations, volumetric properties, glass transition temperature, elastic properties, and fracture properties

8.2 Mechanisms

8.2.1 The Standard Mechanistic Scheme

The oxidation of organic substrates results from a radical chain mechanism established for the first time by Semenov (Nobel Prize 1956) in the 1930s [46]. In the polymer community of Western countries, however, this mechanism remained ignored until the end of World War II where it was rediscovered by a British team [8], often considered the founder of the discipline. A radical chain process involves at least three steps:

Initiation	Non-radical species	→ Radicals
Propagation	One radical	→ One radical
Initiation	Two radicals	→ Non-radical species

The following nomenclature will be used here: P°: alkyl radical (although the existence of aryl radicals is not excluded); PO_2°: peroxy radical (or peroxyl); POOH: hydroperoxide; PH: the substrate.

In the case of radical chain oxidation, the propagation involves two elementary reactions and two radicals: P° and PO_2°. In a saturated substrate:

(II)	P° + O_2	→ PO_2°	(k_2)
(III)	PO_2° + PH	→ POOH + P°	(k_3)

In an unsaturated substrate, the previous mechanism can coexist with the addition of PO_2° to double bonds:

(IIIa)

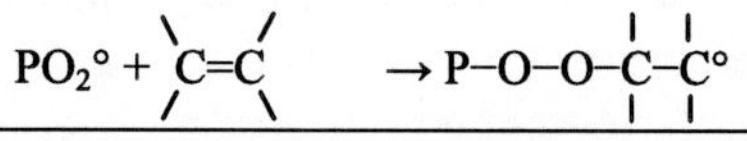

(k_{3a})

This latter reaction can be important in unsaturated polymers such as polydienes [17, 19], but it will not be considered here, because unsaturations are scarce in common composite matrices.

In terminations, radicals deactivate mutually in pairs:

(IV)	P° + P°	→ Inactive species	(k_4)
(V)	P° + PO_2°	→ Inactive species	(k_5)
(VI)	PO_2° + PO_2°	→ Inactive species + O_2	(k_6)

All these terminations can result from radical coupling:

P° + P°	→ P – P	
P° + PO_2°	→ P – O – O – P	
PO_2° + PO_2°	→ P – O – O – O – O – P (very unstable)	→ P – O – O – P + O_2

If, at least, one radical contains one hydrogen in β position, termination can also result from a disproportionation:

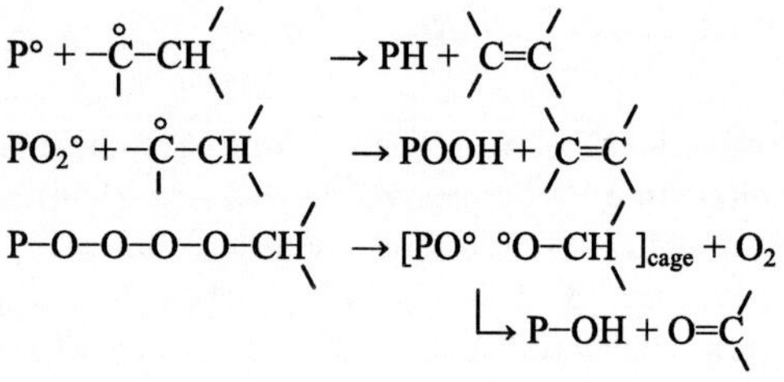

Coupling and disproportionation can coexist.

Three initiation mechanisms can exist, in the absence of any extrinsic initiating agent (irradiation, presence of a radical initiator, etc.).

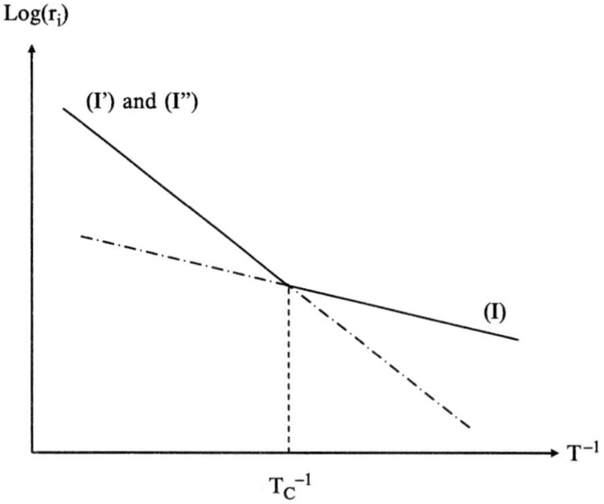

Fig. 8.3 Shape of the Arrhenius plot of initiation rate

- The substrate decomposition:

(I′)	P – P	→ 2P°	(k'_1, r'_1)

- The direct substrate–oxygen reaction:

(I″)	$PH + O_2$	→ $P° + HO_2°$	(k''_1, r''_1)

- The hydroperoxide decomposition:

(I)	δPOOH	→ $\alpha P° + \beta PO_2°$	(k_1, r_1)

The Arrhenius plot of the initiation rate $r_i = d[P°]/dt$ displays generally the shape of Fig. 8.3.

Hydroperoxide decomposition, of which the activation energy ($80\ \mathrm{kJ\ mol^{-1}} \leq H_1 \leq 150\ \mathrm{kJ\ mol^{-1}}$) is noticeably lower than the one of substrate decomposition or hydrogen abstraction by oxygen, tends to predominate at low temperature, provided that the hydroperoxide concentration is high enough. Except in very scarce cases, long-term thermal aging occurs always in the temperature domain where reactions (I′) and (I″) are negligible and the oxidation process generates its own initiator: POOH (degenerate chain branching). The fact that $r_1 >> r'_1$ can be easily checked by an aging test under neutral atmosphere: If r'_1 is negligible, no significant structural change must occur in the timescale of oxidation tests. The fact that $r_1 >> r''_1$ is more difficult to demonstrate. When the temperature and the initial hydroperoxide concentration are low enough, oxidation is autoaccelerated and can even display an induction period. Such autoaccelerated character is due to hydroperoxide accumulation since the reaction rate r_1 is an increasing function of POOH concentration. In contrast, in the case of predominance of initiation by reaction (I″) or (I′), the initiation rate would remain constant (at low conversion) or decrease

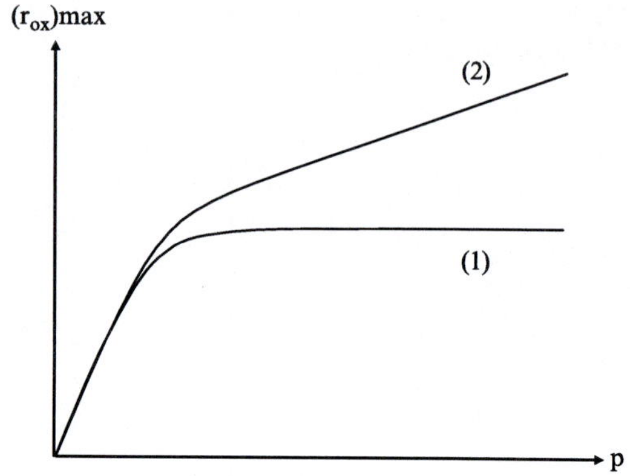

Fig. 8.4 Shape of the pressure dependence of the maximum oxidation rate: (1) reaction (I″) negligible; (2) significant contribution of reaction (I″) to initiation

slowly as the result of substrate consumption. In certain cases, however, when samples are pre-oxidized during processing, or at relatively high temperatures where hydroperoxide decomposition is very fast, the autoaccelerated character can disappear while initiation by POOH decomposition predominates. In such cases, the relative importance of reactions (I″) and (I′) can be recognized from the effect of oxygen pressure on maximum oxidation rate (Fig. 8.4).

Let us consider the theoretical case of a substrate initially free of hydroperoxides. In this case, if oxidation occurs, it is necessarily due to the existence of a process of radical generation other than hydroperoxide decomposition. For a long time, the question of the nature of these initial steps has been considered crucial by the community, and some confusion appeared between "initial steps" and "initiation steps." In general, the nature of eventual "initial steps" is impossible to establish because the corresponding reactive species are in concentrations too low to be measurable. Kinetic modeling brings some light on this almost metaphysical problem: the radicals produced in initial steps are transformed into hydroperoxides of which the concentration increases in such a way as hydroperoxide decomposition becomes rapidly the predominating initiation process, which fully determines the autoaccelerated behavior and, eventually, the length of the induction period [39].

Hydroperoxide decomposition can be unimolecular:	
POOH	$\rightarrow PO^{\circ} + {}^{\circ}OH$
HO° + PH	$\rightarrow P^{\circ} + H_2O$
PO° + PH	$\rightarrow P - OH + P^{\circ}$
PO°	$\rightarrow P^{\circ} + P = O + s$
Balance reaction: POOH	$\rightarrow 2P^{\circ} + H_2O + (1 - \gamma_1)$ $POH + \gamma_1 PO + \gamma_1 s$ (k_{1u})
or bimolecular:	
POOH + POOH	$\rightarrow PO^{\circ} + PO_2{}^{\circ} + H_2O$
PO° + PH	$\rightarrow P - OH + P^{\circ}$
PO°	$\rightarrow P^{\circ} + P = O + s$
Balance: POOH + POOH	$\rightarrow P^{\circ} + PO_2{}^{\circ} + H_2O + (1 - \gamma_1)$ $POH + \gamma_1 P = O + \gamma_1 s$ (k_{1b})

Since HO° and PO° radicals are extremely reactive (they cannot be observed by electron spin resonance (ESR)), the rate-controlling step is the hydroperoxide decomposition and it is licit to consider the balance reaction as a virtual single elementary step.

PO° radicals can react by hydrogen abstraction to give an alcohol or by re-arrangement (β scission):

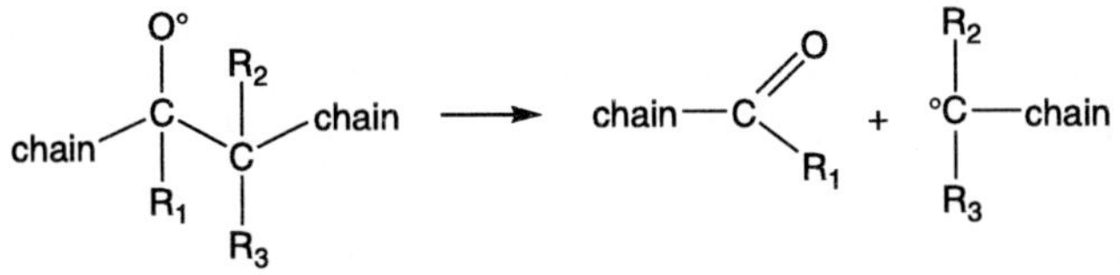

This reaction gives a carbonyl group P = O (absorbing in the 1650–1850 cm^{-1} spectral IR range with a relatively high molar absorptivity: $\varepsilon \approx 150\text{–}600$ L mol^{-1} cm^{-1}) and a chain scission (s). As will be seen, chain scissions play a key role in mechanical property changes.

Let us consider the rates of radical production by both mechanisms:

Unimolecular	$r_{1u} = 2k_{1u}$ [POOH]
Bimolecular	$r_{1b} = 2k_{1b}$ $[POOH]^2$

Both rates are equal when:

$$[POOH] = [POOH]_C = k_{1u}/k_{1b} \tag{8.6}$$

At low POOH concentration, the unimolecular decomposition predominates, whereas at high POOH concentration, the bimolecular decomposition predominates. The activation energy of k_{1u}: $E_{1u} \approx 140$ kJ mol^{-1} is higher than the activation energy of k_{1b}: $E_{1b} = 80\text{–}120$ kJ mol^{-1}. Thus, $[POOH]_C$ is an increasing function of temperature.

To resume: Unimolecular decomposition tends to predominate at high temperature, when hydroperoxides are intrinsically unstable, and when they are in low concentration. Bimolecular decomposition tends to predominate at low temperature, when hydroperoxides are relatively stable, and when they are in relatively high concentration. Indeed, an oxidation process can begin with unimolecular initiation and turn into a bimolecular one when the hydroperoxide concentration becomes higher than the critical one $[POOH]_C$.

8.2.2 Structure–Reactivity Relationships

Even in the most widely studied polymer families, for example, polyolefins and polydienic elastomers, we are far from a consensus about elementary rate constant values. For polymers in glassy state, with heteroatoms and aromatic cycles in the chain, fundamental data are almost nonexistent. In other words, oxidation kinetics in composite matrices is an almost virgin research domain. The knowledge about elementary rate constant values can be summarized as follows.

8.2.2.1 Relative Orders of Magnitude

Let us consider, for instance, the 150–200°C temperature range. Most common rate constant values would generally be in the following intervals:

$10^{-7} \leq k_{1u} \leq 10^{-3}\ s^{-1}$
$10^{7} \leq k_2 \leq 10^{9}\ L\ mol^{-1}\ s^{-1}$
$1 \leq k_3 \leq 10^{2}\ L\ mol^{-1}\ s^{-1}$
$10^{6} \leq k_4$, k_5 and $k_6 \leq 10^{12}\ L\ mol^{-1}\ s^{-1}$

Initiation is, by far, the slowest process. This is the reason why the hydroperoxide accumulation kinetics determines mainly the process autoacceleration.

$k_2 \gg k_3$, so that when oxygen is in excess, the hydrogen abstraction process (III) is the propagation rate-controlling process (but not the whole oxidation rate-controlling process as frequently claimed). Furthermore, since P° radicals are very quickly transformed into $PO_2°$ ones, their probability of reacting by other ways (terminations (IV) and (V)) becomes negligible, chain termination occurs only by reaction (VI), and the whole reaction rate (in steady state) is sharply linked to the ratio k_3^2/k_6 [5]. When oxygen is not in excess, the oxidation rate becomes dependent on the oxygen concentration, and this dependence is sharply linked to the value of a composite rate constant β [12, 14]:

$$\beta = \frac{k_2 k_6}{2 k_5 k_3 [PH]} \tag{8.7}$$

Schematically, oxygen is in excess when its concentration C exceeds a critical value C_C defined by:

$$C_C = q\beta^{-1} \tag{8.8}$$

where q is of the order of 10.

β was found close to 160 $L\ mol^{-1}$ at 200°C, as well in an epoxy network as in a poly(bismaleimide) [12, 14].

8.2.2.2 Rate Constants

Compilations of rate constant values exist in reference books [23, 38, 50], but they are too parceled or too scattered to permit a coherent synthesis, except, eventually, for the propagation rate constant k_3 [31]. According to these authors, the rate constant k_3 at 30°C would be linked to the dissociation energy E_D of the broken CH bond by:

$$\text{Log}\, k_3 = A - 0.048E_D \quad (8.9)$$

where A = 15.4 for a tertiary peroxyl and 16.4 for a secondary peroxyl, k_3 is expressed in L mol^{-1} s^{-1} and E_D in kJ mol^{-1}.

In a similar way, the activation energy E_3 of k_3 would be given by:

$$E_3 = 0.13\,(E_D - 261) \quad (8.10)$$

It can be recalled that $E_D(C-H)$ ranges from a value close to 500 kJ mol^{-1} (for a CH in an aromatic ring) to a value of 300–350 kJ mol^{-1} (for a CH in allylic position, in a benzylic position or in the vicinity of an electronegative atom: –O – or – N<). k_3(30°C) can thus range between 10^{-4}–10^{-3} L mol^{-1} s^{-1} for the most stable aromatic structures, to 0.1–1.0 L mol^{-1} s^{-1} for the most unstable (common) structures present in polymers.

Surprisingly, the hierarchy of polymer stabilities to oxidation is roughly correlated to the hierarchy of E_D values: Polymers having only stable CH bonds, for example, only aromatic groups and methyls such as polyethersulphone, poly(dimethyl siloxane), various aromatic polyimides, etc., are very stable to oxidation. Polymers having very unstable CH bonds, for instance, polybutadiene, polyisoprene, polyamides, polyethers, etc., are especially unstable to oxidation. Polyethylene, in which E_D is intermediary ($E_D \approx 390$ kJ mol^{-1}), displays a medium stability. The existence of this correlation, indicating the key role of k_3, seems to indicate that the other rate constants, especially k_1 and k_6, which are expected to play a role of the same importance as k_3, do not vary much from one polymer to another.

8.2.3 How to Establish Oxidation Mechanisms

For pure molecular substances in liquid or gaseous state, it is possible to build a systematic strategy for the elaboration of an oxidation mechanistic scheme. In such a case, oxidation products can be separated and identified by very powerful methods, especially mass spectrometry (MS) and nuclear magnetic resonance (NMR). For polymers, especially thermosets, this separation is very difficult or even impossible. Furthermore, NMR lacks of sensitivity in the conversion range of practical interest. In conditions of thermal oxidation, electron spin resonance (ESR), which is the tool for radical identification, also lacks of sensitivity owing to the very low steady concentration of radicals (especially P° ones). The remaining available methods are scarce: Infrared spectrophotometry (IR), hydroperoxide chemical titration, and elementary analysis give generally useful information, but these can lead only to a fuzzy image of the structural changes undergone by the polymer.

The most common way to establish oxidation mechanisms cannot be, finally, depicted as a linear, fully logical reasoning. It consists of starting from the assumption that oxidation obeys the standard mechanistic scheme which displays the following characteristics:

1. Hydroperoxides are the unique propagation products
2. Initiation results only from hydroperoxide decomposition
3. All the other products result from initiation, termination, or side reactions

Three important quantitative data are eventually accessible: The whole quantity of oxygen reacted with the polymer (from manometric measurements in a closed cell during exposure or from elementary analysis); the hydroperoxide concentration; and the nature of predominating carbonyl groups (IR). By cross-checking this information with kinetic data (see below), it must be possible to validate the starting assumptions of mechanism or to choose another one. When ambiguities remain, for instance, about the nature of reactive sites and their relative importance, the recourse to studies of model compounds, to which the analytical methods of organic chemistry are fully applicable, can be useful.

8.3 Kinetics

Let us consider the standard mechanistic scheme:

δPOOH	$\rightarrow \alpha P^\circ + \beta PO_2{}^\circ$	(k_1)
$P^\circ + O_2$	$\rightarrow PO_2{}^\circ$	(k_2)
$PO_2{}^\circ$ + PH	$\rightarrow$ POOH + P°	(k_3)
$P^\circ + P^\circ$	$\rightarrow$ Inactive species	(k_4)
$P^\circ + PO_2{}^\circ$	$\rightarrow$ Inactive species	(k_5)
$PO_2{}^\circ + PO_2{}^\circ$	$\rightarrow$ Inactive species + O_2	(k_6)

The following basic principles of chemical kinetics can be applied if oxidation can be considered homogeneous, for example, when the concentrations have a physical sense, which leads to considering local concentrations (in thin thickness layers):

- Concentrations are generally expressed in moles per liter (mol L^{-1}).
- In an elementary step, reaction rates are always expressed as functions of reactants (left side of the arrow). Except in very scarce cases, reaction rates are proportional to reactant concentrations.
- For the whole process, the variation rate of a given species concentration is the algebraic sum of the elementary rates.

Application of these principles to the above scheme leads to the following set of differential equations:

$$\frac{d[P^\circ]}{dt} = \alpha k_1[POOH]^\delta - k_2[O_2][P^\circ] + k_3[PH][PO_2^\circ] - 2k_4[P^\circ]^2 - k_5[P^\circ][PO_2^\circ] \quad (8.11)$$

$$\frac{d[PO_2^\circ]}{dt} = \beta k_1[POOH]^\delta + k_2[O_2][P^\circ] - k_3[PH][PO_2^\circ] - k_5[P^\circ][PO_2^\circ] - 2k_6[PO_2^\circ]^2 \quad (8.12)$$

$$\frac{d[POOH]}{dt} = k_3[PH][PO_2^\circ] - \delta k_1[POOH]^\delta \quad (8.13)$$

$$\frac{d[PH]}{dt} = -k_3[PH][PO_2^\circ] - \gamma k_1[POOH]^\delta \quad (8.14)$$

$$\frac{d[O_2]}{dt} = -k_2[O_2][P^\circ] + k_6[PO_2^\circ]^2 + Diff \quad (8.15)$$

Boundary conditions are:

- At $t = 0$: $[P^\circ] = [PO_2^\circ] = 0$; $[POOH] = [POOH]_0$; $[PH] = [PH]_0$
- At every time: $[O_2] = [O_2]_S$ (equilibrium oxygen concentration) in the sample superficial layer. $[O_2]_S = Sp$, where p is the oxygen pressure and S is the oxygen solubility in the polymer

γ is the number of PH groups consumed in one hydroperoxide decomposition event (generally $1 < \gamma < 2$).

"Diff" is a term expressing oxygen supply by diffusion (see below).

In classical approaches, several simplifying assumptions are made:

1. Low conversions. Since, often, embrittlement occurs at low conversions, it is licit to consider that substrate consumption is negligible and, thus, that $[PH] = [PH]_0 =$ constant. However, in this case, there is no way to simulate the final autoretardation process, which is generally observed in kinetic curves (Fig. 8.5).
2. Constant oxygen concentration: $[O_2] = [O_2]_S$. This assumption is only valid in thin films or in a thin superficial layer of thick samples. In this case: Diff = 0.
3. Oxygen excess (only possible in thin samples and superficial layers). In this case, the terms $k_4\,[P^\circ]^2$ and $k_5\,[P^\circ]\,[PO_2^\circ]$ vanish.

Analytical models exist in this case (using assumption (1) and the assumptions that there is a steady state for whole radical concentration: $d[P^\circ]/dt + d[PO_2^\circ]/dt = 0$, and that a steady state is reached when $d[POOH]/dt = 0$). Two distinct models have been elaborated for unimolecular and bimolecular hydroperoxide decomposition by Tobolsky et al. [51]. They were almost totally ignored or forgotten by the community until the middle of 1990s [4]. Their properties were studied in detail [5], and small improvements were proposed [16].

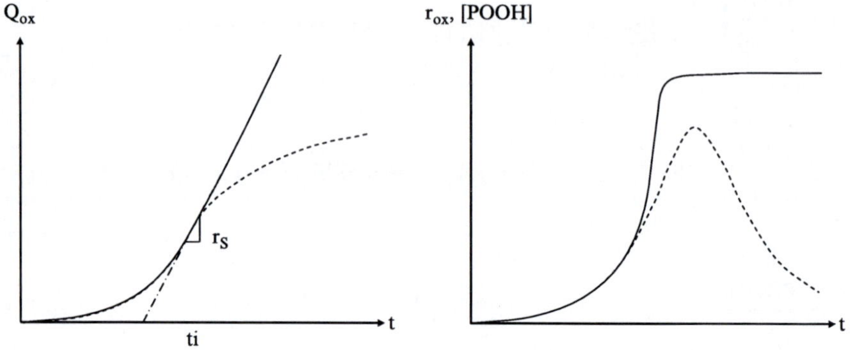

Fig. 8.5 Kinetic curves. *Left*: Oxygen consumption or stable products accumulation (for instance, carbonyls). *Right*: Oxidation rate or POOH formation. *Full line*: Model with $[PH] = [PH]_0 =$ constant. *Dashed line*: Experimental curve

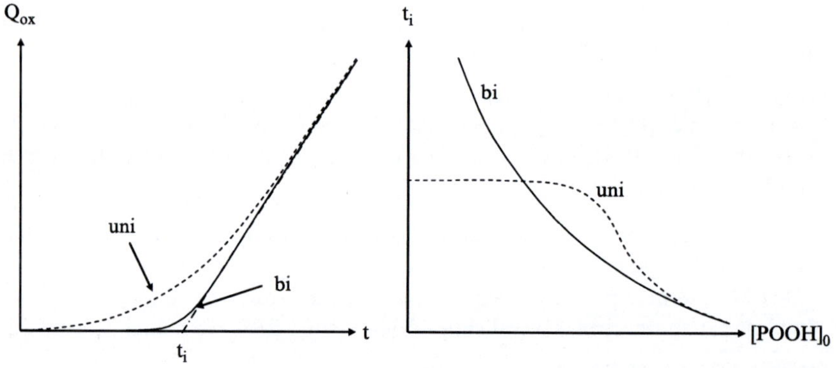

Fig. 8.6 *Left*: Shape of oxidation kinetic curves for unimolecular (*uni*) and bimolecular (*bi*) POOH decomposition. *Right*: Variation of the induction time with the initial hydroperoxide concentration for unimolecular (*uni*) and bimolecular (*bi*) POOH decomposition

Despite their simplicity, these models are very useful in appreciating the trends of kinetic behavior with the exposure parameters and rate constant values or structural variables. Unimolecular and bimolecular hydroperoxide decomposition display two important differences (Fig. 8.6).

Autoacceleration during the induction period is more progressive for unimolecular than for bimolecular decomposition. At low initial POOH decomposition, the induction time tends to be independent of $[POOH]_0$ for low $[POOH]_0$ values in the case of unimolecular decomposition, whereas it increases continuously, when $[POOH]_0$ decreases, in the case of bimolecular decomposition. Indeed, since both initiation modes are competitive, unimolecular decomposition predominates at low $[POOH]_0$ values, whereas bimolecular decomposition predominates at high $[POOH]_0$ values. The best way for modeling consists, thus, in taking into account, simultaneously, both initiation modes, but in this case, we have to renounce to analytical solutions. Analytical expressions of the induction time are very simple [5]. For unimolecular initiation, at low $[POOH]_0$:

$$t_i = \frac{A}{k_{1u}} \tag{8.16}$$

where A = 2.5 or 3.0, depending on the chosen criterion.

For bimolecular initiation:

$$t_i = \frac{\mathrm{Ln}\, b}{K} \tag{8.17}$$

where:

$$b = \frac{[POOH]_\infty}{[POOH]_0}; \quad [POOH]_\infty = \frac{k_3[PH]}{2(k_{1b}k_6)^{1/2}}; \quad \text{and} \quad K = k_3[PH]\left(\frac{k_{1b}}{k_6}\right)^{1/2}$$

A general property of these “closed-loop” schemes, where the reaction generates its own initiator, is that the kinetic chain length: KCL = propagation rate/initiation rate, decreases continuously during the induction period to reach unity in steady state. As a consequence, the apparent activation energy decreases continuously during the induction period. Application of Arrhenius law to predict lifetime appears, therefore, questionable. Since, whatever the initiation mode, the induction time is a decreasing function of temperature, there is a critical temperature above which the induction period vanishes totally. This temperature is as low as initiation is fast and initial hydroperoxide concentration is high.

8.3.1 Case of Oxygen Lack

Let us consider now the case of oxygen lack, first in a thin layer (no thickness gradient). The terms $k_4\ [P°]^2$ and $k_5\ [P°]\ [PO_2°]$ must be reintroduced in the kinetic scheme. Classically, the set of differential equations was solved making the following assumptions:

I : Initiation rate constant (r_i), for example, stable hydroperoxides
S: Steady state for the (whole) radical concentration
T: Relationship between termination rate constants: $k_5^2 = 4k_4k_6$
L: Long kinetic chain: $k_3[PH]/(2r_i\ k_6)^{1/2} \gg 1$

Fig. 8.7 Shape of the curve: oxygen consumption rate versus oxygen concentration: $r_S = k_3[PH]\,(r_i/2k_6)^{1/2}$

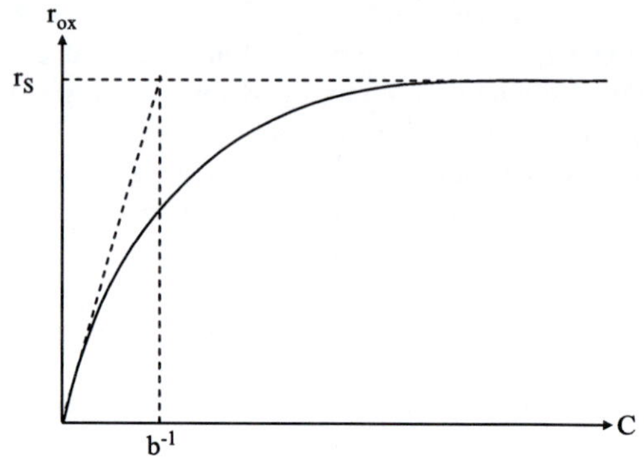

In the frame of this set of assumptions, the oxygen consumption rate r_{ox} is given by a hyperbolic function [21]:

$$r_{oxi} = \frac{aC}{1 + bC} \tag{8.18}$$

where C is the oxygen concentration, $a = k_2\left(\frac{r_i}{2k_4}\right)^{1/2}$ and $b = \frac{k_2}{k_3[PH]}\left(\frac{k_6}{k_4}\right)^{1/2}$.

The shape of the curve $r_{ox} = f(C)$ is shown in Fig. 8.7.

The above set of assumptions is, however, questionable in the case of thermal oxidation where hydroperoxides decompose. Achimsky et al. [2] reconsidered assumption T. For radical mobility reasons [29], it is better to take: $k_5^2 > 4k_4\,k_6$, which leads to a new expression of oxygen consumption rate (in the case of unimolecular initiation):

$$r_{ox} = 2\frac{k_3^2[PH]^2}{k_6}\,\frac{\beta C}{1 + \beta C}\left[1 - \frac{\beta C}{2(1 + \beta C)}\right] \tag{8.19}$$

where:

$$\beta = \frac{k_2 k_6}{2k_5 k_3[PH]}$$

The shape of the curve is the same as Fig. 8.7, but the asymptotic value of r_{ox}: $r_S = k_3^2[PH]^2/k_6$, is different. This model was immediately applied to composite matrices [11]. Despite its improvement, this model remains perfectible because it keeps the assumption of steady state, which is questionable. The next step consists in also suppressing this assumption that imposes solving numerically the system of differential equations constituting the kinetic scheme. This type of resolution was performed, for the first time, in the case of photo-oxidation by Sommersall and Guillet [49], but this tentative remained isolated until the turn of the century where our research group at Arts et Metiers ParisTech, proposed a numerical approach for the thermo-oxidation kinetics of polypropylene [40], polydiene elastomers [41], or poly(bismaleimides) [13]. The great advantage of the numerical approach is that it permits studying the case of variable initiation rate and, thus, having a better simulation of the autoaccelerated phase of oxidation, during the induction period.

8.3.2 Diffusion-Controlled Oxidation

Let us now consider the case of thick samples. Since oxygen is consumed by oxidation, its concentration must be a decreasing function of the layer depth z in the sample thickness. Then a diffusion term must be included in the oxygen balance equation:

$$\frac{\partial[O_2]}{\partial t} = D\,\mathrm{div}\left(\overrightarrow{\mathrm{grad}}([O_2])\right) - r_{ox}([O_2]) \tag{8.20}$$

where $r_{ox}([O_2])$ is the oxygen consumption rate in function of oxygen concentration.

Initially, two assumptions were systematically made, namely, (1) the system is in steady state: $d[O_2]/dt = 0$, and (2) far from sample edges, diffusion is unidirectional:

$$\frac{\partial[O_2]}{\partial t} = D\frac{\partial^2[O_2]}{\partial z^2} - r_{ox}([O_2]) \tag{8.21}$$

Until 2001, only the case of constant oxidation rate (e.g., also constant oxidation profile) was considered. The models differed by the expression of $r_{ox}([O_2])$ [1, 11, 21, 27, 29, 45].

Let us recall that the kinetic analysis leads to defining two kinetic regimes: The oxygen excess regime and the oxygen lack regime, with a critical oxygen concentration $[O_2]_C$ separating both regimes. The shape of the oxidation thickness profile depends on the difference between this critical concentration and the equilibrium oxygen concentration $[O_2]_S$, linked to oxygen pressure p and oxygen solubility in the polymer:

$$[O_2]_S = pS \tag{8.22}$$

where S is the solubility coefficient of O_2 in the polymer.

Two cases can be distinguished (Fig. 8.8).

In the case where $[O_2]_S < [O_2]_C$, all the thickness layers are in oxygen-lack regime, and the oxidation rate and the oxidation conversion decrease continuously with the depth. In the case where $[O_2]_S > [O_2]_C$, all the superficial layers in

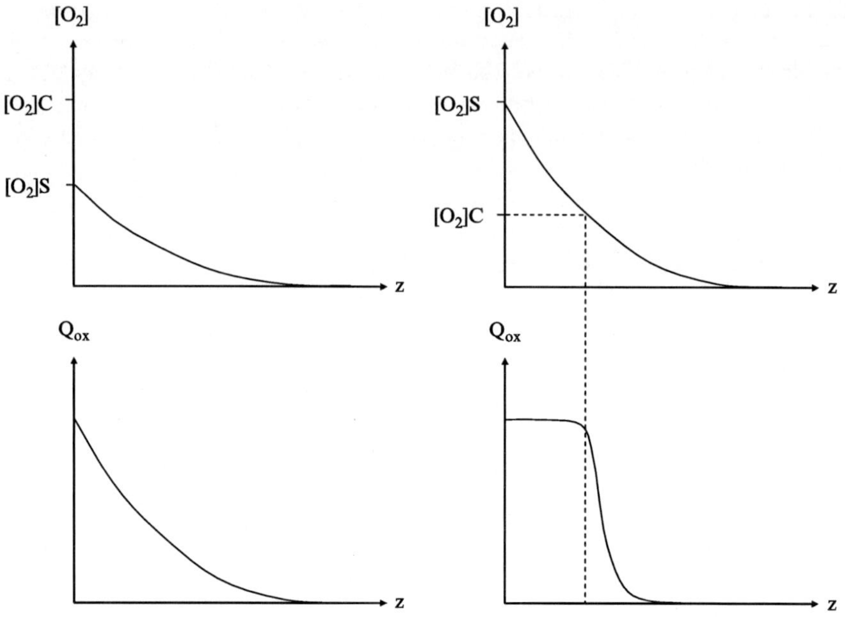

Fig. 8.8 Shape of oxygen concentration profile (*above*) and oxidation conversion profile (*below*). Polymer of type PP (*left*). Polymer of type PE (*right*)

which $[O_2] > [O_2]_C$ are in oxygen-excess regime, in which the oxidation rate is independent of oxygen concentration. The oxidation profile displays, then, a plateau in its superficial layer.

From a simplified approach of the problem, Audouin et al. [3] showed that, in most cases, the thickness of the oxidized layer ℓ can be estimated with a simple scaling law:

$$\ell \approx \left(\frac{D}{K}\right)^{1/2} \tag{8.23}$$

where D is the oxygen diffusion coefficient and K is the pseudo first-order rate constant for oxygen consumption:

$$K = \frac{r_{ox}}{[O_2]} \tag{8.24}$$

This law is useful to predict the trends of variation of ℓ, for instance, in the case of thermo-oxidation, the temperature effect. If D and K are assumed to obey (in a first approximation) Arrhenius law:

$$D = D_0 \exp - \frac{E_D}{RT} \quad \text{and} \quad K = K_0 \exp - \frac{E_K}{RT} \tag{8.25}$$

then:

$$\ell = \ell_0 \exp - \frac{E_\ell}{RT} \tag{8.26}$$

with $\ell_0 = \left(\frac{D_0}{K_0}\right)^{1/2}$ and $E_p = \frac{1}{2}(E_D - E_K)$.

Since, always, $E_D < E_K$, ℓ is expected to be a decreasing function of temperature as experimentally observed.

All the above quoted approaches have in common questionable assumptions, especially the "double steady state": one relative to the radical concentration ($d[P°]/dt + d[PO_2°]/dt = 0$) and the other relative to the diffusion–reaction balance ($\partial[O_2]/\partial t = 0$), which cannot describe the behavior during an autoaccelerated stage. All the simplifying assumptions were suppressed at the turn of the century [40, 41, 13]. The diffusion–reaction equation was integrated to the system of differential equations constituting the kinetic scheme, in the following form:

$$\frac{\partial[O_2]}{\partial t} = D\frac{\partial^2[O_2]}{\partial z^2} - k_2[O_2][P°] + k_6[PO_2°]^2 \tag{8.27}$$

The system of equations is then solved in time (t) and space (z) using the Rosenbrock's semi-implicit method dedicated to stiff problem of chemical kinetics. There is, henceforth, no limit (except a computing capacity one) for the complexity of the kinetic model, as recently shown in the cases of polyisoprene [17] or polybutadiene [19] elastomers, and in the case of chlorine dioxide effect on polyethylene [18] where three diffusion–reaction couplings, respectively relative to oxygen, antioxidant, and chlorine dioxide, are taken into account.

The specific case of composites involves several complications: anisotropy of diffusion, possible interfacial diffusion, modification of oxygen transport due to damage, etc. [14]. These aspects will be treated in another chapter.

8.4 Consequences of Oxidation

8.4.1 Mass Variations

Mass variations are usually expressed in fractions of the initial matrix mass $m = \Delta w/w_0$. A part of mass loss is due to outgassing (sorbed water, residual monomers, and other molecular impurities). The rest of mass variation, which will be analyzed here, is due to oxidation and can be divided into two components: Mass gain due to oxygen grafting to the polymer and mass loss due to the evolution of volatile products, in which one can distinguish water coming from hydroperoxide decomposition and other volatile (organic) molecules coming from rearrangements of chain end radicals or chain scissions near to chain ends. It has been already established that the balance of oxygen grafting to the polymer is given by:

$$-\frac{d[O_2]}{dt} = k_2[O_2][P°] - k_6[PO_2°]^2 \tag{8.28}$$

so that mass gain is given by:

$$\frac{dm^+}{dt} = \frac{32}{\rho_0}(k_2[O_2][P°] - k_6[PO_2°]^2) \tag{8.29}$$

ρ_0 is the specific weight in g L^{-1}.

Mass loss can occur only in initiation or termination reactions. However, since the kinetic regime is always close to steady state, initiation and termination rates are almost equal and it is licit to consider that mass loss occurs only in the initiation step:

$$\delta POOH \rightarrow \alpha P° + \beta PO_2^° + H_2O + \upsilon V$$

where V is a "virtual volatile molecule" other than H_2O, of which the molar mass M_V is the number average molar mass of the real volatile molecules and υ is the yield of volatiles, that is to say, the number of V molecules per initiation event. The, mass loss rate is given by

$$\frac{dm^-}{dt} = -\frac{1}{\rho_0}k_1[POOH]^\delta(18 + \upsilon M_V) \tag{8.30}$$

Then, the whole mass variation is:

$$\frac{dm}{dt} = \frac{dm^+}{dt} + \frac{dm^-}{dt} \tag{8.31}$$

It is noteworthy that mass gain is mainly linked to propagation, whereas mass loss is exclusively linked to initiation or termination. As a consequence, mass gain must predominate in the case of long kinetic chain, whereas mass loss must predominate in the case of short kinetic chain. Two cases can then be distinguished (Fig. 8.9). In case (a), for instance, poly(bismaleimides) [12] or polypropylene [42], oxidation displays an induction period or, at least, an initial autoaccelerated period. The kinetic chain length decreases during exposure and mass gain, which increases during the induction period, is progressively supplanted by mass loss. In case (b),

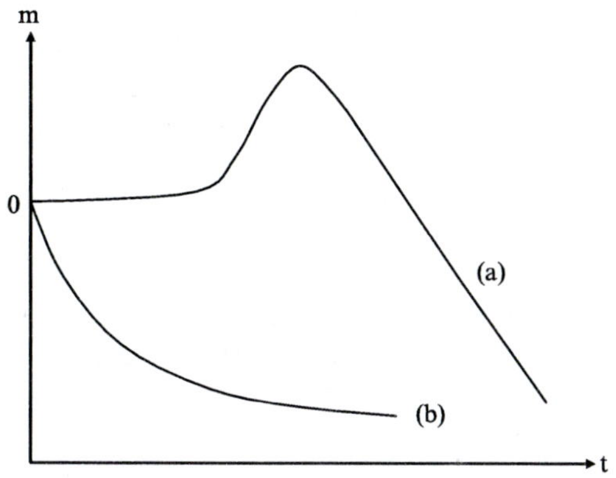

Fig. 8.9 The two kinds of gravimetric curves (see text)

for instance, epoxies [12], hydroperoxides are very unstable, there is no induction period, the kinetic chain is initially short, and mass loss predominates as soon as exposure begins. The ability of the model to simulate the non-monotonic gravimetric behavior in the case, for instance, of poly(bismaleimides), constitutes a good argument in its favor. Indeed, mass loss must be calculated into elementary thickness layers and then, integrated in the whole sample thickness.

8.4.2 Shrinkage

In a given elementary layer, after the end of the induction period, the mass decreases due to volatile loss, and the density increases because light hydrogen atoms are replaced by heavier oxygen atoms. Density variations can be estimated from atomic composition variations using the following method: A constitutive repeat unit (CRU), representative of the whole polymer structure, is defined. Its molar mass is initially M_0 and contains N_0 atoms. One can define its average atomic mass:

$$M_{A0} = \frac{M_0}{N_0} \tag{8.32}$$

Making simple assumptions on oxidation mechanisms, one can determine the new CRU composition after a certain exposure time that leads to a new value of the average atomic mass: M_A. A correlation has been observed between polymer density and average atomic mass [35]:

$$\rho = 350 + 120\,M_A \tag{8.33}$$

ρ is expressed in kg m^{-3} and M_A in g mol^{-1}.

The parameters of this equation can be eventually refined for a given structural series. It is thus possible to predict the local density, at every time, in every location, from the kinetic model.

Then, the volume variation can be calculated from:

$$\frac{dv}{v} = \frac{dm}{m} - \frac{d\rho}{\rho} \tag{8.34}$$

Since m decreases and ρ increases, v decreases. In other words, oxidation induces shrinkage, and this latter induces a stress state. At more or less long term, the latter is responsible for a spontaneous superficial cracking of organic matrix composites [15].

8.4.3 Glass Transition Temperature

Except in the scarce cases where the composite matrix contains double bonds (for instance, diglycidyl ether of bisphenol A-methyltetrahydrophthalic anhydride networks [32]), oxidation induces essentially chain scission and the latter induces a decrease of the glass transition temperature T_g. Chain scissions result from radical rearrangements and can occur only in initiation or termination events, so that the same reasoning as for mass variation can be made: It will be arbitrarily considered that chain scission occurs only in initiation:

$$\frac{dS}{dt} = \frac{1}{\rho_0} \gamma_1 k_1 [POOH]^\delta \tag{8.35}$$

where S is the number of moles of chain scission per mass unit.

In linear polymers, S is linked to number and weight average molar masses M_n and M_W by:

$$S = \frac{1}{M_n} - \frac{1}{M_{n0}} \quad \text{and} \quad S = \frac{2}{M_W} - \frac{2}{M_{W0}} \tag{8.36}$$

In networks (for ideal or quasi-ideal networks at low conversions) by:

$$S = \frac{1}{M_{e0}} - \frac{1}{M_e} \tag{8.37}$$

where M_{e0} and M_e are the respective molar masses of elastically active chains before and after aging.

Thus, for a linear polymer:

$$M_n=\left\{M_{n0}^{-1}+\frac{\gamma_1 k_1}{\rho_0}\int_0^t [POOH]^\delta dt\right\}^{-1} \tag{8.38}$$

and for a tridimensional polymer:

$$M_e=\left\{M_{e0}^{-1}-\frac{\gamma_1 k_1}{\rho_0}\int_0^t [POOH]^\delta dt\right\}^{-1} \tag{8.39}$$

In linear polymers [26]:

$$T_g = T_{g\infty} - \frac{K_{FF}}{M_n} \tag{8.40}$$

where $T_{g\infty}$ and K_{FF} are parameters specific to the polymer chemical structure (mainly chain dynamic stiffness), interrelated by [7]:

$$K_{FF}= 2.715 \times 10^{-3} T_{g\infty}^3 \tag{8.41}$$

The Fox–Flory relationship can be thus transformed into:

$$T_{g0} - T_g = K_{FF} S \tag{8.42}$$

so that,

$$\frac{dT_g}{dS} = -K_{FF} \tag{8.43}$$

The effect of chain scissions is an increasing function of the chain stiffness. Typically, K_{FF} can be of the order of 10 K kg mol^{-1} for flexible chains (rubber) and of the order of 150–200 K kg mol^{-1} for stiff chains (aromatic polymers).

In tridimensional polymers [24]:

$$T_g=\frac{T_{gl}}{1-2F\nu} \tag{8.44}$$

where $\nu = M_e{}^{-1}$ is the concentration of elastically active chains. F and T_{gl}, which are probably interrelated as K_{FF} and $T_{g\infty}$ are, for example, T_{gl} and F increase with the chain stiffness [6]. Thus,

$$\frac{dT_g}{dS}= -\frac{dT_g}{d\nu} = -2F\frac{T_g^2}{T_{gl}} \tag{8.45}$$

Here also, the effect of chain scissions is an increasing function of the initial T_g value.

8.4.4 *Elastic Modulus in Rubbery State for Networks*

For networks in rubbery state (in the absence of post-cure or degradation processes during its measurement), the elastic modulus is linked to the concentration of elastically active chains:

$$G = RT\rho\nu \tag{8.46}$$

where G is the shear modulus.

Since the Poisson's ratio is close to 0.5, the tensile modulus E is

$$E = 3RT\rho\nu \tag{8.47}$$

It becomes:

$$\frac{dG}{dS} = -\frac{dG}{d\nu} = -RT\rho \quad \text{and} \quad \frac{dE}{dS} = -3RT\rho \tag{8.48}$$

$$\frac{dG}{dt} = \frac{dS}{dt}\frac{dG}{dS} = -RT\rho\frac{dS}{dt} \tag{8.49}$$

where dS/dt is directly given by the kinetic model.

8.4.5 *Elastic Modulus in Glassy State*

Elastic modulus in glassy state depends mainly on the following factors:

- For amorphous polymers: cohesion and local mobility (responsible for sub-glass transitions)
- For semi-crystalline polymers: both above factors plus morphological parameters characterizing the crystalline phase

8.4.5.1 In Linear Polymers

Fracture properties fail before elastic properties have undergone significant changes. Semi-crystalline polymers can undergo chemicrystallization when they undergo degradation above T_g of the amorphous phase [25].

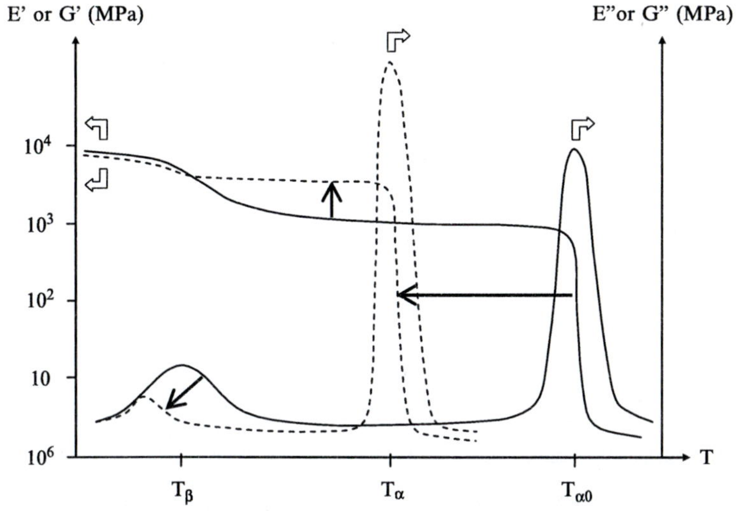

Fig. 8.10 Thermomechanical spectra of a network before (*full lines*) and after (*dashed lines*) degradation

8.4.5.2 In Thermosets

Two families can be distinguished:

1. Polymers having secondary transitions of low activity or overlapping glass transition (for instance, styrene cross-linked polyesters or vinylesters, and anhydride cured epoxies). In these cases, chain scissions have a negligible effect on modulus.
2. Polymers having secondary transitions of high activity and well separated from the glass transition (for instance, amine cross-linked epoxies). In this case, chain scissions induce an internal antiplasticization effect as illustrated by Dynamic Mechanical Analysis (Fig. 8.10).

Chain scission induces a decrease of the glass transition temperature, as expected, but also an increase of the modulus (E' or G') at the glassy plateau between T_β and T_α. This phenomenon is called internal antiplasticization. It results from the inhibition of β motions having the highest degree of cooperativity, and responsible for the high temperature side of the β dissipation band. The bulk modulus K is not affected by the sub-glass transitions. In contrast, the shear (G) and tensile (E) moduli above T_β depend on the activity of the β transition. It can be written as

$$G = G_0\left(1 - \alpha\frac{T}{T_g}\right) - \sum_i \Delta G_i \tag{8.50}$$

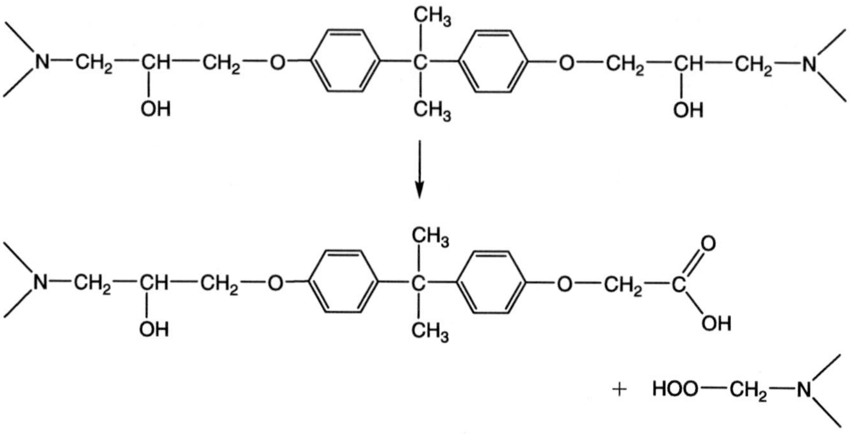

Fig. 8.11 Example of possible chain scission in an aromatic epoxide-amine network

where G_0 is the shear modulus at 0 K, which depends only on cohesive energy density, $\alpha \approx 0.3$–0.5 and ΔG_i is the modulus gap at the ith secondary transition below the temperature T [28].

Antiplasticization affects apparently only skeletal motions of the crankshaft type, so that the above relationship reduces to:

$$G = G_u - \Delta G_\beta \quad (8.51)$$

where G_u is the unrelaxed modulus at the temperature under consideration and ΔG_β is the modulus gap at the β transition. Antiplasticization which results from chain scission, induces a decrease of ΔG_β and, thus, an increase of G. Since K remains constant, this means that the tensile modulus increases:

$$E = E_u - \Delta E_\beta \quad (8.52)$$

and ΔE_β decreases, whereas the Poisson's ratio decreases:

$$\mu = \mu_u + \Delta\mu_\beta \quad (8.53)$$

where $\Delta\mu_B$ decreases.

Quantitative relationships linking modulus changes to cross-link density are not yet established. A possible way to establish these relationships consists in studying model networks of well-defined structure [37]. In the case of aromatic epoxide-amine networks, chain scissions occur only in the isopropanol segment (Fig. 8.11).

It is noteworthy that un-degraded networks based on the same diepoxide and the same diamine, but with an epoxide excess, would contain dangling chains of almost the same structure as those created by oxidative degradation (Fig. 8.12).

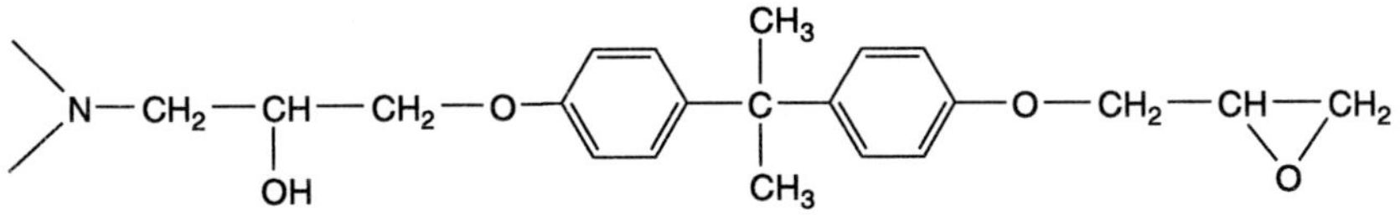

Fig. 8.12 Dangling chain in an epoxide-amine network containing an excess of epoxide

Let us consider the case of a diepoxide (of molar mass M_E)-diamine (of molar mass M_A) network. For un-degraded networks differing by the amine/epoxide functional ratio (a), cross-link density and dangling chain concentration can be estimated as follows: Let us consider the constitutive repeat unit (CRU) based on two diepoxide molecules and a diamine molecule. In the case of epoxide excess, all the nitrogen atoms are fully reacted. For $(1 - a) << 1$, one can consider that all the dangling chains are isolated. The number of cross-link nodes is thus:

N = Total number of nitrogen atoms –
Number of nitrogen atoms connected to dangling chains

so that:

$$N = 2a - (4 - 4a) = 6a - 4 \tag{8.54}$$

and the cross-link density (node concentration) is thus,

$$X = \frac{6a - 4}{2M_E + aM_A} \tag{8.55}$$

so that:

$$\nu = \frac{3}{2}X = \frac{3}{2}\left(\frac{6a - 4}{2M_E + aM_A}\right) \tag{8.56}$$

whereas the concentration of dangling chains is:

$$b = \frac{4 - 4a}{2M_E + aM_A} \tag{8.57}$$

Let us consider a stoechiometric network (a = 1) which has undergone S chain scissions per mass unit. The concentration of elastically active chains is thus, at low conversions:

$$\nu = \nu_0 - 3S = \frac{2}{2M_E + aM_A} - 3S \tag{8.58}$$

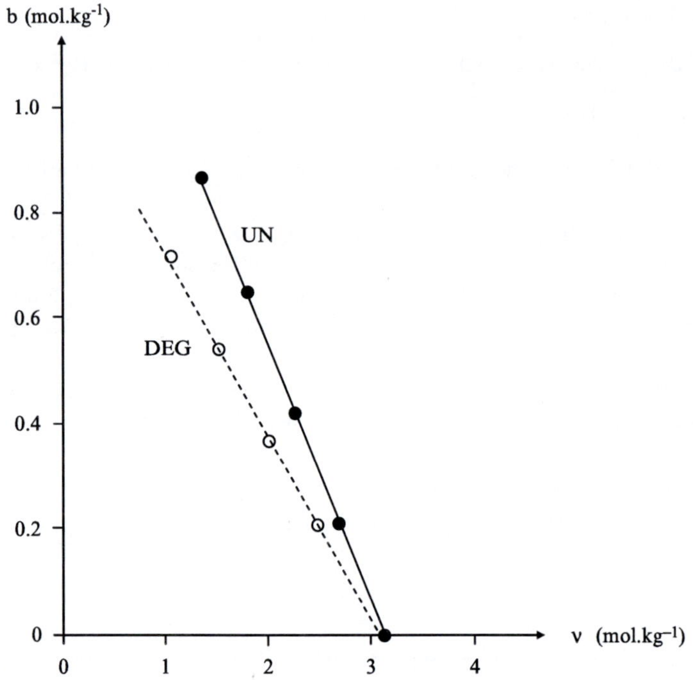

Fig. 8.13 Dangling chain concentration against cross-link density for un-degraded DGEBA-DDS networks containing an epoxide excess (UN) and for a stoichiometric DGEBA-DDS network undergoing chain scission (DEG)

and, since one dangling chain is created per chain scission:

$$b \approx S \tag{8.59}$$

The dangling chain concentration has been plotted against cross-link density for model networks differing by the amine/epoxide functional ratio (a/e $\leq$ 1) and for a degraded network, both structural series being based on the diglycidyl ether of bisphenol A-diamino dyphenylsulphone (DGEBA-DDS) system (Fig. 8.13).

The un-degraded networks contain more dangling chains, for a given cross-link density, than degraded ones, but the concentrations remain of the same order of magnitude, so that networks containing an excess of free epoxides are interesting model networks for degraded ones.

Modulus profiling by nano-indentation on oxidized samples shows that PMR-15 resin behaves as an epoxide-amine one, indicating thus the probable existence of an internal antiplasticization phenomenon [36].

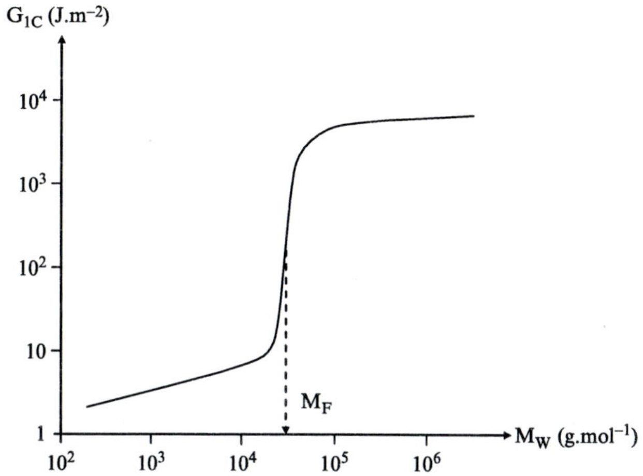

Fig. 8.14 Critical elastic energy release rate against weight average molar mass. General shape of the curve for linear polymers

8.4.6 Toughness

8.4.6.1 Linear Polymers

The toughness of linear polymers depends on molar mass according to a curve which always displays the shape of Fig. 8.14. This curve reveals the existence of two regimes of fracture: The "tough" regime where, typically, $G_{1C} > 1$ kJ m^{-2}, and the "brittle" regime where, typically, $G_{1C} \leq 100$ kJ m^{-2}. A relatively sharp transition between both regimes is located at a critical molar mass M_F, which is always of the order of 10–100 kg mol^{-1}.

For an industrial grade, the initial molar mass M_{W0} is always higher than M_F. Thus, during polymer oxidative degradation, the toughness will remain almost constant until the time where the molar mass value will reach M_F and the behavior will suddenly turn to brittle regime (Fig. 8.15).

It is important to note that toughness varies almost in a discontinuous way while molar mass varies continuously. In initially brittle polymers (polystyrene, poly(methyl methacrylate), etc.), G_{1C} is the most pertinent quantity. In initially ductile polymers (polyolefins, polycarbonate, polysulphones, etc.), G_{1C} can be replaced by J_{1C} or simply, by the ultimate tensile elongation.

It has been tried to compare M_F to the entanglement molar mass M_e. Two categories of polymers can be distinguished, depending on the ratio $q = M_F/M_e$:

1. For amorphous polymers and semi-crystalline polymers having their amorphous phase in glassy state (for instance, poly(ethylene terephthalate), poly(ether ether ketone), polyamides, etc.): $2 \leq q \leq 10$. This means that embrittlement occurs

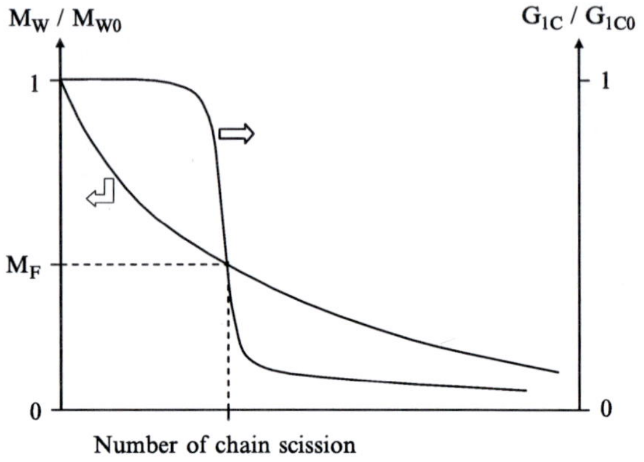

Fig. 8.15 Shape of molar mass and toughness variations during aging

when the entanglement network has been significantly damaged. It is well known that chain drawing, responsible for plastic deformation, can occur only if the polymer is physically cross-linked by entanglements.

2. For semi-crystalline polymers having their amorphous phase in rubbery state (polyethylene, polypropylene, poly(tetrafluoroethylene), etc.), typically $q > 25$. In these polymers, embrittlement results essentially from morphological changes (e.g., decrease of interlamellar amorphous phase thickness) resulting indirectly from chain scission [25].

8.4.6.2 Thermosets

Thermosets are generally brittle or semi-ductile, except eventually in the immediate vicinity of their glass transition. In ideal networks, the toughness depends mainly on the difference $(T_g - T)$, T being the temperature at which fracture properties are measured. This means that, at a fixed measurement temperature T, toughness is, roughly, a decreasing function of T_g [35].

In a given structural series, T_g is generally an increasing function of the cross-link density, so that toughness is found to be a decreasing function of the cross-link density, for example, for an ideal network, an increasing function of the molar mass M_C of elastically active chains. It is tempting to apply these rules to degraded networks; however, in these, the number of elastically active chains decreases, but their length remains constant. As far as model networks are representative not only for elastic properties, but also for fracture properties, chain scission is expected to induce toughness decrease as illustrated in Fig. 8.16.

Unfortunately, very little is known on the quantitative relationships between structure and fracture properties in nonideal networks.

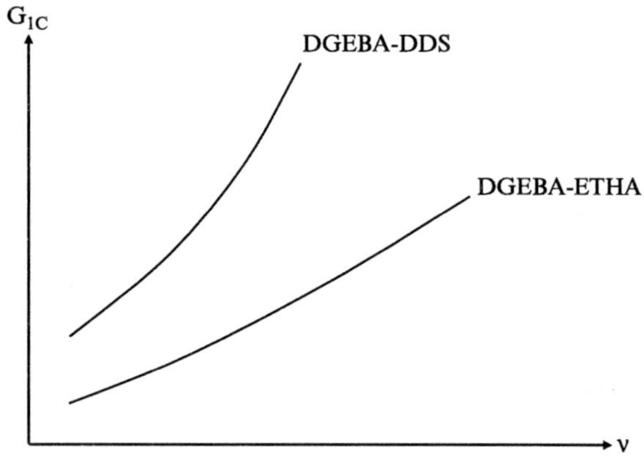

Fig. 8.16 Critical elastic energy release rate against cross-link density for model networks DGEBA-DDS and DGEBA-ETHA [37]

8.5 Conclusion and Prospects

The domain of polymer oxidation kinetics was characterized by a somewhat chaotic progression in the second half of the twentieth century. Among the possible reasons of its relative underdevelopment are, probably,

- A relative mistrust of chemical engineers about mathematical modeling
- The use and abuse of simplifying hypotheses to permit analytical resolution of kinetic schemes

The introduction, at the turn of the century, of numerical methods, allows suppressing of most of the undesired hypotheses and reconstructing kinetic curves from a mechanistic scheme that permit, first, a rigorous validation of the model by comparison to experimental curves and, second, the determination of certain elementary rate constants by an inverse approach.

For bulk polymer samples, it is now possible to accurately predict structural changes at all the pertinent structural scales:

- Local concentrations of oxygen containing groups (carbonyls, hydroxyls, hydroperoxides, etc.)
- Local quantities characterizing the macromolecular structure and directly linked to mechanical properties (molar masses, cross-linking densities, etc.)
- Depth distribution of the above characteristics at the micrometric scale

The ability of the model to predict at the same time complex kinetic behaviors (e.g., non-monotonic gravimetric curves) and distribution thickness profiles, independently of sample geometry, in a wide range of temperatures and oxygen pressures, illustrates well its power.

In the case of composites, many complications appear. The most important one, linked to the anisotropy of oxygen diffusion, is practically resolved now. The problem of interfacial degradation, in contrast, remains largely open. The problem of couplings between chemistry and diffusion with stress remains an almost virgin domain and would need, in future, significant research efforts. The key problem of the relationships between chemical changes and mechanical properties is that they are not totally elucidated and will also need, in the future, a noticeable volume of investigations. Numerical problems linked to tridimensional diffusion are now in the way to be solved, which will permit solving of heterogeneous kinetic schemes, for instance, due to chemical reactions between carbon fibers and reactive species participating in polymer oxidation.

References

1. Abdeljaoued, K.: Study of matrix thermal oxidation in carbon fibers-PMR-15 composites, PhD Thesis, ENSAM Paris, 1999.
2. Achimsky, L., Audouin, L., Verdu, J., Rychla, L., Rychly, J.: The effect of oxygen pressure on the rate of polypropylene oxidation determined by chemiluminescence, Europ. Polym. J. **35**, 557–563 (1999).
3. Audouin, L., Langlois, V., Verdu, J., De Bruijn, J.C.M.: Role of oxygen diffusion in polymer aging. Kinetic and mechanistic aspects, J. Mater. Sci. **29**, 569–583 (1994).
4. Audouin, L., Gueguen, V., Tcharkhtchi, A., Verdu, J.: Close-loop mechanistic schemes for hydrocarbon polymer oxidation, J. Polym. Sci.: Part A: Polym. Chem. **33**, 921–927 (1995).
5. Audouin, L., Achimsky, L., Verdu, J.: Kinetic modelling of low temperature oxidation of hydrocarbon polymers, In: Halim Hamid S. (ed) Handbook of polymer degradation, 2nd edition, chap. 16, pp. 727–763, Marcel Dekker, New York (2000).
6. Bellenger, V., Morel, E., Verdu, J.: Effect of structure on the glass transition temperature of amine crosslinked epoxies, J. Polym. Sci.: Part B: Polym. Phys. **25**, 1219–1234 (1987).
7. Bicerano, J.: Prediction of polymer properties, 3rd edition, chap. 6, p. 212, Marcel Dekker, New York (2002).
8. Bolland, J.L., Gee, G.: Kinetic studies in the chemistry of rubber and related materials. Parts II and III, Trans. Faraday Soc. **42**, 236–252 (1946).
9. Bowles, K.J., Meyers, A.: Specimen geometry effects on graphite/PMR-15 composites during their thermo-oxidative aging, In: Proc of the 31st International SAMPE Symposium and Exhibition, vol. **3**, pp. 1285–1299. Covina, CA, US (1986).
10. Bowles, K.J., Nowak, G.: Thermo-oxidative stability studies of Celion 6000/PMR-15 unidirectional composites. PMR-15 and Celion 6000 fiber, J. Compos. Mater. **22(6)**, 966–985 (1988).
11. Colin, X.: Kinetic modelling of high thermomechanical performances polymer and composite materials, PhD Thesis, ENSAM Paris, 2000.
12. Colin, X., Marais, C., Verdu, J.: Kinetic modelling and simulation of gravimetric curves. Application to the oxidation of bismaleimide and epoxy resins, Polym. Degrad. Stab. **78(3)**, 545–553 (2002).
13. Colin, X., Verdu, J.: Thermal ageing and lifetime prediction for organic matrix composites, Plastics Rubber Compos.: Macromol. Eng. **32(8/9)**, 349–356 (2003).
14. Colin, X., Marais, C., Verdu, J.: Kinetic modelling of the stabilizing effect of carbon fibers on thermal ageing of thermoset matrix composites, Compos. Sci. Technol. **65**, 117–127 (2005a).
15. Colin, X., Mavel, A., Marais, C., Verdu, J.: Interaction between cracking and oxidation in organic matrix composites, J. Compos. Mater. **19(15)**, 1371–1389 (2005b).
16. Colin, X., Fayolle, B., Audouin, L., Verdu J.: The classical model for radical chain oxidation of hydrocarbon substrates initiated by bimolecular hydroperoxide decomposition, Intern. J. Chem. Kin. **38(11)**, 666–676 (2006).

17. Colin, X., Audouin, L., Verdu, J.: Kinetic modelling of the thermal oxidation of polyisoprene elastomers. Parts I, II and III, Polym. Degrad. Stab. **92(5)**, 886–914 (2007).
18. Colin, X., Audouin, L., Verdu, J., Rozental-Evesque, M., Rabaud, B., Martin, F., Bourgine, F.: Aging of polyethylene pipes transporting drinking water disinfected by chlorine dioxide. Parts I and II, Polym. Eng. Sci., **49**(7), 1429–1437 (2009) and **49**(8), 1642–1652 (2009).
19. Coquillat, M., Verdu, J., Colin, X., Audouin, L., Nevière, R.: Thermal oxidation of polybutadiene. Parts II and III, Polym. Degrad. Stab. **92(7)**, 1334–1349 (2007).
20. Crews, L.K., McManus, H.L.: Modelling the high temperature degradation of graphite/epoxy, In: Proc. of the American Society for Composites, 12th Technical Conference on Composite Materials, pp. 1123–1132. Detroit, MI, US (1997).
21. Cunliffe, A.V., Davis, A.: Photo-oxidation of thick polymer samples, Polym. Degrad. Stab. **4(1)**, 17–37 (1982).
22. Cunningham, R.A., McManus, H.L.: Coupled diffusion-reaction models for predicting the distribution of degradation in polymer matrix composites, In: Proc. of the ASME Aerospace Division, ASME International Mechanical Engineering Congress and Exposition, Symposium on Composite Materials, vol. **52**, p. 353. Atlanta, GA, US (1996).
23. Denisov, E.T., Afanas'ev, I.B.: Oxidation and antioxidants in organic chemistry and biology, CRC Taylor and Francis, Boca Raton (2005).
24. Di Marzio, E.A.: On the second-(order transition of a rubber, J. Res. NBS: Section A: Phys. Chem. **68**, 611–617 (1964).
25. Fayolle, B., Richaud, E., Colin, X., Verdu, J.: Review: Degradation-induced embrittlement in semi-crystalline polymers having their amorphous phase in rubbery state, J. Mater. Sci. **43**, 6999–7012 (2008).
26. Fox, T.G., Flory, P.J.: Second-order transition temperatures and related properties of polystyrene. I. Influence of molecular weight, J. Appl. Phys. **21**, 581–591 (1950).
27. Furneaux, G.C., Ledbury K.J., Davis, A.: Photo-oxidation of thick polymer samples. Part I: The variation of photo-oxidation with depth in naturally and artificially weathered low density polyethylene, Polym. Degrad. Stab. **3(6)**, 431–442 (1981).
28. Gilbert, D.G., Ashby, M.F., Beaumont, P.W.R.: Modulus maps for amorphous polymers, J. Mater. Sci. **21**, 3194–3210 (1986).
29. Gillen, K.T., Wise, J., Clough, R.L.: General solution for the basic autoxidation scheme, Polym. Degrad. Stab. **47**, 149–161 (1995).
30. Golike, R.C., Lasoski, S.W.: Kinetics of hydrolysis of polyethylene terephthalate films, J. Phys. Chem. **64**, 895–898 (1960).
31. Korcek, S., Chenier, J.H.B., Howard, J.A., Ingold, K.U.: Absolute rate constants for hydrocarbon autoxidation. XXI: Activation energies for propagation and the correlation of propagation rate constants with carbon-hydrogen bond strengths, Can. J. Chem. **50**, 2285–2297 (1972).
32. Le Huy, M., Bellenger, V., Paris, M., Verdu, J.: Thermal oxidation of anhydride cured epoxies. Part I, II and III, Polym. Degrad. Stab. **35**, 77–86 (1992), **35**, 171–179 (1992) and **41(2)**, 149–156 (1993).
33. Nam, J.D., Seferis, J.C.: Anisotropic thermo-oxidative stability of carbon fiber reinforced polymeric composites, SAMPE Quart. **24(1)**, 10–18 (1992).
34. Nelson, J.B.: Thermal aging of graphite/polyimide composites. In: O'Brien, T.K. (ed) Long term behaviour of composites, ASTM STP **813**, pp. 206–221. American Society for Testing and Materials, Philadelphia, (1983).
35. Pascault, J.-P., Sautereau, H., Verdu, J., Williams, R.J.J.: Thermosetting Polymers. Marcel Dekker, New York (2002).
36. Putthanarat, S., Tandon, G.P., Schoeppner, G.A.: Influence of polishing time on thermooxidation characterization of isothermally aged PMR-15 resin, Polym. Degrad. Stab. **92**, 2110–2120 (2007).
37. Rasoldier, N., Colin, X., Verdu, J., Bocquet, M., Olivier, L., Chocinski-Arnault, L., Lafarie-Frenot, M.-C.: Model systems for thermo-oxidized composites matrices, Composites: Part A: Applied Science and Manufacturing **39**, 1522–1529 (2008).

38. Reich, L., Stivala, S.S.: Autoxidation of hydrocarbons and polyolefins. Kinetics and mechanisms, Marcel Dekker, New York (1969).
39. Richaud, E., Colin, X., Fayolle, B., Verdu, J.: Induction period in the low temperature thermal oxidation of saturated hydrocarbons. Example of polyethylene, Intern. J. Chem. Kin. **40(12)**, 769–777 (2008).
40. Rincon-Rubio, L.M., Fayolle, B., Audouin, L., Verdu, J.: A general solution of closed-loop kinetic scheme for the thermal oxidation of polypropylene, Polym. Degrad. Stab. **74(1)**, 177–188 (2001).
41. Rincon-Rubio, L.M., Colin, X., Audouin, L., Verdu, J.: A theoretical model for the diffusion-limited thermal oxidation of elastomers at medium temperatures, Rubber Chem. Technol. **76(2)**, 460–482 (2003).
42. Rychly, J., Matisova-Rychla, L., Csomorova, K., Achimsky, L., Audouin, L., Tcharkhtchi, A., Verdu, J.: Kinetics of mass changes in oxidation of polypropylene, Polym. Degrad. Stab. **58**, 269–274 (1997).
43. Salin, I.M., Seferis, J.C.: Anisotropic effects in thermogravimetry of polymeric composites, J. Polym. Sci.: Part B: Polym Phys. **31**, 1019–1027 (1993).
44. Seferis, J.C.: Aging analyses of polymer composites through time-temperature equivalence. In: Reifsnider, K.L., Dillard, D.A., Cardon, A. (eds) Progress in durability analysis of composite systems, pp. 193–201, Balkema, Rotterdam (1998).
45. Seguchi, T., Hashimoto, S., Arakawa, K., Hayakawa, N., Kawakami, W., Kuriyama, I.: Radiation induced oxidative degradation of polymers. Part I: Oxidation region in polymeric films irradiated in oxygen under pressure, Radiat. Phys. Chem. **17(4)**, 195–201 (1981).
46. Semenov, N.M.: Chemical kinetics and chain reactions, Oxford University Press, London (1935).
47. Skontorp, A., Wong, M.S., Wang, S.S.: High temperature anisotropic thermal oxidation of carbon fiber reinforced polyimide composites. Theory and experiments, In: Street, K., Poursartip, A. (eds) Proc. of ICCM-10, 10th International Conference on Composite Materials, vol. **4**, pp. 375–382. Whistler, BC, Canada (1995).
48. Smith, Vale: Chimie et Industrie (Paris) **105(18)**, 1179 (1972).
49. Sommersall, A.C., Guillet, J.E.: Computer modelling studies of polymer photooxidation and stabilization. In: Klemchuk, P.P. (ed) Polymer stabilization and degradation, chap. 16, pp. 211–234, American Chemical Society, Washington, DC (1985).
50. Swern, D.: Organic peroxides, vols **1** and **2**, Wiley-Interscience, New York (1971).
51. Tobolsky, A.V., Metz, D.J., Mesrobian, R.B.: Low temperature oxidation of hydrocarbons. The phenomenon of maximum rates, J. Amer. Chem. Soc. **72**, 1942–1952 (1950).

Chapter 9
Characterization of Thermo-Oxidation in Laminated and Textile Composites

G.P. Tandon

Abstract An understanding of the effects of thermo-oxidation in high-temperature PMCs for structural components subjected to arbitrary service environments is critical to life-performance predictions. Durability and degradation mechanisms in composites are fundamentally influenced by the fiber, matrix, and interphase regions that constitute the composite domain. The thermo-oxidative behavior of the composite is significantly different from that of the constituents as the composite microstructure, including the fiber/matrix interphases/interfaces, architecture, and ply layup introduce anisotropy in the diffusion and oxidation behavior. In this work, light microscopy and scanning electron microscopy techniques are used to characterize the oxidative process in laminated and textile carbon-fiber-reinforced polyimide composites. The observed anisotropy in composite oxidation is explained by carefully monitoring the development and growth of damage through the use of fluorescence imaging using dye impregnation. It is shown that alternative pathways for transport of oxygen into the interior of the composite are fiber–matrix debonds and matrix cracks that propagate with the oxidation front. It has been further determined through closer examination of the oxidation front and the crack front for discrete regions of the various composite specimens, that the oxidation front consistently precedes the crack front. This mechanism for accelerated oxidation is an excellent example of the intrinsic coupling of chemical oxidative aging and damage, which needs to be properly represented in predictive models.

G.P. Tandon
Multi-Scale Composites and Polymers Division,
University of Dayton Research Institute, Dayton, OH 45469-0060, USA
e-mail: Gyaneshwar.Tandon.ctr@wpafb.af.mil

K.V. Pochiraju et al. (eds.), *Long-Term Durability of Polymeric Matrix Composites*,
DOI 10.1007/978-1-4419-9308-3_9, © Springer Science+Business Media, LLC 2012

9.1 Introduction

Polymer matrix composites (PMCs) used in aerospace high-temperature applications, such as turbine engines and engine-exhaust-washed structures, are known to have limited life due to environmental degradation. Evaluating the extended service life of composite structures subjected to mechanical loading, high temperature, moisture, and corrosive conditions is challenging [1, 2] due to the complex physical, chemical, and thermo-mechanical mechanisms involved. A valuable resource [3] for understanding the high-temperature behavior of PMCs is the tremendous volume of long-term, high-temperature aging data generated in the NASA High-Speed Research (HSR) program. The HSR effort was a national effort to develop the next-generation supersonic passenger jet designed for a 60,000-h life with temperatures approaching 177°C. An important contribution to understanding the high-temperature performance of the polymer composites evaluated in the HSR program was the use of viscoelastic formulations to model long-term behavior (e.g. [4, 5]). The assumption inherent in the use of these viscoelastic formulations is that the material is chemically static or the original chemical structure is thermally recoverable [6]. Physical aging is a reversible process known to be easily altered with stress and temperature [7]. Earlier studies on the effects of physical aging on polymer composite behavior were conducted by Sullivan [8] and McKenna [9]. Practical considerations of physical aging generally become important when aging temperatures are near the T_g. For high-temperature PMCs that are used at temperatures near the material's T_g, physical aging may dramatically affect the time-dependent mechanical properties (creep and stress relaxation) and rate-dependent failure processes [10, 11]. However, for highly cross-linked polyimide systems, it is difficult to separate the physical aging from chemical-aging effects when conducting tests near the T_g because the aging effects are coupled.

Chemical aging, unlike physical aging, is typically not thermo-reversible. Hydrolysis (the chemical reaction of the polymer with water) and oxidation (the chemical reaction of the polymer with oxygen) are the primary forms of chemical degradation in high-temperature polymer matrix composites (HTPMCs). Although hygrothermal degradation, which is a significant concern for HTPMCs, is not specifically addressed here, the degradation mechanism is somewhat similar. As with oxidation, hygrothermal degradation involves transport and chemical reactions of the polymer with the diffused media. The chemical changes occurring during oxidation include chemical bond breaks that result in a reduction in molecular weight, mechanical response changes, and a local loss of mass associated with outgassing of oxidation by-products. Although the rate of oxidative chemical reactions is, in part, governed by the availability of reactive polymer, the oxygen concentration, and the reaction temperature for high T_g glassy polymers, the rate of oxidation can be greater than the rate of diffusion of oxygen into the polymer. For such circumstances, the oxidative process is diffusion-rate limited. Diffusion-rate-limited oxidation of neat polymer specimens typically results in the development of an oxidative layer [1] or graded oxidative properties near the free surfaces of the specimen. Within the oxidized region of the polymer, it is typical that the tensile strength, strain to failure, flexural strength, density, and toughness decrease while the modulus increases. The effect of oxidation on changes in the T_g is dependent on the specific polymer system. Some polymers initially have a decrease and then an increase in T_g, others may have only a decrease, and still others may only have an increase in T_g. This may be due to competing chemical and physical aging phenomenon or differences in the oxidation reaction mechanisms. Since HTPMCs typically operate at temperatures near their initial design T_g, any changes in the local or global T_g can have detrimental effects on performance.

While the time-dependent physical, chemical, and damage-induced degradation mechanisms have been studied for some resin systems, polymer composite thermal oxidation studies from a mechanistic perspective are nascent. Notable exceptions to this are the recent works of Colin et al. [12, 13] and Tandon et al. [14, 15], and past work by Skontorp et al. [16] and Wang et al. [17]. Equally important are the experimental characterization efforts by such groups as Bowles et al. [18–20], Wong et al. [21], Tsotsis et al. [22–24], Abdeljaoued [25], Bellenger et al. [26], and Whitley et al. [27]. However, most literature is confined to thermal oxidation of unidirectional polymer systems and there are limited studies on laminated and woven composite materials.

Nelson [28] was the first to document the anisotropy of oxidation in PMCs and that the oxidation process was sensitive to the surface area of the test specimens. He found that the dominant degradation mechanism for the graphite/polyimides is oxidation of the matrix at the laminate edges. Additionally, the materials degraded preferentially at the specimen surface perpendicular to the fiber (axial surface) and the rate of oxidation is hastened by microcracks opening on the axial surface, increasing the surface area for oxidation. Subsequently, numerous other investigators (e.g., Bowles et al. [29, 30], Nam and Seferis [31], Salin and Seferis [32], Schoeppner et al. [33], and Tandon et al. [34]) also observed the anisotropic nature of oxidation in high-temperature unidirectional PMCs. In these composites, the diffusion-limited oxidation behavior is predominantly controlled by the properties of the resin and the fiber–matrix interface and the total surface area through which the oxygen can diffuse. Various factors [12–15] may lead to the preferential oxidation along the fiber paths. Resin cure shrinkage and mismatches in the coefficient of thermal expansion of the fibers and matrix during the composite cure process give rise to localized micromechanical residual stresses in the fiber–matrix interphase region. The diffusion process in materials is known [35] to increase with increasing stress levels even becoming nonlinear, as a function of stress, for materials with linear mechanical behavior. Therefore, due to stress-assisted diffusion, and hence stress-assisted oxidation, the highly stressed fiber–matrix interphase regions may tend to oxidize at an accelerated rate compared to the lower stressed bulk polymer phase of the composite. Another factor that may lead to preferential oxidation along the fiber is that the local stoichiometry of the polymer may be altered in the fiber–matrix interphase region due to coupling agents or sizing agents. Woven and braided preforms require the use of a fiber coupling or sizing agents to protect the fibers from damage during the weaving and braiding process. Owing to the significant use of woven fiber prepreg and preforms in HTPMC applications, the proper selection of fiber sizing is of paramount concern. The presence of the sizing can have a strong influence on the fiber–matrix interphase or interfacial properties [36–39] and can ultimately affect the local diffusivity and/or thermal oxidative stability. In addition, severe surface oxidation degradation results in the formation of fiber–matrix disbonds coalescing with transverse surface cracks. These cracks not only reduce strength, but also create enhanced pathways [40–47] for oxygen to penetrate deeper into the composite.

In laminated systems, interlaminar residual stresses (besides the fiber–matrix micromechanical stresses at the ply level) further play an important role in the degradation process. In a study by Bowles et al. [19], both unidirectional and cross-ply (±45) PMR-15 laminates were isothermally aged at 288°C for up to 1,000 h. They found that the weight loss of the (±45) cross-ply laminates was much greater than that of the unidirectional laminates. The primary reason given for the greater weight loss in the (±45) cross-ply laminates was that free-edge cracks, resulting from interlaminar residual stresses, were much more prominent and cause extensive advancement of oxidation into the interior of the composite. Similar observations were made by Tsotsis and Lee [23]. They examined the long-term thermo-oxidative degradation behavior of G30-500/R922-1 and G30-500/R6376 carbon-fiber-reinforced epoxy composites. Their results indicated that residual stresses arising from aging-induced differential resin shrinkage and interaction between plies of different orientations were found to have a strong effect on the degradation process for plies close to the surface and, especially, near free edges. An analytical model has been developed by McManus and Chamis [42] that links matrix material degradation and shrinkage to the ply and laminate behavior that result. Recently, Lafarie-Frenot and Rouquie [48] and Rouquie et al. [49] performed thermal cycling tests on four different carbon/epoxy laminates: $[0]_8$, $[0_3/90_3]_S$, $[45_3/-45_3]_S$ and $[45/0/-45/90]_S$ in nitrogen and air. Microscopic observations and X-radiographs showed an acceleration of matrix cracking and matrix shrinkage due to coupling between oxidation and cyclic thermal stresses. Moreover, damage observations on the polished edges of the samples were highly dependent on the laminate stacking sequence. Thermo-oxidative aging of multi-hole carbon/epoxy laminates has been studied by Ammar-Khodja et al. [50]. They have shown that oxidation and cracking progresses faster on the hole edges than on the mold surfaces resulting in damage of the multi-hole panels bulk, and greater diminution of the perforated laminate properties compared to the plain panels. Obviously, increasing the surface in contact with the aging environment leads to enhancement in oxygen action and resulting degradation.

There is also substantial evidence that for woven composites, the oxidation preferentially advances in the in-plane direction along the fiber paths [51, 52]. The oxidation in the transverse direction (normal to the specimen's top and bottom or tool surfaces) is typically constrained by the presence of the fibers. When testing a T650-35/PMR-15, eight-harness, satin-weave graphite fiber material, Bowles [53] observed two types of surface degradations. Higher temperature aging (288–316°C) results in the formation of a light-colored surface layer that propagates into the material causing voids and microcracks to initiate and grow within the surface layer.

At temperatures lower than 288°C, specimens show the same advancement of voids and microcracks into the surface but the oxidized light band of matrix material is not visible. That is, at lower temperatures oxidation does not cause a visible change in the color of the polymer. Composite oxidation is primarily a surface reaction phenomenon controlled by the diffusion and rate of reaction of oxygen with the material so that surfaces with different microstructural characteristics are expected to exhibit different oxidation behavior due to differences in diffusivity.

While these limited studies demonstrate coupling between material degradation [54–56] and damage development [40–47] associated with thermo-oxidation aging, very little work has been reported on examining and quantifying the oxidation growth behavior in multidirectional laminated and textile composite systems. In this work, we investigate the influence of laminate architecture, layup, thickness, and the neighboring ply effects on oxidation growth within HTPMCs. Light microscopy techniques are used to characterize the oxidative process in laminated and woven carbon-fiber-reinforced polyimide composites. Several different laminates are considered, namely, unidirectional $[0]_{16T}$, quasi-isotropic $[0/\pm45/90]_{2S}$, cross-ply $[0/90]_{4S}$, angle-ply $[\pm45]_{2S}$, plain weave, and 3D textile composites. Isothermal aging is conducted at 177°C. In addition to oxidation growth behavior, we also examine the oxidation-induced damage development and the associated coupling through fluorescence imaging using dye impregnation in conjunction with optical and scanning electron microscopy techniques. Alternately, x-ray micro computed tomography (CT) enables one to examine the details of the microstructure of a material in a nondestructive fashion. A limited comparison of optical microscopy images of crack growth on cut and polished cross sections with CT imaging is therefore made to validate the use of optical microscopy techniques for quantitative assessment of oxidation and crack growth measurements.

9.2 Isothermal Aging of Test Specimens

Rectangular composite specimens measuring approximately 125 mm × 12.5 mm and of varying thicknesses were used to monitor oxidation propagation and damage development along the length and transverse to the fibers. The isothermal aging was accomplished in an air-circulating oven that provided a continuous replenishment of oxygen in the ambient air by convection through the oven inlet. At specified sparse time intervals, specimens were removed from the ovens and samples were dry-sectioned from the aged larger specimen, potted in epoxy, and polished as illustrated in Fig. 9.1. The cross section was cut at a minimum of 6 mm from the exposed surface. This allowed monitoring of four of the exposed edges in the interior of the large specimen. The remainder of the large sample was subsequently placed back in the oven to continue the aging process. Additional details on specimen handling, aging, and vacuum impregnation procedure are included in [34].

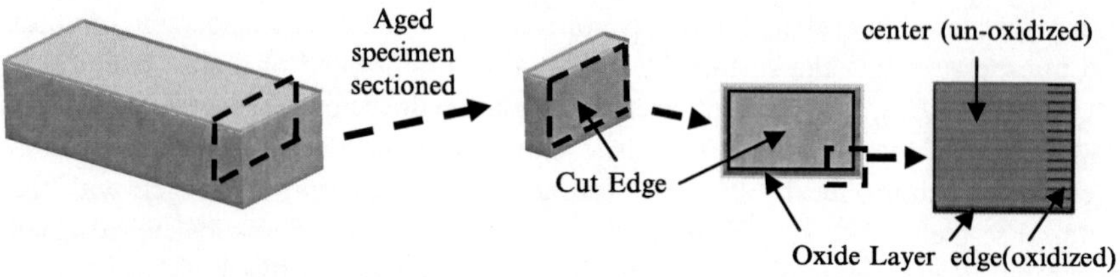

Fig. 9.1 Oxidation measurement procedures

Dark-field microscopy was utilized to monitor the oxidation propagation in $[0]_{16T}$, $[0/\pm45/90]_{2S}$, $[0/90]_{4S}$, $[\pm45]_{2S}$, plain-weave, and 3D resin impregnated composites. The oxidation layer formation near the exposed specimen edges changes the optical characteristics of the material, and the oxidized layer is observed by light microscopy. Specifically, the oxidation layer appears as a frame around the composite cross section, and is clearly seen as the lighter oxidized material using standard light microscopy in the grayscale mode. The extent of oxidation and damage was measured in both axial and transverse directions of the laminates. Note that for the remainder of the chapter, axial and transverse directions refer to the length and thickness of the cross section of the laminate under consideration.

9.3 Thermo-Oxidative Behavior of Laminated Composites

9.3.1 *Unidirectional, $[0]_{16T}$ Composite*

Figure 9.2 shows the extent of oxidation near the laminate edges in a cross section parallel to 0° direction in a $[0]_{16T}$ composite, as a function of aging time. For illustration, only one edge of the cut section is shown displaying oxidation growth from the exposed left edge and top and bottom surfaces. Note that each observation of extent of oxidation is for a unique cross section of the sample since the specimen preparation process, which includes sectioning, impregnating, and polishing specimens at each discrete aging time, precludes tracking oxidation growth on a unique cross section.

It is clearly evident that with increase of aging time, the axial propagation of oxidation is much greater than oxidation in the transverse direction, thus displaying anisotropic oxidation growth. Also, with aging time, the axial oxidation front becomes nonuniform through the laminate thickness, and can clearly be seen following the fiber paths and dominating the oxidation process. On the other hand, small changes are observed in the oxidation thickness in the transverse direction with aging time. Figure 9.3 shows the extent of oxidation near the laminate edges in a cross section perpendicular to 0° direction in a $[0]_{16T}$ composite. In this cross-section orientation, the oxidized region appears as a picture frame near the exposed free

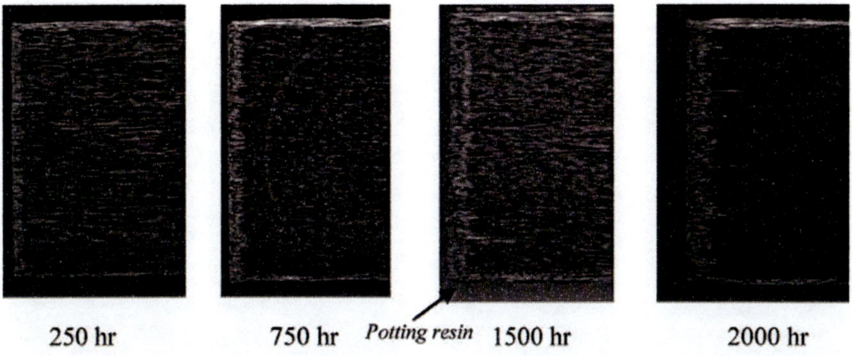

Fig. 9.2 Oxidation growth near the laminate edge in a cross section *parallel* to 0° direction in $[0]_{16T}$ laminate as a function of aging time

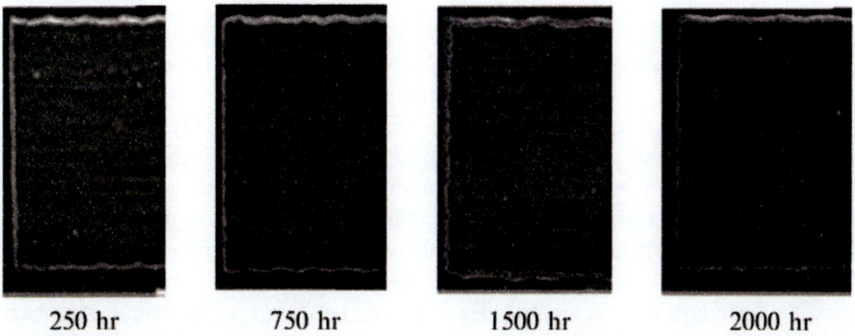

Fig. 9.3 Oxidation growth near the laminate edge in a cross section *perpendicular* to 0° direction in $[0]_{16T}$ laminate as a function of aging time

surfaces of the specimen, and the behavior is similar to that of the neat resin. Moreover, the oxidation thickness is fairly uniform around the entire cross section, and shows only marginal growth, for the aging times considered.

The dye-impregnated potted specimens were also viewed under fluorescent light to document the growth and development of damage. Figures 9.4a, b show the extent of oxidation and damage development near the laminate edges in a cross section parallel to 0° direction in a $[0]_{16T}$ composite aged for 1,000 and 4,996 h, respectively. The images on the left in Figs. 9.4a, b are the dark-field images whereas the images on the right are the optical microscopy images viewed under fluorescent light. The fluorescent image in Fig. 9.4a shows short fiber–matrix debonds along the 0° fiber direction. Experimental evidence suggests that these debond cracks will continue to increase in length, while new cracks continue to develop at the free surfaces of the specimen with further aging. The left image in

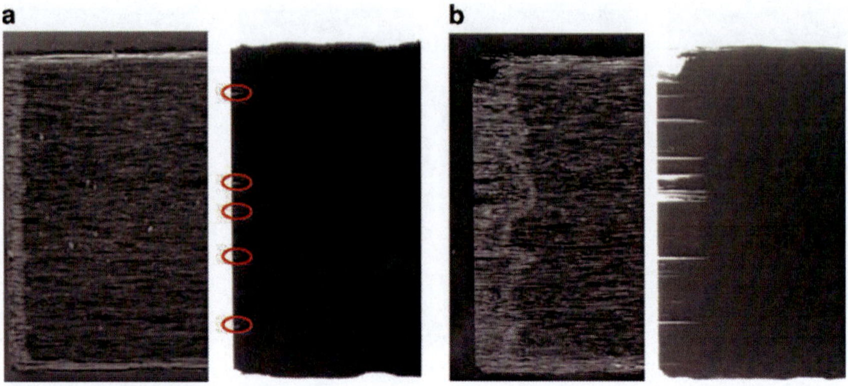

Fig. 9.4 Dark-field image of oxidation and fluorescence imaging of cracks in a cross section *along* fiber direction in $[0]_{16T}$ composite aged for (**a**) 1,000 h and (**b**) 4,996 h

Fig. 9.4b reveals that the oxidation front is nonuniform, and the oxidation has advanced to a greater extent in the region of the specimen corresponding to the location of the debond cracks. For short aging times, prior to damage development in the axial direction, oxidation growth is fairly uniform along the exposed edges. However, with further aging, axial cracks begin to develop near the edges parallel to the fiber in areas where enhanced oxidation is observed. Thus, damage develops as a consequence of oxidation-induced material embrittlement and interfacial and interlaminar stresses induced during curing and resulting from material shrinkage. However, with further increase in aging time, it is these cracks which will then provide pathways for oxygen to diffuse deeper into the specimens and facilitate oxidation deeper into the interior, while new cracks will continue to develop at the exposed free edges. Therefore, for longer aging times, damage growth is the principal driver for enhanced oxidation growth and the increase in the oxidation axial to thickness anisotropy ratio.

9.3.2 Cross-Ply, $[0/90]_{4S}$ Laminate

Figure 9.5 shows the oxidation growth with aging time in a cross section parallel to 0° direction in $[0/90]_{4S}$ laminate. Unlike the relatively uniform oxidation front observed for the unidirectional laminate at short aging times, we observe preferential oxidation growth along the fiber paths for the cross-ply laminate. Moreover, oxidation is no longer uniform through the ply thickness even at short aging times, as individual plies influence oxidation growth in neighboring plies in a multidirectional laminate. The oxidation propagation rate in a ply is strongly influenced by the orientation of its neighboring plies. For example, a 0° ply will tend to increase the oxidation rate in a neighboring off-axis ply as compared to its oxidation rate in a

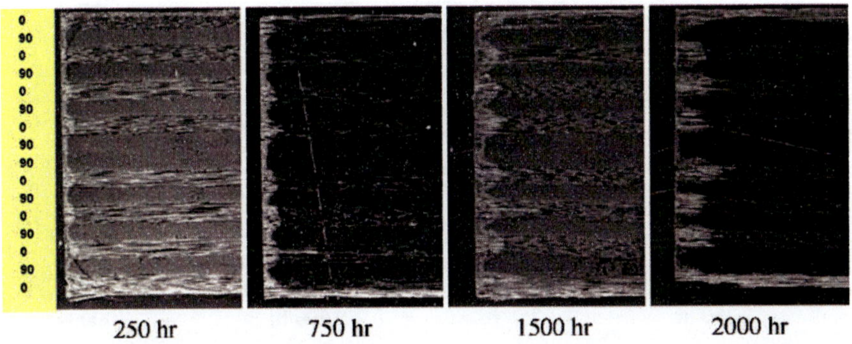

Fig. 9.5 Oxidation growth near the laminate edge in a cross section *parallel* to 0° direction in $[0/90]_{4S}$ laminate as a function of aging time

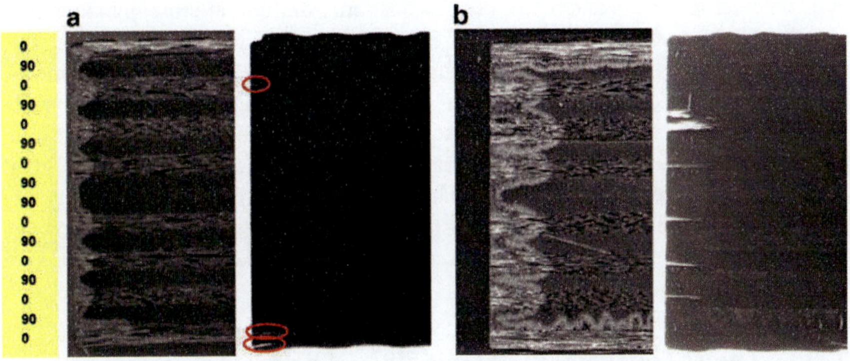

Fig. 9.6 Dark-field image of oxidation and fluorescence imaging of cracks in a cross section *parallel* to 0° direction in $[0/90]_{4S}$ laminate aged for (**a**) 1,000 h and (**b**) 4,996 h

unidirectional composite of the same off-axis orientation. Similarly, a 90° ply will tend to decrease the oxidation rate in neighboring plies of different orientation. In the $[0/90]_{4S}$ cross-ply laminate, every interior ply except for the 90° plies located at the laminate midplane has a symmetric neighbor on either side. Consequently, it is observed that maximum oxidation in the interior 0° plies and minimum oxidation in the 90° plies occur at their respective ply midplanes, if the adjacent top and bottom plies are symmetrically oriented about the midplane of the ply of interest. Obviously, the axial oxidation growth in the 0° surface plies is affected by the oxygen exposure from the transverse surface as well.

Figures 9.6a, b show the oxidation and damage growth in a cross section parallel to 0° direction in a $[0/90]_{4S}$ laminate aged for 1,000 and 4,996 h, respectively. As seen in Fig. 9.6, the nonuniform oxidation front advances and the nonuniformity becomes more pronounced because of the differences in oxidation growth rates for the two-ply

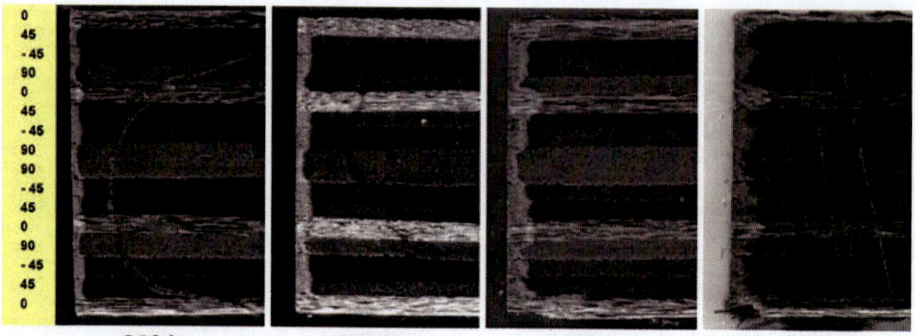

Fig. 9.7 Oxidation growth near the laminate edge in a cross section *parallel* to 0° direction in $[0/\pm45/90]_{2S}$ laminate as a function of aging time

orientations, with increase in aging time. On the other hand, the oxidation thickness perpendicular to the top and bottom surfaces is fairly uniform in cross-ply laminates for the shorter aging time investigated. However, with increasing aging time, the oxidation front from the top and bottom surfaces also tends to become nonuniform, as transverse matrix cracks develop from these exposed edges and coalesce with fiber–matrix debonds. Another thing to observe in the fluorescence images in Figs. 9.6a, b is the formation of axial cracks, mainly in the 0° plies along the fiber direction, in regions corresponding to enhanced oxidation growth demonstrating the synergistic coupled effect between oxidation and damage development and growth.

9.3.3 Quasi-Isotropic, $[0/45/-45/90]_{2S}$ Laminate

Figure 9.7 shows the oxidation growth with aging time in a cross section parallel to 0° direction in a $[0/\pm45/90]_{2S}$ laminate. In this composite layup, none of the individual plies (0°, 45° or 90°) have a symmetric distribution of neighboring plies. Hence, the minima or maxima oxidation extent does not occur at the ply mid-planes. Instead, oxidation growth within individual ply and across the ply interfaces is strongly influenced by the orientation of its neighboring plies. For example, in Fig. 9.7, within the 90° layer of the $[0/\pm45/90]_{2S}$ laminate which is not located at the laminate midplane, the maximum oxidation extent in the axial direction is largest at the 90/0 ply interface, followed by the −45/90 interface and is the least in the interior of the 90° ply. This is because the 0° ply neighbor on one side tends to increase the oxidation growth in the 90° ply at a faster rate compared to the −45° neighboring ply on the other side. Consequently, two gradients in oxidation growth develop across the thickness of this 90° ply. Thus, the oxidation growth profile for this 90° layer has a maxima at the 90/0 interface and a minima which is not in the ply center but closer to the −45/90 interface. As seen in Fig. 9.7, axial oxidation growth is largest in the 0° plies, followed by 45°/−45° and then the 90° plies.

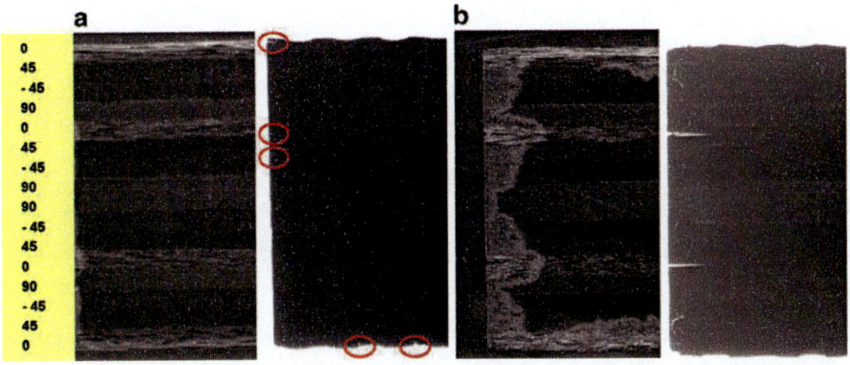

Fig. 9.8 Dark-field image of oxidation and fluorescence imaging of cracks in a cross section *parallel* to 0° direction in $[0/45/-45/90]_{2S}$ composite aged for (**a**) 1,000 h and (**b**) 4,996 h

The minimum oxidation in the cross-ply laminate in the axial direction occurs within the 90° plies at the laminate midplane.

Figures 9.8a, b show the oxidation and damage growth in a cross section parallel to 0° direction in a $[0/\pm45/90]_{2S}$ laminate aged for 1,000 and 4,996 h, respectively. As expected, the differences in the oxidation growth rate of individual plies are accentuated at longer aging time. At 1,000 h of aging, the oxidation thickness perpendicular to the top and bottom surfaces is fairly uniform in a quasi-isotropic laminate. However, similar to the observations in a cross-ply laminate, the oxidation front tends to become nonuniform with further increase in aging time, as damage initiates and propagates with aging time. Close examination of the fluorescence and dark-field images in Figs. 9.8a, b further shows that short edge cracks develop in the regions of enhanced oxidation extent, especially in the 0° plies along the fiber direction at 1,000 h of aging. With further aging, the edge cracks grow in length driving oxidation deeper into the laminate, while new cracks continue to develop at the free surfaces of the specimen.

9.3.4 Angle-Ply, $[45/-45]_{2S}$ Laminate

The oxidation growth as a function of aging time near the laminate edges in a $[\pm45]_{2S}$ laminate is shown in Fig. 9.9. The oxidized region appears as a picture frame near the exposed free surfaces of the specimen for short aging times (~750 h), and the growth is relatively uniform, similar to that observed in a neat resin specimen.

The oxidation and damage growth at 1,000 and 4,996 h of aging near the laminate edges in a $[\pm45]_{2S}$ laminate are shown in Figs. 9.10a, b, respectively. It is observed

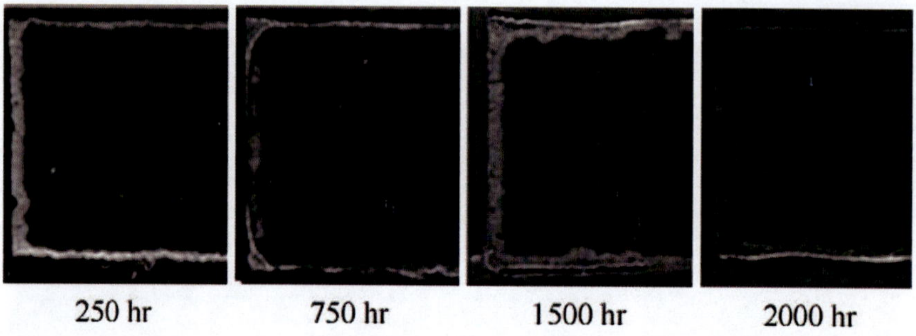

Fig. 9.9 Oxidation growth near the laminate edge in $[\pm45]_{2S}$ laminate as a function of aging time

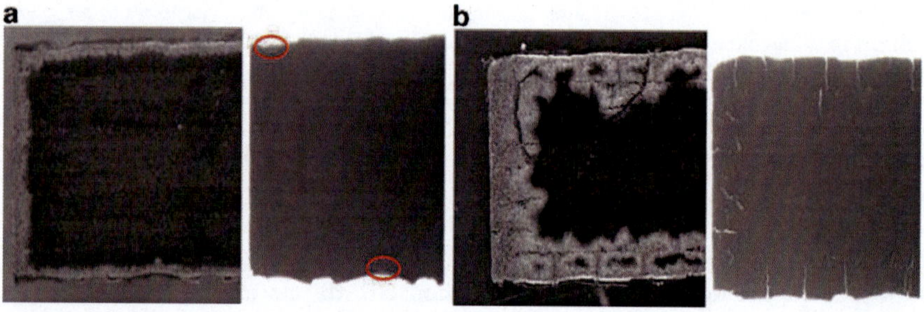

Fig. 9.10 Dark-field image of oxidation and fluorescence imaging of cracks in a $[45/-45]_{2S}$ composite aged for (**a**) 1,000 h and (**b**) 4,996 h

that the oxidation front quickly becomes nonuniform in extent with further increase in aging time, as axial and transverse cracks begin to develop in areas where enhanced oxidation is observed, similar to the observations in $[0/90]_{4S}$ and $[0/\pm45/90]_{2S}$ laminates. A comparison of dark-field and fluorescence images clearly reveals the inherent synergistic coupling between damage development and oxidation growth.

9.4 Quantification of Oxidation and Damage Growth

Quantification of the oxidation growth and damage propagation history is complicated by the fact that the optical microscopy techniques utilized preclude tracking individual oxidation zones and cracks as a function of aging time. Each observation on a polished cross section represents oxidation and damage growth measurements within a unique specimen cross section. Alternately, x-ray micro computed tomography (CT) enables one to examine the details of the microstructure of a material in a nondestructive fashion. Based on oxidation-induced density changes, CT can be used to measure oxidation in some neat resins and laminated composites. Although, composite oxidation is difficult to detect with CT, certainly oxidation-induced

Fig. 9.11 Comparison of (**a**) fluorescence and (**b**) x-ray imaging of cracks in a cross section *perpendicular* to 0° direction in $[0/90]_{4S}$ laminate aged for 4,996 h

cracking can be tracked with CT. Moreover, this technique can be utilized to examine the 3D volume of the material without the need for sectioning, as required for optical microscopy measurements.

Figures 9.11a, b are fluorescence and x-ray imaging of cracks in a cross section perpendicular to 0° direction in $[0/90]_{4S}$ laminate aged for 4996 h. The CT image was taken from the center of the remainder of the large sample from which a section was cut earlier for optical microscopy measurements. Note the presence of axial cracks in the 0° plies near the edges and extensive cracking perpendicular to the top and bottom surfaces of the laminate. The x-ray sample was held from its left edge and therefore the axial cracks emanating from the left edge of the sample in Fig. 9.11b are not seen. There seems to be a general agreement between the observed crack lengths, the crack distribution, and the density of crack growth between the fluorescence and x-ray images shown in Figs. 9.11a, b, respectively.

This comparison of optical microscopy images of crack growth on a cut and polished cross section (Fig. 9.11a) with CT imaging (Fig. 9.11b) shows that the damage observations made on a polished discrete section are generally representative of the damage behavior in the interior of the laminate. Similarly, Figs. 9.12a, b and Figs. 9.13a, b compare the fluorescence and x-ray imaging of cracks in a cross section perpendicular to 0° direction in $[0/45/-45/90]_{2S}$ laminate, and in a cross section of $[45/-45]_{2S}$ laminate, respectively, after aging for 4,996 h. For the quasi-isotropic laminate, axial cracks are mainly confined to the 0° plies, while extensive cracking develops perpendicular to the top and bottom laminate surfaces. On the other hand, the angle-ply laminate has transverse cracks around all the exposed surfaces. Again note that the x-ray sample in Fig. 9.12b was held from its left edge and that in Fig. 9.13b was held from its right edge and therefore the axial cracks emanating from the respective left and right edges of the sample are not seen. Similar to the observations made for the cross-ply laminate, there is very good agreement of damage growth and distribution between the respective fluorescence and x-ray

Fig. 9.12 Comparison of (**a**) fluorescence and (**b**) x-ray imaging of cracks in a cross section *perpendicular* to 0° direction in $[0/45/-45/90]_{2S}$ laminate aged for 4,996 h

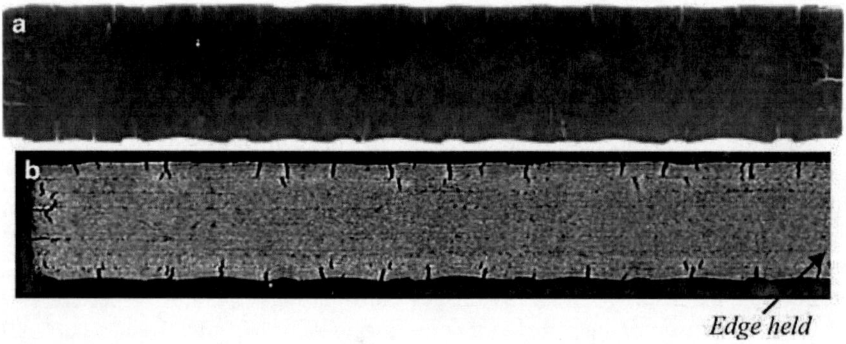

Fig. 9.13 Comparison of (**a**) fluorescence and (**b**) x-ray imaging of cracks in a cross section of $[45/-45]_{2S}$ laminate aged for 4,996 h

micrographs for both the laminates shown. Thus, these comparisons show that the process of sectioning and polishing does not induce any additional damage within the cut section and, therefore, validate the use of optical microscopy techniques for quantitative assessment of oxidation and crack growth in HTPMCs.

9.5 Anisotropy of Oxidation

While the oxidation growth is fairly similar within plies of the same orientation along the exposed edges for short aging times, the front quickly becomes nonuniform as earlier cracks continue to grow while new cracks and oxidation zones continue to

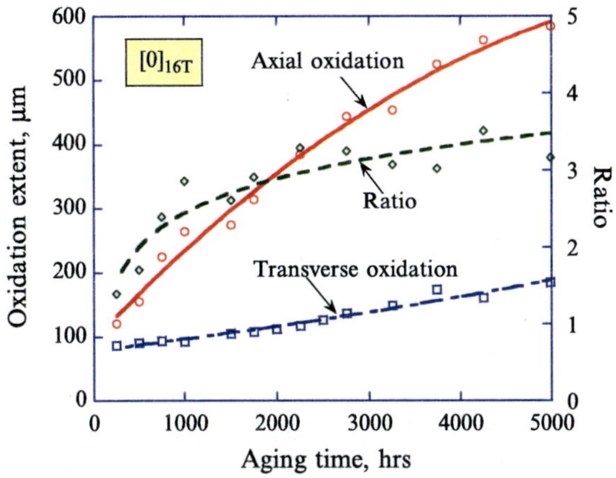

Fig. 9.14 Comparison of oxidation growth in axial and transverse directions in a unidirectional composite with aging

form on the free edges and/or in the interior with further aging. Due to the stochastic nature of the crack formation and the limited data samples at each aging time, two assumptions are made in order to determine the oxidation growth and crack propagation rates. Firstly, it is assumed that each discrete section from which data are obtained is representative of the behavior of the rest of the sample. The comparisons in Figs. 9.11–9.13 are generally supportive of this assumption. Secondly, it is assumed that the initial oxidation zone formed and damage observed at the beginning of aging will lead to maximum oxidation and the largest damage throughout the aging history. This is reasonable to expect and therefore a measurement of the maximum oxidation zone size and maximum crack length within each cross section is assumed to provide a measure of oxidation and crack growth rate, respectively.

9.5.1 *Unidirectional, $[0]_{16T}$ Composite*

As shown previously, it is clearly evident from Figs. 9.2 and 9.3 that with increase of aging time, the axial propagation of oxidation in a unidirectional composite is much greater than oxidation in the transverse direction leading to anisotropy in oxidation growth. Figure 9.14 is a plot of the measured values of maximum oxidation extent in the axial and transverse directions in a $[0]_{16T}$ laminate as a function of aging time. Also plotted in Fig. 9.14 is the ratio of axial to transverse oxidation growth, which is seen to increase rapidly with aging. After 5,000 h of aging, the anisotropy ratio is roughly about 3.5.

The unidirectional composite cross sections were also evaluated for maximum crack length measurements parallel and perpendicular to the fiber direction. A comparison of the maximum oxidation front with the damage front in the respective axial and transverse directions in Fig. 9.15 further reveals that the oxidation front consistently precedes the crack front for all aging times.

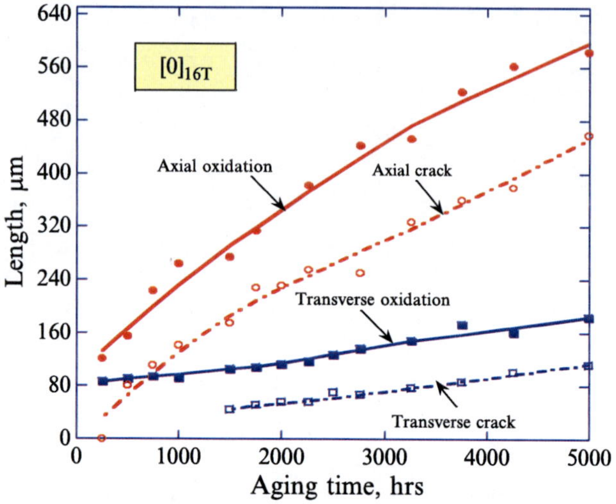

Fig. 9.15 Comparison of oxidation and damage growth in a unidirectional composite with aging

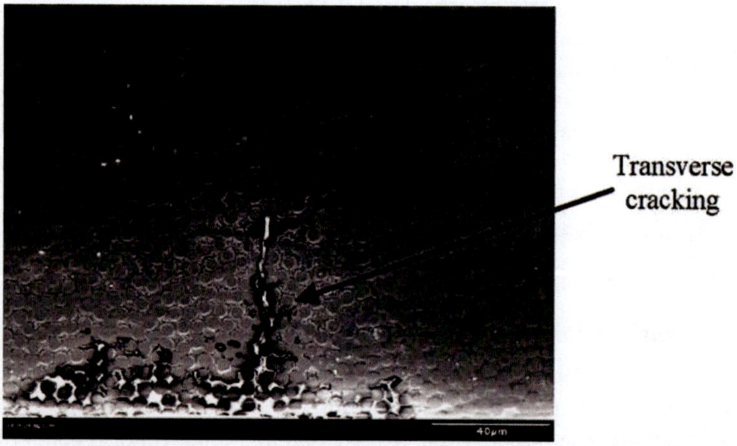

Fig. 9.16 SEM photomicrograph showing transverse matrix cracking within the white oxidized area in a unidirectional composite aged for 3,750 h

Figure 9.16 is an SEM photomicrograph of a composite cross section perpendicular to the fiber length at 3,750 h of aging. The white area near the bottom exposed edge is the region undergoing oxidation. Also seen in the photomicrograph is a transverse crack which initiated at the outer edge and is contained within the oxidized area. Figure 9.17 is a photomicrograph of the cross section in the axial direction at 4,996 h of aging. Similar to Fig. 9.16, one can see the white oxidized area with axial fiber–matrix debonds emanating from the left edge which is exposed to surface

Fig. 9.17 SEM photomicrograph showing axial fiber–matrix debonds within the white oxidized zone in a unidirectional composite aged for 4,996 h

oxidation. Clearly, both Figs. 9.16 and 9.17 are examples of the intrinsic coupling of chemical oxidative aging and damage. Whereas cracks can accelerate oxidation by providing pathways for oxidants, so can oxidation lead to cracking in the resin matrix and at fiber–matrix interfaces. The critical nature of the fiber–matrix interphase on degradation and failure processes in composites signifies the importance of proper representation of its behavior in predictive models. Since failure initiation is typically associated with the fiber–matrix interface/interphase region, modeling the coupling effects of damage and oxidation is important.

9.5.2 Cross-Ply, $[0/90]_{4S}$ Laminate

Figure 9.18 compares the extent of oxidation and damage growth in the axial and transverse directions of a cross-ply composite with aging. As expected, oxidation is largest in the axial direction in the 0° ply while the presence of the fibers slightly retards the oxidation growth in the transverse direction in the axial plane (cross section parallel to 0° direction) in comparison to the transverse oxidation in the perpendicular plane (cross section perpendicular to 0° direction). Damage in the axial direction in the 0° ply initiates after ~500 h, while transverse cracks develop in the perpendicular plane at ~1,000 h, and axial cracking in the 90° ply (i.e., crack in the 90° ply parallel to specimen length) is observed in excess of 4,000 h of aging. Similar to unidirectional composite, it is observed that the oxidation front precedes the respective crack front at all aging times.

Figure 9.19 is an SEM photomicrograph of a cross section perpendicular to 0° direction in a $[0/90]_{4S}$ laminate at 2,250 h of aging. The image shows a transverse crack in the top 0° layer which has stopped just short of the neighboring 90° ply

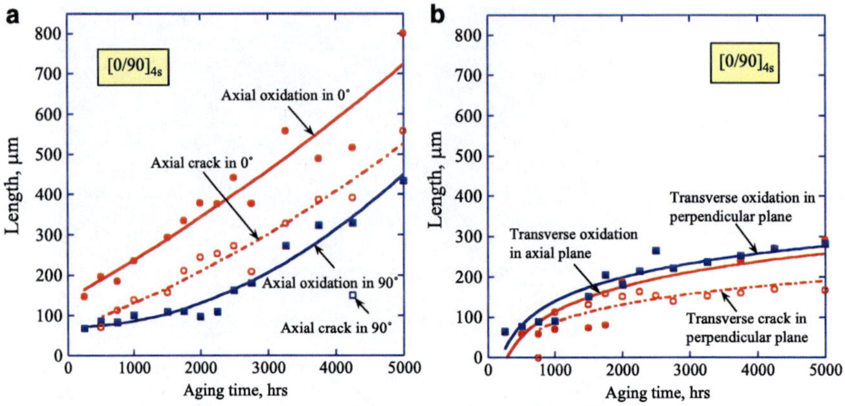

Fig. 9.18 Comparison of oxidation and damage growth in (**a**) axial and (**b**) transverse directions in a cross-ply composite with aging

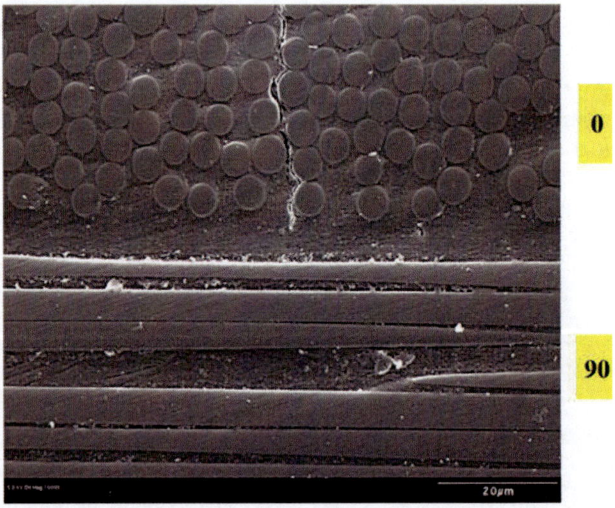

Fig. 9.19 SEM photomicrograph showing transverse matrix cracking in a cross section *perpendicular* to 0° direction in a [0/90]$_{4S}$ lamiante aged for 2,250 h

interface. With further aging, the crack front advances and reaches the ply interface. Two scenarios are now possible. If the interface is "strong," the crack can continue to grow vertically with fiber bridging, as seen in Fig. 9.20a. Alternately, if the fiber–matrix interface ahead of the crack front has deteriorated and the interface is "weak," the crack front advances by delamination along the fiber direction, as seen in Fig. 9.20b. In both situations, the oxidation zone continues to advance with additional pathways created for oxygen to diffuse deeper into the composite laminate.

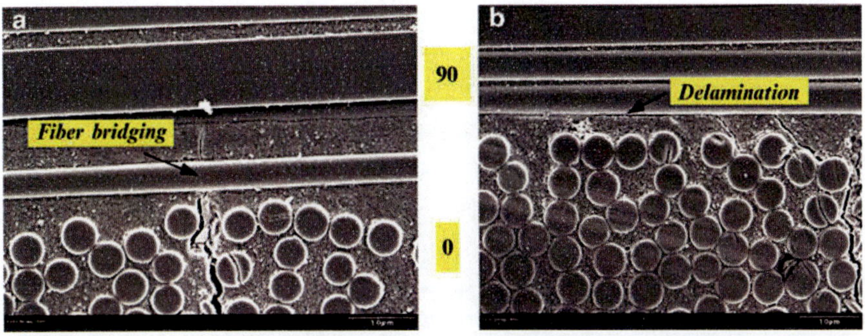

Fig. 9.20 SEM photomicrograph showing (**a**) fiber bridging and (**b**) delamination along ply interface in a cross section *perpendicular* to 0° direction in a $[0/90]_{4S}$ laminate aged for 3,750 h

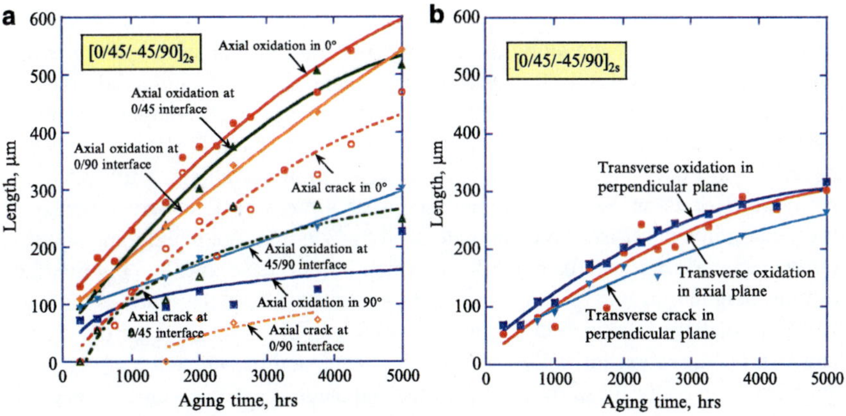

Fig. 9.21 Comparison of oxidation and damage growth in (**a**) axial and (**b**) transverse directions in a quasi-isotropic laminate with aging

9.5.3 *Quasi-Isotropic, $[0/45/-45/90]_{2S}$ Laminate*

The influence of near neighboring ply on oxidation and damage growth in axial and transverse directions within the quasi-isotropic laminate is shown in Fig. 9.21. Similar to the cross-ply composite, maximum oxidation in the axial direction in the $[0/45/-45/90]_{2S}$ laminate is in the 0° plies, while minimum oxidation occurs in the 90° plies, and oxidation along the 0/45, 0/90, and 45/90 ply interfaces is intermediate. The longitudinal fibers in the 0° surface ply are again seen to retard the oxidation growth in the transverse direction in the axial plane (cross section parallel to 0° direction) in comparison to the transverse oxidation in the perpendicular plane (cross section perpendicular to 0° direction). Damage in the axial

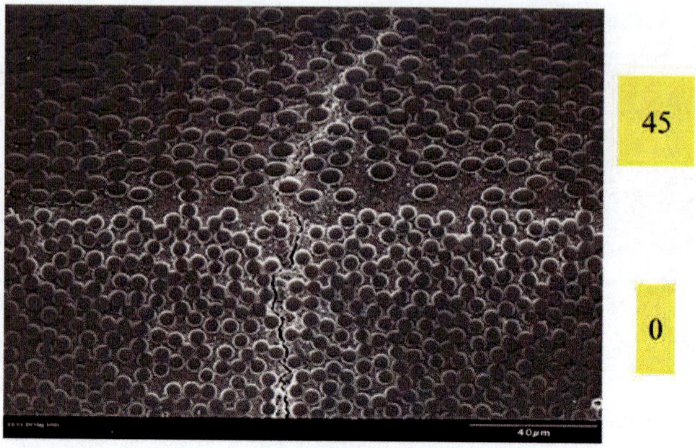

Fig. 9.22 SEM photomicrograph showing a matrix crack deflected at the ply interface in a cross section *perpendicular* to 0° direction in a quasi-isotropic laminate aged for 3,750 h

direction in the 0° ply and along the 0/45 ply interface initiate after ~500 h, while transverse cracks develop in the perpendicular plane at ~750 h, and axial cracking in the 90° ply is observed in excess of 2,000 h of aging. A comparison of the maximum oxidation extent and crack length measurements in Fig. 9.21 shows that the oxidation fronts precede their respective crack fronts at all aging times.

Figure 9.22 is an SEM photomicrograph of a cross section perpendicular to 0° direction in a quasi-isotropic laminate aged for 3,750 h. The photograph shows a transverse matrix crack in the outer 0° ply which is deflected at the 0/45 ply interface into the neighboring 45° ply. The white region surrounding the crack is the area which has undergone oxidation. Figure 9.23 shows an SEM image of a cross section parallel to 0° direction in $[0/45/-45/90]_{2S}$ laminate, and the influence of ply layup and orientation on the anisotropy in oxidation and damage growth is clearly revealed.

9.5.4 Angle-Ply, $[45/-45]_{2S}$ Laminate

The oxidation and damage growth measurements along the length (axial) and thickness (transverse) of the cross section of $[45/-45]_{2S}$ laminate are shown in Fig. 9.24. The comparisons reveal the anisotropy as the axial growth is larger than the growth in the transverse direction. Also, the crack length measurements are smaller compared to their respective oxidation measurements, which is indicative of existence of oxidized zone ahead of the crack tip. These trends are clearly seen in the SEM photomicrograph in Fig. 9.25 at 4,996 h of aging. Additionally, one observes crack deflection on encountering the neighboring ply interface, while oxidation is observed around the damage profiles.

Fig. 9.23 SEM photomicrograph showing the influenec of ply layup and orientation on oxidation and damage growth in a cross section parallel to 0° direction in a quasi-isotropic laminate aged for 4,996 h

Fig. 9.24 Comparison of oxidation and damage growth in $[45/-45]_{2S}$ laminate

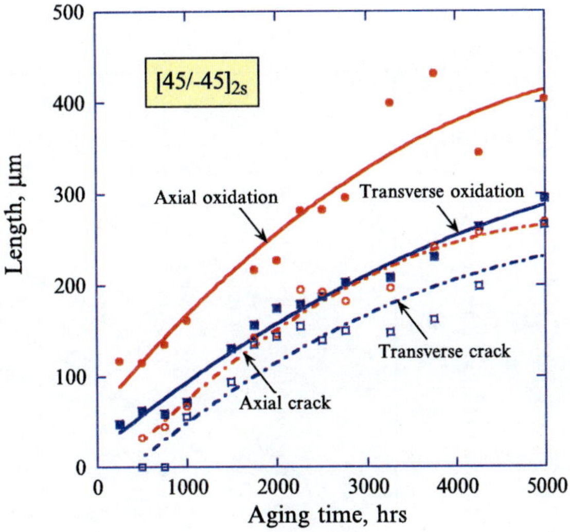

9.6 Influence of Ply Stacking Sequence

Next, we will compare the oxidation and damage growth characteristics within the plies and at their interfaces in the various laminates investigated. Figures 9.26 and 9.27 compare the measurements of the oxidation extent in axial and transverse directions, respectively. It is observed that maximum oxidation in the axial direction is comparable in $[0]_{16T}$ and 0° ply in the $[0/90]_{4S}$ and $[0/45/-45/90]_{2S}$

Fig. 9.25 SEM photomicrograph showing the influenec of ply layup and orientation on oxidation and damage growth in $[45/-45]_{2S}$ laminate aged for 4,996 h

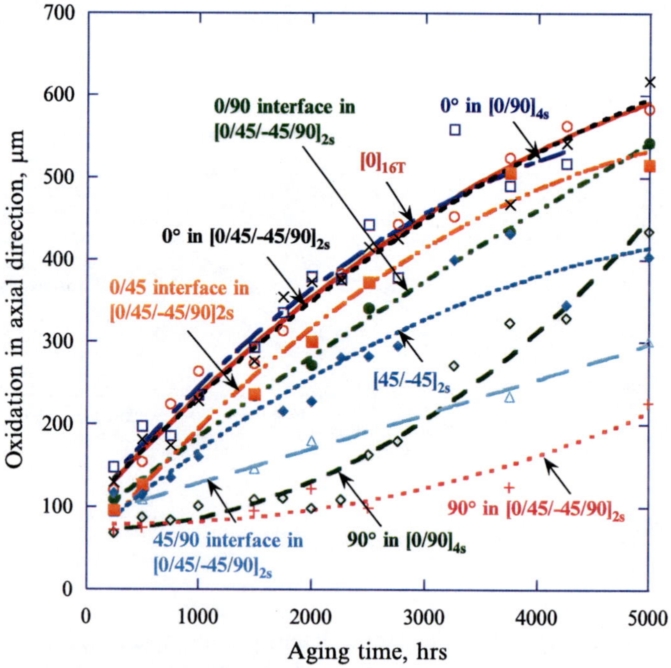

Fig. 9.26 Comparison of oxidation extent in axial direction in $[0]_{16T}$, $[0/90]_{4S}$, $[0/45/-45/90]_{2S}$, and $[45/-45]_{2S}$ laminates as a function of aging time

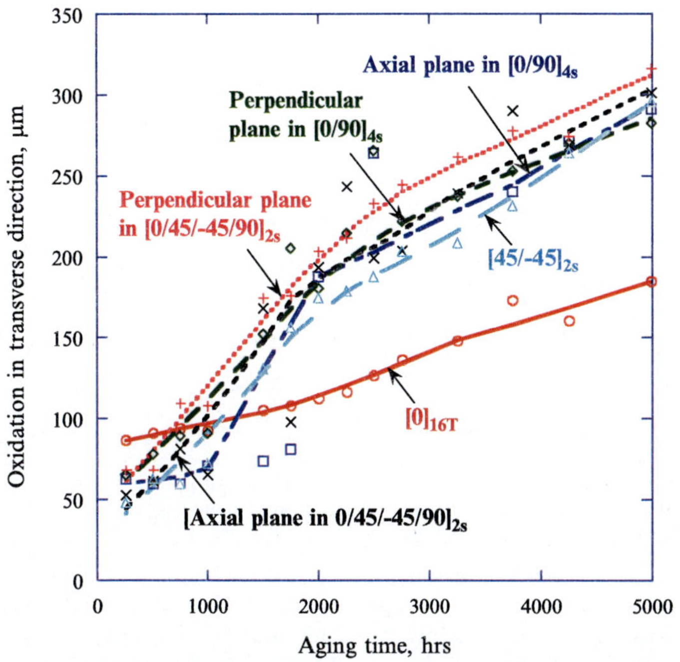

Fig. 9.27 Comparison of oxidation extent in transverse direction in $[0]_{16T}$, $[0/90]_{4S}$, $[0/45/-45/90]_{2S}$, and $[45/-45]_{2S}$ laminates as a function of aging time

laminates, while the axial growth is minimum in the 90° ply in the $[0/45/-45/90]_{2S}$ laminate. These trends are consistent with the direction of high oxygen diffusivity in the axial fiber (i.e., 0°) direction, and low oxygen diffusivity perpendicular to the fiber direction (i.e., 90°). At short aging times (<1,000 h), oxidation behavior of the 90° ply in $[0/90]_{4S}$ laminate is similar to the 90° ply in the $[0/45/-45/90]_{2S}$ laminate, but with further aging, extent of oxidation increases rapidly in the 90° ply in the $[0/90]_{4S}$ laminate. This demonstrates the influence of ply stacking sequence as the neighboring 0° ply on either side accelerates the oxidation growth for the 90° ply in $[0/90]_{4S}$ laminate in comparison to the 0° and 45° neighbors for the 90° ply in the $[0/45/-45/90]_{2S}$ laminate. The oxidation behavior of $[45/-45]_{2S}$ laminate is intermediate to the oxidation growth observed in 0° and 90° plies. On the other hand, oxidation extent in the transverse direction is the largest for the unidirectional composite for <500 h of aging, but increases rapidly with further aging for cross-ply, quasi-isotropic, and angle-ply laminates, and is the smallest for the unidirectional composite at longer aging times (>1,500 h). The rapid increase in oxidation is consistent with initiation and growth of damage in the transverse direction in the laminates investigated. These results in Figs. 9.26 and 9.27 clearly demonstrate the influence of neighboring ply orientation (or stacking sequence effect) on ply oxidation characteristics in laminated composites.

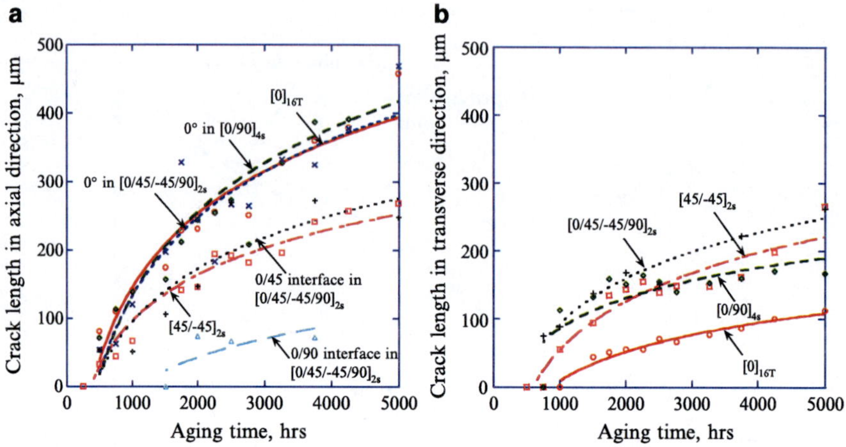

Fig. 9.28 Comparison of crack length in (**a**) axial and (**b**) transverse direction in $[0]_{16T}$, $[0/90]_{4S}$, $[0/45/-45/90]_{2S}$, and $[45/-45]_{2S}$ laminates as a function of aging time

Figures 9.28a, b are the measurements of the maximum crack lengths in axial and transverse directions in the various ply orientations in the unidirectional, cross-ply, quasi-isotropic, and angle-ply laminates considered. Similar to the trend in Fig. 9.26, the axial cracks in unidirectional laminate and the 0° ply in the cross-ply laminate and quasi-isotropic laminate are comparable in size. Further, no axial cracks were observed to develop in the 90° ply of the cross-ply laminate till 4,000 h of aging. In the transverse direction, least amount of damage occurs in the unidirectional composite, and is maximum for the quasi-isotropic composite. A comparison of the maximum oxidation extent and crack lengths in Figs. 9.27 and 9.28 quickly reveals that the oxidation front precedes the crack front in both axial and transverse directions for all laminates considered. Moreover, there seems to be an oxidation zone of approximately constant thickness ahead of the crack front for all laminates studied.

9.7 Thermo-Oxidative Behavior of Textile Composites

9.7.1 Plain-Weave Composites

Thermo-oxidative behavior of plain-weave and 3D textile composites has recently been studied by Tandon et al. [57, 58]. The influence of ply weave orientation and composite laminate thickness on oxidation growth was examined by considering several different plain-weave laminates. For discussion, a four-layer plain-weave laminate is designated by $[PW]_4$, a sixteen-layer plain-weave laminate by $[PW]_{16}$,

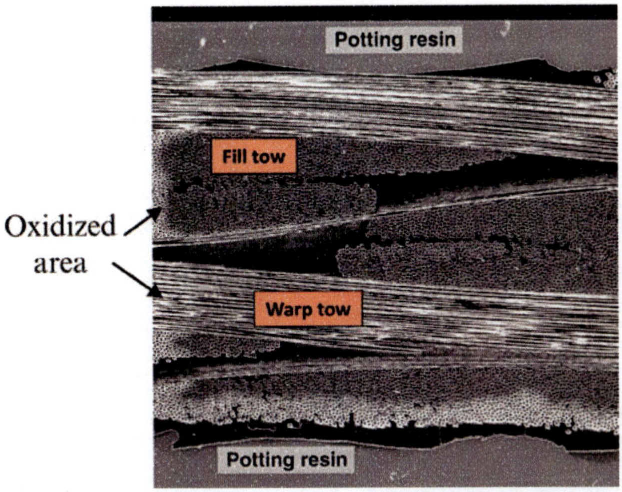

Fig. 9.29 Dark-field image of oxidation growth in a four-layer plain-weave laminate ($[PW]_4$) aged for 500 h

a four-layer plain-weave laminate oriented at 45° by $[PW45]_4$, and an eight-layer plain-weave symmetric laminate with alternating plain-weave and plain-weave layers at 45° by $[PW, PW45]_{2S}$. Further, the fiber tow aligned along the length of the cross-section under view is referred to as the warp tow, while the fiber tow perpendicular to the plane of view is referred to as the fill tow, as depicted in Fig. 9.29. The figure shows the dark-field image of the cross section of a $[PW]_4$ laminate aged for 500 h. It is clearly evident that the axial extent of oxidation is much larger in the warp tow in comparison to the fill tow. Thus there is an anisotropy in oxidation growth between the warp and fill tow orientations, which is very similar in nature to the anisotropy in oxidation growth observed between the 0° and 90° plies in the cross-ply laminate, as discussed earlier. Note the presence of a thin layer of resin rich region along the bottom surface of the $[PW]_4$ laminate. Autoclave processing of the plain-weave prepreg during laminate manufacturing results in this excess resin on one surface for all the plain-weave laminates studied here.

Figure 9.30a compares the dark-field oxidation image of a $[PW]_4$ laminate aged for 3,000 h with the corresponding fluorescence image in Fig. 9.30b. It is clear that oxidation has advanced to a greater extent in the axial direction in the warp tow in regions where short cracks have developed along the edges. Also notice the development of transverse cracks (i.e., damage in the thickness direction) originating in the thin resin-rich layer on the top surface. Figure 9.30c is an SEM image of the transverse cracks which shows the nature of the coupling between oxidation and damage growth. Clearly, oxidation has advanced to a greater depth in regions where transverse cracks have developed and there is an oxidation zone surrounding the transverse crack, which is reasonable to expect since the crack provides a pathway for oxygen to diffuse deeper into the specimen.

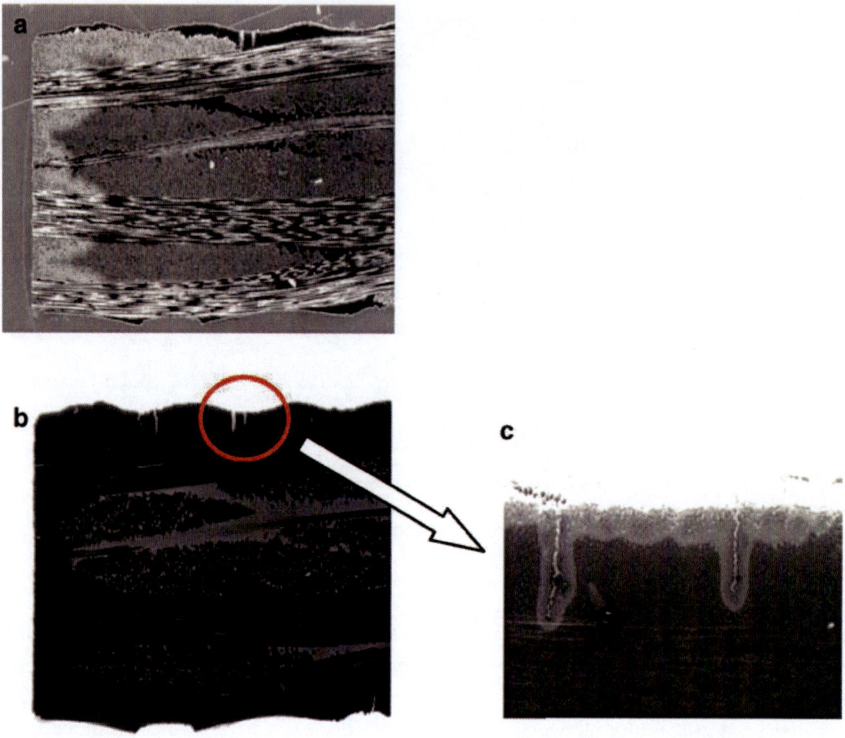

Fig. 9.30 (**a**) Dark-field, (**b**) Fluorescence, and (**c**) SEM imaging of oxidation and damage growth in $[PW]_4$ laminate aged for 3,000 h

Figures 9.31a, b compare the oxidation and damage growth in axial and transverse directions, respectively, in a $[PW]_4$ laminate. The measurements in the transverse direction are made from the laminate surface which does not have the excess resin-rich layer. It is evident from these measurements that oxidation front precedes the damage front at all aging times. In the axial direction, both oxidation and damage develop at a faster rate in the warp tow versus the fill tow. Oxidation-induced damage development in the axial direction is measured in $[PW]_4$ laminate only after 1,500 h of aging. In the thickness direction, measurements were made both in the resin pockets (designated as matrix in Fig. 9.31b) in regions corresponding to warp yarn undulation and also in fill tows (designated as tow in Fig. 9.31b) near the surface. Though the oxidation measurements in the tow and matrix regions are comparable at shorter aging times (<1,000 h), the oxidation values begin to separate at longer aging times, as the presence of the fiber bundles impedes the transverse oxidation growth in the fill tow. A comparison of axial and transverse damage growth in Figs. 9.31a, b further shows that transverse cracks begin to develop much earlier during isothermal aging of $[PW]_4$ laminates.

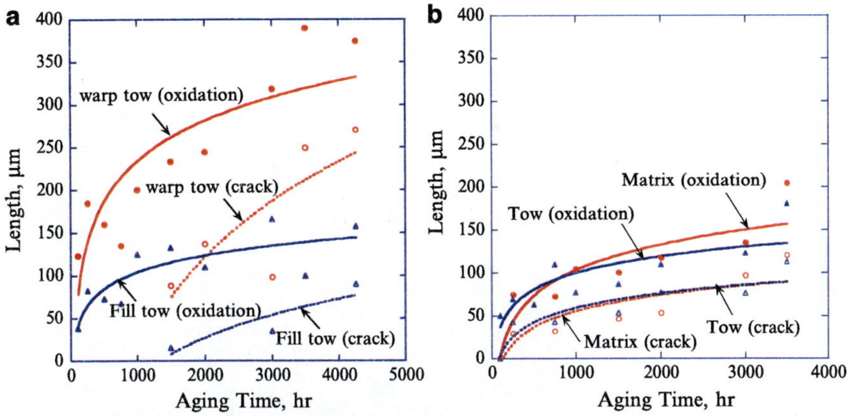

Fig. 9.31 Comparison of oxidation and damage growth in (**a**) axial and (**b**) transverse directions in $[PW]_4$ laminate with aging

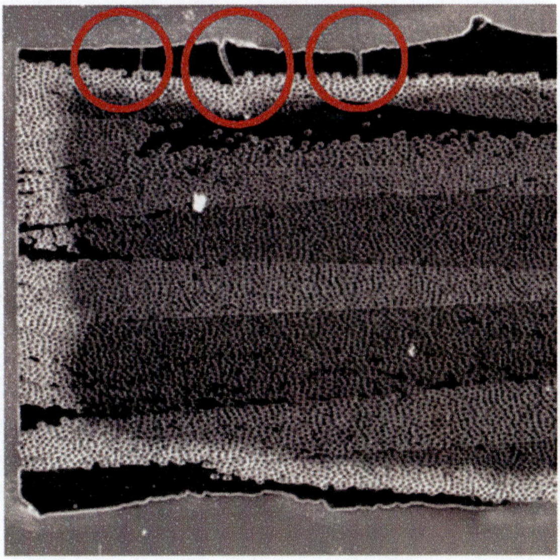

Fig. 9.32 Dark-field image of oxidation growth in a four-layer plain weave laminate cut at 45° ($[PW45]_4$) aged for 500 h

The oxidation characteristics of a four-layer plain-weave laminate cut at 45°, $[PW45]_4$, are shown in Fig. 9.32 after 500 h of aging. The oxidized region appears as a lighter picture frame near the exposed free surfaces of the specimen for short aging times, and the growth is relatively uniform, similar to that observed in $[45/-45]_{2S}$ laminate shown previously. In this laminate orientation, both the warp and fill tows undergo similar oxidation growth in the axial direction. Notice the development of transverse cracks in the resin-rich top surface of the laminate.

Fig. 9.33 Dark-field image of oxidation growth in a 16-layer plain weave laminate ($[PW]_{16}$) aged for 500 h

Figure 9.33 shows the dark-field image of oxidation in a sixteen-layer plain-weave laminate, $[PW]_{16}$, aged for 500 h. The oxidation behavior of the warp and fill tows within an individual layer is strongly dependent on the stacking of the neighboring plies (or layers). For example, the oxidation growth in a fill tow in one layer will be accelerated if the neighboring ply has a warp tow. Similarly, two neighboring warp tows will tend to complement each other's oxidation growth, resulting in a maxima near the tow boundaries. On the other hand, two neighboring fill tows will develop a minima in their oxidation profile at the common boundary. These trends are clearly seen in the photomicrograph in Fig. 9.33.

Finally, Fig. 9.34 shows the coupling of damage and oxidation growth in an eight-layer plain-weave symmetric laminate with alternating plain-weave and plain-weave layers oriented at 45°. In this laminate configuration, oxidation features observed in both $[PW]_4$ and $[PW45]_4$ laminates can be seen. In particular, notice how the transverse cracks drive oxidation deep into the interior of the laminate from the adjoining surface plies.

The influence of laminate thickness and orientation on axial oxidation growth in plain-weave laminates is compared in Fig. 9.35. It is seen that oxidation is largest in the warp tow and is minimum in the fill tow of $[PW]_{16}$ laminate. Reducing the number of plain-weave layers from sixteen to four results in a decrease in the maximum oxidation extent in the warp tow, but is followed with a small increase in the fill tow. Thus, the extent of axial oxidation in the warp tow increases with the thickness of the laminate. This is likely a result of the ply stacking sequence with

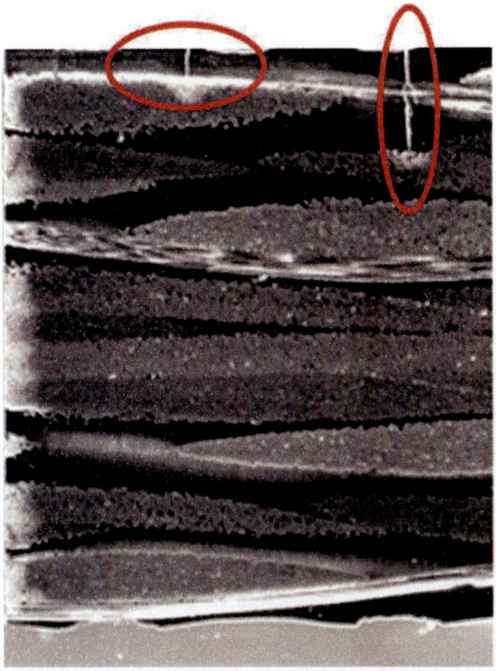

Fig. 9.34 Dark-field image of oxidation in [PW, PW45]$_{2S}$ laminate aged for 500 h

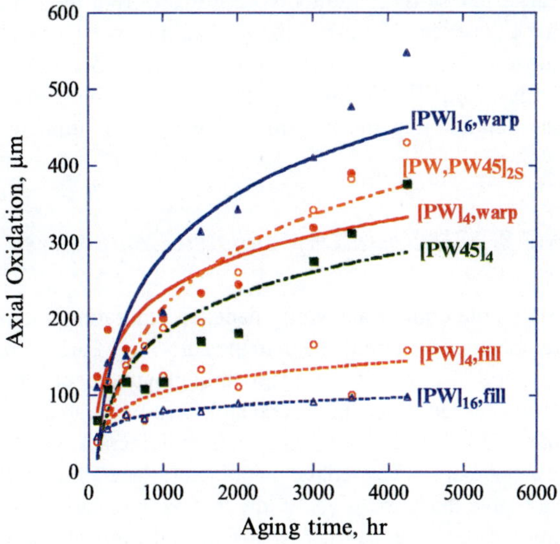

Fig. 9.35 Influence of laminate thickness and orientation on oxidation growth in axial direction in plain-weave laminates

neighboring plies (or layers) positioned to accelerate the growth of oxidation and the probability of these occurring increases with an increase in the number of layers involved. Likewise, maximum oxidation in an eight-layer [PW, PW45]$_{2S}$ laminate is observed to be larger than that measured in a four-layer [PW]$_4$ laminate at longer

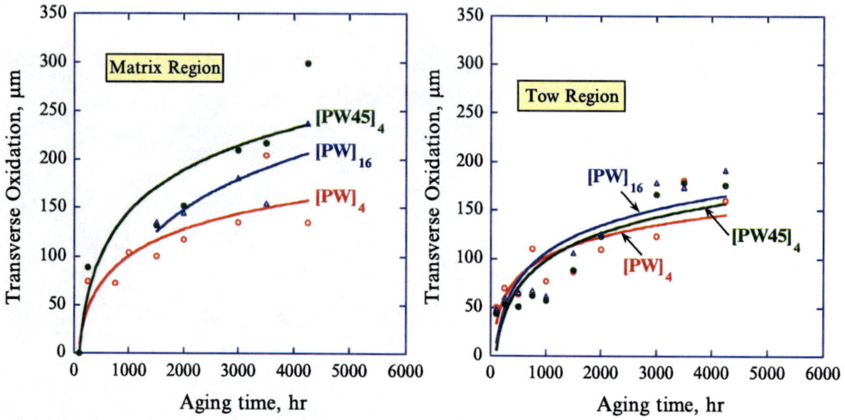

Fig. 9.36 Influence of laminate thickness and orientation on oxidation growth in transverse direction in plain-weave laminates

aging times (>2,000 h). As expected, the oxidation characteristics of $[PW45]_4$ laminate are intermediate to the oxidation behavior of the warp and fill tows in a $[PW]_4$ laminate.

A comparison of the oxidation extent in the thickness direction in the tow and matrix regions of various plain-weave laminates is made in Fig. 9.36. As mentioned earlier, these measurements were made from the laminate surface which does not have the excess resin-rich layer. The transverse oxidation growth is very similar in the tow region for $[PW]_4$, $[PW]_{16}$, and $[PW45]_4$ laminates, but is observed to be the largest for $[PW45]_4$ laminate in the resin pockets in regions corresponding to warp yarn undulation, and is minimum for $[PW]_4$ laminates.

9.7.2 3D Resin Impregnated Composites

3D textile composites were made by infiltrating a carbon-fiber preform with high-temperature polyimide resin in the autoclave. Earlier manufacturing attempts resulted in good wetting of the resin throughout the preform with little porosity in the laminate. However, microcracking was widespread in the resin matrix throughout the composite. In order to reduce the microcracking, slower heating and cooling rates were attempted beyond the manufacturer's recommended cure cycle. The former was to enable the resin to gel at low temperature such that the cure shrinkage strain was mitigated by the thermal expansion as the temperature was raised further during the cure cycle, and to ensure that the resin was completely cured before cool-down began. Slower cooling rates were also employed to minimize the development of residual stresses due to thermal expansion mismatch between the resin and fiber preform. These changes led to a significant reduction in the amount of matrix

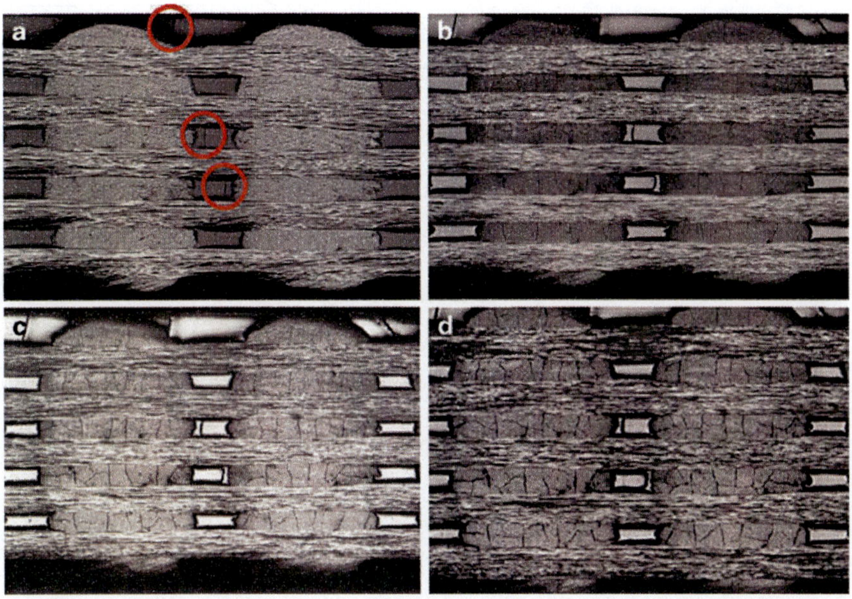

Fig. 9.37 Surface damage characterization in a 3D composite aged for (**a**) 0 h, (**b**) 1,000 h, (**c**) 1,500 h, and (**d**) 2,000 h

cracking. However, microcracks are hard to completely eliminate in a 3D composite due to large resin pocket in 3D weave resulting in large shrinkage in the thickness direction.

Figure 9.37a shows a portion of the polished edge of a 3D laminate in unaged condition. There is some limited cracking within the resin pockets between the fiber tows. However, no cracking was visible within the matrix in the tow bundles during processing. Figures 9.37b, c, and d are the photomicrographs of the same region of the 3D composite after isothermal aging for 1,000, 1,500, and 2,000 h, respectively.

There is little change in the matrix crack density between the fiber tows with aging. However, damage develops mainly in the form of transverse and horizontal cracks within the fill tows (i.e., tow bundles which are perpendicular to the plane of view) on aging. The transverse cracks are seen to span the entire height of the fill tows, while the length of the axial cracks is limited by the transverse crack spacing. In addition, some of the cracks running parallel to the warp tows do not remain completely horizontal as some limited crack branching is observed. The damage in the fill tows is extensive after 1,500 h of aging. As seen in Fig. 9.37d, the cracks within the tow bundles open wider as the crack opening displacement increases with further aging.

Figure 9.38 is a histogram plot of the measured transverse surface crack density within the fiber tows as a function of aging from three different areas of the cross section, namely, near the left edge, right edge, and from the center of the specimen.

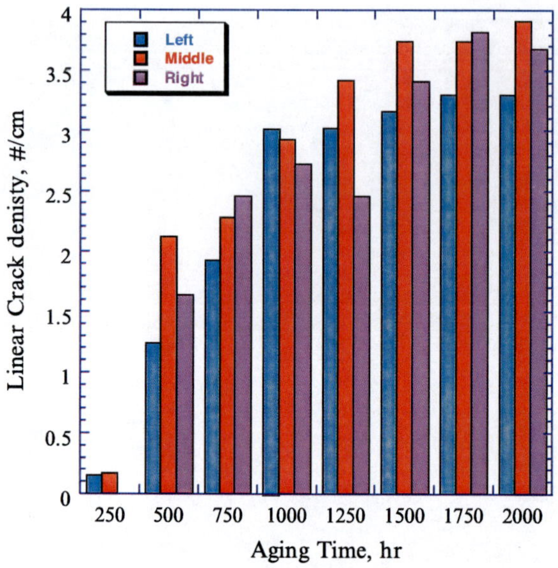

Fig. 9.38 Transverse surface crack density in 3D composite as a function of aging time

It is seen that there is a rapid increase in the transverse crack density during the initial 1,000 h of aging and approaches a steady-state value at 1,500 h of aging. Continued aging results in minor changes only in the transverse crack density but the vertical cracks widen as their crack face separation increases.

Lastly, we will examine the oxidation characteristics of a 3D textile composite. The process of resin infiltration in a 3D fiber preform leaves a thin layer of excess resin near the top and bottom surfaces of the laminate. Three different microscopy techniques were therefore utilized to assess oxidation and damage growth. Figures 9.39a, b, and c are the respective bright-field, dark-field, and fluorescence images of the same cross section of a 3D composite aged for 2,000 h. The bright-field imaging is used to measure resin oxidation near the top and bottom surfaces, as shown in Fig. 9.39a. The dark-field image in Fig. 9.39b reveals the tow oxidation, while fluorescence imaging in Fig. 9.39c shows both the processing-induced as well as oxidation-induced damage within the cross section. Note that the combination of preexisting internal matrix cracks coupled with aging-induced surface cracks provides a penetrating network for oxygen to diffuse deeper resulting in oxidation in the interior tow regions. Near the vertical edges, oxidation advances more in the axial direction in the warp tow in comparison to the fill tow, and the behavior is similar to that observed in a cross-ply laminate, with alternating 0° and 90° layers.

A comparison of the oxidation and damage extent in the axial direction in warp and fill tows near the edges in a 3D composite is shown in Fig. 9.40. Damage in the

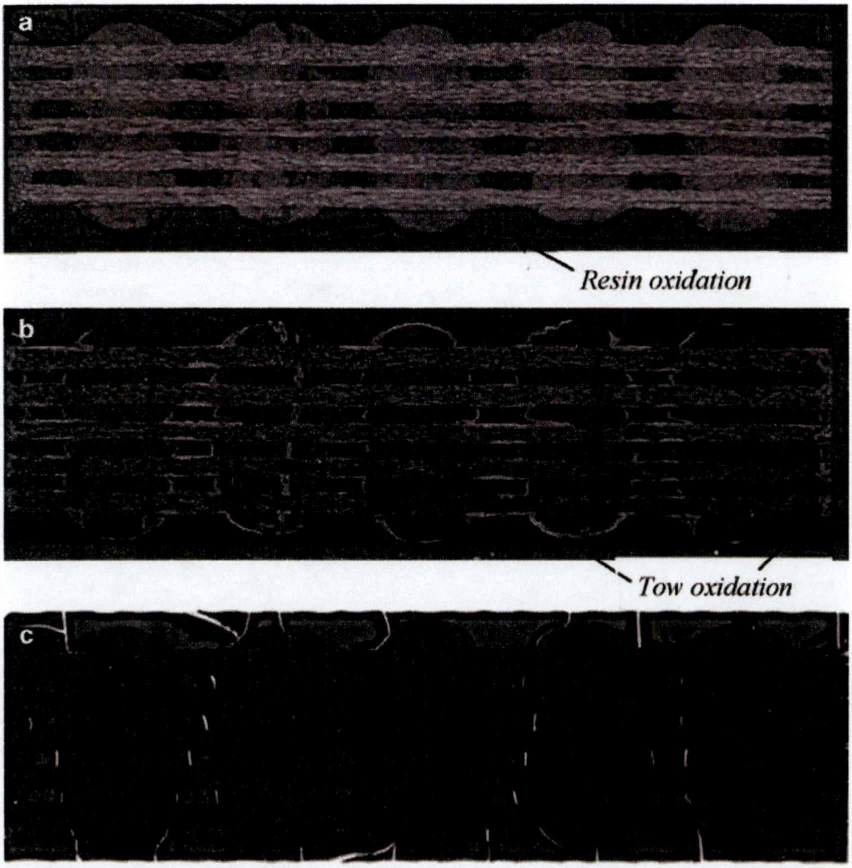

Fig. 9.39 (**a**) Bright-field, (**b**) dark-field, and (**c**) fluorescence imaging of oxidation and damage in a 3D composite aged for 2,000 h

warp tow develops fairly early during aging (~ 400 h) while in the fill tow occurs at longer aging time (~1,200 h). Further, both oxidation and axial cracking are much larger in the warp tow in comparison to the fill tow. Similar to our earlier observations, we find that oxidation precedes damage for the aging times considered, thus there is always an oxidation zone ahead of the crack tip. Figure 9.40 also shows that the resin oxidation in the thickness direction is larger than the oxidation growth in the fill tow, as the presence of fibers impedes transverse oxidation development.

Fig. 9.40 Aging-induced oxidation and damage growth in a 3D composite

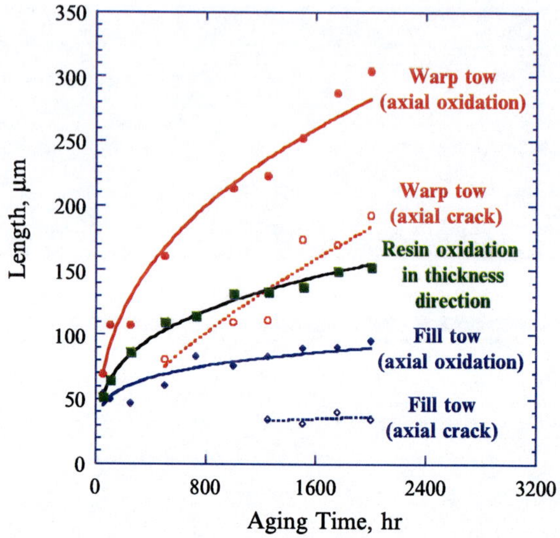

9.8 Concluding Remarks

Thermo-oxidative degradation of HTPMCs is a subject of major importance as designers push the high-temperature endurance limits of polymer composites to improve the performance of aerospace systems. Although the anisotropy in oxidation of unidirectional composites was observed and reported earlier, limited work exists on characterizing thermo-oxidation in laminated and woven composite systems. This effort describes the influence of ply layup, laminate architecture and thickness, and the role of damage on the transport of oxygen into the interior and resulting oxidation growth in laminated and textile composites.

Light microscopy and SEM techniques are used to characterize the oxidative process in laminated and woven carbon-fiber-reinforced polyimide composites. Several different composites are considered, namely, unidirectional, cross-ply, quasi-isotropic, angle-ply, plain-weave, and 3D woven composites. It is shown through comparisons with nondestructive CT imaging that the process of sectioning and polishing does not induce any additional damage within the cut section, thereby, validating the use of optical microscopy techniques for quantitative assessment of thermo-oxidation in HTPMCs. It is shown that the oxidation propagation rate in a ply (or a layer) is strongly influenced by the orientation of its neighboring plies (or layers). Because of the increased diffusivity in the axial direction, a 0° ply will tend to increase the oxidation rate in a neighboring off-axis ply, while the 90° ply will tend to decrease the oxidation growth rate. An interior ply in the laminate, which has plies of two different orientations as its immediate neighbors, will therefore develop oxidation growth gradients across the ply interfaces because of the anisotropy of diffusion. Similar observations are made for anisotropic oxidation growth in warp and fill tows within plain-weave and 3D textile composites.

It is shown that during isothermal aging, damage develops as a consequence of oxidation-induced material embrittlement and interfacial and interlaminar stresses induced during curing and resulting from material shrinkage, although the contribution of residual stresses has not been explicitly evaluated. With further increase in aging time, it is these cracks which then provide pathways for oxygen to diffuse deeper into the specimens and facilitate oxidation deeper into the interior, while new cracks continue to develop at the exposed free edges. Therefore, for longer aging times, damage growth is the principal driver for enhanced oxidation growth and the increase in the oxidation axial to thickness anisotropy ratio. Thus, there is a synergistic coupled effect between oxidation and damage development and growth, which should be properly represented and accounted for in mechanistic simulations of thermo-oxidation in HTPMCs.

In this work, quantification of the oxidation growth and damage propagation history is made by assuming that a measurement of the maximum oxidation zone size and maximum crack length within each cross section of the composite provides a measure of oxidation and crack growth rate, respectively. A comparison of the maximum oxidation extent and crack lengths quickly reveals that the oxidation front precedes the crack front in both axial and transverse directions for all composites considered. Moreover, there seems to be an oxidation zone of approximately constant thickness ahead of the crack front for all composites studied.

Acknowledgments This work was supported by the Air Force Office of Scientific Research (AFOSR) under two programs, with Dr. Charles Lee and Dr. Joycelyn Harrison as the Program Managers, and Dr. Richard Hall as the AFRL program monitor. Funding was provided under contract FA8650-05-D-5052. The author gratefully acknowledges the assistance of Michael Nickell (UDRI) for aging specimens, William Ragland (UDRI) for optical microscopy, and Dr. Sirina Putthanarat (UDRI) for CT imaging.

References

1. Schoeppner, G. A., Tandon, G. P., and Pochiraju, K. V. (2008) "Predicting thermo-oxidative degradation and performance of high temperature polymer matrix composites" in *Multiscale Modeling and Simulation of Composite Materials and Structures*, Y. Kwon, D. Allen and R. Talreja, eds, ISBN: 978-0-387-36318-9, Springer Verlag, pp 359–462.
2. Saff, C., Hahn, G., Griffith, J., Ingle, R. and Nelson, K. (2005) Accelerated insertion of materials – composites. AIAA-2005–2165 46th AIAA/ASME/ASCE/ AHS/ASC Structures, Structural Dynamics and Materials Conference, Austin TX.
3. Accelerated Test Methods for Durability of Composites, High Speed Research Materials Durability (Task 23), High Speed Research Program, March 1998.
4. Brinson, L. C. and Gates, T. S. (1995) Effects of physical aging on long term creep of polymers and polymer matrix composites. Int. J Solids Struct, 32:827–46.
5. Veazie, D. R. and Gates, T. S. (1998) Effects of aging-time reference on the long term behavior of the IM7/K3B composite. 39th AIAA/ASME/ASCE/AHS/ASC Structures, Structural Dynamics, and Materials Conference Paper AIAA 98–1962.

6. Struik, L. C. E. (1978) Physical aging in amorphous polymers and other materials. Elsevier, Amsterdam.
7. Seferis, J. C. (1999) Aging analyses of polymer composites through time-temperature equivalence. J Comp Tech and Research 21:173–179.
8. Sullivan, J. L. (1990) Creep and physical aging of composites. Comp Sci and Tech 39:207–32.
9. McKenna, G. B. (1994) On the physics required for the prediction of long term performance of polymers and their composites. J. Res. NIST, 99:169–189.
10. Scott, D. W., Lai, J. S. and Zureick, A. H. (1995) Creep behavior of fiber reinforced polymeric composite: a review of the technical literature. Journal of Reinforced Plastics and Composites 14:588–617.
11. Lincoln, J. E. (2001) Structure-property-processing relationships and the effects of physical structure on the hygrothermal durability and mechanical response of polyimides. Ph.D. Thesis, Michigan State University.
12. Colin, X. and Verdu, J. (2005). Strategy for studying thermal oxidation of organic matrix composites. Composite Science & Technology, 65:411–419.
13. Colin, X., Marais, C. and Verdu, J. (2005) Kinetic modeling of the stabilizing effects of carbon fibers on thermal aging of thermoset matrix composites. Comp Sci and Tech 65:117–127.
14. Tandon, G. P., Pochiraju, K. V. and Schoeppner, G. A. (2008) Thermo-oxidative behavior of high temperature PMR-15 resin and composites. Materials Science and Engineering – A, 498: 150–161.
15. Tandon, G. P. and Pochiraju, K. V. (2011) "Heterogeneous thermo-oxidative behavior of multidirectional laminated composites," Journal of Composite Materials 45(4):415–435.
16. Skontorp, A., Wong, M. S. and Wang, S. S. (1995), High-temperature anisotropic thermal oxidation of carbon-fiber reinforced polyimide composites. Theory and experiment. Proc of ICCM-10, Whistler B.C., Canada, IV:375.
17. Wang, S. S., Chen, X. and Skontorp, A. (2003) High-temperature mechanics modeling and experiments of thermal oxidation, degradation and damage evolution in carbon/fiber/polyimide composites. Technical Report CEAC-TR-03-0106, University of Houston.
18. Bowles, K. J. and Nowak, G. (1988) Thermo-oxidative stability studies of Celion 6000/PMR-15 unidirectional composites, PMR-15, and Celion 6000 fiber. J Composite Materials 22:966–985.
19. Bowles, K. J., Jayne, D., Leonhardt, T. A. and Bors, D. (1993) Thermal stability relationships between PMR-15 resin and its composites. NASA Technical Memorandum 106285.
20. Bowles, K. J. (1999) Durability of graphite-fiber reinforced PMR-15 composites aged at elevated temperatures. J Comp Tech and Research 21:127–132.
21. Wong, M. S., Skontorp, A. and Wang, S. S. (1994) Thermal oxidation of carbon fibers and carbon-fiber reinforced high-temperature polyimide composite at elevated temperature. In: Proc. of Am. Soc. Comp., 9th Tech. Conf., Technomic Publ. Co., Lancaster, PA., 458–467.
22. Tsotsis, T. K. (1998) Long-term thermo-oxidative aging in composite materials: experimental methods. Journal of Composite Materials 32:1115–1135.
23. Tsotsis, T. K. and Lee, S. M. (1998) Long-term thermo-oxidative aging in composite materials: failure mechanisms. Composites Science and Technology, 58:355–368.
24. Tsotsis, T. K., Keller, S., Lee, K., Bardis, J. and Bish, J. (2001) Aging of polymeric composite specimens for 5000 hours at elevated pressure and temperature. Comp Sci and Tech 61:75–86.
25. Abdeljaoued, K. (1999) Thermal oxidation of PMR-15 polymer used as a matrix in composite materials reinforced with carbon fibers. MS thesis, Ecole Nationale Superieure des Arts et Metiers, Paris.
26. Bellenger, V., Decelle, J. and Huet, N. (2005) Aging of a carbon epoxy composite for aeronautic applications. Composites Part B: Engineering 36:189–194.
27. Whitley, K. S. and Collins, T. J. (2006) Mechanical properties of T650/AFR-PE-4 at elevated temperatures for lightweight aeroshell designs. AIAA-2006-2202 47th AIAA/ASME/ASCE/AHS/ASC Structures, Structural Dynamics, and Materials Conference, Newport, Rhode Island, AIAA-2006-2202.

28. Nelson, J. B. (1983) Thermal aging of graphite polyimide composites," in *Long-Term Behavior of Composites*, T. K. O'Brien, ed., ASTM STP 813: American Society for Testing and Materials, Philadelphia, pp. 206–221.
29. Bowles, K. J., and Meyers, A. (1986) Specimen geometry effects on graphite/PMR-15 composites during thermo-oxidative aging," in Proceedings 31st International SAMPE Symposium, April 7–10, 1986.
30. Bowles, K. (1991) Effect of fiber reinforcements on thermo-oxidative stability and mechanical properties of polymer matrix composites. NASA TM 103648.
31. Nam, J. D. and Seferis, J. C. (1992) Anisotropic thermo-oxidative stability of carbon fiber reinforced polymeric composites. SAMPE Quarterly 24:10–18.
32. Salin, I. M. and Seferis, J. C. (1993) Anisotropic effects in thermogravimetry of polymeric composites. J. Polymer Science 31:1019–1027.
33. Schoeppner, G. A., Tandon, G. P. and Ripberger, E. R. (2007) Anisotropic oxidation and weight loss in PMR-15 composites. Composites Part A: Applied Science and Manufacturing 38:890–904.
34. Tandon, G. P., Ragland, W. R. and Schoeppner, G. A. (2009) Using optical microscopy to monitor anisotropic oxidation growth in high-temperature polymer matrix composites. Journal of Composite Materials, 43:583–603.
35. Weitsman, Y. (1987) Stress assisted diffusion in elastic and viscoelastic materials. J. Mechanics and Physics of Solids, 35(1):73–97.
36. Bowles, K. J., Madhukar, M., Papadopolous, D. S., Inghram, L. and McCorkle, L. (1995) The effects of fiber surface modification and thermal aging on composite toughness and its measurement. NASA TM-106765.
37. Upadhyaya, D. and Tsakiropoulos, P. (1995) Evaluation of the effect of sizing levels on transverse flexural and shear strengths of carbon/epoxy composites. Journal of Materials Processing Technology 54:17–20.
38. Paipetis, A. and Galiotis, C. (1996) Effect of fiber sizing on stress transfer efficiency in carbon/ epoxy model composites. Composites Part A 27A:755–767.
39. Kung, H. K. (2006) Effects of surface roughness on high-temperature oxidation of carbon-fiber-reinforced polyimide composites. Journal of Composite Materials 39:1677–1687.
40. Skontorp, A. and Wang, S. S. (1995) High temperature creep with physical and chemical aging, and associated viscoelastic constitutive equations of polyimide-matrix composites. Recent Advances in Composite Materials, S.R. White. H.T. Hahn, and W.F. Jones, Eds., ASME MD Vol. 56, American Society of Mechanical Engineers, New York, NY, 57–70.
41. Kung, H. K. (1996) High-temperature oxidation, aging and creep in carbon-fiber reinforced polyimide composites during thermal fatigue. Ph.D. Thesis, Univ. of Houston, TX.
42. McManus, H. L., and Chamis, C. C. (1996) Stress and damage in polymer matrix composite materials due to material degradation at high temperatures," NASA Technical Memorandum 4682.
43. Colin, X. and Verdu, J. (2003) Thermal aging and lifetime prediction for organic matrix composites. Plastics, Rubber and Composites 32:349–356.
44. Zhang, Y., Xia, Z. and Ellyin, F. (2005) Nonlinear viscoelastic micromechanical analysis of fibre-reinforced polymer laminates with damage evolution. International Journal of Solids and Structures 42:591–604.
45. Ahci, E. and Talreja, R. (2006) Characterization of viscoelasticity and damage in high temperature polymer matrix composites. Comp Science and Tech. 66:2506–2519.
46. Colin, X., Mavel, A., Marais, C. and Verdu, J. (2005) Interaction between cracking and oxidation in organic matrix composites," Journal of Composite Materials, 39:1371–1389.
47. Pochiraju, K. V. and Tandon, G. P. (2009) Interaction of oxidation and damage in high temperature polymeric matrix composites. Composites Part-A, V 40, Issue 12, pp. 1931–1940.

48. Lafarie-Frenot, M. C., and Rouquie, S. (2004) Influence of oxidative environments on damage in C/epoxy laminates subjected to thermal cycling. Composites Science and Technology, 64:1725–1735.
49. Rouquie, S., Lafarie-Frenot, M. C., Cinquin, J. and Colombaro, A. M. (2005) Thermal cycling of carbon/epoxy laminates in neutral and oxidative environments. Composites Science and Technology, 65: 403–409.
50. Ammar-Khodja, I., Picard, C., Fois, M., Marais, C. and Netchitailo, P. (2009) Preliminary results on thermo-oxidative aging of multi-hole carbon/epoxy composites. Composites Science and Technology, 69:1427–1431.
51. Bowles, K. J., McCorkle, L. and Ingraham, L. (1998) Comparison of graphite fabric reinforced PMR-15 and Avimid N composites after long term isothermal aging at various temperatures. NASA/TM-1998-107529.
52. Bowles, K. J., Tsuji, L., Kamvouris, J. and Roberts, G. D. (2003) Long-term isothermal aging effects on weight loss, compression properties, and dimensions of T650-35 fabric-reinforced PMR-15 composites–data. NASA TM-2003-211870.
53. Bowles, K. J. (1998) Thermal and mechanical durability of graphite-fiber-reinforced PMR-15 composites. NASA TM 113116/REV1.
54. Cunningham, R. A. (1996) High temperature degradation mechanisms in polymer matrix composites. MS Thesis, M.I.T., Cambridge, MA.
55. Chung, K., Seferis, J. C. and Nam, J. D. (2000) Investigation of thermal degradation behavior of polymeric composites: prediction of thermal cycling effect from isothermal data. Composites Part A 31:945–957.
56. Chung, K. and Seferis, J. C. (2001) Evaluation of thermal degradation on carbon fiber/cyanate ester composites. Poly Degradation and Stability 71:425–434.
57. Tandon, G. P., Hall, R. B., Jefferson, G. and Hay, R. (2009) Durability and damage tolerance of complex 3-d fiber composites. Presented at AFOSR Low Density Materials Program Review, Oct. 8–9, Arlington, VA.
58. Tandon, G. P., Ragland, W. R., and Hall, R. B. (2010) Thermo-oxidative response of polyimide textile composites. Presented at 30th High Temple Workshop, Feb 1–4, Destin, FL.

Chapter 10
Modeling Thermo-Oxidative Aging and Degradation of Composites

Kishore V. Pochiraju

Abstract Long-term durability and use-life of polymeric matrix composites operating at elevated temperatures are limited by their thermo-oxidative stability. Although weight loss testing is traditionally performed to characterize the oxidative degradation of composite systems, the results of such tests are neither translatable to other composites architectures with the same constituents nor scalable to longer exposure times or higher temperatures. A comprehensive modeling framework for understanding the morphological changes in the composites and degradation of the mechanical performance is described in this chapter. A thermo-chemo-mechanics model that defines and utilizes an oxidation state parameter for each constituent is formulated. The effect of oxygen diffusion in the fiber and fiber–matrix interphase on the oxidation of the composite is simula-ted. The role of damage in accelerating the oxidation growth along the fiber direction leading to high orthotropy in lamina oxidation is also addressed. The stiffness changes due to oxidation as well as the strains induced due to shrinkage are explicitly modeled leading to a detailed simulation of oxidation growth around discrete cracks. Oxidation growth in laminated composites is predicted using microscale and homogenization techniques. The model is applied to study the long-term thermal oxidation of polyimide composites.

10.1 Introduction

Modeling of coupled chemo-mechanical behavior of composites from constitutive, lamina, and structural scales for high-temperature polymer matrix composite is the focus of this chapter. Complementing the earlier chapters of this book on the

K.V. Pochiraju (✉)
Department of Mechanical Engineering, Design and Manufacturing Institute,
Stevens Institute of Technology, Hoboken, NJ 07030, USA
e-mail: Kishore.Pochiraju@stevens.edu

K.V. Pochiraju et al. (eds.), *Long-Term Durability of Polymeric Matrix Composites*,
DOI 10.1007/978-1-4419-9308-3_10,

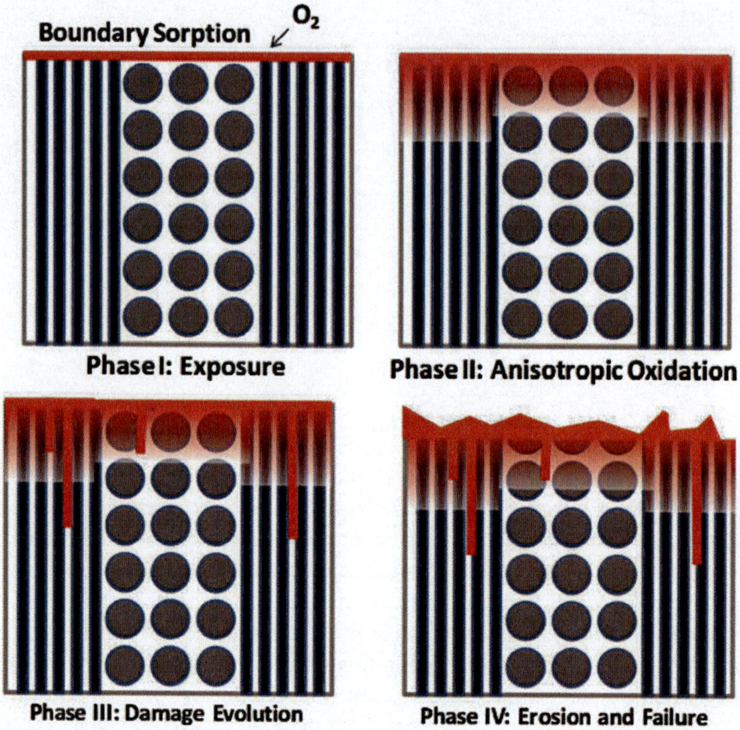

Fig. 10.1 Evolution of oxidation-induced degradation in polymeric composites

chemistry of degradation of the resin (Chap. 8) and experimental characterization of oxidative growth (Chap. 9), this chapter provides a comprehensive view of predictive modeling of oxidative degradation. Age-related degradation of High Temperature Polymeric Matrix Composites (HTPMCs) is a result of complicated interaction of time-dependent physical creep and relaxation processes, temperature-dependent chemical degradation of polymer and interfaces such as chain scission and hydrolysis, and oxidation-induced degradation [1]. Long-term durability and performance of polymer matrix composites in high-temperature environments is typically simulated using physical aging models [2–4], linear viscoelastic, and time-temperature superposition models [2].

Thermo-oxidative stability of the HTPMC is typically determined in practice by the weight loss behavior of the composites specimens. In contrast, Pochiraju et al. [3, 4] consider the chemo-mechanics-based mechanisms for modeling oxidation and damage growth in HTPMCS. In these efforts, oxygen diffusivity, rate of oxidation reaction, and damage evolution kinetics of the materials were determined for high-temperature resins and the oxidation behavior of the composite was simulated from the constituent behavior. Figure 10.1 depicts various mechanisms responsible for oxidative degradation. A composite when subjected to oxidative

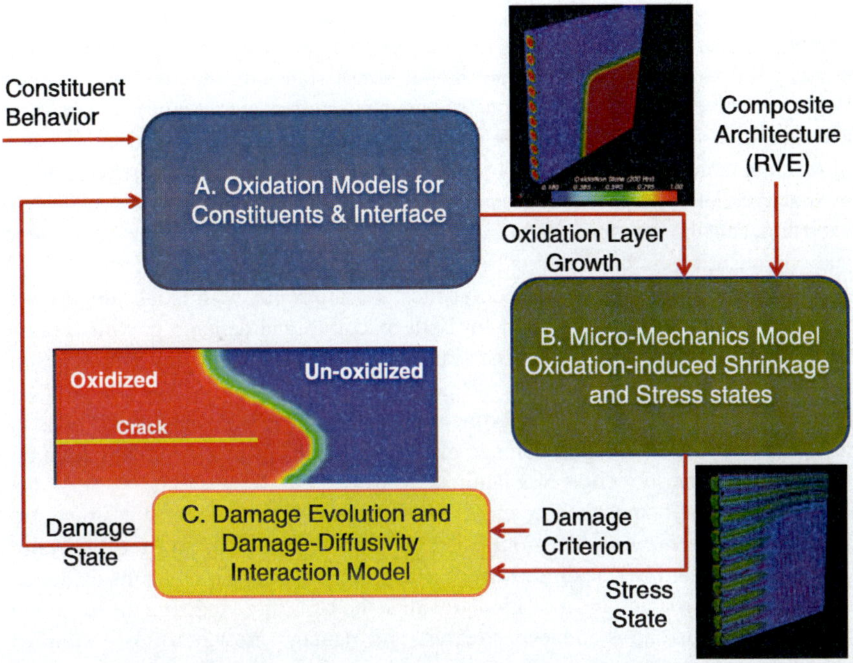

Fig. 10.2 Modeling framework for considering morphological changes in oxidation growth, effects of oxidation-induced stresses and discrete damage

environments absorbs the oxygen at the gas–solid interface and the dissolved oxygen diffuses deeper into the solid. The exposure and absorption at the surface is first of several interacting mechanisms leading to thermo-oxidative degradation. The anisotropic diffusion and the reaction of the dissolved oxygen with the polymer substrate is the mechanism driving morphological changes. The conversion of the polymer into oxidation products is generally accompanied by oxidation-driven strain and damage evolution shown as phase-III of the figure. The discrete damage (crack faces) form new surfaces from which oxygen is further absorbed into the structure. Damage and oxidation layer growth are strongly coupled and lead to accelerated degradation of the material.

In order to develop a frame predictive modeling framework, a multi-scale and multi-physics approach, shown in Fig. 10.2, has been adopted. The material behavior and morphological changes are tracked at constituent, lamina, and laminate scales. Chemo-mechanics modeling of diffusion, oxidation reaction, and substrate conversion into oxidation products is carried out for each constituent. Evolution of the oxide layer in each constituent is discretely simulated over several hundred hours using finite element techniques. Figure 10.2 shows the constitutive scale modeling as the block-A of the modeling framework. The oxidized and unoxidized regions are quantitatively correlated with experimentation for both neat resins and unidirectional (UD) composites. As the chemical composition of the material is changed during oxidation, the physical and mechanical properties changes lead to oxidation-induced strain fields. Block-B denotes the coupled micro-mechanics model used to determine the evolution of stress stated during oxidative aging. Damage evolves in the oxidized zone of composites after several hours of exposure and oxidation is seen to accelerate ahead of the crack tips. The location and rate of discrete damage growth controls the extent of oxidation growth after inception of damage. There is material erosion from the edges typically after 1,000 h of exposure, causing inaccuracies in experimental measurements of oxidation growth and complications in correlations between predictions and experiments.

Oxidation in fiber-reinforced composites is orthotropic, with axial direction of fiber being the preferred direction for both oxidation and damage growth. Transverse oxidation growth at the lamina scale correlates with growth rates observed in neat resins after accounting for the presence of the fibers. Close coupling is observed between discrete crack growth rates and oxidation layer growth rates in the axial direction. Modeling damage evolution is critical for predicting oxidation growth in the fiber direction of a lamina.

For laminated composites or those with complex 3D reinforcement structures, effective homogenization techniques of the sorption, diffusion, and reaction rates play a critical role. The stacking sequence and the neighboring plies alter the oxidation layer diffusion behavior and oxidation growth in the laminate. Modeling the oxidation layer growth, oxidation-induced stresses, and damage growth using a coupled oxidation and micro-mechanical damage model as depicted in Fig. 10.2 is essential for predicting the failure mechanisms and residual strength of composites operating in high-temperature oxidative environments. Each of these building blocks is further discussed in the subsequent sections of this chapter.

10.2 Modeling Oxidation in Constituent Materials

The model presented in this chapter is adapted from the stress-assisted moisture diffusion framework [5] with a chemical potential for a diffusing vapor phase as given in Eq. 10.1 and including the effects of oxygen reaction rate and polymer conversion. The mass transport of the dissolved vapor phase is governed by Eq. 10.2 in which C represents concentration field of dissolved oxygen in the constituent. D_{cc}, D_{cT}, and D_{cE} are the Fickian, thermo-diffusion, and strain-assisted diffusion coefficients, respectively. The equilibrium equation written in terms of the strain, displacement, and concentration and temperature fields is shown in Eq. 10.3 with λ and G as the elastic constants of the material and C_C and C_T as the concentration and thermal coupling coefficients.

$$\tilde{\mu} = A_c(C,T) - C_C(C,T)\varepsilon_{kk} \tag{10.1}$$

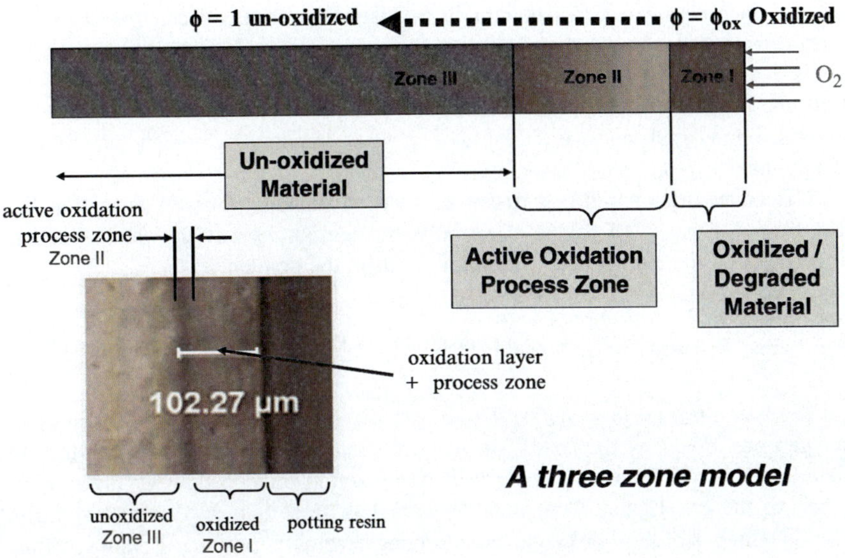

Fig. 10.3 A three-zone model used for oxidation growth prediction in constituent materials

$$\frac{\partial C}{\partial t} = D_{cc}\nabla^2 C + D_{cT}\nabla^2 T + D_{cE}\nabla^2 \varepsilon_{ij} \tag{10.2}$$

$$-C_C C_{,i} - C_T T_{,i} + (\lambda + G)\varepsilon_{,i} + G u_{i,kk} + O(\varepsilon) = 0 \tag{10.3}$$

In the absence of thermo-diffusion (e.g., isothermal aging) and when all the material coefficients are constant, the diffusion equation uncouples from the elastic equilibrium equation as shown in Eq. 10.4. The diffusion coefficients, D_{cc} and D_{cE}, however, can be strain dependent.

$$\frac{\partial C}{\partial t} = \left(D_{cc} + \frac{C_m}{\lambda + 2G} D_{cE} \right) \nabla^2 C \tag{10.4}$$

The model presented in this chapter extends the vapor phase diffusion framework outlined in Eqs. 10.1–10.4 to include reaction of the dissolved vapor phase with the substrate. The reaction rate expressions are adopted from the mechanistic chemistry models. The reaction rate, $R(C)$, is known to be oxygen concentration and temperature dependent. As long-term aging requires tracking reactions over extended periods, the number of polymer sites available for oxidation reaction must be considered. Therefore, an oxidation state parameter (ϕ) is used to denote the progress of the oxidation reaction and its termination. Figure 10.3 shows the three-zone model used for prediction of oxidation layer growth in an oxidizing constituent material. Oxygen diffuses from the exposed end and oxidation reactions take place over an active reaction zone. The active reaction zone (zone-II) moves into the thickness after the polymer completely reacts into its oxidation products. When the oxidation reaction rate is zero only a diffusion zone (zone-I) is present. The oxidizing constituent is treated as a composite material consisting of the oxidized (*ox*) and unoxidized (*un*) phases. The oxidation state parameter value ($\phi = 1$) is chosen to be unity for unoxidized materials and a limit value of ($\phi = \phi_{ox}$) denoting complete oxidation of the polymer.

The concentration field of dissolved oxygen content (molar volume) in the constituent, $C(x,y,z;\ t,T;\ \varepsilon)$, is a spatially and temporally varying field which is dependent on the temperature and strain fields in the domain.

$$\frac{\partial C}{\partial t} = D_{ij}(T,\phi,\varepsilon)\nabla^2 C - R(C,\phi,T) \tag{10.5}$$

The reaction rate term $R(C,\phi,T)$ defines the net rate (molar volume/time) of oxygen consumed by the oxidation reaction. The mechanisms and the rates are determined by the chemistry of the oxidation reaction and depend upon the composition of the constituent. Considerable work has been done by Verdu and Colin [6–11] (Chap. 8 of this book) on characterizing several for polymeric resin systems. The oxidation reaction of carbon fiber has also been studied extensively. However, the oxidation reaction rate of the carbon is negligible [12, 13] if use temperature of the HTPMC remains below 375°C.

The pertinent boundary and interface conditions are determined by the continuity of chemical potential at the boundary ($\vec{x}$) exposed to the ambient vapor with a chemical potential of (μ_A)

$$\tilde{\mu}(C,T,\varepsilon_{ij};\vec{x}) = \mu_A(\vec{x}) \tag{10.6}$$

For a linear elastic solid, saturation concentration (C_s) at the boundary can be assumed to be linear with stress for low concentrations or quadratic as shown in Eq. 10.7 with σ_{ij} as the stress tensor and S_{ij} as its deviatoric component.

$$C_s \cong C_s^0 + C_s^1\sigma_{kk} + C_s^2\sigma_{ii}\sigma_{jj} + C_s^3 S_{ij}S_{ij} \tag{10.7}$$

The saturation concentration without stress dependence at the boundary is related to the partial pressure (P) of the ambient vapor and a solubility parameter (S) via the Henry's law as given in Eq. 10.8.

$$C_s^0 = SP \tag{10.8}$$

When the individual fiber and matrix domains are homogenized into an effective medium, the saturation oxygen content for that medium must be determined. Volume averaging is used to model the sorption behavior and the saturation oxygen concentration of the effective medium.

10.2.1 Resin Behavior

Thermo-oxidative aging of the polymeric material changes the chemical composition of the polymer and hence the physical properties of the material. Therefore, any part of the polymer subjected to thermo-oxidation invariably contains multiple material phases – namely, the unoxidized or virgin polymer, the active reaction zone (where a mix of oxidized and unoxidized polymers exist) and an oxidized polymer phase. In general, the diffusivities for each of these phases will be different, while the diffusivity of the oxidized polymer layer controls the oxygen available to the unoxidized layer. The oxygen diffusivity of the material is, therefore, a function of the oxidation state, $D_{ij}^*\ (\phi)$. The diffusivity is taken to be D_{ij}^{ox} when the resin is fully oxidized and D_{ij}^{un} when completely unoxidized.

$$D_{ij}^*(\varphi) = D_{ij}^{un}\frac{(\varphi - \varphi_{ox})}{(1 - \varphi_{ox})} + D_{ij}^{ox}\frac{(1 - \varphi)}{(1 - \varphi_{ox})} \tag{10.9}$$

The oxygen diffusivity tensor is also temperature dependent and assumed governed by the Arrhenius equation (Eq. 10.10) with a pre-exponent D_{ij}^0, and activation energy (E_a). The pre-exponent and the activation energy will vary with the oxidation state.

$$D_{ij}(\phi) = D_{ij}^0(\phi)\exp\left(\frac{-E_a(\phi)}{RT}\right) \tag{10.10}$$

The oxidation reaction rate, $R(C,\phi,T)$, depends upon the temperature, oxidation state, and the concentration of oxygen. The saturation reaction rate (R_0) is temperature dependent [14]. Typically this dependence is characterized by the Arrhenius-type relationship ($R_0 = R_0 * \exp(-R_a/RT)$), where R_0* and R_a are the scaling and activation constants. While the Arrhenius-type kinetics models are suitable for capturing the temperature dependence, the mechanistic models capture [2, 8, 15–17] the reaction-rate dependence on the dissolved oxygen concentration. Models for concentration dependence of reaction rate investigated by Colin et al. [9] and Abdeljaoued [16] vary with the type of the polymer. Colin et al. [9] investigated epoxy and bismaleimide (BMI) matrices and the latter work was on PMR-15 polyimide resin systems.

In order to determine the saturation reaction rate (R_0), we observe the active reaction zone in Fig. 10.3. Considering the diffusion-reaction equation (Eq.10.5), the proportion of the diffusivity versus reaction rate determines the length of zone over which the oxygen diffuses in and gets consumed by the reaction. This zone is indicative of the saturation reaction rate (R_0) in Eq. 10.11. The lower the reaction rate, the larger the active reaction zone is for a constant diffusion coefficient. The concentration dependence is also shown in Eq. 10.11, which models the case of oxygen starvation (dissolved oxygen concentration is lower than that required for the maximum reaction rate). The concentration-reaction rate dependence shown here follows the analysis presented in Abdeljoued [16].

$$R(C,\varphi) = R_0 2\xi\left(1 - \frac{\xi}{2}\right)\left(\frac{\varphi - \varphi_{ox}}{1 - \varphi_{ox}}\right) \tag{10.11}$$

$$\xi = \frac{\beta C}{1 + \beta C} \tag{10.12}$$

This model has three dependence regimes, namely, a rapid acceleration regime ($0 < \beta C < 3$), followed by a more tempered dependence in which saturation reaction rate is reached ($3 < \beta C < \sim 20$) and then a constant regime ($\beta C > \sim 20$). The oxygen concentration at which the behavior changes from rapid acceleration to constant reaction rate can be easily controlled by varying the value of β. For the polyimide systems considered in this work, an appropriate value of β is determined based on mass loss behavior of the resin when exposed to various oxygen partial pressures. The procedures used for determining these parameters for a resin system are explained in Sect. 10.4.

10.2.1.1 Reaction Termination and Oxidized Layer Growth

While the active zone modeling describes the reaction rate and its dependence on oxygen availability (concentration fields), the diffusion-reaction system will be stationary (does not grow into the polymer thickness) if sufficient amount of polymer is always available for reaction. However, under longer-term aging (tens to hundreds of hours) conditions, the oxidation layer will grow moving the active reaction zone into the polymer, and three zones can be observed along the depth of the polymer, as illustrated in Fig. 10.3. The oxidation state variable (ϕ) indicates the availability of an active polymer site for reaction. As the polymer reacts with oxygen, the oxygen molecules will no longer be able to find active polymer sites for reaction and will prefer to diffuse into the interior of the polymer. We define a cut-off value (ϕ_{ox}) at which the oxygen will predominantly diffuse rather than react with the polymer indicating that the polymer has completely oxidized. The state variable ϕ is parameterized to vary from $\phi_{ox} < \phi < 1$, where ϕ_{ox} denotes completely oxidized polymer and $\phi = 1$ denotes unoxidized polymer. The value of $\phi = \phi_{ox}$ is used instead of $\phi = 0$ to establish its relation to mass loss/gain due to oxidation.

10.2.1.2 Oxidation State, Weight Loss, Reaction Rate Relationships

The oxidation state parameter is related to the mass loss ratio and the oxidation reaction rate as shown in Eq. 10.13.

$$\frac{d\varphi}{dt} = \frac{dW^{ox}}{dt} = -\alpha R(C) \tag{10.13}$$

The oxidation state variable (ϕ) is determined at every time step during the computation using the weight loss relationship shown in Eq. 10.14.

$$\varphi = \max\left\{ \varphi_{ox}, \left(1 - \int_{0}^{t} \alpha(\zeta) R(\zeta) d\zeta \right) \right\} \tag{10.14}$$

In an inert environment, polymers undergo physical and chemical aging only, whereas in an oxidizing environment, specimens will undergo physical, chemical, and oxidative aging.

Note that we define the irreversible surface diffusion response and chemical changes occurring during oxidation of the polymer as oxidative or thermo-oxidative aging of the polymeric matrix. This excludes other mechanisms such as curing, light-initiated degradation, and moisture-induced degradation. Aging experiments are conducted both in inert atmospheres (e.g., argon) and oxidative atmospheres (air), and the weight loss is recorded at various time intervals in order to determine the effect of oxidation on aging. Let the initial specimen dimensions be given by length (L), width (W), and thickness (T). At a particular time interval, if the weight loss fraction (weight loss/original weight) in the inert atmosphere is given as γ and weight loss fraction after the same time period in air is given by ε, then the weight loss fraction due to oxidation alone is ($\varepsilon-\gamma$). Let the observed thickness of the completely oxidized layer and the active oxidation layer be denoted by t_o, and t_a, respectively. The initial volume of the specimen (V), the volume of the fully oxidized zone (V_0), and the volume of the active oxidation zone (V_a) are therefore given by Eq. 10.15:

$$\varphi_{ox} = 1 - \frac{V(\varepsilon - \gamma)}{V_0 + \frac{1}{2} V_a} \tag{10.15}$$

10.2.2 Fiber Oxidation and Interphases

High-temperature oxidation of carbon fibers has been reported in the literature in the context of ceramic matrix and carbon–carbon composites [18–20]. Several papers report the high-temperature oxidation of carbon fibers with a variety of coatings [21–23].

Diffusion of oxygen into pitch at low temperatures (from −25 to 150°C) was studied by Singer and Mitchell [24] using Electron Paramagnetic Resonance (EPR). The study included fibers with diameters ranging from 11 μmto 200 μm for which both oxygen diffusion and sorption and times to reach the maximum oxygen content were determined. The temperature dependence of diffusivity for the characterized fibers is given by

$$-\ln D = \frac{\Delta E}{RT} + a \tag{10.16}$$

The activation energy ΔE was computed to be 8.0 Kcal/mole and $a = 6.0$. The oxygen diffusivity at room temperature (RT) was determined to be $2 \times 10^{-9}\,\text{cm}^2/\text{s}$. A mass diffusion coefficient of $D = 1.5 \times 10^{-7}\,\text{cm}^2/\text{s}$ was observed for a fiber with 200 μm diameter. The time required to achieve 90% sorption, which is fiber diameter dependent, was found to be 2.7 s for 10 μm diameter fibers.

Dhami et al. [12] studied weight loss behavior of pitch-based carbon fibers at temperatures greater than 3,000°C in vacuum and around 500°C in air. The isotropic pitch precursor fiber exhibited most reaction rate at 400°C. The reaction rate is defined in this paper as the rate of change of mass loss with time as given in Eq. 10.17.

$$\frac{\Delta m}{m_0} = K_1 t \tag{10.17}$$

Δm is the mass change at time t from an initial mass of m_0 and K_1 is the reaction rate constant. The lowest temperature at which mass loss was documented in this work is 400°C and for the fiber with the isotropic pitch precursor. The reaction rate constant K_1 has a temperature dependence as shown in Eq. 10.18. This equation is fitted from the temperature-dependent reaction rate data taken from Table 2 of Ref. [12].

$$K_1 = 33814 e^{\frac{-14421}{T}} \tag{10.18}$$

Using Eq. 10.18, the reaction rates can be extrapolated at lower temperatures to the values of $K_1 = 1.94 \times 10^{-9}$/s at 200°C and at 100°C, $K_1 = 1.94 \times 10^{-13}$/s. These rates are several orders of magnitude lower so that mass losses due to reaction rates are expected to be considerably low in carbon fibers for several thousand hours of oxygen exposure. However, mass gains may be expected due to substantial high oxygen diffusivity seen at low temperatures.

Eckstien [13] studied oxidation of carbon fibers made with several precursors in a temperature range of 230–375°C. He studied pan, rayon, pitch, and meso-pitch precursor-based fibers. Several fibers with 99.9% carbon content showed insignificant weight loss behavior after thousands of hours of oxidation. Those with carbon content less than 97% did not survive aging for 1,000 h at 375°C. The weight loss per square meter of exposed surface after 100 h of oxidation at temperatures ranging from 230°C to 375°C are summarized in Table 7 of Ref. [13].

Table 10.1 shows the weight loss rate for selected carbon fibers after 100 h of oxidation. The weight loss per square meter data is taken from Eckstein's work and the equivalent reaction rates as determined by Dhami et al [12]are shown.

From these observations, the assumptions that fibers diffuse oxygen to the same extent or higher than that of typical polymeric matrix composites but have nearly zero reaction rates at use temperatures in 150–200°C temperature range seem quite reasonable.

Table 10.1 Weight loss and equivalent reaction rates for several carbon fibers

Fiber	Fiber dia (μm)	Precursor	% Weight loss/m^2 @ 230°C [13]	Equivalent K_1 of [12] (10^{-9})/s
Thornel-300, WYP, UC-309	6.9	PAN	0.08	2.22
VYB	9.5	Rayon	0.01	0.28
Thornel-25, WPD, UC-307	7.4	Rayon	0.04	1.11
SP-1 (low temp)	10	Pitch	0.01	0.28
VSA-11, UC-304	13	Mesopitch	0.28	7.78
VC-150 cloth	13	Mesopitch	0.06	1.67

10.3 Numerical Simulations

The diffusion-reaction-conversion system described in Eqs. 10.5–10.14 is solved numerically for through-the-thickness oxidation growth in resins with 1D simulations and for orthotropic oxidation growth in lamina and laminate scales using 3D finite element methods.

10.3.1 Through-the-Thickness Oxidation Growth with 1D Simulations

The diffusion-reaction system of equations is solved using numerical solutions to differential algebraic equations in 1D domain. These methods are computationally effective when performing parametric sensitivity analysis. This section describes the solution algorithms used to solve the diffusion reaction system with the nonlinear reaction rate as given in Eq. 10.11. The diffusion reaction system is solved using modified implementation of *ode15s* and *pdepe* solvers in MATLAB™ system. The *ode15s* is a variable order multistep solver based on the numerical differentiation formulas (NDFs) and uses the backward differentiation formulas (BDFs, also known as Gear's method). *Pdepe* solves initial-boundary value problems for systems of parabolic and elliptic partial differential equations (PDEs) in the one space variable x and time t. The ordinary differential equations (ODEs) resulting from discretization in space are integrated to obtain approximate solutions at times specified in t span. The *pdepe* function returns values of the solution on a mesh provided in *xmesh*. *Pdepe* solves PDEs of the form:

$$c\left(x,t,u,\frac{\partial u}{\partial x}\right)\frac{\partial u}{\partial t} = x^{m}\frac{\partial}{\partial x}\left(x^{m}f\left(x,t,u,\frac{\partial u}{\partial x}\right)\right) + s\left(x,t,u,\frac{\partial u}{\partial x}\right) \tag{10.19}$$

The PDE in Eq. 10.19 is solved for a time interval of $t_i \leq t \leq t_f$ and over a spatial domain of $a \leq x \leq b$. The initial time is represented by t_i and the final time of the simulation as t_f and the boundary points of the computational domain are $x = a$ and $x = b$. The interval *[a,b]* must be finite, while m can be 0, 1, or 2, corresponding to slab, cylindrical, or spherical symmetry, respectively. If $m > 0$, then a must be $>= 0$. In the above equation, $f(x, t, u, \frac{\partial u}{\partial x})$ is a flux term, and $s(x, t, u, \frac{\partial u}{\partial x})$ is a source term. The coupling of the partial derivatives with respect to time is restricted to multiplication by a diagonal matrix $c(x, t, u, \frac{\partial u}{\partial x})$. The diagonal elements of this matrix are either identically zero or positive. The diffusion-reaction system given in Eq. 10.5 can be cast into the form shown in Eq. 10.19 by using $m = 0$ and the following functions:

$$u \equiv C; C\left(x, t, u, \frac{\partial u}{\partial x}\right) = 1 \tag{10.20}$$

$$f\left(x, t, u, \frac{\partial u}{\partial x}\right) = D(\varphi)\frac{du}{dx}; \quad s\left(x, t, u, \frac{\partial u}{\partial x}\right) = -R(u) \tag{10.21}$$

For (t–t_0) and all x, the solution components satisfy the initial conditions of the form where $C^S(x)$ is the known initial state of oxygen concentration. For all t and either $x = a$ or $x = b$, the solution components satisfy a boundary condition of the form:

$$p(x, t, u) + q(x, t) f\left(x, t, u, \frac{\partial u}{\partial x}\right) = 0 \tag{10.22}$$

The boundary conditions for the present problem can be written as follows:

$$\begin{aligned} @x = a(= 0) &\begin{cases} p(0, t, u) \equiv u - C^S \\ q(0, t) = 0 \end{cases} \\ @x = b(= L) &\begin{cases} p(L, t, u) = 0 \\ q(L, t) = 1 \end{cases} \end{aligned} \tag{10.23}$$

The oxidation-dependent diffusivity $D_{ij}(\phi)$ is determined at each computational point using the model discussed in Sect. 10.2.1.1, namely, For initial results, the proportionality constant, α relating the weight loss and reaction rate is assumed constant with time (i.e., aging). In later simulations, we also consider the variation of α with time, as discussed in Sect. 10.4.

10.3.2 Oxidation Simulations with 3D Finite Element Methods

The next step is to discretize the domain, Ω, by a union of finite elements as $\Omega = \bigcup_{e=1}^{N_e} \Omega^{(e)}$ and N_e is the total number of elements in Ω. The approximate element solution for the concentration $C(x,y,z,t)$ in the diffusion model is defined as

$$C(x,y,z,t) = [\Phi(x,y,z)]^T\{\tilde{C}(t)\} \tag{10.24}$$

where Φ (x,y,z) are the spatial interpolation functions for the concentration field and $\tilde{C}(t)$ is the vector of time varying nodal concentration values. Using the element-free Galerkin method as described for thermal diffusion problems in [25], the weighted residual for the diffusion-reaction equation (Eq. 10.5) must vanish. Equation 10.26 can be derived after integration by parts and with zero flux boundaries.

$$w = \Phi(x,y,z) \tag{10.25}$$

$$\oint \left(\Phi\Phi^T \frac{\partial\tilde{C}(t)}{\partial t}\right) dV + \oint (\Phi R^*(C)) dV + \oint_v \left(\begin{array}{l} D^*_{11}\left(\dfrac{\partial\Phi}{\partial x}\dfrac{\partial\Phi^T}{\partial x}\tilde{C}(t)\right) + \\ D^*_{22}\left(\dfrac{\partial\Phi}{\partial y}\dfrac{\partial\Phi^T}{\partial y}\tilde{C}(t)\right) + D^*_{33}\left(\dfrac{\partial\Phi}{\partial z}\dfrac{\partial\Phi^T}{\partial z}\tilde{C}(t)\right) \end{array} \right) dV = 0 \tag{10.26}$$

After some mathematical manipulations, Eq. 10.26 can be written in matrix form as

$$[\mathbf{C}]\frac{\partial\tilde{C}(t)}{\partial t} + [\mathbf{B}]\tilde{C}(t) + \{\mathbf{R}\} = 0 \tag{10.27}$$

Assuming orthotropic diffusivity, the matrices [**B**] and [**C**] and vector {**R**} are given by

$$[\mathbf{B}] = \sum_e \oint_{\Omega(e)} \left(D^*_{11}\frac{\partial\Phi}{\partial x}\frac{\partial\Phi^T}{\partial x} + D^*_{22}\frac{\partial\Phi}{\partial y}\frac{\partial\Phi^T}{\partial y} + D^*_{33}\frac{\partial\Phi}{\partial z}\frac{\partial\Phi^T}{\partial z} \right) d\Omega \tag{10.28}$$

$$[\mathbf{C}] = \sum_e \int_{\Omega(e)} \Phi\Phi^T d\Omega \tag{10.29}$$

$$\{\mathbf{R}\} = \sum_e \int_{\Omega(e)} \Phi R^*(C) d\Omega \tag{10.30}$$

$R^*(C)$ is defined in Eq. 10.11. The oxidation-dependent diffusivity D^* is defined as the linear interpolation of unoxidized and oxidized diffusivity as determined by Eq. 10.9. Equation 10.27 is discretized by backward Euler method in time domain, where $\Delta t = t^{n+1} - t^n$ and can be written as a system of algebraic equations:

$$[C]\left\{\frac{\tilde{C}^{n+1} - \tilde{C}^{n}}{\Delta t}\right\} + [B]\{\tilde{C}^{n+1}\} + \{R\} = 0 \tag{10.31}$$

The oxidation state is evaluated from the concentration fields using Eq. 10.14. Assuming that the oxidation state (ϕ) remains invariant over the time step, the reaction rate vector is a nonlinear term.

10.3.2.1 Assumption – Reaction Rate Remains Constant Over the Time Step

Assuming that the reaction rate is unchanged during the time step, $\{R^*(C^n)\}$ is computed at the beginning of the time step and held constant during the iterative solution scheme.

$$\{\mathbf{R}\} = \sum_e \int_{\Omega(e)} \Phi R^*(C^n) d\Omega \tag{10.32}$$

$$[\mathbf{D}]\{\tilde{\mathbf{C}}^{n+1}\} - \{E\} = 0 \tag{10.33}$$

$$[D] = [C] + [B]\Delta t \tag{10.34}$$

$$[E] = [C]\{\tilde{C}^n\} - \{\mathbf{R}\}\Delta t \tag{10.35}$$

10.3.2.2 Assumption – Reaction Rate Assumed to Vary Linearly Over the Time Step

As an alternative, we assume that $\{R^*\}$ varies linearly over the time step, leading to the set of equations given below.

$$[\mathbf{D}^*]\{\tilde{\mathbf{C}}^{n+1}\} - \{E^*\} = 0 \tag{10.36}$$

$$[\mathbf{D}^*] = [C] + [B + G]\Delta t \tag{10.37}$$

$$[E^*] = [C]\{\tilde{C}^n\} - \{\mathbf{R}_0\}\Delta t \tag{10.38}$$

$$\{\mathbf{R}\} = \sum_e \int_{\Omega(e)} \Phi \left[R^*(C^n) + \left.\frac{\partial R^*}{\partial C}\right|_{C^n} (C^{n+1} - C^n) \right] d\Omega \tag{10.39}$$

$$\{\mathbf{R}_0\} = \sum_e \int_{\Omega(e)} \Phi \left[R^*(C^n) - \left.\frac{\partial R^*}{\partial C}\right|_{C^n} \Phi^T \tilde{C}^n \right] d\Omega \tag{10.40}$$

$$\{\mathbf{R}_1\} = [G]\tilde{C}^{n+1}$$
$$[G] = \sum_e \left\{ \int_{\Omega(e)} \left[\Phi \left.\frac{\partial R^*}{\partial C}\right|_{C^n} \Phi^T \right] d\Omega \right\} \tag{10.41}$$

Algorithm #1: *Preconditioned Conjugate Gradient (PCG) method*

$\boldsymbol{x}$ = *initial guess*

$\boldsymbol{r} = \boldsymbol{b} - \boldsymbol{Ax}$

$\mathbf{W} = \boldsymbol{M}^{-1}$

$\boldsymbol{d} = \mathbf{W}\boldsymbol{r}$

For k = *1, 2, . . . max_iterations*

$\alpha = (\boldsymbol{r}^T\mathbf{W}\boldsymbol{r})_{k-1}/\boldsymbol{d}^T\boldsymbol{Ad}$

$\boldsymbol{x} = \boldsymbol{x} + \alpha\boldsymbol{d}$

$\boldsymbol{r} = \boldsymbol{r} - \alpha\boldsymbol{Ad}$

If $||\boldsymbol{r}|| <$ *tolerance, then go to End*

$\beta = (\boldsymbol{r}^T\mathbf{W}\boldsymbol{r})_k/(\boldsymbol{r}^T\mathbf{W}\boldsymbol{r})_{k-1}$

$\boldsymbol{d} = \mathbf{W}\boldsymbol{r} + \beta\boldsymbol{d}$

End

In order to describe the algorithm used for the iterative solution, we cast Eq. 10.33 or Eq. 10.36 into a general form. In order to accelerate the convergence, the residual vector **r** is preconditioned before the iterative solution. Preconditioning entails multiplication of **A** by a matrix $\mathbf{M}^{-1}$ so the condition number of $\mathbf{M}^{-1}\mathbf{A}$ is better than that of **A** [26]. When **A** is a symmetric and positive definite (SPD) matrix, we can use Preconditioned Conjugate Gradient (PCG) method to solve Eq. 10.18. Algorithm #1 presents the pseudo code for a single-core PCG algorithm [26, 27]. Here **r** is the residual vector, ||**r**|| is the norm of the residual, **d** is direction of the descent during the minimization search, α is the length of descent, β is the ratio of values of $\mathbf{r}^T\mathbf{W}\mathbf{r}$ in the current iteration and with that in the previous iteration.

After carrying out several iterations, the accumulated round-off error may be large. This accumulation can be particularly significant for problems with minor nonlinearity. To remedy this, the residual **r** can be recalculated not from $\mathbf{r} - \alpha\mathbf{Ad}$ but from $\mathbf{b} - \mathbf{Ax}$ after a user-selected number of iterations. This update corrects for the reaction rate dependence on the concentration.

Heath [26] discussed the selection of the vector norm for the residual **r**. The value of ||**r**|| depends on the scale of values of the problem. A better stopping criterion is a relative residual, ||**r**||/(||**A**||·||**x**||). If **A** is ill-conditioned, a small ||**r**|| does not necessarily indicate convergence. Typically, the user can select either first or second norm. As the calculations of ||**A**|| (a one-time calculation) and ||**x**|| (calculation at every iteration) are time-consuming, vector **b** can be used in lieu of **Ax**. ||**b**|| is calculated once per time step, ||**r**||/||**b**|| can be treated as the relative residual. The user sets a small value as tolerance to indicate convergence of the solution. An alternative convergence criterion is the use of relative change in ||**r**||. Barrett et al. [28] discussed the residual fluctuation within a small range. This fluctuation not only wastes the computation time but may lead to an inaccurate solution. They suggest stopping the iteration when it no longer makes any progress.

The parallel version of PCG follows the single processor version (Algorithm#1), except that, calculations of vectors **r**, **Wr**, **x**, **d**, are carried out on multiple processor cores. The calculation of inner product of two vectors requires inter-core communications. Suppose there are *p* cores, each core uses a rank number *i* to index:

Algorithm #2: *Parallel PCG (PPCG) method for core of rank i*

x^i = *initial guess*

$r^i = b^i - A^i x^i$ *(needs communication with other cores)*

$W^i = (M^i)^{-1}$

$d^i = W^i r^i$

For k = 1, 2, ... max_iterations

$\alpha = (r^T W r)_{k-1} / d^T A d$ *(needs communication with other cores)*

$x^i = x^i + \alpha d^i$

$r^i = r^i - \alpha A^i d^i$ *(needs communication with other cores to get* $A^i d^i$*)*

If $||r|| <$ *tolerance (needs communication with other cores), then go to End*

$\beta = (r^T W r)_k / (r^T W r)_{k-1}$

$d^i = W^i r^i + \beta d^i$

End

rank 0 represents the first core and rank $p - 1$ represents the last core. A vector (or matrix) followed by a superscript *i* represents the vector (or matrix) is a sub-vector (or sub-matrix) and the core of rank *i* calculates that sub-vector. For example, sub-vectors $\mathbf{r}^0, \mathbf{r}^1 \ldots, \mathbf{r}^{p-1}$ make up the vector **r**. Algorithm #2 describes the PPCG procedure.

We chose MPI as the parallel programming model for the PPCG implementation. MPI uses *rank* to index each core, and rank 0 identifies the first one. In the PPCG, rank 0 acts as the master core which not only performs the same CG computation as other cores but also undertakes the message collection and distribution to all cores. A PPCG library is designed such that user-developed programs call several PPCG library functions which wrap MPI functions and simplify communication interfaces. For each iteration there is data communication among cores. To exchange, send, receive, collect, or broadcast data among cores, the library utilizes a series of MPI functions. Jimack et al. [29] provided some examples in which they utilized MPI_Send and MPI_Recv to perform blocking communication. The programs block if either the data sender or the data receiver is not ready for the task or has not completed its task. Blocking communication may cause deadlocks in certain circumstances and MPI_Isend and MPI_Irecv, which perform non-blocking communications, provide better alternatives. Our experiments show that using of MPI_Isend and MPI_Irecv decreases the running speed and avoids deadlocks. The performance of MPI_Isend and MPI_Irecv depend on the problem and the computer architecture. The implementation uses blocking communication.

Figure 10.4 shows the efficiency of the parallelization. A speedup factor is defined as the ratio of the elapsed times to solve the problem on a single processor over that required on multiple processors. The perfect acceleration scenario is shown as the linear relationship in which the elapsed time proportionally decreases with the number of processors. The actual speedup is shown with the symbols and the efficient number of processors required depends upon the problem size. For a problem with 200 K unknowns, it is seen that using 16 cores is most effective.

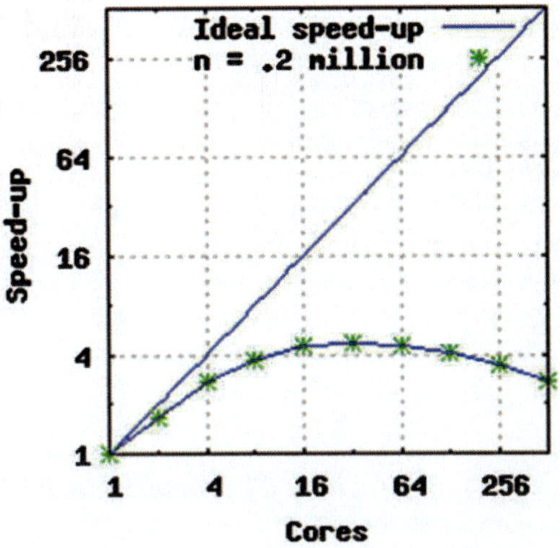

Fig. 10.4 Observed speed up in elapsed time with number of processor cores. The speed up is nonlinear due to communication overhead and depends upon the problem size

2 CPU Cores	PPCG-Iterative Conv. Tol = 10^{-9}	PPCG-Iterative Conv. Tol = 10^{-10}

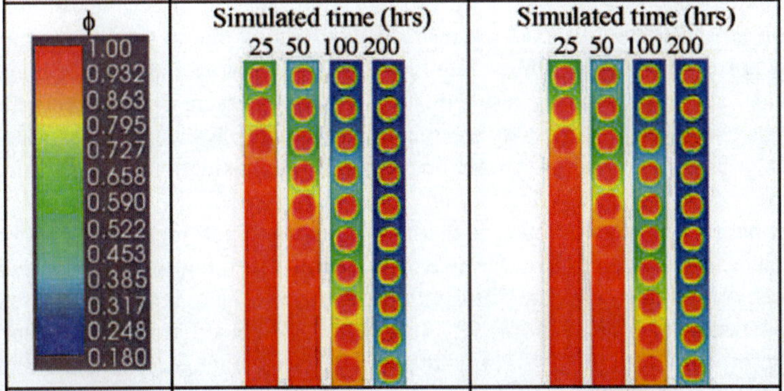

Iterations	10435	23578
Time (sec)	1088	2323

Fig. 10.5 Convergence of oxidation zone size and the number of iterations required to simulate oxidation layer growth in a lamina over 200 h of exposure

Figure 10.5 shows the effect of the tolerance choice on the iterative algorithm. It is seen that choice of 10^{-9} as convergence tolerance in the PPCG scheme produces similar results as the finer 10^{-10} tolerance choice but at about half the computational burden. All the simulations described in this chapter use 10^{-9} as the convergence tolerance. Coarser tolerances (10^{-6}–10^{-7}) produce results must faster but the errors in the concentration fields add up during simulations over long aging times (hundreds of hours).

10.4 Model Parameter Determination

The parameter set required to simulate oxidation growth in homogeneous isotropic materials includes the oxygen solubility parameter (S) or saturation concentration (C_s) diffusivity tensor (D_{ij}), reaction rate (R_0), reaction-concentration dependence parameter (β), substrate conversion parameter (α), and the oxidation state of the material at reaction termination (ϕ_{ox}). Considering the oxidized material as nonreacting but with properties different from that of the unoxidized material, the oxygen diffusivity tensor is required for oxidized material. In this section, the procedures used to characterize these parameters are discussed.

10.4.1 *Diffusivity Characterization*

Oxygen diffusivity data are available for selected unoxidized high-temperature resins. For example, Abdeljaoued [16] determined the diffusion and sorption coefficients for PMR-15; Colin et al. [8] obtained the diffusion coefficients for (Bis)Maleimdes; and Decelle et al. [30] for epoxies. The direct measurement of the oxygen diffusivity in oxidized resins is challenging due to the brittle nature of oxidized resin and the thin films required for permeation testing. The acceleration methods, such as elevating test temperatures and pressures, also accelerate the damage evolution kinetics, oxygen reaction rate and physical/chemical aging of the material. Indirect methods based on weight loss and oxidation layer growth [3, 4] may be used to infer the diffusivity of the oxidized resins.

An observation worth noting is that the diffusivity small molecules in polymers may have some relationship to the size of the molecule. In Fig. 10.6, the relationship between oxygen and moisture diffusivity is examined. The figure shows the temperature-dependent moisture diffusivity for PMR-15 obtained from both absorption and desorption tests [31]. A curve-fit gives the temperature-dependence relation: $D_x = 1.38 \times 10^{-6}\, e^{-4.29(1000/T)}$. From this result, the estimated diffusivity for 25°C is 7.84E-13 m^2/s. The value of the average moisture diffusivity from absorption tests was found to be $D_x = 8.58\text{E-}13\ m^2/s$. The oxygen diffusivity obtained from [3, 4] is also shown for oxidized and unoxidized PMR-15 in this figure.

10.4.2 *Reaction Rate and Concentration Dependence*

The reaction behavior as described by Eq. 10.11 requires the determination ofmaximum reaction rate (R_0) and the concentration dependence parameter (β). Under the assumption that abundant active sites are available for reaction in the polymer, the diffusion reaction behavior leads to an active reaction zone over which oxidation takes place.

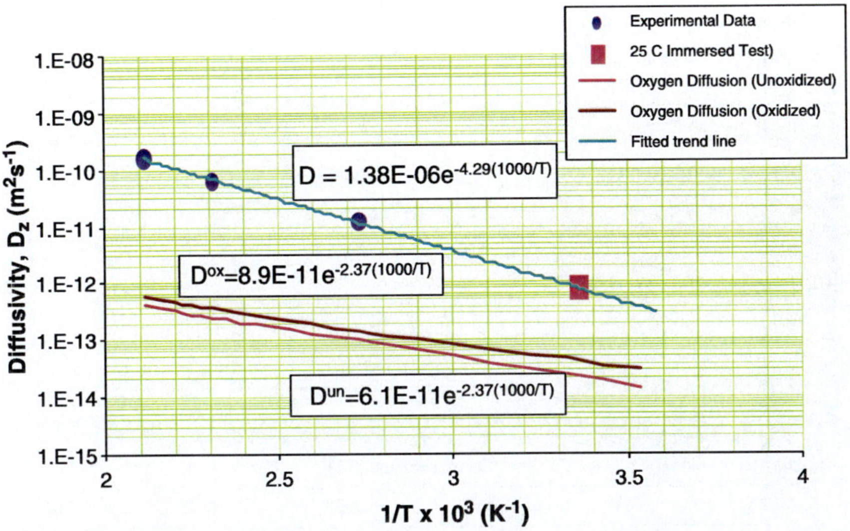

Fig. 10.6 Temperature-dependent moisture and oxygen diffusivity in PMR-15 resin

Figure 10.7 shows the results from parametric simulations showing the active zone size for several values of R_0 at several temperatures. This plot is made for PMR-15 polyimide resin with diffusivity varying with temperature using the unoxidized constant values. Figure 10.8 shows the active reaction zone measurements for PMR-15 polyimide resin at 288°C. This figure shows two distinguishable region sizes – namely, 17 μm before 40 h and 25 μm after 40 h of aging. These active reaction zone sizes correspond to a R_0 value of about 3.5 mol/m^3 min for the first 40 h of aging which then slows down to a reaction rate corresponding to about 2 mol/m^3 min. The reaction rates are determined using a diffusivity of 53.6e-6 mm^2/min evaluated at 288°C. The figure can be used to determine the reaction rate from experimental observation of the active reaction zone or predict the active zone if the reaction rate is known at any temperature. Based on the experimental observations of the reaction zone size of PMR-15 resin [32], we determine the saturation reaction rate R_0 as 5.5 mol/m^3 min at 343°C from the results shown in Fig. 10.8.

The zone in which oxidation reaction is actively in progress is observable in several polymers including PMR-15. The thickness or size of this zone is determined by the relative dominance of oxygen diffusivity (increasing the dissolved oxygen concentration) and the rate of reaction (decreasing the dissolved oxygen concentration). This zone will be proportional to the ratio of the diffused oxygen volume over reacted volume. For the polyimide resin under consideration, the active reaction zone forms quickly (can be seen within 1 h). The optical experimental observations reported by Schoeppner et al. [33] show the active reaction zone between the oxidized and unoxidized regions for PMR-15 resin.

The value of β can be determined from weight loss data obtained at two oxygen partial pressures, typically in pure O_2 and in air. This formulation leads to $R(C) \rightarrow R_0$ when $C \rightarrow$ infinity. That is, when the reaction is not oxygen deprived;

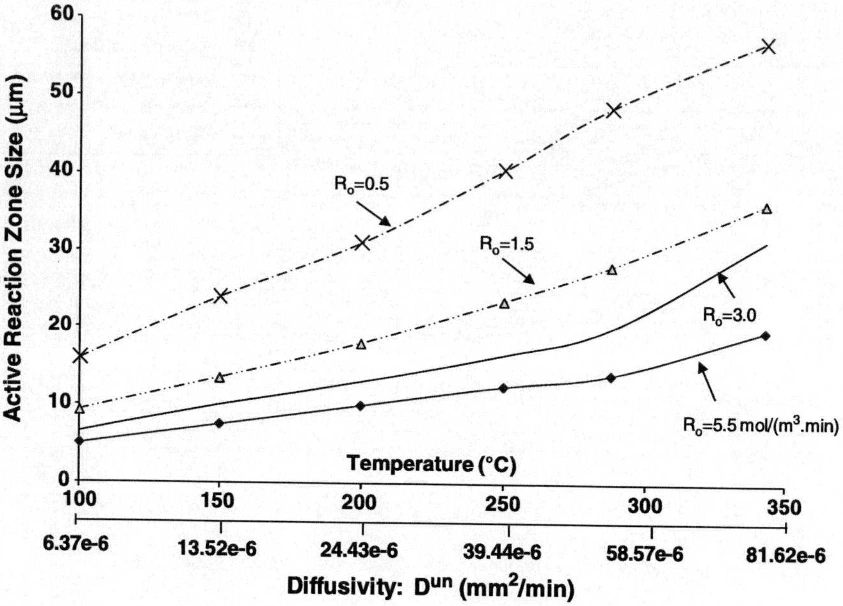

Fig. 10.7 Estimated active reaction zone size for various diffusivities and reaction rates

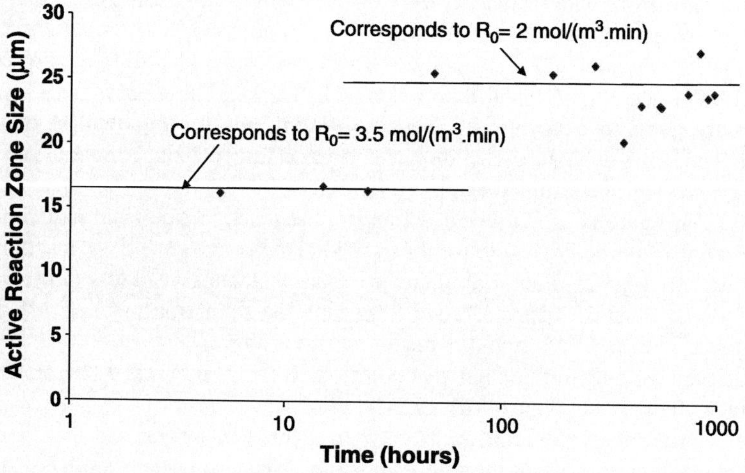

Fig. 10.8 Active reaction zone size measured from experiments [32] and corresponding reaction rate for PMR-15 at 288°C

the reaction rate reaches a maximum saturation rate, R_0. In order to determine this value, we need weight loss measurements at two concentrations, which translate to the reaction rates at those two concentration values. Since the weight loss is taken to be proportional to the oxygen consumption rate, the ratio of the weight loss determined at two concentrations is the same as $R_1(C_1)/R_2(C_2)$. For example, from

Abdeljaoued [16], the ratio of the weight loss between pure O_2 ($C_2 = 3.74$ mol/m^3) and air ($C_1 = 0.79$ mol/m^3) is about 0.7.

$$\frac{2\beta C_1}{1+\beta C_1}\left[1-\frac{\beta C_1}{2(1+\beta C_1)}\right] = 0.7\frac{2\beta C_2}{1+\beta C_2}\left[1-\frac{\beta C_2}{2(1+\beta C_2)}\right] \quad (10.42)$$

Solving Eq. 10.42 produces three roots: [−3.5593, −0.4266, 0.9195] with only one physically feasible value for PMR-15 resin to be $\beta = 0.919$. With all the parameters in the diffusion-reaction equation Eq. 10.1 defined, the simulation of the oxidation layer growth requires the specification of the reaction rate R_0 and a mechanism for reaction completion/cessation. Considering that, the oxidation reaction will stop if it is oxygen-starved, the reaction will also cease when all the active polymer sites are consumed. In the latter case, the oxygen will again diffuse deeper into the material in favor of the reaction. In this section, we describe the case when the diffusing oxygen finds the active polymer reaction sites creating an active oxidation layer followed by advancement of the layer upon the reaction completion.

10.4.3 *Effect of Fiber and Fiber–Matrix Interphase*

The effects of the fiber and fiber–matrix interphase on the oxidation of composites are discussed in this section. The fiber and fiber–matrix interphase regions are treated as a single concentric cylindrical assemblage with effective material behavior. At the use temperatures of polymer matrix composites, carbon and glass fibers are generally thermally stable. However, oxygen diffusivity and solubility in the fiber must be considered. In a recent paper [34], we established a framework to differentiate the effects of diffusivity of fiber and fiber-interphase assemblages. We treated the fiber and fiber–matrix interphase as a single geometric and material domain and referred to it as the fiber–interphase assemblage.

The primary effect observed in the oxidation behavior of the unidirectional composite due to presence of fiber reinforcements is orthotropy in oxidation growth. Although carbon fibers are generally assumed to be inert and do not react with oxygen, the fiber and the fiber–matrix interphase can act as oxygen conduits. The anisotropy in oxidation of the composite will therefore depend upon the architecture of the fiber reinforcement and the nature of the diffusivity tensor and saturation oxygen concentration of the fiber and fiber–matrix interphase assemblage. Direct characterization of the relative diffusivity and the saturation concentration of oxygen in fibers and through the fiber–matrix interphase is complicated. In this paper, we therefore illustrate an indirect methodology to characterize the relative diffusivity in which the oxidation zone sizes are measured in a unidirectional composite in axial and transverse directions in regions where there are no discrete cracks or observable damage. These measurements correspond to the minimum oxidation layer sizes, as explained previously.

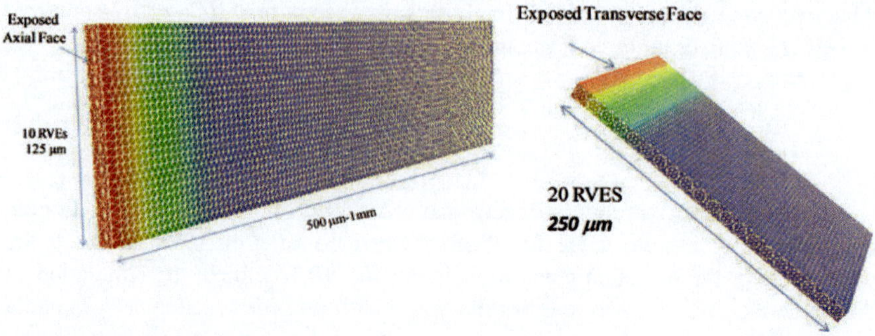

Fig. 10.9 Lamina-scale models used for characterization of fiber diffusivity effects (after [34])

We define a scaled diffusivity parameter for the fiber–interphase diffusivity as $D_{ij}^{f*} = D_{ij}^{f}\frac{C_m^s}{C_f^s}$, where D_{ij}^{f} is the fiber–interphase diffusivity and C_m^s and C_f^s are the saturation oxygen concentrations of the matrix and fiber–interphase assemblage, respectively.

A procedure for establishing the effective diffusivity and solubility of the fiber as a ratio of the matrix diffusivity is determined from oxidation growth observations. A unidirectional composite lamina is modeled and the fiber and matrix domains are explicitly represented as shown in Fig. 10.9. The dimensions of the finite models used for axial oxidation growth (Fig. 10.9a) and the transverse oxidation growth (Fig. 10.9b) are shown in the figure. Properties of PMR-15 resin and carbon fibers are used in the numerical simulations. Experimental observations [35] show that the axial oxidation growth is substantially higher than the transverse growth.

Using homogenization theories in an earlier effort [36], the values of $D_{11}^{f*}$is varied from 1 to 30 times that of the oxidized matrix and the transverse diffusivity D_{22}^{f*} is varied from 0.01 to 10 times that of the oxidized matrix. Simulations are conducted for 200 h at 1 min time increments and the oxygen concentration profiles as well as oxidation state (ϕ) profiles are obtained. The oxidized regions ($\phi = \phi_{ox}$) are measured and plotted with time for each of the assumed values of D_{ij}^{f*}.

Figure 10.10 shows the oxidized layer measurements obtained from simulations in the axial direction and the corresponding power law fits to the data for several assumed values of relative diffusivity parameters. Power law fits are used to extend the predictions to beyond 200 h. The figure shows axial oxidation with various assumptions for the axial diffusivity of the fiber and experimental observations [35] of oxidation layer sizes. The simulations are conducted for 200 h and extrapolated to 4,000 h using power law fits while the experimental data are actual observations at specified hours. The minimum axial oxidation layer size observed at each time is plotted as a filled triangle and corresponds to locations where there is no observable damage. The maximum values typically correspond to locations where discrete cracks are present and these values are shown as filled squares. Comparing the minimum oxidation layer size and the simulation results, the axial diffusivity ratio

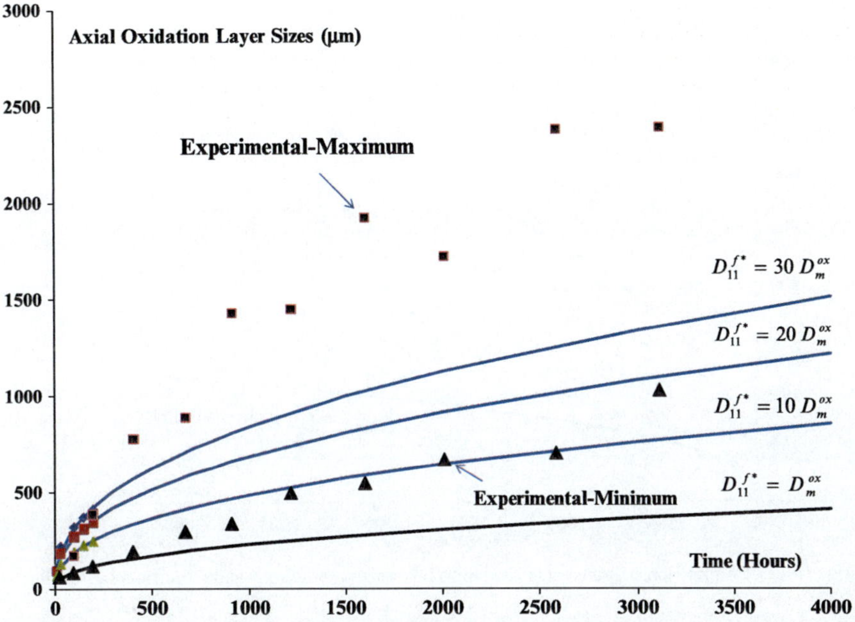

Fig. 10.10 Comparison of experimental and simulation results. This correlation helps estimate the relative D_f^* of the fiber–matrix assemblage

D_{11}^{f*} can be estimated between D_m^{ox} and $10D_m^{ox}$ during the simulation period and to be at $10D_m^{ox}$ after approximately 1,200 h.

In the transverse direction, the model considered has 20 fibers and can only simulate oxidation regions up to 250 μm. The fiber–interphase assemblage diffusivity, D_{22}^{f*} is varied between $0.1D_m^{ox}$ and $10D_m^{ox}$ of the oxidized matrix diffusivity. Figure 10.10 shows the oxidized layer measurements obtained from simulations in the transverse direction and the corresponding power law fits to the data for several assumed values of relative diffusivity parameters. Once again, the power law fits to the simulation results are used to extend the predictions to 4,000 h of aging time and compared with the transverse oxidation layer sizes observed in the experiments. Figure 10.11 shows the correlation of the experimental and the simulated results and it can be seen that the fiber–interphase assemblage diffusivity D_{22}^{f*} equal to $0.1D_m^{ox}$ fits the data well. Therefore we establish that the transverse diffusivity of the fiber–interphase assemblage to be around 78e-7 mm^2/min.

The estimate of the fiber–interphase assembly diffusivities enables simulation and prediction of the oxidation zone sizes for the unidirectional composites in regions where there is no damage present. Figure 10.11 shows the oxidation layer sizes at 50, 100, 150, and 200 h for the case in which the fiber–interphase assemblage has orthotropic diffusivity with $D_{22}^{f*} = D_m^{ox}$ and $D_{22}^{f*} = 0.1D_m^{ox}$. The oxidation layer sizes shown in this figure correlate well with the minimum oxidation zone observations in the experimental analysis. The orthotropic diffusivity

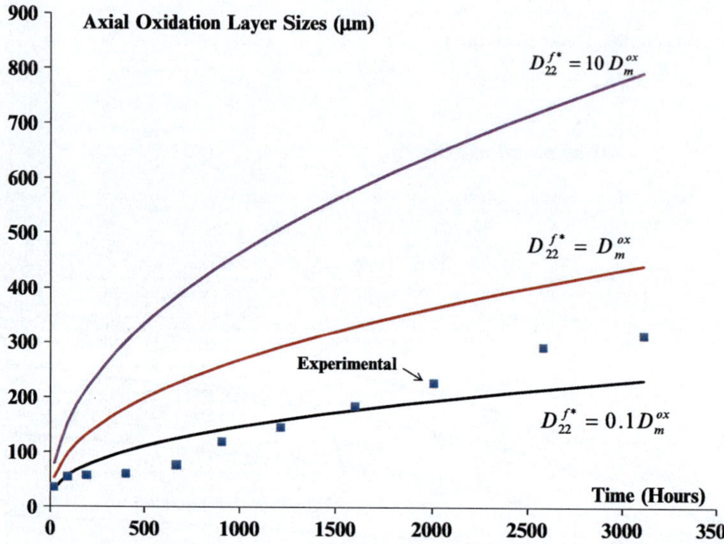

Fig. 10.11 Comparison of transverse oxidation layer size predictions and experimental results

values used for the fiber–interphase assemblage are $D_{22}^{f*} = 78 \times 10^{-6}$ mm^2/min and $D_{22}^{f*} = 78 \times 10^{-7}$ mm^2/min. The axial layer measurements are comparable to that obtained earlier with isotropic fiber–interphase assemblage diffusivities, that is, $D_{11}^{f*} = D_{22}^{f*} = 10D_m^{ox}$ and the transverse measurements compare with simulations performed with isotropic diffusivity values of $D_{11}^{f*} = D_{22}^{f*} = 0.1D_m^{ox}$.

10.5 Predictive Modeling of Oxidation Growth

Oxidation layer size and its growth rate at the constituent, lamina, and laminate scales enable prediction of morphological changes in the composite, evolution of stress, and degradation due to damage progression. In this section, the nature of the oxidation growth and damage evolution is discussed.

10.5.1 Long-Term Oxidation Growth in Resins

In order to understand the thermo-oxidative behavior of the polymer, we compare and correlate the experimental oxidation layer growth and weight loss data obtained at 288°C (550°F) and 343°C (650°F) with parametric analyses obtained from simulations (Fig. 10.12). These comparisons provide quantitative values for various parameters of the model, especially the apparent diffusivity of the oxidized region (D_{ij}^{ox}), the behavior of the proportionality ($\alpha(t)$) between reaction rate and the weight loss, and the temperature-dependent saturation reaction rate (R_0).

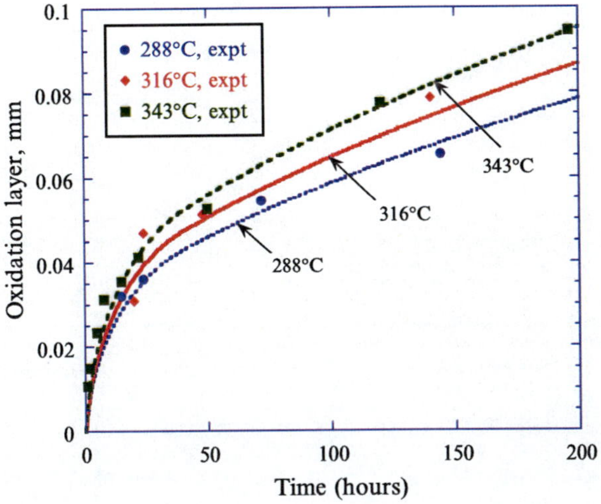

Fig. 10.12 Oxidation layer growth in PMR-15. Experimental data are shown as symbols and predictions are shown as lines

Experimental observations suggest that the behavior of PMR-15 is diffusion controlled, and the reaction rate is high enough to consume the available polymer in observable times (tens of hours). Therefore, growth of oxidation into the polymer is observed with a clear oxidized, active reaction, and unoxidized regions and the growth of the oxidized region with time. Figure 10.13a shows the growth of the oxidized polymer layer in the PMR-15 specimen aged at 288°C measured using optical microscopy, where the thickness of the oxidation layer increases steadily approaching a plateau value, while the thickness of the active "reaction" zone remains nearly constant as the aging time increases. Further, although not shown here, it was observed that the elastic modulus of the oxidized layer in PMR-15 resin remains relatively constant for the aging times considered.

Next, we considered two more high-temperature resin systems. As the specific chemical compositions and commercial names of these resins cannot be disclosed in open literature, the resin with lower use temperature is denoted as Polyimide-A and the higher operating temperature as HT-Resin A. Figure 10.13b shows the oxidation growth behavior of the resin denoted as Polyimide-A when aged at 177°C. Unlike PMR-15, where the thickness of the active "reaction" zone remains nearly constant over the aging time considered, the thickness of the transition region in Polyimide-A resin increases considerably, while the region undergoing complete (uniform) oxidation remains rather small in size. However, the total thickness of oxidation regions (I + II) in Polyimide-A is comparable to PMR-15 for similar aging times considered.

The visual characteristics of HT-Resin-A neat resin do not change during oxidation. Therefore, nano-indentation was used to characterize the change in modulus of the neat resin, and thereby to monitor the extent of oxidation. For HT Resin-A, the total thickness

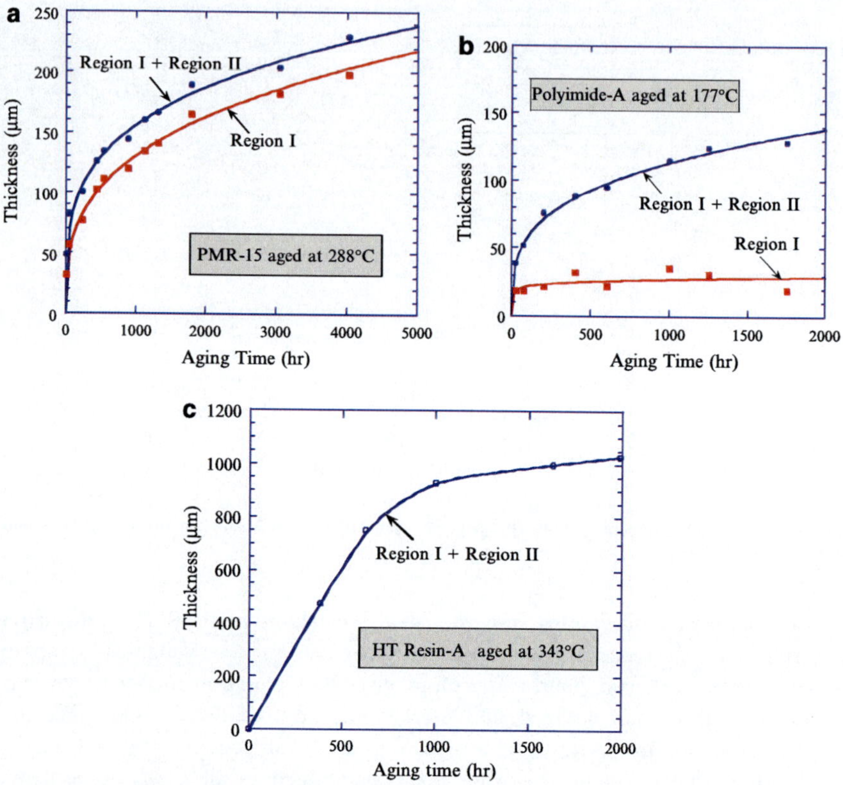

Fig. 10.13 Oxidation behavior over thousands of hours in three different resin systems. (**a**) PMR-15 resin shows oxidation layer growth through the thickness and a small active reaction zone indicating high reaction rate-diffusivity ratio. (**b**) Polyimide-A shows that larger active reaction zone indicating reduced reaction rate-diffusivity ratio. (**c**) HT Resin-A shows a very large active reaction zone size and undetectable oxidized layer. This system also has very large active reaction-zone indicating high oxygen diffusivity

of the oxidized region was estimated from the distance over which the modulus deviates from its average value in the interior of the resin. Figure 10.13c shows that these extrapolated values of depths of oxidation are considerably larger compared to the measured thickness of the oxidation layer in PMR-15 and Polyimide-A resins for the same aging time periods. Further, for HT-Resin-A resin, the modulus value within the oxidized region continually increases up to the maximum aging time of 1,628 h that was examined in this study. These results therefore suggest that during the isothermal aging process, reactions are occurring at significant depths of HT-Resin A resin specimens (~1,000 μm), as compared to PMR-15 and BMI resins for which material changes are isolated to “thin” surface layers. The growth of the oxidation layer is a function of the material chemical composition, cures kinetics, and reaction mechanisms.

10.5.2 Oxidation Growth in Unidirectional Lamina

The oxidation behavior in a unidirectional lamina differs from the neat resin oxidation in several ways. The heterogeneity induced by the presence of fibers in the oxidation behavior is threefold. Firstly, the oxygen diffusivity tensor is strongly orthotropic and the axial oxidation growth rate is higher than the transverse growth rate. Secondly, the interstitial matrix areas between the fibers and interphases are areas where damage can be preferentially onset due to the oxidation-induced stresses. Lastly, the fibers bridge any transverse cracks enhancing the apparent toughness of the oxidized materials. Therefore, the observed oxidation rate in axial direction is substantially higher than that in the transverse and accelerates with damage evolution.

Figure 10.14 shows the simulated oxidation state in a carbon fiber/PMR-15 lamina aged for 200 h with the orthotropic diffusivity of oxygen into the lamina. The predicted oxidation layer sizes can be correlated with experimental observations in regions where there is no damage.

Figure 10.15 shows the correlation of experimentally observed minimum oxidation layer size (discrete symbols) in axial and transverse directions and the results of simulations. The simulations results show that the fiber–interphase assemblage values of $D_{11}^{f*} = D_m^{ox}$ and $D_{22}^{f*} = 0.1D_m^{ox}$ correlate well with the experimental results in both axial and transverse directions. However, after 500 h of aging the minimum oxidation growth in the axial direction accelerates, indicating degradation processes other than observable discrete cracking in the matrix or the fiber–interphase assemblage domains. The degradation processes are perhaps attributable to the fiber–interphase assemblage region as the simulations indicate that increasing the axial diffusivity from D_m^{ox} to $10D_m^{ox}$ enables comparison of experimentally observed oxidation layer sizes with the simulations.

10.5.3 Oxidation Growth in Laminates

In this section, we will describe the modeling framework required for simulating the oxidation growth in a composite laminate by considering the thermo-oxidative behavior of a single unidirectional ply, and the effect of stacking sequence on the oxidative behavior. The oxidation layer growth in a unidirectional ply is inherently orthotropic due to the presence of the fiber and interphase regions, and is further complicated by the damage growth. Experimental measurements taken in the undamaged areas of the unidirectional laminate were then utilized to characterize the effective diffusivity of the fiber-interphase assemblage. Simulations made with an assumed fiber–interphase diffusivity equal to that of oxidized matrix in the axial direction and tenth of the oxidized matrix in the transverse region correlate well

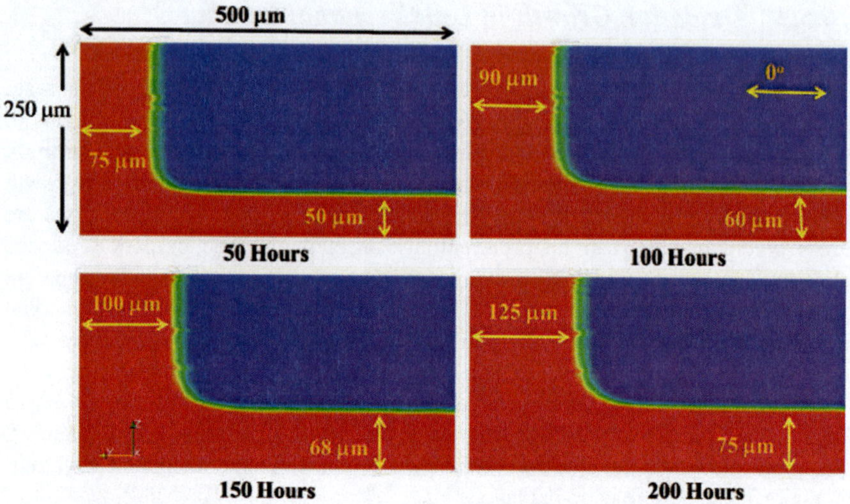

Fig. 10.14 Oxidation growth prediction in a unidirectional composite with orthotropic fiber–interphase assemblage diffusivity. The fiber (0°) direction is horizontal as indicated. Oxidized regions are measured and the sizes are shown (after [34])

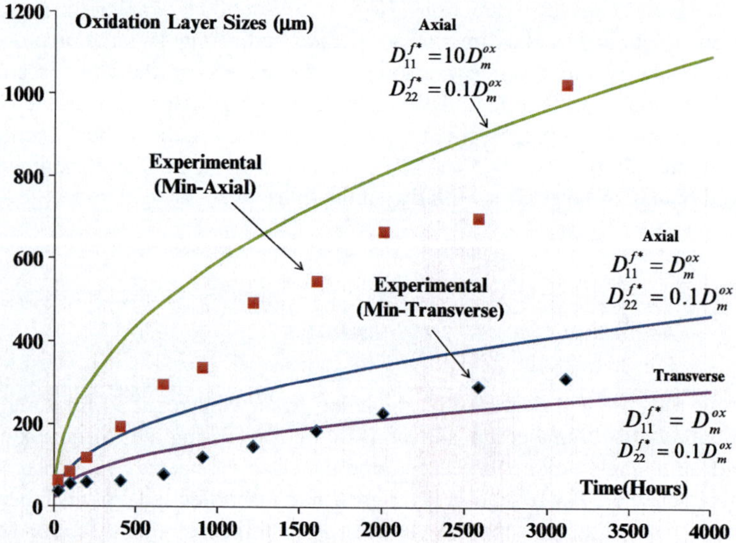

Fig. 10.15 Comparison of experimental and simulated minimum oxidation layer sizes with orthotropic diffusivity for the fiber–interphase assemblage (after [34])

with the experimentally observed values for aging time below 500 h [34]. Once the oxygen diffusion characteristics were estimated for the constituents of the lamina, an effective homogenized orthotropic diffusivity tensor is determined at the lamina scale. Assuming a representative volume element with a fiber fraction of V_f and with the assumptions that fibers–interphase assemblage is diffusive but not reactive to oxygen, we can define the effective diffusivity tensor, reaction rate, and saturation concentration at the lamina scale. The homogenized properties for the composite lamina (shown with the superscript "c") are determined using rule of mixtures for this analysis and are given in Eqs. 10.43–10.47. Equation 10.43 defines the homogenized saturation concentration in the composite lamina and Eq. 10.44 defines the diffusivity parameters.

As the saturation oxygen concentrations are not known at this point for the fiber and matrix, we also assume same solubility for both fiber and matrix. The reaction rate ($R^c(C)$)is homogenized as shown in Eq. 10.45 and the oxidation state parameter (φ^c_{ox})varies from $\varphi^c_{ox} \leq \varphi^C \leq 1$.

$$C^c_s = C^s_m V_m + C^s_f V_f \tag{10.43}$$

$$\begin{aligned} D^c_{11} &= D^m_{11} V_m + D^f_{11} V_f \\ \frac{1}{D^c_{22}} &= \frac{V_m}{D^m_{22}} + \frac{V_f}{D_{22}{}^f} \end{aligned} \tag{10.44}$$

$$R^c(C) = V_m R^m(C) \tag{10.45}$$

$$\frac{d\varphi^c}{dt} = \alpha R^m(C) V_m = \alpha R^c(C) \tag{10.46}$$

$$\begin{aligned} \frac{d\varphi^c}{dt} &= \alpha R^m(C) V_m = \alpha R^c(C) \\ \varphi^c_{ox} &= \varphi^m_{ox} + (1 - \varphi^m_{ox}) V_f \end{aligned} \tag{10.47}$$

where D_{ij} is the diffusivity tensor, C^s is the saturation oxygen concentration, and the superscript and subscripts *m*, *f*, and *c* refer to the stated quantities for the matrix and fiber–interphase assemblage and composite, respectively.

The fiber-induced heterogeneity and anisotropy in the microstructure of the composite creates a preferred direction for diffusion along the axis of the fiber and a more tortuous path transverse to the fibers, assuming that the fibers themselves are less diffusive. Further, the variation in the orientation of the adjacent plies affects the oxygen diffusion and results in through-the-thickness heterogeneity of oxidation state in a laminate. We will examine the relative effects of each of these mechanisms on the oxidation growth using laminate scale simulations and a cross-ply laminate architecture.

The laminate model used for analysis is a 16-ply $[0/90]_4$ composite as shown in Fig. 10.16. Each ply has a thickness of 127 μm and the simulation domain is 318 μm in length along the 0° fiber direction. As our interest is in studying the anisotropy in oxidation growth and neighbor ply effect, only the left edge is exposed to simulate the axial oxidation growth. The exposed surface is assigned a molar concentration of 0.79 mol/m^3 for the dissolved oxygen at the surface [4]. We assume that the

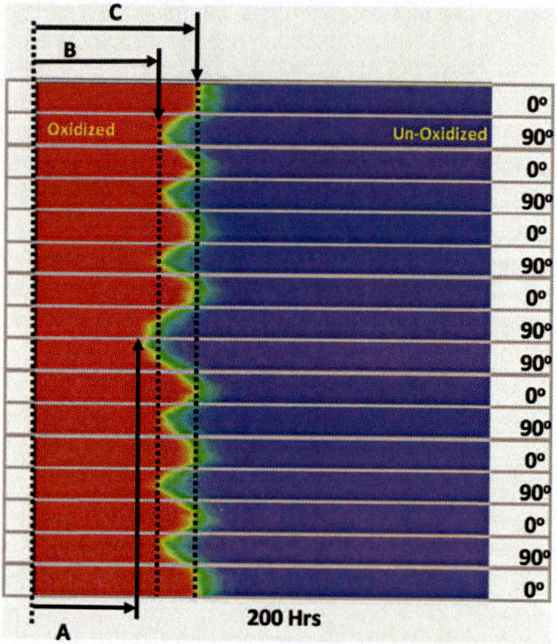

Fig. 10.16 Oxidation growth in $[0/90]_{4s}$ after 200 h of aging. The oxidized zone size in a ply is influenced by the diffusivity (or ply angle) of the neighboring plies

solubility of both the fiber and matrix phases is the same. Although the simulation can account for different saturation concentrations in fiber and matrix, this assumption is made due to the absence of experimental data on oxygen solubility in carbon fibers. The figure shows the through-the-thickness heterogeneity in oxidation growth in the laminate and its evolution with time. The oxidation layer sizes in the 0° plies are considerably higher than the 90° plies. Also noticeable is the effect of the neighboring ply on the oxidation layer sizes. When a ply is 90° flanked by two 90° plies, the oxidation layer size is higher than when the ply is flanked by a 0° and 90°. Figure 10.16 denotes three layer thicknesses, *A*, *B*, and *C* which represent the layer size for a 90° ply flanked by 0° and 90°, for a 90° ply flanked by two 0° plies and for a 0° ply flanked by 90° ply. At 200 h, the smallest oxidation layer size (denoted by length "*A*" in the figure) is in the 90° center ply with a 0° and 90° neighboring ply. The next larger oxidation layer size is in 90° plies that have 0° neighboring plies as denoted by length "*B*." The largest oxidation layer size is for the 0° as denoted by length "*C*." However, the dimension "*C*" is irrelevant due to the growth of transverse oxidation in the top and bottom plies and will not be examined in detail.

Figure 10.17 shows the oxidation layer contours within 0° and 90° plies in the cross-ply laminate. The simulation result shown in Fig. 10.17a corresponds to a 0° layer with two 90° adjacent plies (denoted as the [90/0/90] configuration) and that shown in Fig. 10.17b corresponds to a 90° layer with two 0° adjacent plies ([0/90/0] configuration). The fiber direction for 0° layer is horizontal and the 90° layer is perpendicular to the view. Red depicts complete oxidation state ($\phi = \phi_{ox}$) and blue is the unoxidized region ($\phi = 1$). The measured oxidation region sizes are shown

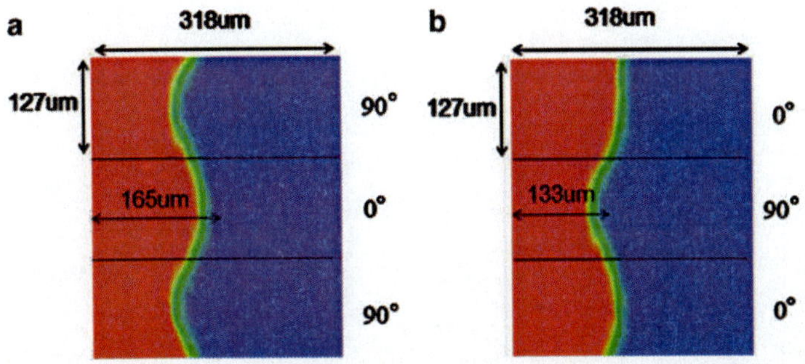

Fig. 10.17 Oxidation layer sizes for [90/0/90] and [0/90/0] stacks in the $[0/90]_{4s}$ composite. The minimum possible oxidation layer growth in a 0° ply occurs for the [90/0/90] ply combination and the maximum possible oxidation growth for a 90° ply occurs for the [0/90/0] combination

for both the 0° and the 90° layers after 200 h of isothermal exposure. The anisotropy in diffusivity between the 0° and 90° layers is evident in the oxidation growth and growth rates. As expected, the extent of oxidation in the center 0° ply is larger compared to the center 90° ply. Similar to our experimental observations reported in Sect. 10.2, the maxima or the minima in oxidation extent occurs at the ply midplane if the ply has symmetrically oriented plies on either side. Figure 10.17 shows the neighbor ply effect for these two cases after 200 h of exposure. For the [0/90/0] configuration, the oxidation layer size in the 90° layer is 133 μm at 200 h, while for the [0/90/90] configuration, the oxidation layer is 108 mm. Thus, the neighboring 0° plies on either side accelerate the oxidation growth for the 90° ply in [0/90/0] configuration compared to 90° ply in [0/90/0] configuration. Thus, for the $[0/90]_{4s}$ laminate, the absolute minimum in oxidation growth occurs at the laminate center in the middle 90° plies and this correlates with experimental observations. For a cross-ply laminate, the [90/0/90] configuration determines the lowest oxidation layer size expected for a 0° ply and the [0/90/0] configuration determines the highest oxidation layer size expected for a 90° ply. The 90° plies that are not at the center of the laminate have two adjacent 0° layers ([0/90/0] configuration), but the 90° layer at the center of the laminate has one 0° layer and one 90° layer adjacent to it ([0/90/90]).

Figure 10.18 shows the adjacent ply effects on oxidation growth in the 90° layer with [0/90/0], [0/90/90] and [90/90/90] stacking sequences for longer aging times. The neighboring ply effect clearly seen in short-term exposure is accentuated for long-term oxidation growth. The trends of these numerical simulations are consistent with the experimental observations.

Considering a quasi-isotropic $[0/\pm45/90]_{2s}$ laminate, the oxidation layer sizes after 200 h exposure are shown in Fig. 10.19. The dependence of oxidation layer size on the fiber orientation is evident in the simulation. Several configuration combinations are present in this structure including [−45/90/0] and [−45/90/90]. The heterogeneity of the oxidation growth for this laminate is seen in the simulation results. The results show highest oxidation layer sizes for the 0 plies and lowest for the center ply with [−45/0/90] configuration.

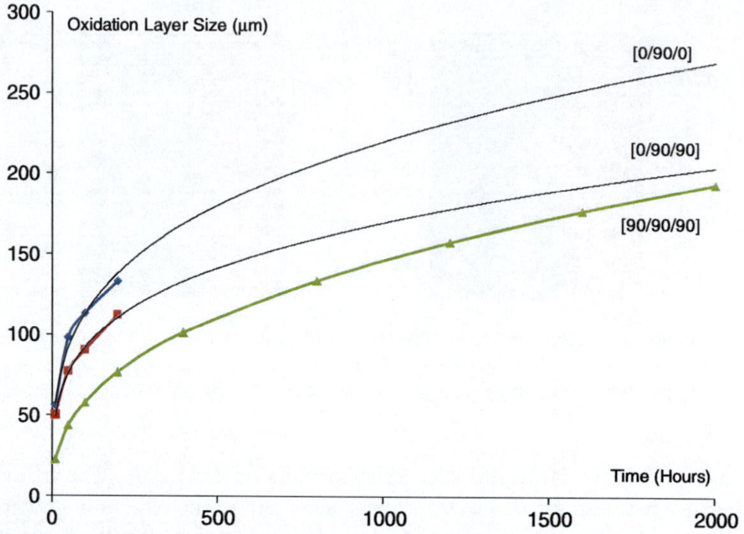

Fig. 10.18 The oxidation predictions for 90° layer with adjacent layer effects

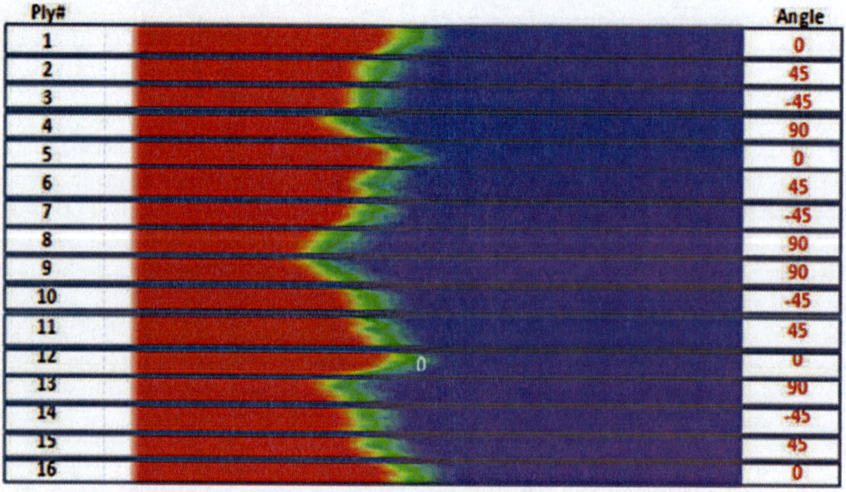

Fig. 10.19 Oxidation growth in $[0/\pm45/90]_{2s}$ laminate

10.6 Oxidation, Stress, and Damage

The evolution of oxidation-induced stress states, oxidation growth near discrete crack tips and the development of discrete damage in oxidized regions are discussed in this section.

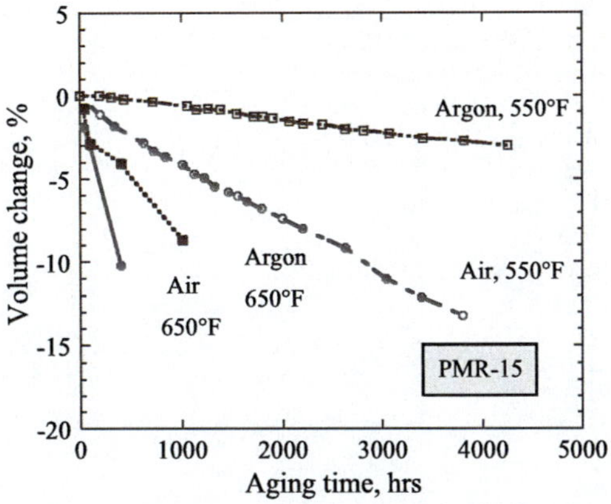

Fig. 10.20 Volumetric shrinkage in PMR-15 in Air and argon environments (after [36])

10.6.1 Evolution of Stress Due to Oxidative Layer Growth

The following three main assumptions were made to establish the shrinkage strain in inert (non-oxidizing) and air (oxidizing) environments. It is first assumed that the shrinkage of the specimen aged in argon is representative of the shrinkage in the interior (unoxidized) region of the specimen undergoing oxidation. Secondly, we assume that the oxidized layer has uniform shrinkage and that it does not depend upon concentration of the absorbed oxygen. Thirdly, we assume that there is no erosion loss of material from the surface of the specimens. It has been documented that for long aging times, the oxidation layer thickness remains constant, that is, the thickness versus aging time reaches some asymptote. This does not mean that the oxidation front is stationary, but it may imply that the rate of erosion from the surface of the specimen is equivalent to the oxidation layer growth. Therefore, the shrinkage model is not applicable for long aging times when erosion becomes significant.

Figure 10.20 shows the volumetric shrinkage of a PMR-15 specimen in argon and oxygen aging environments of 288°C and 343°C. The volumetric shrinkage (ε_v) in air is considerably higher than that seen in argon and increases with an increase in temperature. In order to determine the linear oxidation-induced shrinkage strain, we further assume that the lineal shrinkage (ε_l) is isotropic and that the difference in the volumetric shrinkage strains between argon and oxygen environments is attributed to the shrinkage in the oxidation layer. The rate of volumetric shrinkages computed from the slopes of the shrinkage lines in Fig. 10.20. Subtracting these two volumetric shrinkage rates, the oxidation-caused shrinkage is 0.00254%/h at 288°C. We set the simulation window to 200 h wherein the surface erosion is not significant. The total volumetric shrinkage computed at 200 h for nonoxidized regions are taken as 0.152% and the oxidized region to be 0.66%. The value used for the oxidized zone consists of both the inert shrinkage contribution and the oxidation-induced shrinkage. The isotropy of the shrinkage assumption leads to the definition of the volumetric shrinkage in a unit elemental volume as $\varepsilon_v = (1 + \varepsilon_l)(1 + \varepsilon_l)(1 + \varepsilon_l) - 1 = \varepsilon_l^3 + 3\varepsilon_l^2 + 3\varepsilon_l$, where ε_l is the lineal shrinkage strain.

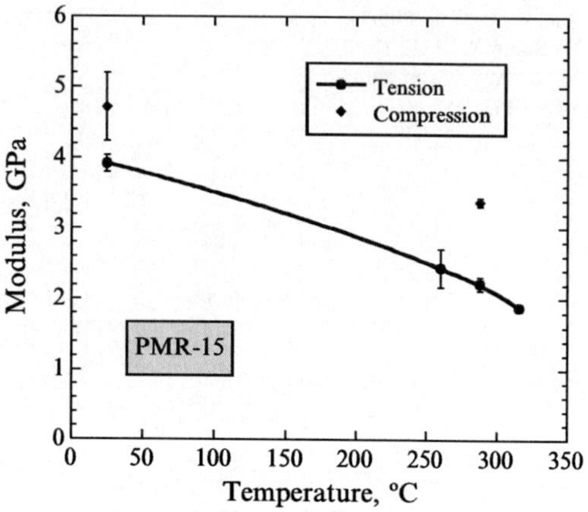

Fig. 10.21 Modulus change with temperature. PMR-15 shows significant drop in modulus with temperature. Matrix property controlled composite behavior such as shear response will be significantly degraded by the temperature (after [36])

Solving for ε_l from known ε_v, we get the lineal shrinkage strains to be 0.0005 (0.05%) and 0.0022 (0.22%) for unoxidized ($\phi = 1$) and oxidized zones ($\phi=\phi_{ox}$), respectively. The oxidation causes an increase of 0.17% strain compared to that induced by inert environments. The total lineal shrinkage strains for a material point with an oxidation state of ϕ are obtained by interpolation of shrinkage strains for oxidized and unoxidized materials.

The modulus of PMR-15 is sensitive to the temperature and oxidation state. The temperature dependence of the unaged resin moduli as obtained from testing bulk PMR-15 specimens in tension and compression are shown in Fig. 10.21. Test data reveal that the resin is stiffer in compression, and that the compressive modulus of unaged PMR-15 resin decreases from 4.72GPa to 3.39 GPa due to temperature rise from 25°C to 288°C.

Nano-indentation was used to determine the room temperature (RT) modulus of the oxidized specimens. The Berkovich indenter tip was calibrated by the method described by Oliver and Pharr [37] using a fused silica standard, the modulus of which is 72 GPa. The values of the modulus obtained from nano-indentation under compressive loading for unoxidized and oxidized resin (see Chap. 9 of this book). The oxidative aging for the data shown in this figure was conducted at an aging temperature of 343°C, and the specimen was cooled to room temperature before the modulus was determined using nano-indentation. Nano-indentation data show that the stiffness of the oxidized resin is higher than that of the unoxidized portion of the sample. The data show that the unoxidized resin has a room temperature modulus of 4.84 GPa and the oxidized resin a modulus of 5.84 GPa. Thus, there is a small discrepancy between the values obtained from the bulk testing (4.72 GPa at RT) and those obtained from the nano-indentation (4.84 GPa) for the unoxidized resin. This small discrepancy could be attributed to some pileup observed in PMR-15 specimens during nano-indentation leading to contact areas that are greater than the cross-sectional area of the indenter at a given depth, and hence a larger apparent modulus measurement.

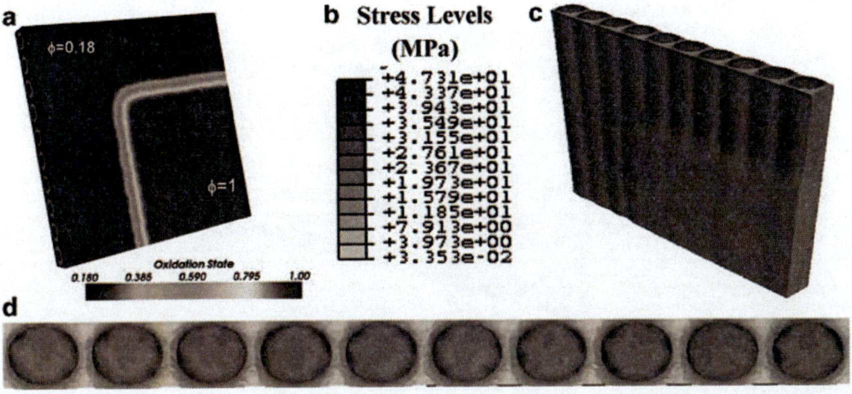

Fig. 10.22 Oxidation layer thickness and effective stress distributions in a unidirectional lamina. (**a**) Oxidation state after 200 hours of aging. $\phi = 1$ denotes the unoxidized region and $\phi = 0.18$ denotes fully oxidized region. (**b**) Stress contours levels (**c**) Oxidation induced stress state in the lamina. (**d**) Cut view showing Stress state in the fiber and at the interface (after [36])

In order to model the temperature (T), oxidation state (ϕ) and time dependence of the modulus, we use the framework suggested by Wise et al. [38]:

$$E(\varphi, T) = E_{un}(T) e^{\left(K_{ox} \frac{1-\varphi}{1-\varphi_{ox}}\right)} e^{\left(K_{nox} \frac{\varphi-\varphi_{ox}}{1-\varphi_{ox}} t\right)} \tag{10.48}$$

where $E_{un}(T)$ is the unaged resin modulus at temperature T. The modulus of the oxidized resin at any temperature is given as in Eq. 10.48. Two constants, K_{ox} and K_{nox}, are assumed to be temperature independent and are obtained from room temperature nano-indentation results. K_{ox} represents the oxidation state (ϕ) dependence and K_{nox} controls the aging time (t)-dependent modulus changes. The modulus of the resin is stable and does not change with the aging time in an inert environment. Therefore, we set K_{nox} to zero and obtain $K_{ox} = 0.1878$.

As the nano-indentation data were not obtained at elevated temperatures, we use the bulk specimen data for the temperature-dependent modulus of the unoxidized resin. Obtaining the modulus of the oxidized resin through bulk mechanical testing is extremely challenging. Therefore, we use a mixed set of parameters to determine the temperature-dependent modulus of the oxidized resin. We determine the K_{ox} value from nano-indentation data and use the same value with the modulus data obtained with bulk mechanical testing at higher temperature. With this assumption, we determine that the modulus of the oxidized resin is 4.09 GPa at 288°C.

Figure 10.22a shows the simulated oxidation state in a G30-500/PMR-15 composite. The picture frame oxidation layer growth is shown after 10 h of exposure. Finite element analysis is conducted to determine the deformation and stress. The model is subjected to symmetry boundary conditions on the unexposed surfaces, and the surfaces exposed to oxygen are specified as stress free. No mechanical loading is applied to the RVE. The oxidation state variable (ϕ) at each material

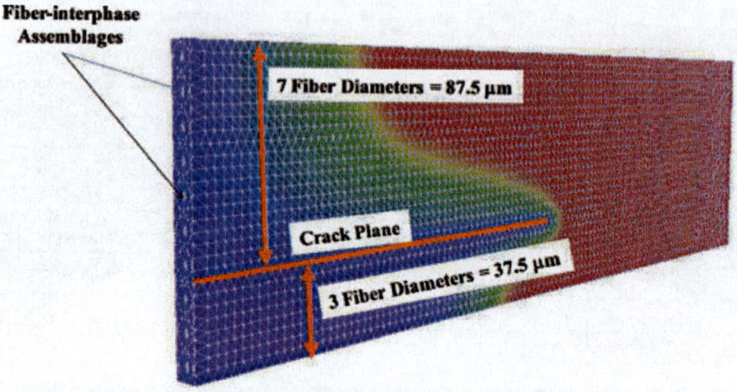

Fig. 10.23 Discrete crack is placed in the 10 fiber model as shown and propagated at 1.3 μm/h (after [34])

integration point in the stress analysis mesh is used to compute the appropriate modulus (Eq. 10.48) and shrinkage strains induced due to thermo-oxidation are determined from Fig. 10.20 and the procedure described earlier. Figure 10.22c, d show the Von-Mises effective stress distribution in the RVE with the corresponding grey-scale bar shown in Fig. 10.22b. The oxidation state and the stress distributions are shown after 200 h of thermo-oxidative aging. The average stress in the fiber increases to 25 MPa, and the interstitial matrix regions have almost 1.2 MPa, while the peak value is 47.3 MPa. Thus, after 200 h of aging, the level of stresses induced due to the shrinkage mismatch is quite high (of the order of the matrix strength of 41 MPa at 288°C) and therefore leads to onset of damage at the fiber–matrix interface/interphase or decohesion of the matrix.

10.6.2 Oxidation Growth Near Discrete Cracks

In this section, we will attempt to predict the oxidation extent in regions in which damage is observed to develop. The effects of discrete cracking on oxidation growth are twofold. The crack surfaces adsorb oxygen bringing higher soluble oxygen into the interior of the specimen. Additionally, the high crack-tip stress fields can increase diffusivity of oxygen near the tip, accelerating the oxidation ahead of the crack tip. In this effort, a discrete axial crack is propagated in the RVE and the oxidation growth monitored around the crack surfaces. Figure 10.23 shows the finite element model used for the analysis. The model consists of ten unit-cells (i.e., fiber–interphase assemblages) and the crack plane is placed between the seventh and the eighth unit cells. This allows determination of the region (as a function of number of fiber diameters) affected by the presence of oxidation. The crack is propagated at a rate of 1.3 μm/h which is the average value estimated from the experimental observations for the aging times considered. The crack growth is simulated by setting the oxygen concentration at the crack face nodes equal to that of boundary sorption at the

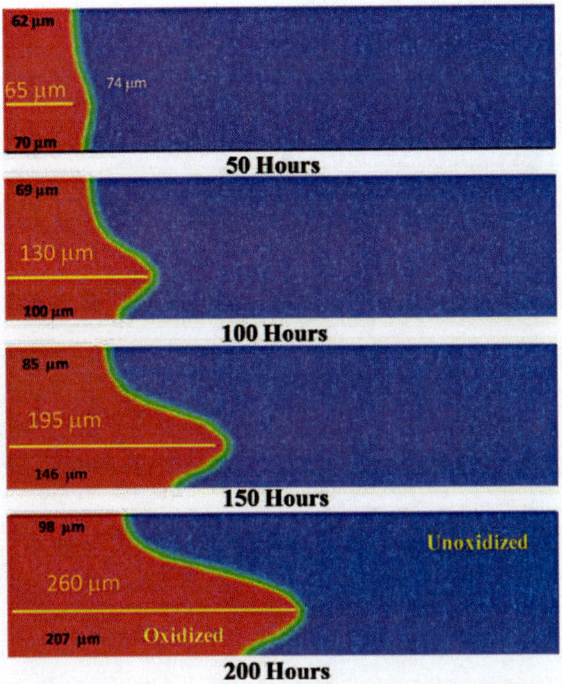

Fig. 10.24 Oxidation growth around a discrete crack propagating at 1.3 μm/h (after [34])

appropriate instant in time when the crack surfaces form. At present, no interactions between crack-tip stress fields and diffusivity are considered in this analysis.

Figure 10.24 shows the oxidation growth around a propagating discrete crack in a unidirectional composite. The diffusivity of the fiber–interphase assemblage is set to orthotropic diffusivities as determined in the previous section. The axial oxidation front is shown at 50, 100, 150, and 200 h. Consistent with experimental observations, the oxidation front is always seen ahead of the crack and the oxidation depth near the crack surface is substantially higher than the regions away from the crack. The bottom picture in Fig. 10.25 shows the detailed measurements of the axial oxidation zone near the discrete crack after 200 h of aging. The crack tip is located at 260 μm from the boundary surface and the oxidation zone is seen ahead of the crack tip. The oxygen concentration field in the top picture of Fig. 10.25 shows that the oxygen concentration gradient is substantially higher near the crack-tip regions compared to the boundary away from the crack tip indicating higher rate of Fickian diffusion near the tip. Since the diffusivity of oxygen is presently assumed to be unaffected by the stress fields, the diffusivity ahead of the crack tip is set to be the same as in any other regions of the model. The oxygen concentration gradients and the material diffusivity near the crack tip contribute to the amount of soluble oxygen, the reaction rate, and the rate of oxidation ahead of the crack tip. In general, the crack propagation rate will be determined by the stress state and the fracture toughness of the oxidized resin or the toughness of the fiber–matrix interphase.

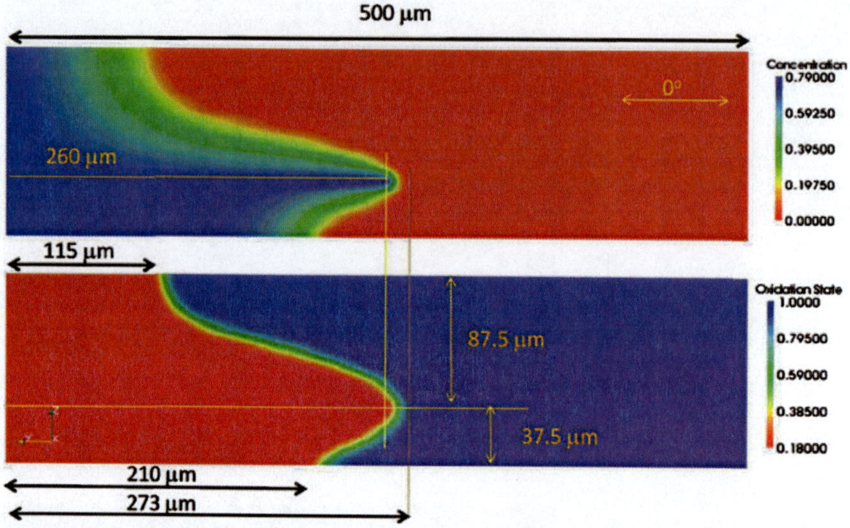

Fig. 10.25 Effect of discrete crack on the oxidation growth. Oxidation state and concentration profiles are shown here for a unidirectional composite after 200 h of aging (after [34])

The measured maximum oxidation zones are typically longer than the associated crack sizes by more than 100 μm. In the simulations however, oxidation front is typically very close to the crack tip (<20 μm). This difference could be due to the fact that stress-assisted diffusivity is not considered in the simulation while the crack-tip stresses are generally considerable. The discrepancy could also be due to damage and degradation processes in the resin ahead of the crack zones, such as micro-porosity or blistering due to escaping volatiles that are not observable due to the resolution of the optical or dye-penetration methods used in this study. It is also possible that micro-cracking, which often precedes a crack front, could occur in the plastic zone ahead of the crack tip. Micro-cracking would not be detectable using the dye-penetrant method. Additionally, the crack lengths reported in this study have been measured through fluorescence imaging using dye impregnation in conjunction with optical microscopy techniques. If the crack faces are fully open, then the dye penetrant will be able to fill the entire length of the crack. However, in case the crack is partially closed, there will be an error in the reported crack length measurement.

10.6.3 Damage Growth in Oxidized Regions

Consider a specimen subjected to uniaxial loading as shown in Fig. 10.26a. The material chosen is PMR-15 resin, the properties of which are known from prior efforts [3, 4]. The specimen is subjected to isothermal aging at 288°C with only the

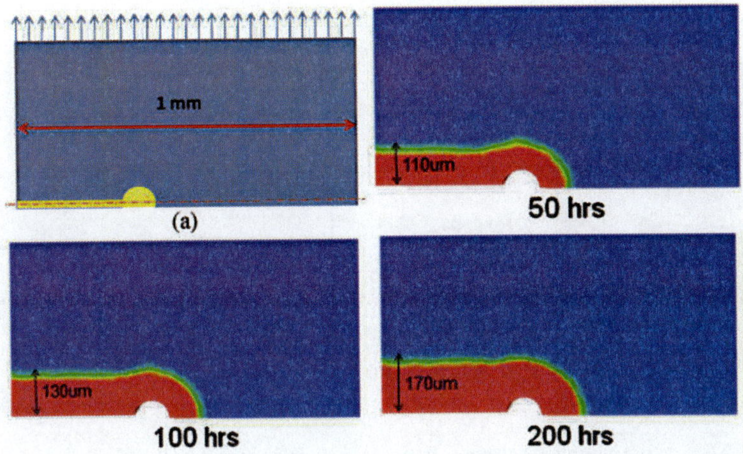

Fig. 10.26 Specimen geometry used for analysis and oxidation growth around the crack

crack surfaces subjected to oxidation. Figure 10.26 also shows the oxidation layer growth near the crack surfaces at selected aging times. The oxidation layer thicknesses predicted from diffusion-reaction simulations are around 90, 110, and 130 μm for 50, 100, and 200 h simulation times, respectively.

We investigated the characteristics of damage propagation into the oxidation zone. We make the fundamental assumption that the oxidized matrix has much lower toughness and critical stress compared to the unoxidized region. The critical stress at which crack propagates is assumed to be 40 MPa for unoxidized resin [36] and 20 MPa for the oxidized resin. Oxidation state dependent material stiffness is mapped to the crack propagation mechanics model. The moduli of oxidized and unoxidized resins are $E_{ox} = 5.84$ GPa and $E_{un} = 4.84$ GPa, respectively. The Poisson's ratio is assumed to be 0.3 for both areas. The information of oxidation layer thickness versus time is mapped to the mechanics model.

Figure 10.27 shows crack growth in the oxidized and unoxidized regions driven by both oxidation-induced stresses and applied loading. Simulation shown here has an oxidized zone corresponding to 50 h (Fig. 10.27a) and 200 h (Fig. 10.27b) of aging. With an oxidized zone after 50 h of aging, the crack initializes at an applied pressure of 0.575 MPa and propagates in a stable mode through the oxidized area. At an applied pressure of 0.865 MPa, the crack reaches the boundary between oxidized and unoxidized regions. An applied pressure of 1.635 MPa is needed for stable propagation of the crack into the unoxidized zone. We assume that the crack is arrested at the interface between oxidized and unoxidized zones and further aging (oxidation) time is necessary for damage to evolve. An oxidized layer corresponding to 200 h of aging is mapped to the model and tensile stress is applied. The crack begins to propagate at a pressure of 0.63 MPa and arrives at the unoxidized region at an applied pressure of 0.672 MPa.

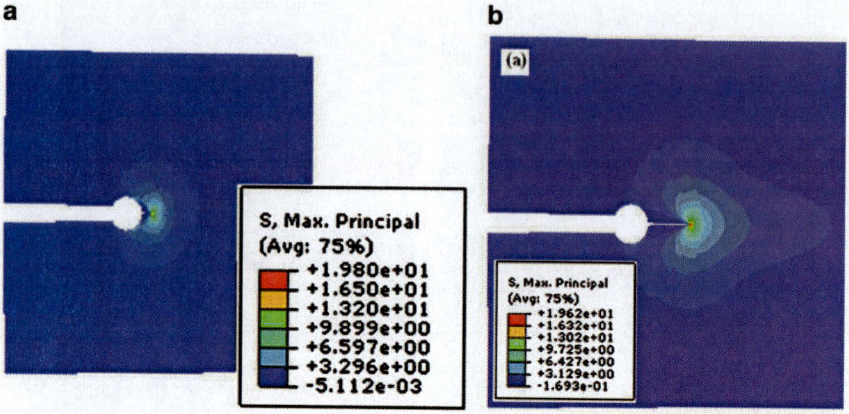

Fig. 10.27 Crack propagation and stress fields: (**a**) oxidized zone after 50 h of aging (**b**) 200 h of aging

Noting that the pressure required to propagate the crack through the entire oxidized zone is only 0.672 MPa at 200 h of aging compared to 0.865 MPa, it can be concluded that for a constant applied pressure when the discrete damage initiates, it continues the pattern of growth and arrest during oxidative aging.

10.7 Concluding Remarks and Future Research

Modeling oxidation-induced damage entails simulation of morphological changes (growth of oxidation layers), oxidation-induced stress states, and damage evolution. The three processes are closely coupled with each mechanism influencing the other. Particularly, oxidation accelerates damage growth due to the presence of a low toughness of oxidized zones and new crack surfaces accelerate the oxidized layer growth.

At the constituent scale, the oxygen diffusion, reaction, and oxide conversion behavior depends upon the chemical composition and molecular structure of the resins and fibers. Three different resin systems studied showed the relative dominance of diffusion and reaction rates on the oxidation layer sizes. The behavior in carbon fiber is diffusion dominant at low temperatures and reaction rate is relatively minimal. Estimates for the oxygen diffusivity and reaction rates for low-temperature oxidation of carbon fibers are summarized from published literature.

At lamina scale, the oxidation growth is divided into time periods and regions where damage is not onset and there are areas with discrete crack growth. In the areas without damage, the oxidation layer growth can be simulated using represented volume elements and constituent behavior models. Indirect methodologies for determining the fiber diffusivity at the aging temperature from the oxidation layer size observations in axial and transverse directions are shown. It is seen that the fiber oxygen diffusivity is orthotropic with higher axial diffusivity than transverse diffusivity. The oxidation growth is controlled by damage growth after the onset of damage. Damage controlled oxidation growth is observed typically after 500 h of aging.

Use of explicit representation of fibers and matrix regions is impractical at the laminate scale and for composites with 3D reinforcements. Homogenization of sorption, diffusivity, and reaction rates is required for numerical simulations of oxidation growth at the lamina scale. We studied the through-the-thickness heterogeneity observed in laminate oxidation and the effect of stacking sequence on the oxidation growth in each ply. The oxidized region sizes for various ply angles and ply sequences. For complex 3D fabric reinforcements, more sophisticated homogenization techniques are necessary.

When predicting the oxidation growth for long aging times, the coupling between oxidation and damage must be rigorously modeled. The diffusion and fracture toughness behavior of oxidized surface regions play a critical role in the protection of the interior regions of the structure. Generally, the oxidized shells exhibit higher diffusivity and lower fracture toughness accelerating the oxidation growth after several hours of oxygen exposure. Consideration of stress-assisted diffusion terms and the accelerated sorption at high-stress regions requires considerably more parameters in the model and experimental analysis to determine the parameter values. Work is in progress on reduction of the parameters required to simulate stress-assisted diffusion and considering stress coupling terms in the sorption. Modeling the acceleration of oxidation growth near the crack tips enables better prediction of maximum oxidization zone sizes (in regions where discrete damage was observed) at the laminate scale. The damage evolution due to oxidation-induced stresses in complex 3D structures also requires careful consideration of the coupling effects.

Acknowledgments A major portion of this work has been funded by Air Force Office of Scientific Research (AFOSR: Drs. Charles Lee and J. Harrison, Program Managers) and Air Force Research Laboratory at Wright Patterson AFB. Collaboration and interactions with Dr. G. P. Tandon, Dr. G. Schoeppner, Dr. R. Hall, and Prof. J. Whitcomb are deeply appreciated. I gratefully acknowledge the efforts of my students, Alan An, Dr. Eva Yu, and Andong Xu, which led to several insights presented in this chapter.

References

1. Schoeppner, G., G. Tandon, and K. Pochiraju, *Predicting Thermooxidative Degradation and Performance of High-Temperature Polymer Matrix Composites*, in *Multiscale Modeling and Simulation of Composite Materials and Structures*, Y. Kwon, D. Allen, and R. Talreja, Editors. 2008, Springer US. p. 359–462.
2. Verdu, S. and J. Verdu, *A new kinetic model for polypropylene thermal oxidation at moderate temperatures*. Macromolecules, 1997. **30**: p. 2262–2267.

3. Tandon, G.P., K.V. Pochiraju, and G.A. Schoeppner, *Modeling of oxidative development in PMR-15 resin*. Polymer Degradation and Stability, 2006. **91**(8): p. 1861–1869.
4. Pochiraju, K.V. and G.P. Tandon, *Modeling thermo-oxidative layer growth in high-temperature resins*. Journal of Engineering Materials and Technology, Transactions of the ASME, 2006. **128**(1): p. 107–116.
5. Weitsman, Y., *Stress assisted diffusion in elastic and viscoelastic materials*. Journal of the Mechanics and Physics of Solids, 1987. **35**(1): p. 73–94.
6. Audouin, L., et al., *Review of role of oxygen diffusion in polymer aging kinetic and mechanical aspects*. Journal of Materials Science, 1994. **29**: p. 569–583.
7. Verdu S, V.J., *A new kinetic model for polypropylene thermal oxidation at moderate temperatures*. Macromolecules, 1997. **30**: p. 2262–2267.
8. Colin, X., C. Marais, and J. Verdu, *Thermal oxidation kinetics for a poly (bismaleimide)*. Journal of Applied Polymer Science, 2001. **82**: p. 3418–3430.
9. Colin, X., C. Marais, and J. Verdu, *A new method for predicting the thermal oxidation of thermoset matrices: application to an amine cross-linked epoxy*. Polymer Testing, 2001. **20**: p. 795–803.
10. Colin, X. and J. Verdu, *Strategy for studying thermal oxidation of organic matrix composites*. Composites Science and Technology, 2005. **65**: p. 411–419.
11. Lafarie-Frenot, M.C., et al., *Thermo-oxidation behaviour of composite materials at high temperatures: A review of research activities carried out within the COMEDI program*. Polymer Degradation and Stability, 2010. **95**(6): p. 965–974.
12. Dhami, T.L., L.M. Manocha, and O.P. Bahl, *Oxidation behaviour of pitch based carbon fibers*. Carbon, 1991. **29**(1): p. 51–60.
13. Eckstein, B.H., *The oxidation of carbon fibers in air between 230 and 375*. Fiber Science and Technology, 1981. **14**: p. 139–156.
14. McManus, H.L., B.J. Foch, and R.A. Cunningham, *Mechanism-based modeling of long-term degradation*. Journal of Composites Technology & Research, 2000. **22**: p. 146–152.
15. Schieffer, A., J.F. Maire, and D. Levique, *A coupled analysis of mechanical behavior and aging for polymer-matrix composites*. Composites Science and Technology, 2002. **62**: p. 543–549.
16. Abdeljaoued, K., *Thermal oxidation of PMR-15 polymer used as a matrix in composite materials reinforced with carbon fibers*. 1999, Ecole Nationale Superieure des Arts et Metiers, Paris.
17. Colin, X., C. Marais, and J. Verdu, *Kinetic modeling and simulation of gravimetric curves: application to the oxidation of bismaleimide and epoxy resins*. Polymer Degradation and Stability, 2002. **78**: p. 545–553.
18. Serp, P. and J.L. Figueiredo, *An Investigation of vapor-grown carbon fiber behavior towards air oxidation*. Carbon, 1997. **35**(5): p. 675–683.
19. Sullivan, R., *A model for the oxidation of carbon silicon carbide composite structures*. Carbon, 2005. **43**(2): p. 275–285.
20. Glime, W.H. and J.D. Cawley, *Oxidation of carbon fibers and films in ceramic matrix composites: A weak link process*. Carbon, 1995. **33**(8): p. 1053–1060.
21. Baker, A.A., et al., *Oxidation of Aluminum-Coated Carbon Fibres and Carbon-Aluminum Composites*. Fiber Science and Technology, 1972. **5**: p. 285.
22. Piquero, T., et al., *Influence of carbide coatings on the oxidation behavior of carbon fibers*. Carbon, 1995. **33**(4): p. 455–467.
23. Wang, H., P. Gao, and Z. Jin, *Preparation and oxidation behavior of three-dimensional braided carbon fiber coated by SiC*. Materials Letters, 2005. **59**(4): p. 486–490.
24. Singer, L.U. and S. Mitchell, *Diffusion of oxygen into pitch*. Carbon, 1997. **35**(5): p. 599–604.
25. Tang, L.Q., et al., *Three-dimensional transient mold cooling analysis based on galerkin finite element formulation with a matrix-free conjugate gradient technique*. International Journal for Numerical Methods in Engineering, 1996. **39**(18): p. 3049–3064.

26. Heath, M.T., *Scientific Computing: An Introductory Survey (2nd edn)*. 2002: McGraw-Hill: New York.
27. Shewchuk, J.R. *An Introduction to the Conjugate Gradient Method without the Agonizing Pain* 1994 [cited 2007 April 30, 2007]; Available from: http://www.cs.cmu.edu/~quake-papers/painless-conjugate-gradient.pdf.
28. Barrett, R., et al., *Templates for the Solution of Linear Systems: Building Blocks for Iterative Methods (2nd edn)*. . 1993, Society for Industrial and Applied Mathematics.
29. Jimack, P.K. and N. Touheed, *Developing parallel finite element software using MPI*. In High Performance Computing for Computational Mechanics. B.H.V. Topping and L. Lammer (editors) 2000: Saxe-Coburg Publications, p. 15–38.
30. Bellenger, V., J. Decelle, and N. Huet, *Ageing of a carbon epoxy composite for aeronautic applications*. Composites Part B: Engineering, 2005. **36**(3): p. 189–194.
31. Yu, Y.T. and K. Pochiraju, *Characterization of temperature-dependent moisture diffusivity in PMR-15 resin*. Journal of Applied Polymer Science, 2007. **106**(2): p. 1281–1290.
32. Ripberger, E.R., G.P. Tandon, and G.A. Schoeppner. *Charaterizing the Oxidative Degradation of PMR-15 Resin*. in *Sampe 2004 Symposium and Exhibition*. 2004. Long Beach, CA: SAMPE.
33. Schoeppner, G.A., G.P. Tandon, and E.R. Ripberger, *Anisotropic oxidation and weight loss in PMR-15 composites*. Composites Part A: Applied Science and Manufacturing, 2007. **38**: p. 890–904.
34. Pochiraju, K. and G.P. Tandon, *Interaction of oxidation and damage in high temperature polymeric matrix composites*. Composites Part A: Applied Science and Manufacturing, 2009. **40**(12): p. 1931–1940.
35. Tandon, G.P., *Characterization of Thermo-oxidation in Laminated and Textile Composites*, in *Long-Term Durability of Polymeric Matrix Composites*, K. Pochiraju, G.P. Tandon, and G.A. Schoeppner, Editors. 2011, Springer Science+Business Media, LLC, New York, NY.
36. Pochiraju, K.V., G.P. Tandon, and G.A. Schoeppner, *Evolution of stress and deformations in high-temperature polymer matrix composites during thermo-oxidative aging*. Mechanics of Time-Dependent Materials, 2008. **12**(1): p. 45–68.
37. Oliver, W.C. and G.M. Pharr, *Measurement of hardness and elastic modulus by instrumented indentation: Advances in understanding and refinements to methodology*. Journal of Materials Research, 2004. **19**(1): p. 3–20
38. Wise, J., K.T. Gillen, and R.L. Clough, *Quantitative model for the time development of diffusion-limited oxidation profiles*. Polymer, 1997. **38**: p. 1929–1944.

Chapter 11
Composite Structures Durability Design and Substantiation

Gregory A. Schoeppner and Matthew S. Tillman

Abstract An overview of the durability design requirements and substantiation approaches, with an emphasis on aerospace applications, is provided in this chapter. The service history of both commercial and military composite aircraft structures has provided numerous examples of both good and bad designs from which current design specification and guidance are derived. Issues to be considered in the design of composites used in primary and secondary composite structures including corrosion prevention measures associated with joining composite and metallic components are described. Design details, material selection, and demonstration that designs meet performance criteria are dependent on a structure's thermal and mechanical loading environments, service and economic life requirements, manufacturing constraints, and inspectability requirements. Based on an assessment of the durability of composite materials at the coupon level through modeling and testing, a description of the building block approach used to validate durability predictions of elements, subcomponents, components, and full-scale structural testing is given. The unique requirements for developing load spectra for accelerated full-scale durability testing for both composite and combined composite/metallic structures are discussed.

11.1 Introduction

The long-term durability of aerospace, maritime, and land-based platforms is directly dependent on material choices and structural details established during design. Much of the current guidance for material selection, design, and manufacturing detail is based on historic lessons learned from what worked in

G.A. Schoeppner (✉)
US Air Force, Air Force Research Laboratory Materials and Manufacturing Directorate, Wright-Patterson AFB, Dayton, OH 45433, USA
e-mail: greg.schoeppner@wpafb.af.mil

K.V. Pochiraju et al. (eds.), *Long-Term Durability of Polymeric Matrix Composites*,
DOI 10.1007/978-1-4419-9308-3_11,

previous designs as well as shortcomings from deficient designs. Opening the design space to include new materials and structural configurations increases the risk of unexpected deficiencies in the designs and/or cost and schedule risks to substantiate the design. This was the case for some of the earliest applications of polymer matrix composites in aerospace structures. In particular, the environmental susceptibility of polymer composite materials to hygrothermal degradation, moisture ingression, and galvanic corrosion as well as their susceptibility to impact damage were some of the leading causes of early composite durability and sustainment problems.

This chapter focuses primarily on the application of composite materials in aerospace structures. However, the design guidance and corrosion prevention measures are applicable to other uses of composites in structural and nonstructural applications. The terms 'durability and damage tolerance' are sometimes used interchangeably and incorrectly; therefore their distinct definitions are given here. *Durability* is the ability of a structure to resist degradation, such as fatigue, environmental deterioration, and accidental damage to the extent that it can be controlled by economically acceptable inspection and maintenance programs [1]. *Damage tolerance* is the attribute of a structure that permits it to safely retain its required residual strength for a period of unrepaired usage after the structure has sustained specific levels of fatigue, corrosion, accidental, and/or discrete source damage [2]. Both of these terms are often confused with *Damage resistance*, which is the inherent ability of a structure to endure typical service without incurring damage requiring maintenance action. Durability ensures that a structure can be economically sustained and damage tolerance ensures structural safety in the presence of undetected flaws or damage. While this chapter focuses on durability considerations for the design and substantiation of polymer composites, it would be amiss to discuss durability without discussing its inherent relationship to damage tolerance.

Design guidance for metallic structures benefits from fatigue work dating back some 170 years, culminating in mature methods for robust stress concentration and fracture mechanics analyses. Due to inherent differences in the behavior of composites and metals, successful application of metallic design and analysis tools to composites is limited. These differences include: fatigue characteristics (scatter factors, fatigue testing approaches, etc.); environmental susceptibility (moisture, heat, ultraviolet radiation, etc.); prediction of stress concentration factors (anisotropic effects, stress and strain gradients, through thickness properties, etc.); and damage susceptibility (low-velocity impact, free-edge delamination, etc.). These fundamental differences in material behavior have dictated the need to develop composite specific design and testing approaches. Furthermore, the lack of robust analysis tools to predict damage initiation and growth is manifested in the durability and damage tolerance design and validation approaches and requirements.

Three aspects of durability of polymer composites are discussed in this chapter. *Design* practices to assure long-term durability must be considered during all design and manufacturing phases. *Qualification* is the process of establishing that a material is suitable for a given application and is of sufficient quality. This involves testing, analysis, and establishment or confirmation of material standards. Manufacturing processes, type of loading, operating environment, and material inspectability must be considered during material selection and integration with structural design. Statistically based material properties developed during qualification are used in conjunction with analytical tools to calculate design allowables that size the structure. *Substantiation* of designs involves testing and analysis to verify that an element, component, or full-scale structure is in compliance with the structural design criteria. This chapter describes several design substantiation methods for aerospace composite structures that are an inherent part of the structural certification process. Due to the differences in metal and composite design approaches, testing requirements differ substantially. Substantiating the design of hybrid structures – structures that include both metallic and composite materials – poses challenges to the certification community. In recent years, hybrid structures have become prevalent in both commercial and military aircraft; however, generally accepted methods to substantiate hybrid structure durability are still a matter of debate.

11.2 Design Drivers

The design of aerospace structures is multifaceted, requiring detailed knowledge of the anticipated service environment. Figure 11.1 identifies ten design criteria for aerospace structures, each of which must be addressed to meet service life requirements. All of the criteria are interdependent, so each criterion cannot be evaluated in isolation. It is this interdependence that makes establishment and substantiation of design requirements difficult. Producibility is included here since design details and tolerances must take into account the material form

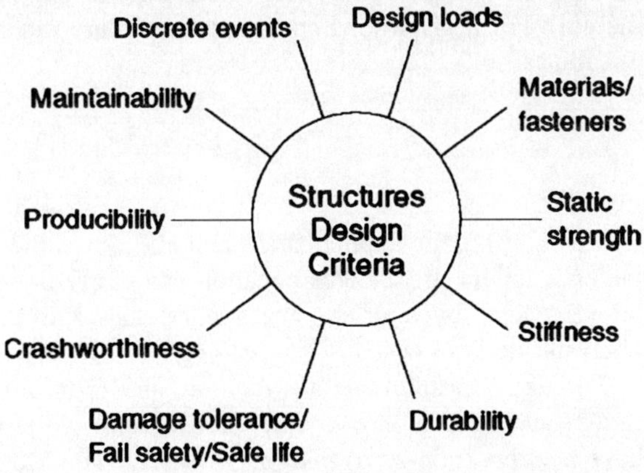

Fig. 11.1 Aerospace structural design criteria [3]

and the type of manufacturing. For instance, composite structures made with unidirectional tape, 2-dimensional weaves, or 3-dimensional textiles all have unique durability properties. In addition, composites manufactured with prepreg, Resin Transfer Molding (RTM), Vacuum Assisted RTM (VARTM), Resin Infusion (RI), and other processes will each have unique durability properties and typically require a separate material qualification effort.

Material qualification is a process designed to produce a quantitative assessment of the ability of a material to meet performance requirements, and the establishment of verification test values to ensure that material properties are controlled and consistent. This process includes the development of statistically based material properties that are temperature and moisture dependent. For metallic safety-of-flight fixed-wing aircraft structures, A-basis allowables (95% confidence that 99% of strengths lie above the allowable value) are required whereas for composites, B-basis allowables (95% confidence that 90% of strengths lie above the allowable value) are often used. For spacecraft structures, A-basis allowables are often used since many spacecraft do not feature redundant load paths to ensure fail safety. Since the strength of composite materials and structures are dependent on the fiber form, lay-up, geometry and type of loading, the B-basis allowables must include these factors [4–6]. Unlike metallic allowables, which are limited to a few key properties, composite design allowables provide validated test data over a range of laminate families (percentages of 0°, +/−45°, and 90° plies), stress concentrations (hole size effects, hole spacing effects, bending–tension coupling, load axis effects), and configuration effects (fasteners, joint configuration, and bearing-bypass load distribution). In total, a complete design allowables data set can require several thousand test specimens per material form.

For aerospace structures, the criteria to meet strength, stiffness, durability, and damage tolerance requirements can be at odds with overarching system requirements such as meeting range or payload requirements that may necessitate minimum weight designs. Commonly, structural sizing is done for a wide variety of parameters and failure modes. In each of these cases, the basis for comparison is the dimensionless parameter referred to as margin of safety (MS, or simply margin). For a given load condition and failure mode, the MS can be calculated as follows:

$$MS = \frac{\varepsilon_{applied}}{\varepsilon_{allowable}} - 1 \tag{11.1}$$

where $\varepsilon_{applied}$ is the applied strain and $\varepsilon_{allowable}$ is the strain allowable for a given material failure mode. This equation can easily be rewritten for stress- or load-based designs by replacing the applied and allowable strains with applied and allowable stresses or loads, respectively.

For each location on a structure, one criterion will typically govern the design and have a zero or near-zero margin while non-governing criteria will have positive (non-zero) margins of safety. The MS are based on analysis, with critical margins validated through test at multiple scales. Designs based on zero or near-zero margins of safety introduce risk associated with the accuracy with which predicted design loads represent actual in-service loads and fidelity with which the material basis values and analysis tools accurately represent the as manufactured material behavior and structural configuration. This is particularly true for composite materials when calculating margins based on prediction of failure initiation or ultimate strength because composite analysis tools are not as mature and reliable as those for predicting onset of yielding (plastic deformation) in metals.

Composite damage tolerance and fatigue performance is strongly dependent on structural design details (e.g., skin laminate stacking sequence, stringer or frame spacing, stiffening element attachment details, damage arrestment features, and structural redundancy) [7]. It is common in the aerospace industry that predictions of aircraft weights are underestimated during the preliminary design phase. As designs mature to include appropriate structural features and details, historically structural weight tends to grows. These "overweight" designs subsequently require costly weight reduction programs late in the design phase. As a result, the system requirements for reduced weight are often achieved at the expense of durability and damage tolerance [1]. Similarly, there are concurrent trade-offs with the other system and structural design criteria such as minimization of cost (typically in manufacturing procurement cost not life cycle costs).

11.2.1 Material Environmental Requirements

Before a material is selected, it is critically important to understand the application loading environment and expected service life. Unfortunately, material selection typically takes place early in the design phase when the design loads and operational environments are not fully defined. However, similar applications of materials or a class of applications can provide foundational information for selecting materials. As an example, cyclic loading requirements will significantly differ amongst applications. A typical number of fatigue cycles in a lifetime for a fixed-wing air vehicle rarely exceed 1 million cycles, whereas dynamic components in rotorcraft tend to be loaded at frequencies that represent 7 million to 70 million cycles per 1,000 operating hours [8]. Exceptions for fixed-wing structures can include buffet-dominated structure such as vertical and horizontal stabilizers. Although the rotorcraft hubs and rotors are subjected to much higher frequency of loading, they experience a much narrower distribution of loads thanequivalent structural members on fixed-wing aircraft. Rotorcraft rotors typically operate in a centrifugal-force tension-dominated stress field while airplane wings and fuselages experience combined tension, compression, and shear induced by structural torsion and bending. In contrast, wind power turbine blades operate at relatively lower frequencies with life requirements of up to 30 years, accumulating on the order of 10 million cycles [9]. The type, magnitude, and frequency of loading, along with cyclic life requirements, need to be primary considerations when selecting the resin and fiber systems and the form of fiber reinforcement.

It is well documented that polymer matrix composites are susceptible to moisture and temperature-induced degradation. For aerospace structures, the stability and properties of resins at the combined humidity and temperature extreme are often controlling factors for composite material selection and vehicle design. Depending on the application, numerous material hygrothermal properties or characteristics may need to be determined including:

• Thermal expansion coefficients	• Hygrothermal diffusivity
• Thermal conductivities	• Thermal oxidative stability
• Heat capacity/specific heat	• Thermal decomposition
• Thermal mechanical stability	• Flammability/fire resistance
• Cure transition (gelation) temperatures	• Combustion products/smoke toxicity
• Glass transition temperature	• Moisture saturation

The design environment should be derived from projected operational use temperatures and moisture conditions ranging from dry to the most critical of the possible moisture conditions and the allowables should include the effects of these critical operational environments [10].

Beyond use temperature and hygrothermal stability, the material selection process should evaluate a material's suitability for the detailed design, producibility, safety, and life expectancy. From the list above, the heat capacity and cure transition temperatures of the material provide information about its producibility by identifying curing characteristics which may have a higher importance for large and thick structures. The Navy has particular safety concerns about flammability and smoke toxicity, so fire resistance and combustion products for ship fires are a primary concern. For composites mechanically attached or bonded to dissimilar materials, the thermal expansion coefficients need to be considered for thermal strain mismatch. Finally, for high-temperature applications, the life expectancy of the material will be dependent on the thermal mechanical and oxidative stability as well as the thermal decomposition temperature.

Heat sources within an aircraft (engines, auxiliary power units, electronics racks) and external to the aircraft (ground soak thermal radiation, aerodynamic heating) must be a major consideration during material selection. As an example, Table 11.1 shows the specific hot temperature values used for designing a light air/ground fighter [10]. These temperatures are based on temperature surveys for a class of aircraft accounting for global temperature data such as those found in MIL-HDBK-310 [11]. If materials are to be used in a new type of application or the environment is not fully defined, a survey of the environment for a similar application may be used or the environment may be derived from projected operational usage. The material selection and design must include provisions for the cumulative effects of the temperature/load history for its planned service life.

Table 11.1 Light fighter hot temperature values [10]

Area	Maximum temperatures		
Description	Soak	Sustained	Intermittent
Forward avionics bay		71°C (160°F)	95°C (203°F)
Gun bay		85°C (185°F)	99°C (210°F)
Nose wheel well		71°C (160°F)	95°C (203°F)
Cockpit area	93°C (200°F)	54°C (130°F)	
Conditioning compartment			102°C (215°F)
External skin		85°C (185°F)	99°C (210°F)
Engine bay and tail cone		121–482°C (250–900°F)	
Drag chute container		85°C (185°F)	121°C (250°F)

To eliminate the need for viscoelastic-based analysis, adequate margin must be maintained between the highest anticipated use temperature and the temperature at which the material properties begin to rapidly degrade. Often this inflection point can be estimated using the wet glass transition temperature (Tg) of the material. Design guidance recommends the maximum sustained use temperature to be a minimum of 28°C (50°F) below the wet glass transition temperature of the material [10]. Materials can be used closer to the wet Tg provided that sufficient statistical data of critical resin-dominated properties such as shear, interlaminar tension, notched compression are generated and the service temperature ceiling is well defined [12]. For the light fighter application in Table 11.1, the external skins will require a material with a 113°C (235°F) wet Tg using the recommended guidance. It should be realized that the worst case environment may not be the same for all structural details, e.g., hot–wet conditions can be critical for some failure modes, while cold–dry conditions may be critical for others [7]. Material properties should be determined for the entire range of expected service temperatures including the lowest expected service temperature. While material testing is conducted at −54°C (−65°F) for typical commercial and military aircraft, very high-altitude subsonic aircraft such as the U-2 and RQ-4 require much lower test temperatures to adequately simulate the cold soak environment at higher altitudes. As a minimum, cold temperature dry (CTD), elevated temperature dry (ETD), and elevated temperature wet (ETW) allowables should be determined. Room temperature dry (RTD) and room temperature wet (RTW) allowables are typically determined as well.

Other application-dependent environmental factors may have a primary influence on the selection of the composite fiber and resin systems. The resistance of composites to the environmental exposures/conditions identified in Fig. 11.2 that the material may be exposed to during maintenance operations or field usage should be determined. Inadvertent exposures to mud, dirt, and other contaminates (including spills of chemicals) inside cargo areas of aircraft, ships, or ground vehicles resulting from loading, carriage, and unloading of cargo may be significant.

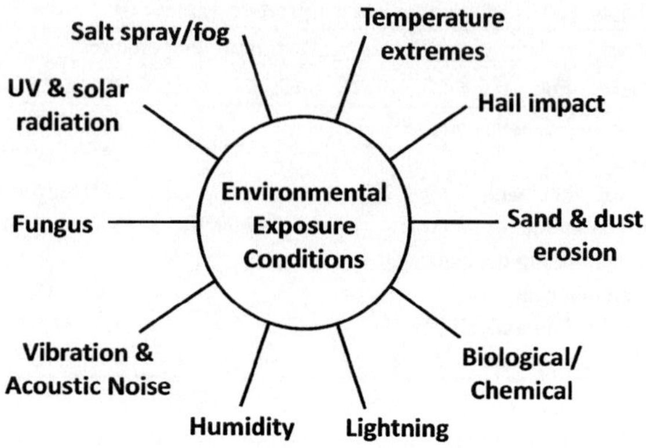

Fig. 11.2 In-service environmental exposures

The resistance of materials to chemicals used to service, maintain, and operate vehicles including solvents, cleaners, fuels, hydraulic fluids, lubricating oils, and anti-icing fluids can be ascertained on small coupon specimens. As part of the material test program, it is important to establish the specific list of fluids and the exposure conditions (temperatures and exposure durations) to be tested. Traditionally, the hot/wet environment has proven to be the most degrading for composite strength, thus designing to the hot/wet allowables usually covers the detrimental effects from other fluids; however, this must be validated through testing. However, some synthetic fibers such as Kevlar® can significantly degrade in hot–wet environments. For most environmental concerns, selecting tests that measure the matrix-dominated response such as shear and notched compression provides the most valuable information.

Environmental degradation often results from long-term exposure that cannot readily be replicated in short duration tests. If accelerated testing methods are used to characterize materials, there must be evidence that the results can be correlated to real-time environmental and load exposure. The only way to obtain such correlation is by conducting exposure tests in the natural environment. Therefore, if verified methods to correlate accelerated tests to real-world exposure do not exist, long lead times are need to characterize long-term exposure degradation [7, 10].

11.2.2 Damage Resistance and Tolerance

Discussions of design criteria warrant definitions of classes of structure. Primary structure is structure that carries the main operational loads and whose failure would reduce structural integrity. In the aviation community, primary structure is also commonly referred to as safety-of-flight structure and flight safety structure. It carries flight, ground, and pressurization loads, and includes specialized features to assure crashworthiness. The remaining structure is referred to as secondary structure and includes components such as the shrouds and fairings. These items may be important for the aerodynamic performance of a vehicle, or may be necessary for a vehicle to execute a particular component of its mission (e.g., pylons for the carriage of external stores), but their failure will not result in immediate impairment of safe flight.

The damage tolerance philosophy currently employed by the Air Force on most aircraft requires that safety-of-flight structures be designed as either (1) slow damage growth structure or (2) fail-safe structure [2]. The requirement for slow damage growth structure is that damage growth is predictable and that it is designed to avoid growth to critical damage size growth for specified periods of unrepaired service usage. The unrepaired service usage time depends on the structure's inspectability. Fail-safe structure is designed such that it retains its required residual strength for a period of unrepaired usage after the failure or partial failure of safety-of-flight structure. Use of the fail-safe criterion implies that failure will eventually occur, will initially be undetected, and therefore must be designed to occur in a manner that is not catastrophic. A fail-safe structure is achieved through either a multiple load path design or a crack arrest design. Multiple load path designs have redundant structures which – in the event of failure of one of the structural members – allow for the redistribution of applied loads to other load-carrying members for a specified period of time. Crack arrest designs include embedded structural features that arrest failure propagation, allowing the structure to retain its required residual strength (not necessarily ultimate load) for a period of undetected damage and unrepaired usage. The ability to accurately predict damage growth in composite structures is still an emerging/developing technology; therefore, it is not typically used as part of certification.

An alternative to the slow damage growth and fail-safe design philosophy is the safe-life design philosophy, used predominantly by the Navy for airworthiness certification. Safe-life philosophy, applicable to safety-of-flight structures, requires that a structure is designed (and subsequently validated) for no failures within a specified design life (including a defined reserve life). At the end of its design life, the component is removed from service. From an operational standpoint, safe-life designs can dramatically reduce the burden of inspections. However, safe-life designs may include overly conservative designs and removal of structures with significant residual life, both of which may be uneconomical. Finally, the success of safe-life designs is intimately linked to the quality of analysis methods, loads prediction, and correct determination of operational usage methods before first flight as well as a robust in-service program for tracking the actual usage of individual aircraft.

Testing and analyses (typically fatigue analysis) are used to provide sufficient confidence that a component will meet its expected life. Since fatigue failure is a stochastic process, a safety or scatter factor (a life-reduction factor) is included in the analysis when estimating how long a structure can be in service before it is expected to fail. Safe-life designs are often employed in structures that are very difficult or uneconomical to inspect or repair. As an example, aircraft landing and arrestment gear components are typically designed using a safe-life approach due to inspection and repair limitations. Prior to the mid-1980s, it was typical that all

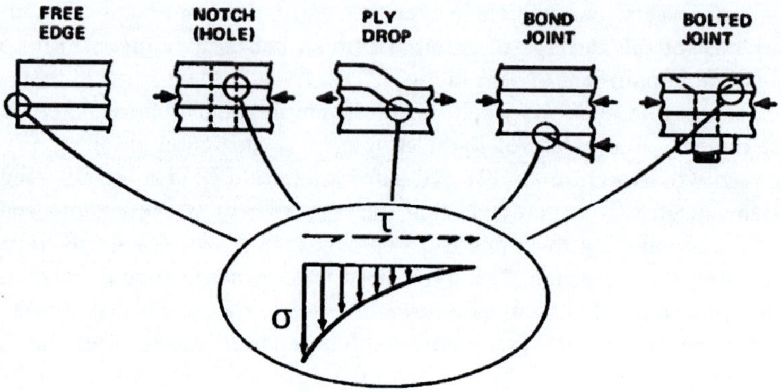

Fig. 11.3 Sources of out-of-plane loads in design [13]

fatigue-critical components of rotorcraft were designed using safe-life approaches. These components were removed from service at or before its defined service life so that the probability of a fatigue failure was remote [13]. Since 1989, amendment 28 to FAA FAR Part 29.571 for civilian transport rotorcraft requires flaw tolerant safe-life or damage tolerance (DT) substantiation for all composite rotor system components [14]. Although the US Army is permitted to use either a safe-life or a fail-safe design [15], safe-life designs and substantiation are used for all rotorcraft dynamic components.

Application of damage tolerance and safe-life design philosophies are typically limited to primary structures since, by definition, failure of secondary structure is not a safety-of-flight issue. However, durability considerations are applicable to both primary and secondary aircraft structures because all structures have economic considerations for operational maintenance. It is important to note that these designations and design drivers are derived from the function of the structure within the air vehicle structural system, rather than being driven by damage modes to which the structure may be susceptible. For instance, composite rotor blades operating in a tension-dominated strain field often exhibit benign and non-catastrophic failure modes such as delaminations and skin cracking that are nonstructural and easily repaired in most cases [14]. This type of damage for rotor blades is a durability issue rather than a damage-tolerant safety issue. In contrast, the case of a composite wing that operates in a tension–compression–shear field having a similar type and size of delamination, could be a safety-of-flight damage tolerance concern since a delamination in a compression field can grow and lead to catastrophic failure.

A primary means of obtaining durable and damage-tolerant designs is to minimize geometric stress concentrations and the effect of inherent material weaknesses on key structural details. It is well documented that laminated polymer composites are intrinsically weak in the out-of-plane direction and susceptible to delamination [16]. Figure 11.3 shows the most common sources of out-of-plane loads. It is impractical to avoid many of these structural details and even under in-plane loads,

Table 11.2 Design ultimate strain levels for graphite/epoxy structures [17]

Considerations	Condition	Results	Ultimate strain level με
Open hole (unreinforced)	Static tension and compression loading	40–60% strength reduction	(0.5 × 9,000) 4,500
90° Ply off-axis loading	Onset of first ply cracking	3,000 μcm/cm	(1.5 × 3,000) 4,500
Fatigue testing (quasi-isotropic laminates)	Tension–compression cycles $R = -0.5$ to -1.0	10^5 to 10^6 cycles at 3,300 μcm/cm	(1.5 × 3,300) 5,000
Non-detectable impact damage	Hail, tool drops, etc.	30–60% strength reduction	(0.55 × 9,000) 5,000

these details give rise to interlaminar normal and shear stresses. However, in some circumstances, actions can be taken to reduce their influence by minimizing loads in the vicinity of the stress concentrations or by designing the critical loads in tension rather than compression (which may result in delamination and local buckling). Tapering the thicknesses of composites using internal ply drops as opposed to external ply drops is one of the design features that may contribute to both durability and damage tolerance. However, mechanical fastening of composites entails unavoidable stress concentrations that often drive limits on design ultimate strains and minimum gage thicknesses. This is evidenced by the simple fact that the vast majority of specimens tested in the development of composite design allowables are those designed to evaluate the open and filled hole tension and compression, fastener bearing, and fastener pullout strengths of the material.

Design ultimate strains for composite structures are restricted by many durability, damage tolerance, and repairability concerns including joints and splices, impact damage, and transverse cracking in 90° plies. For an aerospace grade graphite/epoxy material, typical un-notched quasi-isotropic laminates have failure strains on the order of 9,000 με. Table 11.2 summarizes the strain-level restrictions needed to avoid damage initiation and growth during service. The open hole strength of composites, which is dependent on layup and stacking sequence, is 40–60% less than un-notched strength restricting the design ultimate strain to about 4,500 με. The strain level for cracking of 90° plies varies considerably, depending on curing stresses, moisture content, layup, etc. However, 3,000 με is the order of magnitude for onset of first ply cracking in well-designed and cured laminates. The reduction in strength resulting from impact damage (even below the visible detectable level) can be on the order of 50%. Impact damage can reduce both the tensile and compressive strengths; however, the compressive properties are more adversely affected due to the propensity for local and global buckling. Since damage of this type is likely to occur during the structure's lifetime, the design ultimate strain can be restricted to as much as 50% of that for an unflawed laminate [17].

Overall, the design ultimate strains are limited to about 4,500–5,000 με, and practical working strain levels (limit load levels) are limited to 3,000–3,500 με. The magnitude of Table 11.2 design strain restrictions should not be used as a general rule-of-thumb because specific limitations are material and application dependent. Many of the tests performed in materials qualification programs (notched strength, fastener bearing and pullout, compression strength after impact and un-notched fatigue) provide material-specific design strain limitations. However, it is worth noting the clear penalty paid in structural design for the inclusion of stress concentrations and for the incorporation of damage-tolerant and durable features. Interestingly enough, many of these considerations are not penalties at all. In the design of modern aircraft, many of which have composite skins and metallic substructure, it is important to moderate the design strains of the skins so as not to rapidly accelerate fatigue damage in the substructure. From this perspective, the 3,000–4,000 $\mu\varepsilon$ skin strain region often provides a balance of capability against weight growth in substructure.

The ultimate strain levels, susceptibility to impact damage, tolerance to manufacturing defects, resistance to fatigue damage, manufacturability, and durability are some of the considerations for selecting the minimum gage thicknesses for sandwich skins, fuselages, etc. For sustainment, repairability can also drive minimum design thicknesses to allow for notched strength reductions, fastener bearing, and pull-through for bolted joint repairs [7]. Of all these considerations, the most critical will be used to establish the minimum structural thickness requirements.

11.2.3 Manufacturing and In-Service Inspection

Residual strength substantiation of damage-tolerant designs requires demonstration by either test or analysis supported by test, that safety-of-flight structure is capable of carrying ultimate load when it contains damage or manufacturing flaws up to the threshold of detectability [18]. Inherent in the residual strength requirement is that it must be met when the structure is subjected to the worst case combined service temperature and moisture conditions. Additionally, durability substantiation requires that the structure be designed for two lifetimes of loading with the maximum size of undetectable damage without the damage growing to a critical size. Durability and damage tolerance design requirements are closely linked to the capability and fidelity of the Non-Destructive Inspection (NDI) equipment and procedures that define the range of detectable damage. High-fidelity inspection capability with demonstrated high probability of detecting small-size damage can potentially result in higher design allowables and lighter structure. If the inspection capability is limited to detecting relatively large size damage, the damage tolerance design requirements will lead to lower design ultimate strain allowables resulting in heavier structure. Therefore, structural designs and part geometries conducive to the type of inspection equipment and techniques used for field inspections are desirable to maximize probability of damage detection. Although large unitized structures may be desirable from a manufacturing and cost standpoint, unitized designs must be tempered with the need to inspect critical interfaces and locations

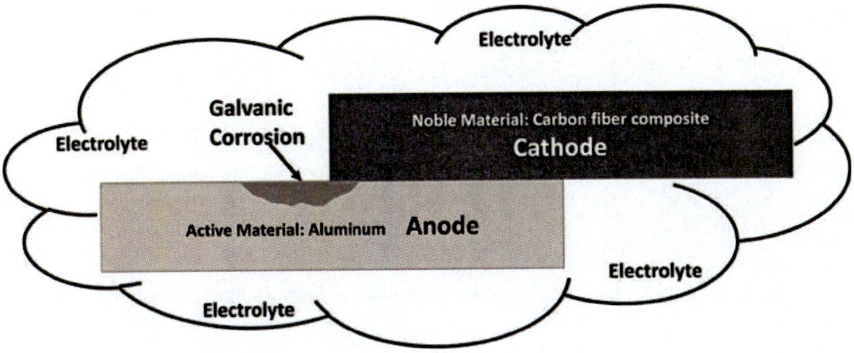

Fig. 11.4 Basic galvanic corrosion

that could have limited inspectability within the structure [19]. In many cases, the need for post-manufacture and in-service inspections of interior structure drives the inclusion of access covers, selection of fastener types, and structural interchangeability requirements – all of which add weight to the structure.

The extent of initially detectable damage used for durability and damage tolerance designs should take into consideration the manufacturing processes and manufacturability of the components. Inherent manufacturing flaw sizes vary with manufacturing processes and defining acceptable flaw sizes less than the expected inherent flaw size may lead to low manufacturing yields and increased actions to disposition parts. Also, the maximum permitted flaw size should be consistent with inspection methods used by trained inspectors during scheduled maintenance in conditions representative of operational maintenance environments. The vehicle system maintenance plan should include an inspection program that establishes inspection intervals, methods, and extent of inspections. Consistent with metallic structures, inspection intervals should be established such that the damage will be reliably detected between the time it initially becomes detectable and the time at which the damage extent reaches the requirement limits for residual strength capability [7]. Data and analysis of the probability of detection and probability of occurrence for different damage sizes and location on the structure helps define this inspection interval.

11.2.4 Corrosion Prevention and Control

Corrosion is the environmental deterioration of materials. Common types of corrosion for metals include pitting, erosion, uniform, galvanic, and crevice corrosion. Galvanic corrosion is the most frequently encountered form of corrosion associated with polymer matrix composites. The basic mechanism of galvanic corrosion is illustrated in Fig. 11.4. Materials with higher electrochemical potential are considered to be more noble, whereas materials with lower potentials are considered to be

More Noble Most cathodic or passive (+) ↑	Platinum Graphite Titanium Stainless Steel (passive) Nickel Copper Brass Tin Lead Stainless Steel (active) Iron Mild Steel Aluminum Alloys Magnesium Alloys
↓ Most anodic or active (-) Less Noble	

Fig. 11.5 Galvanic series of various metals exposed to saltwater

more active. When the materials are electrically connected in the presence of an electrolyte (a solution with capability to conduct electricity), the noble material acts as the cathode and the active material acts as the anode [20]. The anode (i.e., aluminum) is dissolved during this electrochemical process causing it to become brittle and crack. Water typically contains dissolved ionic compounds such as salts that can make it an excellent electrolyte. Therefore, the presence or absence of water and its soluble contents can greatly affect galvanic corrosion.

The galvanic series, such as that shown in Fig. 11.5, is a listing of metals and alloys based on their order and tendency to independently corrode in a particular electrolyte solution. The tendency for corrosion is related to the electric potential of the metal. Metals closely positioned in the series have electric potential nearer one another whereas the greater the difference in potential, the more significant the corrosion effects. Methods such as that found in ASTM G82 can be used to predict the rate of corrosion [21]. Graphite and composites containing graphite fibers behave as noble materials and when in electrical contact with a less noble metal such as aluminum alloys, can cause substantial corrosion damage to the aluminum.

Galvanic corrosion can occur between carbon fiber composites and a metal, even when no fibers are initially exposed on the surface of the composite. Corrosion occurs when moisture is absorbed by the composite in-service, thereby affecting electrolytic resistance of the resin and transporting ions to the graphite fibers. It was estimated that after 200 days of seawater exposure, 15% of the planar area of a graphite/epoxy composite specimen contained seawater-exposed graphite fibers [22]. Galvanic corrosion between dissimilar materials can be controlled by one or more of the following:

- Minimize differences in electrochemical potential between attached materials
- Electrically isolate dissimilar materials
- Eliminate electrolytes from the environment

- Material Selection
  - Avoid materials unsuitable to operational environment if possible
  - Consider material compatibility
  - Isolate dissimilar materials from each other
- Protective Coatings
  - Use to isolate vulnerable materials from the environment
- Design Geometries
  - Avoid crevices when possible
  - Avoid design features that make it difficult for protective coatings to function (sharp corners for instance)
  - Avoid geometries that unnecessarily trap containments/moisture
- Modify the Environment
  - Consider a design that allows for the modification of the environment to which material will be exposed
  - Dehumidification and sheltering can be effective to modify environment

Fig. 11.6 Precautionary design measures [23]

Given the prevalence of aluminum alloys in aircraft structures, graphite/epoxy to aluminum joints are likely in structures with hybrid construction. In addition, many aircraft operate in areas with significant natural electrolyte presence (especially those employed in coastal or naval service). Therefore, the majority of effective galvanic corrosion control measures are aimed at electrically isolating dissimilar materials as a matter of design and manufacturing practice.

Effective design and implementation of corrosion prevention measures depends on an understanding of the conditions and environments that lead to corrosion. The most effective of these control measures are implemented during design and manufacturing. Figure 11.6 identifies four categories of corrosion prevention measures that may be tailored to the operating environment of the system. Galvanic corrosion of polymer matrix composites is fundamentally linked to carbon fibers, and is not a concern for quartz or glass fibers. Due to their insulative properties, quartz and glass can be used as barrier plies to electrically isolate dissimilar materials. For epoxy-based laminates, the glass barrier ply should extend a minimum of 25.4 mm (1 in.) beyond the contact region. For condensation polyimide-based laminates (e.g., bismaleimide and cyanate ester), the glass barrier ply should fully cover the laminate surfaces in contact [23].

In addition to barrier plies on the composite, the fay surfaces of metals should be sealed by painting or coating to insure that they are watertight and resistant to electrical current. It is recommended that the surfaces be sealed with a corrosion inhibiting primer (MIL-P-23377) and/or a corrosion inhibiting sealant (MIL-S-81733 or MIL-S-29574). Fay seals cannot be considered primary seals because they are extremely thin and may be separated due to structural deflections. Therefore, fillet and edge seals, using the same sealant as was used for the fay surface, should be used as the primary seals. Figure 11.7 illustrates bolted joint barrier plies and seals. Sealing can be accomplished using polysulfide sealants that are available

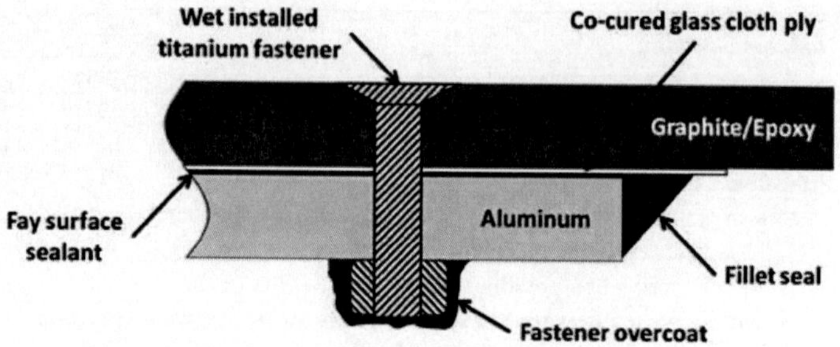

Fig. 11.7 Bolted joint barrier ply and seals

in a variety of product forms with a range of viscosities and cure times. Polysulfides are elastomeric sealants with excellent adhesive characteristics. Polysulfide sealants contain solvents that can degrade some plastics such as acrylic or polycarbonate; however, they contain leachable corrosion-resistant compounds that aid in preventing the aluminum from corroding. Polysulfide sealants are usable in temperatures ranging from −54 to 121°C (−65 to 250°F) with short-term capabilities to 178°C (350°F). If higher temperatures are required, then silicone sealants with temperature capabilities as high as 260°C (500°F) can be used [24].

Conventional aluminum, stainless steel, and cadmium-plated fasteners will corrode when placed in contact with graphite fiber composites. Fastener materials for use in joining one graphite composite structure to another graphite composite structure should be titanium or a corrosion-resistant nickel–copper alloy. Fastener materials for joining a graphite composite structure to an aluminum structure should be titanium alloys or corrosion-resistant steels (CRES) and high-molybdenum stainless steel alloys for high-strength applications. All permanently installed fasteners (particularly, mold-line fasteners) should be wet installed with a corrosion-inhibiting epoxy primer or a corrosion-inhibiting sealant. Wet (liquid sealant coated) installation of fasteners typically involves dipping fasteners in a primer or sealant before installation. Coatings should not be applied to the threaded portion of fasteners if torque requirements were established without the coatings. After assembly, the joint should be overcoated with primer or primer and topcoat corresponding to the finish requirements of the surrounding area [23]. Fasteners should be overcoated after installation to the maximum extent practical. These measures provide good corrosion protection except when load cycling causes relative movement of the fastener and the skin. This movement can compromise the seal allowing moisture to penetrate into the joint and potentially lead to crevice corrosion. In all cases, it is critical to evaluate the propensity of the materials (resin, fiber, fasteners, and metals) to corrode in a realistic assembly configuration.

Consideration should be given to design alternatives that preclude the traditional galvanic corrosion problems created by dissimilar metals for frequently removed items such as doors and panels that cannot be permanently sealed. Removable graphite composite doors/panels fastened to aluminum structure should be avoided if possible, particularly on upper surfaces where moisture/salt spray can potentially migrate through the fastener holes and cause corrosion of the aluminum substructure [10]. Quick release fasteners and removable fasteners penetrating exterior surfaces should be designed and installed so as to provide a seal which prevents moisture or fluids from entering. In some cases, an O-ring rubber seal can be placed on the fastener shank to form the seal [25]. Joints that require separation as part of normal maintenance should use a form-in-place seal with a suitable release agent on one surface substituted for a fay sealant.

Normal operation and maintenance of a structure over time can degrade corrosion control measures. During design, consideration should be given to accessibility for maintenance and corrosion inspection. Particular areas of concern are fasteners, grooves, joints, and removable panels where protection schemes can be compromised. Long-term structural durability can depend on early detection of corrosion before significant damage has occurred. A corrosion prevention and control program [23] should be established as part of the design and manufacturing process and followed through the structure's service life. Additional guidance for corrosion prevention and control can be found in MIL-HDBK-1568 [26] and MIL-HDBK-729 [27].

11.2.5 Minimum Thickness Structures

For lightly loaded structures, the minimum thickness of composite skins is typically governed by durability, damage tolerance, or repairability requirements. Occasionally, supplementary requirements are imposed on thin-skinned sandwich structure to provide additional resistance to damage, thereby alleviating maintenance cost at the expense of weight and manufacturing cost. Sandwich structures with thin face sheets are used in a variety of aircraft structures such as control surfaces, leading edges, fairings, skins, and doors. Sandwich structures, particularly those with low-density core materials, typically have significant maintenance requirements. Largely, this increased maintenance burden is driven by sensitivity to in-service threats such aslow-velocity impact, surface face sheet damage, and moisture intrusion. For example, the level of impact energy which typically must be sustained by honeycomb sandwich control surfaces without requiring repair or allowing moisture ingestion is quite low, e.g., 0.5–0.7 J (4–6 ft lbs) [4]. These components are often designed for minimal weight, resulting in minimal damage resistance. As such, they must be handled carefully to prevent damage during processing, transport, installation, and use.

Sandwich structures with honeycomb and open cell foam cores are very susceptible to water ingress via cracked face sheets, failed sealant, porosity in the face sheet, porous adhesive, and through any intentional penetrations in the structure resulting from modifications. Moisture ingression can lead to face sheet-to-core debonding and bondline degradation and subsequent corrosion of the core material. Fabricating sandwich panels using phosphoric acid anodized corrosion protection for aluminum honeycomb and properly sealing edges has been shown to be successful in preventing corrosion degradation [28]. Prevention of moisture ingression into the core can be facilitated by increasing the face sheet minimum thicknesses to reduce moisture diffusion. The use of surfacing films that can be co-cured with the laminate also should be considered. Paint should not be relied on to provide a moisture barrier as operational experience has shown that paint suffers from erosion and abrasion.

Areas of sandwich panels with thin face sheets located adjacent to supporting fittings are particularly vulnerable to damage during component installation and removal. Consequently, solid laminate construction is commonly used within a reasonable working distance of fittings. The aft 102 mm (4 in.) of trailing edges of control surfaces are vulnerable to damage from ground collisions, handling, lightning strikes, and precipitation-static (p-static) discharges. Repairs in this region can be difficult because both the skins and the trailing edge reinforcing substructure may be involved. A desirable approach for the design is to provide a load carrying member forward of the trailing edge to react loads and have easily repairable material for the trailing edge whose damage will not compromise the structural integrity of the component [4]. The design of sandwich panels is a complex interaction that must seek to balance minimizing weight, recurring production cost, and maintenance burden.

11.3 Damage and Defect Threat Assessment

A durability and damage tolerance design approach recognizes that during normal manufacturing and operation, structures may be subjected to defects and damage from a variety of sources and thus must be designed to operate cost effectively and safely with acceptable levels of damage. A thorough evaluation of the potential manufacturing- and application-specific damage/degradation threats should be considered when establishing structural design details. A damage threat assessment includes categorizing or zoning a region of a structure based on the likelihood that it will be damaged by a particular threat. Potential damage threats including maintenance personnel and tools, runway debris, service equipment, ground hail, in-flight hail, and lightning will have a higher probability of damaging particular regions of aircraft. During initial manufacturing and assembly, composite parts are susceptible to impact damage, scratches, hole drilling damage, etc. Threat assessment should account for part function, location on the airplane, past service data, accidental damage threats, environmental exposure, impact damage resistance, durability of assembled structural details, adjacent system interface (e.g., potential overheating or other threats associated with system failure), and anomalous service or maintenance handling events that can overload or damage the component [7]. The assessment should consider threats ranging from those that cause durability or maintenance concerns to those having a significant effect on residual strength. Understanding the threat and identifying regions susceptible to the threat can prevent the use of overly conservative designs in zones where they are not required.

Table 11.3 Maintenance-induced damage

Damage type	Damage source
Cuts and scratches	Mishandling
Fastener hole wear	Removal/installation of fasteners
Delamination	Tool drop, walk in no-step regions
Edge damage	Mishandling of doors and removable parts
Heat damage	Heating blanket malfunction
Impact damage	Ground equipment impact

11.3.1 Maintenance/Handling Damage

The majority of damage to composite aircraft structures occurs during assembly, routine maintenance, and ground handling [14]. A large percentage of the maintenance-induced composite damage is delamination resulting from impacts and mishandling of components both on and off the aircraft. Delamination damage induced by blunt impact has been reported to reduce the composite laminate tensile strengths by as much as 25% and the compressive strength by amounts in excess of 60% with very little or no external indication of damage [30]. Although the 60% reduction is not typical of commonly used materials and structural configurations, it illustrates the need to account for non-visible damage in design criteria. Since the presence of internal delaminations may not be readily visible, maintenance events that cause damage most often are unreported. In recent years, this has driven several operators to institute on-condition inspections, in which known impact events must be immediately inspected. While this methodology is more likely to catch non-visible damage, it can dramatically increase the time required for maintenance events. Table 11.3 contains a list of common types of damage incurred during routine maintenance handling.

11.3.2 Impact Damage

Service history has established that impact damage is a common threat to the integrity of composite structures. Of primary concern is the effect of undetectable damage on the residual strength. This is particularly true for configurations such as co-cured/co-bonded discretely stiffened structures for which impacts to the external laminate may leave no surface indications of subsurface damage to the stiffeners. The threshold of reliable visual detectability of impact damage is often referred to as barely visible impact damage (BVID). Substantiation that the structure can maintain acceptable strength throughout its service life with BVID involves identification of impact threats often long before any actual service history is available. The impact threat assessment relates the types and severity of discrete events to the probability of their occurrence based on surveys of service history

Table 11.4 Low energy impact (tool impact) [10]

Zone	Damage source	Damage level	Requirements
1. High probability of impact	– 12.7 mm dia. solid impactor – Low velocity – Normal to surface	Impact energy smaller of 8.16 J or visible damage (2.54 mm deep) with min. of 5.44 J	– No functional impairment or structural repair required for two design lifetimes and no water intrusion – No visible damage from a single 5.44 J impact
2. Low probability of impact	Same as Zone 1	Impact energy smaller of 8.16 J or visible damage (2.54 mm deep)	– No functional impairment after two design lifetimes and no water intrusion after field repair if damage is visible

incidents for similar applications. The impact damage criticality for local structural details and the ability to locally inspect for damage must be considered in residual strength substantiation test programs. The impactor size and shape used for residual strength substantiation should be consistent with likely impact damage scenarios that may go undetected for the life of an aircraft [29].

11.3.2.1 Zoning and Impact Survey

Since structures less susceptible to impact do not have to meet the same requirements as those more susceptible to impact, aircraft structures should be zoned according to their probability of impact based on field experience or surveys for similar applications. Damage-susceptible regions of airplanes include the lower fuselage and adjacent fairings, lower surfaces of the inboard flaps, and areas around doors. Often, special attention is given to the regions immediately behind the landing gear due to the possibility of runway debris impact and regions around work stations or built-in work platforms due to the possibility of tool drop. Measures to increase the impact damage resistance for these regions may include additional plies or the use of glass rather than carbon fiber reinforcement. Structure in the wheel well area may require special consideration because of damage susceptibility from explosive tire disintegration (both in flight and on landing). Similarly, structure in the vicinity of the thrust reversers is damage prone due to ice or other debris thrown up from the runway [4].

The US Air Force and Navy in their Joint Service Specification Guidance 2006 (JSSG-2006) [10] implement Table 11.4 to divide structures into two impact zones for durability considerations. Zone 1 contains regions that have a relatively high likelihood of impact damage from maintenance or similar sources. Zone 2 contains regions that have a relatively low probability of the structure being impacted in-service. The damage source, damage level, and specific requirements for these

Table 11.5 Low energy impact (hail and runway debris) [10]

Zone	Damage source	Damage level	Requirements
All vertical and upward facing horizontal surfaces	Hail – 20.3 mm diameter – Sp. Gr. = 0.9 – 27.4 m/s – Normal to horizontal surfaces – 45° angle to vertical surfaces	Uniform density 20.3 mm on center	– No functional impairment or structural repair required for two design lifetimes – No visible damage
Structures in path of debris	Runway debris – 12.7 mm diameter – Sp. Gr. = 3.0 – Velocity appropriate to system	N/A	– No functional impairment after two design lifetimes and no water intrusion after field repair if damage is visible

two zones are used for design evaluation during impact testing. Visible damage as identified in Table 11.4 is damage that is visible to the unaided eye from a distance of 1.5 m (dent depths of 2.54 mm). As part of the impact survey testing, it should be demonstrated that damage not readily visible on the composite surface – less than 2.54 mm (0.1 in.) dent depth – will not result in subsequent degradation of the part, impair function, or require maintenance actions. Although the 2.54-mm dent depth is recommended in JSSG-2006, recent aircraft programs have demonstration that smaller dent depths (on the order of 0.25–1.27 mm) are visible, whereby the design burden to meet durability requirements is lessened. Knowing that dent depth visibility is platform dependent, an assessment of the visibility of damage should be part of the impact survey to define the durability design requirements. The intent is to ensure that costly maintenance will not be incurred due to service exposure. The damage tolerance impact requirements imposed on safety-of-flight structure are much more severe than those identified in Tables 11.4 and 11.5. Additional information can be found in the JSSG-2006 where initial flaws/damage assumptions used to design safety-of-flight structures are identified.

Two additional impact threats that may cause an economic burden or adversely affect safety are ground hail impact and runway debris impact. Table 11.5 identifies the regions of the aircraft that are susceptible to these threats. Design substantiation requires that the structure meets the requirements of Table 11.5 for the defined damage sources and damage levels. Composite aircraft structure should be designed to resist damage caused by typical hail strike energies, thereby minimizing the amount of repair required after a hailstorm. The hailstone size was chosen such that this size or smaller represents 90% of hailstorms [10]. Table 11.5 is not applicable to in-flight hail damage that normally results in significant damage to leading edges and radomes. However, in-flight hail damage to safety-of-flight structure must be accounted for in the design. Methods to account for in-flight hail damage vary amongst manufacturers and are discussed in more detail in the 'Impact Damage of Composites' chapter of this book. The runway debris size was chosen to include most of the potentially damaging objects found in ground operations such as broken pavement, aircraft or ground vehicle parts, and garbage. The velocity of the runway debris is dependent on the category of aircraft.

There is little guidance on the effects of repeated low-level impacts such as hail impact, tool drop, or damage caused by walking on the structure. Although damage caused by a single impact incident may not be apparent, repeated impact for a given area over the structural service life may affect the durability of the structure. If the durability of an area proves to be sensitive to a repeated damage source, consideration should be given to simulating the damage on the full-scale test article to verify the effects of the damage.

An impact survey consists of a series of impact tests representing a wide range of conceivable impacts applied at various locations to a structure. The goal is to define the relationships between impact energy, damage detectability, damage characteristics, and residual strength for representative structure subjected to boundary conditions characteristic of the real structure. Typical coupon level residual strength measurements are based on compression strength after impact tests. Impact surveys with representative structure are required to establish critical damage scenarios and to identify structural elements that are particularly susceptible to impact damage (e.g., stiffener terminations). As mentioned earlier, assessing the impact susceptibility and effects on stiffened structure using representative specimens is key to determining how impacts to the external surfaces that may produce non-visible surface damage can have significant effects on underlying structure. The results of the survey are used to identify structural features crucial to integrity and to establish the impact variables (energy, location, etc.) to be applied to full-scale test articles used to determine residual strength [4].

Due to the large number of material, structural, and extrinsic variables affecting damage, impact surveys have been found to provide the most consequential results when applied to large structural test articles with design detail and boundary conditions representative of production structure. Many different impact scenarios and locations should be considered in the survey, with a goal of identifying the most critical impacts possible (i.e., those causing the most serious damage but are the least detectable). The survey should include blunt and sharp impactors and the entire range of potential impact energy levels. Although large test articles are expensive, such studies are practical because numerous impacts can be applied to a single test article [7], and in many cases, this work can be done at the conclusion of a static or fatigue test with a full-scale test article. Service data collected on structures over time can better define impact threats and design criteria for structures with similar applications, as well as establish more rational inspection intervals and maintenance practices. Impact surveys should be conducted when there is insufficient service experience to make good engineering judgments on energy and impactor variables. In review of service experience databases, it should be understood that the most severe and critical impact damages that are possible

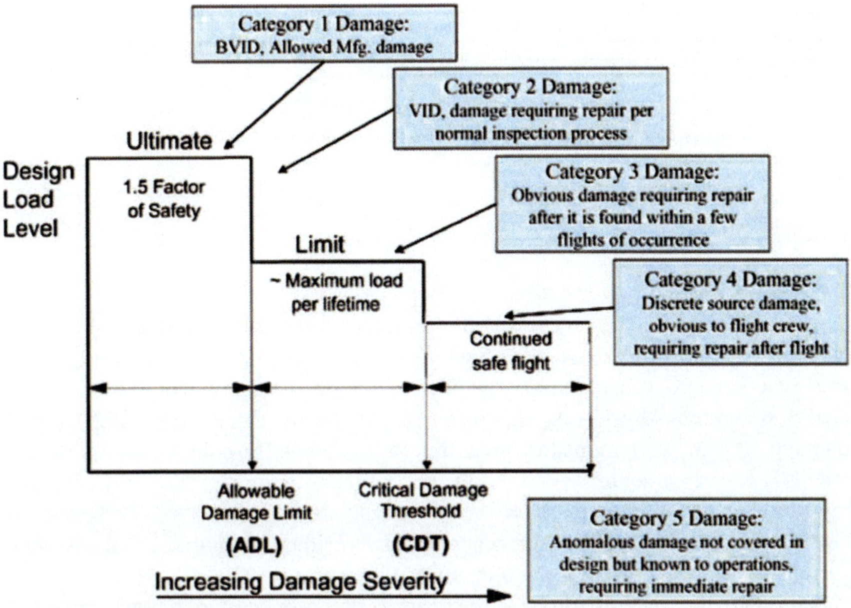

Fig. 11.8 Residual strength requirements vs. categories of damage [7]

may not be part of the database. This consideration is important to the assumptions needed for use of probabilistic damage threat assessments in defining design criteria, inspection methods, and repeat inspection intervals for maintenance.

11.3.2.2 Damage Classification

The damage threat assessment and the impact survey should account for an extensive range of damage levels with varying levels of detectability, ranging from non-visible to clearly evident. The various levels of damage and detectability for the impact survey are critical in defining damage tolerance and residual strength substantiation requirements. Damage classification for safety-of-flight structure can be derived from the impact survey based on different levels of detectability. Detectability or probability of detection of damage is dependent on the type of inspection, inspection equipment, skills of the inspector, structural details, damage size, and the inspection interval. Both the FAA and DOD have categorized damage according to levels of detectability and defined the residual strength requirements for each category of damage.

In Advisory Circular 20-107B released by the FAA in 2009, five categories of damage for safety-of-flight structures are defined [7]. Figure 11.8 shows the damage levels and residual strength requirements for each of the five categories. The damage levels for the first four categories should be based on the threat

Table 11.6 AC 20-107B damage categories

Category	Degree of detectability	Residual strength requirement
1	Undetectable	Retain ultimate load capability for lifetime
2	Reliably detected – scheduled inspection	Retain "above" limit load capability
3	Reliably detected – walk-around visual inspection	Retain limit or near-limit capability
4	Discrete source-known event	Retain continued safe flight capability

assessment and impact survey for the particular platform of interest. The fifth category is based on extensive damage events not covered in the design criteria. Residual strength substantiation by component or subcomponent testing or by analysis supported by testing should be performed for the various categories of damage. It should demonstrate that the environmentally compensated residual strength of the structure will reliably be equal to or greater than the strength required for the specified design loads – considered as ultimate. The statistical significance of reliable subcomponent and detail residual strength assessments may include conservative methods and engineering judgment [7].

The degrees of detectability and the residual strength required for categories 1–4 damage are shown in Table 11.6. For category 2, the residual strength requirement for a given damage depends on the scheduled inspection interval and fidelity of the inspection method. Substantiation of the requirement ensures that the structure will sustain sufficient residual strength until damage is found during scheduled inspections. Category 2 damages should not grow or, if slow or arrested growth occurs, the level of residual strength retained for the inspection interval is sufficiently above limit load capability. Category 3 damage should be such that it can be reliably detected or indications of potential damage are readily detectable during walk-around inspections. The primary difference between categories 2 and 3 damage is that category 3 requires demonstration of limit or near limit load capability for larger damages, albeit for a shorter inspection interval. Category 4 damage is presumed to be caused by either in-flight or ground evident events that limit flight maneuvers. Likewise, category 5 damage results from known events that may include collisions with aircraft ground equipment, maintenance errors, anomalous flight overload conditions, abnormally hard landings, maintenance jacking errors, and loss of aircraft parts in flight.

The JSSG-2006 similarly classifies damage based on detectability, but bases residual strength requirements on the upper bound spectrum loading developed for durability testing [10]. While the Air Force and Navy have used the JSSG-2006 requirements for fixed-wing aircraft, the Army has recently utilized modified AC 20-107B damage categories for rotorcraft. The modifications by the Army are (1) category 2 damage must retain ultimate load capacity and (2) category 3 damage must be repaired before next flight. The JSSG-2006 residual strength requirements are given in the first four columns of Table 11.7. Although a precise correlation of the AC 20-107B damage categories to the damage classification in JSSG-2006

Table 11.7 Residual strength load [10]

P_{XX}[a]	Degree of inspectability	Typical inspection interval	Magnification factor, *M*	AC 20-107B category
P_{FE}	In-flight evident	One flight[b]	100	4
P_{GE}	Ground evident	One day (two flights)[b]	100	3–4
P_{WV}	Walk-around visual	Ten flights[b]	100	3
P_{SV}	Special visual	One year	50	2
P_{DM}	Depot or base level	1/4 lifetime	20	2
P_{LT}	Non-inspectable	One lifetime	20	1

[a] P_{XX} is the maximum average internal member load (without clipping) that will occur once in *M* times the inspection interval. Where P_{DM} and P_{LT} is determined to be less than the design limit load, the design limit load should be the required residual strength load level

[b] Most damaging design mission

is open to interpretation, the fifth column of Table 11.7 shows a correlation based on detectability of the damage.

For DOD fixed-wing manned aircraft, all safety-of-flight structure including doors and door/ramp mechanisms must meet the residual strength requirements given in Table 11.7. The minimum required residual strength is specified in terms of the internal member load P_{XX} which the airframe must be able to sustain with damage present for the specified period of unrepaired service usage. The magnitude of this load is based on the overall degree of damage detectability and is intended to represent the maximum load the internal member might encounter during a specified inspection interval or during a life time for non-inspectable structure. To account for the fact that any individual aircraft may encounter loads considerably in excess of the average during its life, the required residual strength must be equal to or larger than the maximum load expected during a given interval between inspections. This is accomplished by magnifying the inspection interval. For example, the P_{XX} load for in-flight evident damage is the maximum load that could be expected once in 100 flights (see Table 11.7). A direct comparison of residual strength requirements for AC 20-107B and JSSG-2006 for each of the damage categories is difficult since JSSG-2006 requirements are based on the inspection interval and spectrum loading and AC 20-107B is based on limit and ultimate loads.

11.3.3 Non-Impact Damage Sources

A preponderance of service damage concerns are related to impact, however environmental degradation and damage are principal causes for maintenance. Damage is often a result of improper characterization or underestimation of service environments during early design phases leading to the selection of materials with inadequate performance. Damage sources are highly dependent on the application and service environment and vary greatly between sea-, land-, air-, and space-based

DAMAGE		CAUSE OF DAMAGE
Abrasion	→	Rain/sand/grit erosion
Surface Oxidation	→	Lightening strike
Thermal damage/oxidation	→	Overheat
Hole elongation	→	Overload/bearing failure
Delamination	→	Freeze/thaw (moisture expansion)
Blistering	→	Thermal spike (steam formation)
Honeycomb panel disbond	→	Freeze/thaw (moisture expansion)
Core corrosion	→	Moisture penetration into honeycomb
Surface swelling	→	Solvent damage

Fig. 11.9 Typical nonimpact service environment damage [31]

platforms. The varieties of potential degradation and damage sources that should be considered during design include:

- Acidic atmosphere
- Acoustic noise
- Cyclic temperature
- Dust
- Fluid contamination
- Fungus
- Humidity
- Immersion
- Low/high pressure
- Out-gassing
- Lightning strike/P-static discharge
- Rain (erosion)
- Salt fog
- Sand (erosion)
- Shock
- Solar radiation
- Temperature – hot/cold
- Temperature shock
- Vibration

The susceptibility of materials to damage sources should be evaluated on small-scale test articles as part of the material qualification process. When small-scale article tests are not sufficient to demonstrate durability and damage tolerance against these sources, larger scale developmental test articles are required.

Figure 11.9 shows frequently encountered types of in-service damage and their causes. Aircraft leading edges and radomes must endure the erosive effects of sand and rain during flight that can erode standard coating systems and underlying materials. Flexible, durable polyurethane rain erosion-resistant coatings and elastomeric tapes can often be used to alleviate such erosion problems. Polymer composites are particularly susceptible to overheating from hot gas impingement, fire, repair over-temperature, and lightning strike. Over temperature due to exhaust from engines and auxiliary power units is most often a result of poor predictions of the service environment because materials are often selected before enough design detail is established to provide sufficient service temperature predictions. Design temperatures can potentially be derived from operational use temperature from similar platforms. For such cases, the designer could conservatively select a high-temperature composite material, such as a polyimide, that has a service temperature well above that required for most applications. However many polyimides are more expensive and more difficult to process than traditional epoxy systems and may require specialized manufacturing facilities. Therefore, a cost-benefit assessment of the risk associated with selecting a lower temperature material having a threat of overheating with increased cost of higher temperature materials should be conducted.

Metallic aircraft have an inherent resistance to damage caused by the electrical current and electromagnetic forces generated by lightning strikes and p-static discharges because the conductive skins permit the current to remain on theexterior of the structure. Graphite fiber composites are significantly less conductive than metals and fiberglass/quartz composites are minimally conductive. The effects of electromagnetic discharges on composites include delamination and possible burn-through of the laminate in the immediate strike area, potential arcing at nonconductive gaps and burning with delamination at fasteners and connectors. Arcing is a potential ignition source in fuel tanks. Lightning strike zones such as wing tips and trailing edges are regions with high probability of lightning strike occurrence. The SAE Aerospace Recommended Practice publication ARP 5414 provides information for aircraft lightning zoning [32]. Protection of composite structure by conductive materials is required on and around lightning strike zones to provide conductive paths away from the attachment zones [7].

The strategy to prevent lightning strike damage includes providing a continuous conductive path so the electrical current dissipates and remains on the exterior of the aircraft and arcing is prevented by elimination of nonconductive gaps. Even though lightning strike zones may only be on the end of wing tips, all-composite wings may require a conductive layer over their entire surface. A continuous conductive path across the exterior of the aircraft gives the current numerous paths to safely exit the aircraft. Conductive layers may include bonding aluminum foil to the outermost ply, bonding aluminum or copper mesh as an external ply or embedded one ply from the outer mold line, or incorporate strands of conductive material in laminate [33]. Aluminum foils and meshes can readily be used with glass fiber composites, however due to galvanic corrosion concerns, their use is not recommended with graphite fiber composites. The relative locations of graphite and copper in the galvanic series (see Fig. 11.5) make copper less susceptible to galvanic corrosion; however, the density of copper is 3.3 times that of aluminum, resulting in a significant weight penalty.

There is significant interest in the aerospace community in the development of lighter weight lightning strike protection solutions for graphite fiber composites. Owing to the propensity for maintenance and in-service damage for composites, lightning protection systems should be designed to be easily repairable using replaceable conductive materials and designs that ensure that all conductive path attachments are easily accessible. To prevent arcing between metallic fasteners and composite skins, encapsulating fastener nuts with plastic caps or polysulfide coatings should be considered. FAA advisory circulars AC20-53 and AC20-136 as well as MIL-STD-1795, MIL-STD-1757, and MIL-B-5087 provide means of compliance with lightning strike design requirements [34–38].

11.4 Durability Analysis and Testing

Full-scale durability tests are the primary indicators of compliance to the structures durability requirements. However, durability analyses support the design by establishing design stress levels, aiding in definition of structural details, and reducing risk relative to testing [10]. Commercially available finite element programs have successfully been used by the aerospace industry to accurately calculate the linear elastic stress, strain and displacement fields for laminated composites subjected to thermal and in-plane mechanical loading. For potential areas of concern that may involve out-of-plane loading or complex structural geometries, three-dimensional finite element methods can be employed to estimate local stress and strain fields. However, the current composite structural design and analysis methods used by industry are largely semi-empirical due to the fact that accurate predictions of failure modes, ultimate strength, residual strength, and fatigue life are limited [29]. Importantly, it has become common practice in the aerospace industry to design to allowable stress levels in composite structures that have validated confidence level offsets from actual failure data. Unlike metallic designs, this methodology may produce structure that has no reasonable chance of failure (barring dramatic error in method) for a room temperature static or durability test. The inability to predict damage initiation and growth is a major design consideration for addressing durability and damage tolerance. For composites, neither fatigue crack growth and life nor stable damage progression prior to ultimate failure can be rigorously predicted because of the lack of robust generalized failure criteria and the complexity of the interacting non self-similar growth failure modes. The relevance of this statement precedes definition of the broad range of mechanical, hygrothermal, and chemical environments pertinent to durability and damage tolerance concerns.

The prediction of damage initiation and growth is further complicated by additional factors. It is well documented that failure stresses and strains are dependent on localized residual curing stresses and models to predict these residual stresses for complicated geometries and fiber architecture are lacking [16]. Nonlinear structural response characteristics such as buckling, postbuckled strengths, crippling, and pressurized structure deformations are more difficult to predict for composite structures than they are for metallic structures [29]. For particular environments, polymer composites are known to chemically degrade (hydrolysis, oxidation, etc.) throughout their service life. Models to predict environmental degradation of polymer composites and their subsequent material properties for failure analysis are lacking [39]. Since there is a clear lack of ability to predict damage growth and initiation, design methods are based on limited coupon tests and full-size test articles to either substantiate the design or indicate hot spots that need to be reworked. While the initial design process is fairly mature, less progress has been made in understanding the long-term response of composite structures subjected to adverse in-service loading or environmental influences [40].

Repair concepts for composite structures include doubler and scarf patches attached using mechanical fasteners and adhesive bonding. The purpose of repairs is to restore the original strength, stiffness, and/or fatigue capability of the damaged structure. The source data for the original design must be used to design the repair to insure that the repaired structure conforms to the original design requirements. Strict material and process controls for repair installation should be established as part of the structures maintenance plan. Following the discussion above, the capability to predict the long-term response of repairs for composite structures is lacking. Therefore, structural repair substantiation should be part of the overall structural substantiation program that includes sufficient testing to ensure the long-term integrity of the repair. This can be achieved by including structural repairs on the full-scale durability and static strength test articles.

11.4.1 Safe-Life Analysis

Fixed-wing manned aircraft manufacturers have in general adopted damage tolerance design philosophy for both military and commercial aircraft. Historically, helicopter rotor system components and Naval fixed-wing aircraft were designed and qualified using safe-life approach. Beginning in 1989, the FAA has required damage tolerance substantiation for all commercial rotorcraft composite rotors. However, there are still aircraft, rotorcraft, wind turbine, pressure vessel, and composite fly wheel manufacturers using the classical safe-life approach or a modification thereof. One reason for this is that there are no clearly defined damage tolerance design criteria for composite structures operating in high-cycle loading environments due to difficulties with analysis and testing [41]. When properly used, the safe-life design approach places statistically predictable failures at an extremely remote probability of occurrence. However, early rotorcraft operational experience showed that the safe-life approach by itself is not adequate because of unpredictable failures resulting from such sources such as manufacturing defects and service-induced damage [8]. Furthermore, review of this operational experience has shown these unpredictable failures far outnumbered the predictable ones as shown in Fig. 11.10. The classical safe-life analysis assumes a flaw-free structure, whereas modified safe-life approaches (Flaw-tolerant safe-life or Enhanced safe-life) have been developed to account for manufacturing and service-induced damage. The modified safe-life approaches provide an estimate of the life of components that contain defects that may have been introduced during the manufacturing (scratch, ply distortion, embedded material) or generated during operation (impact, pitting) but are not specifically load-induced flaws (such as fatigue cracks or load-induced delaminations) [42].

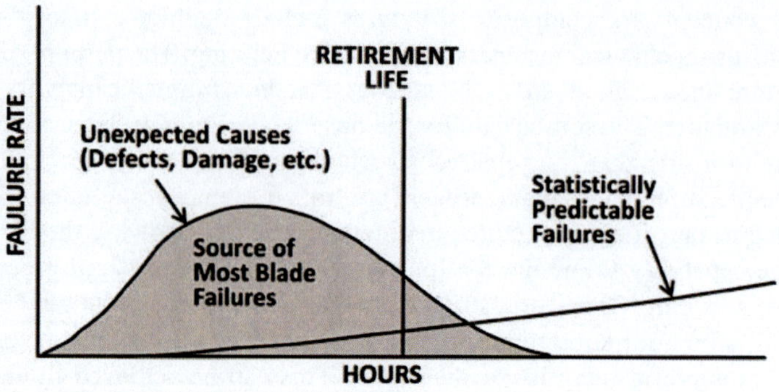

Fig. 11.10 Early operational experience for safe-life rotor designs [8]

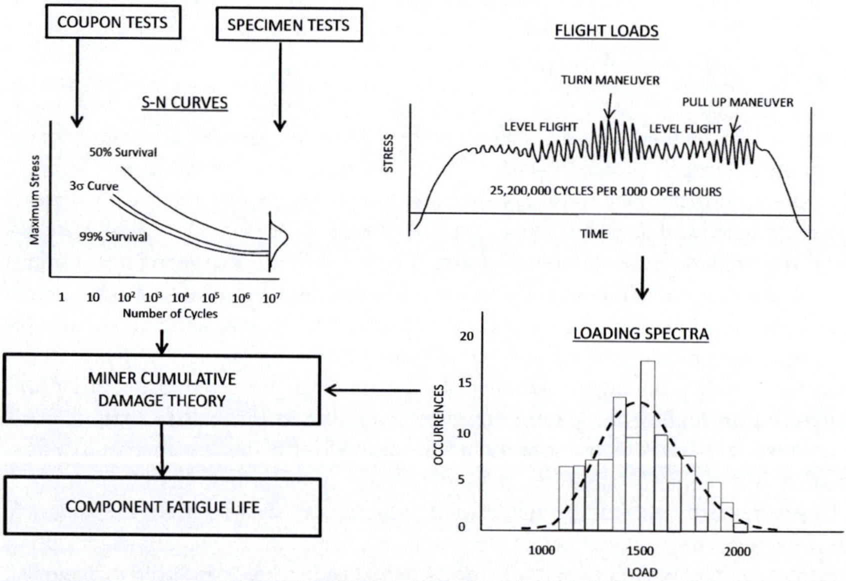

Fig. 11.11 Safe-life approach to fatigue life [8]

The methodology for the fatigue design of rotorcraft primary structural components is shown in Fig. 11.11 [8]. The three basic elements of safe-life design include:

- Working fatigue S-N curve
- Design load spectrum
- Cumulative damage law

The fatigue S-N curve development is based on material coupon tests to determine the curve shape and component tests with constant amplitude loading to construct the mean failure curve. For the case illustrated in Fig. 11.11, the working S-N curve is approximately three standard deviations from the mean. The design load spectrum is developed from flight loads for various maneuvers and flight conditions of the vehicle mission profiles. The cumulative damage law is then used to predict fatigue life for the loading spectrum based on constant amplitude fatigue data. The fatigue life calculations are commonly performed using the linear damage accumulation law according to Palmgren-Miner [43]. However, this technique has not been proven to accurately assess the fatigue behavior of composite materials and structures to the degree that it has for metallics. While the accumulation of fatigue damage in composite is nonlinear and stress-independent, airframe composite structures designed using traditional methods (notched failure strain cutoffs) are typically insensitive to the durability test fatigue cycles. Therefore, application of the Palmgren-Miner law is focused primarily on evaluation of the metallic components. The calculated fatigue life should be divided by relevant scatter factors to ensure that the probability of occurrence of failure is low. A fatigue life of the structure should be demonstrated by extensive structural ground testing including a full-scale fatigue test of all major structural items. In the case where the composite structure is fatigue insensitive, demonstrated via coupon and element tests for all failure modes, this methodology is simplified further. In this situation, the composite structural design can be conducted on a static basis, with fatigue design considerations only required for metallic components.

In cases where the damage tolerance or crack growth analysis approaches may not be applicable or verifiable, such as in the case of composite structures and high strength steel landing gear structure, fatigue safe-life analysis methods can be used. In these cases, it is critical that sufficient development test data is generated to demonstrate compliance with design requirements [10, 44]. For such structures, the limit of detectability of damage may be near that of the critical damage size. Appropriate scatter factors and mandatory replacement times must be applied for parts with safe-life design approach. For the substantiation of rotorcraft dynamic components, a scatter factor of 10 or greater should be used while a scatter factor of three or greater should be used for metallic components [45, 46].

11.4.2 Initial Flaw/Damage Assumptions

Dimensional tolerances can be more critical for composite structures than for metallic structures based on their susceptibility to damage during assembly of parts [29]. Proper tool design to account for part spring back during removal of the part from the tool is critically important in meeting assembly dimensional tolerances. Exceeding tolerances can lead to both short-term labor-intensive part rework (liquid shimming, hand drilling, etc.) and long-term service life reduction resulting from potential fastener hole damage and the addition of residual stresses during part fit-up.

Before assembly and entering service, composite components should undergo quality assurance inspections to detect life-limiting defects that require repair or rejection of a part. As discussed in Sect. 2.3, the maximum acceptable manufacturing defect size is dependent on the fidelity of the NDI equipment and processes and the part geometry. The limitation of the inspection process to detect common manufacturing defects such as delaminations, ply waviness, and voids should be established for each component geometry. Thick components with complex geometries or with limited two-sided access are often difficult to inspect and may be particularly susceptible to marcelling and voids. The capability of a component to meet service life requirements while containing damage below the threshold of detectability should be demonstrated. Acceptable defects may not only be geometry dependent, they may also be load-type dependent. The compression strength of composites is particularly affected by the presence of out-of-plane imperfections such as delamination, fiber waviness, and surface mark off that can trigger local buckling whereas these same defects have little effect on tensile strength.

Flaws induced during manufacturing and in-service (foreign object damage, handling damage, etc.) must be considered when initial flaw size assumptions are established. For both durability and damage tolerance considerations, the capability of composite structures containing initial flaws to meet service life requirements should be demonstrated through testing or analysis supported by testing (accounting for the effects of the appropriate environment). The initial flaw size assumptions for durability are not explicitly identified because they are dependent on the extent of initially detectable damage consistent with the inspection techniques employed during manufacture and in-service, including considerations for the probability of detection. Durability requirements specify that a structure containing damage smaller than the threshold of detectability subjected to fatigue evaluation is expected to retain ultimate load capability for the life of the aircraft structure. These requirements are applicable to the undetectable category 1 damage of Table 11.6 and the non-inspectable damage of Table 11.7.

For damage tolerance, the size of initial damage for safety-of-flight structures is that size which would not be readily detectable in a routine visual inspection. The impact energy level to be assumed in design for each area of the structure should be that level which produces barely perceptible front face damage in the structure. Because the amount of energy necessary to achieve this level of damage is usually a function of the thickness of the structure, an upper bound energy level cutoff should be established for various zones on the structure dependent on the damage threat assessment. The initial manufacturing flaws and in-service damage for composite safety-of-flight structures defined in JSSG-2006 [10] are described in Table 11.8. Initial manufacturing flaw sizes less than that identified in the table can be used if it is demonstrated that there is a B-basis level of probability of detection for the smaller flaw sizes using the inspection techniques employed during manufacturing and service. The 136 J of energy required to cause a dent 2.54 mm deep may be reduced if a damage threat assessment demonstrates an acceptable risk of lowering the maximum impact energy level and a smaller dent depth is determined by testing to be visible.

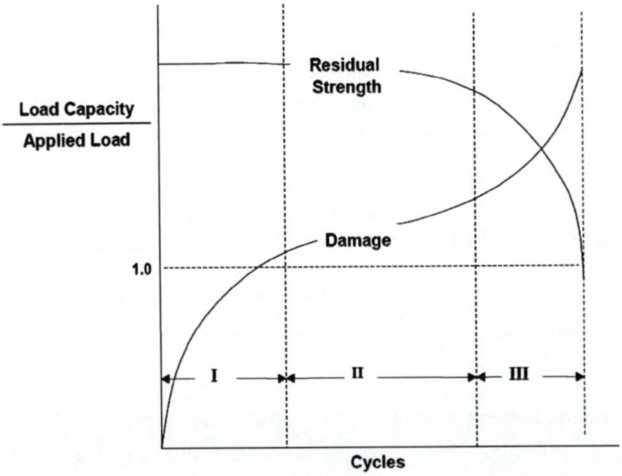

Fig. 11.12 Composite blade damage development [45]

11.4.3 Damage Growth and Inspection Requirements

Figure 11.12 illustrates the typical progression of fatigue degradation and residual strength reduction of a composite structure. Region I represents the initiation of fatigue degradation and damage growth that is typically in the form of microcracks that may have only subtle effects on strength and stiffness. In region II, microcracks coalesce and this eventually leads to steady damage growth to damage levels that decrease stiffness and residual strength. The region III advanced damage state in which large delamination and secondary failure modes advance results in rapid degradation of component stiffness and residual strength [45]. Sufficient fatigue evaluation and testing, including full-scale component substantiation, should demonstrate that region III levels of damage are not reached in service.

Three damage growth approaches are used for slow damage growth and fail-safe structures. They are: (1) no growth; (2) arrested growth; and (3) slow growth or no growth to failure. The growth or no growth evaluation should be performed by analysis supported by test evidence or by tests at the coupon, element, or subcomponent level for critical structure areas [7, 18]. The testing should assess the effect of the environment on the damage growth characteristics and the no-growth validation and should yield conservative and repeatable results. Arrested growth and slow damage growth structures are expected to retain ultimate load capability for the life of the aircraft structure for category 1 (non-inspectable) damage levels. To achieve this, empirically derived strain limits below those that result in damage growth should be used for design ultimate strain levels (see Sect. 2.2).

To account for variability in material properties, manufacturing quality, and inspection reliability, the JSSG-2006 requires that safety-of-flight structure be able to sustain service usage requirements for a period that is longer than the service life.

Table 11.8 Initial flaw damage assumption – applicable to safety-of-flight structure [10]

Flaw/damage type	Flaw/damage size
Scratches	Surface scratch 101.6 mm long and 0.508 mm deep
Delaminations	Interply delamination equivalent to a 50.8 mm diameter circle with dimensions most critical to its location
Impact damage	Damage from a 25.4 mm diameter hemispherical impactor with 136 J of kinetic energy or with that kinetic energy required to cause a dent 2.54 mm deep, whichever is less

Structures must demonstrate positive crack arrestment or no damage growth for the manufacturing and in-service damage/flaws identified in Table 11.8 for two lifetimes [10]. The FAA requires that the number of load cycles used to validate both slow growth and no-growth concepts should be statistically significant, and may be determined by load and/or life considerations and a function of damage size [7]. Unlike metallic structures, for which accurate fatigue crack growth models exist, there is no generally applicable tool for damage growth prediction for composite structures. Therefore, for composites, the no growth or arrested growth approaches are typically favored over the slow damage growth design approach that requires growth predictions.

An inspection program should be developed consisting of the frequency, extent, and methods of inspection for inclusion in the maintenance plan as part of each of the damage growth approaches. Inspection intervals should be established such that damage will have a very high probability of detection between the time it becomes initially detectable and the time at which the damage size reduces the residual strength to below ultimate load, including the effects of environment. For any detected damage size that reduces the residual strength capability below ultimate or causes functional impairment, the component must be either repaired to restore ultimate load capability or replaced [7].

Actual service usage for aircraft rarely matches the design usage spectrum particularly for aircraft featuring new designs, new capabilities, or new mission profiles. Tracking the aircraft fleet to obtain flight-by-flight load or acceleration histories provides a capability to assess fleet usage severity and project its impact on service life. Individual aircraft tracking (IAT) data obtained from a fleet can be instrumental in monitoring structural control points (fatigue-critical locations) and assessing individual aircraft usage to evaluate remaining fatigue life and to make repair decisions. Figure 11.13 illustrates how flaw size can vary with severity of usage. The determination of expended or remaining fatigue is highly dependent on the analytical ability to relate damage size and growth rate to severe, average, or mild usage which is particularly challenging for composite materials.

Equivalent Flight Hours (EFH) are determined by comparing Actual Flight Hours (AFH) to a reference usage defined as the design usage or an updated

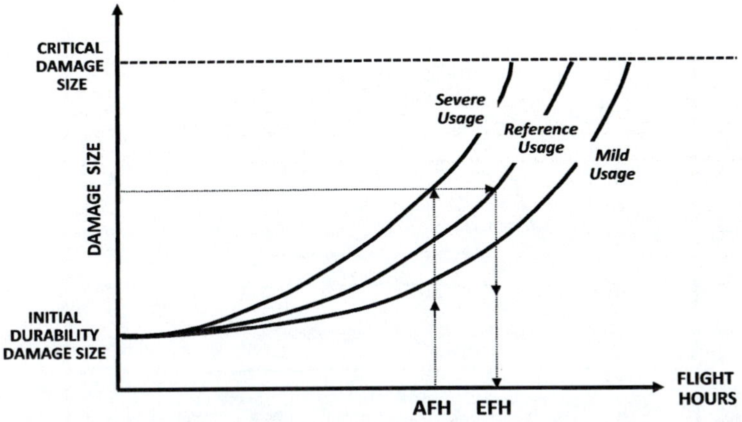

Fig. 11.13 Damage growth dependence on usage severity

baseline usage from fleet usage data. EFH are the AFH of an aircraft adjusted for the usage severity by comparing to the reference usage [2]. Utilizing the severity of usage versus damage size curve and the AFH, the damage size is used to relate AFH to EFH as shown in Fig. 11.13. The severe usage results in EFH being greater than AFH while mild usage results in EFH being less than AFH. Using the EFH approach, usage for all aircraft in a fleet can be compared and the structural health of the fleet assessed. Since composites may be critical for the severe loading cases, care must be exercised to ensure that these high-level load occurrences are properly taken into account in the force management tracking program. Life extension programs are highly reliant on projecting IAT data and projected fleet usage. Extensions in service life should include evidence from component repeated load testing, fleet leader programs (including NDI and destructive tear-down inspections), and appropriate statistical assessments of accidental damage and environmental service data considerations [7]. While demonstration of life requires two lifetimes of durability testing, a third lifetime of testing should be conducted for the purpose of life extension planning and/or modification verifications.

11.4.4 Design Development Tests (Building Block Approach)

Durability analyses and substantiation for composite structures are intimately linked to development tests that consist of "building blocks" ranging from coupons to elements, structural details, subcomponents, and finally components. The range of development tests are used to assess material strength and durability, evaluate producibility and manufacturing defects, determine environmental susceptibility and compensation factors, appraise material and structural variability, resolve static and fatigue-dependent damage modes, and support predictive model development and validation. The tests also address issues of structural discontinuities

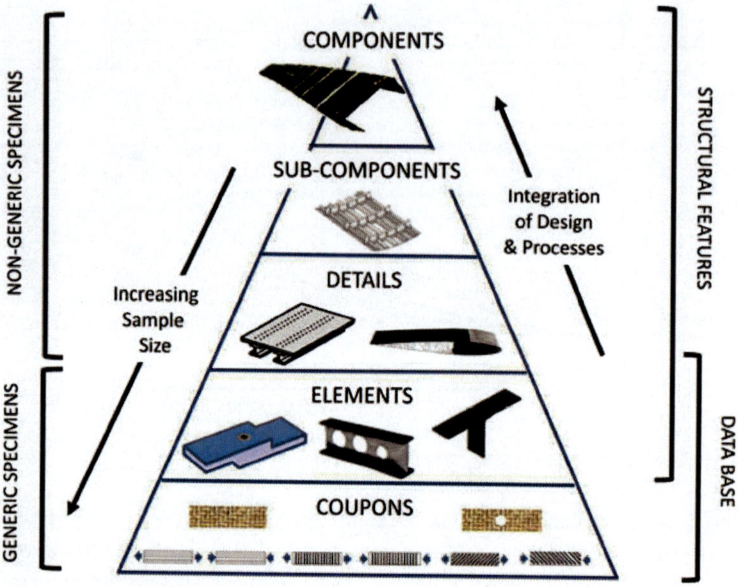

Fig. 11.14 Building block test program

(e.g., joints, cut-outs, or other stress risers), design or process-specific details as well as the effects of local details and internal load paths on structural behavior. A complete building block approach to structural substantiation addresses most of the critical structural issues in the test articles with increasing levels of complexity such that many areas of reliable performance can be demonstrated prior to the full-scale component tests [29]. In addition, the use of a progressive building block is of particular value in composites, because it can avoid the inherent limitations of metal fatigue at lower levels of testing, allowing for a more complete exercising of the composite design.

A schematic of tests typically included in the building block approach for composite structures is shown in Fig. 11.14. Typically, testing progresses from simple coupon specimens to more complex components. The number of specimens needed for the coupon and element tests is commensurate with statistically compensated allowables determination. The coupon and elements tests are used for process evaluation, statistical material property data generation, and environmental effects evaluation. A smaller sample size is used for details and subcomponents tests to validate analytical predictions and establish design allowables. These tests are used to evaluate splices and joints, manufacturing methods, failure modes, and allowable strain levels as well as environmental effects on failure modes and failure strain. The component tests of large assemblies or full-scale components can provide early verification of static strength and evaluation of structural designs of critical areas [10]. Finally, the component tests provide substantiation of the damage tolerance and fatigue performance of the structure.

In addition, each stage of the building block allows for the development and application of engineering compensation factors (commonly referred to as "knock-down" factors). These factors are developed and applied to the structural configuration analysis for two primary reasons. First, in cases where the predictive methodology used in the development of the aircraft structure does not agree with measured test data and the design is too mature to allow for the correction of predictive methodologies, correction factors can be used to modify design allowables or operational load cases. In addition, correction factors can be applied in cases where limited validation data is available for a particular test condition and engineering judgment must be used to establish appropriately conservative limits on design.

Inherent in the serial building block approach is the requirement to scale-up manufacturing processes to produce more complex larger scale test articles. Addressing manufacturing scale-up is critical in processing polymeric composites where it is particularly challenging to accurately scale-up the curing kinetics to full-scale component fabrication [29]. Design and manufacturing lessons learned from smaller scale test articles help avoid costly failures in more complex larger test articles that typically occur in later stages of the design development test program. Successful manufacturing scale-up ensures that the material properties of the full-scale component are reflective of the properties of the smaller scale coupon and element test articles from which material and design allowables are derived. The manufacturing of full-scale test articles should be representative of production manufacturing to account for manufacturing dependent flaws.

Composite materials are known to have environmentally dependent failure modes and these effects must be discovered, characterized, and understood before full-scale component testing. Environmental conditioning of test articles adds considerable cost at all levels of development testing and is likely cost prohibitive for most full-scale aircraft structures. However, substantiation may require full-scale environmental testing when limitations of analysis tools and component tests cannot adequately demonstrate compliance to requirements. Full-scale hot/wet environmental tests have been conducted on rotorcraft components, fuel tanks, and external pods. Additionally, full-scale cold tests have been conducted on aircraft components. To preclude full-scale environmental testing, sufficient development testing is required to validate the failure mode and failure strain levels for the critical design cases with critical temperature and end-of-life moisture. For bonded joints in safety-of-flight structures, environmental testing of full-scale structure adhesive bonds should be conducted to substantiate the design; whereas, for non-safety-of-flight parts, representative coupon testing of adhesive bonds is adequate to validate performance [5]. The development test articles should be strain gaged adequately to obtain data on potentially critical locations and for correlation with the full-scale static strength test. The development testing should provide assurance that the structure is designed such that ambient moisture room temperature failure modes do not change with critical temperature and moisture conditions. If these steps are completed adequately, the full-scale test should be a validation of predicted load paths and failure modes.

To substantiate the full-scale structural static strength and account for the degradation of material properties due to combined temperature and moisture effects, one of the following methods should be used for the test [10].

(a) Environmentally precondition the test article for the worst-case combined temperature-moisture condition and test under these conditions to design ultimate load (DUL).
(b) Test the composite article at room temperature to a load level in excess of DUL to demonstrate the environmental compensation factors for temperature and moisture. The strains measured at DUL in the critical location of the structure must be less than the failure strains in the environmentally conditioned development tests for the same design details and loadings. Development testing must also show that there is no change in failure mode between the room temperature and environmentally conditioned tests. Development testing must also validate the statistically compensated compensation factor.

Since composite structures are often designed to ETW allowables, the margin of safety for a RTD test loaded to DUL is equivalent to

$$MS_{RTD} = \frac{P_{allowable-ETW}}{P_{allowable-RTD}} - 1 \tag{11.2}$$

where $P_{allowable\text{-}RTD}$ is the material allowable at RTD and $P_{allowable\text{-}ETW}$ is the material allowable at ETW. To demonstrate the environmentally compensated ultimate load capability at room temperature, the DUL must be enhanced by the factor $P_{allowable\text{-}RTD}$ divided by $P_{allowable\text{-}ETW}$. For a typical graphite/epoxy composite, the ETW environmental compensation factor is on the order of 1.10–1.15. If CTD or ETD conditions were shown to be critical in development tests, the corresponding compensation factors should be used.

The design development test program plays a key role in the development and validation of the models used to predict static and fatigue behavior accounting for the expected service temperature and moisture extremes. Owing to the lack of damage prediction tools for composites that are based on fundamental principles, empirically based analytical tools are developed and validated throughout the design development test program. This approach provides the data needed for sufficient analysis correlation and the necessary test replicates to quantify variations occurring at the larger structural scales to be economically obtained. At completion of detail and subcomponent testing, failure criteria should be established and the developed analytical tools should be able to account for all potential failure modes for the full-scale component tests. When establishing details of the design development test articles, the types of manufacturing and in-service damage/degradation expected in the structure must be incorporated to satisfy durability and damage tolerance requirements. An example of these progressive steps up the building block is provided in Table 11.9. The design development tests are complete when the failures modes have been identified, the critical failure

Table 11.9 Examples of progressive building block validation of damage tolerant and durable structures

	Coupons	Elements	Subcomponents	Components
Damage tolerance (impact damage)	Fatigue and static testing of compression strength after impact specimens	Damage survey of impacts over stiffener locations on test panels and subsequent residual static and fatigue testing of excised elements	Impact of critical location on multi-stiffener skin panel followed by fatigue test verification	Impact of critical damage location on full-scale fatigue article between first and second lifetime
Durability (fastener hole wear)	Static and fatigue testing of bearing, filled hole and bearing bypass specimens	Fatigue test of production representative joints with manufacturing toleranced holes	Fatigue test of critical fastened joint (such as a rib/spar/skin integration subcomponent)	Post-fatigue test inspection of critical holes for evidence of hole wear

modes in the component tests are judged to be not significantly affected by the nonrepresentative portion of the test structure, and the structural sizing is judged to be adequate to meet the design requirements [10].

11.4.5 Structural Fatigue Testing

The fatigue behavior of composite structures (high fatigue thresholds, high data scatter, and multiple damage mechanisms) differs significantly from that of metallic structures [4]. Because of these differences and the inability to correctly predict the fatigue response of full-scale components, durability of composite structures is assured by testing and the empirical models developed during the design development test program. Figure 11.15 shows a comparison of the fatigue life and static strength scatter distributions for graphite/epoxy composites (based on two-parameter Weibull distribution) [47]. The mean of the Weibull shape parameter for composite fatigue strength is approximately 2.17 while that for static strength is 23.2 indicating that the static strength has a much smaller coefficient of variation. Based on a US Navy data set, Whitehead [47–49] found that

- Compression–compression ($R = -\infty$) fatigue loading produces a significantly higher fatigue life scatter than compression–tension ($R = -1$ and -2) fatigue loading.

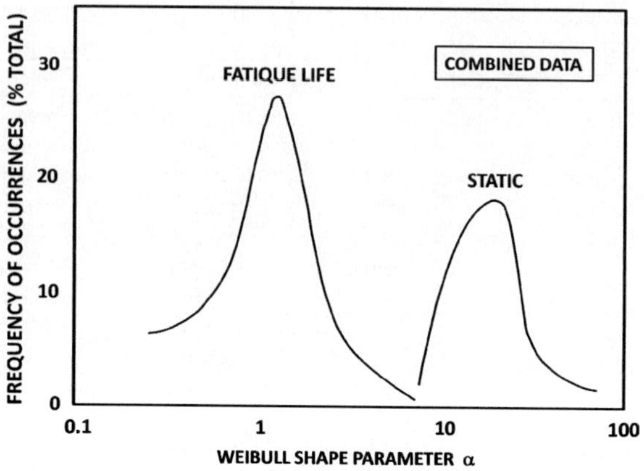

Fig. 11.15 Graphite/epoxy fatigue life and static strength scatter distributions [47]

- There is no significant difference between fatigue life scatter for constant amplitude and spectrum loading.
- Laminate layup does not significantly influence fatigue life scatter.
- The RTD test environment produces significantly lower fatigue life scatter than the ETW test environment.

The fatigue scatter factor is defined as the ratio of the mean life and the allowable life. The factor signifies the relation between the central tendency of a data set (the mean) and the extreme statistics (the allowable). The value of this factor depends upon the shape parameter, the mean, and the sample size. Figure 11.16 shows a typical plot of the scatter factor as a function of shape parameter for a sample size of five [47]. With such large scatter factors for fatigue and the potential of fatigue damage from high stresses, it is difficult to establish a test program that will ensure the durability of composite components [50]. However, it has been demonstrated that the durability performance of composites is generally excellent when the structure is adequate to meet its strength requirements [10, 49–51]. Therefore, the thrust of the full-scale durability test must be to locate and provide early indications of detrimental stress concentration areas that were not found in the static tests to minimize retrofit cost, determine durability life, assess service life capability and economic burden, and establish inspection, repair, and modification needs.

11.4.5.1 Fixed-Wing Aircraft Structures

The full-scale durability test article is used to substantiate the airframe design service life and to identify critical structural details not previously identified by analysis or development tests that may require modifications or special in-service inspections. The loading spectrum development is central to the goal of achieving equivalent damage in the test article to what will occur in the anticipated service usage aircraft. Therefore, the loading spectrum should accurately represent the

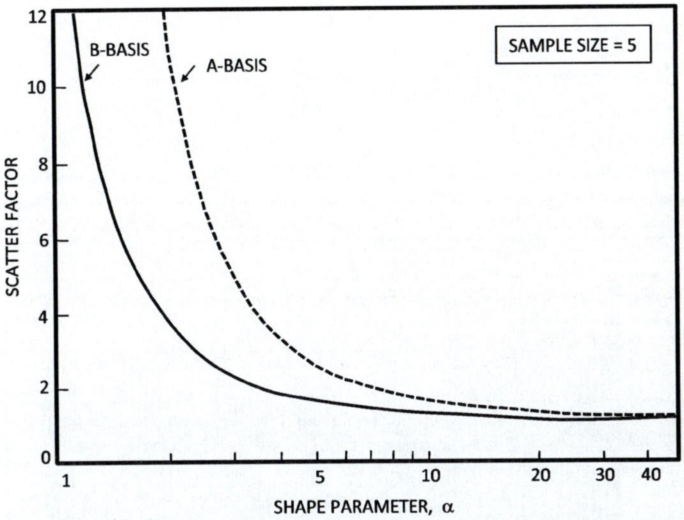

Fig. 11.16 Relationship between shape parameter and scatter factor [47]

design usage while being practical for analysis, testing, and test result evaluation. All of the design flight-by-flight mission profiles including taxiing, climb, cruise, maneuvers, descent, flight times, operational speed, and altitude should be considered in the development of the spectrum. The primary loads that need to be considered are flight loads (gust and maneuver), ground loads (taxiing, landing, turning, engine run-up, braking, thrust reversing, towing, jacking), catapult and arrestment for Naval aircraft, and pressurization loads [52]. The percentage of time that the aircraft is operated in each of the loading regimes should be reflected in the spectrum. When all vehicle loads and frequencies are combined, the resulting load profile is commonly known as the design loads spectrum or the master events spectrum. An assessment of the sensitivity to changes in the baseline design loads spectrum is recommended [10], but this is often much more critical for metallic structure than composites. For design development test articles, the aircraft loads must be translated to coupon, element, detail, and subcomponent stresses to establish stress spectra for each of the critical areas as illustrated in Fig. 11.17.

Typical design spectra for modern fixed aircraft may include hundreds of thousands of loading cycles over the duration of the vehicle design life. Truncation of the design loading spectrum is typically required to reduce excessive testing time (multiple lifetimes of testing) and optimize the durability analyses by eliminating stress cycles that are less than the fatigue threshold and do not contribute to damage growth. Depending on the type of aircraft and the design service usage, truncation of loads from the design spectrum that result in stresses less than the material fatigue threshold can significantly decrease the testing time and the analysis effort. For damage-tolerant designs containing metallic structures, high amplitude loads may be clipped because they can result in the creation of crack-tip plastic zones that

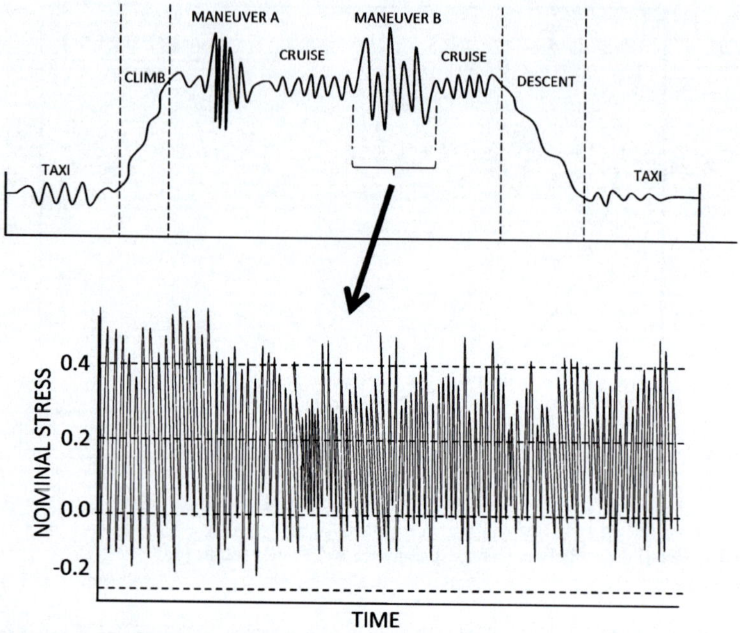

Fig. 11.17 Mission profile translated to component stress spectrum

can retard crack growth and slow crack growth rates. Therefore, clipping of all stress levels above 90% limit load may be required in order to reduce the impact of crack retardation or beneficial residuals that result in a conservative level of crack growth for metallic structures [10]. However, composites are very sensitive to high load applications and to preclude the development of nonconservative analysis spectra, the clipping of high loads is generally not accepted and if used, must be justified by development tests [7, 10, 53]. Therefore, the loading spectrum for damage-tolerant, hybrid designs must be a compromise between crack retardation in metallic components and elimination of high-amplitude load cycles with the potential for damage initiation in composite components. For safe-life design certification, high-amplitude load clipping is seldom used for metals and/or composites because it can have a detrimental effect on damage initiation, thereby reducing the overall conservatism of the test.

Establishment of low-amplitude thresholds for truncation of load cycles requires determination of the load levels that can be shown not to contribute to damage growth in the aircraft lifetime. Stress (strain) levels below the material's fatigue threshold are typically considered to cause no fatigue damage (initiation or progression) and theoretically can be removed from the spectrum without changing the test results. However, in practice, the truncation level is usually a certain percentage of the A- or B-basis fatigue threshold (e.g., 60–70%) due to the large scatter factor for composites [4]. An example of the truncation level determination based on the RTW A basis strength is illustrated in Fig. 11.18. For this example, the truncation

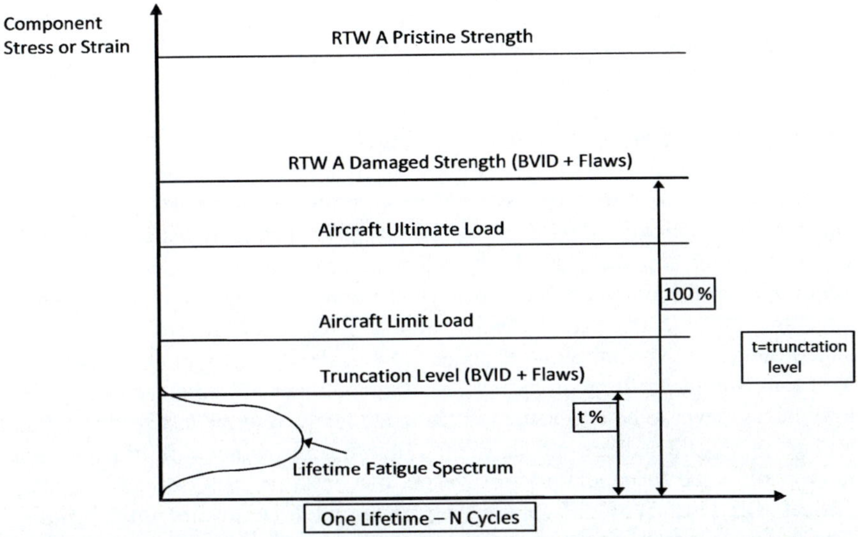

Fig. 11.18 Truncation level determination [4]

level is determined as a ratio of the stress (or strain) corresponding to 108 cycles on the S-N curve to the static room temperature wet A basis strength with damage. The truncation ratio can be shown to depend on the R ratio ($\sigma_{min}/\sigma_{max}$) and damage type. It also depends on the material type and layups used. For this reason, coupon and element tests are needed to establish and validate that the truncation level conservatively covers all cases. In all practical cases, the truncation ratio is largely driven by damage accumulation in metallic substructure, and not by the composites. This is especially true in the case of aluminum, where fatigue growth can occur at relatively low levels of strain once damage sites have initiated.

Fully assembled structures are preferred for durability testing; however, individual major component testing may be used if approved by the certification authority. The testing of individual components such as wings, fuselages, and empennages requires that the load introduction (supporting and loading fixtures) accurately simulates the load and deflection interactions with the adjacent structure not being tested [10]. The test articles should be representative of production airframes to account for inherent process and manufacturing-dependent anomalies. The thrust of testing should be to locate detrimental stress concentration areas that were not found in the design development tests, particularly for metallic parts. Extensive databases have shown that composite structures, which demonstrated adequate static strength, were fatigue insensitive. Whitehead et al. went so far as to suggest that no full-scale durability test is required for all composite structures or mixed composite/metal structures with non-fatigue critical metal parts, provided durability and fatigue life are successfully demonstrated on development test articles and the full-scale static tests are successful [47, 49]. A successful durability test culminates with a demonstration of residual strength. The environmentally compensated residual strength load level requirement depends on the level of damage present (Tables 11.6 and 11.7).

11.4.5.2 Rotor Blades and Dynamic Components

Fatigue spectra for rotorcraft have a very high number of cycles per flight as compared to fixed-wing aircraft. A significant difference in traditional rotorcraft and fixed-wing structural durability substantiation programs is that constant-amplitude loading is often favored over spectrum loading for high-frequency dynamic components. Alternatively, high-frequency cyclic stresses can be superimposed on the mission-dependent low-cycle load providing a spectrum load test. Similar to fix-wing aircraft, the large number of stress cycles requires a truncation level to be established to eliminate loads from the test spectrum that will not propagate damage in the aircraft's lifetime. A modified safe-life approach is typically used to generate stress versus life or cycle (S-N) curves for each critical part [54]. With the impetus of DOD and FAA toward implementing damage tolerance requirements for rotorcraft, the flaw-tolerant or enhanced safe-life approaches can be used to identify a component-based retirement life. The retirement life is then defined as a period of operation of the structure that may be affected by flaws of expected types and sizes (as defined in the threat assessment) with only routine inspections necessary [42].

Traditional rotorcraft fatigue substantiation programs use multiple full-scale specimen fatigue tests (4–6 articles) rather than just one full-scale test for fixed-wing programs [54]. The test articles typically contain anticipated damage levels and as manufactured flaws (ply distortion, embedded material, etc.). Since the rate of accumulation of stress cycles for rotor blades, hubs, and other rotor head components is very high, the goal of testing is to demonstrate that stresses are below the component or material endurance limit. Endurance limits may be based on S-N curves for safe-life evaluations or crack growth curves (*da*/*dN* curves) and delamination growth curves for damage tolerance evaluations. Crack growth curves – crack length versus number of load cycles – and crack growth rate curves are traditionally defined for metallic components for which the crack front and thus crack length are easily identified in test specimens. Their application to composites is complicated by non-self-similar crack growth and ill-defined crack and damage fronts. The fatigue endurance limits are developed with an understanding that the dynamic components include the effect of manufacturing and in-service damage consistent with flaw tolerance safe-life and damage tolerance approaches [45].

11.4.6 Hybrid–Composite/Metallic Structures

Material and design constraints often require a combination of composites and fatigue-sensitive metals in airframe structures. Since truncation and clipping strategies for substantiation test loading spectra development differ for composites and metals, great care should be taken to assure that each major component is exercised as close as practical to its full service life. However, the high loads needed to demonstrate the composite durability and environmentally compensated static strength capability may result in metal fatigue crack retardation or exceedance of the metal's yield or ultimate load capability. If the composite component durability has sufficiently been verified in the design development tests, the loading spectrum for the full-scale durability test should be focused on demonstrating the durability of the fatigue-critical metal parts [49, 51, 55]. Based on past experience, since the durability performance of composites is generally excellent when the structure is adequate to meet its static strength requirements, the full-scale static strength test is essential for verification of the composite structure. The full-scale static test is also essential for substantiation of the metallic components of the structure [51]. This being said, it is important to carefully inspect the composite structures on the full-scale fatigue test article to ensure that no unforeseen failures were induced in test.

Approaches for demonstrating static strength requirements for hybrid composite/metallic structures provide a limited number of options that include load-based (discussed in Sect. 4.4) and strain-based substantiation. The more commonly used method for full-scale tests of hybrid structures is strain-based substantiation which is dependent on the ability to accurately predict the strains measured on the test article up to DUL. For the strain-based approach, critical locations of the structure are instrumented to monitor strains and the structure is loaded at ambient room temperature to the DUL. The strain-based approach requires that the predictions for the strains at the critical location match the measured strains for loads up to DUL. The analysis tools are developed throughout the design development tests and validated at the subcomponent levels for environmentally conditioned test articles. This, along with accurate predictions of the fully assembled structural response for room temperature conditions, implies that the method can accurately predict the response for the assembled structure for worst-case temperature and moisture conditions. The strains measured at DUL in the critical location of the structure must be less than the statistically derived failure strains in the environmentally conditioned development tests for the same design details and loadings. Development testing must also show that there is no change in failure mode between room temperature and environmentally conditioned specimens [10].

11.4.7 Substantiation Methods

11.4.7.1 Life Factor Approach

The use of the Life Factor approach is common when attempting to substantiate a design based on an analytical fatigue life prediction, and has been in use for metallic structures for many years. Commonly used in spectrum fatigue, the Life

Factor approach simply multiplies the load cycle count at a particular load level for the test condition by the fixed life factor, effectively applying multiple lifetimes of the master events spectrum (or truncated test spectrum) to determine any apparent flaws in the design. In most cases, a Life Factor of two is used to substantiate fixed-wing designs and non-dynamic components of rotary wing aircraft.

In general, this approach works very well for metallic structure. The initiation of metallic fatigue damage is fundamentally dependent on the load levels applied during test (or during operation). In most cases, the master events spectrum developed during system design will not accurately represent actual usage, but it approximates the operational spectrum of the aircraft to a certain extent. From a metallic fatigue initiation perspective, one of the key components to this is the accurate prediction/determination of the maximum loads in the spectrum. Once initiated, the propagation of cracks will depend on the number of load cycles applied, and the magnitude and sequencing of those load cycles. In this manner, the Life Factor approach offers a great advantage, as it provides a higher level of damage propagation cycles than will ever be encountered during operational usage. Since the aircraft were either designed to be safe-life or damage tolerant, it is thereby possible to ensure that any damage accumulated during the multiple lifetimes of fatigue test will be more severe than the actual aircraft. So long as the fatigue test is carried out so that the fatigue test article accumulates hours faster than the fleet leading aircraft, the potential for operational danger is very low. In the case of metallic fatigue, the ability to detect design/manufacturing flaws, extend the life of aircraft, and develop and test retrofits are great advantages of the Life Factor approach. These capabilities are only limited by the budget available for testing.

Unfortunately, composite materials do not benefit as much from the Life Factor approach. While metallic fatigue is dependent upon both the load levels and the number of cycles (with damage occurring even at low loads), composite fatigue in structural components is much less dependent on the number of cycles until damage is initiated. In many cases, with the allowables and design methodologies that have been developed for composite materials, there is such a significant margin between operational strains and damage-inducing fatigue strains that the composite materials are significantly below their endurance limit, hence the common use of spectrum truncation. In these cases, all that is learned from Life Factor composite fatigue testing is that the composite is not fatigue sensitive for the spectrum tested. This being said, Whitehead provides a methodology for calculating the improvement in confidence levels demonstrated during test based on additional test duration (at a fixed load level) [47]. As might be expected, the return on additional testing is nonlinear, with diminishing returns as the test duration grows. With the shape of typical fatigue S-N plots shown in Fig. 11.19, this is to be expected.

In a safe-life composite design, it is common to fatigue the components to a fixed life factor, fully inspect the composite components to ensure there is no damage, and then operate the vehicle within the operational limitations set by the metallic fatigue. In a damage tolerant design, it has become more common to induce the maximum tolerable flaw size in the composite prior to test, or in between the first and second lifetimes, fatigue the component, and evaluate the flaw for growth. Both of these

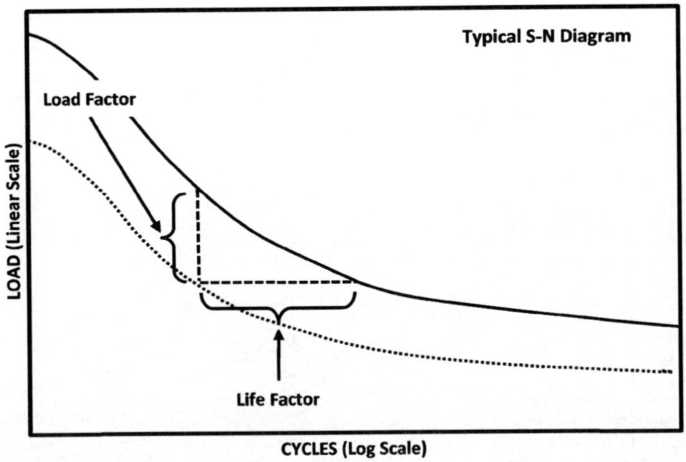

Fig. 11.19 Load and life factors [45]

methods result in a reliable data set that is the basis for vehicle certification. However, they also leave a vexing question: what if there is a significant change in the operational loads (due to additional stores carriage, engine upgrades, changes in mission, or weight growth of the vehicle) over the course of the operational life of the vehicle?

11.4.7.2 Load Enhancement Factor Approach

The development of the Load Enhancement Factor (LEF) approach to composite fatigue began in earnest with a series of phenomenological studies performed by Northrop [47] in the 1980s. As discussed above, the end result of much of this work was a mathematical model that related a typical reliability scatter model (Weibull) to the statistical confidence levels that could be substantiated with a given test profile. The basic equation that is the center of the LEF methodology is given below:

$$LEF = \frac{\mu\Gamma\left(\frac{\alpha_R+1}{\alpha_R}\right)}{\left(\frac{-\ln(p)}{x_\gamma^2(2n)/2n}\right)^{\frac{1}{\alpha_R}}} \tag{11.3}$$

where Γ is the gamma function, and n the number of test replicates (sample size),

$$\mu = \frac{\left[\Gamma\left(\frac{\alpha_L+1}{\alpha_L}\right)\right]^{\frac{\alpha_L}{\alpha_R}}}{\Gamma\left(\frac{\alpha_R+1}{\alpha_R}\right)} \tag{11.4}$$

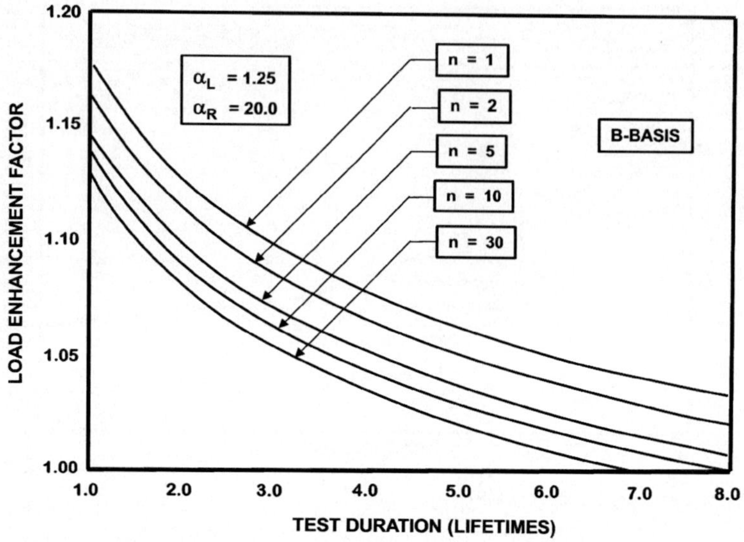

Fig. 11.20 Load and life factors [47]

and the desired probability of survival is

$$p = \exp(\ln(l)N^{\alpha_L}) \tag{11.5}$$

with α_L and α_R being the Weibull shape parameter for fatigue life and residual strength, respectively, l being the required reliability at γ confidence and N representing the test duration [47].

As with the Life Factor approach, the LEF approach is a method for providing statistical confidence in the lack of measured failure in test. In its purest sense, the LEF is derived from the basic S-N curve behavior shown in Fig. 11.20. As such, it is of no value to apply it to constant amplitude fatigue testing. Instead, its value comes when performing spectrum fatigue, as it allows for the enhancement of every load in the spectrum by a fixed value.

While there are advantages to the methodology, and its use has become common within the certification community, there are some notable drawbacks. Perhaps foremost among these is the limitation that metallic structure – including fasteners – places upon load enhancement. Even the slight enhancement (3–5%) commonly used in certification to provide statistical compensation for actual test time pushes the limit of aluminum fatigue in many designs. This is the result of the nonlinear relationship between load and damage (both initiation and propagation) rates in metallic structure. It has been argued in many cases that the use of load enhancement factors artificially penalizes aluminum structure, often leading to the reconsideration of rework and retrofits that are developed. In many cases, full-scale structural testing does not use a LEF for this very reason, choosing instead for the simpler Life

Enhancement approach. However, in cases of all composite subcomponents, it is possible to significantly reduce testing time and the number of test articles required for a given level of statistical confidence by enhancing the load (at times up to 10%).

Philosophically, the LEF approach is commonly paired with the Life Factor approach to provide a complete certification picture for the entire vehicle. For example, if a vehicle has a hybrid structure with composite skins and metallic substructure, it is likely that the component fatigue test would be performed with a Life Factor of two and an LEF of 1.0 (no enhancement). However, if there was an all-composite joint that was a critical load path, it would likely be tested in a subcomponent fatigue test using a Life Factor and LEF combination that would provide B-Basis statistical confidence in a multi-lifetime test (for one lifetime of operation). As shown in Fig. 11.20, this could be achieved using any point along each of the lines, depending on how many test articles (n) were to be tested [47].

Now that some early composite applications are reaching the end of their design services lives, certification agencies are looking for a means of evaluating end-of-life performance of composite structure. In some cases, it has been possible to use the LEF methodology as a means to compare the relative confidence conditions present for various test articles (including operational vehicles) [56, 57]. While these methodologies do provide a comparison of confidences, they cannot ultimately provide more than that, as they are not linked to failure data. Using mechanisms such as this, it is possible to overcome the fundamental problem of operational envelope variation to some extent. By using spectrum comparison methods, which can be derived in several ways from the fundamental mathematics of the LEF, it is possible to recalculate the confidence intervals when comparing a fixed test spectrum to a variable operational spectrum. Unfortunately, unless the initial testing was done with enough foresight to use an artificially elevated LEF or an artificially lengthened test life, these methods always result in confidences less than those obtained for initial certification.

11.4.7.3 Quasi-Static and Ultimate Strength Approaches

While not commonly used for component structural certifications, the Quasi-static methodology is a potent tool for use at lower levels of the building block. The premise of the methodology is simple, and can save considerable time and cost during certification. In an all composite structure – or a structure with non-fatigue critical metallic elements (e.g., oversized titanium) – this approach can be used to validate that the structure is not fatigue sensitive. However, it cannot be used to effectively determine the fatigue life of elements of a structure (as the Life Factor and LEF methodologies can).

For each loading mode, the structure is exercised to a given Life Factor (usually two lifetimes) of constant amplitude fatigue, with the applied load designed to induce stresses in the structure at a fixed reference level. Depending on the level of the building block, the fixed reference level may vary. For example, at lower levels of the building block, it is common to employ the Quasi-static methodology with a reference load level of DLL/DUL multiplied by the relevant design allowable. This effectively demonstrates that a particular failure mode is not fatigue critical (e.g., if the structure is designed to withstand static loading at ultimate load, it can withstand two lifetimes of cycling at limit load).

Alternatively, if a particular coupon or element configuration cannot pass the Quasi-static load level testing, it is possible to repeat the testing at a lower reference load level (iterating until a successful test is achieved). The resulting load level can be used to define a new material or design allowable (depending on the type of test conducted) for that failure mode by multiplying by the factor of uncertainty (usually 1.5). This allowable may then be used for static stress analysis, since it will always be lower than the test validated static allowable for that failure mode.

However, this is generally not possible for higher order building block testing. In structural subcomponents or components where this methodology is employed, the complexities driven by multiple load paths within the structure preclude such simple analyses. As such, the Quasi-static methodology is used in these cases to simulate the stresses seen at the maximum design load for a particular loading case. Naturally, this could be equivalent to the coupon level test for the critical load case at the location of minimum margin. In many cases, this condition is never reached due to the load shedding that arises from multiple load paths present in actual structure.

Alternately, the Ultimate Strength approach provides a means to design using a reduced allowable (or an enhanced factor of uncertainty) to provide additional robustness against fatigue sensitivity. The calculation of the factor required for this is presented elsewhere [47], but the end result is a structure with an analytically insensitive design for fatigue. At this point, there is no theoretical need to verify the fatigue performance of the design. However, from a practical perspective, some level of fatigue validation is always required for certification. At the point at which fatigue testing is done (even on the coupon level), the fundamental differences between the Ultimate Strength approach and the Quasi-static approach are irrelevant.

The key to the robustness of the Ultimate Strength and Quasi-static methodologies lies in the simple fact that operational stresses (that are not of fixed amplitude) will never exceed those placed upon the structure in this manner. Unfortunately, it is limited in its utility beyond simple verification that certain failure modes are not fatigue sensitive. From the simple reason that it cannot be used to establish the fatigue life of components, it has limited value if the test article does not pass. This, more than the complexity of loading, is the primary reason that it is only used at the coupon and element levels of the building block.

11.5 Concluding Remarks

An accurate understanding and characterization of a structure's operating environment – including damage threats – during the design phase has a most significant impact on its long-term performance and economic viability. Durability requirements can drive the design of a significant percentage of aircraft structures. The need to meet platform performance and production costs can sometimes lead to compromises in durability and damage tolerance, often at the expense of decreased margins of safety and increased supportability and life cycle costs. As with all designs, the risk to cost and safety for a particular design must be managed and weighed against its performance benefits. The building block approach and verification methodologies discussed within this chapter provide source data to aid designers in predicting the impact of structural design modifications. The durability and damage tolerance data is also the source data to support damage disposition and repair designs throughout the life of the structure.

Characterizing the sufficiency of a design to meet the requirements for an application should be assessed on the smallest possible test specimens that represent the structural feature being evaluated. Cost can be a strong consideration during material screening; however, the cost is only a small percentage of the investment required to complete the building block program and even a smaller percentage of the cost associated with maturing a design with an inadequate material. Therefore, adequate resources should be expended early in a program to ensure the right material is selected for the application. Models used to predict the linear and nonlinear stress–strain behavior of materials are used throughout the design process to size structures. However, physics-based *strength* prediction models based on fundamental constituent material behavior are typically not used because they are not accurate and robust for the variety of loading conditions seen in aircraft structures (see Sect. 4). For example, strength prediction models do not typically account for processing and manufacturing anomalies resulting from scale-up from coupon to full-scale structures that can adversely affect damage resistance and tolerance. These effects, as well as the material statistical variability, are inherently accounted for in the building block empirical methodology. Consequently, utilization of models that predict *failure* or *strength* is limited in the durability and damage tolerance building block approach because designers rely on test data and empirically based models to establish allowables and to characterize damage resistance and tolerance. Although the number of specimens needed to characterize the material strength can be significant, the risks associated with maturing a design using inaccurate strength predictions far outweighs the cost of empirically characterizing material strength. Thus, for the building block approach, designers rely on empirically based *descriptive models* rather than *predictive models* for composite *strength.* Predictive models for *strength* are better utilized for designing materials, for predicting the suitability of a material for a given application, for disposition of the effects of in-service damage outside the bounds of that tested in the durability and damage tolerance program, and for the fundamental design of an appropriate, cost-effective building block test program. Advancements in analytical modeling, fatigue behavior, crack/damage growth, and the effects of defects (both as manufactured and service induced) are desired in order to refine today's methods and improve predictive capability.

The goal of the durability and damage tolerance design process is to attain a structure that is safe and can be economically maintained for its service life. However, full-scale static strength and fatigue substantiation testing and analyses often reveal critical features of the structure that require additional attention such as modifications, regular inspections, or regular maintenance actions. Monitoring and inspection of these critical features for safety of flight structures – known as control points - must be scheduled to support the ongoing evaluation of structural integrity and to ensure damage never reaches the sizes that can cause catastrophic failures. Managing an aircraft fleet's durability can be achieved through implementation of a Force Structures Maintenance Plan (FSMP) that defines when, where, how, and the estimated cost of modifications, inspections, and maintenance actions [2]. The FSMP should be regularly updated based on usage severity data (IAT data) to modify inspection intervals to reflect differences between the design usage and actual usage. Projected maintenance actions based on analytical condition inspections that evaluate the corrosion, fatigue, and other damage concerns in fleet leading aircraft (aircraft with the highest flight hours or fatigue life expended) should also be included in the updates to the FSMP.

There are numerous topics and shortcomings of current composite durability design and substantiation practices that need further study and attention. In particular, the ability to fully exercise both metal and composite structural components in full-scale tests of hybrid structures is lacking. This topic is of special interest due to recent hybrid aircraft structural weight reduction programs and advancements in design optimization tools that drive design margins-of-safety to near zero for all components. The inability to fully substantiate all aspects of the design could eventually result in safety concerns and prohibitive sustainment costs.

Acknowledgments The authors acknowledge contributions from David Stone from US Army, Aviation Engineering Directorate, Aviation and Missile RDEC for his many discussions on the US Army's substantiation methods for rotorcraft. In addition, the contributions of Kevin Miller, Wayne Koegel, David Quinn, and David Barrett from the Naval Air Systems Command were critical, in particular for their knowledge of design substantiation methods for Naval systems. Finally, the consultations with Jeffery Hendrix, Carl Rousseau (Lockheed Martin), Harlan Ashton (Boeing), and Charles Babish (Air Force, Aeronautical Systems Center) were of tremendous value.

References

1. Ransom JB, Glaessgen EH, Raju IS, Knight NF, and Reeder JR, 2008, 'Lessons Learned from Recent Failure and Incident Investigations of Composite Structures,' Proceedings of the 49th AIAA/ASME/ASCE/AHS/ASC Structures, Structural Dynamics, and Materials Conference, 7 – 10 April 2008, Schaumburg, IL, AIAA 2008–2317.
2. MIL-STD-1530C, 2005, DOD Standard Practice, Aircraft Structural Integrity Program (ASIP).
3. Mohaghegh M, 2004,'Evolution of Structures Design Philosophy and Criteria,' 45th AIAA/ASME/ASCE/AHS/ASC Structures, Structural Dynamics & materials Conference 19–22 April 2004, Palm Springs, CA, AIAA 2004–1785.
4. MIL-HDBK-17F, 2002, DOD Composite Materials Handbook Volume 3. Polymer Matrix Composites Materials Usage, Design, and Analysis.
5. ADS-13F-HDBK, 1997, Aeronautical Design Standard Handbook; Air Vehicle Materials and Processes, United States Army Aviation and Troop Command.
6. Tomblin JS, Ng YC, Raju KS, 2003,'Material Qualification and Equivalency for Polymer Matrix Composite Material Systems: Updated Procedure,' DOT/FAA/AR-03/19
7. AC 20-107B, 2009, DOT Federal Aviation Administration Advisory Circular on Composite Aircraft Structure.
8. Reddick HK, 1983,'Safe-Life and Damage-Tolerant Design Approaches for Helicopter Structures,' In NASA. Langley Research Center Failure Analysis and Mechanics of Failure of Fibrous Composite Structures, pp 129–152.
9. Lieblein S, 1981,'Survey of Long-Term Durability of Fiberglass-Reinforced Plastic Structures,' NASA CR-165320.
10. JSSG-2006, 1998, DOD Joint Service Specification Guidance, Aircraft Structures.
11. MIL-HDBK-310, 1997, DOD Handbook, Global Climatic Data for Developing Military Products.
12. Rousseau C, 2001, Test Program Planning, *ASM Handbook*, Volume 21 Composites, ASM International.
13. Wilkins DJ, 1983, A Preliminary Damage Tolerance Methodology for Composite Structures, In NASA. Langley Research Center Failure Analysis and Mechanics of Failure of Fibrous Composite Structures pp. 67–94.
14. Reddy DJ, 2006, Composites in Rotorcraft Industry & Damage Tolerance Requirements, Presented at the FAA Composites Workshop, Chicago, IL, July 19–21, 2006.
15. ADS-51-HDBK, 1996, Aeronautical Design Standard Handbook, Rotorcraft and Aircraft Qualification (Raq) Handbook.
16. Pagano NJ, Schoeppner GA, 2000, Delamination in Polymer Matrix Composites: Problems and Assessment', *Vol. 2 Polymer Matrix Composites*, in series *Comprehensive Composite Materials*, Elsevier, New York.
17. Davis GW, Sakata IF, 1981, Design Considerations for Composite Fuselage Structure of Commercial Transport Aircraft, NASA CR-159296.
18. FAR PART 23—Airworthiness Standards: Normal, Utility, Acrobatic, And Commuter Category Airplanes, § 23.573 Damage tolerance and fatigue evaluation of structure.
19. Vosteen LF, Hadcock RN, 1994, Composite Chronicles: A Study of the Lessons Learned in the Development, Production, and Service of Composite Structures, NASA CR-4620.
20. MIL-STD-889B, 1993, DOD Standard Practice, Dissimilar Materials.
21. ASTM G82-98, 2009, Standard Guide for Development and Use of a Galvanic Series for Predicting Galvanic Corrosion Performance.
22. Aylor DM and Murray JN, 1992, The Effect of a Seawater Environment on the Galvanic Corrosion Behavior of Graphite/Epoxy Composite Coupled to Metals, Carderock Division Naval Surface Warfare Center, CDNSWC-SME-92/93.

23. 'Corrosion Prevention and Control Planning Guidebook Spiral 3', 2007, Department of Defense, Principle Deputy Under Secretary of Defense, Acquisitions, Tech and Logistics, Washington DC.
24. Campbell FC, 2004, Manufacturing Processes for Advanced Composites, Elsevier Ltd.
25. Hoeckelman LA, 2001, "Environmental Protection and Sealing," *ASM Handbook*, Volume 21 Composites, ASM International.
26. MIL-HDBK-1568, 1996, DOD Standardization Handbook, Materials and Processes for Corrosion Prevention and Control in Aerospace Weapons Systems.
27. MIL-HDBK-729, 1983, DOD Standardization Handbook, Corrosion and Corrosion Prevention Metals.
28. Gintert LA, Hihara LH, 2002, 'Corrosion Considerations for Military Applications of Composite Materials,' CORROSION 2002, NACE International, Houston, Paper 02162.
29. Harris CE, Starnes JH, Shuart MJ, 2001, An Assessment of the State-of-the-Art in the Design and Manufacturing of Large Composite Structures for Aerospace Vehicles, NASA/TM-2001-210844.
30. Demuts E, Whitehead RS, and Deo RB, 1989, 'Assessment of Damage Tolerance in Composites,' *Composite Structures*, vol. 4, pp. 45–58.
31. Baker AA; Jones R; Callinan RJ, 1985, 'Damage Tolerance of Graphite/Epoxy Composites,' *Composite Structures*, vol. 4, pp. 15–44.
32. ARP-5414, 2005, Aircraft Lightning Zoning, Society of Automotive Engineers, Inc. (SAE) Aerospace Recommended Practice.
33. Gardiner, G, 2006, 'Lightning Strike Protection for Composite Structures' *High Performance Composites*, http://www.compositesworld.com/articles/lightning-strike-protection-for-composite-structures, Accessed in 2009.
34. AC 20-53B, 2006DOT Federal Aviation Administration Advisory Circular on Protection of Aircraft Fuel Systems Against Fuel Vapor Ignition Caused by Lightning.
35. AC 20–136, 1990, DOT Federal Aviation Administration Advisory Circular on Protection of Aircraft Electrical/Electronic Systems Against the Indirect Effects of Lightning.
36. MIL-STD-1795A, 1989, DOD Standard Practice, Lightning Protection of Aerospace Vehicles and Hardware.
37. MIL-STD-1757A, 1983, DOD Standard Practice, Lightning Qualification Test Techniques for Aerospace Vehicles and Hardware.
38. MIL-B-5087, 1964, DOD Specification, Bonding, Electrical, and Lightning Protection for Aerospace Systems.
39. Schoeppner GA, Curliss DB, 2002,'Model-Based Design for Composite Materials Life Management,' Proceedings of the 9th AIAA/ISSMO Symposium on Multidisciplinary Analysis and Optimization Conference, 4–6 September 2002, Atlanta, GA, Paper number AIAA-2002-5516.
40. Tomblin J, Lacy T, Smith B, Hooper S, Vizzini A, Lee S, 1999, Review of Damage Tolerance for Composite Sandwich Airframe Structures, DOT/FAA/AR-99/49.
41. Lincoln JW, Yeh HC, 1999, Treatment of High-Cycle Vibratory Stress in Rotorcraft Damage Tolerance Design, RTO AVT Specialists' Meeting on Application of Damage Tolerance Principles for Improved Airworthiness of Rotorcraft, held in Corfu, Greece, 21–22 April 1999, and published in RTO MP-24.
42. Lazzeri L, Mariani U, 2009 'Application of Damage Tolerance Principles to the Design of Helicopters,' *Int J, Fatigue*, vol. 31, pp. 1039–1045.
43. Miner MA, 1945, 'Cumulative Damage in Fatigue,' *J. Appl. Mech.*, pp. A159 – A164.
44. FAA FAR Part 25, Airworthiness Standards: Transport Category Airplanes.
45. AC 35.37-1A, 2001, DOT Federal Aviation Administration Advisory Circular on Guidance Material for Fatigue Limit Tests and Composite Blade Fatigue Substantiation.
46. AC 20-66A, 2001, DOT Federal Aviation Administration Advisory Circular on Vibration and Fatigue Evaluation of Airplane Propellers.

47. Whitehead RS, Kan HP, Cordero R, Saether ES, 1986, 'Certification Testing Methodology for Composite Structures, Volumes I and II,' Report numbers NADC-87042-60 and DOT/FAA/CT-86/39.
48. Whitehead RS, 1991, Lessons learned for composite structures, NASA. Langley Research Center, The First NASA Advanced Composites Technology Conference, Part 1, pp. 399–415.
49. Kan HP, Cordero R, Whitehead RS, 1997, Advanced Certification Methodology for Composite Structures, DOT/FAA/AR-96/111.
50. Lincoln J, 1986, 'Certification of Composites for Aircraft,' Proceedings of the 1986 Aircraft Structural Integrity Conference.
51. Lincoln JW, 2000, 'USAF Experience in the Qualification of Composite Structures,' *Composite Structures: Theory and Practice, ASTM STP 1383*, P. Grant, Ed., American Society for Testing and Materials. West Conshohocken. PA, pp. 1–11.
52. AC 25.571 -1C, 1998, DOT Federal Aviation Administration Advisory Circular on Damage Tolerance and Fatigue Evaluation of Structure.
53. CMH-17-3F, Composite Materials Handbook, Volume 3. Polymer Matrix Composites Materials Usage, Design, and Analysis.
54. AC 29-2C, 2003, DOT Federal Aviation Administration Advisory Circular on Certification of Transport Category Rotorcraft.
55. Rouchon J, 2007, How, Over the Past 30 Years, "Part 25" Composite Structures Have Been Coping with Metal Minded Fatigue and Damage Tolerance Requirements, 24th ICAF Symposium, Naples, Italy, 16 May 2007.
56. Tillman MS, 2009, Addendum to NAWCADPAX/TR-2009/139 – Evaluation of Flight Load Spectrum and Fatigue Test Spectrum Severity, Report No: NAWCADPAX/TM-2009-169, Naval Air Systems Command, Patuxent River, MD.
57. Tillman MS, Tsai HC, Peek M, 2009, An Investigation of the End-of-Life Residual Strength of the F/A-18A-D Inner Wing Step Lap Joint, Report No: NAWCADPAX/TR-2009/139, Naval Air Systems Command, Patuxent River, MD.

Chapter 12
Durability of Structural Joints

Carl Q. Rousseau and Endel V. Iarve

Abstract Structural joints are where the durability, or lack of durability, is usually most evident in composite structural response to combined environmental effects and mechanical loads. The overall durability or degradation of a structural joint is a function of the response of all of the constituents of the joint: individual joined members, fasteners and/or adhesive, shims, sealants, coatings, etc. The objective of this chapter shall be to provide a brief overview of observed durability behavior of composite structural joints, experimental techniques for exploring the durability of joints, analytical predictive methods, and certain empirical case studies, mainly from the aerospace industry. The chapter is logically divided into separate bolted and bonded sections.

12.1 Introduction

Long-term structural durability and degradation of polymeric matrix composite materials is a multifaceted, multidisciplinary issue and is addressed from many different angles in this book. Section C covers various aspects of structural response, including thermomechanical damage and failure, fatigue behavior, impact damage, and – in this chapter – durability of bonded and bolted joints.

The overall durability or degradation of a structural joint is a function of the response of all of the constituents of the joint:

- Individual joined members
- Fasteners and/or adhesive
- Shims
- Sealants
- Coatings

C.Q. Rousseau (✉)
Lockheed Martin Aeronautics Co., 748 MZ 6516, Fort Worth, TX 76101, USA
e-mail: carl.q.rousseau@lmco.com

K.V. Pochiraju et al. (eds.), *Long-Term Durability of Polymeric Matrix Composites*, 483
DOI 10.1007/978-1-4419-9308-3_12, © Springer Science+Business Media, LLC 2012

The basic response of individual joined members, as well as composite shims, to long-term environmental exposure and mechanical loading is discussed in sections A and B. The individual response of fasteners, adhesives, metallic shims, sealants, and coatings to environmental exposure and mechanical loads is beyond the scope of this book, but well covered in several extensive references [1–4]. The combined response of these constituents in bonded and/or bolted joints is the topic of this chapter.

The objective of this chapter shall be to provide a brief overview of:

- Observed durability behavior of composite structural joints
- Experimental techniques for exploring the durability of joints
- Analytical predictive methods

The chapter is logically divided into separate bolted and bonded sections, as well as subsections addressing the above-noted issues.

12.2 Bolted Joint Durability

12.2.1 Overview

The basic components of a bolted joint generally include two or more structural elements, fasteners, shims, sealants, and coatings, as shown in Fig. 12.1. Important design details that affect durability include part and hole tolerances, the bolt torque and if/how it might relax over the life of the structure, thermal and chemical mismatch of dissimilar materials, and fretting/wear due to joint compliance. The following subsections address observations regarding long-term bolted joint behavior, experimental techniques for testing, analytical methods for prediction of joint durability, and a brief summary of key bolted joint durability issues.

12.2.2 Key Observations on Long-Term Behavior

Bolted joints have the potential to exhibit failure mode changes between static loading and long-term thermomechanical fatigue loading. Basic bolted joint failure modes are shown in Fig. 12.2. While a joint might fail statically in bolt bending, after a lifetime of fatigue loading, it might instead become bearing or pull-through-critical. This change in response is due to fatigue-induced and environmental changes in joint load path and/or constituent material properties. The most common of these changes are relaxation of bolt torque, due to through-thickness creep of joined composite elements and/or geometric changes in the fastening system (nut back-off, tail-element deformation, head deformation, etc.); and environmental degradation of shims/gaskets and/or composite structural elements (e.g., galvanic corrosion of fasteners or metallic shims and substrates;

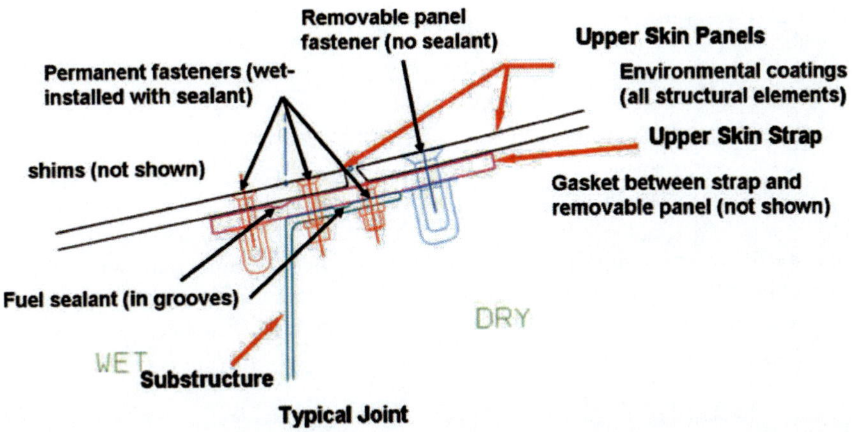

Fig. 12.1 Typical bolted joint configuration

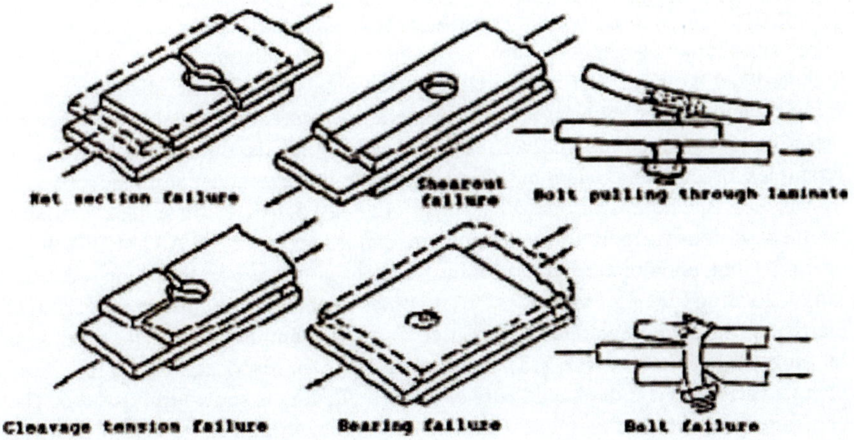

Fig. 12.2 Bolted joint failure modes

embrittlement of composite matrix or plastic shim material; etc.), leading to stiffness and/or strength reductions. Thus, well-designed bolted joints must use materials whose long-term response/degradation is understood, and which are then designed to avoid a critical amount of degradation during the joint's design lifetime.

Given large differences in environment, loading conditions, design life, etc., this leads to very different typical joint concepts and materials in various different applications (e.g., marine, aeronautics, space, infrastructure, electronics, etc.). However, all of these applications share a common design goal of not allowing a mechanically fastened joint to be the weak link in the durability of their structures

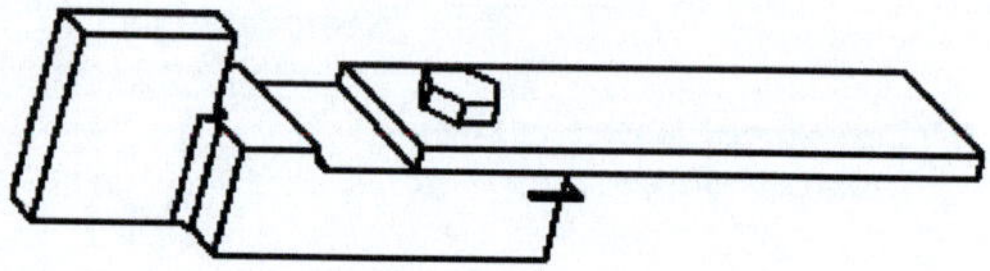

Fig. 12.3 Composite bearing test specimen

(except for a few unique fuse-like applications). For the most part, all of these industries rely on empirical (test-based) design criteria and conservative bolted joint design rules to ensure that their joints have the required durability.

12.2.3 *Experimental Techniques for Bolted Joint Durability Testing*

As noted above, both individual constituent (laminate, bolt, shim, etc.) and overall joint sub-element response are required to be measured experimentally. Standardized coupon test methods for individual constituent stiffness, static strength, and fatigue endurance measurement are generally available in the open literature [5–9] and summarized in several useful guidance documents [1, 10–12]. Examples of composite laminate bearing, pure bypass, and combined bearing/by-pass test specimens are shown in Figs. 12.3–12.5. All of these tests can also be run at various temperatures and moisture conditions (e.g., use ASTM D5229 [5] or equivalent for moisture conditioning), and in constant amplitude or spectrum fatigue loading (per ASTM D6873 [5]), in order to obtain cyclic thermomechanical elastic and strength degradation responses. A good example of such an experimental program is found in Ref. [13]. Fastener constituent response may be measured using a variety of standardized fastener tests [8, 9], e.g. as shown in Fig. 12.6. The constituent response of shim, sealant, and coating material may be measured using standardized metallic, plastic, and/or adhesive standard test methods [6, 7] or, more commonly for aerospace applications, at the sub-element level.

Sub-element-level bolted joint tests are generally nonstandard in nature (developed specifically for a given design application or class of design features), but generic guidance exists [10, 11]. Figures 12.8, 12.10, and 12.11 are examples of such sub-elements. Certain notched plate or beam coupon-level test methods, such as ASTM D5961, D6873, D7248, D7249, and D7332 [5], are easily adapted to sub-element use through slight changes in specimen configuration (to yield a range of failure modes), instrumentation (to isolate the variation in failure mode), and loading parameters. Finally, larger component-level and full-scale tests (all non-standard and design specific) can sometimes be used for bolted joint durability characterization, if taken to failure and if the failure location is (either by design or by chance) in one of the bolted joints of the component.

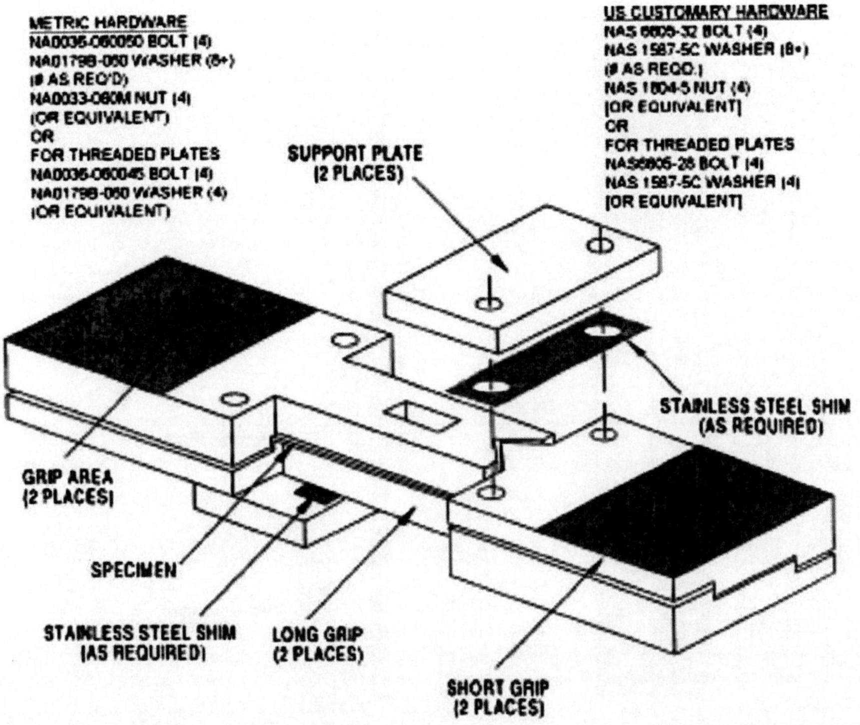

Fig. 12.4 Composite bypass compression test setup

12.2.4 Analytical Methods

Simple empirical models for bolted joint durability include sub-element fatigue-test-based S-N curves combined with Miner's Rule [14]; or elevated-load spectrum testing, from which an endurance limit is derived [15]. A notional example of a curved-beam bolted flange/skin sub-element S-N curve is shown in Fig. 12.7, and the associated test specimen in Fig. 12.8. An example of a normalized wing root bending moment spectrum is shown in Fig. 12.9, and a single-shear fastener sub-element test specimen in Fig. 12.10. Also, in lieu of a durability-based criterion, a fatigue crack growth rate (FCGR)-based delamination onset or growth criterion can occasionally be used for bolted joints, if delamination is a critical, life-limiting failure mode, such as in stiffener run-outs with bolted flange-ends, or bearingless rotor hubs with bolted attach fittings adjacent to ply drops and flexure features. A tension/bending tee/hat sub-element fatigue delamination onset/growth test specimen is shown in Fig. 12.11. In prior certification programs, such a specimen (with no artificial flaw) or a full-scale fatigue test article would be tested at an elevated constant-amplitude load level until a delamination naturally

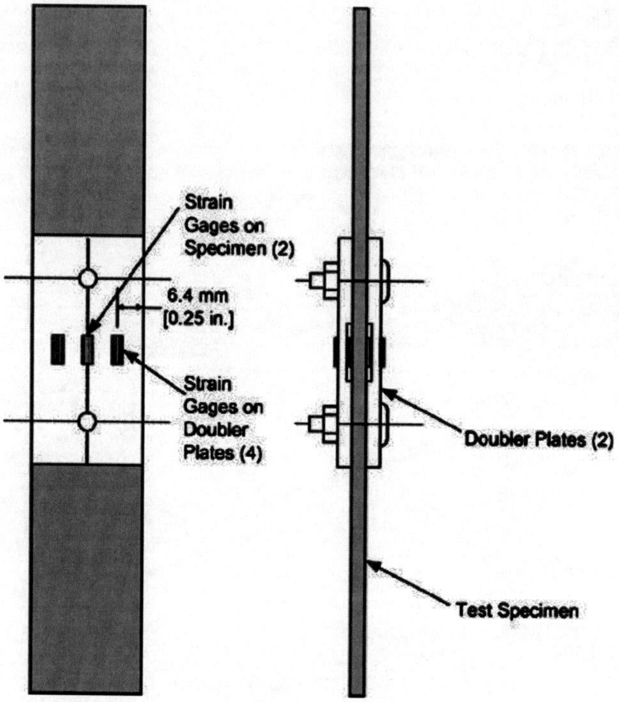

Fig. 12.5 Composite bearing/bypass test configuration

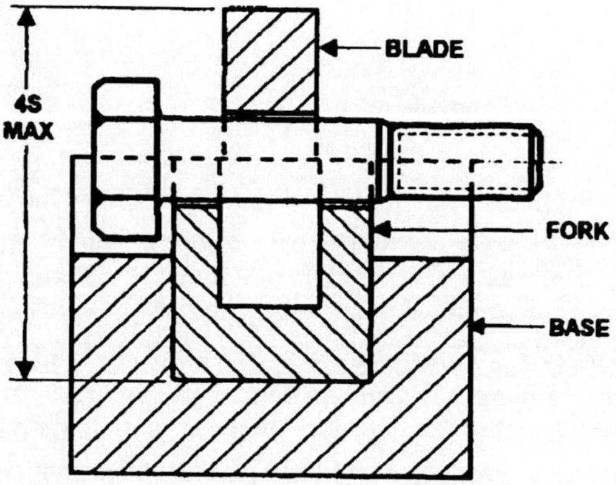

Fig. 12.6 Standard fastener test configuration

developed (defining a point on a durability S-N curve for the same configuration). Then, the specimen would be run under a design spectrum and fatigue delamination growth tracked. After 2–4 lives (depending on certification requirements), the *a* versus N response is plotted, and a FCGR criterion applied to the bolted (or bonded) sub-element feature in order to predict life and/or inspection interval.

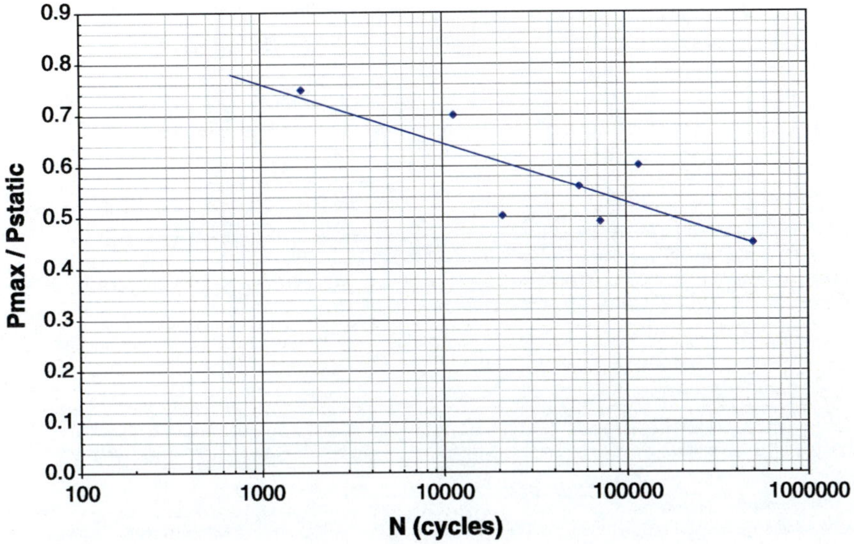

Fig. 12.7 Notional bolted skin-spar S-N curve

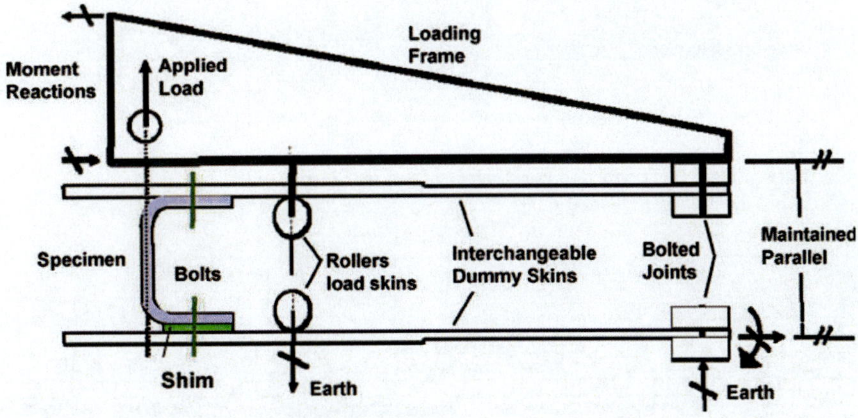

Fig. 12.8 Bolted skin-spar test specimen

The combined effects of thermal and mechanical fatigue are usually neglected in the simple analytical models used in current production airframe certification programs. However, there are occasionally situations in which bolt clamp-up (and loss thereof) is critical. In these cases, sub-element test specimens are tested under combined thermal and mechanical loading spectra. A notional thermal spectrum, over which a mechanical spectrum such as that shown in Fig. 12.9 would be superimposed, is shown in Fig. 12.12. Note that for most aircraft

Sequence	*Min Load*	*Max Load*	*Number of Constant Amplitude Cycles*
1	-0.31	+1.00	1
2	-0.06	+0.65	135
3	-0.2	+0.85	14
4	-0.25	+0.95	3
5	-0.16	+0.75	44
6	-0.22	+0.90	3
7	+0.04	+0.55	300
		Total Cycles (One Block) =	*500*

Fig. 12.9 Normalized wing root bending moment spectrum

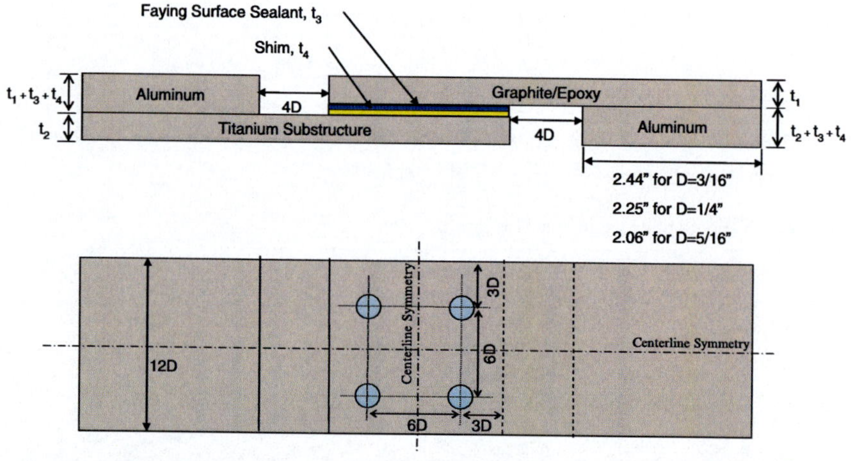

Fig. 12.10 Single-shear bolted joint sub-element test specimen

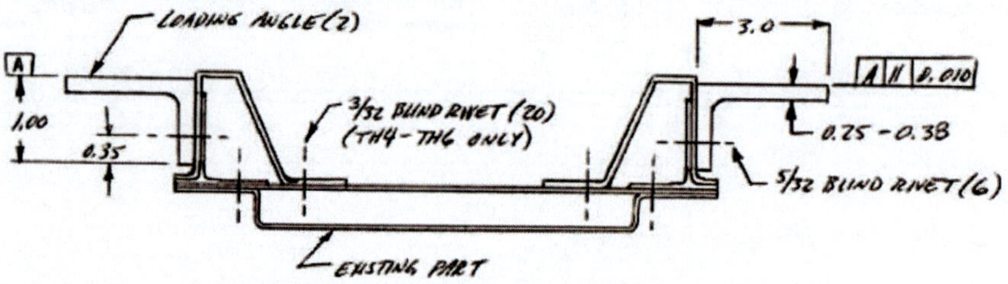

Fig. 12.11 Bolted/bonded tee/hat sub-element fatigue test specimen

airframes, a very low fraction of a part's fatigue life is spent at either very high or very low temperatures.

The state of the art in high-fidelity mechanism-based methods for durability and life prediction in composite bolted joints reflects that of unnotched composites. A considerable amount of work has been performed on static strength prediction

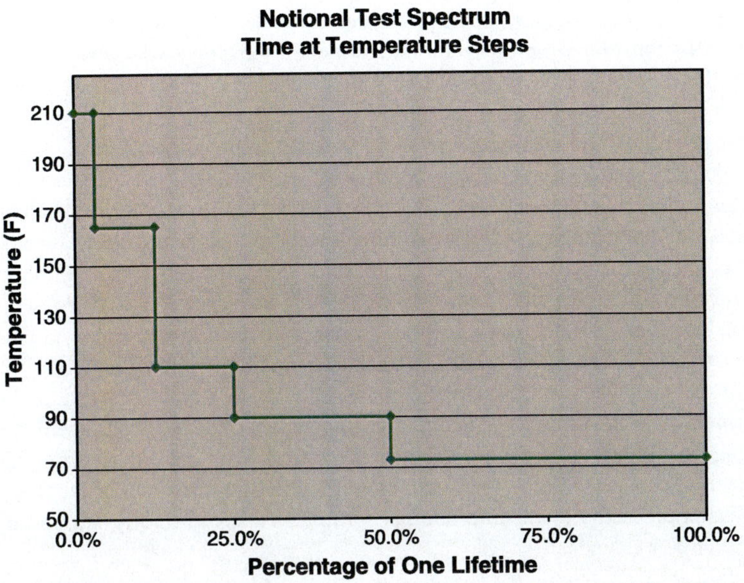

Fig. 12.12 Notional thermal test spectrum

methods in ambient environments, whereas the fatigue loading and environmental degradation such as thermal and oxidative exposure are addressed mostly by empirical models discussed above.

12.2.5 Three-Dimensional Stress Analysis

Stress analysis of composites with elastic fastener holes has evolved through a sequence of stages starting with original works by Lekhnitskii [16] extending the Kolosov-Muskhelishvili complex variable formalism for an anisotropic body.

Using finite element analysis to actually solve for the contact stresses and the contact region in a two-dimensional formulation, Crews et al. [17] showed that the cosine distribution assumption is very accurate for isotropic plates, however it underpredicts the amplitude of the contact stress by about 40% for orthotropic plates with $E_1/E_2 = 13.5$. An extensive review of composite pinned-joint failure models based on linear and/or nonlinear two-dimensional finite element stress analysis combined with various failure criteria and property degradation models was given by Lessard and Shokrieh [18].

A significant effect of stacking sequence on the pin-bearing strength of quasi-isotropic carbon-epoxy laminates has been reported by Hamada et al. [19]. The lamination and stacking sequence effect on pin-bearing strength was also addressed in much earlier papers by Collings [20] and Quinn and Matthews [21].

In engineering practice [22], however, the effect of lamination on bearing strength is generally considered unimportant due to a clamp-up force which suppresses the out-of-plane deformation in the vicinity of the hole. While variation of bearing strength as a function of stacking sequence under high clamp-up torque is small, the damage initiation load may be as low as 60% of the failure load and strongly depend on stacking sequence, according to Smith and Pascoe [23].

A three-dimensional analysis [24–27] is required to properly model the lamination and stacking sequence effects. Although the topic of three-dimensional stress analysis has been addressed for several decades, the amount of work in this area is relatively small.

Marshall et al. [24] utilized a 20-node isoparametric finite element model for 3D stress analysis of pin-loaded and fastener holes in $[0/90]_s$ and $[90/0]_s$ laminates. A rigid fastener was modeled. Inside the contact zone, which was assumed to be uniform through the thickness, the radial displacement of the plate was zero in order to simulate contact with a rigid fastener. The mesh was, reportedly, not suitably refined to accurately predict normal and shear interlaminar stresses at the free edge. The results predicted lower interlaminar normal stresses in $[0/90]_s$ supporting the experimental evidence.

Arnold et al. [25] applied an eight-node linear 3D finite element model to assess the effects of clamp-up force and washer friction on stress distribution in composite bolted joints. Due to computational restraints, homogenized orthotropic laminate properties were used. Clamp-up was modeled by applying a prescribed stress distribution (flexible washers) or displacements (rigid washer) at the plate surface. The clamp-up stress distribution with a maximum at the washer edge resulted in smoother contact stresses at the hole-edge along with higher remote shear stresses under the washer edge. The numerical results showed the friction to be beneficial in reducing the bearing stresses.

Matthews et al. [26] developed a special 20-node isoparametric finite element to represent a laminate. Analogous to higher order plate theory, it implied quadratic variation of all displacement components through the thickness. The integration method actually took into account the stacking sequence, providing results more accurate than utilization of homogenized laminate properties, and allowing a prediction of interlaminar stresses. Perfect-fit rigid pin and frictionless contact were assumed. The clamping was modeled by applying displacement boundary conditions simulating a rigid washer. The clamp-up was found to reduce the interlaminar normal stresses at the hole edge, which is consistent with increased strength observed experimentally.

We find no full-field solution in the literature for the addressed problem with converged stresses and consistent boundary conditions, such as contact stress and displacement continuity between the plate and the fastener in each ply, along with interlaminar stress continuity.

The elastic stresses at the intersection of the hole edge and ply interface exhibit singular behavior. Independent spline approximation of interlaminar stress components and displacements proposed by Iarve [29, 32, 33] along with the curvilinear transformation technique also based on spline approximation, allowed the performance of accurate three-dimensional stress prediction in practical composites containing open holes. Results for rigid fastener hole composites were obtained by using this technique by Iarve [30, 31]. Elastic fastener and countersunk fastener hole problems were addressed in Refs. [35, 36, 38]. The stress singularities [28] were addressed by combined asymptotic numerical solutions by Wang [34], Iarve and Pagano [37] and Whitney et al. [38].

12.2.6 Strength Prediction

To provide accurate strength prediction in notched composites, it is important to take into account the stress redistribution due to damage progression prior to failure. Both two- and three-dimensional models have been developed for the strength prediction of composites with open and filled holes. A review of such models was given by Camanho and Matthews [44]. We shall focus mostly on three-dimensional models where each ply is explicitly modeled as an initially orthotropic material. Our interest in three-dimensional models derives from the fact that only within this framework can the influence of stacking sequence and through-the-thickness loading (e.g., clamp-up) on strength be rigorously addressed.

Three-dimensional, property-degradation based, progressive damage finite element analysis in open-hole composite laminates was first performed about 30 years ago [41]. In Ref. [41], a quasi-isotropic $[0/45/-45/90]_s$ polymer composite laminate was analyzed. Three damage modes were considered: (1) fiber failure, (2) transverse failure or intraply cracking, and (3) delamination. In each case, both the normal and shear stress components were checked in the center of each element to determine local failure. When failure was detected in a particular element, its elastic properties were modified (degraded) in the following manner for each damage mode: (1) all components of the stiffness matrix were set to zero, (2) all components of the stiffness matrix contributing to transverse in-plane stress components were set to zero, and (3) all components of the stiffness matrix contributing to transverse out-of-plane stress components were set to zero. No interface regions between plies were modeled. It is interesting to note that for uniaxial loading, the predicted strength values reported in Ref. [41] were equal to the fiber direction tensile strength value divided by the stress concentration factor in the fiber direction prior to damage. This indicates that the progressive damage algorithm in this case did not lead to significant stress redistribution before final failure. At the time, this type of analysis was at the limit of the available computer capabilities, and no mesh refinement studies were possible.

Recently, other three-dimensional progressive damage models have been developed for strength prediction in open-hole [42, 43] and filled-hole laminates [40, 44, 45]. Three-dimensional property degradation modeling described in Ref. [42] was used in an ABAQUS [54] environment and consisted of a modified Hill failure criteria and associated property degradation rules. Exactly the same damage modes and property degradation rules were used as in Ref. [41].

A significant modification in the model was a resin-rich layer, with the properties of cured matrix, between each ply in which the type (3) stiffness degradation was applied. Due to computational constraints, this layer had the same thickness as the unidirectional plies. Several examples were considered. Axial splits in a 0° ply emanating from a crack-type notch under axial tension were modeled. Good agreement with the experimentally observed split length as a function of load was demonstrated. Symmetric cross-ply laminates with open holes under uniaxial tension were also considered. It was stated that the strength of the laminate was predicted to about 5% accuracy, and that the presence of the resin-rich layer did not significantly affect the prediction. Evaluation of the damage distribution showed qualitative agreement with the radiographic image. Residual stresses were not accounted for in the analysis.

A complex boundary condition contact problem of transverse bending of a composite plate with a countersunk bolt was considered in Ref. [46]. A complete element failure approach was used for property degradation modeling. In this model, the element stiffness was completely degraded if any one of the failure criteria were violated for the average stress component values of the element. In Ref. [46], the maximum stress failure criterion was used for in-plane damage, and Ye's [47] delamination failure criterion was used for damage caused by out-of-plane stresses. Thus, by means of the complete element failure approach, delaminations were included in the model. No residual stresses were considered. The model was shown to produce load deflection curves in good agreement with the experimental results for $[90_8/0_8]_s$ and $[0_8/90_8]_s$ laminates. However, no detailed correlation of the damage progression prediction was made with the experiment.

Complete degradation of the element transverse properties used in the cited works leads to unrealistic stiffness degradation results in unnotched laminates. Indeed, considering an unnotched cross-ply laminate, one will simultaneously find failure in all elements of the 90° ply and obtain the ply discount result for laminate stiffness degradation, which is significantly lower than the experimental values at saturation crack density in the 90° ply (e.g. [48, 49]).

An internal state variable approach to establish property degradation rules was used in Refs. [43, 44]. Two failure mechanisms were incorporated: matrix transverse cracking and fiber tensile breakage. The degraded stiffness matrix due to transverse cracking was represented by

$$C_{ij} = C_{ij}^0 e^{k_{ij} a_f},$$

where C_{ij}^0 was the initial (before damage) stiffness matrix, a_f the damage variable, and k_{ij} the rates of individual stiffness component degradation, established from the self-consistent model. Fiber breakage was detected from the maximum strain failure criterion. The commercial finite element package SAMCEF [55] was chosen to implement the methodology. The numerical results presented were for the same laminates as in the two-dimensional analysis of Tang [46]. It was concluded that more realistic transverse cracking patterns in the off-axis plies were observed.

The loading was modeled only up to about 85% of the failure due to numerical instability. No residual stress or delamination effects were considered.

The internal state variable approach developed in Ref. [46] for a two-dimensional analysis was applied in Ref. [44] in a three-dimensional analysis for the prediction of bearing strength in composite bolted joints. A rigid bolt was modeled, and the ABAQUS finite element package was used to model property degradation. Hashin's failure criterion was used to identify the damage mode in the element. Linear eight-node elements with reduced integration were used for the three-dimensional modeling with one element through the thickness of each ply. Delaminations and residual stresses were neglected. In the bearing failure mode, the analysis was stopped after the damage propagated outside the washer-covered area around the hole. The model proved capable of predicting the mode of failure: tensile, shear-out, or bearing and the initial load drop-off.

The advantage of the CDM approach is a relatively straightforward implementation into the commercial finite element programs, which greatly facilitates transition of such methods to industry as well as allows taking advantage of the advanced nonlinear capabilities of commercial packages. However, a significant drawback of such approaches is the inability to model local interaction between damage modes such as matrix cracking and delamination as well as the stress redistribution in the vicinity of cracks, i.e. fiber direction stress redistribution resulting from ply splitting [39].

A significant advancement in understanding the evolution and interaction of fundamental failure modes such as matrix cracking, delamination, and fiber failure in laminated composites with and without open holes has been achieved in a series of studies devoted to scaling effects in quasi-isotropic laminates by Wisnom and colleagues [56–58]. The numerical component of this work consisted of modeling evolution of a damage pattern consisting of predefined matrix cracks (up to three in each ply) and the delamination between plies. Interface cohesive constitutive formulation was used to model both the evolution of matrix cracking and delaminations. The fiber failure was then predicted by applying Weibull-based tensile criterion. The importance of this work is that a significant amount of experimental data was predicted based on explicit modeling of basic failure mechanisms rather than the effect they have on reduction of effective laminate properties, as in the stiffness reduction progressive damage modeling approach described above.

12.2.7 Advanced Bolted Joint Fatigue Life Modeling Methods

A recent review paper by Thoppul et al. [59] thoroughly addresses several aspects of time-dependent behavior of composite bolted joints. Time and temperature-dependent relaxation effects, including modeling techniques, fatigue and environmental effect on residual strength, were reviewed. In the present paper, only recent developments in and future direction in mechanism-based progressive damage modeling of fatigue and environmental effects will be addressed. Despite a significant amount of work devoted to development of advanced methods of predicting static strength of composite bolted joints, the amount of work on progressive failure modeling of composite bolted joints under fatigue loading is relatively small. Critical components of damage mechanics-based static strength prediction framework consist of a set of failure criteria with respective values of strength parameters and subsequent property degradation rules invoked to model local failure after the failure criteria is met. Conceptual extension of such an approach to model effects of environmental or fatigue degradation is straightforward. Indeed the most common failure criteria such as Tsai-Wu and Hashin were shown to be applicable to residual strength prediction in fatigue by using degraded strength values as functions of fatigue parameters, such as number of cycles, maximum stress, and stress ratio [60–63]. This approach has been applied in industrial practice by Sawicki [64] and Sawicki et al. [65] to show reasonable correlation of maximum shear and fiber direction stress with residual strength under fatigue loading in tension and compression regimes for open and filled hole composites. Practical difficulties in general application of such an approach are the amount of experimental data required to describe various strength values reduction as a function under various stress ratios, which must be performed for each laminate of interest. A significant reduction of such testing is achieved by developing ply-level analysis methods [66, 67]. Authors consider two types of property degradation gradual and catastrophic. Gradual degradation of strength and stiffness as a function of aforementioned fatigue loading parameters is introduced in modified three-dimensional Hashin-type failure criteria with shear nonlinearity. When the failure criterion with degraded strength values is exceeded, macroscopic damage occurs and catastrophic degradation of the respective (to damage mode) stiffness component is introduced. A complete set of fatigue tests required to characterize the strength and stiffness degradation on a ply level was also described. Two load levels in 60–70% and 40–50% of static strength were chosen for characterization. A full spectrum strength and stiffness degradation law was then obtained based on the power law model [68]. This methodology has quite general applications allowing performing analysis in mixed fatigue spectrum loading as well as residual strength prediction by continuing fatigue static loading from an arbitrary point. Although such methodology is a quintessential step forward, it inherits the shortcomings of static stiffness degradation-based strength prediction methods, which were addressed above. These shortcomings transpire from replacing actual damage modes such as matrix cracking and delaminations with low stiffness volumes (elements) and not capturing correctly the stress redistribution in their vicinity.

It appears that methods extending the discrete damage modeling methods proposed for static analysis [50–53], where damage evolution is modeled directly by applying fracture mechanics methodology, holds promise to further the mechanistic, mechanism-based approach to fatigue life prediction in composite bolted joints. In this respect, it is interesting to look back at earlier work by Spearing et al. [69]

where, perhaps for the first time, a discrete damage modeling framework for life prediction in notched composite laminates was demonstrated. At the level of computational resources, at a time one crack in each ply was considered and the shape of the delamination between the load-carrying 0° ply was predefined. Nevertheless, a very clear failure scenario emerged from analysis revealing stress blunting in the 0 ply due to splitting and delamination. Emerging computational techniques based on mesh-independent modeling of a significant amount of matrix cracking in arbitrary locations as well as multiple delaminations and their interaction [52, 70, 71] offer a promising framework for explicit modeling of fatigue failure in composite materials.

12.2.8 Summary

The basic components of a bolted joint generally include two or more structural elements, fasteners, shims, sealants, and coatings. Important design details such as part and hole tolerances, the bolt torque and if/how it might relax over the life of the structure, thermal and chemical mismatch of dissimilar materials (including galvanic corrosion), and fretting/wear due to joint compliance all affect bolted joint durability. Bolted joints may exhibit failure mode changes between static loading and long-term thermomechanical fatigue loading, due to fatigue-induced and environmental changes in joint load path and/or constituent material properties, but they are generally designed to avoid this outcome. While large differences in environment, loading conditions, etc. lead to very different typical joint concepts and materials in various different applications, they all generally share a common design goal of not allowing a mechanically fastened joint to be the weak link in the durability of their structures. Most industries rely on test-based design criteria and conservative bolted joint design rules to ensure that their joints have the required durability.

12.3 Bonded Joint Durability

12.3.1 Overview

The basic components of a bonded joint, as shown in Fig. 12.13, generally include various adhesives, fillers, and adherends. Important design, processing, and material-response details that affect durability include the bond-surface preparation process, individual part and assembly tolerances, thermal mismatch of dissimilar adherends, and long-term environmental response/degradation of adhesives, fillers, and adherends. The following subsections address observations regarding

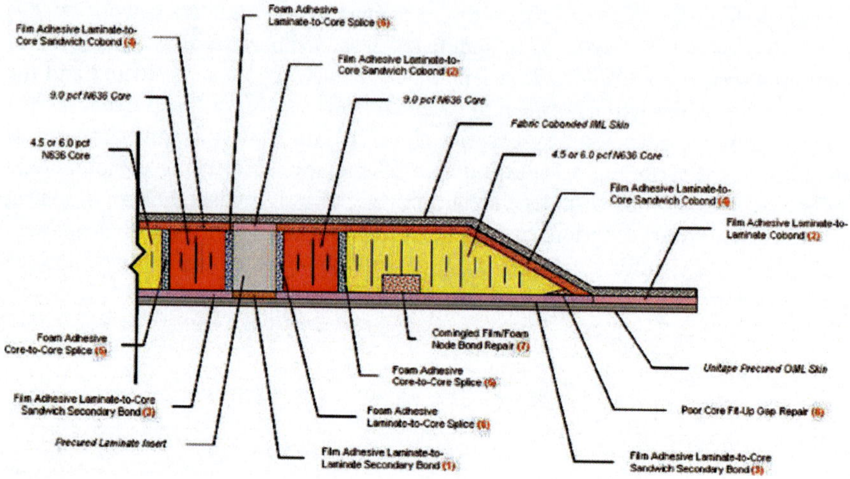

Fig. 12.13 Illustration of various bonded joint details

long-term bonded joint behavior, experimental techniques for testing, analytical methods for prediction of joint durability, and a brief summary of key bonded joint durability issues.

12.3.2 Key Observations on Long-Term Behavior

Bonded joints often exhibit failure mode changes between static loading and long-term thermomechanical fatigue loading. Basic bonded joint failure modes are shown in Fig. 12.14. While a joint might fail statically due to adherend interlaminar or in-plane strength, after a lifetime of fatigue or creep loading, it might instead yield a bondline adhesive failure. This change in response is due to fatigue-induced and environmental changes in constituent material properties, most commonly the adhesive. Thus, well-designed bonded joints must use materials whose long-term response/degradation is understood, and which are then designed to avoid a critical amount of degradation during the joint's design lifetime. A good overview of aerospace-related bonded joint durability, fatigue, environmental effects, and associated test methods is found in Ref. [2].

Given large differences in environment, loading conditions, design life, etc., this leads to very different typical joint concepts and materials in various different applications (e.g., marine, aeronautics, space, infrastructure, electronics, etc.). However, all of these applications share a common design goal of not allowing a bonded joint to be the weak link in the durability of their structures. For the most part, all of these industries rely on empirical (test-based) design criteria and conservative bonded joint design rules to ensure that their joints have the required durability.

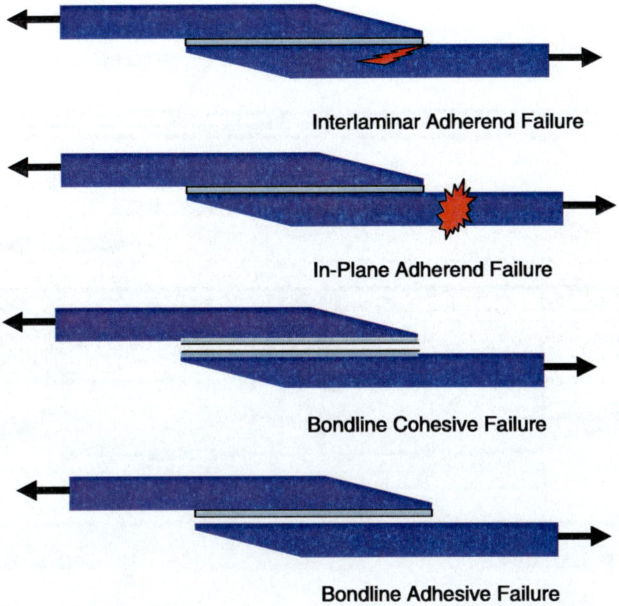

Fig. 12.14 Bonded joint failure modes

12.3.3 Experimental Techniques for Bonded Joint Durability Testing

As noted in the prior section, both individual constituent (laminate, adhesive, etc.) and overall joint sub-element response is required to be measured experimentally. In a manner similar to bolted joint coupon testing, standardized coupon test methods for individual constituent static and fatigue endurance properties are available in the open literature [5–7], as is useful guidance information [1, 10–12]. Examples of adhesive shear and peel tests are shown in Figs. 12.15 and 12.16. Adherends may be either metallic or composite. These tests can also be run at various temperatures and moisture conditions (e.g., use ASTM D5229 [5] or equivalent for moisture conditioning), and in constant amplitude or spectrum fatigue loading (per ASTM D6873 [5]), in order to obtain cyclic thermomechanical elastic and strength degradation responses. A good example of such an experimental program is found in Ref. [72].

As with sub-element-level bolted joint testing, bonded joint sub-element tests are generally nonstandard in nature, but generic guidance exists [10, 11]. Figures 12.17 and 12.18 are examples of bonded joint sub-elements. Also, certain coupon-level adhesive test methods, such as ASTM D1781, D1876, D2294, D3528, D5656, D6415, D6671, and D7250 [5, 7], are easily adapted to sub-element use through slight changes in specimen configuration (to yield a range of failure modes),

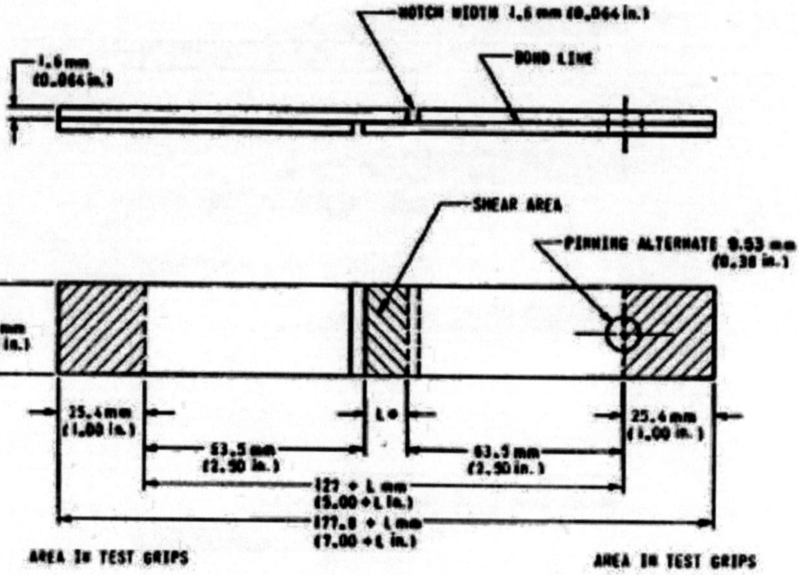

NOTE—*L = length of test area. Length of test area can be varied. Recommended length of lap is 0.50 ± 0.01 in. (12.7 ± 0.3 mm).

Fig. 12.15 Standardized adhesive shear test

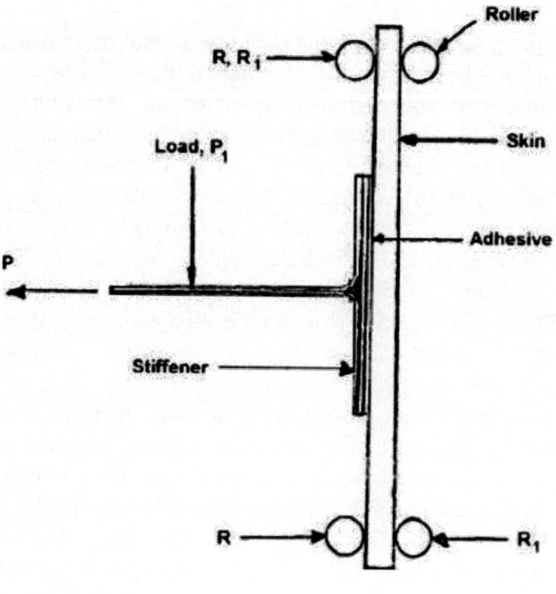

Fig. 12.16 Standardized adhesive peel test

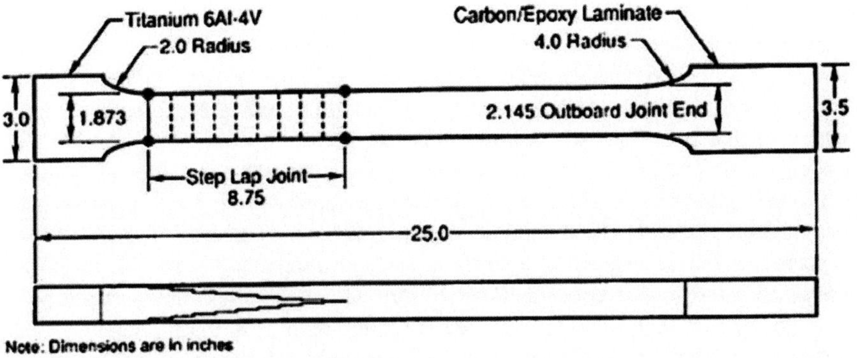

Fig. 12.17 Bonded stepped-lap tensile sub-element test

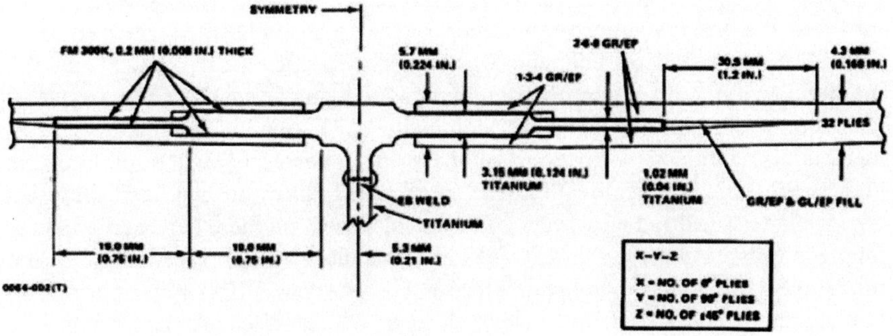

Fig. 12.18 Bonded stepped-lap bending sub-element test

instrumentation (to isolate the variation in failure mode), and loading parameters. Finally, larger component-level and full-scale tests (all nonstandard and design specific) can sometimes be used for bonded joint durability characterization, if taken to failure and if the failure location is in one of the bonded regions of the component.

12.3.4 Analytical Methods

Simple empirical models for bonded joint durability are similar to those for bolted joints in that they rely on either constant-amplitude or spectrum sub-element testing. FCGR criteria are more common, since delamination is a more common failure mode, but delamination onset, rather than subcritical growth, is a more common definition of failure, since bonded joints tend to be more conservatively designed. A thorough review of all aspects of durability of metal and composite bonded joining was recently performed by Baldan [73]. A more recent collection of review articles dealing with modeling aspects of bonded joining with metal and composite adherends can be found in Ref. [74]. Bonded joints are potentially more efficient than bolted or mechanical connections since they do not introduce similar stress concentration and may be ideally designed to sustain greater loads than the adherend materials [75], which corresponds to failure mode 2 in Fig. 12.14. The critical drawbacks of this technology are associated with the difficulties of assuring the bond quality on one hand and known susceptibility of most adhesives to degradation under hot/wet conditions and other environmental factors. These aspects as well as effects of adherends surface treatments, joint shapes (fillet, spew, etc.) on joint strength and durability and related stress analysis methods were addressed in Ref. [73]. The present brief review is mostly dealing with analysis methods, application of which to strength and durability prediction of composite bonded joints and adhesives has emerged in the past several years, although some experimental data will also be discussed. For the past decades, the so-called Boeing Wedge Test (BWT) [77], recently revisited by Adams et al. [76] was used as a primary adhesive durability screening test in hostile service environment. The data reduction technique was extended to calculate the critical value of the strain energy release rate (SERR). In parallel, a double cantilever beam (DCB) technique was evaluated and found, along with BWT, to provide a reliable assessment and quantification in terms of the critical values of SERR of interface preparation and adhesive quality. The interfacial failures on one hand represent nonmechanical and rather process or chemical issues, on the other hand apparent interface failures observed in Ref. [76] and evaluated by using scanning electron microscopy (SEM) and x-ray photoelectron spectroscopy (XPS) appeared to actually represent cohesive failures very close to the interface. Apparent interface failures were found under combined stress and environmental loading in earlier work [78] devoted to evaluation of cyclic stress testing of lap joints in hostile environments. A general conclusion made by authors was that the durability of adhesives under cyclic and constant loading under the same environmental exposure was different. It appears that this issue is still not sufficiently addressed.

In recent years, the modeling efforts have shifted from detailed stress analysis in the bonded joints toward progressive failure analysis. Delamination is the major failure mode in composite bonded joints, where its location may also be within the adhesive layer as in Fig. 12.14. The delamination onset and propagation investigation in composite laminates has been a critical research topic for several decades and is the subject of several reviews, e.g. Refs. [79, 80]. Several numerical techniques are used for fracture mechanics analysis, among them the J-integral technique [81], stiffness derivative method [82], the cohesion-type interface technique [83], and virtual crack closure technique (VCCT) [84]. Significant achievements in practical application of the VCCT both in static and fatigue regimes to delamination propagation were recently reported in Refs. [85, 86]. The advantage of the first three techniques is a relative insensitivity to the crack tip stress computation fidelity; however, these methods do not allow for mode separation in the mixed-mode crack propagation conditions. The cohesive interface element is a crack propagation technology especially suited for finite element implementation. Recent work [87, 88] demonstrates the ability to use it for mixed-mode and fatigue delamination propagation.

Several of these methods were applied to simulation of delamination initiation and propagation in composite bonded joints. Bogdanovich et al. [89, 90] considered double-lap joints with unidirectional adherends. Initial starter cracks in three critical locations, at the adhesive adherent interface, in the middle of the adhesive layer, and in the unidirectional adherend close to the interface, were modeled based on stress analysis in an uncracked joint. Progressive failure analysis was based on SERR calculation. Engelstad et al. [94] applied the VCCT technique for modeling of progressive failure in complex composite bonded elements. A complex joint consisting of a vertical spar bonded to a laminate resembling the stiffener termination area on the fuselage (skin) was considered. The analysis consisted of two parts. Stress analysis was first performed in a pristine joint to determine the failure initiation site. A micromechanical failure criterion called Strain Invariant Failure Theory (SIFT) [95] was applied both in the adhesive and composite. In all cases, the failure initiation site was in the first ply of the skin. Progressive failure analysis by using interface fracture elements [85, 86] was then performed for an assumed initial flaw at the critical location. The decohesive delamination modeling methodology, which incorporates both the crack initiation and propagation capabilities, has been applied to failure analysis of bonded joints in several forms. In its original form [83, 87, 88], this technique is used to model interface failure. The thickness of the decohesive element is zero, and its initial response (before damage) is nearly rigid, so that the interfaces which it connects are deforming together. In view of the small thickness of the adhesive layer, it is tempting to replace it with a decohesive interface altogether. De Moura and Chousal [93] replaced the adhesive in a double cantilever metal joint with a traditional initially stiff decohesive interface and compared the results with predictions based on the continuum damage model (CDM). The CDM constitutive equations were defined based on fracture mechanics relations similar to the decohesive interface formulation, which was achieved through introducing a characteristic length equal to the length of influence of a Gauss point in the given direction. Authors compared the stress distribution in the initial undamaged state and found that overall the CDM model predicted the failure more accurately due to better account of the shear deformation in the adhesive and allowing failure to occur in nonsymmetric fashion with respect to the adhesive midsurface. In the context of bonded joining, the decohesive interface model can also be modified to reflect the low modulus of adhesive by choosing the initial penalty parameter on the order of E/h, where E is the adhesive elastic modulus and h is the thickness. However, observations similar to [93] were made in an unpublished study by E. Iarve, where such a "soft" decohesive model was applied to composite repair. Composite scarf repair was considered and the stress distributions in the adherends showed significant difference between the adhesive modeled as a decohesive interface versus a three-dimensional solid element modeling. Based on these studies, a three-dimensional nonlinear continuum model for adhesive used to examine and optimize scarf repair in quasi-isotropic laminates was considered [101]. Considerable success in matching the experimentally measured strength of repaired laminates was achieved based on independently characterized constitutive properties of the adhesive. Campiho et al. [102] also developed a "soft" decohesive model for composite repair simulation. They, however, concluded that despite differences in initial response, consistent with previous remarks, the adhesive failure prediction is promising. Application of the decohesive interface modeling to composite bonded joints as an equivalent adhesive model can still offer promise. The constitutive relation in the decohesive law can express more complexity in physical behavior than linear elastic fracture mechanics, i.e. it can be modified to capture propagation effects such as fiber bridging [91, 92, 98]. DCB specimens with random reinforced composite adherends were considered, and it was concluded that such representation is computationally efficient and allows one to predict the change of the mode of failure from adhesive to failure in adherend arms as a function of arm thickness.

Fatigue and environmental effects in the context of progressive damage and failure modeling in bonded joints were also addressed by several researchers. The decohesive interface modeling technique has found numerous applications in this area as well. Liljehadi et al. [96] used this method to predict the residual strength of single-lap aluminum and composite joints in a wet environment. Two-parameter constitutive law was used in the cohesive formulation. The strength and SERR values were characterized in independent tests. Coupled stress-moisture loading of aluminum joints was considered in Ref. [103]. The CDM approach was taken to adhesive modeling. The constitutive equation for the adhesive consisted of three parts: linear, elasto-plastic with hardening and damage accumulation, characterized by damage parameter evolution. The damage parameter was also a function of moisture content determined within a fully coupled diffusion–mechanical problem statement. The stiffness degradation was performed based on an equivalent displacement gap value by introducing a characteristic distance parameter to avoid mesh dependence. The fatigue thresholds in composite bonded joints at various temperature and moisture exposure were predicted by using CDM in Ref. [100].

Fatigue performance of composite bonded joints depends both on the behavior of the adherends and the adhesive. The fatigue life of laminated composites was addressed both in the previous section of this chapter as well as in other chapters of this book. In the context of bonded joints, however, additional considerations were investigated in Ref. [104]. The authors conducted experiments to determine residual fatigue strength of a composite laminate subject to normal loading in the thickness direction. Such interest is warranted due to significant out-of-plane stress magnitudes in bonded lap joints as compared with stress distributions in other applications. A specimen consisting of at least 25 bonded stacked 2-mm unidirectional laminates was prepared and bonded to aluminum end pieces. Tensile testing with $R = 0.1$ and 5 Hz frequency was performed. The transverse strength both in static and fatigue regimes was less than the in-plane strength in the direction perpendicular to fibers by 10% and 26% (10^6 cycles), respectively. It was concluded, however, that the transverse strength specimen preparation may have been a factor in the observed trends. The in-plane versus normal (through the thickness) load direction fatigue resistance of composite laminates is still not characterized fully.

Fatigue loading is usually associated with constant and/or variable amplitude sinusoidal loading. With this regard, it is especially interesting to consider other types of loading, [106] e.g. impact fatigue [105]. Authors considered axial tensile loading of strap–lap joint under standard (sinusoidal) fatigue loading with load ratio (minimum to maximum load) of $R = 0.1$ and frequency of 5 Hz and maximum load at 60% of the quasi-static failure load; impact loading was applied with initial potential energy of 1.07 J and impact velocity of 1.9 m/s (resulting in maximum load of 11% of static failure load) at 15 s. intervals; the third loading regime was combined standard fatigue with intermittent impact events. Three damage propagation stages were identified in all cases. The first stage of relatively slow crack growth was identified when the crack propagated inside the adhesive, the next stage was identified when the crack was propagating in the adhesive and partially in the first ply of the composite adherent, and finally the third stage when the crack was mostly propagating in the adherend. What was different for the three loading cases was the crack growth rate per cycle and the number of cycles at each stage observed for the three loading cases. It appeared that the impact fatigue-induced crack growth rates were an order of magnitude higher than the ones observed for standard fatigue. Similarly, the intermittent impact events seemingly accelerated the fatigue crack growth rate under standard fatigue. The authors also observed that the crack growth in the adhesive under impact fatigue was characterized by a lack of rubber particle cavitation, thus reducing the effectiveness of the particle toughened adhesive. Understanding and modeling of such phenomena requires high-fidelity mechanism-based composite and adhesive modeling capabilities. As observed by Quaresimin and Ricota [97, 107], the crack (damage) initiation phase can be a very significant portion (from 20% to 70%) of the total fatigue life of the bonded joint. They proceeded into modeling the crack initiation and propagation in separate stages by using the Paris law in the propagation phase and the singular term coefficient at the adhesive and adherents interface intersection with the free edge for crack initiation prediction. In the propagation phase, the joint life was predicted integrating the Paris law equation for crack growth [99]

$$N_p = \int_{a_0}^{a_f} \frac{da}{K|\Delta G(a)|^n},$$

where K and n are Paris law parameters, and a_0 and a_f are the rack length in the beginning of the propagation phase and the crack length, which causes static failure at the maximum load, respectively. The SERR for a given length of the crack was obtained by using finite element analysis. Mixed-mode crack propagation in the bonded joint has been approximated by (1) taking into account only Mode I contribution and (2) by using the total SERR. The latter approach was shown to give more realistic fatigue life prediction results. The Paris law parameters were obtained from DCB fatigue tests. The crack initiation stage was defined as the time required for $a_0 = 0.3$ mm crack origination. This particular value of a_0 was chosen as a reliable size to be detected by optical microscopy. The crack initiation was addressed by analyzing the stress singularities at the edge of the joint at the adhesive adherent interface. The authors established experimentally that $\Delta H_0 \sim -\log N_0$, where ΔH_0 – is the singular stress term coefficient range and N_0 number of cycles before crack initiation. Note that the dimension of H_0 is (stress)(length)$^{\lambda}$, where

$$\sigma \sim H_0 r^{-\lambda}, \quad r \to 0$$

and r is the distance from the singular point. The power of singularity λ is, however, different for different edge angles and adhesive/adherend stiffness combinations. In the case of the fillet edge and straight edge and unidirectional adherends, $\lambda = 0.197$ and $\lambda = 0.4285$, respectively. This means that the predictive capability of this approach is limited to a particular system with a particular edge treatment only.

Application of fracture mechanics-type approaches outlined above is natural in the bonded joints, where failure is associated with a dominant delamination. On the other hand, the stress fields in typical composite lap joints exhibit nonuniformity in-plane [108] making the application of *one-dimensional* Paris law insufficient. In addition to that, the crack propagation in the transverse cross section is also not occurring along a straight line. Thus the fracture mechanics approach requires additional work in order to build a general bonded joint life prediction capability. Especially challenging appears generalization of the crack initiation stage. One possibility to achieve this is by using the cohesive formulations discussed above for static failure prediction by replacing the static constituent formulation of the decohesive law with a relationship capable of modeling fatigue cracking such as [88].

Several techniques such as cohesive zone models, CDM and VCCT-based methods for crack propagation have been proposed for durability prediction in composite bonded joints in recent years. Validation of these methods and their transition into an industrial environment offers high payoffs in term of design improvement and cost savings.

The basic components of a bonded joint generally include various adhesives, fillers, and adherends. Important design, processing, and material-response details that affect durability include the bond-surface preparation process, individual-part and assembly tolerances, thermal mismatch of dissimilar adherends, and long-term environmental response/degradation of adhesives, fillers, and adherends. While bonded joints can exhibit failure mode changes between static loading and long-term thermomechanical fatigue loading, this is not a desirable design condition. Thus, well-designed bonded joints must use materials whose long-term response/degradation is understood, and which are then designed to avoid a critical amount of degradation during the joint's design lifetime. While large differences in environment, loading conditions, etc. lead to different typical joint concepts and materials in various different applications, all share a common design goal of not allowing a bonded joint to be the weak link in the durability of their structures. For the most part, industry relies on empirical design criteria and conservative bonded joint design rules to ensure that their joints have the required durability.

12.4 Conclusions and Recommendations

Structural joints are where the durability is usually most evident in composite structural response to combined environmental effects and mechanical loads. The overall durability or degradation of a structural joint is a function of the response of all of the constituents of the joint: individual joined members, fasteners, and/or adhesive, shims, sealants, coatings, etc. The objective of this chapter was to provide a brief overview of observed durability behavior of composite structural joints, experimental techniques for exploring the durability of joints, analytical predictive methods, and certain empirical case studies, mainly from the aerospace industry. It was shown that, for both bolted and bonded joints:

- Joints can exhibit failure mode changes between static loading and long-term thermomechanical fatigue loading.
- Important design, processing, and constituent material-response details affect durability.
- Well-designed joints must use constituent materials whose long-term response/ degradation is understood, and which are then designed to avoid a critical amount of degradation during the joint's design lifetime.
- Advanced analytical methods for prediction of joint durability are in their infancy; and therefore,
- Industry generally relies on empirical design criteria and conservative joint design rules to ensure that their joints have the required durability.

Thus, the following recommendations can be made regarding joint durability:

- Further research into the synergistic relationship between constituent thermomechanical response and overall joint behavior is required if industry is to move beyond expensive test-based design criteria.
- Regulatory authorities should not be overly concerned with the current joint design state-of-the-art with respect to durability, as long as the empirical design criteria are well adapted to realistic requirements, and the design and manufacturing rigor of using well-characterized and controlled materials and processes is followed.

References

1. J. R. Davis, ed., *ASM Metals Handbook, Desk Edition*, 2nd Ed., ASM Int'l, Metals Park, OH (1998).
2. Kinloch, A., *Developments in Adhesives – 2*, Applied Science Publishers (1981).
3. Kulak, Geoffry L. et al., *Guide to Design Criteria for Bolted and Riveted Joints*, American Institute of Steel Construction, Chicago, IL (2001).
4. Bickford, John H., *An Introduction to the Design and Behavior of Bolted Joints*, Third Edition, Marcel Dekker, Inc., New York, NY (1995).

5. *Annual Book of ASTM Standards*, "Space Simulation; Aerospace and Aircraft; High Modulus Fibers and Composites," Vol 15.03, ASTM International, 100 Barr Harbor Dr, West Conshohocken, PA (revision issued annually).
6. *Annual Book of ASTM Standards*, "Plastics," Vols 08.01 – 08.03, ASTM International, 100 Barr Harbor Dr, West Conshohocken, PA (revision issued annually).
7. *Annual Book of ASTM Standards*, "Adhesives," Vol 15.06, ASTM International, 100 Barr Harbor Dr, West Conshohocken, PA (revision issued annually).
8. *Annual Book of ASTM Standards*, "Fasteners," Vol 15.08, ASTM International, 100 Barr Harbor Dr, West Conshohocken, PA (revision issued annually).
9. *Fastener Test Methods*, "NASM1312 Standard Practice," Aerospace Industries Assoc., Washington, DC (2002).
10. *DoD Composite Materials Handbook*, "Vol. 1 Polymer Matrix Composites Guidelines for Characterization of Structural Materials," MIL-HDBK-17-1 F (17 June, 2002).
11. *DoD Composite Materials Handbook*, "Vol. 3 Materials, Usage, Design, and Analysis," MIL-HDBK-17-3 F (12 Dec, 2001).
12. *Metallic Material Properties Development and Standardization*, MMPDS02, Battelle Memorial Institute, Columbus, OH (April, 2005).
13. Pinnell, M., 1995, "A Parametric Study of the Factors Affecting the Open and Filled Hole Performance of Fiber Reinforced Composite Materials," Ph. D Dissertation, University of Dayton, Dayton OH, p 141.
14. Altman, L. K., Reddy, D. J., and Moore, H., 2000, "Fail-Safe Approach for the V-22 Composite Proprotor Yoke," *Composite Structures: Theory and Practice, ASTM STP 1383*, P. Grant and C. Q. Rousseau, eds., American Society of Testing Materials, West Conshohocken, PA, pp. 131–139.
15. Eisenmann, J., Rousseau, C. "Practical Fatigue Analysis of Shear-Loaded Fasteners," *Proc. 9th Int'l Fatigue Conf.*, Atlanta, GA (14 – 19 May, 2006).
16. Lekhnitskii, S. G., 1954, "Stress Distribution in an Anisotropic Plate with an Elliptic Elastic Core (plane problem)," *Inzhenernyy sbornic*, v. XIX., Moscow.
17. Crews, J. H. Jr., Hong, C. S. and Raju, I. S., 1981, "Stress Concentration Factors for Finite Orthotropic Laminates with a Pin-Loaded Hole," *NASA Technical Paper 1862*.
18. Lessard, B. L. and Shokrieh, M. M., 1995, "Two-Dimensional Modeling of Composite Pinned-Joint Failure," *J. of Composite Materials*, Vol. 29, No. 5, pp. 671–697.
19. Hamada, H., Haruna, K. and Maekawa, Z.-I., 1995, "Effect of Stacking Sequences on Mechanically Fastened Joint Strength in Quasi-Isotropic Carbon-Epoxy Laminates," *J. of Composite Technology & Research*, July, Vol. 17 (3), pp. 249–259.
20. Collings, T.A., 1977," The Strength of Bolted Joints in Multi-Directional CFRP Laminates," *Composites*, January, Vol. 8 (1), pp. 43–54.
21. Quinn, W. J. and Matthews, F. L., 1977, "The Effect of Stacking Sequence on the Pin-bearing Strength in Glass Fiber Reinforced Plastics," *Journal of Composite Materials*, April, Vol. 11 (2), pp. 139–145.
22. Shyprykevich, P., 1995, "Characterization of Bolted Joint Behavior: MIL-HDBK-17 Accomplishments at Standardization," *J. of Composite Technology & Research*, July, Vol. 17 (3), pp. 260–270.
23. Smith, P. A. and Pascoe, K. J., 1986, "The Effect of Stacking Sequence on the Bearing Strengths of Quasi-isotropic Composite Laminates," *Composite Structures*, Vol. 6, pp. 1–20.
24. Marshall, I. H., Arnold, W. S., Wood, J. and Mousley, R. F., 1989, "Observations on Bolted Connections in Composite Structures," *Composite Structures*, Vol. 13, pp. 133–151.
25. Arnold, W. S., Marshall, I. H. and Wood, J., 1989, "A Finite Element Assessment of Clamping Effects in Composite Bolted Joints," *J. of Aerospace Engineering, Proc Instn Mech Engrs*, Vol. 203, pp. 39–45.
26. Matthews, F. L., Wong, C. M. and Chryssafits, S., 1982, "Stress Distribution Around a Single Bolt in Fibre-Reinforced Plastic," *Composites*, July 1982, pp. 316–322.

27. Chen, W.-H., Lee, S.-S. and Yeh, J.-T., 1995, "Three Dimensional Contact Stress Analysis of a Composite Laminate with Bolted Joint," *Composite Structures*, Vol. 30, pp. 287–297.
28. Folias, E. S., 1989, "On the Interlaminar Stresses of a Composite Plate Around the Neighborhood of a Hole," *Int. J. Solids Structures*, Vol. 25, pp. 1193–1200.
29. Iarve, E. V., 1996, "Spline Variational Three Dimensional Stress Analysis of Laminated Composite Plates with Open Holes," *Int. J. of Solids and Structures*, Vol. 33, No. 14, pp. 2095–2117.
30. Iarve, E. V., 1995a, "Three Dimensional Stress Analysis of Fastener Hole Composites," *Proceedings of the ASME Materials Division*, MD-Vol. 69–1, 1995 IMECE, ASME.
31. Iarve, E. V., 1995b, "Stress Analysis in Laminated Composites with Fastener Holes," *Proceedings of the ASC Tenth Technical Conference*, Santa Monica, CA, October 18–20, Technomic Publishing Company, pp. 408–419.
32. Iarve, E. V., 1994, "On the Accuracy of Interlaminar Stress Calculation in Vicinity of an Open Hole in Composite by Use of the Spline Variational Technique," *Proceedings of the ASC Ninth Technical Conference*, Technomic Publishing Company, DE, September 20–22, pp. 797–806.
33. Iarve, E. V., 1993, "Interlaminar Stress Analysis in Compression Loaded Composite Plates Containing Open Hole," *Proceedings of the ASC Eighth Technical Conference*, Technomic Publishing Company, Cleveland OH, October 19–21, pp. 1025–1034.
34. Wang, S. S. and Lu, X., 1993, "Three-Dimensional Asymptotic Solutions for Interlaminar Stresses Around Cutouts in Fiber Composite Laminates," *Mechanics of Thick Composites*, AMD-Vol. 1962, pp. 41–50.
35. Iarve, E. V., 1997, "Three-Dimensional Stress Analysis in Laminated Composites with Fasteners Based on the B-spline Approximation," *Composites Part A*, **28**, pp. 559–571.
36. Sihn, S., Iarve, E. V. and Roy, A. K., 2006, "Asymptotic Analysis of Laminated Composites with Countersunk Open- and Fastened-Holes," *Composites Science and Technology*, **66**(14), pp. 2479–2490.
37. Iarve, E. V. and Pagano, N. J., 2001, "Singular Full-Field Stresses in Composite Laminates with Open Holes," *Int. J. Solids Struct.*, **38**(1) pp. 1–28.
38. Whitney, T. J., Iarve, E. V. and Brockman, R. A., 2004, "Singular Stress Fields Near Contact Boundaries in a Composite Bolted Joint," *Int J Solids Struct.*, **41**(7), pp. 1893–1909.
39. Iarve, E.V., Mollenhauer, D., Kim, R., 2005, "Theoretical and Experimental Investigation of Stress Redistribution in Open Hole Composite Laminates Due to Damage Accumulation," *Composites Part A*, **36**(2), pp. 163–171.
40. Camanho, P. P. and Matthews, F. L., 1999, "Stress Analysis and Strength Prediction of Mechanically Fastened Joints in FRP: A Review," *Composites Part A 28A*, pp. 529–547.
41. Lee, J., 1983, "Three-Dimensional Finite Element Analysis of Damage Accumulation in Laminated Composite," In *Strength and Fracture in Composite Materials* (175–186). G. C. Sih and V. P. Tamuz, eds. Latvia: Zinatne.
42. Gamble, K., Pilling, M. Pilling, and Wilson, A., 1995, "An Automated Finite Element Analysis of the Initiation and Growth of Damage in Carbon Fiber Composite Materials," *Composite Structures 32*, pp. 275–285.
43. Nguyen, B. N., 1997, "Three-Dimensional Modeling of Damage in Laminated Composites Containing a Central Hole," *J. Composite Materials 31*(17), pp. 1672–1694.
44. Camanho, P. P. and Matthews, F. L., 1999, "A Progressive Damage Model for Mechanically Fastened Joints in Composite Laminates," *J. Composite Materials* 33, pp. 2248–2280.
45. Chen, W. H. and Lee, S. S., 1995, "Numerical and Experimental Failure Analysis of Composite Laminates with Bolted Joints under Bending Loads," *J. Composite Materials 29* (1), pp. 15–35.
46. Tang, S. C., 1991, "A Progressive Failure Model for Composite Laminates Containing Openings," *J. Composite Materials 25*, pp. 556–557.
47. Ye, L., 1988, "Role of Matrix Resin in Delamination Onset and Growth in Composite Laminates," *Composites Science and Technology 33*, pp. 257–277.

48. Pagano, N. J., Schoeppner, G. A., Kim, R. Y., and Abrams, F. L., 1998, "Steady-State Cracking and Edge Effects in Thermomechanical Transverse Cracking of Cross-Ply Laminates," *Composite Science and Technology 58*, pp. 1811–1825.
49. Schoeppner, G. A. and Pagano, N. J., 1999, "3-D Thermoelastic Moduli and Saturation Crack Density for Cross-Ply Laminates with Transverse Cracks," *Int. J. of Damage Mechanics.*
50. Iarve, E. V., 2003, "Mesh Independent Modeling of Cracks by Using Higher Order Shape Functions," *Int. J. Numer. Meth. Engng 56*, pp. 869–882.
51. Moes, N., Dolbow, J., and Belytschko, T., 1999, "A Finite Element Method for Crack Growth without Remeshing," *Int. J. Numerical Methods Engineering 46*, pp. 131–150.
52. Ling, D. S., Yang, Q. D., and Cox, B. N., 2009, "An Augmented Finite Element Method for Modeling Arbitrary Discontinuities in Composite Materials," *Int. J. of Fracture*, **156**, pp. 53–73.
53. Huynh, D. B. P. and Belytschko, T., 2009, "The Extended Finite Element Method for Fracture in Composite Materials," *Int. J. Numer. Meth. Engng.*, **77**, pp. 214–239.
54. ABAQUS, User Manual, Hibbitt, Karlson and Sorensen, Inc.
55. SAMCEF, User Manual, M5-V.5.1-94 Samtech S.A. (Belgium).
56. Green, B. G., Wisnom, M. R., Hallett, S. R., 2007 "An experimental investigation into the tensile strength scaling of notched composites," *Composites Part A*, **38**(3), pp. 867–878.
57. Wisnom, M. R. and Hallett, S. R., 2009, "The Role of Delamination in Strength, Failure Mechanism and Hole Size Effect in Open Hole Tensile Tests on Quasi-Isotropic Laminates," *Composites: Part A* **40**, pp. 335–342.
58. Hallet, S. R., Green, B. G., Jiang, W. J., and Wisnom, M. R., 2009, "An Experimental and Numerical Investigation into the Damage Mechanisms in Notched Composites," *Composites: Part A*, **40**, pp. 613–624.
59. Thoppul, S. D., Finegan, J. and Gibson, R., 2009, "Mechanics of Mechanically Fastened Joints in Polymer-Matrix Composite Structure – A Review," *Comp. Science and Techn.*, **69**, pp. 301–329.
60. Hasin, Z. and Rotem, A., 1973, "A Fatigue Failure Criterion for Fiber Reinforced Materials," *J. Composite Mater.*, (7), pp. 448–468.
61. Philippidis, T. P. and Vassilopoulos, A. P., 2002, "Complex Stress State Effect on Fatigue Life of GRP Laminates, Part I, Experimental," *Int J Fatigue* 24:8, pp. 13–30.
62. Philippidis, T. P., Vassilopoulos, A. P., 2002, "Complex Stress State Effect on Fatigue Life of GRP Laminates, Part II, Theoretical Formulation," *Int J. Fatigue* 24:8, pp. 13–30.
63. Tserpes, K. I., Papanikos, P., and Labeas, G., 2004, "Fatigue Damage Accumulation and Residual Strength Assessment of CFRP Laminates," *Compos Struct* 63:2, pp. 19–30.
64. Sawicki, A. J., 2004, "Development of Compression Design Allow-ables for Composite Bolted Joints Using ASTM Standard D 6742," in *Joining and Repair of Composite Structures, ASTM STP 1455*, K.T. Edward and H. Kim Eds., ASTM International, West Conshohocken, PA.
65. Sawicki, A. and Minguet, P., 2001, "Comparison of Fatigue Behavior for Composite Laminates Containing Open and Filled Holes," *Proceedings of the American Society for Composites 16th Annual Technical Conference*, Blacksburg, VA.
66. Shokrieh, M. M. and Lessard, L. B., 2000, "Progressive Fatigue Damage Modeling of Composite Materials, Part I: Modeling," *J Compos Mater* 34(13), pp. 1056–116.
67. Shokrieh, M. M. and Lessard, L. B., 2000, "Progressive Fatigue Damage Modeling of Composite Materials, Part II: Material Characterization and Model Verification. *J Compos Mater*, 34(13), pp. 1056–116.
68. Adam, T., Dickson, R. F., Jones, C. J., Reiter, H. and Harris, B, 1986, "A Power Law Fatigue Life Damage Model for Fiber Reinforced Plastic Laminates," *Proceedings of Institution of Mechanical Engineers, Part C: Mechanical Engineering Science*, **200**(C3), pp. 155–166.
69. Spearing, S. M., Beaumont, P. E. R. and Kortshot, M. T., 1992, "The Fatigue Damage Mechanics of Notched Carbon Fibre/PEEK Laminates," *Composites*, **23**(5), pp. 305–311.

70. Van der Meer, F. P. and Sluys, L. J., 2009, "A Phantom Node Formulation with Mixed Mode Cohesive Law for Splitting in Laminates," *Int J Fracture* DOI 10.1007/s10704-009-9344-5 (online).
71. Iarve, E. V., Gurvich, M. R., Mollenhauer, D. H., Rose, C. A. and Dávila, C. G., 2011, "Mesh-Independent Matrix Cracking and Delamination Modeling in Laminated Composites," *Int. J. Numer. Meth. Engng*, published online DOI: 10.1002/nme.3195.
72. Rousseau, C., 2000, "Primary Structure Composite Joining – Final Report," USAAMCOM TR-01-D-03.
73. Baldan, A., 2004, "Review. Adhesively-Bolted Joints in Metallic Alloys, Polymers and Composite Materials: Mechanical and Environmental Durability Performance," *J. Material Science*, **39**, pp. 4729–4797.
74. daSilva, L. F. M. and Ochsner, A., Eds, 2008, "Modeling of Adhesively Bonded Joints," Springer-Verlag Berlin Heidelberg, 335p.
75. Hart-Smith, L. J., 1973, "Adhesive Single Lap Joints," NASA CR 112236.
76. Adams, R. D., Cowap, J. W., Farquharson, G., Margary, G. M. and Vaughn, D., 2009, "The Relative Merits of the Boeing Wedge Test and the Double Cantilever Beam Test for Assessing the Durability of Adhesively Bonded Joints, with Particular Reference to the Use of Fracture Mechanics," *Int J of Adhesion & Adhesives*, **29**, pp. 609–620.
77. Marceau, J. A., Moji, Y., McMillian, J. C., 1977, "A Wedge Test for Evaluating Adhesive-Bonded Surface Durability," *Adhesives Age*, pp. 29–35.
78. Briskham, P. and Smith, G., 2000, "Cyclic Stress Durability Testing of Lap Shear Joints Exposed to Hot-Wet Conditions," *Int J of Adhesion & Adhesives*, **20**, pp. 33–38.
79. Tay, T., 2002, "Characterization and Analysis of Delamination Fracture in Composites: An Overview of Developments from 1990 to 2001," *Appl. Mech. Rev.*, **55**(6) (270 ref).
80. Pagano, N. J. and Schoeppner, G. A., 2000, "Delamination of Polymer Matrix Composites: Problems and Assessment," In *Comprehensive Composite Materials, 2*, A Kelly and C. Zweben (eds.), Elsevier Science, Oxford, pp. 433–528.
81. Rice, J. R., 1968, "A Path Independent Integral and the Approximate Analysis of Strain Concentration by Notches and Cracks," *J. Applied Mechanics, 35*, pp. 379–386.
82. Parks, D. M., 1974, "A Stiffness Derivative Finite Element Technique for Determination of Crack Tip Stress Intensity Factors," *Int. J. Fracture, 10* (4), pp. 487–502.
83. Mi, Y., Crisfield, M. A., and Davis, A. O., 1998, "Progressive Delamination Using Interface Elements," *J. of Composite Materials, 32* (14), pp. 1246–1270.
84. Krueger, R., 2002, "The Virtual Crack Closure Technique: History, Approach and Applications," *NASA/CR-2002-211628*.
85. Mabson, G., 2003, "Fracture Interface Elements,", Mil Handbook 17 Committee Mtg, Charleston, SC.
86. Deobald, L. R., Mabson, G. E., Dopker, B., Hoyt, D. M., Baylor, J. and Greasser, D., 2007, "Interlaminar Fatigue Elements for Crack Growth Based on Virtual Crack Closure Technique," *48th AIAA/ASME/ASCE/AHS/ASC Structures, Structural Dynamics, and Materials Conference*, Honolulu, HI.
87. Turon, A., Camanho, P. P., Costa, J., and Dávila, C. G., 2006, "A Damage Model for the Simulation of Delamination in Advanced Composites Under Variable-Mode Loading," *Mechanics of Materials*, **38**, pp. 1072–1089.
88. Turon, A., Camanho, P. P., Costa, J., and Dávila, C. G., 2007, "Simulation of Delamination in Composites Under High-Cycle Fatigue," *Composites: Part A*, doi:10.1016/j.compositesa.2006.11.009.
89. Bogdanovich, A. E. and Yushanov, S. P., 1998, "3-D Progressive Failure Analysis of Bonded Composite Joints," In: *AIAA/ASME/ASCE/AHS/ASC Structures, Structural Dynamics and Materials Conference*, Vol. 2, New York: AIAA, pp. 1616–26.
90. Bogdanovich, A. E. and Yushanov, S. P., 1999, "Progressive Failure Analysis of Adhesive Bonded Joints with Laminated Composite Adherends," *Journal of Reinforced Plastics and Composites*, Vol. 18, No. 18, pp. 1689–1707.

91. Gustafson, P.A. and Waas, A.M., 2009, "The Influence of Adhesive Constitutive Parameters in Cohesive Zone Finite Element Models of Adhesively Bonded Joints," *International Journal of Solids and Structures*, In Press.
92. Li, S., Thouless, M. D., Waas, A. M., Schroeder, J. A., and Zavattieri, P. D., 2005, "Use of Mode-I Cohesive-Zone Models to Describe the Fracture of an Adhesively-Bonded Polymer-Matrix Composite," *Composites Science and Technology* 65, pp. 281–93.
93. De Moura, M. F. S. R. and Chousal, J. A. G., 2006, "Cohesive and Continuum Damage Models Applied to Fracture Characterization of Bonded Joints," *Int. J. Mechanical Sciences*, **48**, pp. 493–503.
94. Engelstad, S. P., Berry, O. T., Renieri, G. D., Deobald, L. R., Mabson, G. E., Dopker, B., Notterf, E. W., and Clay, S. B., 2005, "A High Fidelity Composite Bonded Joint Analysis Validation Study - Part I: Analysis," *46th AIAA/ASME/ASCE/ASC Structure, Structural Dynamics & Materials Conference*, Austin, TX, AIAA, pp. 21–66.
95. Wollschlager, J. A., Gosse, J. H., Christensen, S., and Engelstad, S. P., 2002, "Current Developments in Composite Analysis and Failure Advocating an Advanced Composite Design Methodology," *Proceedings of SAMPE*, Long Beach, CA.
96. Liljehadi, C. D. M., Crocombe, A. D., Wahab, M. A. and Ashcroft, I. A., 2007, "Modeling the Environmental Degradation of Adhesively Bonded Aluminum and Composite Joints Using a CZM Approach," *Int J Adhesion and Adhesives*, **27**(6), pp. 505–518.
97. Quaresimin, M. and Ricotta, M., 2006, "Life Prediction of Bonded Joints in Composite Materials," **28**, pp. 1166–1176.
98. Schom, J, Nyman, T, Blom, A. and Ansell, H., 2000, "A Numerical and Experimental Investigation of Delamination Behavior in the DCB Specimen," *Composites Sci and Technology*, **60**, pp. 173–184.
99. Wahab, M. M. A., Ashcroft, I. A., Crocombe, A. D., Smith, P. A., "Finite Element Prediction of Fatigue Crack Propagation Lifetime in Composite Bonded Joints," *Composites Part A*, **35**, pp. 212–222.
100. Wahab, M. M., Ashcroft, I. A., Crocombe, A. D., Hughes, D. J. and Shaw, S. J., 2001, "The Effect of Environment on the Fatigue of Bonded Composite Joints. Part 2: Fatigue Threshold Prediction," *Composites Part A*, **32**, pp. 59–69.
101. Breitzman, T. D., Iarve, E. V., Cook, B. M., Schoeppner, G. A. and Lipton, R. P., 2009, "Optimization of a Composite Scarf Repair Patch Under Tensile Loading," *Composites Part A*, **40**(12), pp. 1921–1930.
102. Campilho, R. D. S. G., Moura, M. F. S. F. and Domingues, L.2007, "Using a Cohesive Damage Model to Predict the Tensile Behavior of CFRP Single-Strap Repairs," *Int. J. Solids and Structures*, **45**, pp. 1497–1512.
103. Hua, Y., Crocombe, A. D., Wahab, M. A. and Ashcroft, I. A., 2008, "Continuum Damage Modeling of Environmental Degradation in Joints Bonded with EA9321 Epoxy Adhesive," *Int. J of Adhesion and Adhesives*, **28**, pp. 302–311.
104. Martin, R. H. and Sage, G. N., 1986, "Prediction of the Fatigue Strength of Bonded Joints Between Multi-Directional Laminates of CFRP," *Composite Structures*, **6**, pp. 141–146.
105. Casas-Rodriguez, J. P., Ashcroft, I. A. and Silbershmidt, V. V., 2008, "Delamination in Adhesively Bonded CFRP Joints: Standard Fatigue, Impact-Fatigue and Intermittent Impact," *Composites Sci. and Techn.*, **68**, pp. 2401–2409.
106. Erpolat, S., Ashcroft, I. A., Crocombe, A. D. and Abdel-Wahab, 2004, "Fatigue Crack Growth Acceleration Due to Intermittent Overstressing in Adhesively Bonded CFRP Joints," *Composites: Part A* **35**, pp. 1175–1183.
107. Quaresimin, M. and Ricota, M., 2006, "Fatigue Behavior and Damage Evolution of Single Lap Bonded Joints in Composite Materials," *Composite Science and Technology*, **66**, pp. 176–187.
108. Panigrahi, S. K. and Pradhan, B., 2008, "Onset and Growth of Adhesion Failure and Delamination Induced Damages in Double Lap Joint of Laminated FRP Composites," *Composites Structures*, **85**, pp. 326–336.

Chapter 13
Durability and Aging of Composite Aircraft Structures

John Tomblin, Lamia Salah, and Dan Hoffman

Abstract The use of fiber-reinforced composites in aircraft structural components has significantly increased in the past few decades due to their improved specific strength and stiffness and superior resistance to both corrosion and fatigue with respect to their metal counterparts. Furthermore, current economic conditions require the use of most military and commercial aircraft beyond their original design service objectives; therefore, it is necessary to understand composite aging and in-service-induced damage to ensure the airworthiness and structural integrity of these airframes. Most aging aircraft studies conducted thus far have focused on metallic structures; however, as more composite components are being certified and used on aircraft structural components, it is necessary to address this aging concern for composite components as well.

The primary objective of this chapter is to summarize the findings of the teardown conducted on a B-737 CFRP composite stabilizer after 18 years of service. The B-737-200 stabilizer was developed by Boeing as part of the NASA ACEE program initiated in July 1977. Five B-737 horizontal stabilizer shipsets were manufactured and certified in August 1982.

As of March 2011, three shipsets have been retired from service, one has been sold to a foreign carrier, and one is owned by a commercial aircraft part supplier and has been reported for sale. This chapter provides highlights of the ACEE program, a summary of the B-737 horizontal stabilizer teardown activities, and results of the aging study. Results found indicate that the composite structure maintained its structural integrity over its service life and did not show significant degradation or detrimental signs of aging.

J. Tomblin (✉)
National Institute for Aviation Research, 1845 Fairmount, Wichita, KS 67260-0093, USA
e-mail: john.tomblin@wichita.edu

K.V. Pochiraju et al. (eds.), *Long-Term Durability of Polymeric Matrix Composites*,
DOI 10.1007/978-1-4419-9308-3_13,

13.1 Boeing 737 CFRP Horizontal Stabilizer RH Teardown Overview

As more commercial and military airplanes are required to maintain operational capability beyond their original design life objectives, it has become necessary to answer the questions of their continued airworthiness and structural integrity. Most research conducted on aging structures thus far has focused on metallic components; however, with the increasing use of composite materials in primary and secondary structures, it has become necessary to address the long-term structural health of aging composite components as well.

For advanced composite materials to be used in aircraft primary structures, it is necessary to demonstrate equivalent levels of safety, durability, and damage tolerance with respect to metal structures. These materials also improve profitability to the operators in terms of lower operating and maintenance costs. Composites offer great advantages over metals: improved specific strength and stiffness, the ability to be tailored to design requirements in various directions, enhanced manufacturing cost advantages (especially assembly), operating and maintenance costs and the potential of significant weight savings, which is directly related to fuel savings. This chapter examines the structural integrity of a Boeing 737-200 composite horizontal stabilizer after 18 years of service and presents data that can be used to substantiate aging of composite structures.

The B-737-200 right-hand (R/H) horizontal stabilizer, which is the subject of the current investigation, is the oldest commercial medium primary structure built using composite materials. The structure was built as part of the National Aeronautics and Space Administration (NASA) Aircraft Energy Efficiency (ACEE) program using carbon-reinforced graphite/epoxy composites and was certified by the Federal of Aviation Administration (FAA) and entered service in August 1984. The structure was in service for 18 years and was retired in 2002 after completing 48,000 flights and 52,000 h.

In order to characterize the structural integrity of the horizontal stabilizer after 18 years of service, the National Institute for Aviation Research (NIAR) acquired the aged structure and conducted several nondestructive and destructive tests to evaluate its structural integrity. Generated data can be used to understand aging mechanisms on composite parts currently in service and to reveal the main differences between damage mechanisms and damage accumulation in metallic versus composite components. This data could aid in future inspection and maintenance plans for composite structures to ensure their continued airworthiness and safety.

Nondestructive inspection (NDI) methods were conducted on the structure, prior to disassembly. The objective was to use current state-of-the-art inspection methods and evaluate their accuracy in detecting flaws in the structure. All inspections prior to teardown were conducted at the Sandia National Laboratories and at The Boeing Company. Methods used included thermography, Rapidscan™, laser UT, and Boeing NDI. NDI inspections conducted after teardown included TTU (through transmission ultrasonics) and PE (pulse echo) ultrasonics. Destructive evaluation included thermal analysis, image analysis, and physical and mechanical tests according to the appropriate ASTM standards and/or the standards used to generate the baseline data as summarized in the NASA report [1].

Thermal analysis was conducted using both Dynamic Mechanical Analysis (DMA) and Differential Scanning Calorimetry (DSC) to determine the aged material's glass transition temperature (Tg) as well as its degree of cure. Image analysis was conducted to characterize the state of the structure at the microscopic level and to detect possible flaws induced during manufacture or service. Physical tests were conducted to establish moisture and porosity levels in the composite structure and compare them to the design values. Mechanical tests were performed using the same standards applied to generate the allowable values. Strength and stiffness values of coupons/elements extracted from the aged structure were compared to those obtained during the design phase.

The ultimate goal of the investigation was to assess the overall structural health of the composite stabilizer after 18 years of service, to identify possible changes in the material properties due to environmental effects and/or flight service, to provide data to help understand aging mechanisms in composite structures, and to gain confidence in the long-term durability of composite materials.

13.2 ACEE Program Overview

The B-737-200 graphite/epoxy horizontal stabilizer used for this investigation was built as part of the NASA ACEE program, which was initiated in late 1975. This comprehensive program was motivated by the escalation of jet fuel prices and, thus, the necessity of developing new technology concepts in designing and building more fuel efficient commercial aircraft components.

The ACEE comprehensive program was subdivided into four development areas: laminar flow systems, advanced aerodynamics, flight controls, and composite structures [2]. The ACEE composite program focused on using advanced materials in existing aircraft structural components thus offering a high potential for weight savings and thereby operating costs of commercial aircraft. A building block approach was adopted where composite structure development would start with lightly loaded secondary components followed by medium primary components, and finally conclude with wing and fuselage design. Although considered highly successful, the program was terminated before the implementation of composite materials in wing or fuselage primary structures.

Under the ACEE program, The Boeing Commercial Aircraft Company, Douglas Aircraft Company, and Lockheed Corporation contracted to develop the following secondary and medium primary components: the upper aft rudder of the DC-10, the inboard ailerons of the L-1011, the elevators of the B-727, the vertical stabilizers of the L-1011 and DC-10, and the horizontal stabilizer of the B-737-200 as shown in Figs. 13.1 and 13.2. All of these components yielded weight savings of at least 21.6%.

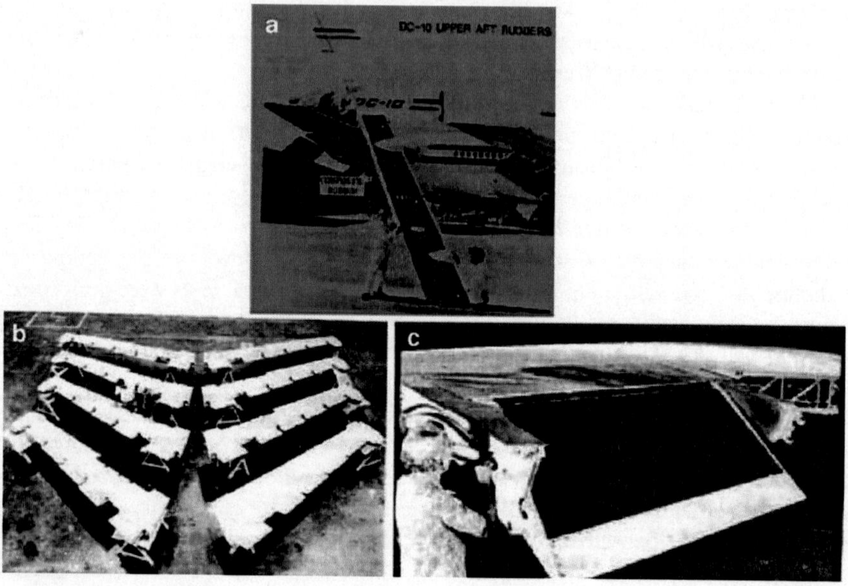

Fig. 13.1 Composite secondary components developed under the ACEE composites program: (a) DC-10 Upper aft rudder, (b) Boeing 727 elevators, (c) Lockheed L-1011 inboard aileron [2]

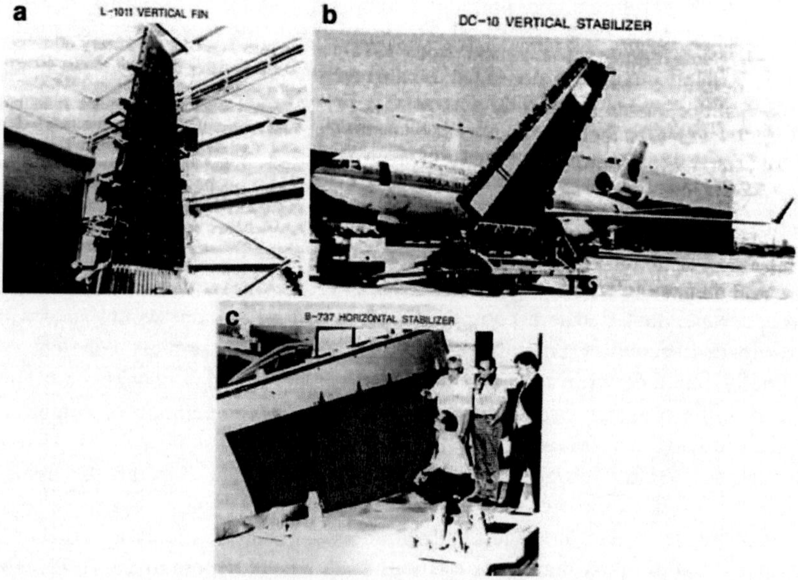

Fig. 13.2 Composite medium primary components developed under the ACEE composites program: (a) Lockheed L-1011 vertical fin, (b) Douglas DC-10 vertical stabilizer, (c) Boeing 737 horizontal stabilizer [2]

Specific objectives of the B-737-200 composite horizontal stabilizer program included achieving a minimum of 20% weight reduction with respect to the metal structure, fabricating at least 40% by weight of its components using composite materials, demonstrating cost-effectiveness of the structure, obtaining FAA certification for the structure, and closely monitoring its performance in service.

As part of the program, Boeing redesigned and manufactured five shipsets of the B-737-200 horizontal stabilizer using graphite/epoxy composites. The Original Equipment Manufacturer (OEM) adopted a fail-safe damage-tolerant approach to certify the structure where the structure's capability to sustain ultimate loads was validated by analysis supported by appropriate test substantiation [3]. This complies with FAA Part 25 Certification Requirements for Commercial Transport Aircraft.

Analysis was conducted to establish the most critical environment in which the stabilizer would be operating. It was found that a combined moisture level of 1.1% with temperature extremes of 180°F and −65°F would simulate the worst environmental conditions to which the structure would be exposed [1]. Allowable laminate, element, and subcomponent data were generated at these extreme conditions. Subsequent statistical reduction factors were implemented and provided high-confidence design values to use in the substantiating analysis.

Certification was completed in August 1982 and all five shipsets were deployed in 1984. The OEM closely monitored the performance of the stabilizers for 7 years. Outstanding performance was demonstrated with no in-service incidents attributed to the composite structure.

13.3 Composite Stabilizer Description

The B-737 horizontal stabilizer shown in Fig. 13.3 consists of an aluminum structural box, and leading edge, a fiberglass laminate tip, fiberglass honeycomb trailing edge and ribs, an elevator and body gap covers. Each stabilizer is secured to a metal center structure with three lugs at the rear spar and two lugs at the front spar. The structural box was redesigned using graphite/epoxy composites such that maximum commonality is achieved with the existing metal configuration and that both structures are interchangeable in terms of geometry and aerodynamic shape. The bending and torsional stiffness of the composite stabilizer as well as its aerodynamic shape and planform were made comparable to the metal stabilizer to meet control effectiveness and flutter requirements [1]. Furthermore, the composite stabilizer was designed to be damage tolerant and its strength, durability, inspectability, and serviceability equivalent to that of the metal structure.

Structural details of the existing metal and composite stabilizer designs are shown in Figs. 13.4 and 13.5. Several composite structural box arrangements were evaluated during the design phase of the composite stabilizer: a multiple rib concept, a honeycomb concept, and a stiffened skin concept.

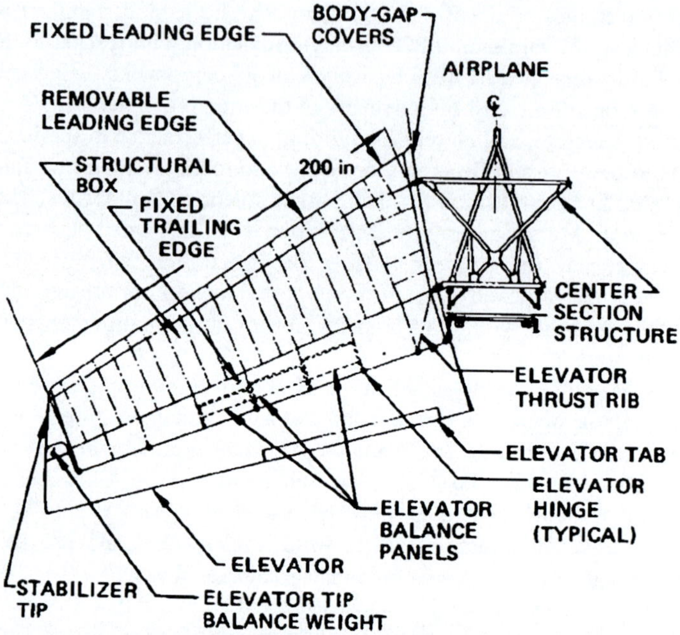

Fig. 13.3 B-737 horizontal stabilizer general arrangement [1]

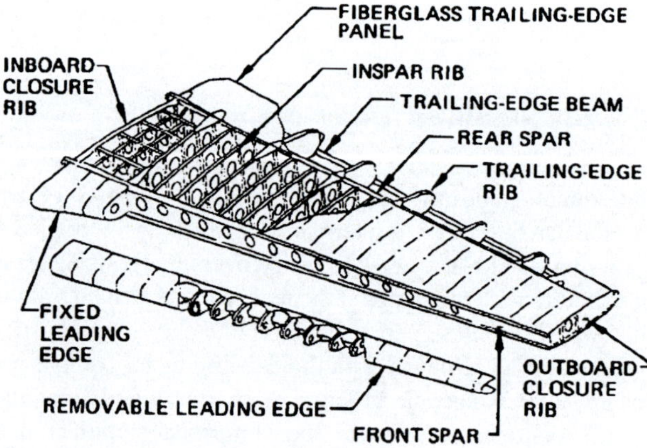

Fig. 13.4 Metal stabilizer structural arrangement [1]

The stiffened skin concept as shown in Fig. 13.5 was chosen due to its structural efficiency and minimal cost [1].

The structural arrangement consists of a single co-cured skin/I stiffener combination, 191 in long and 50.5 in wide at the inboard end with stringers spaced 3.85 in apart. Details of the stiffened skin and ribs are shown in Fig. 13.6. Mechanical

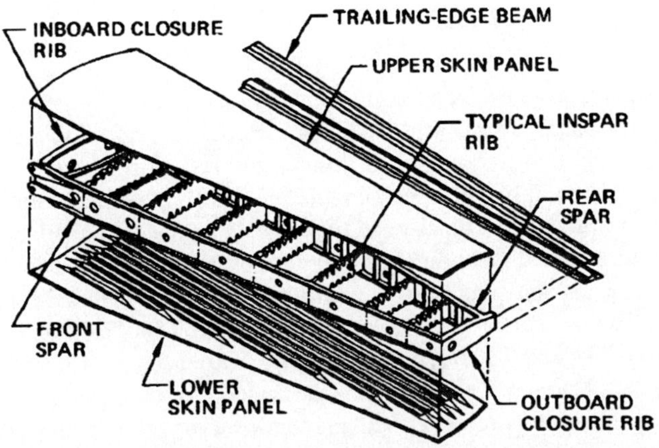

Fig. 13.5 Composite stabilizer structural arrangement [1]

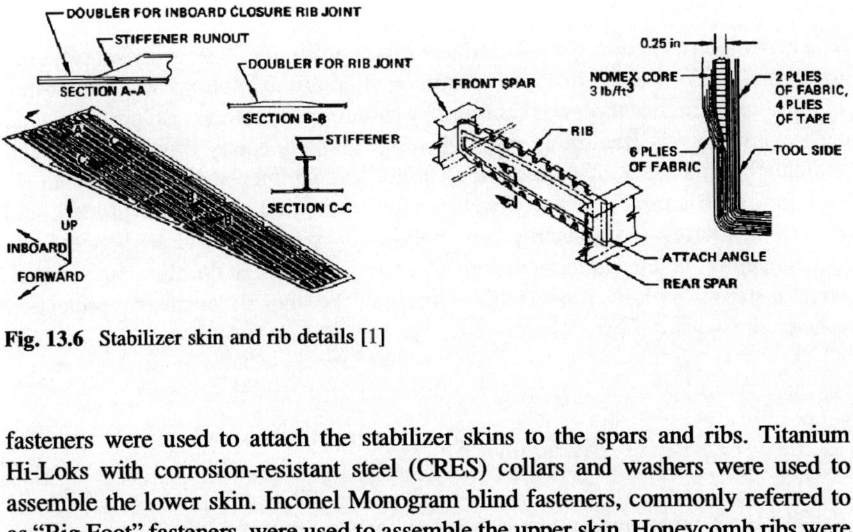

Fig. 13.6 Stabilizer skin and rib details [1]

fasteners were used to attach the stabilizer skins to the spars and ribs. Titanium Hi-Loks with corrosion-resistant steel (CRES) collars and washers were used to assemble the lower skin. Inconel Monogram blind fasteners, commonly referred to as "Big Foot" fasteners, were used to assemble the upper skin. Honeycomb ribs were used because of the simplicity of the concept in terms of tooling, fabrication, and cost and were fastened to the skins using graphite/epoxy shear ties. The spars were I beams consisting of two precured C-channels and two precured caps subsequently bonded together using film adhesive. Bolted steel spar lugs were used to attach the structural box to the fuselage center section. The spar lugs used two steel plates bonded and bolted externally to a graphite/epoxy chord. The composite design yielded 21.6% weight savings with respect to the metal configuration with a final weight of approximately 206 lbs.

13.3.1 Material Selection

Various materials were evaluated and manufacturing producibility was assessed by fabricating test panels from each candidate material to determine factors such as drape, tack, work time, and degree of difficulty in layup. In addition, all materials were expected to comply with specific requirements and tolerances on prepreg and cured laminate properties. Finally, additional factors such as available industrial database, demonstrated resin durability in various environments, supplier production experience, capacity and control, ability to produce all material forms were all considered. The systems evaluated included Narmco T300/5208 and T300/5235, Fiberite T300/934 and T300/976, Hercules AS/3501-5A, and Hexcel T300/F263 and T300/F288. Narmco T300/5208 was the material of choice because it satisfied most of the evaluation criteria [1]. The predominant material form was fabric with selected use of tape. The structural details used hand-layup procedures throughout.

13.3.2 Corrosion Protection Scheme

The corrosion protection system used on the composite stabilizer aimed at isolating the graphite/epoxy areas from the aluminum structure to avoid galvanic corrosion of the aluminum. Isolation was achieved by co-curing a fiberglass isolation ply onto the graphite/epoxy structure in areas where the graphite/epoxy structure interfaced the aluminum surface or painting the surface with primer and epoxy enamel. In addition, all aluminum surfaces were anodized or alodine treated, primed, and enameled. During the assembly, wet polysulfide sealant was used between the graphite/epoxy and the aluminum parts. All fasteners through the aluminum surface were installed with wet polysulfide sealant. The overall corrosion protection scheme is shown in Fig. 13.7.

13.3.3 Lightning Protection Scheme

The lightning protection system developed for the advanced composite stabilizer provided an electrical path around the entire perimeter of the structural box by means of bonding straps connecting the aluminum leading edge, the aluminum rib cap of the outboard closure rib, and the aluminum elevator spar. All these components were electrically grounded to the fuselage by means of the spar lugs and the leading and trailing edge ribs.

An aluminum flame spray was applied to the composite stabilizer's critical strike zone. This was determined to be the outboard 18 in on the upper and lower skins as shown in Fig. 13.8. These outboard skin panels were insulated using a layer of fiberglass co-cured to the skin panels. The flame spray was applied to the outboard

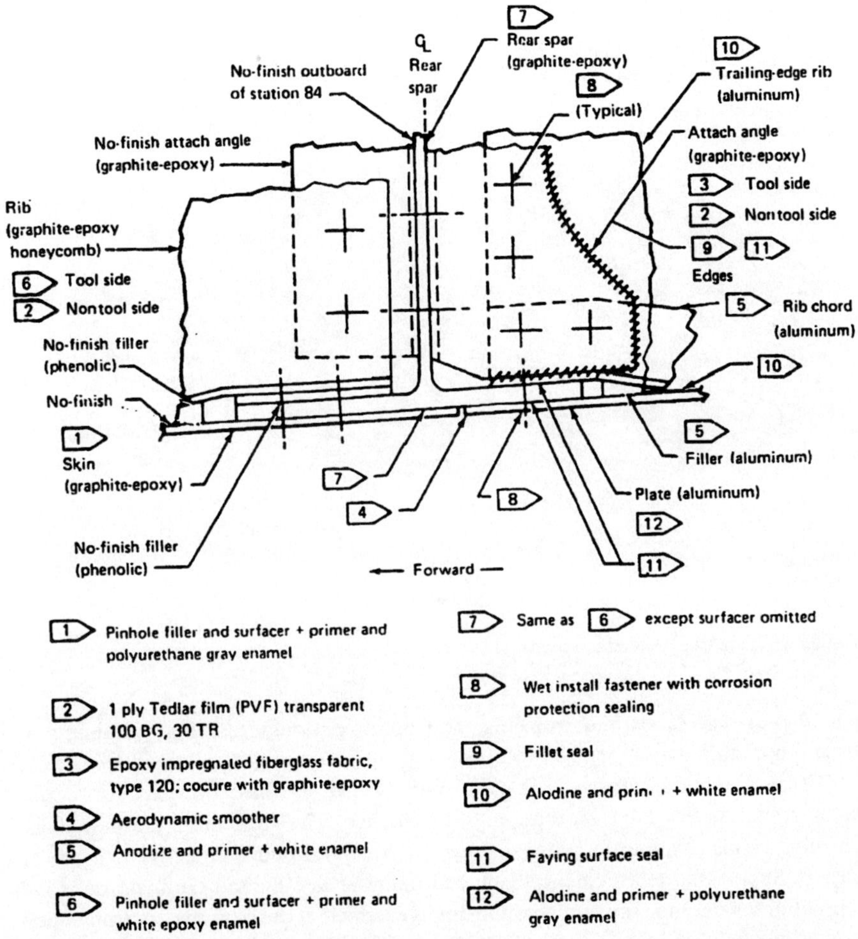

Fig. 13.7 Corrosion protection scheme [1]

sections of the skin after fabrication. It was then electrically connected to the metal cap of the outboard closure rib using four mechanical fasteners. The conductive surface was then alodine coated, primed, and painted.

13.4 B-737 Horizontal Stabilizer Certification

Certification requirements for commercial transport airframes are defined in the FAA regulations. Compliance items of prime importance to composite structures include requirements pertaining to material and process specifications, material properties, proof of structure static strength, proof of structure fatigue

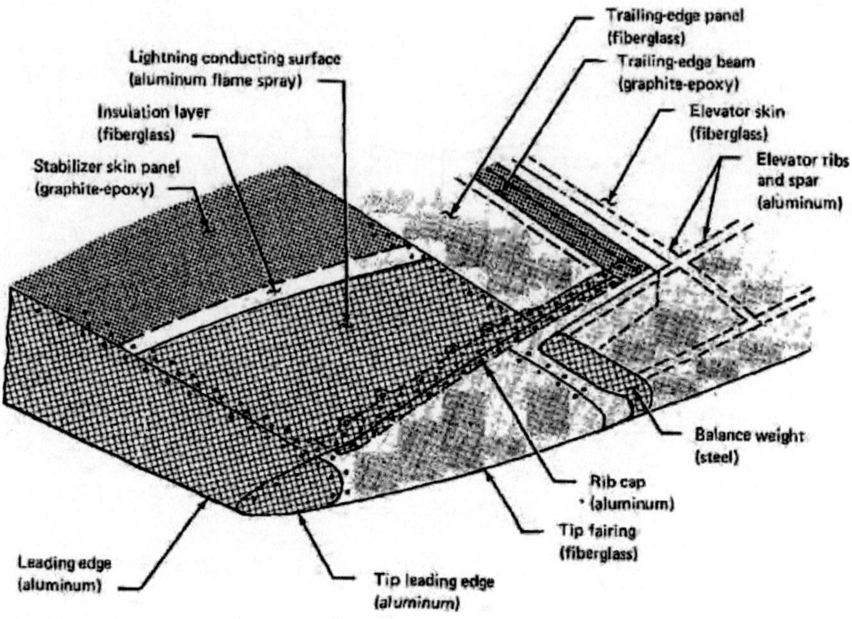

Fig. 13.8 Lightning protection scheme [1]

and damage-tolerance requirements, and additional requirements pertaining to lightning protection, structure protection, and structure repair and maintenance.

The certification approach adopted by the OEM was based on analytical substantiation supported by appropriate test evidence. The structure's ultimate load-carrying capability was demonstrated by analytically produced positive margins of safety. Supporting test evidence included material lamina and laminate data and structural detail, and subcomponent testing generated at the extreme environmental conditions to which the structure would be subjected.

Extensive static, dynamic, sonic environments, electrodynamic effects and environment analyses were conducted to substantiate the structure for airline flight use but also to provide the necessary data for certification. A detailed finite element analysis was developed for the stabilizer and the most critical load cases were considered.

Ultimate strains were obtained by combining the mechanical load strains and the thermal and moisture analysis strains. Margins of safety were obtained by comparing the calculated ultimate strains with the allowable strains for the different environments.

A thorough test program was carried out to support the data generation necessary to certify the structure. A building block approach was followed where coupons, elements, large panels, and test boxes were tested [1, 4]. Coupons and elements were used to generate mechanical properties, interlaminar properties, identify stress concentration effects, environmental factors, and to characterize impact damage properties and demonstrate material durability. Large panels and test boxes were used to validate design concepts, verify analysis methods, and correlate test and analytical data.

13.5 Maintenance Planning Program

A maintenance program was developed by the OEM to help the operators in defining inspection procedures and schedules to ensure the airworthiness and safety of the horizontal stabilizers [5]. The plan was accepted by the FAA and is presented in Table 13.1. In addition, the OEM agreed to perform an early Structural inspection, as described in the table, for the first two horizontal

Table 13.1 Maintenance planning schedule [5]

Check	Inspection interval (Flight Hr)	Description
Preflight-transit		Walk around
A	75	Visual inspection of exterior surface, from ground level
B	300	Visual inspection of external surfaces
C	1,200	External visual inspection
		Exposed rear spar area
		Exposed hinge fittings and thermal linkage
	2,400	Front and rear spar to center section attachment lugs
		Inboard edge of rear spar web
		Trailing edge cavities
Structural	14,000	External visual inspection
		NDT Inspection upper and lower skin from the rear spar forward to stringer 3 between the side-of-body and the rib at stabilizer station 111.1
		Front and rear spar attachment lugs, pins, bushings, and fittings
		Internal trailing-edge structure
		Internal structure, spars, stiffeners, closure ribs; access by removing gap covers, access hole covers, removable leading edge, removable lower trailing edge panels, and removable tip

stabilizer shipsets to reach 7,000 flights. Both inspections required 17-h and 24-h downtime, respectively, which is very consistent with the downtime required to inspect the metal stabilizers. No structural problem or in-service deterioration was reported [5].

13.6 B-737 Horizontal Stabilizer Fleet Status

Table 13.2 summarizes the status of all five composite stabilizer shipsets as of March 2011. As shown in the table, four shipsets have been removed from service after accumulating 47,000, 19,300, 55,000, and 48,000 flights, respectively. Shipset 1 has been sold and is being operated by a foreign carrier. Shipset 2 has been reported for sale by an aircraft commercial part supplier.

Shipset 3 became available for a teardown in 1990 after a crash of the aircraft on landing approach that led to minimal damage on the L/H stabilizer. The OEM acquired the stabilizer shipset and conducted a partial teardown on the structure. The inspection of the areas that were not damaged during the crash revealed no structural damage or delaminations, no corrosion on the interface of the aluminum/graphite interfaces, no evidence of visual degradation, cracking, deterioration or fretting [5].

Shipsets 4 and 5 were removed from service after skin/stringer delaminations were discovered at the inboard end of the upper skin. A design oversight omitted a "No Step" designation in this area. Airline personnel routinely stood on the horizontal stabilizer's upper skin while servicing the vertical stabilizer. The repeated flexing of the upper surface eventually resulted in the delaminations.

Table 13.2 B-737 composite stabilizer fleet status as of March 2011

Shipset/production line #	Entry into service	Carrier	Status as of March, 2011
1/1003	2 May 1984	A and E	Sold to a foreign carrier (60,000 h, 45,000 flights)
2/1012	21 March 1984	A	Reported for sale by a commercial part supplier (62,000 h, 47,000 flights)
3/1025	11 May 1984	B	Damaged beyond repair 1990; partial teardown completed in 1991 (17,300 h, 19,300 flights)
4/1036	17 July 1984	B and C	Stabilizers removed from service 2002 (approx. 39,000 h, 55,000 flights); partial teardown of R/H unit at Boeing
5/1042	14 August 1984	B and D	Stabilizers removed from service 2002 (approx. 52,000 h, 48,000 flights); teardown of L/H unit at Boeing; teardown of R/H unit at NIAR, WSU

When the damage was initially discovered, the OEM ordered inspections on the other three aircraft remaining in service. All three had similar delaminations. As a result, a "No Step" zone was added in this area, shipsets 4 and 5 were retired from service, and shipsets 1 and 2 were repaired and returned to service.

13.7 B-737 Horizontal Stabilizer Teardown Results

The objective of this teardown was to evaluate the aging effects on the structural integrity of the composite structure after 18 years of service. The main goal was to evaluate the structural health of the aged composite component after 48,000 flights equivalent to about 2/3 of its design service objective.

To accomplish this task, the research was subdivided into several nondestructive and destructive subtasks. The goals of the nondestructive inspections were to characterize the state of the structure after 18 years of service and to investigate the existence/extent of flaws introduced during manufacture or service using current methods used in the field but also more sophisticated methods including thermography, Laser UT, Rapidscan™, etc. The objective of the destructive inspection/evaluation was to confirm the existence of flaws detected using NDI, to conduct mechanical tests, thermal analysis, physical tests, image analysis, and to compare the data to that generated during the design phase.

13.7.1 Visual Inspection/Nondestructive Inspection Prior to Teardown

The general visual inspection conducted on the outside of the structure prior to teardown revealed numerous cracked and peeled paint areas in the upper and lower skins. In most fastener areas, the paint was cracked and loose. The surface blemishes were most probably caused by excessive paint/surfacer, aging, and exposure to weather.

Nondestructive Inspection scans were conducted at the Sandia National Laboratories using pulse echo ultrasonics (Rapidscan™) as shown in Fig. 13.9. The upper-skin scans showed extensive regions of porosity. The Rapidscan™ inspections also confirmed all known upper-skin stringer disbonds as shown in the figure. The areas circled in red showed excessive porosity from the Rapidscan pulse echo inspection but were identified as disbonds after teardown.

The Rapidscan™ inspections also showed a very porous circular repair, applied in the field, between rib stations 2 and 3 and stringers 5 and 8. Because of the excessive porosity in the repair, most of the ultrasonic signal was dissipated through the air pockets and no useful signal could be read. The lower-skin scans showed less porosity in comparison to the upper skin but a few areas were identified as disbonds, at the inboard section of the lower skin at stringer 1 runout, at the

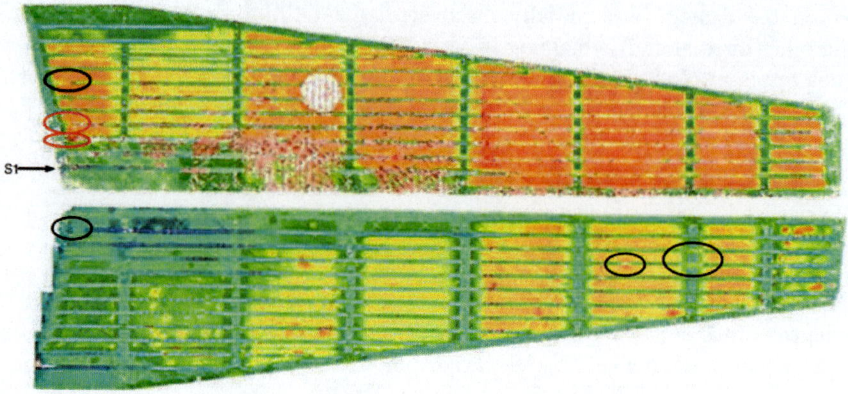

Fig. 13.9 Rapidscan™ analysis (pulse echo time of flight data) of the R/H of the B-737 stabilizer (Courtesy of Sandia National laboratories and NDT Solutions Ltd., UK)

outboard section of the skin at rib 6, stringers 2 and 3, and between ribs 5 and 6, at stringer 3 as shown in Fig. 13.9.

The main findings of the NDI inspection were the excessive porosity levels found in the upper skin and the stringer disbonds in the upper and lower skins.

13.7.2 Disassembly

The R/H B-737 horizontal stabilizer as shown in Fig. 13.10 was disassembled as follows: the upper skin was first demated by drilling out the "Big Foot" blindfasteners. A combination of carbide and cobalt drill bits was used to drill out the "big foot" blindfastener head as shown in Fig. 13.11. Once the fastener's head was drilled out, the fastener's shank was accessible, and could be easily driven out of the structure.

As shown in Fig. 13.12, all "Big Foot" blind fasteners attaching the upper skin to the spars and the ribs were carefully drilled out in order not to introduce additional damage to the structure. The dismantled upper skin is shown in Fig. 13.13. The lower skin was dismantled next at the Titanium Hi-Lok attachment points to the spars and ribs. The lower skin, after disassembly is shown in Fig. 13.14.

As shown in Fig. 13.13, the R/H upper skin appeared to be flat with no evidence of curvature or residual strains. Furthermore both skins appeared to be in extremely good condition with no evidence of detrimental aging or deterioration.

Once both skins were demated, the structural box, shown in Fig. 13.15 was further disassembled by demating the spars from the ribs. The front and rear spars are shown in Fig. 13.16.

Fig. 13.10 B-737 CRFP R/H horizontal stabilizer prior to disassembly

Fig. 13.11 "Big foot blind fastener removal"

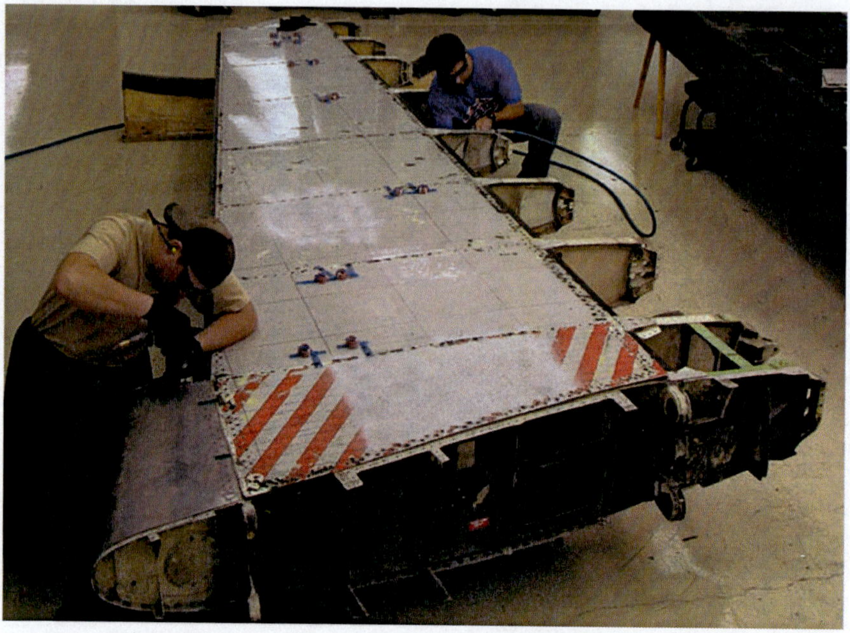

Fig. 13.12 Upper-skin disassembly

Fig. 13.13 B-737 R/H composite stabilizer upper skin, after disassembly

13.7.3 Visual Inspection Findings

As shown in Figs. 13.13–13.16, the first significant observation that could be made after disassembly is that the stabilizer appeared to be in really good condition with no evidence of pitting or corrosion, as would be observed in a metal structure with

Fig. 13.14 B-737 R/H composite stabilizer lower skin, after disassembly

Fig. 13.15 R/H B-737 composite stabilizer structural box

similar service history. All fasteners seemed to be in really good condition with a few exceptions as shown in Fig. 13.17.

Visual inspection also showed evidence of Tedlar® degradation due to environmental exposure as seen in Fig. 13.18. Tedlar® was used as a moisture protective film and was co-cured to parts of the structure, mainly ribs and spars and used as a protection/barrier from moisture ingression.

The disassembled structure revealed the presence of both liquid and phenolic solid shims as shown in Figs. 13.19 and 13.20. These shims were woed to seol the

Fig. 13.16 R/H B-737 composite stabilizer spars – front spar (*top*), rear spar (*bottom*)

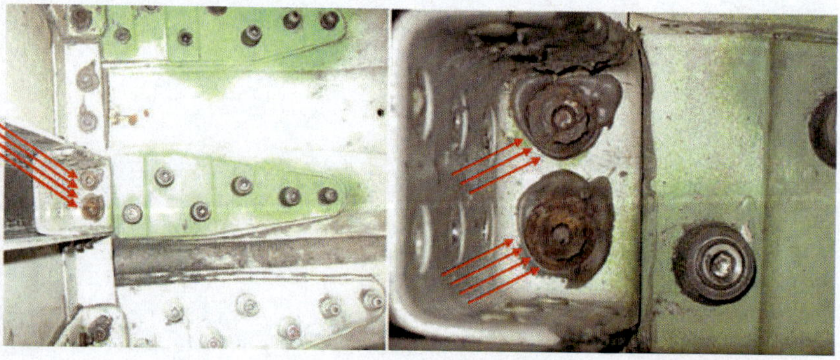

Fig. 13.17 Corroded fasteners due to sealant deterioration

gaps between the stabilizer's skins and ribs. The shims used seemed to be very effective and held up very well in service. The bonding of the liquid shim was effective yet the release agent used on one side, allowed separation at disassembly. There were no signs of the liquid shim breaking up, softening, deteriorating or cracking.

The visual inspection/findings of the B-737 R/H stabilizer teardown showed that the composite structure held extremely well with minor signs of degradation or deterioration due to environmental exposure.

Fig. 13.18 Degradation of Tedlar®, moisture barrier film

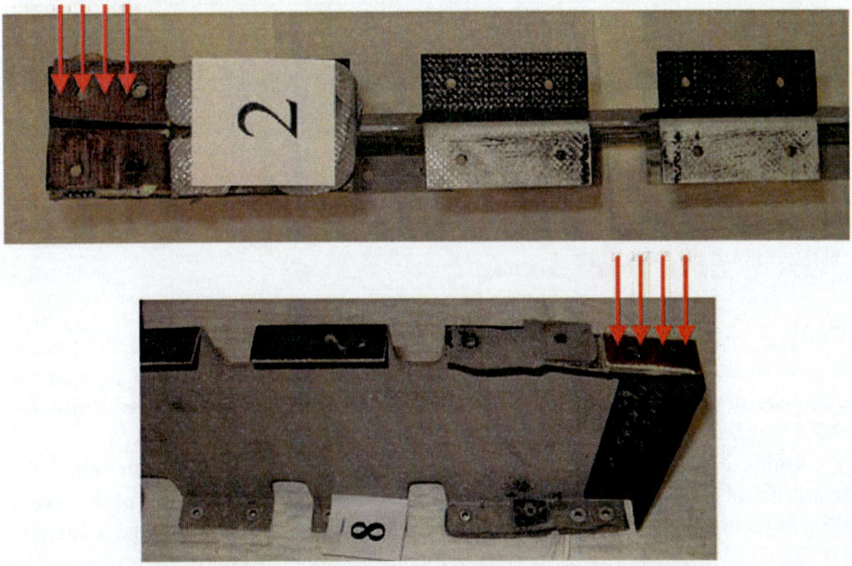

Fig. 13.19 Phenolic shims used to fill the gaps between the upper skin and ribs 2 and 8

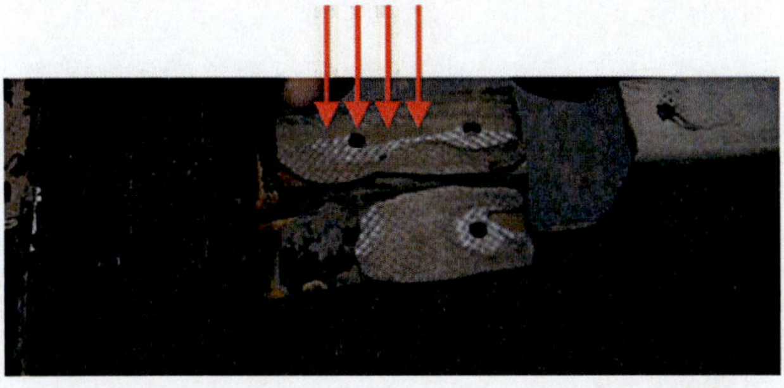

Fig. 13.20 Liquid shims used to fill gaps between the upper skin and the stabilizer ribs

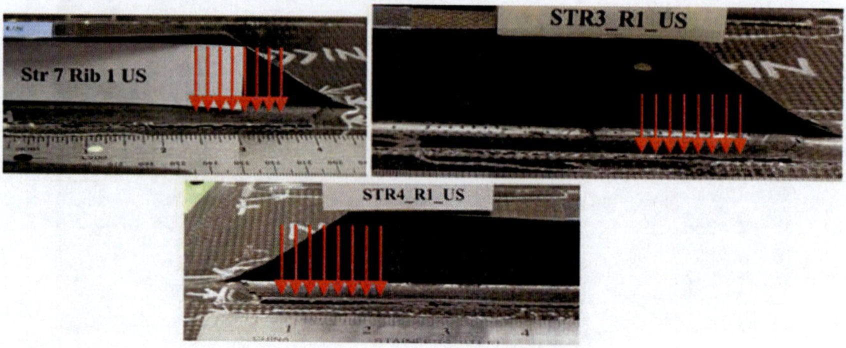

Fig. 13.21 Upper skin delaminations at stringer 3, 4, and 7 runouts

Tedlar® and fastener sealant degradation were the only signs of deterioration due to environmental exposure found on the structure. Finally, a visual inspection of the upper-skin repair found between rib stations 2 and 3 confirmed that the repair was a surface repair only and did not extend to the stringers.

13.7.4 Visual Inspection/Destructive Evaluation of Disbonded Areas

Visual inspection confirmed the existence of the disbonded stringers at the inboard section of the upper skin at stringer 7 runout. Visual inspection also showed the existence of additional disbonds at the upper-skin stringer runouts for stringers 3 and 4 and 6 as shown in Fig. 13.21.

Similar to the delaminations found in the inboard section of the upper skin at the stringer runouts, a delamination was detected in the inboard section of the lower skin at stringer 1 runout and was confirmed after disassembling the structure. Destructive inspection was conducted on all other disbonds identified using Rapidscan™ in the lower skin and all were confirmed. Some of those disbonded stringers are illustrated in Fig. 13.22.

13.7.5 Nondestructive Inspection after Teardown

Pulse echo and through-transmission nondestructive inspection methods were used to assess the stabilizer's damage state after 18 years of service. Pulse echo results using today's equipment capabilities showed the increased accuracy/sensitivity of the current inspection methods compared to those used in the 1970s. In accordance

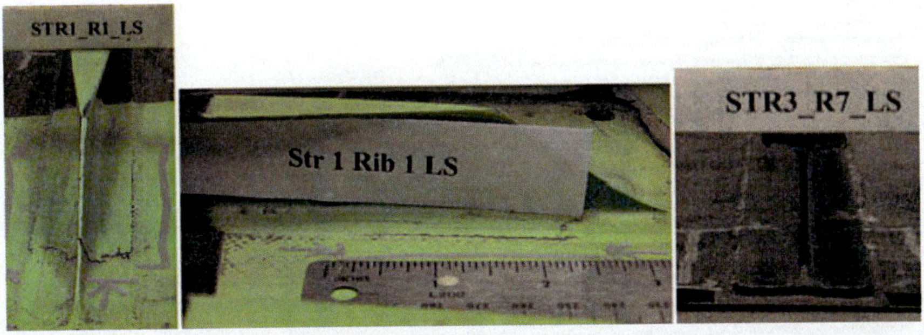

Fig. 13.22 Lower-skin disbonds confirmed after teardown

Fig. 13.23 1980s ultrasonic inspection sensitivity versus today's equipment capabilities. (**a**) 1970's capability (**b**) current capability

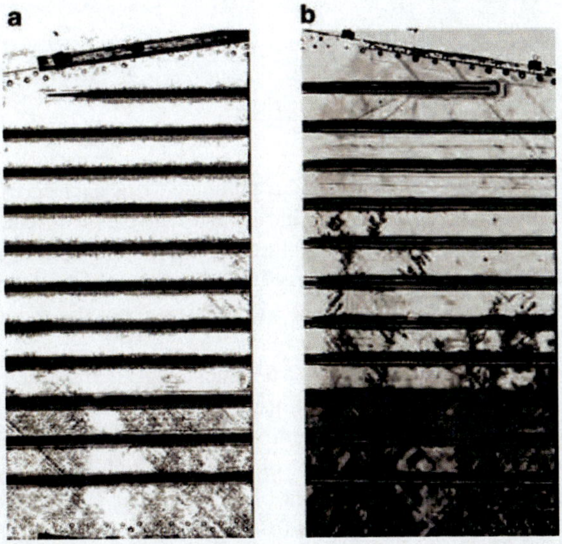

with current inspection requirements, 5 MHz transducers were used to inspect the aged structure; 1 MHz transducer was used to obtain a scan according to the 1970s requirements. This is shown in Fig. 13.23.

Pulse echo scans conducted confirmed the large amounts of porosity in the upper skin as previously identified in the Rapidscan™ inspection results. Results summary is shown in Fig. 13.24. The field skin repair is also shown in the same figure, in the upper skin between rib stations 2 and 3 and stringers 1 and 8.

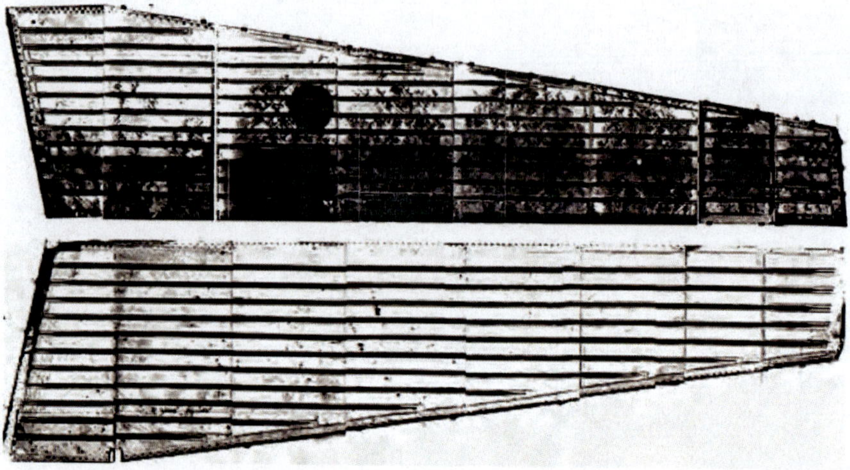

Fig. 13.24 NIAR pulse echo scans of the B-737 RH horizontal stabilizer

13.7.6 Thermal Analysis

Thermal analysis was conducted on coupons excised from both the upper and lower skins to investigate possible changes in the resin chemistry of the material. The thermal tests conducted included Dynamic Mechanical Analysis (DMA) and Differential Scanning Calorimetry (DSC). DMA tests measure the response of a material to a periodic stress and provide information about the modulus and the damping of the material.

DMA curves provide two values of glass transition temperature, a value based on the onset storage modulus or material fiber stiffness loss and a value based on material damping/maximum viscosity which is the peak of Tan Delta. The storage modulus is a characteristic of the material fiber stiffness whereas the damping is a characteristic of the matrix of the material. The glass transition temperature (Tg) value based on the onset of storage modulus is always more conservative than the value obtained using the peak of Tan Delta.

All DMA curves obtained for coupons extracted from both skins exhibited a shallow storage modulus transition and a narrow Tan Delta which is an indication of a highly cross-linked material – therefore, a material fully cured, as shown in Fig. 13.25.

DSC was also conducted on specimens extracted from both skins to evaluate the degree of cure of the material. As shown in Fig. 13.26, the low heat of reaction values support the results obtained using DMA, that is, that the material is fully cross-linked and fully cured. DMA was conducted using the parameters specified in the ASTM standard E1640 but also according to the Boeing method and both sets of results are summarized in Figs. 13.27–13.29. All values are compared against the baseline dry Tg obtained using TMA supplied by NARMCO.

Figures 13.27 and 13.28 summarize DMA data obtained for the stabilizer upper skin following the ASTM standard and the Boeing method, respectively.

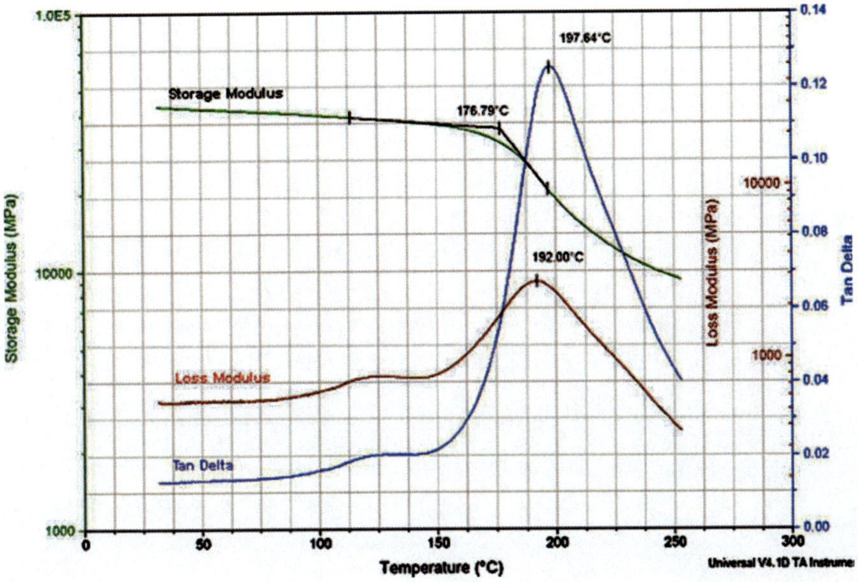

Fig. 13.25 DMA analysis of a coupon extracted from the B-737 upper skin (rib specimen, tested as extracted)

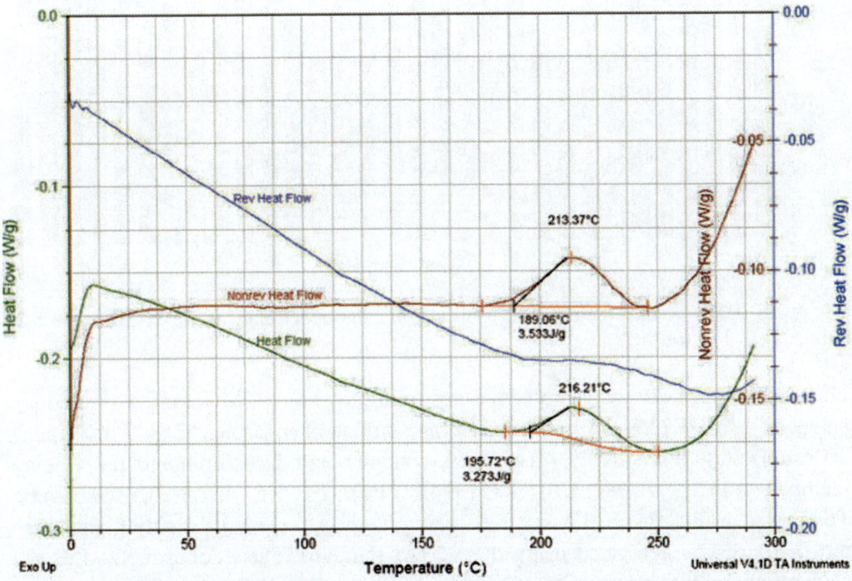

Fig. 13.26 DSC of a coupon extracted from the B-737 upper skin

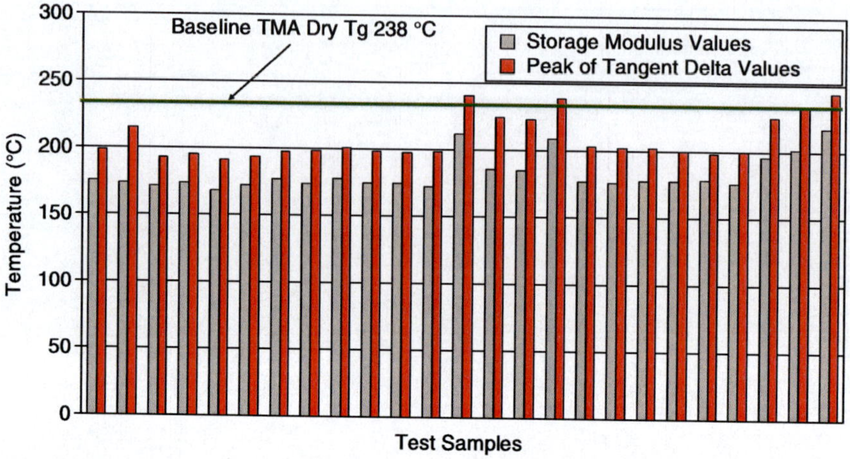

Fig. 13.27 DMA results for coupons excised from the upper skin of the B-737 horizontal stabilizer (ASTM standard)

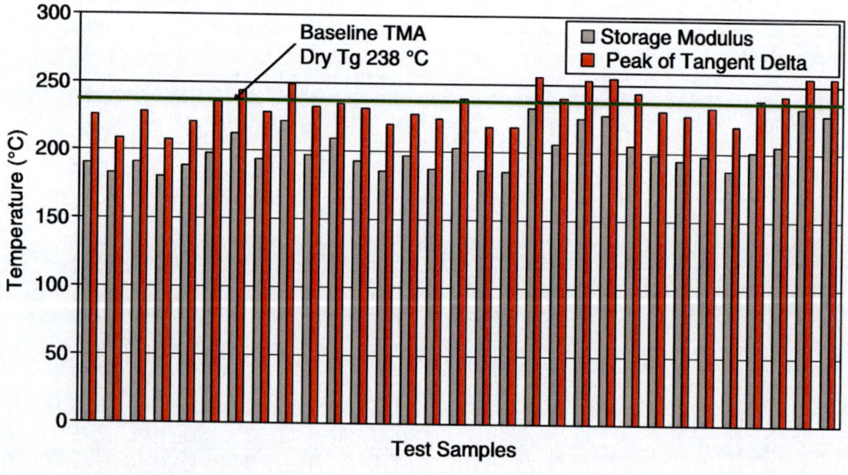

Fig. 13.28 DMA results for coupons excised from the upper skin of the B-737 horizontal stabilizer (Boeing method)

The DMA upper-skin coupons tested according to the ASTM standard yielded Tg values of 182.2°F (359.9°F)(onset of storage modulus) and 208.5°C (407.3°F) (peak of Tan Delta), respectively. The DMA coupons tested according to the Boeing method yielded Tg values of 201.1°C (394.0°F) and 233.5°C (452.3°F), respectively. The DMA results obtained using the Boeing method, were consistently higher than the DMA results generated using the ASTM standard. The Boeing method uses a slower temperature ramp-up rate than the ASTM standard allowing the specimen to dry during the DMA test and thus yielding a higher Tg for a given coupon. Tg values

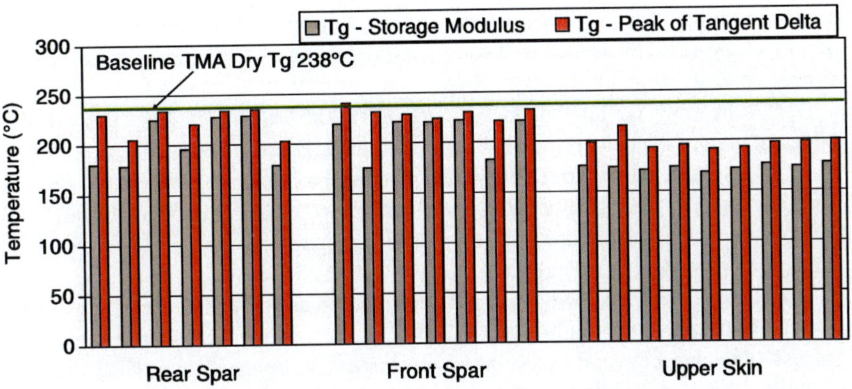

Fig. 13.29 DMA results comparison for coupons extracted from the upper skin and the front and rear spars of the B-737 horizontal stabilizer

obtained were consistent with the values obtained for the L/H stabilizer [6], are consistent with the Tg values of a 350°F cure material system and are significantly higher than the maximum temperature that the structure was designed for.

Similarly, DMA lower-skin coupons tested according to the ASTM standard yielded Tg values of 170.2°C (onset of storage modulus) and 198.4°C (407.3°F) (peak of Tan Delta), respectively. The DMA coupons tested according to the Boeing method yielded Tg values of 196.9°C and 230.9°C (452.3°F), respectively.

When comparing the Tg values obtained for the front and rear spars to the upper-skin Tg, it was found that the glass transition temperature obtained from coupons extracted from the spars was at least 9°C higher than the upper-skin Tg values as shown in Fig. 13.29. The increased Tg is most probably due to the additional curing that occurred during the secondary bonding of the spars. The skins were only subjected to a single cure whereas the spars were initially cured as separate C-channels and secondarily bonded using film adhesive.

Similarly, when comparing the Tg values obtained for the front and rear spars to the lower-skin Tg, it was found that the glass transition temperature obtained from coupons extracted from the spars was at least 21.3°C higher than the lower-skin Tg values, which is consistent with the results obtained for the upper skin.

Throughout the teardown investigation, baseline material capability was not available for all tests and configurations considered. In order to establish baseline material data for all tests and laminate configurations, current Cytec T300/5208 material was acquired. Laminates of different configurations were manufactured and tested. The intent was to investigate the thermal, physical, and mechanical properties of the new system and possibly establish baseline capability when the information was not available.

DMA tests were conducted on the new material system and yielded an average Tg of 193°C (onset of storage modulus) and 225°C (peak of Tan Delta), respectively. The current material glass transition temperature correlated very well with the average dry-skin Tg. The average upper-skin "dry" Tg was 205°C (onset of storage modulus) and 224°C (peak of Tan Delta), respectively. The average lower-skin "dry" Tg was 199°C (onset of storage modulus) and 221°C (peak of Tan Delta) respectively.

DSC tests were conducted on the current T300/5208 material to compare the degree of cure of the new material and the aged material extracted from the skins and spars. A DSC test on uncured T300/5208 was used to generate the heat of reaction value necessary to yield a fully cured part. Heat of reaction values obtained from DSC tests on spar and skin coupons were then used to generate a cure conversion percent, indicative of the degree of cure of the part.

DSC data generated showed that spars reached an almost fully cured status (99% cured), as additional post-curing most likely occurred during the secondary bonding process (4% cure conversion increase with respect to the new material). The skin coupons showed a slightly lower degree of cure as they were only subjected to one cure cycle. The average heat of reaction values obtained from the skin tests showed that the upper skin is about 97% cured whereas the lower skin is about 95% cured. This is most likely due to a higher thermal exposure of the upper skin during service. All coupons extracted from the aged structure showed an overall higher degree of cure than the new material as additional curing/aging has occurred during the life span of the structure.

13.7.7 Moisture Content Evaluation

Moisture content in the aged structure was quantified using ASTM D5229. Figures 13.30 and 13.31 show the moisture content distribution in both skins based on coupons excised from both locations and subsequently dried. All moisture levels obtained were lower than the design moisture level of 1.1%.

13.7.8 Void Content Evaluation

Physical tests were conducted per ASTM D3171 and ASTM D2734 to determine the resin and fiber volume fractions in the structure but also to quantify the porosity levels in the structure.

As shown in Fig. 13.32, the maximum void content in the upper skin was in the order of 7%, whereas the maximum void content found in the lower skin was in the order of 3%.

13.7.9 Microscopy/Image Analysis

Image analysis was conducted on samples extracted from both the upper and lower skins to inspect the structure at the microscopic level for voids, micro-cracks, or any evidence of aging or material degradation. Extracted samples were

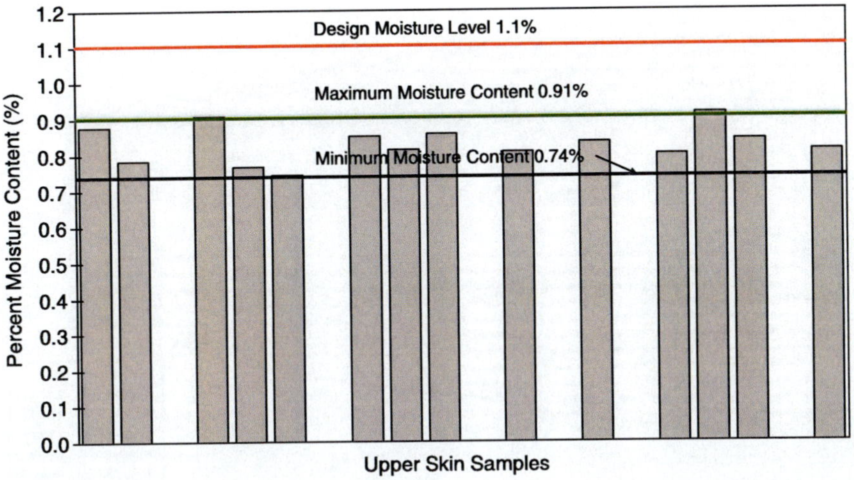

Fig. 13.30 Moisture distribution in the upper skin

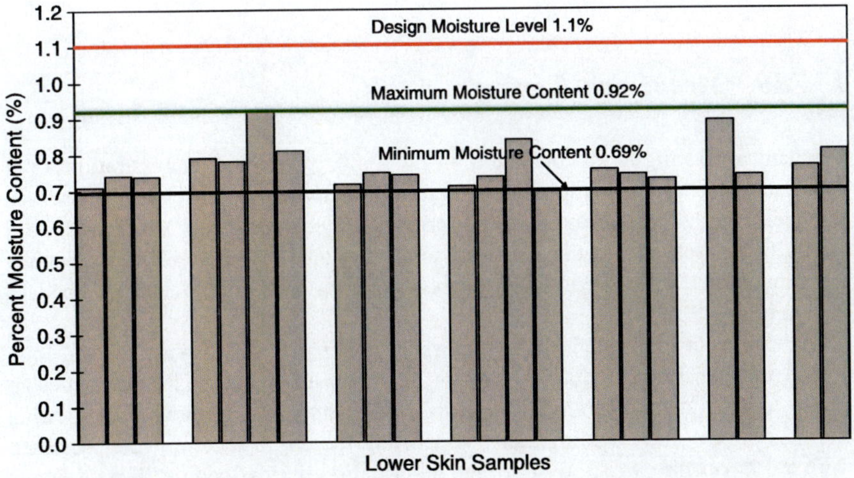

Fig. 13.31 Moisture distribution in the lower skin

potted, polished, and viewed under the microscope to detect any possible evidence of material aging. Figures 13.33 and 13.34 show evidence of porosity but also micro-cracking emanating from the voids/porous areas. The photos show micro-cracks; however, the exact source of micro-cracks could be resin shrinkage, in service degradation, or machining.

Fig. 13.32 Void content summary

13.7.10 Mechanical Testing

Mechanical testing was conducted using 1980s specimen configurations and standards. Because of layup schedule variation between the different rib stations, thickness variation, curvature in the structure, mechanical test specimens were carefully extracted from the structure, and individually tabbed when applicable, then machined and tested. All tension specimens were individually tabbed as shown in Fig. 13.35.

As shown in Fig. 13.36, the compression specimens are 1″ wide by 5.5″ long rectangular specimens. This is the coupon geometry that was used during the design phase, to generate allowables for the stabilizer structure. The fixture used for testing consists of two anti-buckling plates supporting the entire length of the specimen with a 0.2″ opening to allow specimen deformation during loading. The compressive loads are transferred to the specimens through end loading primarily. Two edge extensometers were used to generate stiffness and strain values for the different configurations tested.

A 22-kip servo hydraulic load frame was used for loading. The test equipment was calibrated and verified according to ASTM E4 standard to ensure the accuracy of the load and displacement readings. The coupons were loaded at a rate of 0.05in/min.

Figure 13.37 summarizes the compression results obtained for coupons extracted from the upper skin, tested with no drying. The figure presents measured and normalized RTA compression strength for coupons extracted from different

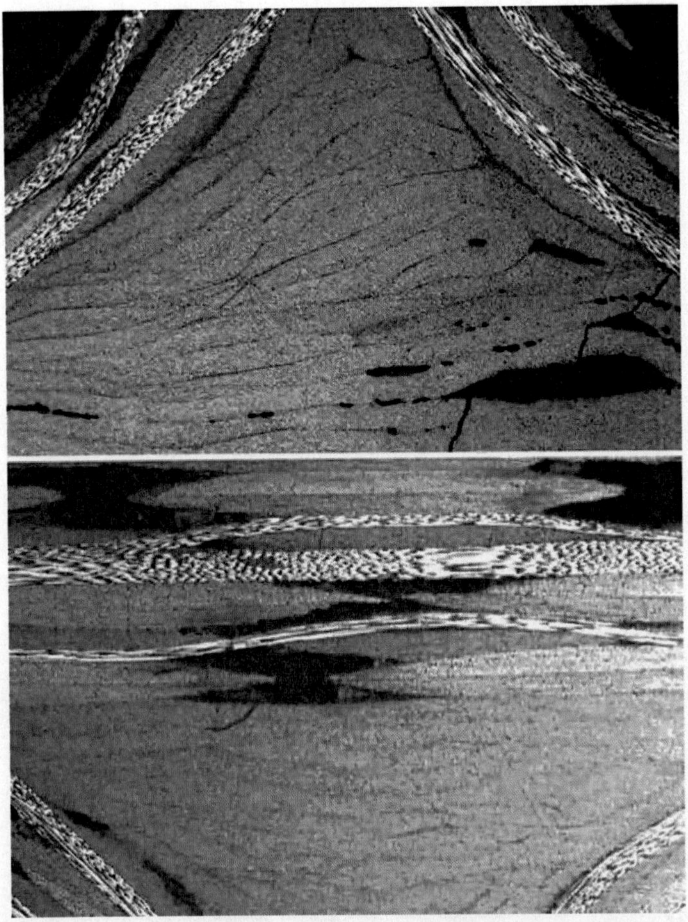

Fig. 13.33 Cross section of stringer flanges 2 and 3, rib station 2 at a magnification of 50× (*upper skin*)

locations on the upper skin. It should be noted that only specimens extracted from the skin around rib 1 had the same stacking sequence as the baseline 102-1 and 102-2. As illustrated in the figure, normalized compression strength for the three specimens correlated very well with the baseline compression strength.

Similar to the upper-skin values, the lower-skin normalized data obtained from the coupons extracted from rib 4 correlated very well with the baseline data as shown in Fig. 13.38. It should be noted that specimens excised the skin around rib 4 had the same layup schedule as the two baseline samples 102-1 and 102-2. The RTA compression data did not show any evidence of material degradation that can be attributed to the aging of the stabilizer.

Fig. 13.34 Cross sections from two different stringer webs at a magnification of 50×. Both cross sections show evidence of porosity between the laminate plies

Fig. 13.35 Tabbing fixture for the tension specimens

Tension testing of extracted coupons from the aging stabilizer was also conducted per the 1980s requirements and specifications. As shown in Fig. 13.39, the coupon has a tapered cross section with a 1″ width at the gage section and a 1.5″ width at the ends.

Fig. 13.36 Compression test fixture and test specimens

A 22-kip servohydraulic load frame was used for loading. The coupons were loaded at a rate of 0.05 in./min.

Figure 13.40 summarizes the tension results obtained for coupons extracted from the upper skin, tested with no drying. The figure presents measured and normalized RTA tension strength for coupons extracted from different locations on the upper skin. It should be noted that only skin specimens extracted around rib 1 and rib 3 had the same stacking sequence as the baseline 102-1 and 102-2. As illustrated in the figure normalized tension strength data for the three specimens correlated very well with the baseline tension strength. The tension strength data did not show any evidence of degradation that can be attributed to the aging of the stabilizer.

Three-stringer crippling panel tests were conducted on a few elements extracted from the upper and lower skins. The test setup is shown in Fig. 13.41.

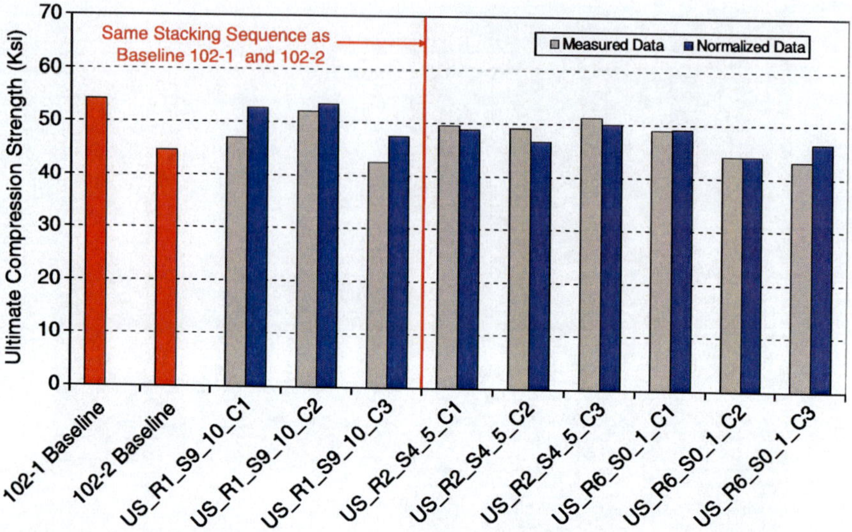

Fig. 13.37 Upper-skin compression strength test data, as extracted

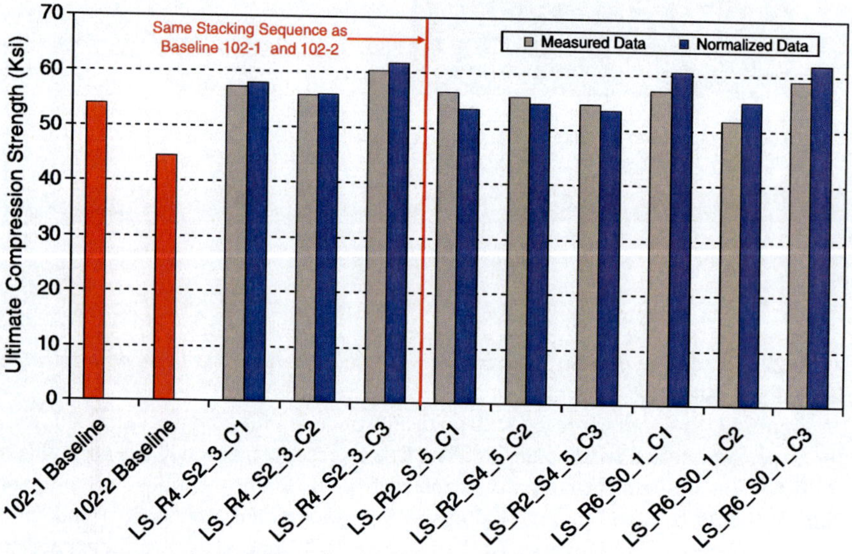

Fig. 13.38 Lower-skin compression strength data, as extracted

A 55-kip servohydraulic load frame was used for loading. The elements were loaded at a rate of 0.05 in./min.

A picture of a failed three-stringer crippling element is shown in Fig. 13.42. Failure load of the stiffened skin panel, though slightly lower, correlated very well

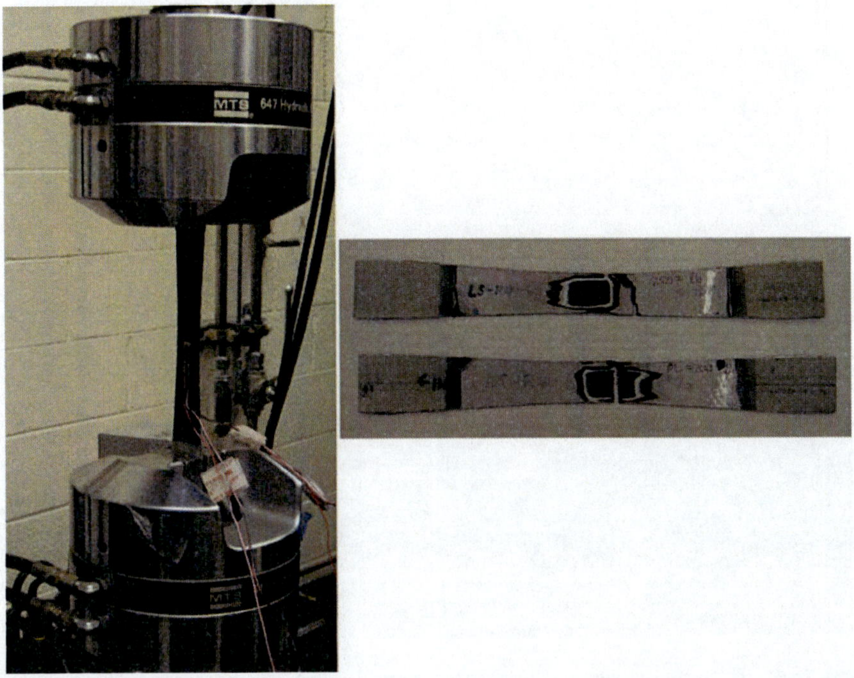

Fig. 13.39 Tension test setup and specimen configuration

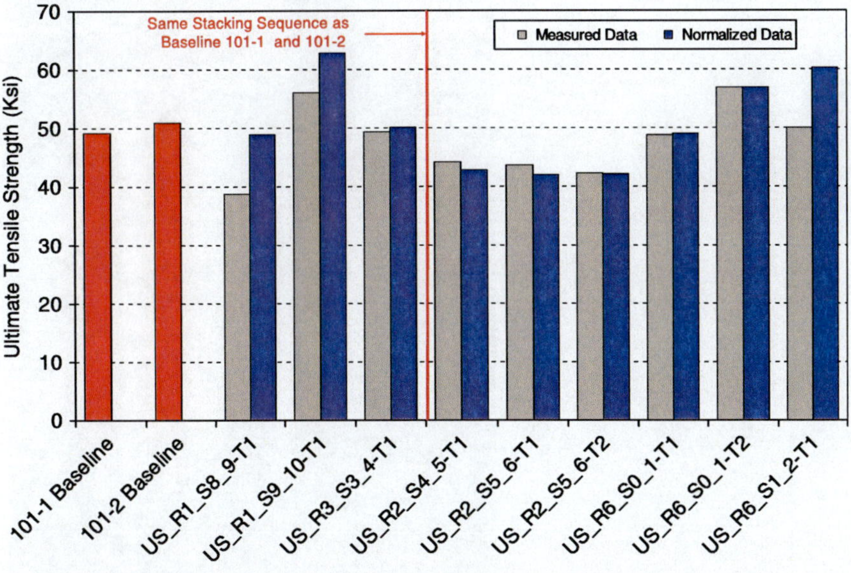

Fig. 13.40 Upper skin tension strength data, as extracted

Fig. 13.41 Three-stringer element crippling test setup

Fig. 13.42 Post-test picture of a three-stringer element

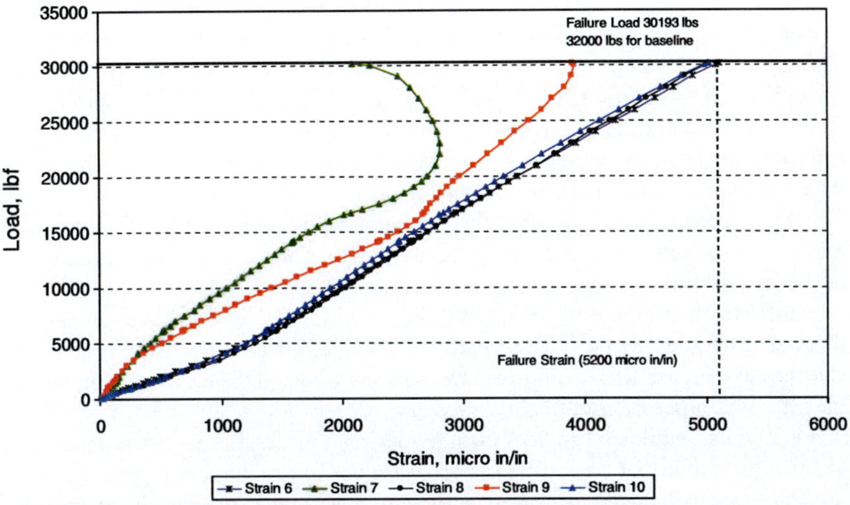

Fig. 13.43 Crippling element load versus strain data

with the failure load of the equivalent panel tested during the design/certification phase. Load versus strain data are shown in Fig. 13.43. It should be noted that all coupons/elements tested during the design phase were flat, whereas the specimens extracted from the aged stabilizer had a slight curvature characteristic of the shape of the structure. The failure strain achieved, 5,200 microstrains, correlated very well with the strains to failure of the equivalent panel tested during certification.

13.8 Conclusions

The teardown of the R/H Boeing-737-200 stabilizer revealed a composite structure that held up extremely well after 18 years of service with no obvious signs of aging to the naked eye, such as pitting and corrosion, as would a metal structure with a similar service history exhibit. The teardown also demonstrated good agreement between the results obtained from the R/H and L/H of shipset 5.

Physical test results showed moisture levels in the structure after 18 years of service as predicted during the design phase. Thermal analysis results were very consistent with those obtained for the left-hand stabilizer and with the baseline data provided by the supplier.

Thermal analysis showed that the degree of cure of the spars is close to 100%, that additional curing may have occurred during the secondary bonding of the spars. Furthermore, additional curing may have occurred in the upper skin due to environmental exposure (overall at least 95% cure was achieved in the structure).

For mechanical tests where baseline data were available (1970s certification and test data), the residual strengths met or exceeded the baseline values. No significant degradation was noted in any of the tests.

The teardown also showed that significant improvements in composite manufacturing processes and nondestructive inspection methods have been made in the last 30 years.

Furthermore, the current cytec T300/5208 material appeared to have comparable thermal properties to the 1970s material; that is, comparable resin system but significantly higher tensile and compressive properties indicative of an improvement in fiber processing in the last 30 years.

Finally, the teardown provides closure to a very successful NASA program and affirms the viability of composite materials for use in structural components. From all data generated, the margins were sufficient to warrant a "no significant degradation" conclusion.

Acknowledgments The authors would like to thank and acknowledge the invaluable guidance, support, and contributions of the Boeing team, our FAA technical monitor, and NIAR team members from the Aging Aircraft, Composites, and Mechanical Test Laboratories. This research program was sponsored by the Federal Aviation Administration William Hughes Technical Center and the National Institute for Aviation Research and has been presented at the Aging Aircraft Conference and will also be available to the public in an FAA report.

References

1. Aniversario, R. B., et al., "Design, Ancillary Testing, Analysis, and Fabrication Data for the Advanced Composite Stabilizer for Boeing 737 Aircraft." December 1982.
2. Dow, Marvin B., "The ACEE Program and Basic Composites Research at Langley Research Center (1975 to 1986)."
3. McCarty, J.E., "737 Graphite Epoxy Horizontal Stabilizer Certification." Boeing Commercial Airplane Company, Seattle, Washington, AIAA No 82-0745, May 1982.
4. Ekvall, J.C. and Griffin, C.F., "Design Allowables for T300/5208 Graphite-Epoxy Composite Materials." Presented at the AIAA/ASCE/AHS 22nd Structures, Structural Dynamics and Materials Conference, 1981.
5. Quinlivian, J.T., Kent J.A., and Wilson, D.R., "NASA-ACEE/BOEING 737 Graphite-Epoxy Horizontal Stabilizer Service, NASA DOCUMENT 95N28489, 1991.
6. Tomblin, J., Salah, L., Davies, C., Miller, M., Hoffman, D., Kollgaard, J., "Aging Effects evaluation of a decommissioned Boeing CFRP 737–200 Horizontal Stabilizer," presented at the 8th Joint FAA/DoD/NASA Aging Aircraft Conference – Jan. 31st – Feb. 3rd, 2005.

Chapter 14
Durability of Polymer Composites in Power and Propulsion Applications

John C. Thesken, Cheryl L. Bowman, and James K. Sutter

Abstract Advancements in polymer matrix composites have made them attractive to developers of power and propulsion equipment for spaceflight and aeronautic applications. However, many of these applications have very unique operational environments that are not easily found in the available design databases. Rapid insertion of these materials through prototype development and concept demonstration programs are hampered by the absence of relevant design data. In such cases, development programs conducted by the NASA Glenn Research Center have found it beneficial to employ pathfinder experimental methods designed to focus on the specific application and operational environment at hand. This chapter describes specialized experimental investigations of composite durability for applications that include flywheel energy storage, combustion chamber, and fan case structures. The experiments were designed to investigate complex thermomechanical and hygrothermal environments posed by these technologies. Beyond cycles-to-failure and residual strength, dimensional stability and stiffness degradation are key, if not primary, concerns in preserving functional capability for the cited applications.

14.1 Introduction

14.1.1 Background

Power and propulsion technology for spaceflight and aeronautic applications is a central element of the NASA Glenn Research Center's research and development portfolio. As the continuing push for lightweight structures is resulting in improved thrust and power-capacity-to-weight ratios, polymer matrix composites (PMCs) are

J.C. Thesken (✉)
NASA Glenn Research Center, Cleveland, OH 44135, USA
e-mail: john.c.thesken@nasa.gov

K.V. Pochiraju et al. (eds.), *Long-Term Durability of Polymeric Matrix Composites*,
DOI 10.1007/978-1-4419-9308-3_14, © Springer Science+Business Media, LLC 2012

finding usages in the relatively cooler segments of these lightweight systems. Still, operating composites in even moderate elevated temperatures with moisture can diminish the crucial glass-transition temperature safety margin. This adversely affects stiffness and dimensional stability while enhancing damage evolution processes associated with durability performance.

Technology development and concept feasibility demonstration programs rarely have access to design data for the precise operation environments associated with the new application. Pathfinder experimental programs designed with a unique focus on the specific environment in question can play a crucial role in the program's success. As a consequence of the focused nature of such experiments, statistical rigor in the quantity of repeated test conditions is often traded in favor of documenting complex synergistic phenomena at play in the operation environment.

Theoretical mechanics of composite durability are well covered elsewhere in this book; this chapter reviews and highlights three specific examples of experimental pathfinder investigations associated with power and propulsion applications. The applications include composites for flywheel rotors [1], high-temperature polyimide composites for reusable launch vehicle propulsion systems [2], and hygrothermal and mechanical fatigue of aero engine components [3].

The first example concerning flywheel rotor applications [1] might more formally be considered as an isothermal experimental investigation of PMC linear viscoelasticity. However, since the notion of linear hereditary response is so fundamental to the complexity of PMC mechanical behavior, this study provides a clear first example of pathfinder experimentation. The study contained a targeted group of isothermal creep and stress-relaxation tests, made in both compression and tension, to highlight weaknesses in the rigorous application linear viscoelasticity and the corresponding time-temperature superposition principle (TTSP). This complex synergy was further born out in the paradoxical response of the PMC during a prototypical flywheel operation cycle.

High-temperature polyimide composites applied to reusable launch vehicle applications is considered in the second example [2]. Here the transient phase of ignition and launch dominates the operation cycle for the mission environment. While abundant isothermal data were available, the evaluation of possible synergistic effects of rapid heating and loading were considered critical to the success of composites in this component. Application of a unique thermal heating and cooling system made this evaluation possible.

The final example concerns the hygrothermal and mechanical fatigue of a PMC to mimic an aero engine environment [3]. The goal was to simulate the moisture absorption–desorption, for instance an aircraft on the tarmac in a humid locale, followed by subsequent drying in flight. Here, the entire operations history was separated from the flight loads in order to design a tractable research program. Years of hygrothermal fatigue were followed by a course of mechanical fatigue cycles. The study reviews the techniques and results from one such investigation.

14.1.2 Time-Temperature Response of Composites for Flywheel Rotors

Section 2 of the following will describe pathfinder experiments for composite flywheel rotors. Flywheel energy storage offers a possible alternative to the electrochemical battery systems currently used for space applications such as the International Space Station. Rotor designs utilizing the load carrying capacity of carbon fibers wound in the hoop direction are capable of high operational speeds and specific energies. Despite having the primary load in the hoop wound fiber, the long-term durability of such rotors can be limited by the time-temperature dependent behavior of the epoxy matrix, which dominates the shear properties and stiffness transverse to the fibers. This test program employed a carefully chosen combination of isothermal creep and stress relaxation tests for the prototypical rotor material, IM7/8552 [1]. Flat filament wound panels were made using the same processing route as for composite rotors. Coupon specimens, sectioned normal and parallel to the winding axis, were tested in compression and tension, at room temperature (RT), 95°C and 135°C for strain rates from 5×10^{-6} to 5×10^{-3}/s. Creep and stress relaxation testing ran 72 h followed by a 72-h recovery. Time-, temperature-, and load sign-dependent effects were significant transverse to the fiber. At a fixed deformation of −0.5% strain applied for 72 h, compressive stresses relaxed 16.4% at 135°C and 13% at 95°C. Tensile stresses relaxed only 7% in 72 h at 135°C for a comparable 0.5% strain. The postulate of a linear hereditary material response and the application of Boltzmann's principle of superposition to describe the behavior observed here was found to be problematic if not intractable. Within the scope of these experiments, uniaxial compressive stress relaxation data could be used to bound the amount of relaxation with time of radial preload stresses in flywheel rotors.

14.1.3 Thermomechanical Fatigue (TMF) of Polyimide Composites for Reusable Propulsion Systems

Section 3 describes the application of thermomechanical fatigue (TMF) to a polyimide composite support structure for a concept reusable launch vehicle engine. Reusable propulsion systems for manned space flight face stringent durability and safety requirements. As Rocketdyne and NASA-Glenn Research Center (GRC) sought to introduce polyimide composites to the harsh environmental and loads familiar to space launch [2], serious consideration was being given to the synergistic effects of temperature and mechanical loads. The candidate component would operate at 316°C, which is about 80% of the polyimide's glass transition temperature (Tg). Earlier, TMF investigations of chopped fiber polyimide composites tested near to Tg showed that cyclic temperature and stress promoted excessive creep damage and strain accumulation. In the proposed propulsion application, stiffness and deformation tolerances were operationally critical and it was crucial to verify that the continuous-fiber, cross-ply laminates maintained these requirements.

The study reviewed here examines the isothermal fatigue and TMF of stitched and unstitched cross-ply laminates of M40J carbon fiber-reinforced polyimide. The maximum operation temperature, 316°C, was used for isothermal tests and as the maximum temperature in TMF tests. The mechanical load cycle was identical in both isothermal and TMF testing. The load magnitudes were guided by design analysis of the actual component. Test waveforms for thermal and mechanical load cycles were in-phase and were chosen to be representative of an engine operation cycle.

Deformation and stiffness degradation due to fatigue loading was monitored periodically during each fatigue test. Isothermal and TMF performance was compared to deduce the presence of synergistic effects. The influence of stitching on the long-term durability of the material was studied and the combined results were used in design guidelines for the development program.

14.1.4 Durability of Hygrothermally Conditioned Composites for Aero Engine Applications

Section 4 describes pathfinder testing to investigate the influence of hygrothermal conditioning on composites for fan case applications. The experimental investigation was conducted to characterize the elevated-temperature, fully reversed fatigue response of a hygrothermally conditioned carbon fiber-reinforced epoxy composite. The material was characterized by its fully reversed fatigue strength, residual compression strength, and modulus degradation both as function of hygrothermal conditioning and number of elevated temperature fatigue cycles. The composite material, AS4/PR500, was a five-harness satin weave in a high-temperature epoxy matrix. Elastic modulus and residual compressive strength were recorded as a function of fatigue cycle. Unconditioned composite test samples were compared to samples after exposure to 12,000 h of hygrothermal conditioning. The 12,000 h of hygrothermal conditioning (12 k h HC) was designed to simulate the life cycle of a commercial gas turbine component. The HC mission cycle consisted of a 90-min hold at 121°C followed by a 22.5-h soak at 30°C with 85% $\pm$ 3% relative humidity. The achievement of fully reversed fatigue and compressive residual strength data at elevated temperature is somewhat unique and is due to the chosen specimen geometry. Specimens were machined into a dogbone geometry that was developed and verified for fully reversed fatigue and static compression testing without anti-buckling supports. Before applying fatigue loading, the compressive strength was compared for the 12 k h HC material and the unconditioned material finding an 11% strength degradation. Fully reversed fatigue data revealed a deterministic stress-life response. Residual compressive strength measurements were made at specified compressive modulus degradation levels as measured throughout the test. As expected, the HC material trends lower in performance capability as made clear by graphing residual compressive strength as a function of compression modules.

14.2 Time–Temperature-Dependent Response of Filament Wound Composites for Flywheel Rotors

Successful space flight operations require onboard power management systems that reliably achieve mission objectives for a minimal launch weight. Flywheel energy storage is a viable space power storage option. Flywheels can attain relatively greater power and energy densities compared to electrochemical batteries with no fall-off in capacity after repeated charge–discharge cycles. However, space power flywheel design implementation requires the safe-life qualification of the technology-enabling rotor material: carbon fiber-reinforced polymer composite. Specifics of flywheel design and the evolution of composites for rotor materials may be traced through a number of comprehensive state-of-the-art reviews [4–9].

Solutions to this problem have evolved rotor/hub designs that induce compressive radial stresses acting transverse to the fiber. Analytical work by Arnold et al. [9] has shown the importance of the interference misfit for these designs. Clearly, the performance and long-term durability of a carbon fiber polymer rotor depends upon multiaxial performance and not just stability in the winding-filament direction. Thus time–temperature-dependent material behavior may have a significant, life-limiting impact. Loss of preload or a substantial change in the pre-induced strain state could lead to disk separation and to loss of system function. Concerns with the time–temperature-dependent behavior of composite flywheel rotors have been raised by several previous researchers [10–13] and relevant data has been modeled within a viscoelastic framework [13]. While viscoelasticity may be a fundamentally useful representation of the matrix-dominated properties, it may not be entirely accurate. The objective of this experimental investigation was to provided a more accurate assessment of the relevant constitutive characteristics for the flywheel-system operating conditions.

The work was initiated within the Rotor Safe-Life Program in support of the Flywheel Energy Storage System (FESS) evaluation of pre-loaded, concentric ring rotors made of wound IM7 carbon fiber, 8552 epoxy, prepreg tapes. Rotor design, development work, and element-level testing for this program was made by the University of Texas Center for Electromechanics (UTCEM) [14].

14.2.1 Experimental Approach

The time–temperature-dependent nature of polymers and their composites, e.g. carbon fiber-reinforced epoxy, have been widely investigated [15–22]. Studies that examine the applicability of viscoelastic and viscoplastic theoretical frameworks may be found by Raghavan and Meshi [23] and Al-Haik and coworkers [24]. These works and the experimental methods applied by Arnold et al. [25] guided this study. The concepts that were tested focused on the notion of linear hereditary constitutive behavior transverse to the fiber and the issue of a recoverable, initial stress/strain state.

Linearity holds when the time-dependent relaxation modulus or the creep-compliance data are independent of the applied deformation or stress amplitude.

The eventual recovery of the initial stress/strain state, when the applied deformation or stresses are returned to zero, indicates the absence of permanent strain and/or damage. The proof of existence and the definition of applicable bounds for such linear hereditary response are necessary for design engineers seeking to use Boltzmann's principle of superposition to predict the long-term, uniaxial stress/strain response.

The temperature reflects the energy available for molecular motion of the polymer and, like time, is a parameter of the constitutive response. Higher temperatures ease the rearrangement of the polymer molecular chain and thereby manifest greater compliance and shorter relaxation times. Indeed, relaxation times are often found to scale with temperature, giving rise to the time–temperature superposition principle (TTSP). Master relaxation-modulus curves and master creep-compliance curves may be constructed over a wide time range by shifting multiple temperature data to a specific reference temperature. Note that at the glass-transition temperature, Tg, a significant change in the degree of molecular motion occurs as the polymer transforms from the glassy state to the rubbery state. Hence, the nature of the shift factor also changes. Necessarily, the operation temperatures for carbon fiber/epoxy rotors are expected to be less than Tg. Dynamic Mechanical Analysis (DMA) was made to determine the glass-transition temperature and all mechanical testing was made at temperatures lower than Tg.

The test program consisted primarily of isothermal, monotonic-loading tests and isothermal, stress relaxation tests and creep tests. Mechanical response to compressive and tensile loading was examined to determine the constitutive-response complexity, which effects the number and type of stress invariants in the constitutive-model object function. Coupons were tested at different monotonic-loading strain rates to gauge the time–temperature-dependent, stress–strain response. Temperature and either strain or stress are held constant in the stress relaxation tests or creep tests, respectively. While none of these variables is expected to be constant during normal operation, prototypical experiments were made to identify possible synergistic effects due to cyclic loads.

14.2.1.1 Flat Coupon Specimen Manufacturing

Hexcel IM7/8552 in the form of pre-impregnated tow was used to wind flat specimen panels. The 8552 resin system is an amine-cured, toughened epoxy. Neat resin has a dry Tg of 200°C when cured at 177°C. Properties of the 8552 neat resin and IM7 carbon fibers are given in Table 14.1.

Flat panels were manufactured on a 35.56-cm square steel hollow mandrel, with a wall thickness of about 0.95 cm. An individual, pre-impregnated tow, under about 35 N tension, was wound onto the mandrel surface to create a square cylinder. Tow advance was about 89.8°, resulting in a slight tow-to-tow overlap. Specifications for the IM7/8552 tow included a bandwidth of about 0.32 cm with an approximate resin content of 60%. After achieving the finished thickness, the part was prepared for

Table 14.1 Composite constituent properties

	Density (g/cc)	Tensile modulus (kN/mm^2)	Tensile strength (N/mm^2)	Tensile strain-to-failure (%)
8,552	1.30	4.67	120	1.7
IM7	1.77	276	5,379	1.8

curing using caul plates on the outside surface of the material and conventional vacuum bagging procedures. Placed in an autoclave, the part was cured at 177°C using the cure cycle specified by the material supplier [26].

The four-sided, cured part was then machined into separate panels using diamond cutoff tooling. Efforts were made to preserve the filament winding orientation, i.e. sectioned panels reflect tow advance. Residual strains are present in the composite panels, in the direction of the carbon fiber. This residual strain is thought to be associated with the tension induced in the fiber direction due to mandrel thermal expansion during cure and the winding tension applied to the tow. During mandrel expansion, prior to resin staging, the fibers closest to the mandrel surface carry more of the mandrel expansion load. After cure, this strain profile, with the relative tensile fiber strain maximum at the inner panel surface and decreasing toward the outer surface, is believed to be responsible for a slight convex warp of the panels.

Prior to specimen machining, panels were subjected to ultrasonic examination to produce C-scan representations of the material quality. The same nondestructive evaluation methods were used as those being used by Baaklini and coworkers [27] for flywheel certification. Compaction attributes of flat panel and hoop wound rotors, such as void content and fiber volume fraction, were found to be similar. Coupon specimens were sectioned from the flat panels using diamond saw. Specimen edges were then lightly sanded with fine grit paper to remove any cutting marks. Specimen geometries were in accordance with ASTM Test Method for Tensile Properties of Polymer Matrix Composite Materials (D3039).

The plan geometry for each specimen was 25.4 × 190.6 mm. The nominal specimen thickness was about 2.82 mm. After machining, specimens were dried in vacuum at 115°C for 48 h and stored in a desiccator prior to testing. Pull tabs were cut from prefabricated sheets of GRFP. These were then tapered and bonded to the ends of the specimen using the 3M structural adhesive AF 163-2L.

The epoxy adhesive was cured following manufacturers guideline. Strain gauges were applied as suggested in ASTM D 3039 and in ASTM Test Method for Compressive Properties of Polymer Matrix Composite Materials with Unsupported Gage Section by Shear Loading (D3410) so that specimen bending could be monitored. Sullivan [18] provides some useful guidelines on the use of strain gages in polymer–composites creep testing but special considerations are still needed for long-term, elevated-temperature testing. This is particularly so with respect to tests made transverse to the reinforcement fiber. Here the polymeric-film backing material may undergo creep and the metallic foil may have a local reinforcement effect.

These issues were discussed with Prof. J. Raghavan who used MicroMeasurement WK-type gauges in creep tests of AS4/3501-6 carbon fiber epoxy at temperatures up to 200°C [23]. In that work, the stability of the WK-type gauge was verified by bonding them to stainless steel samples and subjecting them to elevated-temperature loading over the same duration as the planned creep tests. Perry [28, 29] suggests that 7 kMN/mm^2 is the low-modulus limit for foil strain gauge application to low-modulus materials in order to minimize errors caused by strain-gauge reinforcement. Similarly, Sullivan [18] found that the error is about 6% for materials with a modulus of 4 kN/mm^2 and that this error will decrease with increasing modulus of the base material. Therefore, strain gauge reinforcement effects are probably negligible for IM7/8552 which is expected to have a transverse modulus greater than 7 kMN/mm^2. With this in mind, WK-type gauges were selected for use in these elevated-temperature experiments. Room-temperature tests were made with CEA-type gauges. Gauge bonding practices followed manufacturer's recommendations.

Thermal-stress effects and elevated-temperature stability were examined by heating unconstrained, gauged specimens and by observing the stress-free, gauge output as a function of time and temperature. During these tests, elongation was measured also on one edge of the specimen using an axial extensometer.

14.2.1.2 Test Setup

Elevated temperatures were achieved using a custom-made, radiant mini-furnace. Closed-loop temperature control was provided through a reference thermocouple attached to a fixed position in the furnace. Calibration of the specimen temperature and verification of thermal gradients was accomplished by heating a specimen instrumented with a series of vertical and horizontal embedded thermocouples. Calibration factors were determined for each test temperature. Through careful adjustments of oven position and placement of insulation, it was possible to reduce the thermal gradients in the center measurement region of the specimen to be less than 0.3°C/mm.

Mechanical testing was performed on a computer-controlled, MTS servo-hydraulic test frame. In addition to the back-to-back foil strain gauges used to monitor bending strains, an edge-mounted extensometer with a 12.7-mm gauge length measured deformation. Previous examples of compressive creep or stress relaxation tests transverse to the fiber direction have used special test fixtures and anti-buckling guides [15, 16, 19]. However, these tests were performed in MTS wedge-style grips aligned to within $\pm 20 \times 10^{-6}$strain, without additional support. Based on the specimen geometry and the distance between grips, Euler buckling calculations were made to determine the maximum loads or deformations that could be applied without the occurrence of instability. For a grip edge to grip-edge distance of 54.9 mm and a specimen thickness of 2.82 mm, buckling instability would be expected to occur at a compressive strain of −0.87%. This calculation assumes that the grips may be approximated by fixed-end conditions.

A room temperature, compressive test was made to verify the specimen stability. At an axial strain of −0.88%, the calculated bending strains were measured to be only 6.0% of the axial value; buckling instability did not occur. On subsequent reloading to cause compressive failure, bending strains began to exceed 10% when the compressive axial strain measured −0.81%. The specimen, however, sustained a maximum compressive strain of −1.65% before the loading capacity dropped to zero due to buckling. The discrepancy between measurements and Euler's buckling limit may be due to the tabbing material that supported the specimens about an extra 7.5 mm on either end of the gripped edges. Note that the Euler equation for a length of 40-mm length gives a buckling strain of about 1.62%. This information was then used to limit the maximum compressive strain and stress levels that could be applied in the tests.

14.2.2 *Experimental Results*

14.2.2.1 Material Characterization

Fiber volume fraction and void contents were determined in accordance with ASTM Test Method for Fiber Content of Resin Matrix Composites by Matrix Digestion (D3171) and ASTM Test Method for Void Content of Reinforced Plastics (D2734). A typical range of values for a panel of this material are: fiber volume fraction in the range of 56.6–57.15% and void content in the range of 1.4–0.3%.

The dynamic shear moduli were measured following the ASTM Test Method for Assignment of the Glass Transition Temperature by Dynamic Mechanical Analysis (E 1640). The tests were made using an ARES unit manufactured by Rheometrics Scientific As-received material and a sample aged for 72 h at 135°C were tested to ascertain the chemical-aging potential during extended testing at elevated temperature. The specimen size was 29.6 × 10.3 × 2.8 mm with the fibers oriented in the direction transverse to the axis of rotation. The specimen was deformed in torsion at a fixed frequency of 10 radians per second to a strain of 0.15%. The applied deformation function is sinusoidal so the maximum strain rate for the experiment is 1.5%/s, which is the product of frequency and strain amplitude. The shear storage modulus (G') and shear loss modulus (G'') were recorded as the specimen was heated to 300°C at a rate of 5°C/min. The shear moduli $G'$and G'' are shown as a function of temperature (Fig. 14.1); two vertical dashed lines delineate the glassy, transitory, and rubbery regions at these test conditions.

The Tg was determined to be approximately the same for both samples: 202°C, using the standard intercept approach on a linear scale. The measured Tg agrees well with the manufacturers' reported data for this cure/post-cure fabrication process. Note that the transition region would occur at lower temperatures for slower deformation rates. Further testing would be required to determine if the regime falls as low as 95°C or 135°C. The agreement between the Tg of the

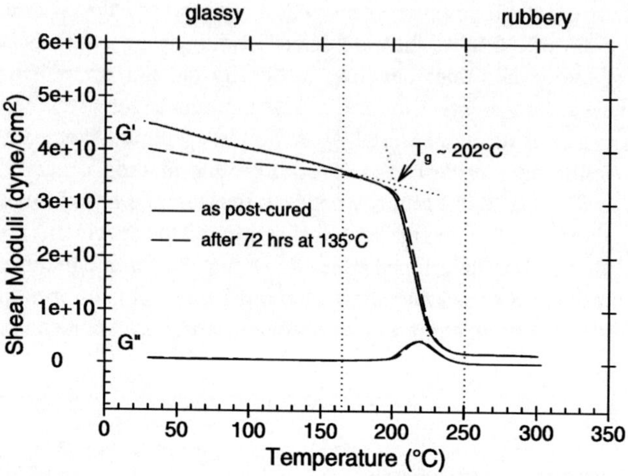

Fig. 14.1 The storage (G′) and loss (G″)moduli as a function of temperature measured by dynamic torsion for a specimen with the fibers oriented transverse to the rotation axis

as-received sample and the Tg of the 72 h-aged sample indicates that, with respect to the times and temperatures evaluated here, the composite may be considered fully cured. Thus chemical aging is not expected to play a role in the time–temperature-dependent material behavior reported here.

14.2.2.2 Monotonic Testing

Longitudinal Properties

The tensile elastic properties in the fiber direction are reported since they are relevant to subsequent design investigations. The longitudinal modulus, E_L, was measured at room temperature for eight different specimens tested at 10^{-4}/s strain rate. The average E_L value measured for room temperature was 151 kN/mm^2 ±4.0%. Tests made at 135°C for strain rates of 5×10^{-6}/s and 5×10^{-3}/s yielded E_L values of 155 kN/mm^2 and 150 kN/mm^2, respectively. Clearly, any variation due to strain rate at this temperature is smaller than the specimen-to-specimen variability measured at room temperature. The Rule-of-Mixtures predicts values for E_L of 158–160 kN/mm^2, This is 6% greater than the mean room-temperature values. Note that if the matrix was completely discounted, the Rule-of-Mixtures prediction would be 156.2 and 157.7 kN/mm^2, still greater than measured values. The tow-winding angle of 89.8° is too close to 90° to account for the 6% difference between measurements and prediction. Indeed, tensor transformation of the

Table 14.2 Transverse material properties[a]

Compression E_T (kN/mm^2)	Tensile E_T (kN/mm^2)	ν_{TL}	Tensile strength (N/mm^2)	Tensile strain-to-failure (%)
9.69 ± 0.78	9.07 ± 0.44	0.021 ± 0.002	69.2 ± 13.4	0.80 ± 0.12
–	8.274	0.017	62.7	0.84

[a]First row data are mean room temperature values; the second row is data for a single 135°C test

orthotropic stiffness coefficients suggests that the winding angle could be decreased to 86° without causing a difference greater than 1%. Therefore, the discrepancy between modulus data and the prediction is likely related to the accuracy of the volume fraction and/or the assumed constituent properties. The average Poisson's ratio corresponding to an applied stress in the fiber direction, ν_{LT}, was measured to be 0.30 ± 0.02 at room temperature.

Transverse Properties

The average room-temperature mechanical properties transverse to the fiber direction were determined from five tests (Table 14.2) and are listed in the first row. The transverse modulus is designated ET and the Poisson's ratio corresponding to a deformation in the fiber direction due to a load in the transverse direction is designateds ν_{TL}. Tests were made at a displacement rate of 0.0254 mm/s. A single tensile test was made at 135°C to determine the reduction in tensile strength due to elevated temperature. This result was used to guide load levels in tensile creep tests. Note also that the longitudinal and transverse data are consistent with the requirement that the ratios ν_{LT}/E_L and ν_{TL}/E_T be equal. The composite transverse tensile strength at room temperature is about 58% of the neat resin tensile strength indicating a considerable tensile-strength knock down due to the presence of reinforcing fibers and fiber/matrix interfaces.

The influence of strain rate on E_T was studied in a series of monotonic compression and tension tests. The experiments were made to a maximum strain of 0.3% in strain control at a range of strain rates from 5×10^{-6}/s to 5×10^{-3}/s. Test temperatures included room temperature (RT), 95°C and 135°C. Figure 14.2 shows E_T as a function of strain rate for the three test temperatures. There were too few tensile E_T/strain rate data to reveal a distinction between compression and tension.

These data give the first indication of the time–temperature dependence of the transverse behavior of IM7/8552. E_T is seen to increase about 5% over three decades of increasing strain rate from 5×10^{-6}/s. The data lie along curved lines of constant temperature that appear self-similar. The room-temperature data are at least 10% greater than the data measured at 135°C. It is proposed that the time scale for more significant changes in modulus is quite long and becomes apparent at slower strain rates. This hypothesis is illustrated by plotting the time-dependent modulus for compressive creep data at 135°C as a function of the average strain rate which was calculated up to that instant in time. The resulting series of calculated

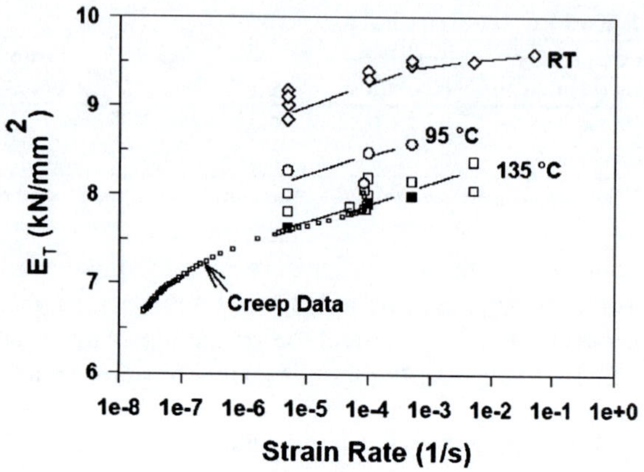

Fig. 14.2 Elastic modulus as a function of strain rate, *open symbols* are compression data, *filled boxes* are tensile, *small squares* are converted compressive creep data

modulus/strain rate data is shown in Fig. 14.2 as the arc of small open squares. This data extends two decades less than 5×10^{-6}/s and, over these two decades, the series of E_T values decreases 13%. Good agreement is found in the region where the calculated, creep-based, transverse modulus data overlaps the traditionally measured, 135°C modulus data. As the total strains in these tests are of the order 0.5%, this shows that the time scales for time-dependent behavior at 135°C are on the order of thousands of seconds or more.

14.2.3 Stress Relaxation and Creep Tests

As described in the experimental procedure, bending strains were closely monitored and all tests reported were within the limits suggested in ASTM D3039 and ASTM D3410. For compression, ASTM D3039 recommends that bending strain, as calculated from back-to-back strain gauges, be less than 10% of the axial strain. A typical plot of the percent bending as a function of axial strain for a long-term compression test is given in Fig. 14.3; the bending strains were less than ±5% of the axial strain level.

The applied strain for the tensile and compressive relaxation tests was 0.5%. The applied stress for the tensile and compression creep tests was 41.4 N/mm^2. Generally, the duration of the applied stress/strain period and the unstressed/unstrained recovery periods were both approximately 2.6×10^5 s (72 h). Instances where these periods differed are apparent in the data plots. The overall results of the stress relaxation and creep tests are represented in Figs. 14.4–14.6. The plots show examples of the respective stress and strain responses for each test type as a function of time. For convenient comparison of the tensile and compressive responses, the stresses are presented as absolute magnitudes. The reader is reminded that the sense of stress for the recovery portion of the stress relaxation plots changes as the control strain approaches zero. This is annotated by the use of filled data symbols for a

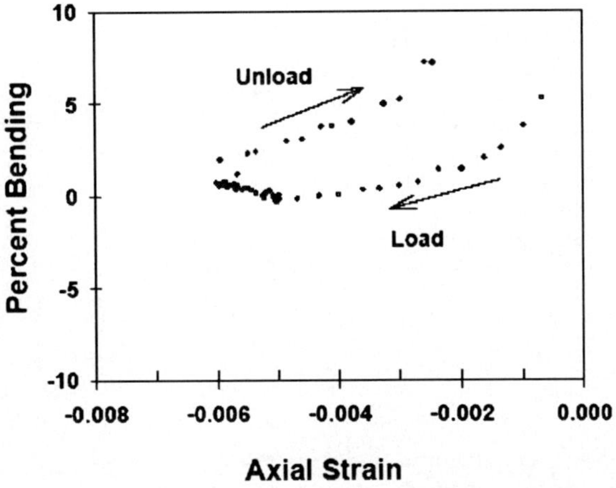

Fig. 14.3 Typical percent bending strain with respect to axial strain for a creep test

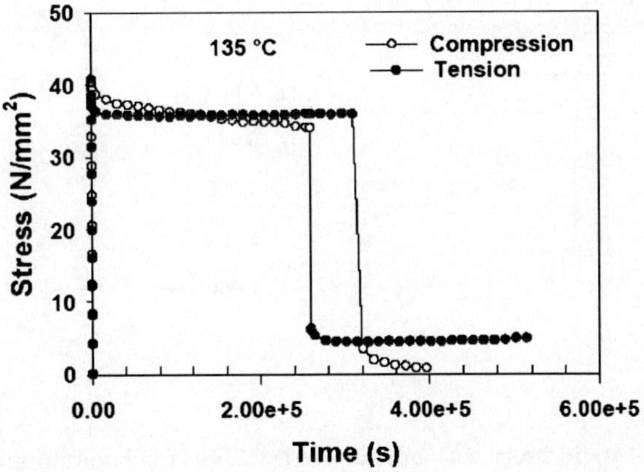

Fig. 14.4 Tensile and compressive stress relaxation at 135°C as a function of time. Stress is plotted as absolute value, note that the sense of stress transposes during the recovery portion of the test so that *filled symbols* are tensile

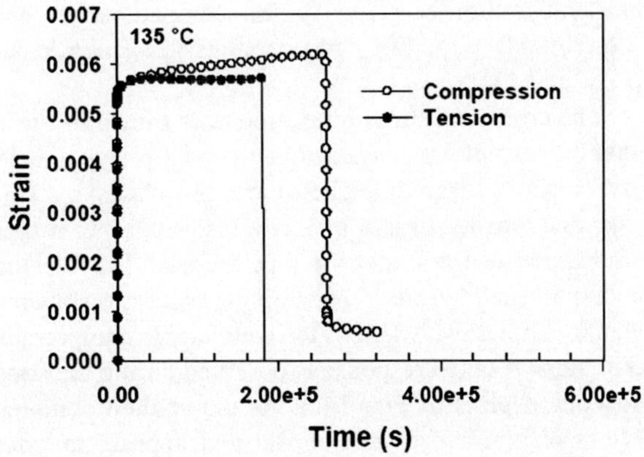

Fig. 14.5 Tensile and compressive creep at 135°C as a function of time. Strain is plotted as absolute value. The tensile test failed after 52 h

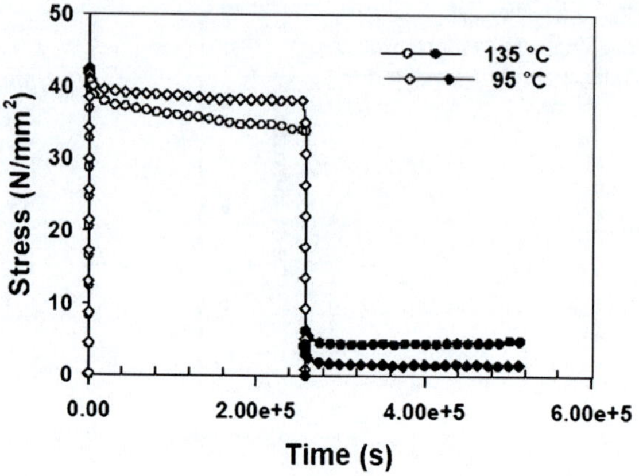

Fig. 14.6 Compressive stress relaxation at 95°C and 135°C as a function of time. Stress is plotted as absolute value, note that the sense of stress transposes during the recovery portion of the test so that filled symbols represent tensile stresses

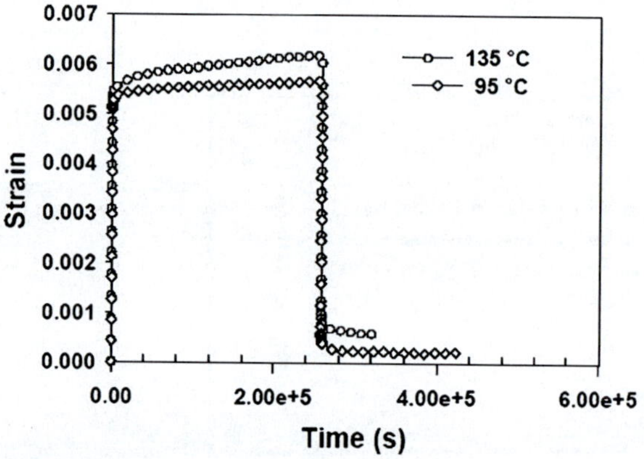

Fig. 14.7 Compressive creep at 95 and 135°C as a function of time. Strain is plotted as absolute value

tensile stress state and unfilled symbols for a compressive stress state. The influence of tensile and compressive states on stress relaxation response (Fig. 14.4) and creep response (Fig. 14.5) is contrasted for tests made at 135°C. The influence of temperature is given in the comparison of compression–stress relaxation experiments (Fig. 14.6) and compression–creep experiments (Fig. 14.7) made at 95°C and 135°C.

The creep and stress relaxation tests shown in Figs. 14.4–14.5 clearly indicated that the response magnitude due to compressive stresses or strains was greater than the response magnitude due to tensile stresses and strains. For example, compressive and tensile stresses relaxed 16.4% and 7%, respectively after 72 h at a strain magnitude of 0.5% and a temperature of 135°C. Qualitatively, the tensile stress relaxation and tensile creep straining processes shown in Figs. 14.4 and 14.5 appear to have shut down whereas the comparable compression curves are still increasing. The same relative response is observed during the stress recovery part of the stress relaxation plots of Fig. 14.4. As the applied compressive strain is removed, the stress response becomes tensile and appears to arrest after a small amount of

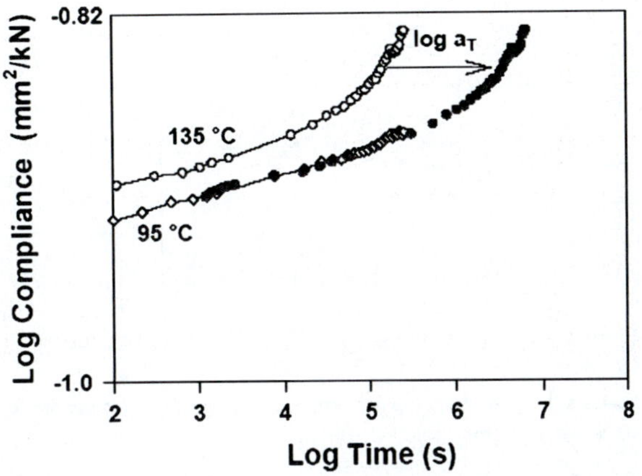

Fig. 14.8 Application of the time–temperature superposition principle to compressive compliance data at 95°C and 135°C. Log aT = 1.4, filled symbols are the shifted 135°C data

recovery. Alternatively, the tensile stress relaxation response that appears to have shut down during the applied tensile strain, yet the compressive stress is seen recovering toward the zero stress state when the applied strain is removed.

The creep data shown in Fig. 14.5 indicate that tensile creep rates are much slower than compressive creep rates and that tensile creep tests were also prone to creep rupture. The creep rupture could be expected due to the presence of weak fiber/matrix interfaces already apparent in the 58% knock down in tensile strength with respect to neat resin. Therefore, recovery data for the unloaded tensile creep specimens is unavailable. Creep recovery in the compressive creep tests were easier to monitor and the rate of recovery was relatively slow with respect to the initial phase of the test.

As expected, reducing the temperature from 135°C to 95°C decreased the stress relaxation and creep response magnitude. For example, compressive stress relaxation was only 13% at 95°C as compared to 16.4% at 135°C. The response curves shown in Figs. 14.6 and 14.7 are classic examples of the time–temperature-dependent material response. Converting the data to time-dependent compliance, shown in Fig. 14.8, it is possible to investigate the applicability of the time–temperature–superposition principle (TTSP). This is accomplished by shifting the 135°C data to the right so that a master curve covering a longer time range is created at 95°C. The horizontal distance, in Fig. 14.8, that the 135°C curve is translated to overlap the 95°C curve is expressed as log aT. The term aT is defined as the shift factor, the ratio of the time scale of the desired master curve to the time scale at an elevated temperature. In this case, the value for log aT was found to be 1.4. Confidence in the result is given by the apparent continuity of the overlapping shifted data (filled symbols) with the underlying 95°C data. Thus the TTSP appears to be applicable in the time and temperature regime of these experiments.

The observation of stress/strain sign-dependent material response brings into question the assumption of linearity and of a recoverable zero stress/strain state. Turner [30] points out that the most practical definition for the term *nonlinear* is

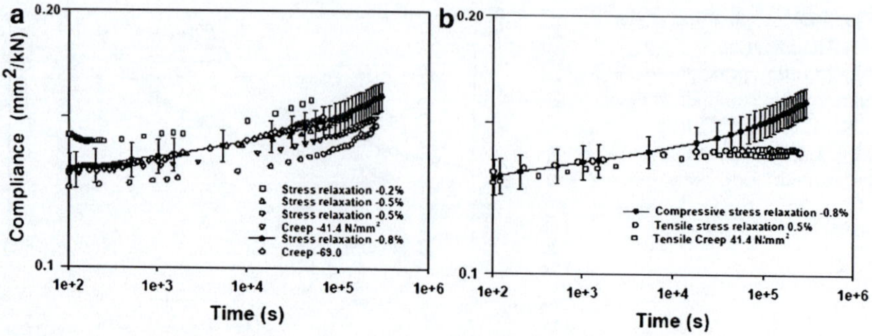

Fig. 14.9 Compliance from both creep and stress relaxation as a function of time on logarithmic axes for (**a**) compression and (**b**) compression and tension

that the stress at any particular time is not proportional to strain. Thus time-dependent compliance curves from tests made at different stress or strain levels should be identical for the material response to be linear.

Further compressive stress relaxation and creep tests were made at 135°C to examine the load-level influence and further explore the linearity assumption. The additional stress relaxation tests were made at −0.2% and −0.8% strain and the additional creep test was made at −68.9 N/mm^2. The results of these tests combined with the previous compressive data at 135°C are converted to time-dependent compliance and plotted versus logarithmic time (Fig. 14.9a). Time-dependent compliance is defined here as the instantaneous ratio of strain over stress for both relaxation and creep tests. Note that the compliance curve determined by the inverse of the relaxation modulus, as this definition implies, is not identical to the creep compliance; in the framework of a linear hereditary material, relaxation modulus and creep compliance are related by a convolution integral [21]. However, for small strains they are approximately equal [31], as seen in Fig. 14.9.

The mean and standard deviation of compliance data as measured from the loading phase of these tests was determined to be 0.128 mm^2/kN and 0.006 mm^2/kN, respectively. The specimen-to-specimen variation in the initial compliance gives a sense of the expected variability of the time-dependent compliance curves. This is illustrated in Fig. 14.9a by attaching error bars corresponding to 0.006 mm^2/kN to a compliance curve near the median of the data. Note that the variation on a replicate −0.5% strain, stress relaxation test is approximately the same magnitude as the variation between the −0.2% and the −0.8% strain, stress relaxation tests. The trend of increasing compressive compliance occurs in all the compressive tests. Figure 14.9b combines the single compressive compliance curve with the error bars, representing the data from Fig. 14.9a, with the compliance data calculated from the tensile creep and tensile stress relaxation tests previously shown in Fig. 14.4 and 14.5. The tensile compliance data shown here has also been compared to tensile compliance data provided by Raghavan and Meshii [23] for AS4/3501-6 carbon fiber-reinforced epoxy. Despite an offset due to a difference in constituent properties and volume fraction, the slopes for the two materials were found to be similar over the common time frame of the two data sets.

Figure 14.9a shows that for a particular range of time, these compression compliance data could be considered to be within the realm of linear behavior when accounting for specimen-to-specimen variability. Note however that acceptance tolerances for distinguishing between linear and nonlinear responses when analyzing hereditary data do not seem to be established (see ASTM Standard Test Methods for Tensile, Compressive and Flexural Creep and Creep-Rupture of Plastics D-2990). In Fig. 14.9b, the two tensile compliance plots coincide closely with each other but begin to diverge from the representative compressive compliance after 10^4 s. As care was taken so that all specimens were prepared identically, their state of physical aging may be presumed to be identical and therefore does not account for this behavior. Based on this observation, the use of the definition of a linear hereditary material and Boltzmann's principle of superposition for IM7/8552 becomes problematic if not intractable. This may be demonstrated further by an attempt to use Boltzmann's principle of superposition to determine the recovery phase of the data given in Fig. 14.5. Since the tensile relaxation response is nearly constant after 10^4 s, superposition of it on to the compressive relaxation response would predict near recovery of the zero stress state for both experiments. This result, however, is only observed for the tensile relaxation test with a compressive recovery phase.

Clearly, further investigation is needed to resolve this apparent paradox but it is believed that complex interactions on the microstructural level involving processing residual stresses can be an explanation. Micro-mechanical analysis and additional experiments are needed to verify this proposition. From an engineering perspective, a sense of the weight of these issues may be given by reviewing the results of the prototypical fatigue test.

14.2.3.1 Prototypical Operation Loads

A prototypical compressive fatigue cycle was defined for the flywheel application as having a mean stress of -41.4 N/mm^2, a stress amplitude of 13.8 N/mm^2, and a triangular waveform. The period of each cycle was 60 s; the corresponding frequency was 0.017 Hz. Each experiment was 2.6×10^5 s (72 h) in duration. Samples were tested at 95°C and at 135°C. The data were processed to provide extreme values of strain and the mean strain for each sample cycle. These maximum, mean, and minimum strains are plotted as a function of time in Fig. 14.10 for each test temperature. Also included in Fig. 14.10 are the creep strain curves generated with -41.4 N/mm^2 applied stress for the respective test temperatures. The apparent mean strain of the fatigue response for both plots in Figs. 14.10a and 14.10b closely matches the corresponding creep response.

It appears that the relationship between the average strain response and the average applied fatigue stress is the same as that between the creep strain response at a constant compressive stress. Using the different tensile and compressive

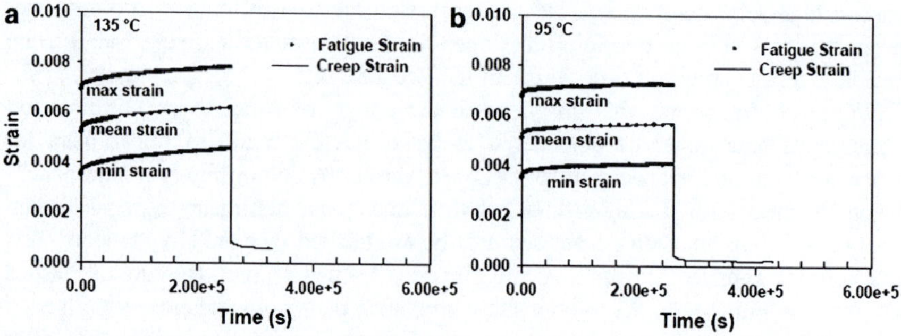

Fig. 14.10 Strain response in compressive fatigue compared to compressive creep data for (**a**) test temperature 135°C and (**b**) test temperature 95°C

compliance curves in Fig. 14.9, it would be impossible to predict this result using Boltzmann's principle of superposition. For instance, the complince curves of Fig. 14.9 would not predict zero creep strain for a fully reversed fatigue experiment. This would be a necessary result for the observations in Fig. 14.10 to be consistent with the definition of linear hereditary behavior.

14.2.3.2 Design Implications

There are several design ramifications of these experimental results for PMC flywheel rotors manufactured with a preload or strain-matching design. Experimentally, this work examines the transverse response of the composite due to a controlled stress or strain boundary condition on uniaxial specimens. Under a controlled strain of −0.5%, the stress relaxation was observed to be 16.4% after 72 h at 135°C in uniaxial compression experiments. Fundamentally, within the strain, time, and temperature regime of these tests, these compressive stress relaxation results can be viewed as an upper bound for calculating the amount of radial preload relaxation exhibited in a composite rotor. There are two primary reasons for this. First, some degree of radial compression will be maintained by the radial deflection of an elastic hub. Second, the relaxation response of a thin uniaxial specimen cannot be less than that of a rotor having any kind of lateral constraint. Obviously, these two features can be controlled to minimize the impact of stress relaxation on radial preload. However, it is apparent from these experiments that any kind of design optimization or durability-limit analysis requires having design data for the appropriate loading mode. The use of IM7/8552 tensile constitutive data would have been nonconservative in any design analysis.

The difference between compressive and tensile time-temperature dependent material behavior could be problematic for designers wishing to assume a linear hereditary response. Although it may be possible to predict pressure losses in press-fit rings fairly well (see, e.g., Emerson and Bakis [13]), extending the analysis to cyclic loading situations would require a consideration for the possibility of sign-dependent behavior in the composite. While comparisons of tensile and compressive time–temperature-dependent behavior for composites are rare, those existing have reported differences between the two responses [16, 21]. These investigations, along with the experimental observations reported here, point out the need for a more rigorous theoretical and experimental treatment of the time–temperature-dependent behavior of PMCs.

Whenever designers face the need for long-term durability data, there is an interest in accelerating the testing. Based on the available data, there is an indication that TTSP may be used by the designer to generate long-term master curves using short-term data at elevated temperatures. Note that the study conducted here does not investigate the phenomenon of physical aging which acts to slow the relaxation rate with time. Accordingly, the master curves developed here would overpredict the long-term relaxation. Struik [32] describes the appropriate test procedures and data-reduction methods to account for physical aging; Gates and coworkers [21, 22] have demonstrated these methods for PMCs.

14.2.4 Summary

A series of experiments were made to appraise the time–temperature response of filament-wound, IM7/8552, carbon fiber/epoxy composite tested transverse to the reinforcing fiber. The polymer matrix composite is considered to be a technology-enabling rotor material in flywheel energy storage systems for space flight applications. Rotor designs utilizing the load carrying capacity of carbon fibers wound in the hoop direction are capable of high operational speeds and specific energies. Several design strategies obviate the problem of tensile radial stresses in the matrix-dominated direction through the introduction of compressive stress. Analytical work by Arnold and coworkers indicates that several transverse geometric parameters control the performance of these rotors. Therefore, the safe-life qualification of PMCs for flywheel rotor applications may be limited by the time–temperature-dependent behavior of the epoxy matrix in compression.

Filament-wound specimen panels were made using the same manufacturing process as for polymer composite rotors. Flat coupon specimens were then cut from the panels parallel to and perpendicular to the winding axis so that fibers were transverse and longitudinal to the loading direction. Compression and tension tests were performed at room temperature, 95°C, and 135°C. The influence of strain rates from 5×10^{-6} to 5×10^{-3}/s was examined on the initial stress/strain response. Creep and stress relaxation responses were studied over a 72-h period. As anticipated, time-dependent effects are negligible when the load was applied parallel to the primary fiber direction. Loading perpendicular to the carbon fiber direction resulted in responses that were dependant on both the sense of load as well as the test temperature. At 135°C, compressive stresses relaxed 16.4% in 72 h under a fixed deformation of 0.5% strain. Lowering test temperature to 95°C reduced the extent of compressive stress relaxation to 13%. At a fixed tensile strain of 0.5%, stresses relaxed only 7% in 72 h at 135°C. Designers of composite flywheel rotors may treat this experimentally determined compressive stress relaxation data as an upper bound to the rate of preload relaxation for this material system.

The time–temperature superposition principle (TTSP) appears to be applicable for the time and temperature regimes explored in this test program. The shift factor to create a master curve at 95°C was log aT = 1.4. The short time-range data at 135°C can be used to predict longer-term response at 95°C. These curves do not account for physical aging, so relaxation rates at longer times are likely to be overestimated [29]. Prototypical compression–compression fatigue tests exhibit a mean strain response that closely tracks the creep strain response for an equivalent creep test; the mean compression stress of the fatigue test was identical to the compression stress of the creep test.

Time-dependent compliance data for tests made at different applied levels of strain or stress were found to nearly coincide if specimen-to-specimen variation is considered. However, compression and tensile compliance data does not coincide. Consequently, the application of the definition of a linear hereditary material response and Boltzmann's principle of superposition to describe the behavior of IM7/8552 in the regime of these experiments is problematic if not intractable. While limited, other comparisons of tensile and compressive time–temperature-dependent data for composites report differences [13, 18]. These findings support the need for more rigorous treatment of the influence sign dependency in a theoretical framework of PMC constitutive models.

14.3 Durability of Polyimide Composites Subjected to Thermomechanical Fatigue(TMF) in Reusable Propulsion Systems

Preserving the ideal flow geometry in thin-shell combustion enclosures is crucial to advanced hypersonic propulsion systems frequently considered for reusable launch vehicles. High-temperature polyimide composites are of particular interest in such lightweight support structures. While the continuous carbon fiber enables a cross-ply laminated skin of high specific stiffness; the polyimide matrix materials ensure that the rigidity and durability is maintained at operation temperatures of 316°C, which is 80% of the polyimide's glass-transition temperature, Tg. Design analysis of prototype composite support structures indicated that the chamber's stiffness requirements could be met while achieving significant weight savings over existing metallic supports [33]. The challenge was to ensure that the stiffness and strength of the composite support would be maintained over the expected design life. Available creep data for similar woven polyimide composites [34] suggest that deflections could increase by about 10% in 1 h at temperatures near the Tg. The concern that themomechanical fatigue (TMF) might accelerate such deformation motivated the experimental investigated reviewed in this section.

This TMF work supported a collaboration between Rocketdyne and NASA-GRC that sought to apply polyimide composites to the harsh environments and loads familiar to space launch propulsion systems [35–37]. Life requirements were only on the order of 200 cycles, but stringent durability and safety requirements for reusable launch systems require that the synergistic effects of cyclic temperature and mechanical loads in polymer matrix composites (PMCs) be understood.

TMF studies of PMCs are not widely available in the open literature. The most relevant work by Castelli and coworkers [38] evaluated the TMF response of chopped fiber-reinforced PMR-15, a polyimide composite in the form of sheet molding compounds (SMCs). Tests made at temperatures up to 80% of Tg showed that cyclic temperature and stress promoted excessive creep damage and strain accumulation [38]. After 120 mission cycles, strain offsets of the order 0.05–0.1% were measured at zero load. From a design standpoint, this corresponds to an increase in deflection at operating load of about 8–16%. In propulsion applications where stiffness and deformation tolerances are operations critical, it was important to verify that the deformation response of cross-ply polyimide laminates subjected to TMF was within design limits.

Physical and mechanical properties of polyimide composites have been studied over a range of hygrothermal states [35, 36, 39, 40]. Particular investigations address degradation due to steam-induced blistering in high-temperature PMCs and point to the usefulness of laminate stitching as a preventive measure [35, 39, 40]. Rate of transverse expansion as a function of temperature is reduced in moisture-conditioned, stitched specimens and is comparable to unstitched vacuum-dried laminates [40]. However, the improved blistering resistance by stitching is offset by a reduction of in-plane properties; the dry ultimate tensile strength of the stitched laminate at elevated temperature was reduced by 40% with respect to unstitched laminate. While the other laminate variables were not strictly controlled in this comparison, the underlying trend is consistent with other studies on the influence of stitching [35, 41]. The reduction of in-plane strength is likely due to misalignment, distortion, and damage of the primary reinforcing fibers accompanied by resin-rich regions, porosity, and resin cracking. The existence of such stitching-induced defects would likely be sources for nonrecoverable strains in TMF. Since stitching has been shown to reduce the in-plane mechanical performance of composite laminates [35, 40, 41], an evaluation of the influence of stitching on TMF response was an important element of this investigation.

The objective was to determine the isothermal fatigue and TMF response of stitched and unstitched cross-ply laminates of M40J carbon fiber in an polyimide matrix. Here the 6F-polyimide matrix made from the ester of 6F-dianhydride (HFDE) and p-phenylenediamine with the phenylethynyl endcap, designated as HFPE is used [42]. The stitched laminate was manufactured using proprietary stitching and resin infusion methods. The exploratory test plan aimed to quickly establish performance limitations and provide design guidelines for the intended application of the composite. While the TMF cycle was synthesized from Rocketdyne's expected operations cycle, an effort was made to use a sufficiently generic test cycle to be of scientific value. However, the nature of the applied TMF cycle and the test material combined to present some challenges in test method development. These difficulties are reviewed and discussed along with the experimental results for the stitched and unstitched laminates.

It would be preferable to have TMF data reflecting the true nature of the operation cycle but such efforts might be unnecessary if the isothermal fatigue degradation can be shown to be a conservative bound on the TMF degradation and the added weight of a conservative design is acceptable. Here a number of baseline isothermal fatigue tests have been made using identical loading waveforms to determine the relative influence of TMF testing.

The maximum loads selected for these tests are guided by analysis of the expected operations stresses in the PMC support using analytical and finite element methods [33]. Due to the relatively long, 480 s duration of a cycle, the number of loading cycles did not exceed about 2,000 on any one specimen so the duration of any test was rarely sufficient to cause failure. Nevertheless, the fatigue load levels and duration of then data are in a regime having factors of safety two or greater relative to the expected operations envelop for the support structure.

14.3.1 *Experimental Approach*

14.3.1.1 Materials

A detailed description of the matrix materials being evaluated in this NASA-GRC/Rocketdyne collaboration, including HFPE-II-52 polyimide resin used here, is given in Ref. [42]. The M40JB carbon fiber is manufactured by Toray Carbon Fibers America, Inc., Santa Ana, CA having a tensile modulus of 377 GPa, strain to failure of 1.2%, and a density of 1.77 g/cm^3. Uni-fabrics with 110 gm/m^2 fiber area weight (FAW) from M fibers were fabricated at Sigmatex High Technology Fabrics, Inc., Benicia, CA. The 24-ply uni-tape panels were fabricated using a 12-ply symmetric, cross-ply layup as in reference [35]. Dry performs were stitched, infused, and cured by using proprietary and/or restricted methods. Autoclave curing reaches temperatures of 371°C and pressures of 1.6 MPa. After manufacturing, ultrasonic scans were made to document laminate quality.

14.3.1.2 Test Methods

The static and fatigue test methods were guided by the Standard Test Method for Tensile Properties of Polymer Matrix Composite Materials (ASTM D3039) and the Standard Test Method for Tension–Tension Fatigue of Polymer Matrix Composites (ASTM D3479) The TMF test methods described here are an extension of those presented in Ref. [38].

The coupon specimen is the dogbone geometry shown in Fig. 14.11 that has been used successfully in several test programs for mat, woven, and cross-ply PMCs

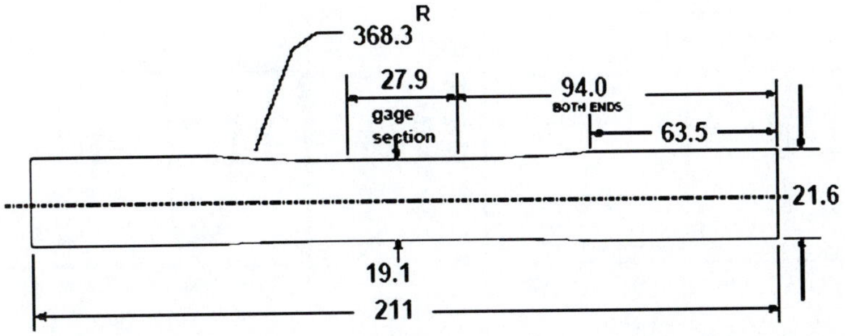

Fig. 14.11 PMC dogbone specimen design for static tension and tension/tension fatigue (lengths in mm)

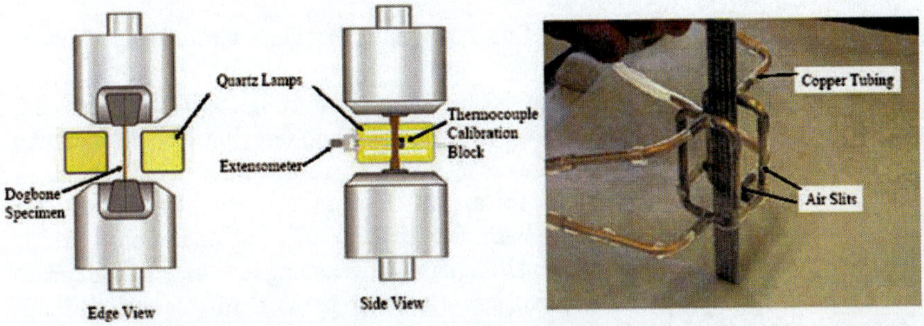

Fig. 14.12 Edge and side view schematic of the experimental setup; TMF cooling cage (far right)

[3, 34, 38]. Specimens were fabricated by water-jet machining. Cut specimens were dried for 48 h at 150°C and 760 mm Hg vacuum and then stored in a desiccator until testing.

The large curvature of the dogbone enables successful testing without tabbing. The width of the specimen in the gauge section is 19.05 mm. Specimens were typically about 2.92 mm thick with a variability of about ±0.13 mm. Stitched specimens were sectioned so that loads are applied parallel to the stitch direction. As the stitch rows are spaced about 5–6 mm, care was taken to ensure that four rows of stitching were centered on each specimen.

Mechanical loads were applied to the specimen using a closed-loop, servo-hydraulic load frame manufactured by MTS™ having a load capacity of 89 kN. Specimens were gripped using hydraulic actuated and SurfAlloy™ coated wedge grips. Axial strains were measured over a 12.7-mm gauge length by an MTS™ air-cooled extensometer. A CVD™ powered bank of quartz lamps was used to heat the specimen during isothermal and TMF testing. Radiant heating by quartz lamps is necessary to meet the dynamic heating requirements of these TMF tests. Edge and side views of the specimen setup are shown in Fig. 14.12. Rapid cooling during the TMF cycle is facilitated by impinging compressed air on the specimen through air slits in a copper

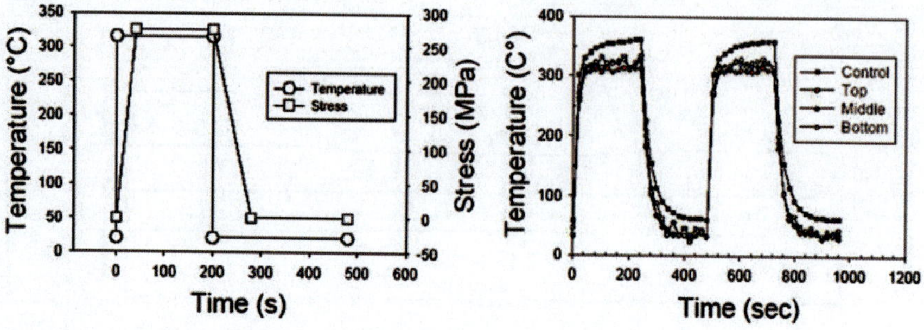

Fig. 14.13 Desired TMF cycle and resulting thermal response of the control block and the Top, Middle, and Bottom thermocouples embedded in a calibration specimen

tubing cooling cage. A photograph of the dismounted cooling cage and a specimen is also shown in Fig. 14.12.

Test control and data acquisition was accomplished using the Test Express 4.0 software and hardware manufactured by McGaw Technology, Inc. The desired load and temperature control waveforms were synthesized from a prototypical operations cycle and are shown in Fig. 14.13 for a max stress of 276 MPa. Start points for thermal and loading ramps are in-phase and the period for the operations cycle is 480 s. The loading waveform is a tensile square wave having load and unload ramps of 40 s and dwell times of 200 s at max stress and zero stress. During launch and shut down, the PMC support structure would reach operating temperature extremes in a few seconds. The most representative thermal cycle would be a square wave form having a Tmax of 316°C and a Tmin of room temperature (RT), each having a dwell time slightly less than 240 s. Details of the actual thermal wave form for these TMF tests will be described further below. Throughout the test program, isothermal fatigue and TMF control cycles were programmed having identical loading waveforms. Based on analysis of the composite supports [33], the maximum operation stress in the prototype composite support is about 138 MPa. Full-scale designs might achieve greater maximum applied stresses so the stresses examined in these experiments ranged from 138 to 483 MPa and could be changed between specific blocks of fatigue cycles. A typical test block was 500 cycles. The isothermal test temperature and maximum TMF temperature (Tmax) were both 316 C. The entire period for a TMF cycle was 480 s.

Temperature measurement and control during testing of PMCs is encumbered by the inability to attach thermocouples (TCs) to each test specimen. Following techniques developed and applied in several fatigue testing programs [3, 34, 38, 40, 43], temperatures in the position of the test gauge section are measured using a calibration specimen with an array of thermocouples. These were correlated to a control TC in a block of the PMC that is positioned adjacent to the specimen (see Fig. 14.12 side view). The calibration specimen is prepared by drilling holes at specific locations and embedding K-type TCs into these holes. Similarly, a control block of PMC is prepared with its own embedded K-type TC. Axial and transverse

Table 14.3 Isothermal static test results 316°C

	E (GPa)	UTS (MPa)	Strain-to-failure (%)
Stitched	63.1	588	0.8
Stitched	72.4	545	0.75
Unstitched	98.0	876	0.95
Unstitched	96.5	834	0.98
Knockdown	30%	33%	20%

temperature variations were then measured in the gage section and correlated to the control TC. Adjustments of the quartz lamps and cooling cage were made to minimize thermal gradients in the specimen gage section. A correlation of the control block TC and the specimen gage length were made for both isothermal and cyclic thermal cases. In the case of isothermal tests, the specimen was allowed to reach an equilibrium stage which makes this correlation constant with time. For the isothermal testing, this correlation was repeatable to within ±2°C.

For the TMF testing, the specimen does not have time to reach an equilibrium state. Since the thermal boundary conditions on the specimen and control block are different, an iterative procedure was used to develop a control waveform that would yield the desired temperature field in the specimen gage length. The control block temperature and specimen temperature response for the best simulation of this operation cycle is shown in Fig. 14.13. The top, middle, and bottom specimen temperatures are from positions at the edges and middle of a 25.4-mm segment of the specimen gauge length. The initial heating rate for the specimen was found to be 12°C/s for the first 25 s of the cycle. The initial cooling rate was about 6°C/s for the first 25 s of the cool down part of the cycle. A slight positive increase in the control block temperature must occur during the hold period to maintain the temperature in the middle of the specimen close to 316°C. The radiant heat flux to the specimen must increase slightly to compensate for nonequilibrium heat loss to the grips. The high power levels needed for rapid heating create a noisy environment for the TC. It is important to note that the variability in the temperature measurement for this TMF cycle is 5–10 times greater relative to the isothermal temperature measurements. The scatter band for the TC data is about ±10 to ±20°C. When the lamps are turned off, this variability decreases down to ±2°C. The minimum temperature, T_{min}, for these tests was about 31–33°C.

14.3.2 Experimental Results

14.3.2.1 Isothermal Static and Fatigue Tests

Two static tests each of the stitched and unstitched specimens were made to determine the elastic modulus (E), ultimate tensile strength (UTS) and strain to failure (Table 14.3). Note that there was a significant knockdown in properties due

Table 14.4 Isothermal fatigue test matrix: stitched at 316°C

Specimen	Peak stress (MPa)	Cycle count
A	276	1–2,000
B	138	1–500
B	207	501–1,000
B	276	1,001–1,500
C	414	1–1,015
C	483	1,015–2,015
D	483	1–200

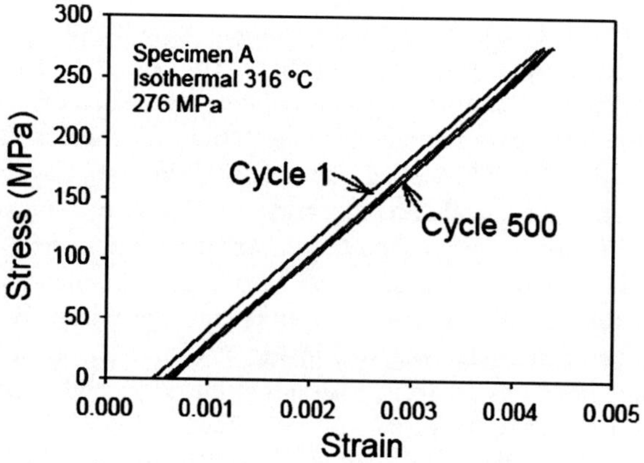

Fig. 14.14 Stitched specimen A stress–strain response at 316°C

to the stitching which was similar to aforementioned studies. A summary of the isothermal fatigue tests at 316°C is given in Table 14.4; only stitched specimens were tested and presented. Unless otherwise mentioned in the discussion, stress values are the maximum values for the fatigue cycle.

Figure 14.14 gives the isothermal stress–strain response for cycle 1 and cycle 500 of specimen A fatigued at a peak stress of 276 MPa. The hysteresis loop for cycle 500 has translated to the right indicating an accumulative strain of 0.02%. This strain-ratcheting phenomenon was observed for all of the isothermal tests and the amount of strain accumulation is plotted as a function of fatigue cycle in Fig. 14.15. Specimen B was initially fatigued at a peak stress of 138 MPa to see if there was a threshold for this behavior above the expected operating stress. It appears in Fig. 14.15 that the strain accumulation for specimen B is fairly similar to specimen A up to 1,000 cycles. At 276 MPa, the strain accumulation in specimen B is slightly greater than in specimen A. The remaining two specimens were fatigued at stress levels up to 80–88% of the composite's UTS. Specimen C acquired 0.13% strain in 1,000 cycles at 414 MPa. The accumulated strain in the specimen increased to 0.25% after 1,000 cycles at 483 MPa. Specimen D began at a maximum stress of 483 MPa accumulated over 0.1% strain in 200 cycles. Figure 14.15 shows that the initial strain-accumulation

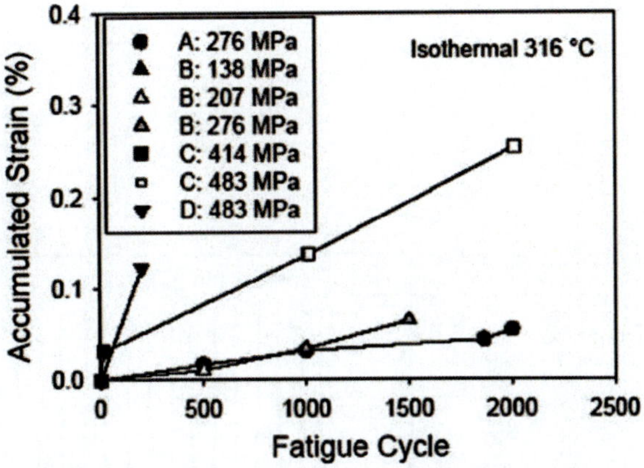

Fig. 14.15 Accumulated offset strain as a function of fatigue cycle: Stitched

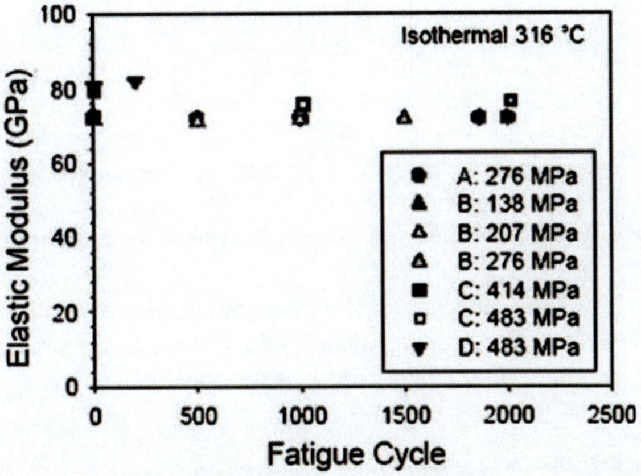

Fig. 14.16 Elastic modulus as a function of fatigue cycle

rate for maximum stress above 80% UTS is significantly greater than at maximum stresses around 50% UTS, which is twice the operating stress.

Changes in the measured elastic modulus are often regarded as a global measure of the composite material's internal damage state. Despite the translation of the hysteresis loops to the right with accumulated strain, as observed in Fig. 14.14, there was no remarkable change in the slope of the stress–strain response. The control software was programmed to make periodic regressions on the load and unload ramps to compute elastic modulus and these results are given in Fig. 14.16 as a function of fatigue cycle. While certain specimen-to-specimen variations are apparent, the elastic modulus values of any one specimen are not found to significantly degrade during the test. This behavior was consistent with observations reported by Castelli et al. [38] for SMC composites.

Apparently, the primary, load-bearing fibers are not being significantly degraded by these short-duration fatigue tests. It was suspected that the presence of a variety

Table 14.5 TMF test matrix

Specimen	Peak stress (MPa)	Cycle count
E	276	1–1,623
F	483	1–500
G	483	1–500

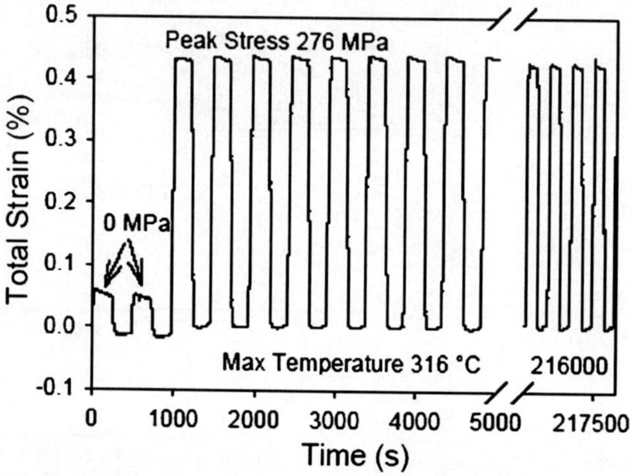

Fig. 14.17 TMF strain response as a function of time: stitched specimen E

of stitching defects provides sites for permanent relative deformations to take place. For example, broken fibers could slip relative to each other without appreciably changing the global stiffness of the specimen.

14.3.2.2 TMF Results – Stitched Material

The TMF test cases that have been completed and evaluated thus far are summarized in Table 14.5. It should be mentioned that progress with TMF testing has been impeded by fatigue failures of the quartz lamps after about 500–800 cycles. Bulb replacement was time consuming due to the necessary temperature recalibration procedures. Increasing cooling air flow to the lamps helped to extend bulb life.

Each TMF test began with a pair of thermal cycles without load to evaluate the thermal expansion response of the specimen. Figure 14.17 shows the strain measurement as a function of time for the first block of 450 cycles that were applied to specimen E. The thermal expansion in the first two cycles was seen to have a maximum value of about 0.05% strain but this was not constant. The initial thermal cycles without load show the extensometer to be measuring an initial rapid thermal expansion followed by shrinkage once maximum temperature had been reached. On cool down, the reverse effect was observed; the extensometer measures a rapid

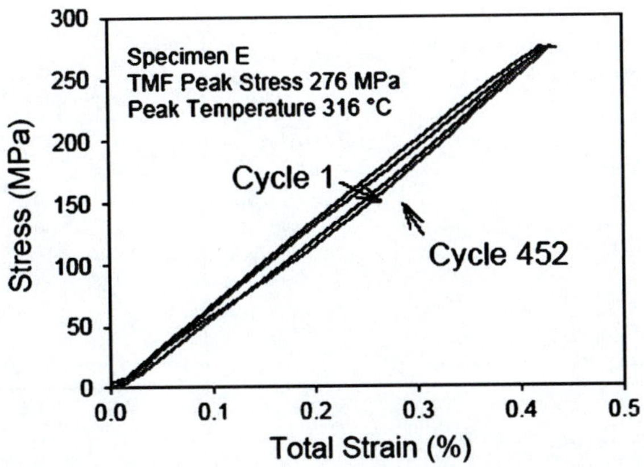

Fig. 14.18 TMF stress–strain response of stitched specimen E

contraction followed by a thermal expansion. The qualitative form of the shrinkage and expansion occurring at the extreme temperatures appears to be of exponential decay as a function of time. Thermal trials with the calibration specimen showed that holding the temperature constant at the peak value allows the strain measurement to eventually reach a constant value.

The reversal of deformation at temperature extremes is counterintuitive to the expected first-order response of the specimen. Indeed the cross-head displacements continue to increase during the heating cycle to maintain zero applied load. Reducing the cooling air pressure was found to retard the initial expansion response and the subsequent amount of shrinkage was smaller. Thermocouple measurements at the extensometer heat shield showed that the heat shield experienced a change in temperature of about 7°C between maximum and minimum temperature. Concerned that the extensometer may be experiencing thermal fluctuations, a validation test was made using a stainless steel specimen. Placing a stainless steel specimen into the grips and subjecting it to the same heating cycle produced a strain measurement that was at all times consistent with the sign of the temperature change. The measurement was not being affected by the cyclic variations in temperature, so there was confidence that the measurement reflects the local behavior of the composite specimen in the gauge length.

Until the cause of this behavior is fully understood, it is not possible to accurately separate the thermal and mechanical strains in the current data set. Nevertheless, certain important observations can be made about the data collected. Figure 14.18 shows the TMF stress-total strain response for cycle 1 and cycle 450 of specimen E. The hysteresis loops coincide with no strain offset. The absence of strain accumulation persisted throughout this experiment until the specimen failed at 1,623 cycles. There was no forewarning of impending failure from the slope of the stress-total strain curve for this specimen. The "apparent stiffness" of specimen E actually increased over the duration of the test by about 14%, from 63 GPa to 71 GPa, and this was not perceived as a harbinger of failure.

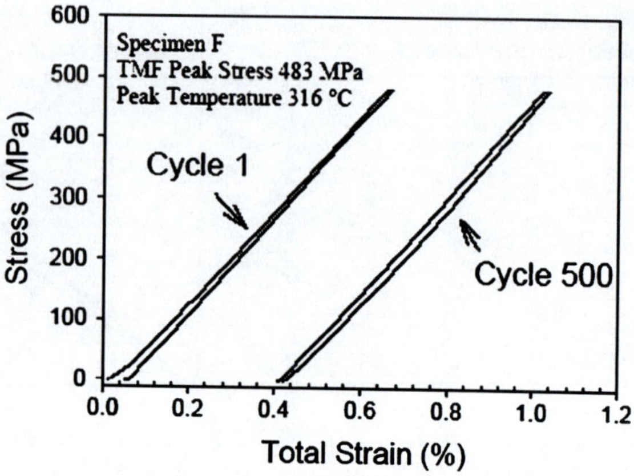

Fig. 14.19 TMF stress–strain response of specimen F

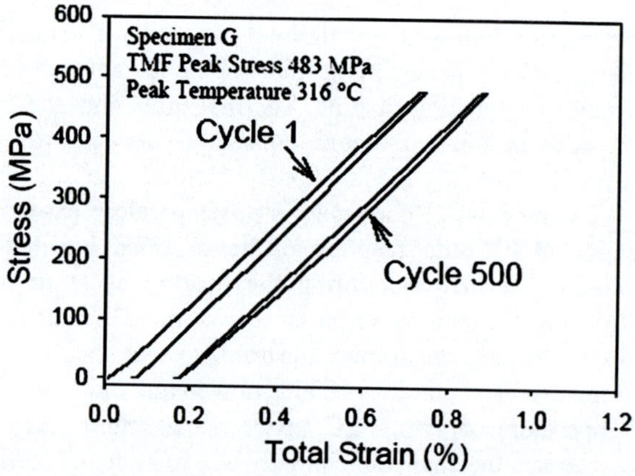

Fig. 14.20 TMF stress–strain response of stitched specimen G

The failure of specimen E at 1,623 cycles prompted an increase in TMF peak stress to 483 MPa or about 88% of the UTS for specimen F and specimen G. Figures 14.19 and 14.20 show the stress-total strain response for cycle 1 and cycle 500. Here a significant amount of strain offset is observed in both curves: 0.38% for specimen F and 0.16% for specimen G. For these tests, changes in the apparent stiffness of the specimen were of the order ±3%. The TMF and isothermal elastic modulus and accumulated strain results for specimens tested at 483 MPa have been summarized in Fig. 14.21. It should be noted that the elastic modulus values for the TMF tests are based on the stress-total strain slope which includes the influence of thermal expansion. The modulus as a function of TMF cycle shown in Fig. 14.21a are within the range of specimen-to-specimen variation seen in Fig. 14.16. This was reasonable since thermal strain was a smaller percentage of the total strain at higher loads. If the maximum of thermal strain was about 0.05% strain for a given TMF cycle, then the influence on total strain at 483 MPa was less than 10%.

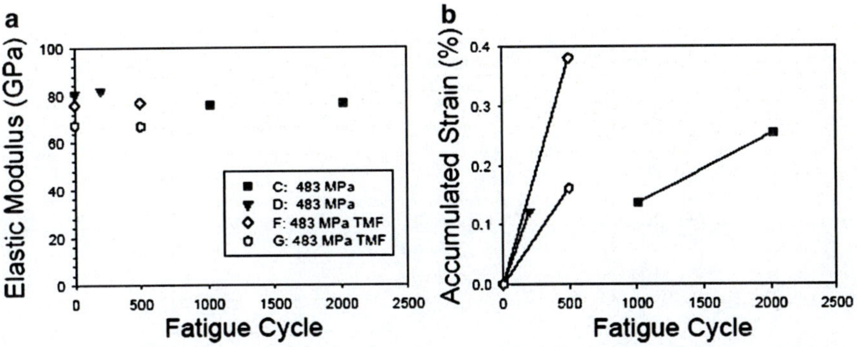

Fig. 14.21 Comparison Isothermal and TMF results for specimens tests at 483 MPa (**a**) elastic modulus and (**b**) accumulated strain as a function of fatigue cycle, symbol table applies to both figures

Table 14.6 TMF test matrix: unstitched material

Specimen	Peak stress (MPa)	Cycle count
H	483	1–2,000
I	724	1–247[a]
J	662	1–58[a]
K	662	1–93[a]

[a]Fatigue failures

TMF was not found to degrade stiffness any differently than isothermal fatigue testing. There was also some indication that strain accumulation does not occur at lower loads in TMF testing. However, Fig. 14.21b seems to indicate that TMF may accelerate strain ratcheting in the higher loads. The manufacturing-induced residual stresses are expected to be the principle mechanism that would accelerate damage and deformation in TMF testing of this polyimide composites. These stresses are out-of-phase with the applied mechanical loads, i.e. the residual stresses are low at elevated temperature and increase as the specimen cools down to 32°C. In this way, the material was continuously subjected to internal and external stresses. However, the absence of accumulated strain in TMF response of specimen E and the presence of accumulated strain at elevated stresses seem to indicate that time and external load at temperature are the significant factors. A telling creep experiment made by Castelli et al. [38] support this speculation. By holding the load on the SMC composite constant and quasi-statically increasing temperature, it was shown that the onset of creep strain occurred at lower temperatures for higher static loads [38].

14.3.2.3 TMF – Unstitched Material

The complete TMF test matrix of unstitched specimens is given in Table 14.6.

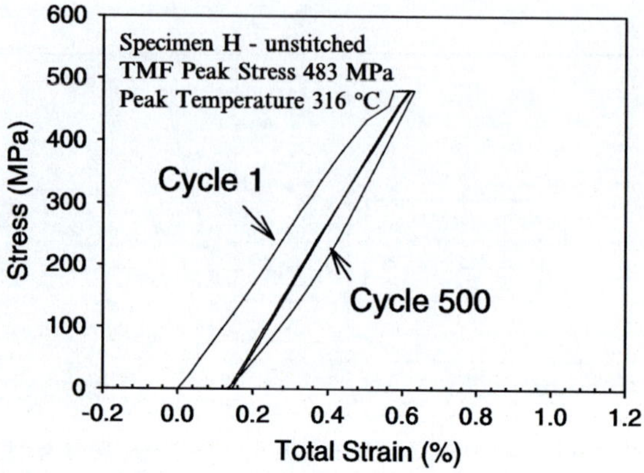

Fig. 14.22 TMF stress–strain response of unstitched specimen H

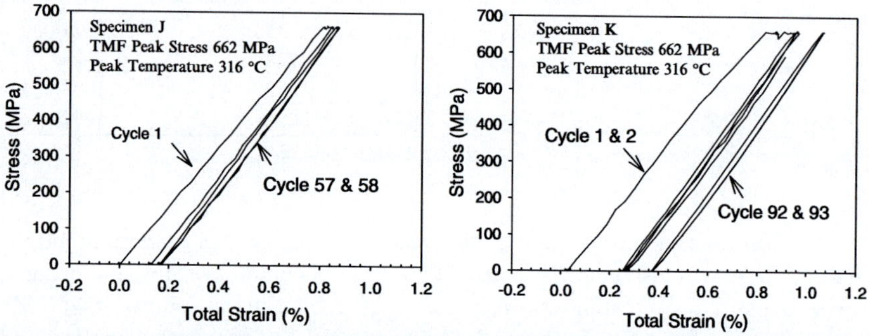

Fig. 14.23 Unstitched repeat test of specimen J and K

Since the UTS for the unstitched material was 33% greater than the stitched material, specimen H was tested at 483 MPa. A comparison of cycle 1 and cycle 500 stress-total strain response for specimen H is shown in Fig. 14.22. All of the strain accumulation occurs in the first cycle and is similar in magnitude to the accumulated strain for the stitched specimen G tested at the same load level. The absence or reduced rate of strain ratcheting in subsequent TMF fatigue cycles is the most notable characteristic in the results of all the unstitched TMF tests. Figure 14.23 shows the first and final cycles of specimens J and K tested at a peak stress of 662 MPa. The major share of strain accumulation occurs on the first cycle in these tests with relatively smaller amounts accruing up to specimen failure.

A comparison of the stitched and unstitched TMF results is shown in Figs. 14.24a, b. Again, the elastic modulus values in Fig. 14.24a appear fairly stable over test duration. The reason for the initial low values for the first load cycle of the unstitched material is the indication of significant damage occurring on that load cycle. Subsequent tangent modulus measurements were quite stable. The

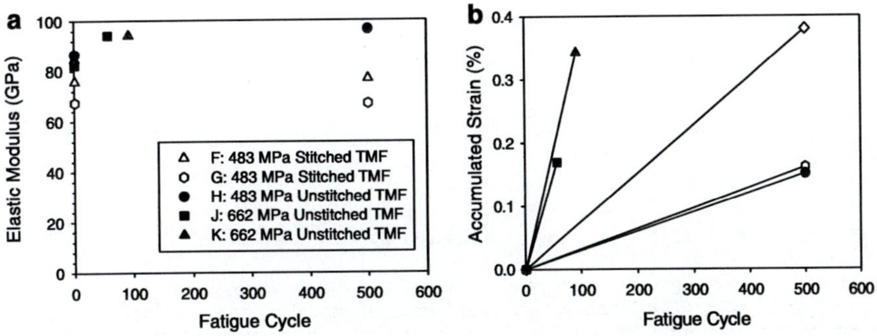

Fig. 14.24 Comparison of stitched and unstitched TMF results for tests at 483 MPa and 662 MPa (**a**) elastic modulus based on total strain (**b**) accumulated strain as a function of fatigue cycle, symbol legend shown in (**a**) applies to both plots

difference between stitched and unstitched modulus is a reflection of the knockdowns caused by the micromechanics of the stitching. The summary of accumulated strain results in Fig. 14.24b shows that similar amounts of strain accumulated at 483 MPa for both stitched and unstitched materials.

Fatigue failures only occurred during TMF tests, not isothermal fatigue, and were more prevalent for the unstitched materials. All three TMF tests made in the range of 79–87% of the minimum unstitched UTS (834 MPa) failed in less than 250 cycles. Stitched materials TMF tested at 88% of the minimum stitched UTS (545 MPA) survived 500 cycles without failure. Unstitched materials are capable of sustaining greater tensile fatigue loads but significant degradation appears on the first cycle, leading to relatively shorter fatigue lives.

A more insightful representation of the TMF and isothermal fatigue strength of the stitched and unstitched materials tested thus far are given in Fig. 14.25. Here the peak applied stress levels are all normalized with respect to the minimum stitched UTS of 545 MPa. Also placed on the curve in bold lines is the tensile fatigue durability requirement based on design operating stresses and life requirements for the component. Wide safety margins are available in both life and load.

14.3.3 Summary

This study compared the isothermal and thermomechanical fatigue (TMF) of stitched and unstitched cross-ply laminates of M40J carbon fiber-reinforced HFPE-II-52 resin. The isothermal test temperature and the maximum temperature of the TMF tests was the maximum operation temperature for the composite structure, 316°C. The square waveforms for thermal and mechanical load cycles are in-phase and designed to be representative of an engine operation cycle. The load waveform is identical in both isothermal fatigue and TMF testing. Peak applied stresses were guided by the results of design analysis of the component [33] and ranged from the component operating stresses of 138–724 MPa. Tensile

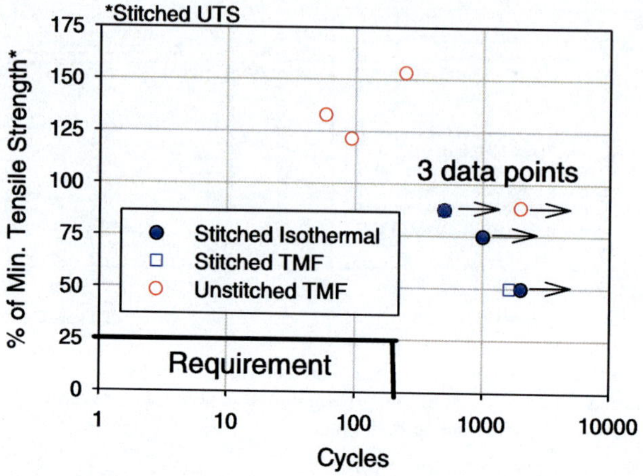

Fig. 14.25 Fatigue strength summary with maximum applied stress in percent of the minimum stitched tensile strength. The outlined bold outlined box represents the possible design window. Three repeated of tests of stitched isothermal case overlap and are annotated

tests for stitched and unstitched materials were conducted at 316°C; the minimum UTS for the stitched composite was 545 MPa, 33% less than the unstitched composite UTS of 834 MPa. Elastic modulus and strain to failure of the stitched material were, respectively, 30% and 20% lower than the unstitched composite.

Due to experimental challenges, examples of TMF testing of polymer composites are not widely found in the literature. The TMF methods used here are reviewed following those reported in [38]. The rapid heating and cooling rates, 12°C/s and 6°C/s respectively, produced an unexpected thermal strain response in the material. The thermal strain rate changed sign at the two temperature extremes. This was observed as relatively small contractions and expansions at maximum and minimum temperature. A variety of experiments demonstrated that the measurement represented the physical behavior of the material. The unusual behavior points to the complex role that heat and mass transfer, creep, stress relaxation, residual stresses, and damage must play in thermal deformation processes for the composite. These mechanisms are inactive during isothermal fatigue testing, so their participation in the fatigue damage of the real component would go unnoticed without TMF investigations. A careful array of experiments is being planned to study the phenomenon. Pending the results of this study, the strain response for the TMF testing is reported as a total strain.

Elastic modulus was monitored throughout the test program but did not degrade with fatigue cycle for any of the tests. However, fatigue strain ratcheting was observed which increased in proportion to test duration and applied load. Accumulated strains ranged in value from 0.01% for 500 cycles at 138 MPa peak stress to nearly 0.4% for 93 cycles at 662 MPa peak stress. Strain accumulation did not occur in the TMF test made at 276 MPa for 1,623 cycles but 0.05% strain did accumulate after 2,000 cycles of isothermal fatigue at the same load level. Therefore, the stress threshold for strain ratcheting in TMF tests appears greater than for isothermal testing. Once strain accumulation is initiated in TMF testing, the trend is toward a higher rate of accumulation than in isothermal tests. Unstitched TMF results were found to exhibit a high initial rate of strain accumulation with most strain accumulation occurring in the first load cycle of the fatigue test.

The following design guidelines can be stated from the results. Isothermal testing did not induce any fatigue failures. Only one stitched specimen failed due to TMF testing. This was at twice the expected operating load of the support structure and after about eight times the required service life of 200 cycles. Stitched specimens TMF tested at a peak load 88% of the UTS were found to survive about 2.5 times the required service life without failure. Unstitched specimens TMF tested at 79–87% of the unstitched UTS failed in tensile fatigue about ten times sooner. Despite these fatigue failures, there appears to be a good margin of safety relative to the life requirement at operational loads. Isothermal fatigue at the operational load level of 138 MPa caused 0.01% strain accumulation in 500 cycles which would be equivalent to an increase in structural compliance of 7% at that load level. Similarly, the 0.05% accumulative strain acquired at 276 MPa, twice the operations load, would cause an increase in structural compliance of about 15% after 2,000 cycles. The stitched and unstitched specimens tested in TMF at 3.5 times the design load would increase the structural compliance by 33% after 2.5 times the design life. These results have guided the laminate dimensions for the support structure to meet the end-of-life stiffness requirements.

14.4 Fatigue Damage and Compressive Residual Strength of Hygrothermal Conditioned (HC) Composites for Aero Engine Applications

Advances in fully automated composite fabrication techniques for production of 2D and 3D woven polymer matrix composite (PMC) have led to increasing use of these materials in aerospace applications. However, predicting the long-term fatigue–damage progression and durability of PMC materials under prototypical environments remains a challenge. Complex environments, such as combinations of elevated temperatures and ambient moisture levels, can lead to enhanced degradation rates [44]. It is well known that PMC will absorb moisture at normal, ambient humidity and that the moisture is removed by elevated temperature excursions. PMCs used in commercial aviation could be exposed to thousands of thermal cycles over years of mission life. However, the repeated cycles of moisture absorption followed by subsequent drying in flight under applied loads are difficult to duplicate experimentally. Yet, these issues are of particular concern for cases where the potential applications are primary load-bearing structures. Compensating for the lack of accurate predictive methods in the area of fatigue durability has resulted in overconservative designs that often mitigate the advantages of using PMCs in primary components.

This section reviews an alternative experimental approach of applying an entire operations history of hygrothermal conditioning (HC) without load followed in series by a course of mechanical fatigue testing. The investigation was made for a five-harness satin, carbon fiber-reinforced epoxy composite, AS4/PR500. Two data sets will be reviewed here. The first is the baseline data for the material behavior without environmental conditioning, which will be labeled as 0 h. The second data set consisted of material that underwent 12,000 h (12 kh) of HC. The elevated temperature fatigue behavior and damage progression was tracked through the measurement of the real-time tensile and compressive modulus. One unique feature of the work was to quantify the severity of fatigue damage via residual compressive strength measurements made at various degrees of progressive compressive modulus degradation. Specifically, samples were subjected to residual compressive strength tests after fatigue cycling led to approximately 0%, 2.5%, 5%, and 7.5% reductions in the compressive modulus. These data illustrate a relationship between fatigue–damage progression as measured by the material's effective modulus and the material's damage tolerance while avoiding all assumptions with respect to the percentage of cyclic life. This detailed examination was completed on material with and without hygrothermal conditioning and then the data trends were compared.

14.4.1 *Experimental Approach*

All material was fabricated by a resin transfer molding (RTM) process at Dow U.T. (USA) from a single batch of resin and fiber mat into thick laminates (16 ply, 6.1 mm) with panel dimensions of 63.5 × 63.5 cm. The HC conditioning, illustrated in Fig 14.26, was a nominal 85/30 (85% ±3% relative humidity at 30°C) exposure combined with a daily "mission cycle" consisting of a 90-min exposure to 121°C in a forced-air, convection oven. The cycle was selected as representative of a static component in the propulsor region of a gas turbine engine. The 12,000 h conditioning was conducted on the material in large-panel form prior to cutting the test coupons to avoid edge effects influencing the mechanical properties. The weight gain, shown in Fig. 14.27, was monitored using a traveler panel with a width-to-thickness ratio of 25:1. Note that the 90-min cycle at 121°C was sufficient to prevent the material from reaching a moisture equilibrium state, even after 12 kh. Test coupons were machined subsequently using a diamond wheel grinding process lubricated with water. After conditioning and machining of the coupons, all material was vacuum dried for 48 h at 105°C and stored in a dessicator until immediately prior to testing. Mechanical testing was performed on a computer-controlled, servo-hydraulic test frame. A radiant mini-furnace was used for specimen heating and strains were measured using an edge-mounted extensometer with a 12.7-mm gauge length. All mechanical testing was performed at 121°C in a laboratory air environment.

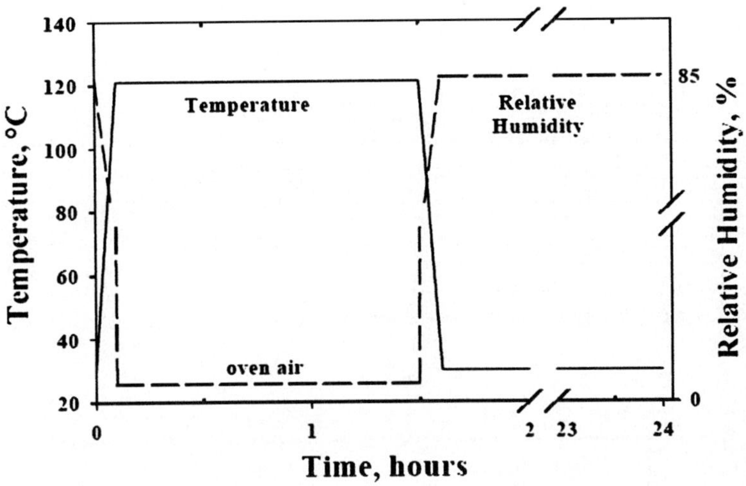

Fig. 14.26 Hygrothermal conditioning (HC) cycle

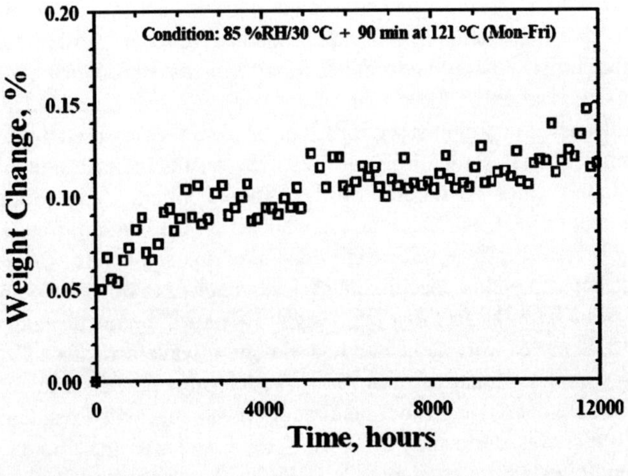

Fig. 14.27 Weight gain during HC conditioning of traveler panels measured before the oven bake-out portion of each cycle

All fatigue tests were conducted under fully reversed loading conditions. Fully reversed fatigue testing of composites is relatively uncommon due to buckling instability issues. Anti-buckling guides may be used, but problems like frictional wear and thermal influences at elevated temperatures are also formidable [45]. The current research used thick laminates with a refined dogbone-specimen design (Fig. 14.28) and an experimental setup that significantly reduced specimen instabilities. The design viability was verified by comparing the compressive strength results to those obtained using the standardized Celanese fixture (ASTM Standard

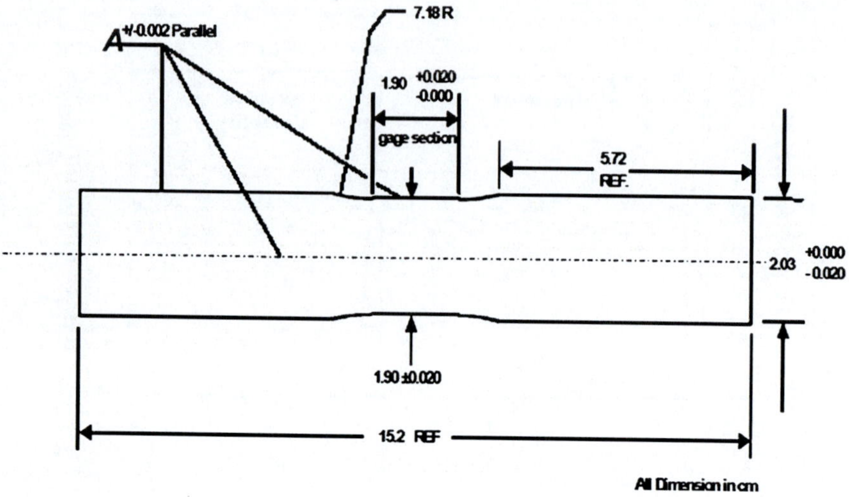

Fig. 14.28 Dogbone geometry

D3410). Ultimate strength results showed good agreement [46] between specimen designs. The specimen stability and experimental setup were further evaluated by examining the compressive behavior while monitoring the strains using back-to-back strain gauges. Deviation of these two strain measurements is a good indicator of bending, as discussed in previous sections. The axial strain measured by strain gauges and the extensometer is shown in Fig. 14.29. The results indicate minimal bending until immediately prior to failure. Further, the maximum fatigue strain ranges examined do not exceed ±0.7%, which is well within the bending-free regime.

The fully reversed fatigue tests were conducted in load-control ($R_\sigma = -1$) with load levels selected to yield a specific initial strain range (±0.4%, ±0.5% and ±0.6%). This is consistent with the fact that the design criteria for the intended application is strain limited. Samples were subjected to a triangular waveform at a 2 Hz frequency either until complete fracture or until a predetermined reduction in modulus. In the cases where the desired modulus reduction was reached, a residual compressive strength test was performed after the fatigue test was interrupted. All static residual strength tests were conducted at 121°C with a controlled displacement rate of 8.5×10^{-3} mm/s.

14.4.2 Experimental Results

14.4.2.1 Fatigue Life and Modulus Degradation

The fatigue strength as a function of the number of cycles to complete failure is presented in Figs. 14.30 and 14.31. Figure 14.30 shows life in terms of the initial

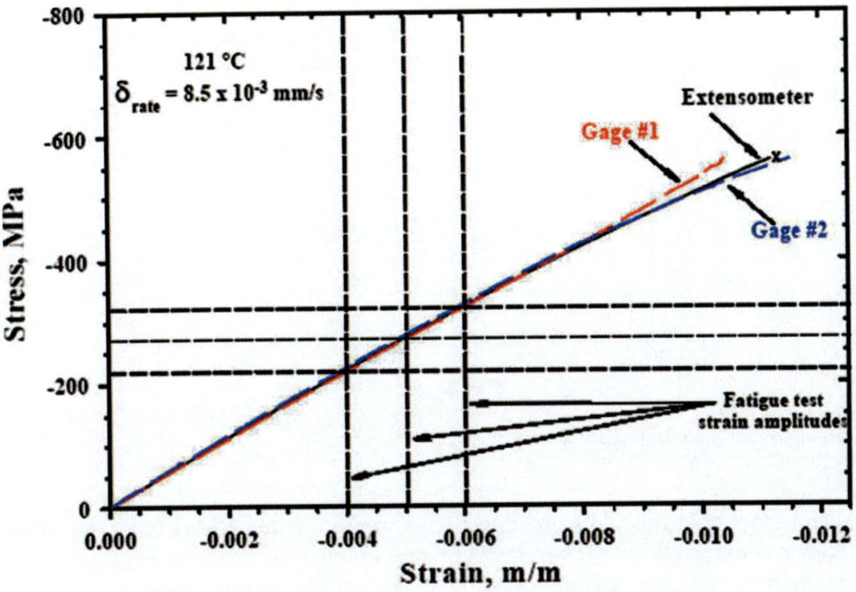

Fig. 14.29 Static stress–strain response determined by extensometry and back-to-back strain gauges indicating insignificant bending at fatigue-testing strain ranges

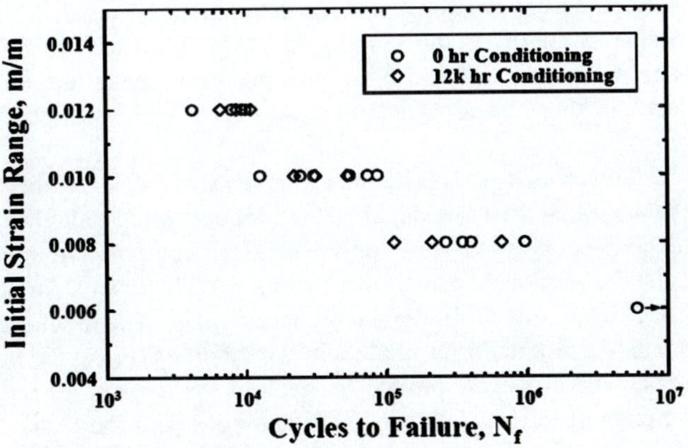

Fig. 14.30 Fatigue life plot based on initial strain range

applied cyclic strain range and Fig. 14.31 shows life in terms of the applied cyclic-stress range. Data are shown at the ±0.4%, ±0.5%, and ±0.6%, strain range, in addition to a datum at ±0.3% where failure did not occur and the test was terminated at approximately six million cycles. The data indicate that HC conditioning does not have a strong influence on the axial fatigue life though some life

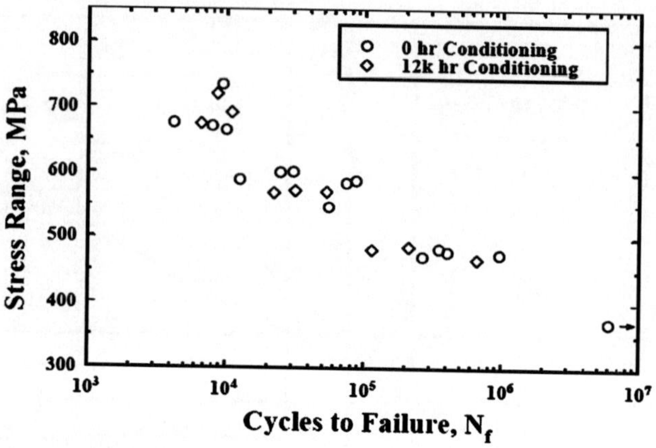

Fig. 14.31 Fatigue plot based upon stress range

deficit may be suggested at the lower load ranges. There appears to be increased scatter at the longer life/lower strain or stress ranges, as might be expected. One should note that the appearance of the data does not correspond with the three-regime format generally observed in the results of PMC axial fatigue [47]. The first high-stress/high-scatter regime, traditionally associated with the probabilistic failure of the fiber, is not present in these results. Due to the fully reversed loading conditions, all of the samples ultimately failed in compression. Compressive loads drive the phenomenon of buckling-induced delamination growth which controls the compressive strength of the composite [48]. Note that there were clear distinctions between the 0 and 12 k h samples regarding the failure morphologies and appearance of these buckled delaminations. This will be discussed in more detail later.

During the fatigue tests, the elastic modulus (E) was calculated periodically to determine its degradation or residual value. Moduli were calculated for both tension and compression. Representative residual compressive modulus data plotted against fatigue cycle are shown in Fig. 14.32. Results for the 0 h and 12 k h material are compared for the ±0.4% strain range. The modulus values are normalized with respect to each sample's initial modulus (E_o) at the beginning of the fatigue test. The last cycle plotted for each of the two curves represents the point of complete specimen fracture. As illustrated, both the 0 and 12 k h HC conditioned materials tended to maintain their initial modulus values over the first several thousand cycles. In general, the modulus of the HC conditioned material degraded earlier and to a greater extent. A number of samples experienced a slight "stiffening" trend prior to the measurable degradation, as illustrated by the 0 h data. This may be a result of the woven fabric "locking-up" with the accumulation of deformation.

All observed trends in the modulus response were consistent between tension and compression with the difference being that the compression modulus

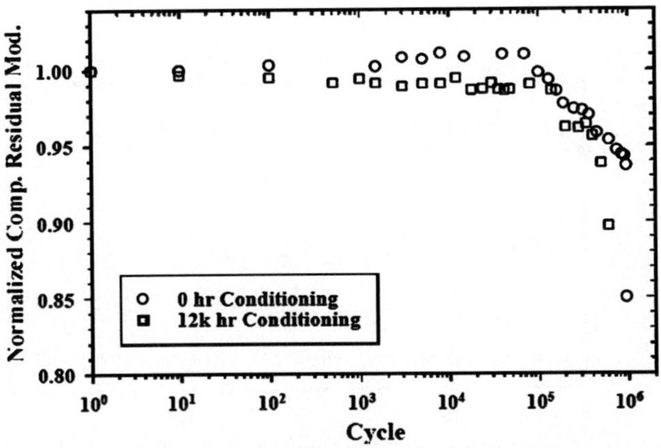

Fig. 14.32 Representative compressive residual modulus(normalized) during fatigue at the ±0.4% range

measurements were generally 2–5% lower for any given sample. In the range of approximately 50,000 cycles and up, the material begins to experience a gradual degradation of modulus. At the ±0.4% strain level, the material consistently experienced a minimum of 15% loss in modulus prior to complete fracture. The following discussion of residual strength and modulus relationships focuses on the ±0.4% initial strain range fatigue data, but the trends were the same for the other strain ranges.

14.4.2.2 Residual Strength/Modulus Relationships

There are several ways to view the material's response to the accumulation of the cyclic hygrothermal and mechanical damage and the damage tolerance under an ultimate compressive event. Three different permutations of the residual modulus, strength, and accumulated cycles data are presented in Figs. 14.33–14.35 for the 0 and 12 kh HC conditioned material tested at 121°C with the initial strain range of ±0.4%. Figure 14.33 shows the residual compressive modulus for each specimen as a function of its total number of accumulated fatigue cycles. As expected, the residual compressive modulus decreased in magnitude for specimens with a greater accumulation of fatigue cycles. While this trend is consistent with normalized modulus data, Fig. 14.32, the degree of degradation is masked in Fig. 14.33 by significant sample-to-sample modulus variation. Prior to fatigue, the compressive modulus of unconditioned material varied 7.6%, from 53.6 to 57.7 GPa, and that of the HC conditioned material varied 5.8%, from 54.4 to 57.6 GPa. This variation is common for RTM-fabricated, woven composites since variations in local fiber orientation and degree of waviness lead to local properties variations. The average initial compressive modulus of all the samples was 55.4 GPa and 55.6 GPa for the 0 h and the 12 kh data sets, respectively. This suggests that the degradation

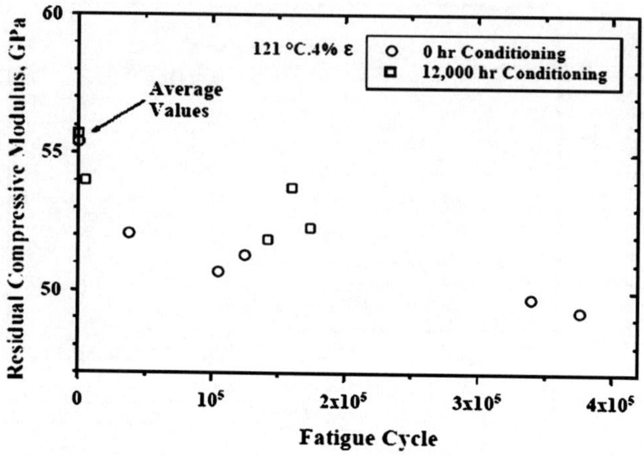

Fig. 14.33 Compressive modulus as a function of accumulated cycles for 0.4% strain range

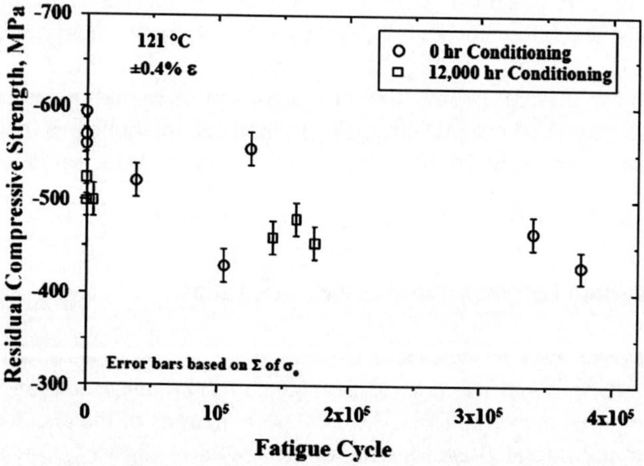

Fig. 14.34 Residual strength as a function of accumulated cycles for 0.4% strain range

associated with the 12 k h of HC conditioning had little to no influence on modulus. Likewise, subsequent fatigue cycling does not cause any observable difference in the rate of compressive modulus degradation between the two data sets.

Figure 14.34 illustrates the residual compressive strength as a function of accumulated fatigue cycles. Again, as expected, this property reveals a downward trend with cumulative fatigue cycling. The trends are relatively linear in cycles, suggesting more regular "deterministic" response when the damage tolerance is examined as a function of residual compressive strength, as opposed to the more commonly explored tensile strength. When the residual tensile strength properties of composites are tracked as a function of tension–tension fatigue, modest degrees of

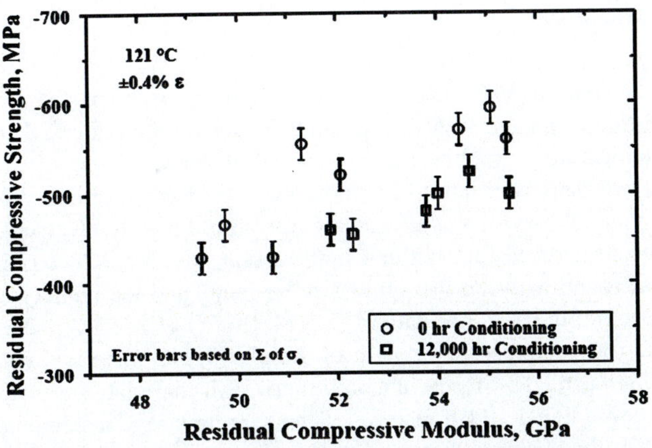

Fig. 14.35 Residual compressive strength vs. residual compressive modulus

damage are ascertained due to the dominating fiber effecting its probabilistic failure characteristics [49]. In contrast, residual compressive strength is more likely controlled by the unstable growth of a critically-sized, buckling-induced delamination [48].

Unlike the residual compressive modulus, the residual compressive strength reveals an 11% on average knockdown due to HC conditioning alone. Damage associated with the HC conditioning may have a greater influence on the extreme values of a defect population (size, spacing) effecting strength, while having an indistinguishable influence on the average distribution of flaw characteristics effecting modulus. Indeed Lacy, McDowell, Willice, and Talreja [50] show that different damage distributions may exhibit identical stiffness properties while having markedly different damage-evolution rates and damage tolerance. The residual compressive strength of the 0 h and 12 kh data sets are intermingled with the accumulation of mechanical fatigue. This is in spite of clear differences in the observed failure morphologies for the two data sets. Upon static failure, the 0 h samples generally revealed clean transverse fractures, with only modest ply delamination at the most aggressive states of fatigue damage. In contrast, the 12 kh HC conditioned material revealed extensive delamination, both at the ply and bundle interfaces. This strongly suggests that fiber/matrix interface damage was induced by the "mission-like" HC cycles.

Figure 14.35 takes the final step of comparing the compressive residual strengths with the corresponding residual compressive modulus values. Presentation of the fatigue data in this manner is believed to be relatively unique; empirically it infers a proportional relationship between residual strength and modulus. First, it can be stated that each of the data sets independently suggests a consistent relationship between these two properties. Furthermore, the 0 h data lie above the 12 kh data with a consistent increase. While the 0 h and 12 kh material may exhibit similar compressive modulus response, the residual compressive strength of the 0 h material tends to be greater.

14.4.3 Summary

This experimental investigation was conducted to examine the potential for increased fatigue–damage progression due to long-term hygrothermal conditioning of a high-temperature graphite/epoxy PMC. PMC panels were conditioned in a prototypic, commercial aviation, hygrothermal cycle. A specimen geometry was identified that could be used at elevated temperature for fully reversed fatigue and compressive strength testing without anti-buckling guides. A deterministic relationship was found between the applied stress range and the number of cycles to failure; the resulting stress-life data in Fig. 14.31 may be used to make design life predictions. The compressive strength of the 12 k h HC material experienced an 11% degradation relative to the unconditioned, 0 h material. No distinction was found between 0 h and 12 k h material when examining the residual compression modulus or residual compression strength as a function of fatigue cycles. Designers may find that the correlation of residual compressive strength and modulus in Fig. 14.35, gives a clearer representation of compressive strength degradation due to HC conditioning. For a given state of stiffness, the HC material had lower residual compressive strength.

14.5 Conclusion

The durability of polymer matrix composites remains a complex issue when applied in power and propulsion technology for spaceflight and aeronautic applications. The harsh environments of interest are rarely well characterized in available design databases, so the insertion of PMCs into these applications is hampered. Design development and proof-of-concept programs have a better opportunity to succeed if they are coupled with effective experimental pathfinder efforts that can quickly illuminate the design drivers for a particular combination of material and application conditions.

This approach was illustrated with the review of three such pathfinder experimental programs for development applications in flywheel energy storage, reusable launch vehicle propulsion structures, and aero engine composites. Through carefully targeted experiments, important issues could be quickly brought forward to enable more efficient and detailed experimental programs in follow-on phases.

The value of interrogating key assumptions was demonstrated in the investigation of time–temperature response of flywheel rotor composites. Through judicious selection of isothermal experiments, the validity of linear viscoelasticity for this application could be refuted. The dependency of deformation response on the sign of stress was an important finding. Designs based on the more conveniently acquired tensile data for component application would have been unconservative. These findings support the need for rigorous treatment of sign dependency in a theoretical framework for PMC constitutive models.

TMF test methods were reviewed in the second example where transient thermal effects dominate the operations cycle for reusable launch vehicle propulsion systems. An appreciation for the complexity of such experimental methods can be gained from the methods employed to calibrate a repeatable thermal cycle in the gauge section of the specimen. The effort highlighted the potential for TMF to accelerate strain ratcheting at higher load levels relative to isothermal testing even though static strength and fatigue life requirements could easily be met for this application. Since the dimensional stability was crucial to the design concept for the support structure, adjustments could be made to account for enhanced ratcheting potential. Unexpected deformation responses were observed in the program that were verified and not artifacts of the experiment. These mechanisms are inactive during isothermal fatigue, so their effect on fatigue damage of the real component would not have been known without TMF testing.

The final example reviewed hygrothermal conditioning (HC) and mechanical fatigue of composites applicable to aero engine applications. Long-term hygrothermal conditioning followed by isothermal fatigue was used to interrogate the material. The experiments reviewed here are also unique in the implementation of a specimen geometry that enabled fully reversed fatigue and residual compression testing at elevated temperature without anti-buckling guides. The test program also used interrupted fatigue tests at prescribed modulus degradation levels to characterize residual compressive strength and modulus as a function of fatigue cycle. The residual compressive strength as a function of residual modulus shows the degrading trend caused by HC conditioning and highlights the sensitivity of compressive properties.

Acknowledgments The three reviewed experimental investigations benefited greatly from the support and expertise of numerous colleagues both within and outside NASA. The mechanical testing was facilitated by Dr. Bradley Lerch, Director of GRC Fatigue and Fracture Laboratory, with the technical assistance of Mr. Christopher Burke. Physical materials characterization was made with the technical assistance of Mr. Daniel Scheiman and Ms. Linda McCorckle. The following were key to specific test programs: Flywheel – FESS/Rotor Safe Life Program with Dr. Steven M. Arnold/GRC and Dr. Richard C. Thompson/UTCEM; Hypersonic Engines – the Aeropropulsion Program Office: Higher Operating Temperature Propulsion Components Project with Dr. E. Eugene Shin/OAI and Mr. Jeffery Fink/Rocketdyne; Aeroengine Composites – Cooperative grant NCC3-821 with Mr. Michael G. Castelli/OAI and Dr J. Rodney Ellis/GRC. Preparation of this chapter was made possible with the support and advice of Ms. Mei-Hwa Liao/ GRC-Chief, Applied Structure Mechanics Branch.

References

1. J. C. Thesken, C. L. Bowman, S. M. Arnold, and R. C. Thompson, (2003) "Time-Temperature Dependent Response of Filament Wound Composites for Flywheel Rotors," Composite Materials: Testing and Design Fourteenth Volume, ASTM STP 1436, C.E. Bakis, Ed., ASTM International, West Conshohocken, PA.

2. J. C. Thesken, E. E. Shin, J. K. Sutter, C. Burke, (2003) "Durability of Polyimide Composites subjected to thermo-mechanical fatigue in reusable propulsion systems", ICCM-14, July 14–18, San Diego, CA.
3. J. C. Thesken, M. G. Castelli and J.R. Ellis (2002) "Hygrothermal and Mechanical Fatigue of Polymer Matrix Composites for Propulsion Applications" Eighth International Fatigue Congress, Stockholm, Sweden.
4. Saleeb, A.F., Arnold, S.M., and Al-Zoubi, N.R., A Study of Time Dependent and Anisotropic Effects on the Deformation Response of Two Flywheel Rotor Designs. *Composite Materials: Testing and Design Fourteenth Volume, ASTM STP 1436*, C.E. Bakis, Ed., ASTM International, West Conshohocken, PA, 2003.
5. Genta, G., Kinetic Energy Storage: Theory and Practice of Advanced Flywheel Systems, Butterworths, 1985.
6. Keckler, C.R., Bechtel, R.T., and Groom, N.J., An Assessment of Integrated Flywheel System Technology. NASA CP 2346, 1984.
7. DeTeresa, S.J., Materials for Advanced Flywheel Energy Storage Devices., MRS Bulletin, 1999.
8. Bitterly, J.G., Flywheel Technology. IEEE AES Systems Magazine, 1998.
9. Arnold, S.M., A.F. Saleeb, and Al-Zoubi, N.R., Deformation and Life of Composite Flywheel Disk and Multi-Disk Systems. NASA/TM.2001-210578.
10. Tzeng, J.T., Viscoelastic Analysis of Composite Flywheels for Energy Storage. ARL-TR-2610, U.S. Army Research Laboratory, Aberdeen Proving Ground, MD 21005.
11. Tzeng, J.T., Viscoelastic Behavior of Composite Rotors at Elevated Temperatures. IEEE Transactions on Magnetics, Vol. 33, No. 1, 1997, pp. 413–418.
12. Emerson, R.P. and Bakis C.E., Relaxation of Press-Fit Interference Pressure in Composite Flywheel Assemblies. Proc. 43th SAMPE Symposium and Exhibition, 31 May.June 4 1998, Long Beach, CA, SAMPE, Covina,CA 1998.
13. Emerson, R.P. and Bakis C.E., Viscoelastic Behavior of Composite Flywheels. Proc. 45th SAMPE Symposium and Exhibition, 21.25 May 2000, Long Beach, CA, SAMPE, Covina,CA 2000.
14. Thompson, R.C., Pak, T.T., and Rech, B.M., Hydroburst Test Methodology for Evaluation of Composite Structures. Composite Materials: Testing and Design *Fourteenth Volume, ASTM STP 1436*, C.E. Bakis, Ed., ASTM International, West Conshohocken, PA, 2003.
15. Irion, M.N., and Adams, D.F., "Compression Creep Testing of Unidirectional Composite Materials," Composites, Vol. 12, No. 2, April 1981, pp. 117–123.
16. Mohan, R., and Adams, D.F., "Nonlinear Creep-Recovery Response of a Polymer Matrix and Its Composites," Experimental Mechanics, Vol. 25, No. 3, September 1985, pp. 262–271.
17. Tuttle, M.E. and Brinson, H.F., Prediction of the Long-Term Creep Compliance of General Composite Laminates. Experimental Mechanics March 1986, pp. 89–102.
18. Sullivan, J.L., Measurement of Composite Creep. Experimental Techniques Sept./Oct. 1991, pp. 32–37002E.
19. Gates, T.S., Experimental Characterization of Nonlinear, Rate-dependent Behavior in Advanced Polymer Composites. Experimental Mechanics March 1992, vol. 31, pp. 68–73.
20. Gates, T.S. and Sun, C.T., Elastic/Viscoplastic Constitutive Model for Fiber Reinforced Thermoplastic Composites. AIAA Journal, vol. 29, No. 3, 1991 pp. 457–463.
21. Gates, T.S., Veazie, D.R. and Brinson, L.C.A. Comparisson of Tension and Compression Creep in Polymeric Composite and the Effects of Physical Ageing on Creep. NASA TM 110273 August 1996, NASA Langley Research Center.
22. Brinson, L.C. and Gates, T.S., Chapter 10.Viscoelasticity and Aging of Polymer Matrix Composites. Comprehensive Composite Materials. Editors in Chief, Kelly, A. and Zweben, C.H., Vol 2. 2000, pp. 333–368, Elsevier Science LTD, Oxford, UK.
23. Raghavan, J. and Meshii, M., Creep of Polymer Composites. Composites Science and Technology 57, 1997, pp. 1673–1688.
24. Al-Haik, M., Vaghar, M.R. Garmestani, H., and Shahawy, M., Viscoplastic analysis of structural polymer composites using stress relaxation and creep data. Composites Part B: engineering, 32, 2001, pp. 165–170.

25. Arnold, S.M., Saleeb, A.F., and Castelli, M.G., A General Time Dependent Constitutive Model: Part II. Application to a Titanium Alloy., Journal of Engineering Materials and Technology, 122, pp. 1–9.
26. Hexcel Advanced Composite Materials Technical Literature, "Hexcel 8552 Epoxy Matrix - Product Data," Hexcel, 5794 West Las Positas Blvd., P.O. Box 8181, Pleasanton, CA 94588-8781, August, 1998.
27. Baaklini, G.Y., Kono, K.K., Martin, R.E., and Thompson, R., NDE Methodologies for Composite Flywheels Certification. SAE Conference, 2000 Power Systems, San Diego October-November 2000. (also NASA/TM.2000-210473, October 2000).
28. Perry, C.C., Strain-Gage Reinforcement Effects on Low Modulus Materials-Section IIID. Manual on Experimental Methods for Mechanical Testing of *Composites*, Ed. M.E. Tuttle and R.L. Pendleton, SEM (1989).
29. Perry, C.C., Strain-Gage Reinforcement Effects on Orthotropic Materials-Section IIID. Manual on Experimental Methods for Mechanical Testing of Composites, Ed. M.E. Tuttle and R.L. Pendleton, SEM (1989).
30. Turner, S. Chapter 4 Creep in Glassy Polymers. in *The Physics of Glassy Polymers* ed. R.N. Haward, Applied Science Publishers LTD, London, 1973
31. Suh, N.S. and Turner, A.P.L., Elements of the Mechanical Behavior of Solids, 1974, Scripta Book Co., Washington D.C. p. 314.
32. Struik, L.C.E. Physical Aging in Amorphous Polymers and Other Materials, 1978, Elsevier Scientific Publishing Co. Amsterdam, Netherlands.
33. J. C. Thesken, M. Melis, E. Shin, J. Sutter and J. Fink (2003) "Design analysis of a Combustion Chamber Stiffened by Polymer Composite Sandwich" *Presented at ICCM-14*, San Diego, CA. July 14–18.
34. C.L. Bowman, J.K. Sutter, J.C. Thesken, and B.P. Rice, "Characterization of Graphite Fiber/ Polyimide Composites for RLV Applications" in Proceedings of the 46th International SAMPE Symposium and Exhibition, Vol 46, Book 2, (SAMPE, Covina CA, 2001) 1515–1529.
35. E. Shin et al (2003): "Effects of Fiber Reinforcement Architecture on the Hygrothermal-Mechanical Performance of Polyimide Matrix Composites for Aeropropulsion applications" *presented at ICCM-14*, San Diego, CA. July 14–18.
36. E. E. Shin et al.(2002) "Design and Fabrication Issues of High Temperature PMCs for Aerospace Propulsion Applications" *SAMPE 2002 Long Beach.*
37. J. E. Fink, B. Shapey, E. Eugene Shin, James K. Sutter, John Thesken, and Demetrios Papadopoulos, Gary Wonacott and Chris Benzie (2002) "Initial Assessment of High Temperature Polymer Matrix Carbon Fiber Composites for 3 rd Generation Liquid Propellant Rocket Engine Applications" *High Temple Workshop 22, Sante Fe, NM.*
38. M. G. Castelli, J. K. Sutter, and D. Benson, "Durability and Damage Tolerance of Polyimide Chopped Fiber Composite Subjected to Thermomechanical Fatigue Missions and Creep Loadings," *Time Dependent and Nonlinear Effects in Polymers and Composites*, ASTM ASTP 1357, R. A Schapery and C. T. Sun, Eds., American Society for Testing and Materials, West Conshohocken, PA, 2000, pp. 285–309.
39. B. P. Rice and C.W. Lee, "Study of Blister Initiation and Growth in a High-Temperature Polyimide," SAMPE International Technical Conference, 29, (1997) pp.675–685.
40. C.L. Bowman, J.C. Thesken, K.C. Chuang, and C.P. Arendt, "Graphite Fiber/Polyimide Composites Subjected To Moisture And Rapid Heating" in *Proceedings of the 47th* International SAMPE Symposium and Exhibition, Vol 47, (SAMPE, Covina CA, 2002).
41. M.B. Dow and D.L. Smith, *SAMPE International Technical Conference*, 21, (1989) pp. 595–605.
42. K.C. Chuang and J.E. Waters, "Effects of Endcaps on the Properties of Polyimide/Carbon Fiber Composites", *SAMPE International Symposium*, 40, (1995) pp. 1113–1123.
43. A.L. Gyekenyesi, et al., "A Study of Elevated Temperature Testing Techniques for the Fatigue Behavior of PMCs: Application to T650-35/AMB21", NASA TM 106927, July 1995

44. Cornelia., D.(1994). In: 39th International SAMPE Symposium, pp. 917–929.
45. Bakis, C.E. et al. (1989). In: Composites (2nd Vol), ASTM STP 1003, pp.180–193.
46. Gyekenyesi, A.L. (1998). ASME-98-GT-106.
47. Talreja, R. (1981). In: Proceedings of the Royal Society of London, A378, pp. 461–475.
48. Nilsson, K. F., Thesken, J. C., Sindelar, P., Giannakopoulus, A. E. and Storåkers, B. (1993) J. Mech. Phys. Solids 41, 749–782.
49. Reifsnider, K. and Stinchcomb, W.W. (1986). In: Composite Materials: Fatigue and Fracture, ASTM STP 907, pp. 298–313.
50. Lacy, T.E., McDowell, D.L. Willice, P. A. & Talreja, R. (1997) International Journal of Damage Mechanics, vol. 6, pp.62–95.

Chapter 15
Durability of Composites in Aircraft Engine Applications

Rajiv A. Naik

Abstract Aircraft engine components are subjected to high temperatures and complex stress states, thus requiring a much more rigorous assessment of durability. This chapter provides a brief introduction to durability assessment of polymer–matrix composites in gas turbine engine applications. The engine environment exposes the composite material to combined missions of hygrothermal exposure, mechanical loading, and fatigue. The complexity of the actual engine conditions is not easily duplicated in the laboratory and simplified simulated exposure missions are often used for durability characterization. Full-term durability testing to determine end-of-life properties is also not practical and accelerated testing and modeling is often considered as an alternative. Thus, a combined experimental and analytical approach is typically used to characterize durability of engine composites. A specific example of a woven composite material is described in this chapter to demonstrate how simple mechanics-based modeling can be used to study the evolution of the different damage mechanisms under fatigue and to predict fatigue life. Finally, the building block approach used to characterize the durability of engine components is discussed along with a specific example of its application.

15.1 Introduction

Today's turbofan aircraft engines are designed to be far more efficient, economical, and environmentally friendly than the turbojet engines of the 1960s. The turbofan engine is a modified turbojet, featuring a large fan section to allow a portion of the incoming air to "bypass" the engine's core. This "bypass" airflow provides a second

R.A. Naik (✉)
Pratt & Whitney, United Technologies Corporation, 400 Main Street,
East Hartford, CT 06108, USA
e-mail: rajiv.naik@pw.utc.com

K.V. Pochiraju et al. (eds.), *Long-Term Durability of Polymeric Matrix Composites*,
DOI 10.1007/978-1-4419-9308-3_15, © Springer Science+Business Media, LLC 2012

source of thrust – similar to that provided by the propeller on a turboprop engine. Generally speaking, the greater the volume of air that is bypassed, the more efficient is the turbofan engine. It is not surprising that the bypass ratios (defined as the ratio of the mass flow rate of air that bypasses the engine core to the mass flow rate passing through the engine core) of new turbofan engine designs have steadily increased over the past three decades. Higher bypass ratios lead to engines with larger and heavier fan sections. Engine designs for the next generation of single-aisle aircraft are expected to have bypass ratios from 10:1 to as high as 12:1, requiring proportionally larger fan sections that could account for about 35% of the engine's total weight [14]. This trend to larger and heavier fan sections is a key driver for the insertion of polymeric–composites in aircraft engines.

The front section of the engine, also known as the "cold section," is ideally suited for the use of polymer–matrix composites due to lower operating temperatures that are usually less than 290°C (550°F). Some of the engine components that make use of polymer–matrix composites are: nacelles, thrust reversers, fan blades, fan containment cases, ducts, fan exit guide vanes, low compressor stators, etc. The cooler front section components (such as fan blades, fan containment cases, nacelles, and fan exit guide vanes) can be made using epoxy resins. However, as the temperatures increase from the front fan section to the compressor section, engine components need to make use of higher temperature resins such as bismaleimides and polyimides.

Polymer–matrix composites on an aircraft engine have to withstand engine operating conditions of combined hygrothermal cycling, mechanical loading, and fatigue. Engine components are typically subjected to high temperatures, pressures, and multi-axial stress states. Durability assessment and the development of end-of-life properties for these components is, therefore, more challenging than other airframe applications. Apart from the technical challenges, durability is also an economic and business concern. An inadequate assessment of the durability of a structure has the potential for significant financial loss to the engine company [5]. The inability of the structure to maintain adequate strength and stiffness over the service life of the engine could lead to excessive warranty, maintenance, repair or redesign, and modification costs. It is, therefore critical to develop both analytical and experimental techniques that will lead to a reliable assessment of durability.

This chapter provides a brief introduction to durability assessment of polymer–matrix composites for gas turbine engine applications. The high temperatures and complex stress states that are experienced by an aircraft engine component necessitate a rigorous assessment of durability. The combined missions of hygrothermal, mechanical, and fatigue loading experienced in an engine environment can be a challenge to reproduce in the laboratory for the direct assessment of durability. Long-term durability characterization is also an issue and accelerated testing is often used together with analytical modeling to assess durability. Such a combined experimental and analytical approach will be demonstrated here using a specific example of a woven composite material. Simple mechanics-based models are developed for the fatigue damage mechanisms observed in the woven composite material. Such modeling can also be used as part of the so-called building block approach [3] by which the durability of engine components is assessed by conducting testing on test configurations with increasing complexity, such as specimens, sub-elements, and components.

The following sections describe first the engine operating conditions of hygrothermal and mechanical loading and how these are simulated in the laboratory. Next, the characterization of the stiffness and strength degradation and the corresponding damage mechanisms under fatigue loading are described for the specific case of a woven composite material. Next, simple analytical models, specific to the woven composite case, are presented to evaluate the damage mechanisms and their evolution under fatigue loading. Finally, the process used to assess durability of an engine component is described along with an example of how it is applied to a specific engine component.

15.2 Durability Characterization

The first step in characterizing durability is to examine the operating and environmental conditions that the engine component will be exposed to during its service life. For example, a typical commercial aircraft engine will be exposed to the temperature and humidity at the airport location while it is on the ground. During every flight, the engine component will be exposed to a gradient of temperature and mechanical loading during takeoff and landing. And at the cruising altitude, it will experience a steady temperature together with combined steady and vibratory loads. Figure 15.1 depicts a typical combined hygrothermal and mechanical loading mission that would be experienced by an engine component. The actual levels of humidity, temperature, mechanical loading, and fatigue will be unique to the type of airplane (for, e.g., commuter, single aisle, wide body, etc.), the type of flight mission (short haul, long range, etc.) and the geographic locations where the airplane is being flown.

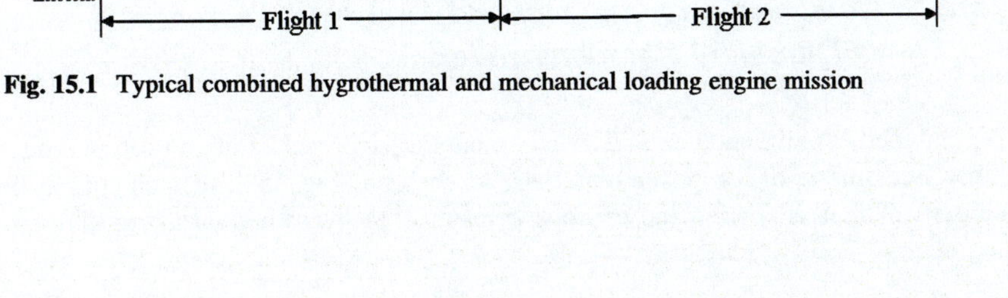

Fig. 15.1 Typical combined hygrothermal and mechanical loading engine mission

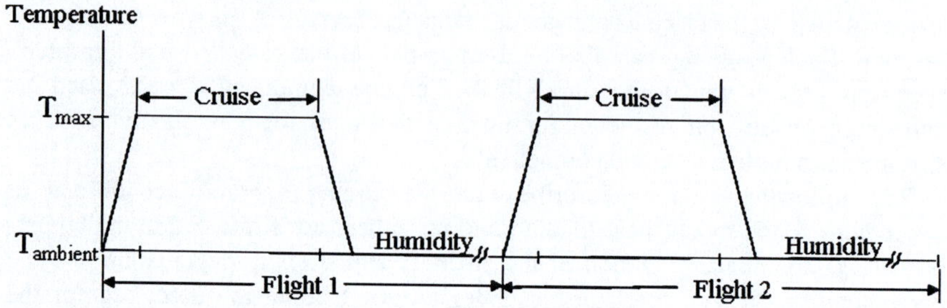

Fig. 15.2 Simulated hygrothermal aging engine mission

15.2.1 Aircraft Engine Hygrothermal Missions

The actual engine mission shown in Fig. 15.1 is not easily replicated in the laboratory due to limitations of test equipment which usually cannot apply a combined hygrothermal and mechanical loading mission. Thus, one has to often settle for a "simulated" engine mission which decouples the hygrothermal mission from the mechanical loading mission. Figure 15.2 shows a typical simulated engine mission. The assessment of durability in this case is performed by first subjecting the composite material to the simulated hygrothermal mission and then applying the mechanical fatigue loading to the hygrothermally aged material. Typically, hygrothermally aged composite panels are removed from the hygrothermal chamber at regular intervals. Test specimens are then extracted from the hygrothermally aged panels for mechanical fatigue testing and residual stiffness and strength measurements.

A commercial engine is typically exposed to several thousand flight cycles in its lifetime. Simulation of this long-term exposure in the laboratory is often not practical. Durability assessment and the estimation of end-of-life properties have to therefore rely on the use of accelerated testing methods and analytical modeling. Mission exposure can be accelerated by aging the material at a higher temperature and/or relative humidity. Modeling can be used to predict the amount of damage in the composite and the subsequent residual stiffness and strength properties. It is important to confirm that the accelerated mission produces the same damage mechanisms as the full-term exposure. For this reason, limited real-time exposure testing is performed to compare the damage mechanisms. This damage mechanism evaluation is also useful in the verification and validation of the durability modeling approach.

15.2.2 Combined Experimental and Analytical Approach to Durability Characterization

A combined experimental and analytical modeling approach is often used in durability assessments of composite materials. As shown in Fig. 15.3, such an approach consists of first subjecting the composite material to hygrothermal aging under a

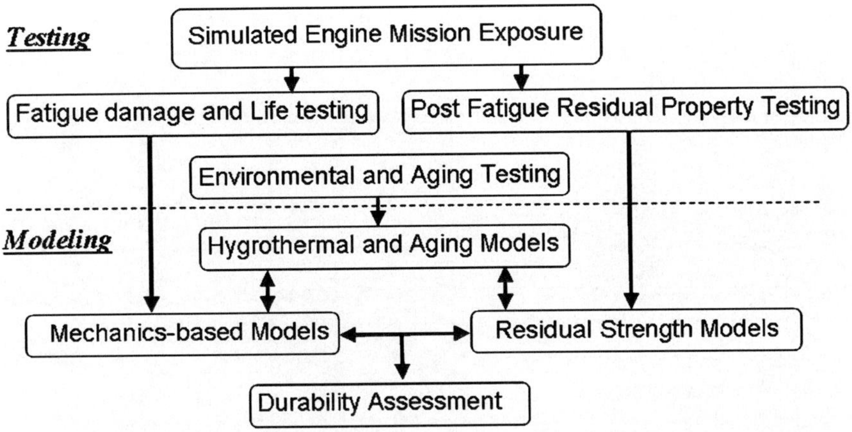

Fig. 15.3 Combined experimental and analytical approach for durability assessment

simulated engine mission. This aged material is then tested under static and fatigue loading. Residual property testing is then performed to establish material durability. Separate environmental and aging testing is also conducted to develop an understanding of hygrothermal and aging effects and to develop appropriate models. The models for hygrothermal and aging effects can be integrated with mechanics-based and residual strength-based models. Mechanics-based models can be used to model the observed fatigue damage mechanisms. Residual strength models can be built based on experimental residual strength results and integrated with the mechanics models. Such a combined experimental and analytical approach that makes use of an integrated modeling strategy can lead to durability assessments that capture some of the synergistic effects of hygrothermal and mechanical loading and fatigue.

15.2.3 Durability Characterization Example

The semi-empirical, mechanics-based modeling approach described above will be demonstrated here for a two-dimensionally (2D) woven, five-harness satin (5HS) fabric composite material consisting of 6 K (6,000 filaments) AS4 graphite fiber yarns (woven at a spacing of 4.724 yarns/cm (12 yarns/in.)) and the PR500 epoxy resin [9]. The laminated panels are fabricated by infiltrating eight warp-aligned layers of the dry 5HS woven fabric with the PR500 epoxy resin using the resin transfer molding (RTM) process. The nominal thickness of the panels is 2.896 mm (0.114 in.) and the composite fiber volume fraction is 55%.

The panels are subjected to a simulated hygrothermal mission before fatigue testing. The hygrothermal mission consists of 24 hour blocks. Each mission block consists of (1) a temperature ramp to 121°C (250°F) followed by (2) a 90 min dwell at 121°C (250°F) followed by (3) a cooldown to 29.4°C (85°F), and then (4) a

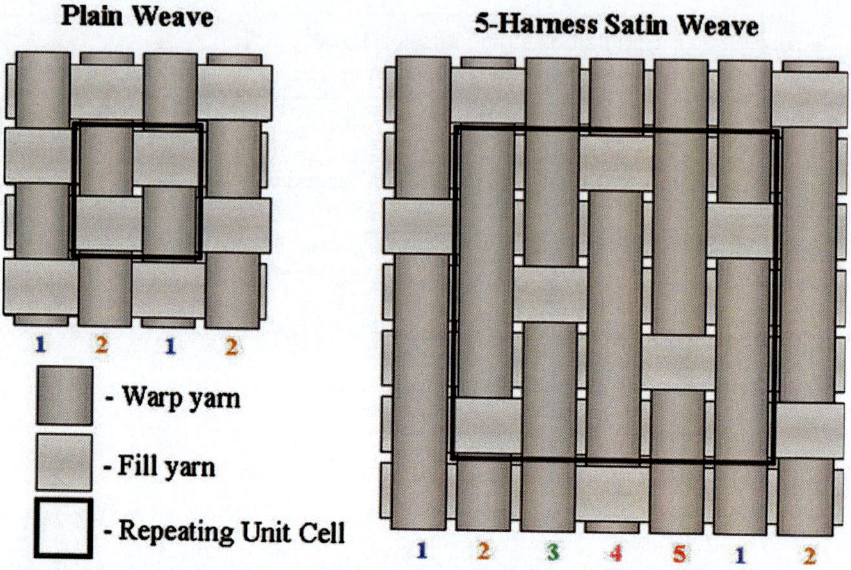

Fig. 15.4 Fiber architecture and repeating unit cell of a plain weave and a 5HS weave

22.5 hour dwell at 29.4°C (85°F) and 85% relative humidity. The panels are hygrothermally conditioned for a total of 12,000 hour (or 500 mission blocks) before testing under constant amplitude fatigue loading [9].

After the hygrothermally conditioned panels are removed from the simulated engine mission, specimens of size 20.32 × 2.54 cm (8 × 1 in.) are machined for fatigue testing. Tension–tension fatigue testing is performed under load control at an *R*-ratio of 0.1 and a frequency of 10 Hz. Testing is conducted at room temperature (RT) 22.2°C (72°F) and at 121°C (250°F). The fatigue stress levels used are at 65%, 70%, 75%, and 85% of the ultimate tensile strength (UTS). The dynamic stiffness of the specimens is monitored during fatigue by using cross-head displacement (or stroke) data [12, 13].

15.2.4 Weave Architecture and Fabrication

Before examining the fatigue damage mechanisms for this composite material, it is instructive to take a closer look at the 5HS weave fiber architecture and to understand how this fabric is woven on a weaving loom. Figure 15.4 depicts the fiber architecture for a plain weave and a 5HS weave fabric. A basic plain weave fabric is created on a weaving loom in which two sets of yarns–warp and fill are interlaced at 90° to each other [1]. Fabricating a plain weave fabric on a loom requires two harnesses to control the motions of the warp yarns as they are being drawn through the reed on the weaving loom. One harness is used to control the motion of all "1" warp yarns (in Fig. 15.4) and the second harness controls all the "2" warp yarns. To form a plain weave, all the "1" warp yarns are lowered and all the "2" warp yarns are raised, forming an opening (called a "shed") through which a fill yarn is inserted (at 90° to the warp yarn direction). In the second step, all the "2" warp yarns are lowered and all the "1" warp yarns are raised before inserting the fill yarn. This leads to the plain weave fiber architecture shown in Fig. 15.4.

A 5HS weave fabric is formed in a similar fashion on a weaving loom. Instead of two harnesses, a 5HS weave requires (as the name suggests) five harnesses. Each harness is used to raise/lower one set of warp yarns when the fill yarn is inserted. The motion of all the warp yarns numbered "1" in Fig. 15.4, for example, is controlled by the same harness. Similarly, the motion of all the warp yarns numbered "2" is controlled by another harness on the loom. If, for example, the harness that controls the motion of all warp yarns numbered "4" in Fig. 15.4 is lowered, and at the same time the four other harnesses that control the motions of the "1," "2," "3," and "5" yarns are raised before the insertion of a fill yarn, this will result in the yarn interlacing pattern depicted, in Fig. 15.4, by the fill yarn in the bottom-most row. The 5HS weave fiber architecture is thus created by raising and lowering a different set of warp yarns at each step of the weaving process using the five harnesses on the loom. The resulting weave pattern is shown in Fig. 15.4 and it can be seen that the repetitive motions on the loom produce a fabric with a repeating yarn interlacing pattern. A repeating unit cell (RUC) as shown in Fig. 15.4 can thus be isolated for the purpose of analysis and modeling of the 5HS woven composite material.

15.2.5 Fatigue Damage Mechanisms

For both the as-fabricated (also referred to as "unaged") material and the 12,000 hour hygrothermally aged (also referred to as "aged") material, the major initial damage mechanisms [12, 13] are transverse yarn cracking followed by debonding at the yarn crossover regions (see Fig. 15.5). This inter-yarn debonding is also known as "meta-delamination" [12, 13]. The locations and the extent of this inter-yarn debonding or meta-delamination can be directly related to the weaving pattern of the 5HS weave depicted in Fig. 15.4. This initial damage is then followed by delamination between the layers of the laminate. Delaminations initiate due to the high stress concentrations at the tips of the transverse yarn cracks that have propagated to the interface of two layers. Delaminations also initiate and grow from the edges of the specimen.

The evolution of these damage mechanisms is also reflected in the dynamic stiffness loss curves that are recorded during fatigue testing. Figure 15.6 shows a typical dynamic stiffness loss curve plotted on a semi-log scale. It is possible to identify three different regions [9] along this normalized stiffness (stiffness after N cycles/initial stiffness, or, E_N/E_o) curve. The first Region I is dominated by the evolution of the initial transverse yarn cracking. This failure mode leads to only

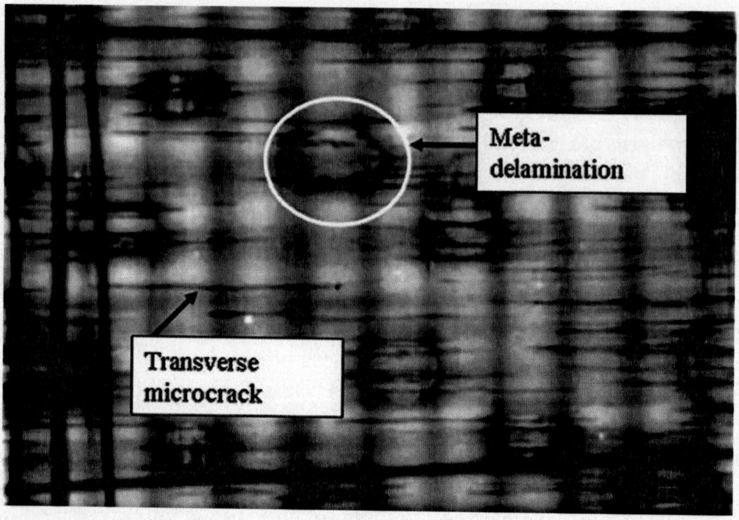

Fig. 15.5 X-radiograph showing transverse yarn cracking and meta-delamination in a 5HS woven composite [9]

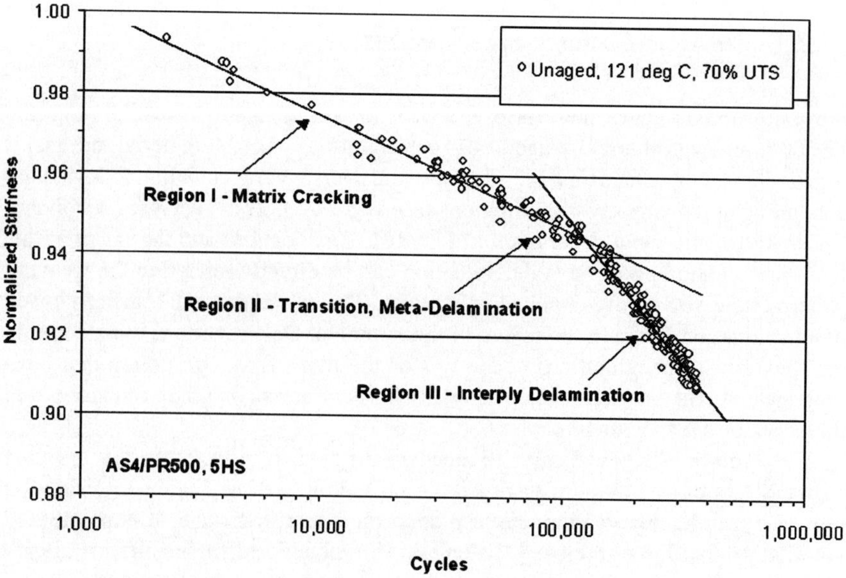

Fig. 15.6 Typical dynamic stiffness loss curve showing three regions of fatigue damage [9]

a minor change in the longitudinal stiffness of the composite. The next Region II is a transition region between Regions I and III. The Region II is characterized by the formation of delaminations at the tips of the transverse yarn cracks and also the formation of inter-yarn debonding. The Region III is dominated by the propagation of edge delaminations. This region leads to the most rapid and the largest loss in composite longitudinal stiffness.

The dynamic stiffness loss measurements in Fig. 15.6 and the observation of the three different regions of damage evolution can be used to develop models for the matrix cracking and the delamination failure mechanisms in the unaged and the aged specimens at both room temperature and at 121°C (250°F).

15.3 Damage Mechanism Modeling

It is possible to develop appropriate mechanics-based models for each of the major damage mechanisms observed during durability testing. In this section, simple mechanics-based models are described for the three fatigue-damage mechanisms exhibited by the woven composite material. The stiffness degradation curves for the unaged and the aged material are used together with mechanics-based models to first study the evolution of matrix cracking and delamination under fatigue loading. Next, these models are further extended to make predictions of fatigue life.

As shown in Fig. 15.6, a linear stiffness degradation relationship can be assumed for each of the major damage mechanism regions (I and III). The transverse yarn cracking in the 5HS woven composite (Fig. 15.7) can be modeled by considering the repeating unit cell (RUC) for the 5HS weave architecture (Fig. 15.8). This RUC can be further idealized as a [0/90] laminate to simplify the modeling of the transverse yarn cracking damage mechanism. The inter-yarn debonding failure mechanism can be modeled by representing the debonded regions in the repeating pattern of the 5HS weave by an "equivalent" debonded region (Fig. 15.8). This equivalent inter-yarn debonding model is used to compute the stiffness loss due to this failure mechanism. The edge delamination observed

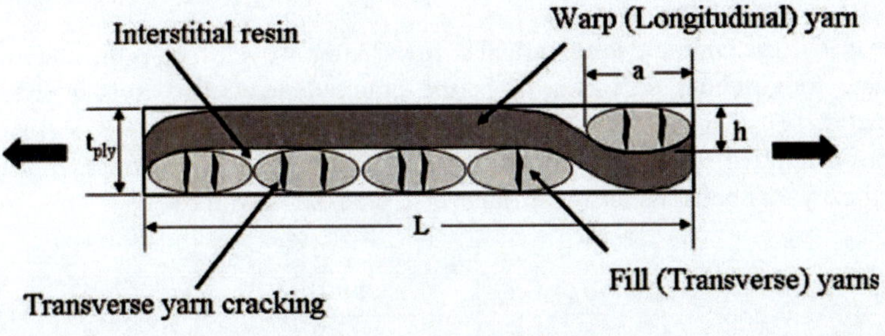

Fig. 15.7 Transverse yarn cracking model showing repeating unit cell of a 5HS layer [9]

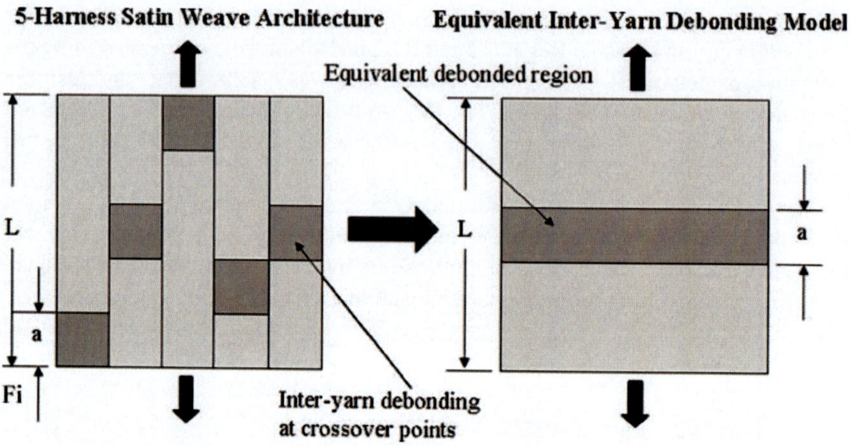

Fig. 15.8 Equivalent inter-yarn debonding model for a 5HS composite layer [9]

during fatigue can be modeled using the classic O'Brien model [11] with appropriate extensions to a woven composite laminate.

15.3.1 Transverse Yarn Cracking Model

The transverse yarn cracking model in Fig. 15.8 is idealized as a [0/90] laminate in which the stiffness, E_1, of the longitudinal, 0° layer is calculated as the "effective" stiffness of the warp yarns after accounting for the yarn undulations. The stiffness of the transverse, 90° layer, E_2, is computed as the "effective" combined stiffness of the fill yarns and the interstitial matrix. The normalized stiffness, E_{mN}/E_o, for a [0/90] laminate with transverse cracks is given by [4]

$$\frac{E_{mN}}{E_o} = \frac{1}{1 + \frac{2\rho}{\alpha}\frac{E_2}{E_1}\tanh\left(\frac{\alpha}{2\rho}\right)} \tag{15.1}$$

where E_{mN} is the laminate longitudinal stiffness after N cycles, E_o is the undamaged laminate longitudinal stiffness, ρ, is the crack density, and α is a shear-lag parameter.

For an applied 0° axial stress, σ_a, the maximum transverse stress, σ_{tr}, in the transverse yarns between adjacent transverse cracks is given by

$$\frac{\sigma_{tr}}{\sigma_a} = \frac{E_2}{E_o}\left[1 - \frac{1}{\cosh\left(\frac{\alpha}{2\rho}\right)}\right] \tag{15.2}$$

Under static loading, as the load is increased, the stress, σ_{tr}, between two adjacent transverse cracks will increase until it reaches a critical value. At this point a new transverse crack will be formed between each pair of adjacent transverse cracks. Thus, the crack density will increase with increasing load until a "saturation crack density" damage state is reached. Once the saturation crack density is attained, further increases in the applied stress, σ_a, result in peak transverse stress, σ_{tr}, values that are lower than the critical transverse cracking stress. Thus, the crack density does not increase further with increasing load. Under fatigue loading, the critical stress value for the formation of transverse cracks is a function of fatigue cycles. This dependence of the critical value on fatigue cycles can be deduced for the unaged and aged AS4/PR500, 5HS weave material from the experimentally measured stiffness degradation (Region I) curves.

Using the first two terms in the Taylor series expansions of the "tanh" and the "cosh" terms in Eqs. 15.1 and 15.2, respectively, it is possible to express the peak transverse stress, σ_{tr}, as a function of the normalized stiffness, (E_{mN}/E_o), as [9]:

$$\frac{\sigma_{tr}}{\sigma_a} = \frac{\left[\frac{E_{mN}}{E_o}\left(\frac{E_2}{E_1}+1\right)-1\right]}{\left[\frac{E_{mN}}{E_o}\left(\frac{5}{3}\frac{E_o}{E_1}+\frac{E_o}{E_2}\right)-\frac{E_o}{E_2}\right]} \tag{15.3}$$

For each of the material conditions (unaged and aged) and each of the temperatures tested, the normalized stiffness, (E_{mN}/E_o), as a function of fatigue cycles, N, can be determined from Region I of the measured dynamic stiffness loss curves (see Fig. 15.6). A linear relationship can be assumed in the form: $(E_{mN}/E_o) = A_m \ln(N) + B_m$. The constants A_m and B_m are determined using linear regression. The expression in Eq. 15.3 is then used to compute the ratio, (σ_{tr}/σ_a), as a function of fatigue cycles, N, for the unaged and the aged material and for each test temperature. For different applied stresses, σ_a, the variation of, (σ_{tr}/σ_a), as a function of fatigue cycles can be represented by a single "master curve." This master curve is approximated using a power law relationship given by [9]

$$\frac{\sigma_{tr}}{\sigma_a} = AN^{-d} \tag{15.4}$$

The variation of crack density with fatigue cycles, $\rho(N)$, for any applied stress level, σ_a, is determined using Eqs. 15.2 and 15.4 as [9]

$$\rho(N) = \frac{\alpha}{2}\sqrt{\frac{1}{2}\left(\frac{E_2}{E_o}\frac{1}{AN^{-d}}-1\right)} \tag{15.5}$$

The variation of crack density for the unaged and the aged material can thus be calculated using Eq. 15.5. Note that the power law master curve relationship of Eq. 15.4 and the variation of crack density with fatigue cycles (Eq. 15.5) are a function of the material condition (unaged or aged) and the test temperature.

15.3.2 Inter-Yarn Debonding or "Meta-Delamination" Model

As shown in Fig. 15.5, the woven composite exhibits inter-yarn debonding at the yarn crossover regions. This damage mechanism is unique to woven-fabric-reinforced composites. A simple model can be used [9] to account for the longitudinal stiffness loss resulting from this damage mechanism. Figure 15.8 depicts an equivalent inter-yarn debonding model that can be used to represent the debonding at the crossover regions in the repeating unit cell (of length, *L*) of a five-harness satin woven composite. By assuming that the composite displacements are the sum of the displacements in the undamaged region (of length, *L-a*) and the locally debonded region (of length, *a*), an expression can be derived for the stiffness, E_{md}, of the locally debonded composite [11] as

$$\frac{1}{E_{md}} = \frac{(1 - a/L)}{E_o} + \frac{(a/L)}{E_{LD}}$$

That is,

$$\left(\frac{E_{md}}{E_o}\right) = \frac{(E_{LD}/E_o)}{\frac{1}{n} + \frac{E_{LD}}{E_o}\left(1 - \frac{1}{n}\right)} \tag{15.6}$$

where E_o is the stiffness of the "undamaged" composite, and E_{LD} is the stiffness of the locally delaminated region. For simplicity, E_{LD}, can be assumed to be $E_1/2$, where E_1 is the equivalent longitudinal stiffness of the yarns [9]. For a woven composite with an "$n \times n$" repeating weave pattern, where *n* fill yarns interlace with *n* warp yarns (in the repeating unit cell), the ratio (*a*/*L*) is given by (1/*n*). For the five-harness satin weave, $(a/L) = (1/5)$. For simplicity, E_o, can be used as the "undamaged" laminate stiffness in this analysis. However, since the "undamaged" region of the weave would have matrix cracks, a more accurate value for E_o can be estimated using Eq. 15.1 if the crack density or the stiffness loss curve is available.

15.3.3 Edge Delamination Model for a Woven Composite

The edge delamination damage mechanism, which is dominant in Region III of the stiffness loss curve (see Fig. 15.6), can be modeled using an analysis similar to O'Brien's [11] edge delamination model for laminated composites. The longitudinal stiffness of the edge delaminated laminate in Fig. 15.9, can be derived by assuming that the total load on the laminate is the sum of the loads in the edge delaminated region (of width, *c*) and the rest of the laminate (of width, 2*b*–2*c*). The longitudinal laminate stiffness, E_{ed}, can be derived as [9]

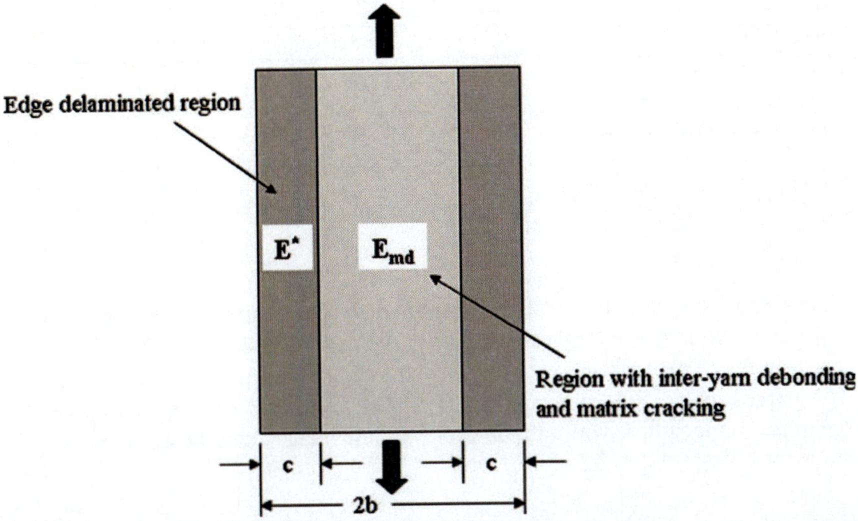

Fig. 15.9 Edge delamination model for a woven composite laminate [9]

$$E_{ed} = A_D E^* + (1 - A_D) E_{md}; \; A_D = (m\,c/b) \tag{15.7}$$

where A_D is the delaminated area fraction defined as the ratio of edge delaminated area to the total surface area of the laminate. This model assumes that edge delaminations initiate and grow equally at all ply interfaces in the woven composite. The symbol, m, represents the number of sublaminates formed by these delaminations. For the eight-ply laminate considered here, $m = 7$. The delamination length, c, represents the length of the delaminations at each of the, m, ply interfaces. This model further assumes that the delamination lengths, c, on both edges of the laminate are equal. E^* is the stiffness of the delaminated region. For simplicity, E^* can be assumed to be equal to $E_1/2$, where E_1 is the equivalent longitudinal stiffness of the yarns. E_{md} is the stiffness of the laminate which has no edge delamination, and is given by Eq. 15.6. This region contains damage in the form of inter-yarn debonding and matrix cracking. For simplicity, the effect of matrix cracking on E_{md} is neglected in this analysis [9].

The edge delamination length, c, for a woven composite laminate made by laying up woven plies with an "$n \times n$" weave pattern can be expressed as a function of the normalized laminate stiffness, E_{ed}/E_o, using Eq. 15.7 as [9]

$$c = \frac{b}{m}\left[\frac{(E_{ed}/E_o) - (E_{md}/E_o)}{(E^*/E_o) - (E_{md}/E_o)}\right] \tag{15.8}$$

where (E_{md}/E_o) is calculated using Eq. 15.6. For all the materials and conditions tested, the normalized stiffness, E_{ed}/E_o, as a function of fatigue cycles, N, is

Table 15.1 Composite properties for as-fabricated AS4/PR500, 5HS at RT [9]

Property	Symbol	GPa	msi
Undamaged laminate stiffness	E_o	64.12	9.30
Yarn equivalent axial stiffness	E_1	119.28	17.30
Yarn and interstitial resin transverse stiffness	E_2	9.03	1.31
Laminate inplane shear modulus	G_{12}	7.58	1.10
Laminate transverse shear modulus	G_{23}	3.79	0.55

determined from Region III of the measured dynamic stiffness loss curves (see Fig. 15.6). A linear relationship can be assumed in the form: $(E_{ed}/E_o) = A_{ed}$ ln $(N) + B_{ed}$. The constants A_{ed} and B_{ed} are determined using linear regression. The expression in Eq. 15.8 can then be used to compare the evolution of the delamination length, c, with fatigue cycles, in the unaged and the aged material.

15.4 Damage Mechanism Comparisons for Aged and Unaged Material

The analytical models described in the previous section can be used to characterize the matrix cracking and the delamination failure mechanisms for the unaged and aged AS4/PR500, 5HS composite. This section examines the results for the matrix cracking, both at 22°C (72°F) and 121°C (250°F), followed by results for the delamination characteristics.

The composite properties used in the calculations are tabulated in Table 15.1. The effective axial stiffness of the undulating yarns, E_1, is calculated using the rule of mixtures and the values of E_o and E_2. E_2 is estimated using constituent graphite fiber and epoxy matrix properties along with micromechanics and textile mechanics analyses described in Refs. [6–8]. The Young's modulus, E_o, for the AS4/PR500 material is reported in Refs. [12, 13] while the other properties are estimated based on typical graphite/epoxy properties.

15.4.1 Transverse Yarn Cracking

The expression in Eq. 15.3 can be used to compute the ratio (σ_{tr}/σ_a), as a function of fatigue cycles, N, (Eq. 15.4) for the unaged and the aged material. The variation of crack density can be calculated using Eq. 15.5 and compared with the measured

Table 15.2 Stiffness loss and transverse yarn cracking master curve parameters during fatigue [9]

Material	Test temperature	Stress, %UTS[a]	A_m	B_m	A	d
Unaged	22°C (72 F)	70	−0.00128	1.00592	0.0883	0.0083
Unaged	121°C (250 F)	70	−0.01164	1.05360	0.1340	0.0842
12,000 h. aged	22°C (72 F)	70	−0.00392	1.01806	0.0967	0.0278

[a]UTS = 744.6 MPa (108 ksi)

crack densities. The shear-lag parameter, α, in Eq. 15.5 is calculated using the following expression [4] for the unaged AS4/PR500 material:

$$\alpha = \frac{1}{h}\sqrt{\frac{6E_oG_{12}G_{23}}{E_1E_2(G_{12}+G_{23})}} \tag{15.9}$$

where G_{12} and G_{23} are the longitudinal and transverse shear moduli of the 5HS woven composite (see Table 15.1).

The yarn thickness, h, (Fig. 15.7) can be estimated as half the ply thickness ($h = 0.181$ mm). For the unaged material at RT (22°C), the value of, α, is calculated as 5.24 mm^{-1} (133.0 in^{-1}). The parameters, A_m, and B_m, in the expression, $(E_{mN}/E_o) = A_m \ln (N) + B_m$, used to define the relationship of the normalized stiffness E_{mN}/E_o with fatigue cycles, N, can be determined from linear regression fits to the measured dynamic stiffness loss data [9]. These parameters are listed in Table 15.2 for the three different cases analyzed.

In order to first verify the analytical model, the variation of the transverse yarn cracking with fatigue cycles needs to be calculated using Eq. 15.5. The "master curve" power law (Eq. 15.4) parameters, A and d, (listed in Table 15.2) can be determined using the measured dynamic stiffness loss for the AS4/PR500, 5HS material tested under the three different conditions (Table 15.2). The predicted crack density evolution with fatigue cycles is compared with the measured crack density evolution [12, 13] in Fig. 15.10. The comparison between the predicted trend and the measured crack density variation with fatigue cycles is reasonably good, thus, verifying the present analytical approach [9].

The shear-lag parameter, α, for the unaged material tested at 121°C and also for the 12,000 hour aged material tested at RT is determined in a slightly different way (instead of using Eq. 15.9) due to a lack of shear stiffness values at these conditions. Thus, α, for these two cases is determined as a best-fit parameter that gives a good correlation with measured crack density data. Note that the value of α merely shifts the trend predicted by Eq. 15.5 either upward or downward. Figure 15.11 shows the predicted trend in the crack density, using Eq. 15.5, for the unaged material tested at 121°C. A value of, α, equal to, 3.35 mm^{-1} (85.0 in.$^{-1}$) is found to correlate well with the measured crack density variation with fatigue cycles [9]. For the 12,000 hour aged material tested at RT, a value of α equal to 2.95 mm^{-1} (75.0 in.$^{-1}$) gives a good correlation of the predicted trend using Eq. 15.5 with the measured crack density [9], as shown in Fig. 15.12.

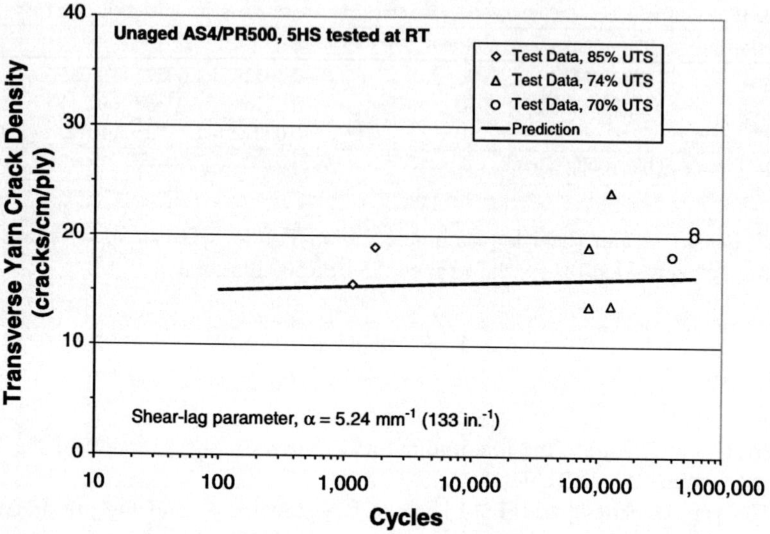

Fig. 15.10 Comparison of measured and predicted transverse yarn crack density for unaged laminate tested at RT [9]

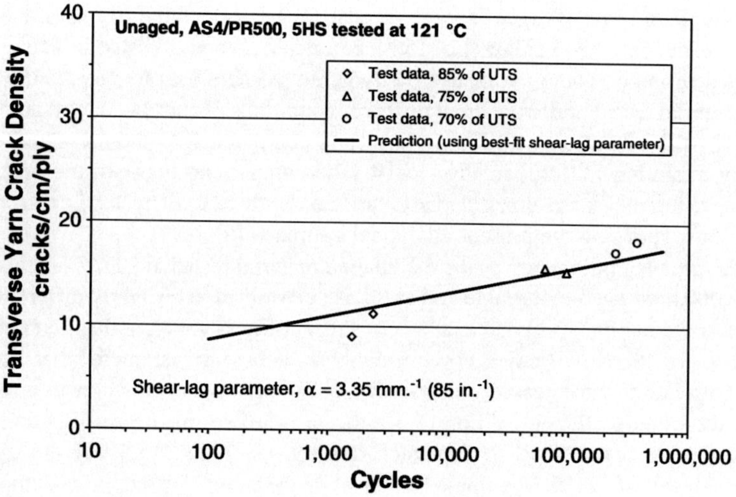

Fig. 15.11 Comparison of measured and predicted transverse yarn crack density for unaged laminate tested at 121°C [9]

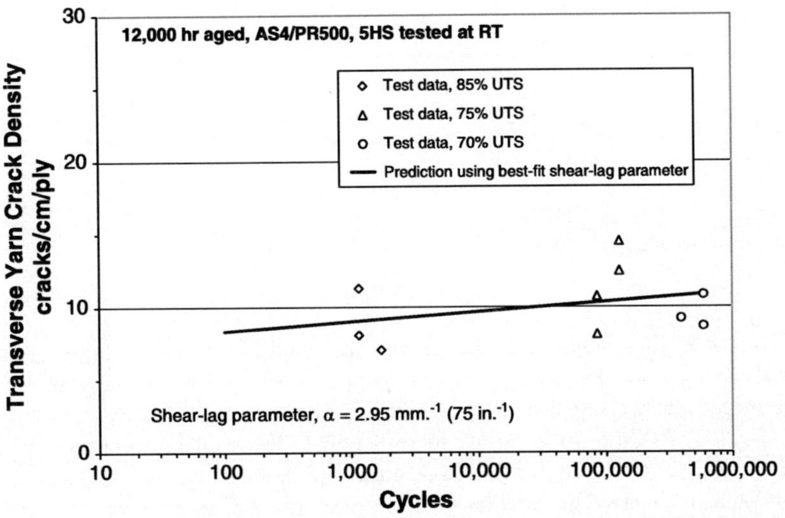

Fig. 15.12 Comparison of measured and predicted transverse crack density for 12,000 hour aged laminate tested at RT [9]

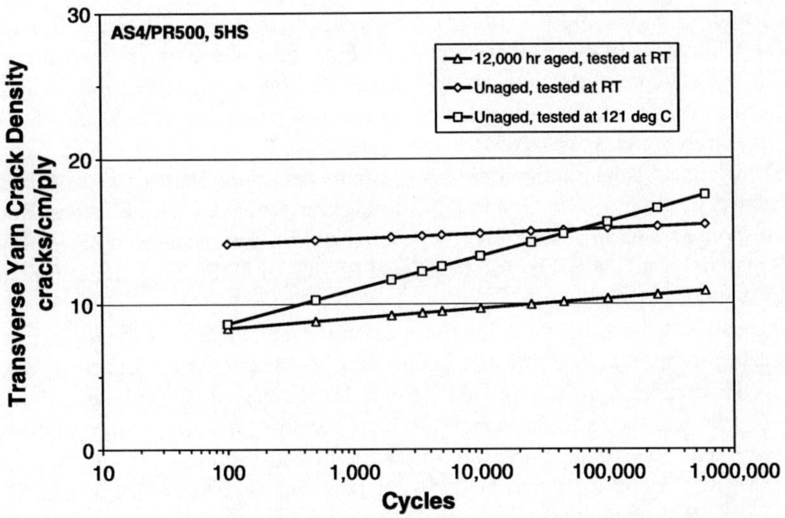

Fig. 15.13 Comparison of transverse yarn crack density for unaged and aged material [9]

Figure 15.13 compares the predicted trends in the variation of crack density with fatigue cycles for the three cases discussed above. It is interesting to note that both, the 12,000 hour aged material and the unaged material tested at 121°C, exhibit low levels of crack density in the early stages of fatigue but high crack density growth rates with

Table 15.3 Stiffness loss parameters for edge delamination (Region III) during fatigue [9]

Material	Test temperature	Stress, %UTS[a]	A_{ed}	B_{ed}
Unaged	22°C (72°F)	70	−0.0164	1.1757
Unaged	121°C (250°F)	65	−0.0845	2.0243
Unaged	121°C (250°F)	70	−0.0337	1.3406
Unaged	121°C (250°F)	75	−0.0178	1.1399
12,000 h aged	22°C (72°F)	70	−0.0282	1.2560
12,000 h aged	121°C (250°F)	70	−0.0489	1.3000

[a]UTS = 744.6 MPa (108 ksi)

fatigue cycles. In contrast, the unaged material tested at RT exhibits high levels of crack density at low fatigue cycles and a low crack density growth rate with cycles. The unaged material tested at 121°C has the highest rate of crack density increase with cycles. At 121°C, the crack density increases 620% faster than the unaged material tested at RT. Crack density for the 12,000 hour aged material increases at a rate of 95% higher than the unaged material tested at RT [9].

15.4.2 Edge Delamination

The parameters, A_{ed}, and B_{ed}, in the expression, $(E_{ed}/E_o) = A_{ed} \ln(N) + B_{ed}$, used to define the relationship (for Region III in Fig. 15.6) of the normalized stiffness E_{ed}/E_o with fatigue cycles, N, can be determined from linear regression fits to the measured dynamic stiffness loss data. These parameters are listed in Table 15.3 for the six different cases analyzed [9].

The cycles to delamination onset, N_{do}, can be determined using Eq. 15.8. N_{do} is calculated by setting, $c = 0$, and substituting the measured normalized stiffness loss response for E_{ed}/E_o (see Table 15.3), into Eq. 15.8. For the unaged AS4/PR500, 5HS material tested at RT and a stress level of 70% of UTS, N_{do}, is determined as 110,884 cycles [9].

Figure 15.14 shows the variation of the delamination length, c, with fatigue cycles calculated using Eq. 15.8 and the parameters in Table 15.3 for the unaged and the aged material. For the same condition (e.g., 121°C, unaged), delaminations initiate at higher fatigue cycles and also reach higher delamination levels, as the stress level is decreased. Note that, since the specimens tested had a width of 25.4 mm, the maximum value for c can be 12.7 mm. For the same stress level (e.g., 70% UTS), elevated temperature and aging lead to a more rapidly increasing and a larger delamination length. This result is in good agreement with experiment [12, 13]. Figure 15.14 indicates that the aged material at 121°C has a very low threshold for delamination, which initiates below 1,000 cycles. After 300,000 cycles, the delamination length at 121°C is 256% larger than that at 22°C for the aged material. The delamination length after 300,000 cycles, at 22°C was 420% larger for the aged material than the unaged material tested at 70% UTS [9].

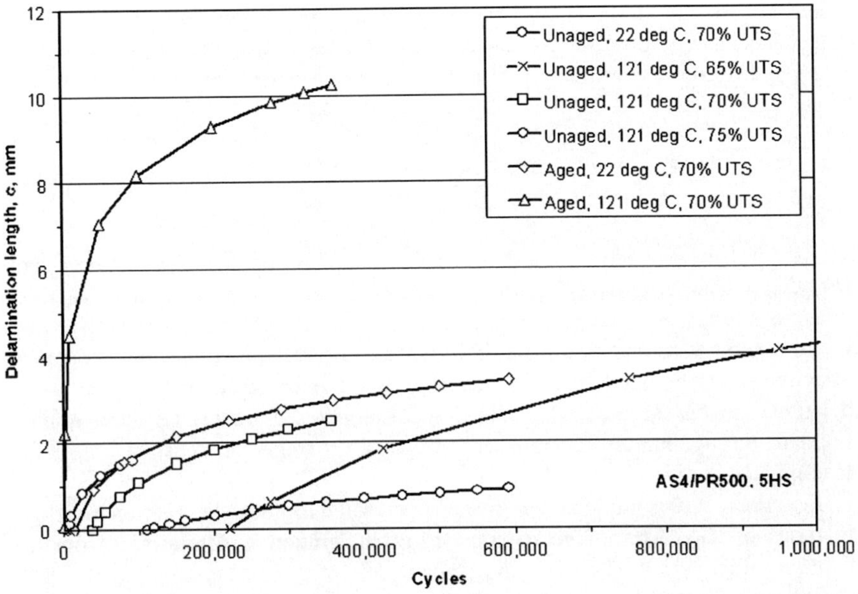

Fig. 15.14 Comparison of delamination length with cycles for unaged and aged material [9]

15.4.3 Summary of Damage Mechanism Modeling

The models developed in the preceding sections provide a simple demonstration of how it is possible to derive information about fatigue damage mechanisms (which are often difficult to characterize), using stiffness loss measurements, which are quite easy to make. These simple mechanics-based models also predict the trends in the matrix crack evolution and the delamination growth reasonably well. For the AS4/PR500, 5HS material, these models, along with the experimentally measured stiffness loss curves under fatigue, are useful in comparing the matrix cracking and delamination characteristics of the unaged and aged composite material at room temperature and also at 121°C.

These models also have the potential of being implemented into a mechanism-based fatigue life prediction analysis for woven composites as demonstrated in the following section.

15.4.4 Fatigue Life Modeling

The modeling of fatigue life can be performed using an empirical growth law for the delamination growth rate. The delamination growth rate, $(dc/dN)_{ed}$, is assumed to follow a power law relationship given by [10]:

$$\left(\frac{dc}{dN}\right)_{ed} = BG_{ed}^{\beta} \tag{15.10}$$

where G_{ed} is the strain energy release rate and the parameters, B and β are determined using standard delamination growth tests. The strain energy release rate G_{ed} for edge delamination is derived using an extension of O'Brien's [11] model as

$$G_{ed} = \frac{t\sigma_a^2}{2mE_o}\left(\frac{E_{md}}{E_o} - \frac{E^*}{E_o}\right) \tag{15.11}$$

where t is the laminate thickness, σ_a is the maximum applied cyclic stress, and (E_{md}/E_o) is computed using Eq. 15.6. Note that Eq. 15.11 is a function of (E_{md}/E_o) which is a function of the evolution of transverse yarn cracking. If we assume that the transverse yarn cracking essentially reaches a plateau and does not propagate after the onset of delamination, then (E_{md}/E_o) can be replaced by its value at delamination onset, $(E_{md}/E_o)_{do}$. Thus, according to Eq. 15.11, G_{ed} will remain constant throughout the delamination propagation stage for constant amplitude loading [10].

Equation 15.10 is integrated to give an expression for the delamination length, c, in terms of, G_{ed}. The unknown constant of integration is determined from the condition that $c \approx 0$ when $N = N_{do}$, where N_{do} is the cycles required for edge delamination onset. This results in the following expression for the delamination length, c [10]:

$$c = BG_{ed}^{\beta}(N - N_{do}) \tag{15.12}$$

The variation of the laminate normalized stiffness, (E_{ed}/E_o), with fatigue cycles can then be expressed using Eqs. 15.7 and 15.12 as [10]:

$$\frac{E_{ed}}{E_o} = \frac{mBG_{ed}^{\beta}(N - N_{do})}{b}\left(\frac{E^*}{E_o} - \left(\frac{E_{md}}{E_o}\right)_{do}\right) + \left(\frac{E_{md}}{E_o}\right)_{do} \tag{15.13}$$

Note that Eq. 15.13 includes the effects of both transverse yarn cracking (until delamination onset) and inter-yarn debonding.

A simple criterion is used for fatigue failure based on the maximum cyclic strain, ε_a, in the laminate [10]:

$$\varepsilon_a = \frac{\sigma_a}{E_{ed}} \geq \varepsilon_{cr} \tag{15.14}$$

where, ε_{cr}, is the critical strain under fatigue. Using Eqs. 15.13 and 15.14 the fatigue life of a woven composite can be expressed as [10]:

$$N = N_{do} + \frac{b}{mBG_{ed}^{\beta}\left((E^*/E_o) - (E_{md}/E_o)_{do}\right)}\left(\frac{\sigma_a}{\varepsilon_{cr}E_o} - \left(\frac{E_{md}}{E_o}\right)_{do}\right) \tag{15.15}$$

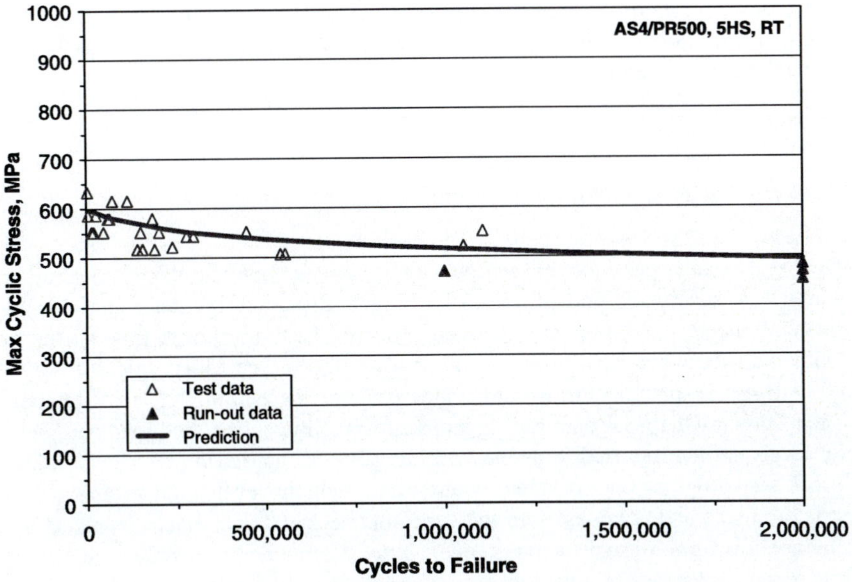

Fig. 15.15 Comparison of predicted fatigue life S–N curve with test data [10]

For the AS4/PR500, 5HS composite, the RT fatigue life can be predicted using Eq. 15.15 and the values in Tabels 15.1 and 15.2 for, E_1, E_o, E_2, A, and d. The parameter, $\beta = 5.25$, in Eq. 15.10 is obtained from Vinciquerra et al. [15] who tested a similar RTM graphite/epoxy composite under mode II fracture conditions. The parameter, $B = 8.3 \times 10^{-18}$, in Eq. 15.10, is selected to match the average measured stiffness degradation rate at 70% of ultimate tension strength (UTS). The value of, N_{do}, is estimated using an initial crack size of 0.3 mm in Eq. 15.10. The critical maximum cyclic strain, ε_{cr}, under fatigue is estimated using $N = N_{do}$ in Eq. 15.15 for a maximum fatigue cyclic stress of 80% of UTS. It is assumed that at this high stress level, delaminations would become unstable as soon as they are initiated. Figure 15.15 shows that the predicted fatigue life (S–N curve) compares reasonably well with test data [10].

15.5 Durability Assessment of an Engine Component

The assessment of durability of a composite engine component generally consists of a mix of testing and analysis. Testing alone can be prohibitively expensive because of the number of specimens needed to verify critical geometric features, loadings, environments, and failure modes. A combined testing and analysis approach can help reduce the cost and the overall effort of assessing durability.

The following sections describe, first, the basic approach used to assess durability of engine components in general and then provide an example of how it is applied to a specific engine component.

15.5.1 *Building Block Approach*

Testing is usually conducted at various levels of structural complexity, beginning with small specimens and progressing through structural elements and details, subcomponents, and finally full-scale components. Each level of testing is used to build and validate analytical models. Each level of test element complexity is also used to assess manufacturing and nondestructive inspection (NDI) techniques. Such a process that uses both testing and analysis of increasing specimen complexity levels is often referred to as the "Building Block" approach [3]. The Building Block test program begins at the coupon and element levels, and progresses in structural size and complexity to the subcomponent and component levels as the engine structural design matures over time. Components are then tested in simulated engine rigs to assess their performance in an assembled engine configuration. In the final phase of the durability evaluation, the component is placed on a ground test engine and/or a flight test engine, and removed at regular intervals to measure residual properties.

The Building Block (BB) approach, thus, starts at the basic material level and progresses to the assessment of structural capability and durability. This process of testing progressively complex sub-elements also reduces the overall risk in the development of a new component. In an effort to contain cost, only a small number of full-scale components are usually tested for durability assessment. Due to limitations in the testing equipment, the component testing often has to be load or life enhanced to account for material scatter, manufacturing process variations, and environmental effects. The load or life enhancement factors for variability and environmental effects are derived from specimen tests that can be performed using a much larger statistically significant population of samples.

In the application of the BB approach to engine components, special consideration needs to be given to thermal and out-of-plane loads that often result in matrix-dominated failure modes, such as interlaminar tension and interlaminar shear [5]. Engine components typically have complex shapes and curvatures that lead to multi-axial stress states in the composite material. It is, therefore, critical to develop experimental and analytical methods that incorporate the effects of thermal and multi-axial stresses.

As mentioned in Sect. 15.2.1, long-term durability on a component level is often assessed using accelerated test methods and modeling. It is important to validate the accelerated test methods and analytical techniques using limited full-term testing. Further confidence in the durability assessment is often established by subjecting preproduction components to actual engine operating conditions by placing them on developmental and flight test engines and removing them at selected intervals to measure residual stiffness and strength properties.

15.5.2 Certification of Durability for Airworthiness

Durability substantiation for airworthiness is accomplished by using the combined experimental and analytical approach discussed earlier. Durability tests need to account for the effects of the appropriate engine environment. The test articles need to be fabricated and assembled in accordance with production specifications and processes so that the test articles are representative of the actual production parts. Sufficient component, subcomponent, element, or coupon tests need to be performed to establish material and testing variability and environmental effects. The component durability testing is also performed with manufacturing and other embedded defects at critical locations. The durability tests have to demonstrate that the residual stiffness and strength properties remain within acceptable design limits. At each stage of the testing, appropriate analytical methods are developed and calibrated in order to help validate the design. The results of the durability tests are also used to establish component replacement and inspection schedules.

For flight safety critical components, the structural integrity and durability have to be demonstrated with a reliability that is at least as good as the "A" basis of a normal distribution. That is, a reliability of 99%, with a 95% confidence level.

15.5.3 Specific Example of Durability of an Engine Component

The engine component described here is the Fan Exit Guide Vane (FEGV) that was being considered for the Pratt & Whitney JT9D engine model [2]. FEGVs are located downstream of the fan blades in the fan bypass duct (Fig. 15.16). In this position, the vanes redirect the flow of the air from the fan blades along the axial engine direction in order to reduce aerodynamic drag and improve engine operating efficiency. The vanes are designed to withstand aerodynamic forces. They do not experience any significant engine structural loads. Failure or loss of a vane during engine operation has minimal effect on the reliability of the engine. The quantity and size of the vanes are determined by aerodynamic and noise requirements.

Vane strength requirements [2] are established by static and dynamic aerodynamic engine stresses. Static stress is primarily due to gas bending loads exerted on the pressure side of the airfoil, whereas dynamic stress is a result of high-frequency vibration of the vane during engine operation. The vanes are designed to have sufficient bending and torsional stiffness to avoid excessive vibratory stress in either the bending or torsional modes. It is also necessary to maintain sufficient

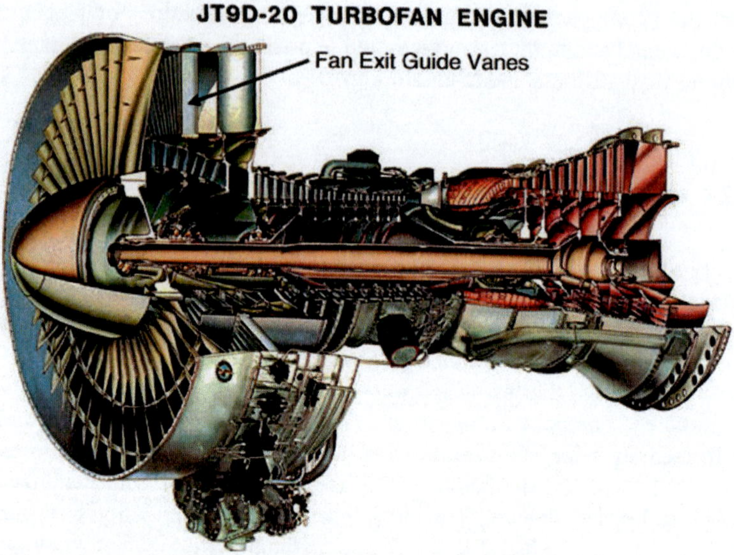

Fig. 15.16 A cutaway section of the Pratt & Whitney JT9D turbofan engine

vane natural frequency differences between first bending and torsion to avoid bending–torsion coupling. Besides these mechanical stress requirements, two basic environmental factors are of concern in establishing the durability of the composite FEGVs that typically operate between −12°C (10°F) and 88°C (190°F). These are long-term hygrothermal exposure and the influence of the absorbed moisture on vane fatigue strength at elevated temperature.

A study was conducted [2] to establish realistic temperature, percent relative humidity, and exposure times for the composite FEGVs operating in the JT9D engine environment. Worst-case average temperature and percent relative humidity conditions at airports around the world were determined to be 24°C (75°F) and 87% relative humidity. Vane moisture absorption and drying cycles were determined based on airline service data for JT9D powered aircraft with the lowest utilization rate during the past several years. The lowest utilization rate would result in the highest level of moisture absorption by the vane since the airplane would be on the ground for the longest cumulative time. This would, therefore, provide the most conservative exposure for determining vane property debits. The lowest utilization time, for one airline operator, was determined to be approximately 20% flight time. When converted to a daily cycle, this utilization was equivalent to ground humidity exposure for approximately 19 hours of every 24 hours, and flight drying conditions for the remaining 5 hours. Typical drying conditions for this 5 hours period would be approximately 15°C (60°F) and 0% relative humidity at air pressures encountered at an altitude of 10,668 m (35,000 ft). A one-dimensional Fickian moisture diffusion analysis was used to determine the required equilibrium moisture content

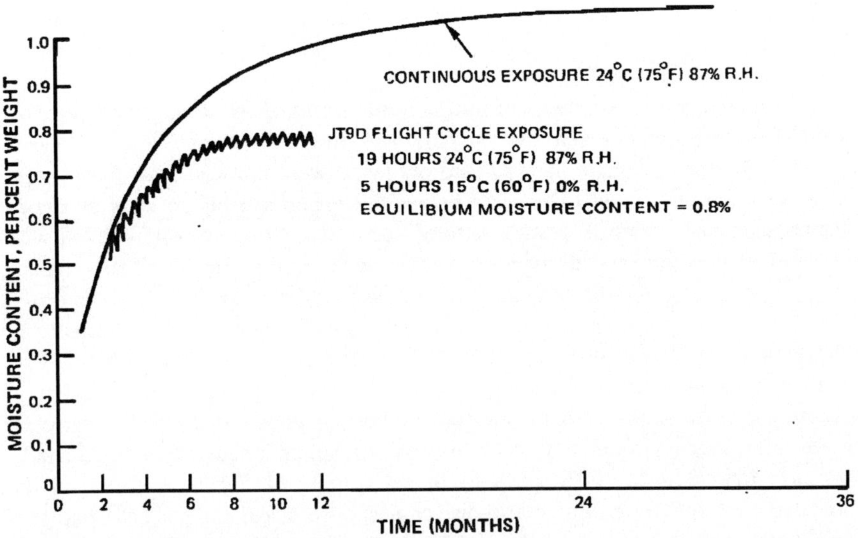

Fig. 15.17 Composite FEGV trailing edge moisture content [2]

in the thinnest trailing edge section of the vane which was 2.03 mm (0.080 in) thick. Moisture content in the trailing edge was considered to be the controlling factor since this airfoil section was also most highly stressed. Calculations for a typical graphite/epoxy (3501-6 Hercules resin) system indicated that the equilibrium moisture content in the 2.03 mm trailing edge section of the vane will be approximately 1.0 wt% after 18 months of zero flight utilization and continuous exposure at 24°C (75°F) and 87% relative humidity. However, for the lowest aircraft utilization factor of 20%, the equilibrium moisture level after 9 months of exposure was calculated to be 0.8% in the graphite/epoxy vane (Fig. 15.17).

The Building Block approach was used to characterize the durability of the FEGVs. Basic specimen testing was first performed to determine static material properties such as flexure modulus and strength and short beam (interlaminar shear) strength. The effects of moisture and temperature on these material properties were also determined. Finally, comparisons were made to material properties measured from specimens removed from actual vanes to ensure that the manufacturing process used for the contoured airfoil shape resulted in similar composite properties [2].

The moisture conditioning of the test specimens was accelerated by exposing the specimens at a higher temperature and a higher relative humidity. Mechanical property test specimens of similar thickness to the vane trailing edge were conditioned for 3–4 weeks at 60°C (140°F) and 95% relative humidity. This accelerated exposure led to the same moisture absorption level as the composite vane exposed to 9 months of normal operation (at the 20% flight utilization rate) in an engine environment. Moisture absorbed by the specimens during this period ranged from 0.4% to 0.9%.

This wet conditioning led to a reduction of short beam shear strength of approximately 10% and no significant reduction in flexure strength or flexure modulus for the graphite/epoxy material [2].

Progressing to the next level of testing in the Building Block approach, FEGVs were fabricated and tested under static and fatigue loading. Thirteen vanes representing the candidate material and ply layup were fabricated for this testing. The first vane was used to check whether the manufacturing process produced vanes that were within dimensional tolerances. Two vanes were fabricated with intentional flaws of various types and sizes at different locations. These flawed vanes were subjected to a variety of nondestructive inspection techniques to determine inspection sensitivity levels. Ten good vanes (with no intentional flaws) were fabricated in two groups of five for static and fatigue testing. Each good vane was tested sequentially under static bending and torsion loading to determine stiffness and processing quality. Subsequent to stiffness testing, each vane was either environmentally conditioned or retained in the as-fabricated condition for fatigue testing. Three vanes of each group of five were tested in the as-fabricated condition and two of five vanes were tested after accelerated hot/wet conditioning at 60°C (140°F)/95% relative humidity for 3 and 4 weeks, respectively. As discussed earlier, this accelerated conditioning would result in an equilibrium moisture level in the trailing edge section of the vane that would be equivalent to that after 9 months of engine operation [2].

Vanes were prepared for fatigue testing by mounting each vane at both ends into aluminum fixtures (Fig. 15.18). Strain gages were strategically mounted to measure surface strains at several locations. Prior to fatigue testing, the aluminum end fixtures on the vane were rotated to provide an airfoil static bending load to simulate the magnitude of aerodynamic loads experienced in the engine. This static load was increased until the desired trailing edge radial static strain was reached. The vane end fixtures were then locked in position for the duration of the fatigue test to maintain the steady static load. Vanes that were tested at 60°C (140°F) in the "wet" condition were first heated to temperature before the static load was applied. No provisions were made to control the moisture content during fatigue testing [2].

Vibrational loading was applied in the first bending mode using an electrodynamic exciter. This exciter was used to vibrate the vane at an increasing frequency level until the first bending resonant frequency was achieved. Fatigue endurance results for all the vanes were then statistically treated to establish a minimum fatigue strain capability. The durability of the FEGVs was established by demonstrating that there was a 25–45% fatigue margin when this minimum strain capability was compared to the anticipated dynamic strain during engine operation. Throughout the test program, finite element modeling was performed and calibrated using measured strains from both the static and fatigue tests. The locations of the observed failure modes were found to correlate reasonably well with the numerical predictions [2].

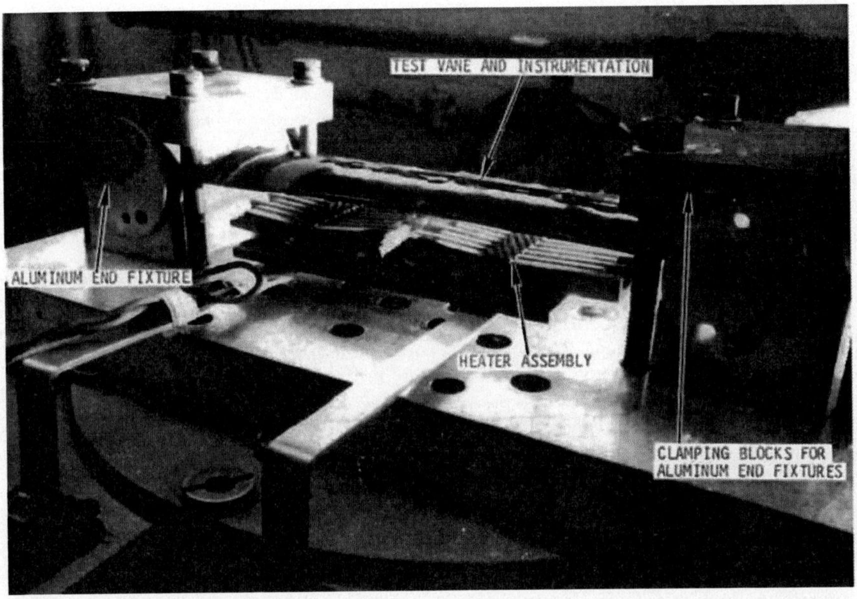

Fig. 15.18 FEGV fatigue test fixture and heater assembly [2]

15.6 Concluding Remarks

This chapter has provided a brief introduction to durability assessment of polymer–matrix composites for gas turbine engine applications. The high temperatures and complex stress states that are experienced by an aircraft engine component, necessitate a much more rigorous assessment of durability. The combined missions of hygrothermal, mechanical, and fatigue loading experienced in an engine environment need to be simplified from a practicality standpoint and simulated engine missions are often used to assess durability of composite materials. Full-term testing is also a challenge and accelerated testing is often used together with analytical modeling to assess durability. Such a combined experimental and analytical approach has been demonstrated here using a specific example of a woven composite material. The good correlation with experimental results demonstrates the feasibility of using simple mechanics-based modeling to study the evolution of the different damage mechanisms under fatigue and to predict fatigue life. Such mechanics-based modeling also augments the use of the building block approach to characterize durability of engine components. The building block approach, as it is applied today, is more focused on experimentation, but as we move into the future, an emphasis on mechanics-based modeling will enable durability assessments that are more robust, less time consuming, and also less expensive.

References

1. Adanur, S., *Handbook of Weaving*, CRC Press, Boca Raton, FL, 2001.
2. Blecherman, S. S., *Design, Durability and Low Cost Processing Technology For Composite Fan Exit Guide Vanes*, NASA-CR 159677, August 1979.
3. Composite Materials Handbook – 17, Mil-HDBK-17-3 F, Volume 3: Polymer Matrix Composites Materials Usage, Design and Analysis, June 17, 2002.
4. Lee, J-W. and Daniel, I. M., *Progressive Transverse Cracking of Crossply Composite Laminates*, Journal of Composite Materials, 1990. **24**(11): p. 1225–1243.
5. Miller, R. J., *Application of Composites in Aircraft Gas Turbine Engines – Critical Issues*. Proceedings of the ASME Aerospace Division, 1996. **52:** p. 79–81.
6. Naik, R. A. and Crews, J. H., *Micromechanical Analysis of Fiber-Matrix Interface Stresses Under Thermomechanical Loadings*, Composite Materials: Testing and Design (Eleventh Volume), ASTM STP 1206, E. T. Camponeschi, Ed., 1993. p. 205–219.
7. Naik, R. A., *Failure Analysis of Woven and Braided Fabric-Reinforced Composites*, Journal of Composite Materials, 1995. **29**(17): p. 2334–2363.
8. Naik, R. A., *Analysis of Woven and Braided Fabric-Reinforced Composites*, Composite Materials: Testing and Design (Twelfth Volume), ASTM STP 1274, R. B. Deo and C. R. Saff, Eds., 1996. p. 239–263.
9. Naik, R. A., Patel, S. R., and Case, S. W., *Fatigue Damage Mechanism Characterization and Modeling of a Graphite/Epoxy Woven Composite*, Journal of Thermoplastic Composite Materials, 2001. **14**(9): p. 404–420.
10. Naik, R. A., *Tension Fatigue Analysis of Woven Composite Laminates*, Proceedings of the 10th International Congress of Fracture (ICF10), Honolulu, Hawaii, Dec. 3–7, 2001, ICF100949OR.
11. O'Brien, T. K., *Analysis of Local Delaminations and Their Influence on Composite Laminate Behavior*, Delamination and Debonding of Materials, ASTM STP 876, W. S. Johnson, Ed., 1985. p. 282–297.
12. Patel, S. R. and Case, S. W., *Durability of a Graphite/Epoxy Woven Composite Under Combined Hygrothermal Conditions*. International Journal of Fatigue, 2000. **22**(9): p. 809–820.
13. Patel, S. R. and Case, S. W., *Durability of Hygrothermally Aged Graphite/Epoxy Woven Composite Under Combined Hygrothermal Conditions*. International Journal of Fatigue, 2002, **24**: p. 1295–1301.
14. Red, C., *Aviation Outlook: Composites in commercial aircraft jet engines*, High-Performance Composites, September 2008.
15. Vinciquerra, A. J., Davidson, B. D., Schaff, J. R., and Smith, S. L., *An Improved Methodology for the Determination of Mode II Fatigue Delamination Growth Rates Using the End-Notched Flexure Test*, Proceedings of the American Society for Composites, 15th Technical Conference, College Station, Texas, Sept. 25–27, 2000. p. 999–1007.

Chapter 16
Composite Durability and Aging in Civil Infrastructures

Dean C. Foster

Abstract As composite materials are finding increasing use in infrastructure projects, their durability over the life span of a structure needs to be quantified. The design life for a typical bridge is 50 years or more. As infrastructure demand is outstripping resources, the life of a bridge can exceed 75 years and sometimes approach 100 years. Additionally with the introduction of the American Concrete Institute (ACI) *Guide for the Design and Construction of Externally Bonded FRP Systems for Strengthening Concrete Structures*, ACI 440.2R-02, allowable stress knockdown factors are incorporated into the design guidelines for exposure to the environment. Consequently, in order to have high confidence that composite material systems, utilized as either structural component parts or for the strengthening of structures, will last the remaining life of the structure, the degradation mechanisms due to exposure to the natural environment must be known. Thus, effects due to the freeze–thaw cycles, de-icing chemicals, humidity, UV radiation, and erosion on the material behaviors need to be studied in order to establish proper design methodologies to account for these conditions over the life of the structure. This chapter will highlight the recent research for understanding composite material behavior when subjected to the naturally occurring environmental conditions. It will also highlight design methodologies suggested in the literature to properly account for these exposures over the structures' lifetime.

D.C. Foster (✉)
Composite & Hybrid Materials Branch, Materials and Manufacturing Directorate, Air Force Research Laboratory, Wright-Patterson Air Force Base, Dayton, OH 45433-7750, USA
e-mail: Dean.Foster@wpafb.af.mil

K.V. Pochiraju et al. (eds.), *Long-Term Durability of Polymeric Matrix Composites*,
DOI 10.1007/978-1-4419-9308-3_16,

16.1 Introduction

Beginning in the 1980s to early 1990s the use of composite materials for civil engineering applications was starting to gain world wide attention. The inherent strength to weight ratios, specifically of carbon fiber reinforced polymers (CFRP), the ability to with stand corrosion that plague typical materials used in civil engineering applications and their prefabricated or ease of installation uniqueness have made these material systems worthy of investigation to use on all types of civil structures. In addition to the inherent composite material qualities previously mentioned, structural components have the added benefit of being lighter in weight thus requiring less heavy equipment for installation. These attributes are quite appealing when adding to an existing structure as the foundation may not need modification. However, the most benefit from composite structural components may be in their application to the highway bridge sector of civil engineering applications. Especially in the United States where bridge infrastructure funding is inadequate to fix the aging structure demand, composite material structural components such as bridge decks and beams offer much faster construction times. Faster construction times equate to less cost to the traveling public in detour routing and also less disruption to first responder times in emergency situations. Furthermore, other composite components such as reinforcing bars add longevity to the structure reducing maintenance costs and disruption to the traveling public.

16.2 Background

The most notable early research work is that of the Swiss Federal Laboratories for Materials Testing and Research's (EMPA) Dr. Urs Meier, on the repair of concrete structures using bonded CFRP plates [1]. Commencing in 1984, CFRP sheets were bonded to reinforced concrete beams having spans up to 7 m. Testing of these specimens involved both static and fatigue loadings at temperatures between −25°C and 40°C and for some specimens in 95% relative humidity. The results of this research at EMPA under the direction of Dr. Meier resulted in the first reinforced concrete bridge strengthened using CRFP plates in 1991. Also in 1991, the Gossau St. Gall Switzerland, City Hall building and a timber bridge near Sins Switzerland were strengthened thus, demonstrating the utility of composite materials for civil structural applications.

In 1993 partially due the world wide interest in utilizing composite materials for infrastructure applications, the US Air Force Materials and Manufacturing Directorate at Wright-Patterson Air Force Base convened a meeting with state and local road and bridge officials on the benefits of composite material application to highway bridges. Discussed at this meeting was the concrete bridge repair application in Switzerland as well as structural bridge components. The results of this meeting lead to the Butler County Engineers Office (BCEO) of Butler County Ohio

Fig. 16.1 Kansas Structural Composite Inc. No-Name Creek Bridge. Note: Steel structure under bridge is for bridge deflection measurement equipment

to investigate the plate bonding techniques for the rehabilitation of deteriorating reinforced concrete bridges. Through a cooperative research and development agreement between the Air Force Research Laboratory (AFRL) at Wright-Patterson Air Force Base (WPAFB) and BCEO culminated in a field demonstration of the plate bonding technique in 1995. Concurrent to these activities were applications of utilizing composite components for bridge applications. Lockheed Missiles and Space Company of Sunnyvale, California in 1995 undertook the notion of lowest cost design to focus on applying composite materials for bridge applications [2]. This independent research work resulted in the design and fabrication of a 30′ (9.2 m) span by 18′ (5.8 m) wide glass-fiber-reinforced polymer (GFRP) bridge. The superstructure comprised three hat-shaped hand-laid beams, 3.0′ (0.91 m) deep, topped by a deck section of bonded pultruded tubes sandwiched by face sheets, resulting in a 5.5″ (139.7 mm) deep deck section. The entire superstructure weighed 23,000 lbs (10,433 kg). The American Association of State Highway Transportation Officials (AASHTO) design specification were used to obtain all loading requirements, although no load reduction intensity factor was used and impact loading was not considered. The bridge was load tested to 130,000 lbs, 180% of the required AASHTO design specifications, which equated to only 8% of the first ply failure criteria used in the design. Constructed as a demonstration project this bridge was never placed in service on a public road.

The first US application of an all composite vehicle bridge opened to traffic in Russell County, Kansas on November 9, 1996 [3]. Manufactured by Kansas Structural Composites Incorporated, the single-span structure crossing No-Name Creek 3 miles west of Russell, Kansas is shown in Fig. 16.1. It consists of three sandwich panels that when bonded together make the bridge a total of 27.75′ (8.45 m) wide with a span of 23.25′ (7.08 m). Each panel had a 0.75″ (19.1 mm) wearing surface applied prior to installation. Constructed of glass-fiber-reinforced polymer composites, the sandwich panels are comprised of corrugated pattern honeycomb core and face skins having a total depth of 22.5″ (571.5 mm). Designed to AASHTO specifications, the center panel was proof tested prior to installation at No-Name Creek. The maximum deflection was 0.442″ (11.23 mm) at a loading of 60.58 kips (269.5 kN) giving a L/d of 1,165 much greater than the 800 requirement. Set on neoprene bearing pads atop reinforced concrete abutments, the total superstructure installation time was accomplished in 10 hours, spread over 2 days commencing on November 7, 1996. The outer panels had the guardrail assemblies installed prior to field construction. Once the adhesively bonded joints and subsequent placement of the wearing course over the joints was allowed to cure, the bridge was opened to traffic. The total construction time was much shorter using a composite superstructure than using conventional materials and techniques.

Load tests were conducted on November 19, 1996, 10 days after opening to traffic, and again on February 20, March 13, April 3, and May 8 in 1997 [4]. The mid-span deflections from the 40 kip (177.9 kN) rear axle load of a fully loaded dump truck showed no difference between the dates the tests were made. Although being in service only 6 months, the time in service was during the winter months under the freezing and thawing conditions of a Kansas winter. Furthermore, the bridge was subjected to the typical snow and ice removal techniques employed by the local Department of Transportation. Thus, the trend of this relatively short service time indicates that the fiberglass material system and fabrication techniques used to manufacture the structure are well suited for civil bridge applications. In addition, since its deployment routine inspections has revealed that the bridge has been used for target practice by the local hunters. Visual inspections indicate no lasting damage from the gun shots. However, no additional data have been published since 1997 but according to Kansas Department of Transportation officials, the bridge has been inspected routinely since installation and is performing well [5].

The second application of an all composite vehicle bridge in the USA was in Lewis County, West Virginia in May 1997. This bridge over Laurel Lick is comprised of a pultruded modular composite deck on FRP wide flange beams. Spanning 20′ (6.1 m) and being 16′ (4.88 m) wide, the superstructure sits on top of a reinforced concrete capped pile abutment. The piling supporting the abutment is also made of FRP materials. The pultruded modular deck is Creative Pultrusions Superdeck™ product made from glass fibers in a vinyl–ester matrix. Consisting of truss and hexagonal profiles, 8″ (203.2 mm) deep, bonded together, it can be made into sections as an increment of the span length with 8′ (2.44 m) wide sections being the maximum width easily transportable to the bridge site [6]. Due to the pultrusion method of fabrication, the length of the profiles is equal to the width of the bridge. Thus, the deck sections were installed transverse to the direction of the superstructure beams. The superstructure beams are a pultruded Creative Pultrusions Pultrex® wide flange section product consisting of glass fibers in a vinyl–ester matrix spaced 2.5′ (0.76 m) apart. The Superdeck™ is field bonded to the superstructure beams using a high strength adhesive. Although this project was a bridge replacement, construction of the superstructure took considerably less time than it would have if conventional bridge building materials had been used. In addition, less required construction equipment was needed due to the lightweight nature of the composite materials used.

The Laurel Lick bridge was designed to AASHTO specifications and prior to construction, the deck was fatigue tested to two million cycles [7]. Test results showed that after two million cycles under a fatigue load range of 33 kip (147 kN), the difference in failure load for the fatigued deck section was 4% lower and the non-cycled deck section. The measured deflection at center span under failure load was 10% greater for the fatigued deck than the non-fatigued deck. Ultimate failure of the non-fatigued deck section occurred at 129.66 kips (577 kN), whereas the post-failure ultimate load was 35.06 kips (156 kN). The AASHTO design live load was 16.3 kips (72.5 kN). Both fatigue tests and ultimate failure load tests used a simply supported configuration having a span of 36″ (915 mm) with the load applied at center span. All applied loads were distributed over a 10″ (254 mm) by 20″ (508 mm) area to simulate wheel loading. The end supports were rollers so not to induce end constraints that would artificially enhance the loading effects. The primary observed failure mechanisms was longitudinal shear failure along the entire deck section in the joint between pultruded profiles [7]. Post-failure analysis indicated that failure initiation was not due to the adhesive but rather interlaminar shear in the webs of the profiles [7]. Although all tests were conducted at room temperature and the residual load capacity after failure was much greater than the live load wheel loading, the durability of this material system due to environmental exposure factors was not evaluated prior to installation. It is not known whether this bridge is still in service since this project served only as a remedial solution.

On July 25, 1997 the third all composite vehicle bridge constructed in the USA was opened to traffic. Dubbed the Tech 21 Bridge for materials technology for the twenty-first century, it is located on Smith Road in the southwest section of the city of Hamilton, Ohio. Manufactured of E-glass fibers in an isothalic-based polyester (iso-polyester) resin, this structure is the second-generation design of the Lockheed Missile System bridge described previously. It is a 33′ (10.06 m) single span by 24′ (7.32 m) wide structure crossing a small urban stream. The superstructure is comprised of three trapezoidal box beams, bonded to a pultruded deck configuration. Each beam has a segment of deck bonded to it prior to construction of the bridge as shown in Fig. 16.2. The bridge rests on neoprene bearing pads atop reinforced concrete abutments. Figure 16.3 shows all three of the sections installed on the abutments. The entire structure came in three pieces and arrived at the construction site on one flat bed tractor trailer as shown in Fig. 16.2. It was erected in less than 1 day, each section being field bonded together [8].

The entire bridge is coated with a tan gel coat to guard against degradation from UV exposure. The wearing surface on the bridge is a typical hot-mix asphalt that has a temperature of approximately 325°F (163°C) when applied. Although the asphalt cools rapidly after placement and during compaction rolling, the application temperature is close to the glass transition temperature of the matrix material. This structure was the first to use a typical hot-mix asphalt wearing surface.

Fig. 16.2 Tech 21 Bridge delivered on flat bed truck

Fig. 16.3 All three sections of the Tech 21 Bridge installed

All others up to this point have either used a polymer modified concrete, cured under ambient conditions, or a polymer modified asphalt, cured at a much lower temperature than hot-mix asphalt, as a wearing surface. Thus, to ensure no material degradation due to the temperature of the asphalt application on the deck, sacrificial plies of the glass fiber composite were placed on the top surface of the deck.

Fig. 16.4 Load testing using two trucks

Designed to AASHTO specifications and based on previous experience, the highest stresses are located in the web of the beams where the Tsai-Hill failure criterion is 0.12 and the demand capacity ratio using a factor of safety of 4 is 0.48 [9]. On September 18, 1997 a load test was conducted as a baseline for long term health monitoring [10]. Twenty-eight strain transducers and four linear varying displacement transducers were placed at center span. Two trucks, one tandem rear axle having a total loaded weight of 47.9 kips (213.1 kN) and a single rear axle having a total loaded weight of 30.8 kips (137 kN) were used. The maximum deflection measured from the initial load test was 0.302″ (7.67 mm). Based on analysis calibrated from the load test results the bridge satisfies all AASHTO deflection criteria load limits. Subsequent load tests were conducted on January 28, 1998 and on August 11, 1998. For the January tests the instrument plan was similar to the one used on September 18 and one tandem rear axle truck having a total loaded weight of 43.9 kips (195.3 kN) was used. Results were the same in that the structure satisfied all AASHTO deflection criteria load limits. Although the January test was specifically scheduled for that time to capture any effects due to weather, it was a relative mild day with the ambient temperature of 52°F (11°C), providing only a difference of 28°F (−2°C) with the initial load test. Thus, the test results and analysis show no effects due to weather 6 months after installation. For the August load test two tandem rear axle trucks having a total loaded weight of 72.0 kips (320.3 kN) and 70.65 kips (314.3 kN) were used as shown in Fig. 16.4. The maximum mid-span deflection measured when both trucks were on the bridge was 0.48″ (12.19 mm). The AASHTO specification maximum allowed deflection for a 72 kip design vehicle is span/800 equating to 0.49″ (12.45 mm). Thus, with almost twice the design load, the maximum deflection limit criterion was not exceeded. Even though the Tech 21

Bridge was designed to AASHTO specification and load tested three times over the course of its first year in service, the effects of the environment on the material system is still not understood nor is it properly taken account of during the design process. However, recent visual inspection of the structure does not indicate that any deterioration is occurring after 13 years in service.

It is apparent from the successful demonstration projects utilizing FRP materials involving both repair of bridges and entire bridge superstructures and the numerous research projects being conducted by academia occurring from the early to mid-1990s, the need for understanding environmental degradation is crucial. Out of this need, an ad hoc review committee was formed in the late 1990s from the efforts of the American Society of Civil Engineers (ASCE), Civil Engineering Research Foundation (CERF), and the FRP composite industry's Market Development Alliance (MDA). This ad hoc committee was comprised of academia, government agencies, and private sector engineers and began addressing the need to understand the long-term effects of the environment on composite materials used in civil engineering applications. The results of their efforts were the *Gap Analysis for Durability of Fiber Reinforced Polymer (FRP) Composites in Civil Infrastructure* report edited by Vistasp M. Karbhari from the University of California, San Diego; Joennie W. Chen from the National Institute of Standards and Technology (NIST); and David Reynaud from the Civil Engineering Research Foundation published in 2001 [11]. This ad hoc group set out to identify the critical needs necessary for composite materials to be readily accepted, just as conventional materials are, for use in civil engineering infrastructure applications. Through a series of workshops led by the end users, the goal was to reach consensus as to the priority of applications and impediments from both a technical and nontechnical nature for the widespread acceptance of composite materials for civil applications. As a starting point, applications were broadly classified into bridges and bridge decks, concrete and masonry reinforcement, and primary and secondary structural members [11]. The workshops identified rebar, external reinforcement of concrete structures, seismic retrofit of concrete and masonry structures, replacement and new bridge deck systems, and wall panels and profiles as priorities for the near-term resolution of durability concerns [11]. To further refine the scope, the applications were limited to those identified as having the highest near-term market acceptance. Thus, the following products were studied for the gap analysis and thus would provide a clear path for future research, assessment, and data collection [11]:

1. FRP rebar alternative to steel reinforcement in concrete
2. FRP jackets/wrap for columns, piers, and walls seismic retrofit
3. External repair and strengthening for beams, slabs, and columns
4. FRP bridge decks
    a. On conventional material beams
    b. On composite material beams
5. FRP structural wall panels and profiles

In addition, the ad hoc committee also addresses the definition of the term "durability" for the purposes of the gap analysis. Thus, the durability of a composite material product shall be defined as the following: "[the] ability to resist cracking, oxidation, chemical degradation, delamination, wear and/or the effects of foreign object damage for a specified period of time, under the appropriate load conditions, under specified environmental conditions" [11].

The gap analysis attempted to conduct a thorough literature search of information relating to the durability in the aforementioned focused applications. Through the evaluation of the available data, the study attempted to identify the critical gaps and prioritize the research needs, thus formulating a strategy to satisfy the gaps. The research areas that the ad hoc committee prioritized for each application are divided into the following seven areas:

1. Effects of moisture/solution on composite properties
2. Effects of an alkaline environment on composite properties
3. Thermal effects from materials processing and due to service life
4. Effects from fatigue loading
5. Material creep and relaxation effects
6. Ultraviolet (UV) radiation effects
7. Effects of fire

Another collection of transportation officials spanning across the federal, state, and local Department of Transportation (DOT) agencies and including academia is the Transportation Research Board (TRB) AFF80 – Structural Fiber Reinforced Polymer Committee. The TRB AFF80 committee began in the late 1990s to address many of the same issues by the ASCE CERF – MDA ad hoc durability committee. Furthermore, the TRB AFF80 committee also began investigating issues due to the lack of design specifications and guidelines. This committee provided a forum for transportation officials, academia, and private enterprise to discuss and prioritize projects for the use of composite materials for highway structures. The prioritized projects were then used to solicit funding from various sources having the backing of the committee. They also sponsored several sessions to present research findings at the annual TRB meetings held in Washington D.C. every January. Many of the projects from the committee were funded through the National Cooperative Highway Research Program (NCHRP). In addition, the annual committee meeting provided collaboration opportunities and was a source of the latest research being conducted by the various agencies. It also provided status updates on the development of the necessary design specifications being developed by the FHWA and the American Concrete Institute (ACI). NCHRP Report 503: Application of Fiber Reinforced Polymer Composites to Highway Infrastructure attempted to identify a path for wide acceptance for FRP utilization for highway structures [12]. It identified the major barriers as being the following:

1. Lack of knowledge of composite material systems by the bridge engineers
2. High cost of materials
3. Lack of bridge-specific material specifications

4. Lack of bridge design guidelines
5. Lack of easy and reliable inspection techniques
6. Lack of repair procedures
7. Lack of support or acceptance from governing government agencies

Through literature review, a questionnaire sent throughout the USA to federal, state, and local Department of Transportation's and through the authors own experience within the civil highway infrastructure and composites communities, the report develops a strategic plan addressing the seven barriers. Recommendations for advancing the acceptance of FRP material systems for highway bridges are the retrofitting of concrete and steel bridge components, seismic retrofitting of bridge piers, special application bridge decks, and internal reinforcement for concrete structures.

Thus, it can be seen that starting in the late 1980s and continuing through the 1990s, a plethora of demonstration projects and research by academia has proven that composite material systems can be successfully applied to highway bridge structures. Through these successful projects, momentum was gaining for continued application within the civil infrastructure industry. However, this emerging material system for bridge structures needed further research and understanding to develop the required specifications and design guidelines necessary for widespread acceptance by bridge owners. This hurdle was recognized by the CERF-MDA Gap Analysis and in the NCHRP Report 503. In addition, cost played a major factor. Because the demonstration projects were unique single applications, it was understood that cost reductions through economies of scale would be realized. However, would that be enough to offset the high cost of the raw materials when compared to conventional materials used in highway structures? Thus, would benefit–cost analysis per project be required to justify the high initial cost of reinforced composite material systems that in some cases was more than two or three times that of conventional material systems? Consequently, one prevailing thought through the civil infrastructure arena was that without adequate specifications and design guidelines, composite material applications for bridges would lose the momentum gained by the successful demonstration project illustrated above.

16.3 Current State of Affairs

Due to the momentum for the application of composite material systems afforded by the successful demonstrations projects, the applications were becoming focused into repair and bridge elements. This was due in part to cost of the composite material systems being utilized but mainly due to risk reduction. Since the repair of concrete structures started the movement of the application of composite material systems to civil infrastructure, the understanding of how these material systems restored the strength to the deteriorated structure or increased the strength due to usage change was becoming better understood than the application of bridge components. So much so that in October 2002 the ACI issued the *Guide for the Design and Construction of Externally Bonded FRP Systems for Strengthening Concrete Structures, ACI 440.2R-02* [13]. Prior to ACI 440.2R-02, the Canadian Network of Centers of Excellence on Intelligent Sensing for Innovative Structures (ISIS) issued Design Manual No. 4, *Strengthening Reinforced Concrete Structures with Externally-Bonded Fibre Reinforced Polymers* [14]. In March of 2001, the International Federation for Structural Concrete (fib) issued its progress report on the Design and Use of Externally Bonded FRP Reinforcement (FRP EBR) for Reinforced Concrete Structures [15]. Each of these documents represents guidelines on the design and use of externally bonded composite material systems for the purpose of strengthening a reinforced concrete structure. They were developed out of the current state-of-the-art research and demonstration projects that have progressed over many years dating as far back as the late 1970s. Each guideline is a living document covering the constituent material systems of the composite material utilized for the strengthening to include mechanical properties, composite element fabrication and installation techniques; construction requirements such as shipping, storage, handling, quality control, acceptance, maintenance and repair; design recommendations including flexural, shear and axial compression strengthening; drawing details and specifications, and design examples. Each guideline also addresses the durability of the material system considered for the strengthening project. Thus, the use of environmental knock-down factors taking into account the durability issues whereby limiting the ultimate strength of the material system utilized is generally accepted by the practicing engineer. Most of the strengthening materials are in the form of pre-cured laminate or rod products fabricated in a tightly controlled manufacturing environment. The most common fiber in composite materials used in strengthening projects is most likely carbon due to its high strength and known high durability. The resin system is usually a very durable epoxy matrix. Thus, combining the selection of materials with known high resistance to environmental factors with a tightly controlled manufacturing system leads to a very durable product. Due to these factors, a large segment of the latest research on durability is focused on the bond line degradation mechanisms. However, new developments in anchoring techniques have made the bond line almost irreverent [16]. Although some strengthening systems use prepreg composite material systems applied to the deteriorated section of the structure and cured in ambient conditions, NCHRP Report 514 Bonded Repair and Retrofit of Concrete Structures Using FRP Composites provide recommendations for construction specifications and process controls to ensure that the repair and/or retrofit performs as designed that addresses wet layup as well as pre-cured laminate materials [17]. NCHRP Report 609 Recommended Construction Specifications and Process Control Manual for Repair and Retrofit of Concrete Structures Using Bonded FRP Composites further investigates surface preparation, FRP defects, and the range of conditions that could possibility be encountered during installation of the bonded composite repair that were discussed in NCHRP 514[18]. NCHRP Report 609 supplements NCHRP Report 514.

Consequently, the civil infrastructure repair and retrofit using composite material systems has defined a set of specifications and design guidelines addressing the durability of FRP material systems for these applications. It should be noted that an emerging repair use for carbon-fiber composite systems is for deteriorated steel structures but more research is needed to understand material compatibility issues as well as degradation mechanisms of the host and repair materials. Thus, this application lags behind that for concrete structures in the development of appropriate design guidelines and specifications.

Composite material civil infrastructure elements comprised mainly of bridge decks and reinforcing bars or grids were identified, via the Gap Analysis, as a composite materials product that had the highest potential for high utilization in the civil infrastructure arena [11]. These products are predominately fabricated from glass-fiber-reinforced polyester or vinyl–ester resin systems (GFRP). Although manufactured in tightly controlled processes resulting in minimal internal flaws, the constituent materials are known to have questionable durability characteristics. However, with all engineered material systems the proper selection of resin matrix materials can withstand the harsh environments attacking the material system. Although bridge decks can be isolated from contact with concrete through neoprene bearing pads and expansion joints, they are still exposed to ambient environmental conditions. Whereas, GFRP rebars are embedded in the concrete where they are exposed to moisture for longer periods of time as the concrete cures and over the lifetime of the structure. Furthermore, the chemicals contained in the moisture embedded in the concrete are more severe than exposure to ambient conditions. Even though the ambient conditions can be harsh such as freeze–thaw, de-icing salts, UV exposure, fatigue loading from applied loads, as well as temperature fluctuations through seasonal changes, these materials still offer tremendous advantages over conventional materials with high strength-to-weight ratios, higher resistance to corrosion that affect conventional materials and their prefabricated or ease of installation uniqueness. It is the understanding of the environmental degradation mechanisms and the corresponding design techniques to compensate for environmental exposure that are not established ensuring the material's performance over its expected lifetime.

Consequently, the remainder of this chapter will focus on the environmental durability of FRP products that have the potential of high utilization within the civil engineering infrastructure arena. Unlike FRP materials used for repair of concrete structures that have several design guidelines and specifications, only GFRP reinforcement has a specification guideline in that the *AASHTO LRFD Bridge Design Guide Specification for GFRP-Reinforced Concrete Bridge Decks and Traffic Railings* was recently published in 2009. Still, the constituent materials comprising the GFRP products are complex systems where the degradation mechanisms are not well understood especially in the infrastructure environment. Testing techniques or methodologies are also not completely standardized to address the long-term effects to environmental exposure. For example, GFRP rebars embedded in concrete are immersed in moisture as the concrete cures and through moisture migration through ambient conditions over the lifetime of the structure. The moisture within the concrete has a high alkalinity contact. Thus to test the long-term effects of this moisture exposure, how long is the GFRP bar immersed in the alkaline solution and at what pH should the solution be? A worst-case scenario would be to soak the bars in a high pH solution and study the degradation of the material system over time. However, is this methodology actually occurring in the field? Data analysis and interpretation is also an issue, although there appears to be a significant amount contained in the literature due to the high interest in the topic. Putting this amount of data into a form readily usable for the practicing civil engineer would be a monumental task unto itself. Testing under combined effects of loading and the environment has not been widely done. In addition, the assessment and characterization of the products as a function of their processing need to be taken into account. That is, what effects does the degree of cure have on the environmental degradation mechanisms subjected to the combined effects of environmental and applied loading? Therefore, it is under these conditions that a review of the relative research conducted over the last 5–6 years, in an attempt to provide additional insight and details for the practicing engineer, is conducted. The goal is that the literature review provided below will point the engineer to more detailed information that will enable the design utilizing FRP material in civil infrastructure projects to achieve the intended design lifetime. The literature review will cover FRP reinforcement and highway bridge deck applications. It will also contain a review of a life-cycle model that would aide in the decision-making process for the application of FRP materials.

16.4 Composite Material Products Durability

The products can be classified into reinforcement for concrete structures and bridge decking, although the bridge decking can be applied to parking structures. FRP reinforcement products are mainly reinforcing bars and or grids that are embedded in concrete structures providing either the main load-carrying reinforcement or temperature reinforcement. The structures can be, either, highway bridges, buildings, retaining walls, foundations, or cladding for buildings. Many of these reinforced concrete structures are exposed to or constructed in severe environments, thus the degradation mechanisms are aggressive. In most of these situations, the structure, using conventional materials, does not achieve its life expectancy before repair, rehabilitation, or even reconstruction occurs. In most cases the root cause is the corrosion of the reinforcing steel. Thus, due to the noncorrosion characteristics of GFRP reinforcing bars, structures utilizing this material system for their reinforcement would appear to achieve their design life expectancy. However, GFRP material systems fail due to degradation caused by the same forces attacking the steel reinforcement. In order to achieve the structure's service life, understanding of the degradation mechanisms and the proper design accounting the degradation and all loading cases must be performed. To this end, is a review of a collection of research work studying the durability of GFRP reinforcing bars. This literature collection is not meant to be the end all but to illustrate the ongoing work in this research area and to give the design engineer good sources to account for the degradation of GFRP material system used for reinforcement in reinforced concrete structures. Furthermore, this collection of recent research work will give the design engineer a better understanding of the data behind the *AASHTO LFRB Bridge Design Guide Specification for GFRP-Reinforced Concrete Bridge Decks and Traffic Railings.*

16.4.1 Durability of GFRP Reinforcing Bars

In *Durability of GFRP Bars: A Critical Review of the Literature*, Nkurunziza et al., identified the durability of GFRP reinforcements as the most critical topic of research and the biggest obstacle to their broad acceptance in the civil infrastructure arena [19]. In this review, Nkurunziza et al. point out that prior studies have used almost two distinctive different reinforcing bar products with respect to their material properties. They contend that due to improvements in manufacturing techniques, the products produced today are far superior than those made 5–10 years prior. Thus, the results shown in the literature must be interpreted with respect to the quality of product tested. They also discuss the design codes in North America and Japan with respect to the knockdown factors used for environmental conditions under load. They conclude that in some cases the reduction factors make achieving the superior properties of FRP products practically impossible and suggest that these factors may be too conservative. They cite one reason for the conservativeness of the reduction factors as being that the tests were done on poor quality specimens due to the time frame in which the research was conducted. In addition, the need exists for developing accelerated test methods that simulate long-term environmental exposure conditions within a practical time frame suited for laboratory experimentation. Thus, tests performed on GFRP rods may not accurately reflect field conditions and in most cases are more aggressive than what occurs in the field. Therefore, data from these studies have a tendency to be more conservative. To support this reasoning, Nkurunziza et al. cite a study in 2004 by ISIS Canada to study the performance of GFRP bars used in demonstration projects. Results of this effort showed no visible degradation of the GFRP reinforcement used in these structures having been exposed to natural conditions over 5–8 years. Conclusions from the ISIS Canada study indicated that under tension GFRP reinforcement is durable and highly compatible with concrete. Consequently, the study results led to an increase in the performance factors used by the Canadian Highway Bridge Design Code (CHBDC).

Nkurunziza et al. [19], next report on the degradation mechanics of the material system identifying the matrix, the fiber, and the fiber–matrix interface as the three main entry paths of deterioration to the GFRP bar. Most critical degradation, Nkurunziza et al. report, is the presence of alkaline solutions which lead to loss of tensile strength and elastic modulus. For the glass fiber in a GFRP bar, resistance to alkalis depends upon the composition of the fiber. By adding an optimum amount of zirconium dioxide (ZrO_2) to the glass fiber, its resistance to alkalis increases. From a systematic standpoint in order for the glass fiber to experience deterioration, the attacking agent has to pass through the matrix. Thus, an impervious matrix such as a high-performance vinyl–ester resin system can greatly improve the resistance to alkaline attack. The fiber–matrix interface is also susceptible to degradation and usually involves a combination of chemical attack and high mechanical loading. Resistance to fiber–matrix deterioration is usually accomplished by a high-strength interface through the use of coupling agents applied to the fiber surface. Therefore, it can be seen that through the use of a high-strength coupling agent and high-performance vinyl–ester resin system, GFRP bars can have a high resistance to alkaline attack commonly found when used in concrete structures.

Thermal effects also contribute to the degradation of the GFRP bars as the temperature near the T_g of the matrix. Prolonged thermal aging near this temperature, in combination with sustained loading, can cause reduced mechanical properties of the matrix. However, Nkurunziza et al. report that temperatures elevated between 30°C and 40°C do not significantly affect the strength or elastic modulus of the majority of glass fibers. Thus, applications where the maximum sustained temperature is near the T_g of the matrix system should be avoided in all cases.

Creep and relaxation of GFRP bars need to be properly accounted for to determine the long-term performance of the material system. Nkurunziza et al. report that most prior experimentation was done on a polyester resin system not a more durable vinyl–ester resin. However, Nkurunziza et al. report that a PhD thesis at the University of Sherbrooke Quebec, Canada in 2005 conducted 330 accelerated aging tests on 9.5 mm, 12.7 mm, and 15.9 mm diameter E-glass/vinyl–ester reinforcing bars. These bars were conditioned in simulated concrete pore water having a pH between 12.6 and 12.8 and distilled water at 23°C, 40°C, and 60°C for 150 and 300 days. Residual tensile and flexural properties were evaluated after 150 and 300 days. Test results show that the residual flexural strength showed a dependence on both duration and temperature conditioning and that change in the flexural modulus was not significant. However, all 300 day specimens aged in the alkaline solution and distilled water at the 60°C temperature exhibited fiber degradation and matrix cracks. Nkurunziza et al. did note that these tests are not indicative of real-world conditions and that the degradation will occur at a much lower rate in the field.

Freeze–thaw cycles also contribute to the degradation of the GFRP bars. Nkurunziza et al. reported a study of concrete beams using GFRP bars subjected to a sustained four-point bending load at 1.4 times the cracking moment of the beam and freeze–thaw cycling between −20°C and +20°C with 50% relative humidity. After conditioning, the reinforced concrete beams were tested to failure in four-point bending. After this study it was concluded that the freeze–thaw cycling has no affect on the flexure behavior of the beams after 100, 200, and 300 cycles. Furthermore, the high sustained loading did not show any significant effect on the flexure strength of the beams after 7, 13, and 26 weeks.

Nkurunziza et al. also list the knockdown factors on the design tensile strength due to environmental conditions suggested by the ACI, Canadian, and Japanese reinforced concrete design codes. Across these three codes, the reduction due to environmental degradation only ranges from 0.7 to 0.08. The Canadian code is the only one to suggest a reduction due to sustained loading being 0.8–1.0. All three codes suggest reducing the guaranteed tensile strength of GFRP bars 70% and 80% due to the combined effects of sustained load and environmental conditions. All three codes also address the creep of the material system by limiting the design tensile stress.

Nkurunziza et al.'s critical review of the literature concerning the durability of GFRP reinforcing bars offers a substantial amount of useful information to the design engineer. The authors do a good job of explaining the degradation mechanism, listing the causes and advances in technology to combat the deterioration of GFRP bars used in reinforced concrete. In the concluding remarks, Nkurunziza et al. recognize that the durability tests cited in the review on the latest generation of GFRP bars subjected to stress higher than design limits, combined with aggressive mediums at elevated temperatures, have concluded that the strength reduction factors adopted by the current codes and guidelines are conservative. The factors adopted by the current codes and guidelines are based on few test results carried out on early generations of GFRP bars that have substantially evolved. Furthermore, accelerated testing techniques are very conservative and that tests more representative of actual field conditions are needed to accurately predict the long-term durability of the GFRP bars.

Another study investigating the long-term durability of GFRP bars was conducted in 2004 by Almusallam et al. [20] in which reinforced concrete beams were subjected to sustained loading under moist environments at constant temperature of 40°C. A total of 36 100 mm square by 2,000 mm long concrete beams were made using a single 10 mm diameter GFRP bar placed at the cross-sectional center 25 mm from the extreme tensile face of the beam. No stirrups were used as the concrete was designed to prevent shear failure during experimentation and stress the GFRP bar 20–25% of ultimate capacity under a reasonable load. In the center 700 mm of the beam, high-alkali cement past was used around the GFRP bar. The cement paste increased the GFRP exposure to the alkalinity normally contained in concrete and also made extraction of the bar easier. One-third of the beams were submerged in tap water and one-third in seawater both at a constant temperature of 40°C. The last third of the beams were subjected to wet–dry conditions where they were submerged in seawater at 40°C for 2 weeks then removed and allowed to dry for 2 weeks. These conditions were kept over three time periods of 4, 8, and 16 months. Half of the beams in all three conditions had sustained four-point bending loads applied over the three time periods causing the GFRP bars to be loaded at 20–25% of their ultimate capacity. After conditioning, the GFRP bars were removed from the beams and tested in tension until failure. Unconditioned bars straight from the manufacturer were also tested and used as control specimens.

Test results from Almusallam et al. show that the tensile strength decreases over time due to the environmental conditions as well as sustained load as shown in

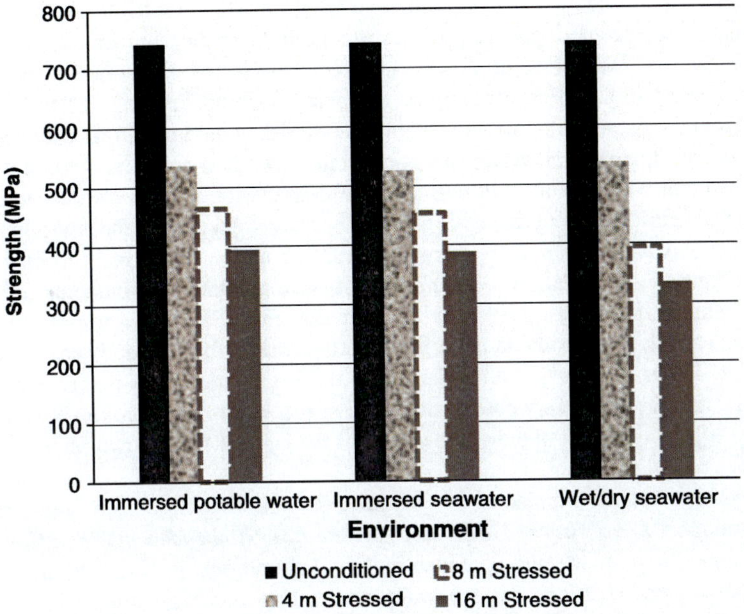

Fig. 16.5 Tensile strength of stressed GFRP bars in RC beams [20]

Table 16.1 Flexure test results of beams under the three exposure conditions after 8 months [20]

		Ultimate loading		Deflection @ ultimate loading	
Condition	Beam designation	Pu (kN)	Decrease (%)	δ (mm)	Decrease (%)
Laboratory @ RT	Control (unstressed)	7.2	N/A	76	N/A
Tap water @ 40°C	Unstressed	6.3	12.5	52	31.6
	Stressed	5.0	30.6	43	43.4
Seawater @ 40°C	Unstressed	5.3	26.4	51	32.9
	Stressed	5.4	25.0	41	46.1
Seawater wet/dry @ 40°C	Unstressed	5.7	20.8	59	22.4
	Stressed	4.8	33.3	46	39.5

Fig. 16.5. Flexure tests were done on the beams after 8 months of exposure and sustained load. The results are tabulated in Table 16.1, where the ultimate strength was reduced 12.5–26.4% for the beams not subjected to sustained load. For the beams under sustained loading, their ultimate strength was reduced 25–33.3%.

The work by Almusallam et al. demonstrates large reductions of mechanical properties of the GFRP bars when subjected to severe but unrealistic environmental conditions. The test conditions in this work will almost never occur in the field and only demonstrate accelerated degradation mechanisms of alkaline agents on GFRP products. Although the data in Almusallam et al. paper are interesting and contribute to the overall understanding of the deterioration of E-glass fibers in a vinyl–ester polymer when subjected to a high alkaline environment, they point out the lack of standards or accepted practices of testing GFRP products under realistic field conditions. In Nkurunziza et al.'s [19] critical review of the literature concerning the durability of GFRP reinforcing bars, it is pointed out that due to the lack of a standard or accepted practice, interpretation of the data obtained from testing where the composite material system is conditioned by total immersion in a severe environment will lead to very conservative degradation results. Furthermore, the Canadian reinforced concrete design guidelines have reevaluated the environmental knockdown factors due to the very conservative nature of testing carried out to understand the degradation mechanisms in severe alkaline environments. In addition, Nkurunziza et al. [19] also pointed out that GFRP materials and their manufacturing into products have improved so much that research conducted in the late 1990s and early 2000s leading to the establishment of design guideline knockdown factors is very conservative. More recent work has proven Nkurunziza et al. [19] to be correct in their assessment of the current knockdown factors due to environmental exposure.

In 2009, a study conducted by Mathieu Robert, Patrice Cousin, and Brahim Benmokrane, *Durability of GFRP Reinforcing Bars Embedded in Moist Concrete* [21], aimed to simulate and approach real field conditions. Robert et al. immersed mortar-wrapped GFRP bars in tap water, contending that this condition is closer to actual field conditions since the FRP material is embedded in concrete as seen in actual applications. Their main objective of this work is to show that past research experimentation that accelerated the aging of GFRP bars in alkaline solution is too severe. Therefore, Robert et al. contend that the past work prematurely degrades the GFRP materials indicating limited life expectancies and too conservative predictions. The test plan involved using sand-coated 12.7 mm diameter GFRP bars comprised of E-glass fiber in a vinyl–ester matrix in a concrete mortar having a pH of 12.15. The 1,440 mm long bars were centered in a 48 mm square concrete mortar section casted in the center third of the bar. Thus, the bar has 18 mm of concrete mortar cover. The exposed ends of the bar were then used as grips for tensile testing after conditioning. The embedded third of the bar was immersed in tap water such that the water could circulate around the mortar wrapping. Water levels were kept constant to avoid an increase in pH due to concentration of the alkaline ions. Three water temperatures were used, 23°C, 40°C, and 50°C, in the conditioning of the mortar-wrapped GFRP bars to accelerate aging but at these temperatures no thermal degradation mechanisms were present. Four time periods were used for this study, 60, 120, 180, and 240 days. All bars were tested under tension to failure and were instrumented with linear variable differential transformers (LVDTs) to capture elongation during testing. After conditioning, the mortar wrap was carefully removed prior to testing.

Robert et al. concluded from their work that the aging of the mortar-wrapped GFRP bars in tap water is significantly less than previous studies trying to simulate GFRP bars in concrete pore solution. In fact, tensile test results show only a slight decrease in strength, 9–16%, over the time and temperature conditions as shown in Fig. 16.6 [21]. In addition, the elastic modulus shows negligible losses over

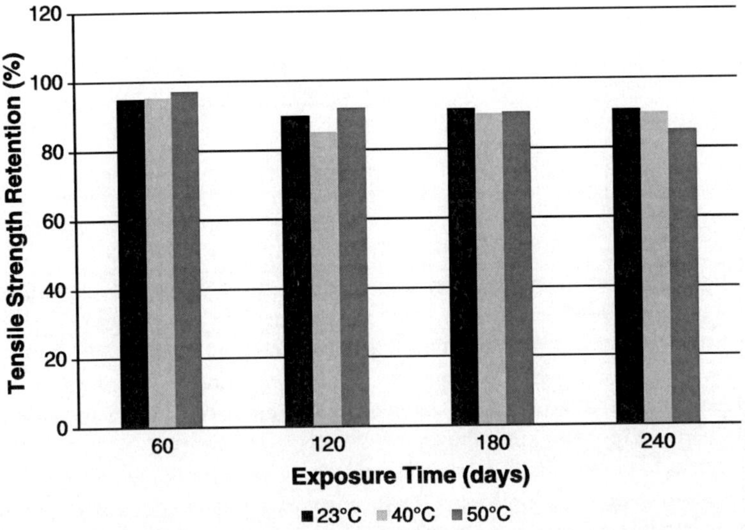

Fig. 16.6 Tensile strength retention of conditioned GFRP bars aged in moist concrete at 23°C, 40°C, and 50°C [21] (With permission from ASCE)

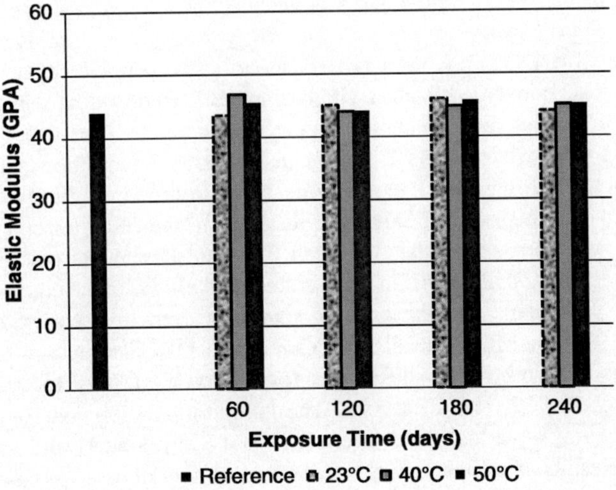

Fig. 16.7 Elastic modulus of conditioned GFRP bars aged in moist concrete at 23°C, 40°C, and 50°C [21] (With permission from ASCE)

the time and temperature conditions as shown in Fig. 16.7 [21]. Robert et al. found that no significant microstructural changes occurred at the concrete–bar interface and at the fiber–matrix interface after being conditioned for 240 days at 50°C. Furthermore, differential scanning calorimetry (DSC) analysis showed practically

Table 16.2 Results of differential scanning calorimetry analysis [21] (With permission from ASCE)

Conditioning	T (°C)	Time (days)	T_{g1} (°C)	T_{g2} (°C)
Unconditioned	–	–	105	134
Embedded in concrete and submerged in water	50	240	104	129

no change in T_g of the vinyl–ester matrix after 240 days at 50°C as depicted in Table 16.2 [21]. Based on no significant microstructural changes and no change in the matrix T_g, it appears that the matrix is not affected by moisture absorption over time and at relatively high temperatures. The experimental results of this durability study led Robert et al. to conclude that many previous studies conducted early when GFRP bars were being introduced in the civil engineering arena as possible replacement to conventional steel reinforcing bars were far too aggressive, leading to conservative degradation losses. These early experimental results in turn led to the incorporation of very conservative environmental knockdown factors used in current design guidelines.

Additional evidence to the conservative nature of previous durability research can be found in a 1999 publication from the Market Development Alliance of the Composite Institute [22]. In the MDA published *Fiber-Reinforced Polymer (FRP) Composite Dowel Bars...a 15 year durability study*, the Ohio Department of Transportation (ODOT) and the Civil Engineering Department of The Ohio State University (OSU) installed E-glass vinyl–ester (GFRP) dowel bars in the contraction joints of concrete pavement sections on Interstate-77 (I-77) in Guernsey County and Ohio State Route-7 (SR-7) in Belmont County between 1983 and 1987; 1 inch diameter GFRP dowel bars were installed in the I-77 pavement sections and 1.25 inch dowel bars were installed in the SR-7 pavement sections. This study was born out of the American Concrete Paving Association (ACPA) research concluding that corrosion of mild steel dowel bars in the contraction joints of concrete pavements was the primary source of premature concrete pavement failure. In 1998, the MDA Dowel Bar Team, under ODOT direction, removed the test sections containing the GFRP bars for laboratory testing to evaluate their performance after 15 years of service. The laboratory testing consisted of tensile strength, flexural modulus, shear strength, surface appearance, and load transfer efficiency. Results of the laboratory test showed no sign of deterioration other than mechanical erosion caused by excessive joint movement after 15 years of use subjected to heavy truck traffic, high pH alkaline exposure, and extensive use of de-icing salts. Although not used for structural reinforcement, the GFRP bars extracted from concrete pavements after 15 years of actual field use had no loss of mechanical properties. This report again highlights the conservative nature of accelerated testing to simulate actual field conditions for civil engineering structures leading to overly conservative knockdown factors.

Table 16.3 Glass transition temperature (T_g) [23]

Structure location	FRP product	Specimen	T_{g1} (°C)	T_{g2} (°C)
Hall's Harbour, wharf	GFRP IsoRod (16 mm dia)	Control	105	125
		Structure	123	125
Joffre Bridge, barrier wall	GFRP C-bar (16 mm dia)	Control	107	108
		Structure	107	108
	GFRP C-bar (9 mm dia)	Control	123	126
		Structure	127	128
Crowchild Trail Bridge, barrier wall	GFRP C-bar	Structure	126	129
Chatham Bridge, barrier wall	NEFMAC grid	Structure	98	116
Waterloo Creek Bridge, barrier wall	NEFMAC grid	Structure	78	117

Another refereed journal article on the durability of GFRP products used in actual structures is the *Field Study of Glass-Fiber-Reinforced Polymer Durability in Concrete* by A.A. Mufti, M. Onofrei, B. Benmokrane, N. Banthia, M. Boulfiza, J.P. Newhook, B. Bakht, G.S. Tadros, and P. Brett [23]. In this paper, GFRP reinforced components were studied after 5–8 years of service by removing cores from various structures across Canada. The research was conducted at several Canadian universities and used several experimental methods to study the alkalis attack on the GFRP in the concrete and compare the composition of the in-service GFRP cores to that of control specimens not exposed to the environment. Five structures that contained GFRP reinforcement throughout Canada were selected for this study. These structures were the 5-year-old Hall's Harbour located in the Bay of Fundy in Nova Scotia, subjected to saltwater and frequent wet–dry freeze–thaw cycles of +35°C to −35°C; the 7-year-old Joffre Bridge located in Sherbrooke, Quebec, subjected to frequent wet–dry freeze–thaw cycles of +35°C to −35°C and de-icing salts; the 8-year-old Chatham Bridge located in Chatham, Ontario subjected to frequent wet–dry freeze–thaw cycles of +30°C to −24°C and sprayed with deicing salts; the 8-year-old Crowchild Trail Bridge located in Calgary, Alberta, subjected to frequent freeze–thaw cycles of +23°C to −15°C and sprayed with de-icing salts; and the 6-year-old Waterloo Creek Bridge located in Vancouver Island, British Columbia, subjected to frequent freeze–thaw cycles of +23°C to 0°C and de-icing salts frequently used on the bridge deck. The GFRP was comprised of E-glass fibers in a vinyl–ester matrix. The GFRP components were typically the deck, barrier walls, and sidewalks of the bridge structures, and deck slab panels and pile cap beams of the wharf. The experimental techniques used were optical microscopy, scanning electron microscopy, energy-dispersive x-ray analysis, DSC, and Fourier transform infrared spectroscopy.

Conclusions from Mufti et al. study show no degradation of the GFRP products in the concrete cored from the five structures previously mentioned after 5–8 years of service. There was no debonding detected through optical microscopy analysis between the concrete and GFRP in any specimens [23]. In addition, optical microscopy, DSC, and Fourier transform infrared spectroscopy show that the matrix of the GFRP products was unaltered from its original state as indicated in Table 16.3 [23].

These results indicate that neither hydrolysis nor significant changes in the T_g of the matrix took place after exposure to the in-service environmental conditions [23]. Energy-dispersive x-ray analysis showed that no alkali ingress was found in the GFRP products from the concrete pore water solution due to the in-service environmental conditions [23]. Furthermore, carbonation and pH testing was done on the GFRP-concrete cores. These results show that the pH of the concrete dropped with time. Thus, it is reasonable to expect that the glass fibers in the GFRP products, protected by the impermeable vinyl–ester matrix, will not experience degradation due to alkali attack in the future, based on in-service environmental conditions. As a result of the Mufti et al. study, the Canadian Highway Bridge Design Code now allows GFRP bars for primary reinforcement or prestressing tendons in concrete applications.

Summarizing the research on the durability of GFRP reinforcements, it appears that the accelerated laboratory testing is far more severe that what actually occurs in the field. This highlights the need to develop acceptable test procedures or standards that best mimic the field conditions. Thus, the data obtained from a strict laboratory experimentation study may be too conservative. However, the benefits from these types of studies to understanding the mechanisms of material degradation when exposed to simulated field conditions is very important for material suppliers and developers for improvements in the material systems to make more durable GFRP reinforcements. For the practicing engineer, *AASHTO LRFD Bridge Design Guide Specifications for GFRP-reinforced Concrete Bridge Decks and Traffic Railings* is now available to use. However, reports and journal articles utilizing test articles from actual structures exposed to the ambient environment over a relatively long period of time, say greater than 5 years, will have data most applicable to deduce appropriate knockdown factors if the AASHTO specification does not address the specific application or situation.

16.4.2 Durability of FRP Bridge Decks

Another fairly larger sector utilizing FRP materials to increase design life of civil infrastructure is in bridge deck applications. In the vast majority of cases the bridge deck provides protection of the main structural members of the bridge while supporting the traveling public, transferring the traffic loads to the structural members. Typically, the conventional material bridge deck is repaired or rehabilitated several times before the main structural members need attention due to exposure to the ambient environment. These material systems are reinforced concrete decks or galvanized steel decking and only span between the main structural members. The cost of repair or rehabilitation of these bridge decks can be on the order of a quarter to a third of superstructure replacement. In some cases the life span of a steel decking system is only 10–15 years and can be replaced four or five times over a 50-year life span of a bridge. However, with shrinking budgets useful life expectancy of a typical highway bridge is approaching 75 years and over. Thus, composite material bridge decks have the potential to offer increased life expectancy of the bridge deck system and also of the main structural load-carrying members.

The prototype structure to the Tech 21 Bridge, designed by and constructed for Lockheed Martin Missile Advanced Technology Center in 1994, was dismantled and shipped to the Idaho National Engineering and Environmental Laboratory (INEEL) in 1997 for durability testing [24]. Fabricated of the same material systems as the Tech 21 Bridge, it had a span of 30 ft and a width of 14 ft. Initial static load testing completed at Lockheed Martin's facility in California [2] showed that this prototype satisfied the length/800 maximum deflection limits [24]. At INEEL the bridge was instrumented to monitor strain in the web, beams, deck, and bolts that attached the deck plates and beams together. The bolts were used for disassembly and removal of the bridge at the conclusion of environmental testing. The instrumentation also measures the vertical deflection at the supports, center of span and longitudinal along the bridge length during static and dynamic loading. In plane displacements, longitudinally and transverse to the span was also measured via the instrumentation plan during static and dynamic loading. Sensors also captured the temperature of the wearing surface, deck and beams, as well as the environmental conditions such as ambient temperature, humidity, solar and ultra violet radiation. Visual inspection were also carried out to supplement the sensor data and evaluated the presence and extent of aging of the deck, wearing surface, impact/snow plow damage, connection integrity, and any environmentally induced damaged not captured by the sensors. Loaded dump trucks and belly dump tractor trailers were used for the static load testing of the bridge.

Accelerated aging of the structure material system was conducted on 3 × 5 inch coupons with and without a UV protective coating. A Q-panel QUV accelerated weathering tester provided UV exposure and a Q-Fog Cyclic Corrosion Tester 1100 provided the salt spray exposure. Test parameters are given in the referenced paper [24]. The QUV meets the requirements of ASTM G53-95 whereas the Q-Fog Cyclic Corrosion Tester meets a General Motors Engineering Standard GM9540, Method B, *Accelerated Corrosion Test* [24]. The coupons were handled in accordance with ASTM G147-96. Prior to the QUV and Q-fog exposure, some panels were subjected to sandblasting and rock hits simulating wind erosion and rock hits from traffic, respectively. After QUV and Q-fog exposure, some panels were subjected to sandblasting and rock hits as the material behavior will most likely be different in these cases.

Freeze–thaw testing was also conducted on 3 × 11 inch coupons of the bridge's material system. Coated and uncoated specimens were tested at two specimen thicknesses, 0.75 and 1.6 inches. These tests were done in accordance with ASTM C666-92, *Standard test Method for Resistance of Concrete to Rapid Freezing and Thawing (Procedure A-Freezing and Thawing in Water)*. A measure of coupon durability in freeze–thaw cycling is the relative dynamic modulus of elasticity. The dynamic modulus of elasticity is a ration of the fundamental transverse frequency at 0 cycles of freeze–thaw to the fundamental transverse frequency after *n* cycles of freeze–thaw. Internal degradation shows a drop in the fundamental transverse frequency. For concrete, when the coupon reaches 60% of its initial modulus after 300 cycles the test is terminated.

Results of the load testing show that the FRP bridge behaves much like a steel and concrete structure subjected to similar conditions, loads, and restraint conditions [24]. Accelerated aging results show that an interaction between a coating, QUV, Q-fog, sandblast, and rock hits was statistically significant. It appears that urethane coatings perform better than gel coating over almost all treatment conditions [24]. Results of the freeze–thaw testing showed no drop in fundamental transverse frequency after 300 cycles indicating no internal deterioration due to freeze–thaw on the FRP material system. Thus, sensor and durability data for this prototype bridge FRP material system indicate that the design is suitable for use in the transportation infrastructure [24].

In 2006, a study was undertaken by Wu et al. to investigate the Durability of FRP-Concrete Composite Bridge Deck Materials under Freeze–Thaw and Low Temperature Conditions [25]. In this study, Wu et al.'s goal was to develop a reliable framework for durability assessment for FRP decks using laboratory testing procedures and finite element modeling. The test program consisted of exposing typical composite deck material systems to freeze–thaw cycling and constant freeze conditions in dry air, distilled water, and in saltwater. Wu contended that bridge deck systems are stiffness critical and thus the environmental conditions having the greatest effect on the material's stiffness are of greatest interest. Wu et al. further state that FRP decks typically are of sandwich construction; therefore, the flexural strength deck depends on the stiffness of the face skins and their location with respect to the neutral axis of the sandwich panel. In addition, the adhesive strength or its shear transfer capability between the face skin and core also affect the composite deck stiffness. Wu et al. focused on the face skins of the composite deck to measure degradation after exposure to various environmental conditions. The materials used for this study were an E-glass fiber and vinyl–ester matrix. Twelve plates 38.1 × 81.3 × 3.175 mm were fabricated of two layers of fabric. Each layer of fabric was made of four plies of unidirectional fiber bonded and stitched to a randomly oriented E-grass scrim. The fiber volume fraction of the skin was approximately 50%. Coupons measuring 25.4 × 254 mm were cut via a diamond saw and the average thickness of the specimens was 3.556 mm. The cut edges of the specimens were coated with a thin layer of epoxy to minimize moisture ingress through these edges. The specimens were immersed in distilled or saltwater for 2 weeks prior to freeze–thaw cycling. Wu simulated realistic bridge deck loading conditions by applying a continuous bending strain while freeze–thaw cycling the face skin specimens. The ASTM C666 test standard was used where the coupons were cycled from 4.4°C to −17.8°C. The ASTM standard requires that distilled water be used. However, a salt solution having 10% sodium chloride by mass was also utilized to simulate de-icing conditions found in the northern USA in addition to distilled water required by the standard. Wu carried out flexural testing using ASTM D790 where the span was 63.5 mm in a three-point bending setup. The flexural strength was measured by the maximum stress in the outermost fiber of the specimen at the instant of fracture. The material's storage modulus and loss factor were also measure by Wu et al. as an indication of degradation due to environmental conditioning. Modal vibration response measurements were taken in determining the storage modulus and loss factor. This nondestructive test can track the accumulation of damage over time from the environmental exposure. The flexural tests and nondestructive vibration tests were conducted after 16, 40, 100, 250, and 625 freeze–thaw cycles.

Test results from the experimentation conducted by Wu et al. shown in Fig. 16.8 indicate that practically no change in flexural strength and storage modulus occurred in the specimens conditioned in dry air, distilled water, and saltwater and subjected to freeze–thaw cycling between 4.4°C and −17.8°C. A slight increase in loss factor was observed for all cycling conditions in saltwater indicating a small level of microdamage, but this did not have a statistically significant effect on the flexural strength or storage modulus. No increases in loss factor occurred in the specimens conditioned in distilled water or dry air.

The New York State Department of Transportation sponsored a study on hybrid FRP-concrete bridge deck systems resulting in two reports detailing the study findings dated 2006 and 2009. In the 2006 report, Aref and Alnahhal described the analysis and design optimization procedures used for development of a hybrid FRP-concrete bridge deck system in the report entitled *Hybrid FRP-Concrete Bridge Deck Systems: Development and System Performance Validation* [26]. The report included tests on the bridge models to assess performance under service conditions. The hybrid deck system developed by Aref and Alnahhal consisted of seven glass fiber vinyl–ester matrix trapezoidal-shaped tubes sandwiched by glass fiber vinyl–ester matrix laminates forming a panel. The composite panel is all bonded together having no mechanical connections. The trapezoidal-shaped tubes were fabricated using an inner and outer tube bonded together. The inner tube height is approximately 20 mm less than the outer tube's internal height dimension. This arrangement provides a 20 mm hollow section across the top flange of the trapezoidal tube. The hollow section is to be filled with concrete, increasing the tube's structural capacity by enhancing the compressive zone of the element. Rectangular shear keys made of the same GFRP material are placed in the hollow sections bonded to the top surface of the inner tube and the bottom surface of the outer tube. The shear keys are spaced at 508 mm along the tube and result in a staggered pattern when the tubes are bonded together forming the hybrid FRP-concrete panels. In October 2009, the second report entitled *Long Term Performance of Hybrid FRP-Concrete Bridge Deck System* by Aref and Warn was made to the New York State Department of Transportation [27]. In this work by Aref et al., they made a three-fourth scale model of the prototype hybrid FRP-concrete deck developed in [26] for assessing its long-term structural performance. The three-fourth scale hybrid FRP-concrete deck bridge consisted of three GFRP deck panels having a total length of 3,657 mm, spanning transverse across three W300X96 steel beams. The main load-carrying steel beams were spaced 1,219 mm apart and have a total length of 4,876.8 mm. Each GFRP deck panel consisted of seven trapezoidal-shaped tubes sandwiched between laminates. Due to

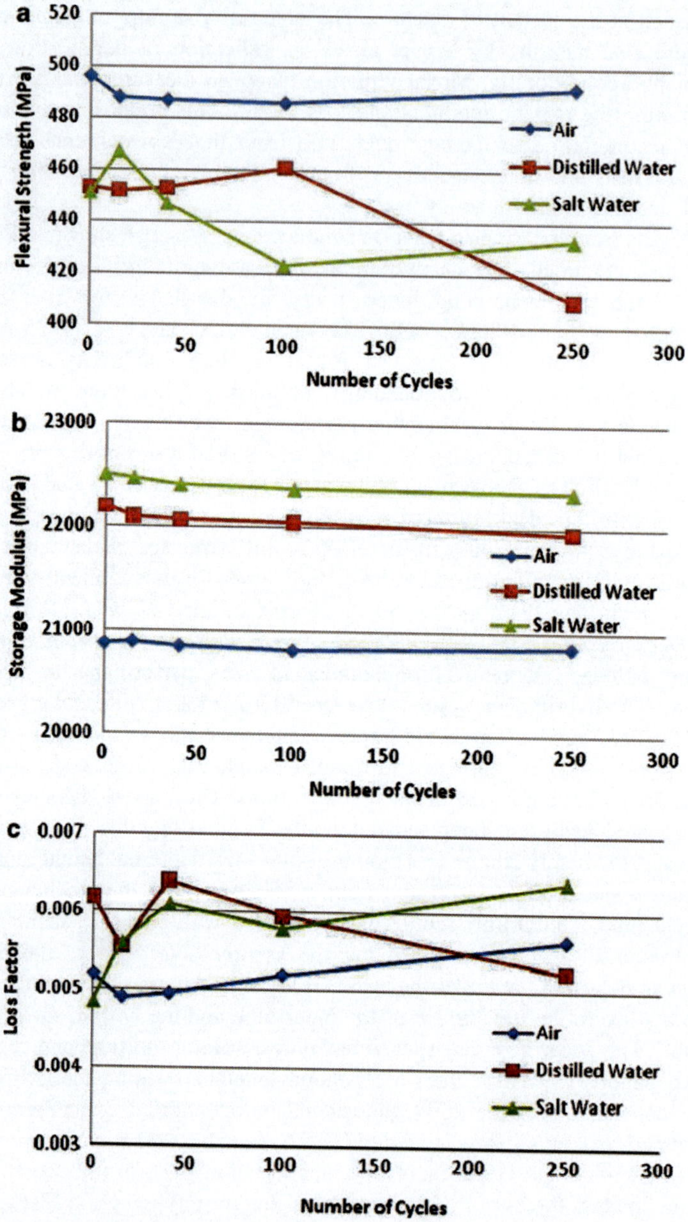

Fig. 16.8 (**a**) Flexural strength 5 h freeze–thaw cycle with pre-straining [25] (With permission from ASCE). (**b**) Storage modulus 5 h freeze–thaw cycle with pre-straining [25] (With permission from ASCE). (**c**) Loss factor 5 h freeze–thaw cycle with pre-straining [25] (With permission from ASCE)

the trapezoidal shape the outer GFRP deck sections had a width of 1,257 mm while the center section had a width of 1,143 mm. The hybrid FRP-concrete deck was attached to the steel beams by three rows of four shear studs welded to the top flange of the steel beams. The shear studs are spaced such that the end and center tubes of the outer panels have shear studs, whereas the third tube and end tube of the center panel have shear studs.

Aref et al. tested their three-fourth scale model of the prototype hybrid FRP-concrete deck for creep, fatigue, and ultimate strength. Creep testing consisted of applying eight ±50 mm linear variable displacement transducers (LVDT) and two thermocouples to the hybrid deck system. The hybrid FRP-concrete deck is attached to three W360×101 wide flange steel beams which are supported by two W760×284 wide flange steel beams sitting on the floor. Four LVDTs are placed under the composite deck and four are placed under the steel beams. The loading consisted of three 383 × 290 mm steel plates weighing 38 kN each for a total loading of 114 kN. The steel plates rested on four steel-rubber bearing pads placed on the top of the composite deck. Data was collected at five samples per second for 30 sec per day over the 73 days of the creep test. Data was collected for the 51 days the deck was loaded and 22 days after the load removal. Temperature readings were also taken above the deck and below the deck during the creep testing. Results shown in Fig. 16.9 indicate that the hybrid FRP-concrete deck displacement increases over time with the constant load applied. The creep displacements of the hybrid FRP-concrete deck were 15–20% of its initial displacement, while displacements under the steel beams remained constant. This suggests that the hybrid FRP-concrete deck had negligible effect on the overall creep of the bridge section. Upon removal of the load, the hybrid FRP-concrete deck recovered 82–84% of its initial displacement while the steel beams recovered 95% of their initial displacement. The average daily temperature ranged from 19°C to 22°C on the top surface of the deck while the bottom surface temperature ranged from 18°C to 21°C throughout the creep test. The temperature differences are a result of the ambient laboratory conditions and had no effect on the material system to contribute to an increase or decrease in creep.

Fatigue testing used the same setup as the creep testing except that the steel load plates were replaced by a vertical loading having amplitude of 165 kN, which is 1.33 times the scaled tandem truck loading. The load configuration was set up to simulate a two-axel truck in accordance with the 1998 AASHTO LFRD Bridge Design Specifications scaled to the three-fourth model experimental setup. The test subjected the bridge system to two million cycles. The fatigue test used eight ±50 mm LVDTs, two ±540 mm string potentiometers, four ±25 mm linear potentiometers, eight uni-axial strain gages and three thermocouples. The instrumentation was placed under the hybrid FRP-concrete deck and the main load-carrying beams to measure the fatigue response of the composite system and bridge system. Results of the fatigue testing show that the bridge system did not degrade through the duration of the testing. Applied load versus vertical displacements plots at zero cycles and at increments of approximately 400,000 cycles throughout the fatigue test show the same force–displacement response. Fatigue results of the composite panels

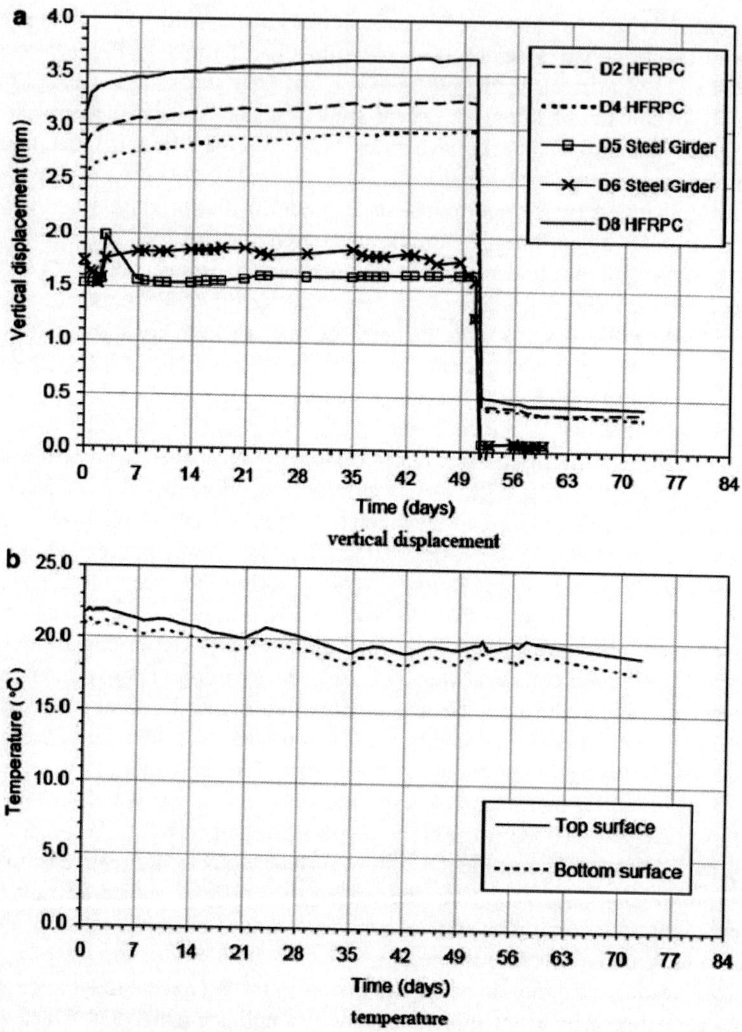

Fig. 16.9 Results of creep testing [27]

show less than a 5% change in the local stiffness of the GFRP material for the duration of the test. Composite action between the hybrid composite panels and the main supporting steel beams was also measured during the fatigue tests. This was accomplished by using the position of the bending neutral axis at two locations. Strain profiles at the exterior and interior beam locations for load levels of 0.4, 0.8, and 1.2 times the scaled applied loading at zero cycles, 800,000 cycles and at two million cycles were plotted with respect to the normalized section height. The strain profiles are practically linear, passing through zero strain at the normalized height of 0.57 of the section depth. Since the zero strain location is greater than half of the section depth, composite action exists between the hybrid FRP-concrete deck and supporting bridge beams. This indicates that the shear stud connectors sufficiently resist fatigue-induced degradation.

Aref et al. next conducted ultimate strength testing on the three-fourth scale prototype bridge section. The same bridge test configuration for creep and fatigue was utilized for the ultimate strength determination except for the application of the loading. The loading for ultimate strength test placed the bridge section in four-point bending. A total of 37 gages were used in this test: six ±50 mm LVDT, two ±540 mm string potentiometers, four ±25 mm linear potentiometers, 13 Krypton light emitting diodes and 12 uni-axial strain gages. Test results show that the prototype bridge section sustained a vertical load of 2.7 times the scaled truck loading without damage. Between 2.7 times and 6.3 times the scaled truck load, the composite deck emitted audible cracking noises and at 6.3 times the scaled truck load, a web of the GFRP deck was crushed. Test results also showed that composite action between the hybrid FRP-concrete deck and main load-carrying steel beams was maintained throughout the ultimate strength test.

Aref et al. offered these conclusions based on the observed test results. Aref et al. contend that creep performance is mainly influenced by the concrete portion of the deck sections and is fully recovered after load removal in 21 days. Since the nature of the applied loads to the hybrid FRP-concrete deck is transient, the observed creep is not a limiting factor to the application of this deck concept as a replacement to typical reinforced concrete decks. Furthermore, the hybrid FRP-concrete deck exhibited no degradation to stiffness and the three-fourth model bridge sections retained significant strength after fatigue loading. In addition, composite action between the hybrid FRP-concrete deck and the main load-carrying steel beams was not altered after two million cycles. This indicates that the concept of clustering shear studs provides a valid connection method between FRP decks to steel beams. Also, the hybrid FRP-concrete deck exceeded design expectations reaching approximately three times the truck loading without any observed damage. This result clearly indicates the robustness of the hybrid FRP-concrete deck concept and its validity to be used in actual bridges to replace existing reinforced concrete decks.

Although the FRP decks are not as numerous as the FRP bars, the research related to their durability shows that it is a very robust material system much more than the conventional materials used today. All indications show that FRP materials are capable of withstanding environmental exposure more severe than typically seen for highway bridge use. Due to their inherent material properties, composite materials will most always be stiffness critical, based on the allowable deflection limits set by the applicable design criteria. Thus, in most bridge deck applications bridge design guideline deflection limits will control. To satisfy these deflection limits, the design will be many times stronger than the stresses imposed by the design loading cases. Consequently, composite materials offer greater service life expectancy over conventional materials. Furthermore, due to their modularity, composite decks offer substantial construction time savings.

16.5 Load and Resistance Factor Design (LFRD) Criteria for FRP Material Structural Components and Systems

In an attempt to facilitate FRP material use in civil infrastructure, the National Institute of Standards and Technology (NIST) commissioned a report to assess the feasibility of implementing practical probability-based LFRD criteria for structural components and systems using FRP composite materials. This study finished in June of 2000 and entitled *Load and Resistance Factor Design (LFRD) For Structures Using Fiber-Reinforced Polymer (FRP) Composites* [28], written by Bruce Ellingwood from the Department of Civil Engineering, Johns Hopkins University, includes a critical appraisal of current approaches to structural design with composite construction; an explanation of reliability fundamentals necessary for probability-based code development; an illustration of the application of reliability principals in LFRD to composite structures through simple examples and development of recommended tasks for implementing LFRD criteria for composite construction. The focus is on utilizing commonly produced composite material structural pultruded shapes either of GFRP or CFRP in a structural system within the civil infrastructure arena. This document identifies and summarizes the reliability bases for LFRD criteria. It also identifies the barriers that need to be addressed for LFRD criteria to be implemented. However, it does not develop specific LFRD provisions as these are developed and approved by the professional standards communities. Ellingwood concludes that it is feasible to develop LFRD criteria for pultruded composite structural shapes. However, many impediments exist primary of which are that the structural engineers lack confidence in using composite structural shapes because of a lack of education, supporting standards, and specifications. Nonetheless, this document gives the structural engineer an understanding of the background of applying LFRD criteria to composite structural shapes. It will also serve as a source of further information on mechanical properties of composite material systems and the factors and combination of factors for probability-based limit states design. Although over 10 years old, this document could form the basis for a design utilizing composite structural shapes along the likes of the Manual for Steel Construction.

16.6 Conclusions

Much of the body of research conducted on the durability of composite materials for civil infrastructure application has focused mainly of reinforcing bars. This is due in part to the high initial cost of larger FRP components such as bridge decks and the cost of testing these components. The composite rebar body of durability research work has a unique distinction relative to the products used in the research. Due to advances in the manufacturing techniques of composite materials, newer GFRP products are far superior that those produced 7–10 years ago with respect to mechanical properties and durability. In addition, durability test techniques lack standardization, thus subjecting the material system to environmental conditions far more extreme than will ever be encountered under actual field conditions. Therefore, earlier durability research serves to understand the degradation mechanisms while recent research gives more meaningful material behavior results under realistic field conditions. However, the *AASHTO LRFD Bridge Design Guide Specifications for GFRP-Reinforced Concrete Bridge Decks and Traffic Railings*, First Edition is the primary reference to be used for GFRP reinforcing bars applications.

With this in mind, FRP composite products offer the civil engineer more tools to counteract the degradation suffered by conventional material systems. Although the initial cost of these products is higher than conventional materials, the benefit/cost ratio due to the composite material's noncorrosive characteristic and excellent long-term durability will be much higher over the design life of the civil structure. Even though applicable standards and design guidelines for FRP products are slow to develop, many examples highlighted above will give the practicing engineer sources of information and confidence to offer design solutions using FRP composite materials.

References

1. Meier U, Deuring M, Meier H, Schwegler G, (1993) Strengthening of structures with advanced composites. In Clarke J (ed) Alternative Materials for the Reinforcement and Prestressing of Concrete, 1st ed. Blackie Academic and Professional, Glasgow
2. Dumlao C, Lauraitis K, Abrahamson B, Hurlbut B, Jacobe M, Miller A, Thomas A, (1996) Demonstration low-cost modular bridge. In Saadatmanesh H, Ehsani M (ed) Fiber Composites in Infrastructure, Proceedings of the First International Conference on Composites in Infrastructure, Tucson, Arizona, January 1996
3. Plunkett J, (1997), Fiber-reinforced polymer honeycomb short span bridge for rapid installation, Idea Project Final Report, Contract NCHRP-96-ID030, Transportation Research Board. http://www.ksci.com. Accessed April 10, 2010
4. Held J, (1997), Fiberglass Bridge Testing Report, http://www.ksci.com. Accessed April 10, 2010
5. Black S, (2003) How are composite bridges performing? http://www.compositesworld.com/articles/how-are-composite-bridges-performing. Accessed April 10, 2010
6. Lopez-Anido R, Troutman D, Busel J, (1998) Fabrication and installation of modular FRP composite beidge deck. In Proceedings of the 1998 International Composites Expo, Nashville, Tennessee, January 1998
7. Lopez-Anido R, Howdyshell P, Stephenson L, (1998) Durability of modular FRP composite bridge decks under cyclic loading. In Benmokrane B, Rahman H (ed) Durability of fiber reinforced polymer (FRP) composites for construction, Proceeding of the first International Conference (CDCC'98), Ottawa, Ontario, Canada
8. Foster D, Richards D, Bogner B, (2000) Design and installation of fiber-reinforced polymer composite bridge. J Comp Const 4 (1): 33–37
9. Abrahamson T, (1998), Tech 21 fiberglass composite bridge structural analysis report. Butler County Engineer's Office internal report

10. Fabrication, installation and health monitoring of the Tech 21 composite material highway bridge Butler County, Ohio. Butler County Engineer's Office internal report
11. Civil Engineering Research Foundation, (2001) Gap analysis for the durability of fiber reinforced polymer composites in civil infrastructure. ASCE, New York
12. Mertz M, Chajes M, Gillespie J, Kulich D, Sabol S, Hawkins N, Aquino W, Deen T, (2003) Application of Fiber Reinforced Polymer Composites to the Highway Infrastructure Cooperative Highway Research Program (NCHRP) Report 503. Transportation Research Board (TRB) Washington D.C.
13. American Concrete Institute (ACI) (2002) Guide for the design and construction of externally bonded FRP systems for strengthening concrete structures. ACI Farmington Hills, Michigan
14. Neale K (2001) Strengthening reinforced concrete structures with externally-bonded fibre reinforced polymers. Intelligent Sensing for Innovative Structures (ISIS) Winnipeg, Manitoba Canada
15. Triantafillou T (2001) Design and use of externally bonded FRP reinforcement (FRP EBR) for reinforced concrete structures. International Federation for Structural Concrete (fib) Lausanne Switzerland
16. Schwegler G, Berset T, (2002) The use of prestressed CFRP-laminates as post strengthening. In Proceedings of the 16th Congress of IABSE September 18–21, 2002, Luzern Switzerland
17. Mirmiran A, Shahawy M, Nanni A, Karbhari V, (2004) Bonded repair ad retrofit of concrete structures using FRP composites, NCHRP Report 514. TRB Washington D.C.
18. Mirmiran A, Shahawy M, Nanni A, Karbhari V, Yalim B, Kalayci A (2008) Recommended construction specifications and process control manual for repair and retrofit of concrete structures using bonded FRP composites, NCHRP Report 514. TRB Washington D.C.
19. Nkurunziza N, Debaiky A, Cousin P, Benmokrane B (2005), Durability of GFRP bars: A critical review of the literature. Prog. Struct. Engng Mater. 7:194–209
20. Almusallam T, Al-Salloum Y (2005), Durability of GFRP Rebars in Concrete Beams under Sustained Loads at Severe Environments, J of Composite Materials Vol 40 7:623–637
21. Robert M, Cousin P, Benmokrane B (2009), Durability of GFRP Reinforcing Bars Embedded in Moist Concrete, J of Composites for Construction Vol 13 2:66–73
22. Market Development Alliance (MDA) FRP Dowel Bar Team (1999), Fiber-Reinforced Polymer (FRP) Composite Dowel Bars. . .a 15-year Durability Study, Composite Institute, Harrison, NY
23. Mufti A, Onofrei M, Benmokrane B, Banthia N, Boulfiza M, Newhook J, Bakht B, Tadros G, Brett P (2007), Field study of glass-fibre-reinforced polymer durability in concrete, Can. J. Civ Eng, 34:355–366
24. Carlson N. Blackwood L, Toores L, Rodriguez J, Yoder S, (1999), Accelerated aging of polymer composite bridge materials, SPIE 6th International Symposium on Smart Structures and Materials, Newport Beach, California
25. Wu H, Fu G, Gibson R, Yan A, Warnemuende K, Anumandla V, (2006), Durability of FRP Composite Bridge Deck Materials under Freeze-Thaw and Low Temperature Conditions, Journal of Bridge Engineering 11:443–451
26. Aref A, Alnahhal W, (2006), Hybrid FRP-Concrete Bridge Deck Systems: Development and System Performance Validation, Technical No. C-02-07, New York State Department of Transportation, Albany New York
27. Aref A, Warn G, (2009), Hybrid FRP-Concrete Bridge Deck Systems: Long Term Performance of Hybrid FRP-Concrete Bridge Deck System, Technical No. C-02-07, New York State Department of Transportation, Albany New York
28. Ellingwood B, (2000), Load and Resistance Factor Design (LFRD) For Structures Using Fiber-Reinforced Polymer (FRP) Composites, National Institute of Standards and Technology, Technology Administration, U.S. Department of Commerce, Gaithersburg, MD

Index

A

AASHTO. *See* American Association of State Highway Transportation Officials
Active protection
  engineering solutions, 47
  NDE, 47
Active reaction zone
  estimation, 401, 402
  measurements, 401, 402
  oxidation behavior, 409
  thickness, 407
  zone-I and II, 387
Additives
  fiber-reinforced composite, 49
  flame retardant
    compounds, 51
    condensed phase char formation, 52–53
    endothermic cooling, 51–52
    fire-safety regulations, 50
    vapor phase radical inhibition, 51
  thermoplastic materials, 49
  UV stabilizers and anti-oxidants
    blooming, 49–50
    effectiveness, 50
    initiation, propagation and termination reactions, 49, 50
    metals, 49
Aging composites
  fiber-reinforced
    Boltzmann superposition integral, 291
    down-jump experiment, 290
    hydrostatic compression, 290–291
    KAHR, 291
    shift factor, 290
    S_{ijkl} tensor, 291
  fibers/particles, complex interaction, 289
  importance, 290
  particle-reinforced polymers
    black-filled and bound rubber, 298
    hydrostatic tension, 297
    residual stress, 296
    semicrystalline polymers, 297–298
    thermoviscoelastic comparison, 297
  shifts, viscoelastic response, 289
  thermal residual stresses, 290
  time-aging time superposition
    aging time shift factor, 292
    calorimetric methods, 296
    compression, effective time theory predictions, 294
    deformation geometries, 294
    vs. effective time theory predictions, 293
    horizontal shift factors, 295
    IM7/K3B fiber-reinforced response, 292
    MMCs, 292
    vs. neat resin and IM7/K3B, 293
    property gradient, 296
    shift rate, K3B resin, 294
    shift rates, temperature and compression, 295
    viscoelastic response, enthalpy, 295
Aging, physical
  composite materials, 289–298
  glassy polymers, 256–289
Aging yield
  failure-related properties, 276–277
  semiempirical model
    data comparison, 270–274
    failure, 265–270
  thermoviscoelastic model, 274–276
Aircraft Energy Efficiency (ACEE) program. *See* Aircraft structures, durability and aging B–737

DOI 10.1007/978-1-4419-9308-3,

Aircraft engines
bypass ratios, 598
"cold section", defined, 598
damage mechanism modeling
edge delamination, woven composite, 608–610
inter-yarn debonding/"meta-delamination" model, 608
transverse yarn cracking model, 606–607
durability assessment, engine component, 617–623
durability characterization
example, 601–602
experimental and analytical approach, 600–601
fatigue damage mechanisms, 603–605
hygrothermal and mechanical loading engine mission, 599
hygrothermal missions, 600
weave architecture and fabrication, 602–603
end-of-life properties, 598
long-term durability characterization, 598
stiffness and strength degradation, 599
turbofan, 597
Aircraft structures, durability and aging B–737
ACEE program
analysis, 517
four development areas, 515
monitoring and certification, 517
motivation, 515
primary and secondary components, 516
redesign and manufacture, 517
use B-737-200 graphite stabilizer, 515
baseline data, 548
CFRP horizontal stabilizer
NASA and ACEE, 514
NDI and NIAR, 514
primary structure and health, 514
requirement, commercial and military airplanes, 514
thermal analysis, DMA and DSC, 515
transmission and PE ultrasonic, 514–515
used advanced materials, 514
composite stabilizer description
'big foot' fasteners, 519
bolted steel spar lugs, 519
corrosion protection scheme, 520
lightning protection scheme, 520–522
material selection, 520
metal center structure, 517, 518
single co-cured skin, 518–519
spars and rib details, 519
stiffened skin concept, 519
structural box arrangements, 517
Titanium Hi-Loks, 519
degree of cure, 548
fleet status, 524–525
horizontal stabilizer certification
coupons and elements, 522–523
data generation and strains, 522
requirements, 521–522
supporting test evidence, 522
improvements, 548
physical test results, 547
planning program maintenance, 523–524
teardown results
destructive inspection, disbonded areas, 532
disassembly, 526–528
findings, visual inspection, 528–530
mechanical testing, 540–547
microscopy/image, 539
moisture content evaluation, 538–539
nondestructive inspection, 532–534
objectives, 525
Rapidscan™, 526
research, 525
thermal analysis, 534–538
visual/nondestructive inspection, 525–526
void content evaluation, 539, 540
T300/5208 material, 548
American Association of State Highway Transportation Officials (AASHTO)
Bridge Design Guide Specification
acceleration, laboratory testing, 646
constituent materials, 636
creep testing, 646
center panel, composite vehicle bridge, 628
deflection criteria load limits, 631
description, 627
Laurel Lick bridge, 629
American Society for Testing and Materials (ASTM), 536
Analytical methods
bolted joints
bonded tee/hat sub-element fatigue test, 490
bypass test, 488
fatigue crack growth rate (FCGR), 487
full-scale fatigue test, 487–488
high-fidelity mechanism, 490–491

Miner's rule, 487
notional skin spar, 489
single-shear joint, 490
S-N curve, notional skin spar, 489
specimen, skin-spar test, 489
standard fastener test, 488
thermal and mechanical fatigue, 489–490
thermal test spectrum, 491
wing root spectrum, 490
bonded joints
adhesives, fillers and adherends, 506
advantages, 502–503
application, fracture mechanics-type, 506
axial tensile loading, 505
Boeing Wedge Test (BWT), 502
continuum damage model, 503
crack growth, 505–506
delamination initiation and propagation, 503
failure mode 2, 502
fatigue and environmental effects, 504
modeling aspects and efforts, 502
residual fatigue strength, 504
SIFT, 503
spectrum sub-element testing, 501
strain energy release rate (SERR), 502
strength and SERR values, 504
stress distribution, 503–504
techniques, 506
VCCT, 502
Angle-ply, $[45/-45]_{2S}$ laminate
lay up and orientation, 364, 366
oxidation and damage growth, 364–365
Anisotropy
angle-ply, $[45/-45]_{2S}$ laminate, 364–366
assumptions, 358–359
cross-ply, $[0/90]_{4S}$ laminate
fiber bridging and delamination, 362–363
oxidation *vs*. damage growth, 361, 362
transverse matrix cracking, 361–362
quasi-isotropic, $[0/45/-45/90]_{2S}$ laminate
oxidation *vs*. damage growth, 363–364
ply lay-up and orientation, 364, 365
transverse matrix cracking, 364
unidirectional, $[0]_{16T}$ composite
axial fiber-matrix debonding, 360–361
oxidation growth, axial and transverse directions, 359, 360
transverse matrix cracking, 360
Antiplasticization, 335–336
ASTM. *See* American Society for Testing and Materials (ASTM)

B

Barely visible impact damage (BVID), 445
B–737 horizontal stabilizer
destructive evaluation, 532
disassembly, R/H skin
big foot blind fastener removal, 527
composition, structural box, 526, 529
CRFP, 527
demating spars and ribs, 526, 530
flat skin, curvature/residual strain, 526, 528
Titanium Hi-Lok attachments, lower skin, 526, 529
upper-skin, 528
nondestructive inspection
lower-skin disbands, 533
NIAR pulse echo scans, 533–534
pulse echo, 532–533
ultrasonic inspection, 1980's, 533
observations, 528–529
thermal analysis
ASTM standard, 536
Boeing method, 536–537
coupon extracted, DSC, 535
description, 534
DMA curve Tan Delta, 534
rib specimen extracted test, 535
upperskin, front, rear spars, 536–537
visual inspection
fasteners, sealant deterioration, 530
front and rear spar, 530
liquid shims, 529, 531
lower skin, 529
phenolic shims, 529, 531
service history, 529
structural box, 529
Tedlar® degradation, 529, 531
upper skin repair, ribs 2 and 3, 532
use shims, 530
Bird strike
damage, C–130 nose cone, 160
experiment and stimulation correlation
damage produced, composite structure, 161, 163
projectile deformation, 161, 162
impact test, helicopter cockpit, 161, 164
penetration damage, Boeing 737, 161, 163
transport Canada, 163–164

Bird strike (*cont.*)
U.S. and foreign bird weights cumulative distributions, 158–159
Bismaleimides (BMIs)
Diels-Alder-type extension and Diallyl bisphenol A, 12
matrix resins, classes, 2
mechanical and time-dependent effects, 16
Michael addition, 12
moisture and solvent effects, 15
monomeric structure, 11, 12
numerical approach, ENSAM, 326–327
polyimide-based lamination, 441
temperature resin, 598
thermo-oxidative degradation, 14–15
BMIs. *See* Bismaleimides
Boeing Wedge Test (BWT), 502
Bolted joints
analytical methods (*see* Analytical methods)
application concepts and materials, 485–486
bypass test setup, 487
components, 484
concepts and materials, 485–486
configuration, 485
design, 485
3D stress analysis
apply eight-node, 492
description, 491
effects, 491–492
elastic fastener, 492–493
formulation, 491
20-node isoparametric use, 492
plate theory and interlaminar stresses, 492
failure modes, 484, 485
fatigue and environmental changes, 484–485
fatigue life modeling, 495–497
strength prediction, 493–495
techniques, 486–487
test-based design, 486
Bonded joints
analytical methods (*see* Analytical methods)
basic components, 497
experimental techniques (*see* Experimental methods)
long-term behavior (*see* Long-term behavior)
observations, 497–498
Building block approach
component tests, 462
correction factors, 463
coupon and elements tests, 461
development tests, 461
DUL, 464
environmental conditioning, 463
graphite/epoxy composite, 464, 466
manufacturing scale-up, 463
margin of safety, RTD test, 464
progressive building block validation, 465
test program, composite structures, 462, 464
Building block (BB) approach, 618–619

C

Carbon fiber reinforced polymers (CFRP), 626. *See also* Aircraft structures, durability and aging B–737
CDM. *See* Continuum damage mechanics
Certification, airworthiness durability, 619
CFRP. *See* Carbon fiber reinforced polymers
Chain lengths
interphase modulus, 102
surface interaction, 99
thermodynamic model, 97
Characterization
diffusivity, 400
Lamina-scale models, 404
Characterization, durability
aircraft engine hygrothermal missions, 600
combined experimental and analytical approach
assessments, composite materials, 600
hygrothermal and aging effects, 601
example
2D woven and 5HS fabric material, 601
fiber architecture and unit cell, 602
hygrothermal mission block, 601
tension–tension fatigue testing, 602
fatigue damage mechanisms
dynamic stiffness loss curves, 603, 604
matrix cracking and delamination failure, 605
meta-delamination, 603
X-radiograph, transverse yarn cracking and meta-delamination, 604
hygrothermal and mechanical loading mission, 599
weave architecture and fabrication
5HS weave fiber, 602
RUC isolation, 603
warp yarns, harness, 602–603

Civil infrastructures
AASHTO design, 627
ad-hoc committee
definition, durability, 633
description, 632
gap analysis, 632–633
research areas, 633
workshops, 632
bridge components, composite material application, 626–627
CFRP and EMPA, 626
composite material products durability
description, structures, 637
fiber reinforced polymer (FRP) bridge decks, 646–653
GFRP reinforcing bars, 638–646
Structure's service life, 637–638
composite vehicle bridge
AASHTO specification, 631–632
E-glass fibers manufacture, 629
fabrication, pultrusion method, 628–629
fatigue tests and ultimate failure load tests, 629
fiberglass material system and fabrication techniques, 628
first, load tests, 628
hot-mix asphalt, 629–630
pultruded modular deck, 628
sandwich panels, 627–628
third, load tests, 631
current state of affairs
applications, 634–635
data analysis and interpretation, 637
environmental durability, FRP products, 636
externally bonded FRP reinforcement (FRP EBR), 635
gap analysis, 636
GFRP, 636
goal, 637
guideline, 635
NCHRP report 514 and 609, 635–636
strengthening concrete structures, 635
LFRD criteria, 654
plate bonding techniques, 626–627
structural components, composite material, 626
TRB AFF80 committee
barriers, FRP composites, 633–634
demonstration projects, 634
description and NCHRP, 633
Classical homogenization, 127
COD. *See* Crack opening displacement
Composite materials and damage
applications, 121
CDM, 122
characterization
CSD, 131
'damage entity', 128–129
dyadic product, 129
mode tensor, 130
physical significance, 129
ply cracks, 129–130
RVE size, 129
Talreja, 128
constitutive framework, 131–138
internal surfaces, 121–122
mechanisms
crack multiplication rate, 123
damage mechanics, 126–128
fiber bundles, 124
formation and progression, damage, 122
internal surfaces, 126
laminate plies, 125
multiple intralaminar cracking, 125
"shear-lag" process, 123
transverse cracking, 122–123, 124
transverse crack multiplication process, 124
use, 126
x-ray radiograph, 122, 123
SDM, 138–140
Compression after impact (CAI), 149
Compressive residual strength
designers, 592
elastic modulus, 552
vs. residual compressive modulus, 590, 591
test program, fatigue, 593
Constitutive framework
CDM, 131
Clausius–Duhem inequality, 132
composite materials, 131
evaluation procedure, 136–137
experimental data, 133–134
"forces", 133
Helmholz free-energy function, 132, 134
parameter *K*, 138
polynomial function, 134–135
response functions and associated restrictions, 133
stress–strain curve, 134
temperature gradient, 132–133
tensor components, 135
thermodynamics, 131–132
transverse cracking mode, 137
Truesdell's principle, 132

Constitutive framework (*cont.*)
  Voigt notation, 135–136
Continuum damage mechanics (CDM)
  definition, 122
  description, 131
  scope, 122
Corrosion prevention
  bolted joint barrier plies and seals, 441, 442
  dissimilar materials, 440
  electric potential, metal, 440
  fastener materials, 442
  galvanic corrosion, 439–440
  moisture, 440–441
  precautionary design measures, 441
  prevention and control program, 443
  quartz/glass fibers, 441
  saltwater exposed metals, galvanic series, 440
  sealing, 441–442
Crack opening displacement (COD), 138–139
Crack sliding displacement (CSD), 131
Creep and stress relaxation test, 560–568
Cross-ply, $[0/90]_{4S}$ laminate, 352–354, 361–363
CSD. *See* Crack sliding displacement

D

Damage. *See also* Composite materials and damage
  definition
    entity, 128–129
    mode tensor, 129–130
  electrical
    circuit failure, 45
    vs. heat, 45
    insulators, 45
    lightning strikes, 45
  heat and fire
    charring and non-charring polymers, 43, 44
    crosslinking and char formation, 43
    fiber reinforcement, 43
    glass transition temperature (T_g), 42
    mass transport application, 43
    matrix resin temperature, 42
    mechanisms and pathways, 42
    mitigate flame propagation, 43
    thermal decomposition, 43
  oxidation, 46
  wear and abrasion
    chemical changes, 46
    fiber reinforcement, 46
    prevention, 46
    rain erosion, 46
Damage classification
  Advisory Circular 20–107B, 449–450
  detectability, 450
  JSSG–2006, 450–451
  residual strength
    load, 451
    requirements *vs.* damage categories, 449
Damage entity, 128–129
Damage growth
  HTPMCS, 384
  and oxidation, 386
  oxidized regions, 420–422
Damage impact
  aviation, 143, 144
  composite structures long-term durability, 145
  development, testing methodology, 145–146
  engineering structures and airframes, 144
  influence
    assessment, 175
    "Build Block" testing, 175
    composite structures design, 174
    discrete threats management, 176
  threat sources
    bird strike, 158–164
    GSE, 155–158
    hail ice, 165–174
    tire rupture, 164–165
    tool drop and FOD, 151–155
  tolerance methodology, 146–149
Damage mechanism, 603–605
Damage mode tensor, 129–130
Damage propagation, 356–358
Damage resistance
  composite rotor blades, 436
  definition, 434
  fail-safe structure, 435
  primary structure, 434
  reparability, 438
  safe-life designs, 435
  sources, out-of-plane loads, 436–437
  strain-level restrictions, 437–438
Damage threat sources
  bird strike, 158–164
  GSE, 155–158
  hail ice, 165–174
  impact damage and affected locations, 149, 150
  status matrix, 150, 151
  tire rupture

FAA, 164
fragment impact experiment *vs.* simulation, 165, 166
tolerance approach, 150
tool drop and FOD, 151–155
Damage threshold
description, 146
use, 146
Damage threshold load (DTL), 147
Damage tolerance (DT)
commercial rotorcraft composite rotors, 455
definition, 428
goal, 478
growth curves, 470
philosophy, Air Force, 435
safety-of-flight structure, 447
Damage tolerance methodology
CAI test results, 147, 148
conceptual damage size *vs.* KE/contact force, 146, 147
DTL prediction relationship, 146–147
falling weight impact testing, 146
impact damage size and CAI strength reduction, 148
threshold conditions, 149
3D composite
microcracking, 374–375
oxidation and damage growth
aging induced, 376–378
imaging, 376–377
surface damage characterization, 375
transverse surface crack density, 375–376
Defect threat assessment
categorizing/zoning, 444
impact damage
BVID, 445
classification, 449–451
non-impact damage sources, 451–453
zoning and impact survey, 446–449
maintenance/handling damage
delamination, 445
types, 445
Degradation. *See also* Durability, structural joints
and fatigue life, 588–590
fiber-matrix interphase, 361
hygrothermal, 346
interlaminar residual stresses, 348
surface oxidation, 347–348
Degradation parameters, 227
Design development tests. *See* Building block approach
Design drivers
aerospace structural design criteria, 429–430
composite damage tolerance and fatigue performance, 431
composite structures, 430
corrosion prevention and control, 439–443
damage resistance and tolerance, 434–438
manufacturing and in-service inspection, 438–439
margin of safety, 430–431
material environmental requirements
degradation, 434
fatigue cycles, 431
heat sources, 432
in-service exposures, 434
light fighter hot temperature values, 433
polymer matrix composites, 432
wet glass transition temperature (Tg), 433
material qualification, 430
minimum thickness structures
lightly loaded, 443
moisture ingression, 443–444
sandwich, 443
solid laminate construction, 444
"overweight" designs, 431
Design, implications, 566–567
3D finite element methods, oxidation simulations
diffusion model, 394–395
Euler method, 395–396
MPI_Isend and MPI_Irecv, 398
orthotropic diffusivity, 395
parallel version, PCG, 397–398
PCG method, 397
PPCG library, 398
speedup factor, 398, 399
time step and held constants, 396
tolerance choice, 399
vector norm, 397
DGEBA-DDS system. *See* Diglycidyl ether of bisphenol A-diamino dyphenylsulphone
Differential Scanning Calorimetry (DSC), 515
Diffusivity
anisotropy, 413
axial fibers, 404
characterization, 400
characterized fibers, 391–392
fiber effects, 404
fiber–interphase assemblage, 405–406, 419
isotropic values, 406

Diffusivity (*cont.*)
matrix ratio, 404
orthotropic, 395–396
oxygen tensor, 389
parameter scale, 404
PMR–15 resin, 401
vs. reaction rate, 389
and reaction rates, 402
saturation oxygen concentration, 403
and solubility, 404
stress-assisted, 420
tensor, 400
Diglycidyl ether of bisphenol A-diamino dyphenylsulphone (DGEBA-DDS) system, 338
Dimensional stability
ratcheting potential enhancement, 593
and stiffness, durability, 550
DMA. *See* Dynamic Mechanical Analysis (DMA)
DSC. *See* Differential Scanning Calorimetry (DSC)
DTL. *See* Damage threshold load
Durability design. *See* Durability testing
Durability, structural joints
aerospace industry, techniques, 507
environmental effects and mechanical loads, 507
function, 483
long-term and degradation, 483
objectives, 484
recommendations, 507
responses, individual joint, 484
Durability testing
composites, 454
damage growth and inspection requirements
composite blade damage development, 459
composite structures, 460
dependence, usage severity, 460, 461
EFH and AFH, 461
IAT data, 460
inspection intervals, 460
region I, II and III, 459
slow damage growth and fail-safe structures, 459
design development tests (*see* Building block approach)
doubler and scarf patches, 455
finite element programs, 454
hybrid, composite/metallic structures, 470–471
initial flaw/damage assumptions
dimensional tolerances, 457
JSSG–2006, 459–460
quality assurance inspections, 458
nonlinear structural response characteristics, 454
safe-life analysis
elements, 456
fatigue S-N curve development, 456–457
modified approaches, 455
operational experience, rotor designs, 455, 456
scatter factors, 457
structural fatigue testing (*see* Fatigue testing, structural)
structural repair substantiation, 455
substantiation methods
LEF approach, 471–473
Life Factor approach, 471–473
quasi-static and ultimate strength approaches, 475–476
Dynamic Mechanical Analysis (DMA), 515

E

End-of-life properties
durability assessment, 598
testing methods and analytical modeling, 600
Environmental effects, 507
Epoxies
amine reactions, 6
carbon/epoxy laminate
delamination failure, 229–231
hygrothermal aging, HWU and HWS, 231
DDS, 6
DGEBA, 4
diffusion, 86
interphase concentration profiles, 92, 94
mass balance, 84, 86
MDA, 5, 6
moisture and solvent effects
toughness, 10
water, 9
moisture diffusion, PMC, 182
novolac, 4, 5
polyether formation, homopolymerization, 7
radiation effect, 8–9
strain assisted diffusion (*see* Strain assisted diffusion)
stress and time, 11

TGMDA, 4, 5
thermal aging, 7–8
thermodynamic equilibrium conditions, 81
trifunctional, 4, 5
Equipresence, 132
Experimental methods
bonded joints
adhesive peel and shear test, 499, 500
coupon test methods, 499
stepped-lap bending test, 501
stepped-lap tensile test, 499, 501
HCs
fatigue life and modulus degradation, 588–590
residual strength/modulus relationships, 590–592
time–temperature-dependent response
flat coupon specimen manufacturing, 554–556
material characterization, 557–558
monotonic testing, 558–560
test setup, 556–557
TMF
isothermal static and fatigue tests, 573–576
materials, 570
stitched material, 576–579
test methods, 570–573
unstitched material, 579–582
Exposure
radiation, 4
thermo-oxidative, 2–3
water and solvents
antiplasticization, 3
case II sorption, 3

F

FAA. *See* Federal Aviation Administration
Failure threshold energy (FTE)
damage thresholds, 146
defined, 169
Fatigue
damage progression and durability, PMC, 583
induced biaxial cracks, 189
life and modulus degradation, 588–590
life plot, strain range, 587
plot, stress range, 587
prototypical compression–compression, 568
prototypical compressive, defined, 565
reversed, strength, 552
static compression testing, 552
static stress–strain response, 586
strain response, 566
uniaxial, 195
Fatigue damage, 603–605
Fatigue life modeling, bolted joints
computational techniques, 497
damage modeling methods, 496–497
environmental effects, 496
mesh-independent model, 497
ply-level analysis methods, 496
static analysis, 496–497
strength and stiffness degradation law, 496
time and temperature effects, 495–496
Fatigue testing, structural
composite and metallic structures, 470–471
durability test, 469
fatigue life and static strength scatter distributions, 465, 466
fixed wing aircraft
clipping, 468
design loads spectrum/master events spectrum, 467
fully assembled structures, 469–470
high loads, 468
loading spectrum development, 467
low amplitude thresholds, 469
stress spectra, 468
truncation level determination, 469
truncation, loads, 468
rotor blades and dynamic components
endurance limits, 470
high frequency components, 470
retirement life, 470
scatter factor, definition, 466, 467
shape parameter and scatter factor, 467
Federal Aviation Administration (FAA)
access, bird-strike data, 158
impact energy, 157
reader, regulations, 162
Federal of Aviation Administration (FAA), 514
Fiber-matrix interphase
assemblage, 403
characterization, 403
diffusivities, fiber–interphase assembly, 405–406
fiber–interphase diffusivity, 404
homogenization theory, 404
Lamina-scale models, 404
oxidation composites, 403
power law fits, 404
primary effect, 403
procedure, 404

Fiber-matrix interphase (*cont.*)
simulation *vs*. experimental results, 405
transverse direction, 405
Fiber reinforced polymer (FRP), 201, 632, 644, 654. *See also* Civil infrastructures
Fickian diffusion, 232, 419
Flame retardants
compounds, 51
condensed phase char formation
materials and metals, 52
polymer decomposition, 52–53
endothermic cooling, 51–52
fire-safety regulations, 50
vapor phase radical inhibition, 51
Flywheel. *See* Time–temperature-dependent response
Foreign object damage (FOD) and tool drop
description, 151
low-velocity impacts, 152, 153
static identification and low velocity impact, 153, 154
static *vs*. dynamic response, 153, 154
weight and pendulum impact tests, 151–152
FRP. *See* Fiber reinforced polymer
FRP bridge decks, civil infrastructures
analysis and design optimization procedures, 649
ASTM C666 test standard, 648
creep testing, 651, 652
deck concept, 653
fatigue testing, 651–652
fiber volume fraction, 648
flexural strength and storage modulus, 649, 650
freeze/thaw testing, 647–648
inherent material properties, 653
instrumentation, 647
load testing, 648
prototype structure, Tech 21 Bridge, 647
Q-panel QUV accelerated weathering tester, 647
repaired/rehabilitated, 646
scale model, 649, 651
test program, 648
ultimate strength testing, 653
visual inspection, 647
FTE. *See* Failure threshold energy

G

Gelation cure, 81
GFRP. *See* Glass-fiber-reinforced polymer
GFRP reinforcing bars, civil infrastructures
alkaline agents, degradation mechanisms, 641–642
carbonation and pH testing, 646
components, 645
concrete beams, 640
creep and relaxation, 639
degradation mechanism, 640
description, 638
differential scanning calorimetry (DSC) analysis, 644
E-glass fiber, 642
elastic modulus, 643–644
fiber-matrix interface, 638–639
Flexure tests, 641
freeze thaw cycles, 639–640
FRP composite dowel bars, 644
knockdown factors, design tensile strength, 640
laboratory testing, 646
linear variable differential transformers (LVDTs), 642
MDA dowel bar team, 644
mortar wrapped, 642–643
optical microscopy analysis, 645–646
reduction factors, 638
significance, 639
standards/accepted practices, 642
tensile strength, 641
thermal effects, 639
wet-dry freeze-thaw cycles, 645
Glass-fiber-reinforced polymer (GFRP), 627. *See also* GFRP reinforcing bars, civil infrastructures
Glass transition
asymmetry approach, 243–244
composite materials, physical aging, 289–298
enthalpy measurements, 245–246
intrinsic isotherms, 242–243
KAHR and TNM models, 247–251
memory, 244–245
phenomenology
fictive temperature, defined, 241
glass transition determination, 242
glass transition temperature, 240
heat capacity measurement, 241
kinectics, 242
logarithm, viscosity, 239
molecular mobility, 240
Vogel–Fulcher–Tammann equation, 239
volume, dilatometric experiments, 240
Williams–Landel–Ferry equation, 239

physical aging, glassy polymers
environmental effects and structural aging, 277–282
nanoscale behavior, 282–289
non-linear response and engineering, 261–277
viscoelastic, 256–261
thermoviscoelastic model, 251–256
Ground service equipment (GSE)
belt loader and Boeing 747 approach, 155, 156
Boeing 737, belt loader and truck, 155, 156
major and minor damage, 155
TUG 660E belt loader velocity, 157–158
GSE. *See* Ground service equipment

H

Hail ice damage
all-composite commercial aircraft, 165
composite panel modes, 168, 170
cumulative probability, diameter, 167
global stiffening and local response phenomenal, 172, 173
growth mechanisms, 167
high-speed video stills, 172, 173
impact-side facesheet, carbon/epoxy sandwich panel, 172, 174
in-flight and "ground hail" events, 167
normalized failure threshold energy, 171, 172
progression, modes, 168, 170
size *vs.* projectile kinetic energy, 169, 171
still images, 168, 169
Hydroperoxide decomposition
autoacceleration, 317–318
balance reaction and PO° radicals, 319
description, 317
kinetic modelling, 318
pressure dependence, oxidation rate, 318
unimolecular and bimolecular, 319
Hygrothermal conditioned (HC), aero engines
anti-buckling guides, 585
compressive residual modulus, 588
cycle, 584
2D and 3D woven, PMC, 583
design viability, 585–586
dogbone geometry, 586
fatigue life and modulus degradation
applied cyclic strain range, 588
compressive modulus, 589
elastic modulus (E), fatigue test, 589
residual compressive strength *vs.* modulus, 590
residual strength, 590
fatigue life plot
initial strain range, 587
stress range, 587
mechanical fatigue testing, 584
residual strength/modulus relationships
permutations, 590
tension–tension fatigue, 591
static stress–strain response, 586
test coupons, 585
weight gain, 585
Hygrothermal testing
least-squares curve-fit procedure, 195–198
maximum moisture uptake, 195, 196
model verification, 198–200

I

Inspection, in-service
flaw sizes, 439
residual strength substantiation, 438
vehicle system maintenance plan, 439
Integrated protective coatings
nanocomposite approach and protection scheme, 56–567
protective layer creation, 57–58
TGA data, 58–63
Integrity bases, 135
Internal delaminations, 125
Interphase
composition prediction, 116
cross-linking polymerization reactions, 77
elastic shear lag analysis, 80
empirical relations, 80
fabrication, 77
fiber surface modifications, 79
formation, 78
formation models
kinetics model, 82–98
thermodynamic model, 98–100
modeling interphase formation, 81
molecular-level mechanisms, 80
overall composite properties, 79
physics based description, 81–82
processing interphase
macroscale thermal models, 110–112
multiscale thermochemical models, 112–116
property relationships
chain/chain interaction, 105
finite element analysis, 103, 104

Interphase (*cont.*)
interphase thickness and modulus, 102
micromechanical analyses, 100
rectangular fiber arrangement, 104, 105
relative fiber arrangement, 106
representative volume elements (RVEs), 103
stress concentration factor, 106–109
Young's modulus variation, 100–101
range, length scales, 78
stress concentration factor, 106–107
Isothermal aging, 349–350
Isothermal fatigue
degradation, 570
load waveform identity, 582
long-term hygrothermal conditioning, 593
stitched and unstitched, tests, 581
test matrix, 574
and TMF, 552

K

KCL. *See* Kinetic chain length
Kinetic chain length (KCL), 325
Kinetics model
activation energy, adsorption, 85
BET theory, 83
binary epoxy–amine mixture, 87
DGEBA/PACM20/Aluminum system, 89–90
diffusivity, 88
electron energy-loss spectroscopy (EELS), 89
fiber arrangement, composite, 82–83
fiber-epoxy-amine system, 83–84
interphase
composition profile comparison, 90–91
concentration profiles, 92–94
thickness, 95, 96
mass conservation analysis, 86
mass transfer processes, 93
nondimensional parameters, 89

L

Lamina oxidation, unidirectional growth, 408–409
Laminated composites
angle-ply, $[45/-45]_{2S}$
cracks and oxidation, imaging, 355–356
oxidation growth, aging time, 355–356
cross-ply, $[0/90]_{4S}$
cracks and oxidation, imaging, 353–354
oxidation growth, aging time, 352–353
quasi-isotropic, $[0/45/-45/90]_{2S}$
cracks and oxidation, imaging, 355
oxidation growth, aging time, 354
unidirectional, $[0]_{16T}$
cracks and oxidation, imaging, 351–352
oxidation growth, aging time, 350–351
Laminates oxidation growth, 409–414
LEF approach. *See* Load enhancement factor approach
LFRD. *See* Load and Resistance Factor Design
Life factor approach
load and life factors, 473
metallic structure, 472
operational and damage-inducing fatigue strains, 472
safe-life composite design, 472–473
spectrum fatigue, 471–472
Load and Resistance Factor Design (LFRD), 654
Load enhancement factor (LEF) approach
confidence intervals, 475
desired probability, survival, 474
life factor approach, 475
load and life factors, 474
reliability scatter model, 473
Load sources, out-of-plane, 436
Long-term behavior
concepts and materials, 498
design rules, 498
failure mode, 498, 499

M

Macro-damage mechanics (MADM), 131
Market development alliance (MDA)
CERF-MDA gap analysis, 634
dowel bar team, 644
FRP composite industries, 632
TRB AFF80 committee, 633
MDA. *See* Market development alliance
MDM. *See* Micro-damage mechanics
Mechanical testing
crippling element load *vs.* strain data, 547
geometry coupon, 540
55-kip MTS load frame, 544
22-kip MTS test, 540
strength data, upper and lower skin, 543, 544
stringer flanges and rib station, 540–541
stringer webs, 542
tabbing fixture and tension testing, 542
tension strength data, upper skin, 545
tension test and specimens, 543, 545

three-stringer crippling panel tests, 543, 546
Meta-delamination model, 608
Micro-damage mechanics (MDM), 122, 125
Model
KAHR and TNM
Boltzmann–superposition formalism, 247
correlations, 250
and enthalpy recovery, comparison, 249
heterogeneity, 251
material parameters, 248
memory effect, 250
temperature dependence, 247
volume recovery, comparison, 249
thermoviscoelastic, 251–256
Modeling. *See* Thermo-oxidative aging and degradation, modeling
Moisture induced degradation
absorption behavior, types, 182–183
diffusion modelling, PMC
biaxial microcracks, 186
boundary condition and maximum saturation level, 192–194
chemical potential, 187–188
closed form solution, laminate thickness, 194–195
damage entity, 188–189
fatigue induced biaxial cracks, 189–190
and heat exposure, 184
hygrothermal testing and data characterization, 195–200
irreducible integrity bases, orthotropic materials, 185
material, 184–185
non-Fickian, 183–184
normalized stress, 185
orthotropic medium, 191–192
quadratic function, crack density, 190–191
transverse crack density, 190
uniaxial loading and Gibbs potential, 186–187
mechanism-based modeling, hygrothermal aging
ISV, 227–229
molecular concentration, water, 226
parameter calculation, 227
strength, 229–231
strain assisted diffusion (*see* Strain assisted diffusion)
Multiscale, 112–116
Multiscale synergistic methodology, 139
Multi-wall nanotubes (MWNT), 67–68

N
Nanocomposites
damage
abrasion and electrical, 55
heat and fire, 54
nanofillers cost, 55
UV and oxidation, 54–55
EMI shield, 53–54
nanoscale science and fillers, 53
reinforcing component, 53
Nanoscale behavior
aging rate *vs*. temperature size, 289
enthalpy recovery, Tf-Ta, 284
evolution
fictive temperature, 288
nitrogen permeability, 286
extension, TNM and KAHR, 283
fictive temperature *vs*. aging time, 287–288
fluorescent probe intensity, 284
implosion/densification event, 287
measurements, ellipsometry, 284
"micro"/fiber-reinforced composites, 282
o-TP, structural and enthalpy recovery, 282–283
reduction, glass transition, 287
"structural recovery", PMMA layer, 285
ultrathin polymer films, 284
vapor-deposited samples, 286
National Aeronautics Space Administration (NASA), 514
National Cooperative Highway Research Program (NCHRP)
application, fiber reinforced polymer composites, 633
bonded repair and retrofit of concrete structures, 635–636
and CERF-MDA gap analysis, 634
National Institute for Aviation Research (NIAR), 414, 514, 524, 534
NCHRP. *See* National Cooperative Highway Research Program
NDE. *See* Non-destructive evaluation
Non-destructive evaluation (NDE), 47, 48
Nondestructive inspection (NDI), 414–415, 514
Non-Fickian
coefficient, diffusion, 219
diffusion process, PMC, 182
hygrothermal effects, 201
material behaviour, 232
temperature, stress and/or strain gradients, 184

Non-impact damage sources
arcing, 453
in-service, 452
metallic aircraft, 453
potential degradation, 452
Non-linear diffusion coefficients
applied strain, 219
diffusion-governing equation, 218–219
least-squares technique, 221–222
mechanical strain, 220–221
non-Fickian, 219
normalized temperature, 219–220

O

Organic matrix thermooxidation, mechanisms and kinetics
autoacceleration, 324–325
boundary conditions, 323
composite mechanical properties, 313
constant oxygen concentration, 323
decomposition, unimolecular and bimolecular hydroperoxide, 324
description, oxidation process, 314, 315
diffusion controlled oxidation
balance equation, 327
differential equations system, 329
double steady state, 329
excess and lack, regime, 327
rate expression, 327
scaling law, 328
solubility coefficient, 327, 328
thermooxidation, 328–329
diffusion controlled process, 313
elementary rate constants, 314
glass transition temperature change, 312, 313
KCL, 325
kinetic modelling and expression, 314
low conversions, 323
oxidation consequences
elastic modulus, rubbery and glassy states, 334–339
glass transition temperature, 332–334
mass variations, 329–331
shrinkage, 331–332
toughness, 339–341
oxidation kinetic, curve, 323, 324
oxidation mechanisms
characteristics and quantitative data, 322
electron spin resonance (ESR), 321
infrared spectrophotometry (IR), 321
separation and identification, MS and NMR, 321
oxygen excess, 323
oxygen lack case
assumptions, differential equations, 325–326
consumption rate *vs.* concentration, 326
photooxidation, 326–327
principles and equations, chemical kinetics, 322–323
residual strength, 312–313
residual strength-mass loss, curve, 311–312
sections, matrix oxidation mechanisms and kinetics, 315
standard mechanistic scheme
hydroperoxide decomposition, 317–319
initiation mechanisms, 316–317
initiation rate, Arrhenius plot, 317
radical chain oxidation, 316
steps, radical chain process, 315
terminations, radicals, 316
structure-reactivity relationships
magnitude, 320
polyolefins and polydienic elastomers, 319–320
rate constants, 320–321
tools, 311
unimolecular and bimolecular initiation, 325
Oxidation and damage growth quantification
CT, 356–357
fluorescence *vs.* x-ray imaging, cracks, 357–358
Oxidation consequences, organic matrix thermooxidation
domain, polymer oxidation kinetics, 341
glass transition temperature
chain scission, 332
Fox–Flory relationship, 333
linear and tridimensional polymers, 332–333
glassy state, elastic modulus
amorphous and semi-crystalline polymers, 334
antiplasticization, 335–336
aromatic epoxide-amine networks, 336
crosslink density and dangling chain concentration, 337
dangling chains, 336, 337
DGEBA-DDS system, 338
dynamic mechanical analysis, 335
high activity and glass transition, 335
internal antiplasticization, 335

linear polymers, 334
low activity/overlapping transition, 335
nano-indentation, 338
shear and tensile modulus, 335–336
mass variations
balance, oxygen grafting, 330
description, 329
gain *vs.* loss, 330–331
gravimetric curves, 330, 331
loss, initiation step and rate, 330
variation, 330
rubbery state, elastic modulus, 334
shrinkage
correlation and volume variation, 331–332
density variations and CRU, 331
structural scales, 341
toughness
amorphous and semi-crystalline polymers, 339–340
brittle and ductile polymers, 339
description, thermosets, 340
energy release rate, 340, 341
ideal network, thermosets, 340
linear polymers curve, 339
molar mass and variations, 339, 340
polymer oxidative degradation, 339
"tough" and "brittle" regime, 339
Oxidation growth
discrete cracks, 418–420
1D simulations, 393–394
lamina scale, 422–423
predictive modeling
laminates, 409–414
resins, 406–408
unidirectional lamina, 408–409
specimen geometry, 421
three-zone model, 387
Oxidation layer growth. *See* Thermo-oxidation

P

Passive protection
anti-oxidants and UV stabilizers, 47–48
engineering solutions, 47
NDE, 48
PETI. *See* Phenylethynyl-terminated imides
Phenylethynyl-terminated imides (PETI)
aromatic polyimides, 20
moisture and solvent effects, 22
phenylethynyl endcap, 21
prepreg materials, 20
radiation, 22–23
thermooxidative aging, 21–22
time and stress, 19–20
Physical aging, glassy polymers
environmental effects and structural aging
"humidity-jump", 277
hypothesis, 278–279
recovery signatures, 279–282
strains, 278
nanoscale behavior, 282–289
non-linear response and engineering
failure-related properties, 276–277
viscoelastic, 261–262
yield-failure and aging, 265–276
yield response, 262–265
viscoelastic, 256–261
Plane weave, oxidation growth
laminate thickness and orientation, 372–373
$[PW]_4$, 368–371
$[PW]_{16}$, 372
$[PW45]_4$, 371
$[PW, PW45]_{2S}$, 372, 373
PMR and related thermosets
cure and thermooxidative aging, 17–18
moisture and solvent effects, 18–19
PMR–15 oligomer, 16
radiation, 19
time and stress effect, 19–20
Polyhedral oligomeric silsesquioxanes (POSS), 68
Polyimide composites, TMF. *See* Thermomechanical fatigue (TMF)
Polymer composites
active and passive protection, 47–48, 64
additive approaches, 48–53
"buckypapers", 55–56
corrosion resistance, 40
damage
electrical, 45
heat and fire, 42–45
wear and abrasion, 45–46
degradation *vs.* decomposition, 41–42
integrated protective coatings, 56–63
interfacial issues
alkyl ammonium treatment, layered silicates, 66–67
carbon nanotubes and nanofibers, 67–68
engineered structure, 68–69
MWNT, 67–68
nanoparticle benefits and POSS, 68
life cycle and environmental challenges, 69
multi-component systems
char formation, 65

Polymer composites (*cont.*)
degradation and decomposition, 65
modeling, 65
multifunctional materials
EMI shield, 65
nanotubes and nanofibers, cost issue, 64–65
polymer + carbon nanotube, 64
nanocomposites, 53–55
trends
environmental resistance, 40
fuel savings, 40
metals and ceramics, 40–41
Polymer composites durability, power and propulsion
absorption–desorption stimulation, 550
applications, 550
creep and stress-relaxation tests, 550
HC, aero engine, 552, 583–592
pathfinder experimental programs, 550
PMCs, 549–550
time-temperature response, 551, 553–568
TMF, reusable propulsion, 551–552, 568–583
POSS. *See* Polyhedral oligomeric silsesquioxanes
Processing interphase property
macroscale thermal model
mathematical expressions, 111
one-dimensional heat transfer, 110
thermosetting matrix composites, 110–111
multiscale thermochemical model
composite transverse modulus, 115–116
interphase formation parameters:, 113–114
macroscale composite domain, 111, 112
scaling factor, 113
Property
finite element analysis, 103, 104
interphase thickness and modulus, 102
micromechanical analyses, 100
Young's modulus variation, 100–101
Propulsion. *See* Thermomechanical fatigue (TMF)

Q
Qualification
definition, 428–429
material
definition, 430
tests, 430
Quasi-isotropic $[0/45/-45/90]_{2S}$ laminate, 354–355, 363–364
Quasi-static methodology
composite structure, 475
maximum design load, 476
vs. ultimate strength approach, 476

R
Radiation
effects, exposure, 4
epoxies, 8
PETI, 22–23
PMR and thermosets, 19
thermoplastics, 27
Reaction rates
active zone modeling, 390
assumption, 396–399
carbon fibers, 393
concentration dependence, 400–403
defined, 388
mechanistic chemistry models, 387
oxidation state and weight loss relationships, 390–391
saturation, 389
Recovery signatures
aging time shift factors *vs.* volume departure, 282
asymmetry approach, 280
characteristic time, viscoelastic, 281
comparison, temperature and RH jumps, 280, 281
intrinsic isopiestics, 280
RH-jump, 279
two-step RH-experiments, memory effect, 281
Representative volume element (RVE) and homogenization
"classical", 127
"evolving microstructure", 127
generic point P, 127–128
heterogeneous solid damage, 127
internal surfaces, 126–127
materials, 126
notion, 127
scalars, 128
Requirements, material environmental
degradation, 434
fatigue cycles, 431
heat sources, aircraft, 432
in-service exposures, 434
light fighter hot temperature values, 433
polymer matrix composites, 432
Residual strength, 590–592

Resin
behavior
Arrhenius-type kinetics models, 389
diffusivities, 389
oxidation state, weight loss, reaction rate relationships, 390–391
oxygen diffusivity tensor, 389
reaction termination and oxidized layer growth, 390
saturation reaction rate, 389
BMI, 11–16
classes, 2
epoxies (*see* Epoxies)
mechanical loads, 4
PETI, 20–23
PMR and related thermosets, 16–20
radiation, 4
thermo-oxidative exposure, 2–3
thermoplastics, 23–27
water and solvents, 3
RVE. *See* Representative volume element

S
Safe-life analysis
elements, design, 456
fatigue life, 456
fixed-wing manned aircraft manufacturers, 455
operational experience, rotor designs, 456
Palmgren-Miner law, 457
scatter factors, 457
S-N curve, 456–457
Saturation concentration, 206
SDM. *See* Synergistic damage mechanics
"Shear-lag" process, 123
Shrinkage
isotropy, 415–416
mismatch, 418
strain, 415
volumetric, 415
Simulated hail ice (SHI), 168
Solvents
BMIs, 15
epoxies, 9–11
PETI, 22
PMR and thermosets, 18–19
thermoplastics, 26–27
and water exposure, 3
Stacking sequence, $[0]_{16T}$, $[0/90]_{4S}$, $[0/45/-45/90]_{2S}$ and $[45/-45]_{2S}$
comparison, crack length, 365, 368
oxidation extent comparison, 365–367
Stepped-lap bending test, 501
Stepped-lap tensile test, 499, 501
Stiffness properties, 136
Strain assisted diffusion
adhesive bonded laminates, 203
boundary condition and saturation concentration, 206
continuum mechanics and thermodynamics, 202–203
diffusion coefficient and saturation mass uptake, 216–218
diffusivity and moisture coefficients, 223–225
environmental cracking, 202
experiment design
beam bending theory, 206–207
force equilibrium, beam cross section, 207–208
longitudinal, 208
objectives, 207
tensile, 209
Waco Mbrace primer, 207
FRP, 201–202
mass saturation and diffusivity, 218, 224
mechanical strain calculation, primer
correction factor, 216
moisture uptake *vs*. sqrt, 211–215
relationship, total and mechanical, 210
through-thickness strain gradient, 210–211
moisture saturation coefficients, nonlinear, 222–224
nonlinear coefficients
least-square error, 221–222
non-Fickian, 219
normalized temperature, 219–220
tensile strain gradient, 218–219
variables separation, 220–221
temperature effect, specimen ring radius, 209–210
total strain influence
diffusivity, 224, 225
saturation concentration, 224, 226
two dimensional equation, 203–206
Strain energy release rate (SERR), 502
Strain Invariant Failure Theory (SIFT), 503
Strength approach, 475–476
Strength prediction, bolted joints
advantages, evolution and interaction, 495
damage modes, 493–494
2D analysis, 495
matrix transverse crack and fiber breakage, 494–495

Strength prediction (*cont.*)
open-hole laminates, 493
property degradation modeling, 494
stress redistribution, 493
Stress-assisted diffusion, 182, 202
Stress relaxation and creep tests, 560–567
Structural joints. *See* Durability, structural joints
Structural recovery
asymmetry approach
comparison, up-jump and down-jump condition, 243
equilibrium, 244
temperature histories, 244
enthalpy measurements
DSC, 242
enthalpy departure, 243
heat capacity, scanning calorimetry, 243
intrinsic isotherms
"geological" ages, 242
thermal history, 243
volumetric strain, 243
memory
description, 244
thermal histories, 245
models
KAHR and TNM, 247–251
thermoviscoelastic, 251–256
Structure-reactivity relationships, organic matrix thermooxidation
magnitude
hydrogen abstraction process, 320
intervals, rate constant values, 320
polyolefins and polydienic elastomers, 319–320
rate constants
coherent synthesis, 320–321
dissociation and activation energy, 321
polymer stabilities, 321
Substantiation approaches
definition, 428
LEF (*see* Load enhancement factor (LEF) approach)
life factor approach, 471–473
quasi-static and ultimate strength approaches, 475–476
Superposition, time-aging time, 291–296
Surface potential
chain/surface interaction, 99
effects, 108
fiber–matrix modulus ratio, 105, 106
functions, 102
thermodynamic model, 97
Synergistic damage mechanics (SDM)
approach, 122
COD, 138–139
definition, 140
E_1 and ν_{12} changes, 138
K-parameter, 139
methods, 139
RVE, 138
viscoelastic, 139–140

T

Termination and oxidized layer growth, 390
Textile composites
3D resin impregnated, 374–378
plain-weave, 368–374
TGA data. *See* Thermogravimetric analysis data
Thermochemical model
composite transverse modulus, 115–116
interphase formation parameters:, 113–114
macroscale composite domain, 111, 112
scaling factor, 113
Thermodynamic model
concentration profiles, 97, 99
interphase formation, 98
Thermodynamics, CDM
first and second laws, 131–132
forces, 133
framework, 131
Truesdell's principle equipresence, 132
Thermogravimetric analysis (TGA) data
polyimide + inorganic systems
alumina (Al_2O_3), 59, 60
autoclave cure, 59
nanofiller, 59, 61
polyimide + protective coating
aging, 60, 61
"protective laminate side up", 60
resin viscosity, process work, 62
thicker top layers, 62
Thermomechanical fatigue (TMF)
collaboration, Rocketdyne and NASA–652 GRC, 569
degradation, 570
elastic modulus, 582
HFDE and p-phenylenediamine, 569
high-temperature polyimide composites, 568
isothermal static and fatigue tests
elastic modulus, 575
hysteresis loop, 575
stitched and unstitched, 573
stitched matrix, 574

stitched, offset strain, 574
stitched, stress–strain response, 574, 578
materials, 570
SMCs, 569
stitched material
"apparent stiffness", 577
isothermal, comparison, 578
strain ratcheting, 579
stress–strain response, 577
stress-total strain slope, 579
thermal expansion response evaluation, 576
thermocouple measurement, 577
tensile test, 582
test methods
axial strains measurement, 571
control and data acquisition, 572
cooling cage, views, 571
cycle and thermal response, 572
dogbone geometry, 570–571
isothermal static test, 573
K-type TCs, 572–573
quartz lamps and cooling cage adjustment, 573
static and fatigue, 570
transverse expansion rate, 569
unstitched material
fatigue strength, 581
micromechanics, 581
repeat test, 580
and stitched, comparison, 580
strain accumulation, 581
stress–strain response, 580
UTS, 583
Thermo-oxidation
anisotropic nature, 347
anisptrophy, 358–365
fiber-matrix interphase region, 347–348
growth behavior, 349
interlaminar residual stresses, 348
isothermal aging, 349–350
laminated composites, 350–356
oxidation and damage growth quantification, 356–358
physical and chemical aging, 346–347
ply stacking sequence, 365–368
surface degradations, 348–349
textile composites, 368–378
Thermo-oxidative aging and degradation, modeling
constituent materials
fiber oxidation and interphases, 391–393
Henry's law, 388
reaction rate, 387–388
resin behavior, 389–391
saturation concentration, 388
three-zone model, 387
vapor phase diffusion, 386–387
fiber-reinforced composites, 386
frame predictive modeling framework, 384
HTPMCs
determination, 384
oxidative degradation, 383–384
laminated composites, 386
mechanisms, 384–385
model parameter
diffusivity characterization, 400
discussion, 400
fiber and fiber-matrix interphase, 403–406
reaction rate and concentration dependence, 400–403
numerical simulations
3D finite element methods (*see* 3D finite element methods)
through-the-thickness oxidation growth, resins 1D, 393–394
oxidation growth, prediction
laminates, 409–414
resins, 406–408
unidirectional lamina, 408–409
oxidized and unoxidized regions, 385–386
stress, oxidation and damage
discrete cracks, 418–420
oxidative layer growth, evolution, 415–418
oxidized regions, 420–422
Thermoplastics
amine end groups, 6
Avimid, K3B, 26
epoxies, 25
fatigue thresholds, 10
generic aromatic polymer, 23, 24
glassy polymer, 2
IM8/ITX, Tg, 25
interlaminar fracture, 20
internal stress, 27
linear polyimide, generic, 23–25
maleimide-capped sulfone polymers, 13
melt viscosities, 23–24
moisture and solvent effects, 22, 26–27
radiation effects, 27
stress and time effects, 11
thermooxidative aging
crosslinking, 22
weight loss, 25–26

Thermovisco elastic model
classical thermodynamics, 251
comparison, calorimetric experiments, 255–256
conventional measurement methods, 254
entropy symbol, 253
Green-Lagrange and Hencky strain measure, 252–253
nonlinear deformation and flow behavior, 252
predictions, 255
shift factor, defined, 254
temperature and strain expansion, 252
Time–temperature-dependent response
FESS evaluation, 553
flat coupon specimen manufacture
composite properties, 555
panels, 554
residual strains, 555
strain gauges, 555
ultrasonic examination, 555
WK-type gauges, 556
flywheel energy, 553
linear hereditary response, 554
master relaxation-modulus and creep-compliance curves, 554
material characterization
ASTM, 557
storage (G') and loss (G") moduli, 558
monotonic testing
longitudinal, 558–559
transverse, 559–560
relaxation modulus/creep compliance data, 553
stress relaxation and creep tests
application, compliance data, 563
Boltzmann's principle superposition, 565
compliance, 564
compression–compression fatigue test, 568
compressive creep, 562
compressive stress, 562
D3039 and D3410, ASTM, 560
data plots and comparison, 560
defined, 564
design implications, 566–567
fiber/matrix interface, 563
filament-wound specimen panels, 567
percent bending strain, 561
prototypical operation load, 565–566
rotor design utilization, 567
specimen-to-specimen variation, 564
tensile and compressive creep, 561
tensile and compressive stress relaxation, 561
TTSP, 563
test programs, 554
test setup
calibration factors, 556
compressive, specimen stability verification, 557
mechanical testing, 556
viscoelasticity, 553
Toughness
behavior, 423
and critical stress, 421
Transportation Research Board (TRB) AFF80 committee
barriers, FRP composites, 633–634
demonstration projects, 634
description and NCHRP, 633
TRB. *See* Transportation Research Board

U

Ultrasonic
pulse echo, 514
transmission, 514
Unidirectional composite
fiber–interphase assemblage, 419
oxidation *vs.* damage growth, 360
prediction, oxidation growth, 410
$[0]_{16T}$, 359–361
UV stabilizers
blooming, 49–50
effectiveness, 50
initiation, propagation and termination reactions, 49, 50
metals, 49

V

Virtual crack technique (VCCT), 502
Viscoelastic, physical aging
aging time shift factors and volume, 260
creep compliance, PVC glass, 257
KWW function, 257
power-law shifting and relaxation/creep curves, 259
shift factors
vs. aging/elapsed time, 260
vs. volume, 261
stress relaxation response, 258
thermodynamic state, 256
time-aging time
shift factors, 257, 258
superposition, 259

W

Water
    concentration profile, 10
    exposure effects, 3
    exposure to radiation, 4
    immersion, K3B, 27
    mechanical and time-dependent effects, 16
    moisture and solvent effects, 22
    permeation, 7
    retention of, 18
    solvent sensitivity, 11
    stress and time effects, 11
    thermoplastic composites, effects, 26
    threshold, 19
    uptake, DDS, 9
Weave architecture and fabrication, 602–603

Y

Yarn cracking model, 606–607
Yield response
    aging time shift factor *vs*. aging time, 263
    compressive stress–strain curves, 264
    stress
        vs. aging time, 265
        vs. strain rate logarithm, 264

Z

Zoning and impact survey
    low energy impact
        hail and runway debris, 447
        repeated, 448
        tool, 446
    residual strength requirements *vs*. damage categories, 449
    visibility, damage, 447
    zones, 446–447